21st Workshop on Biomedical Language Processing (BioNLP 2022)

Dublin, Ireland
26 May 2022

ISBN: 978-1-7138-6728-9

Printed from e-media with permission by:

Curran Associates, Inc.
57 Morehouse Lane
Red Hook, NY 12571

Some format issues inherent in the e-media version may also appear in this print version.

Printed with permission by Curran Associates, Inc. (2023)

Association for Computational Linguistics
209 N. Eighth Street
Stroudsburg, Pennsylvania 18360

Phone: 1-570-476-8006
Fax: 1-570-476-0860

acl@aclweb.org

Additional copies of this publication are available from:

Curran Associates, Inc.
57 Morehouse Lane
Red Hook, NY 12571 USA
Phone: 845-758-0400
Fax: 845-758-2633
Email: curran@proceedings.com
Web: www.proceedings.com

BioNLP 2022 @ ACL 2022

Proceedings of the 21st Workshop on Biomedical Language Processing

May 26, 2022

Reflectively looking into the future of biomedical language processing

Dina Demner-Fushman, Sophia Ananiadou, Kevin Bretonnel Cohen, Junichi Tsujii

The 2022 meeting of the Biomedical Natural Language Processing workshop at the Association for Computational Linguistics conference reminds us of the first such workshop at the 2002 conference. The twenty years that have passed since then have seen enormous growth in the BioNLP community, and now seems like a good time to take stock of where we have come over the course of those two decades.

Interest in scientific natural language processing started soon after the launch of the Sputnik satellite in 1957, when the Anglophone scientific world realized that there was quite a bit of good research being published in Russian that it hadn't been reading. Interest in, first, clinical and then more general biomedical language processing started in the 1960s, and biomedical language processing interest groups soon formed within the clinical, and later the bioinformatics, communities. The Association for Computational Linguistics BioNLP community came together in 2002 to answer the needs of a deeply interdisciplinary area of research focused on natural language processing and text mining methods applied to biomedical text. The field spread quickly, but the events and publication venues for computational linguists interested in the biomedical sublanguage were dispersed across a range of disciplines and conferences. Clinical natural language processing had a natural home in the Association for Medical Informatics, and biologically oriented language processing focused on the rapidly growing scientific literature was well-housed in the International Society for Molecular Biology and Pacific Symposium for Biocomputing publication venues; the Association for Computational Linguistics seemed like a natural home for research that focused around the linguistic nature of our field, rather than being oriented around its clinical and biological applications.

To bring together the passion for the domain and the benefits of belonging to the ACL community, SIGBioMed was formed as an ACL SIG in 2007. SIGBioMed is celebrating 15 years this summer. From the beginning, the SIG strove to be inclusive in terms of the topics of interest, languages studied, and researchers invited for presentations and keynotes. As can be seen in the work presented in this 2022 workshop, SIGBioMed continues that policy of diversity, equity and inclusion. Borrowing from the New York Times, SIGBioMed's (unofficial) motto is "All the Work That's Fit to Print—as long as it broadly applies to the biomedical and clinical domains."

Biomedical language processing started with rigorous text mining research that helped advancing understanding of biomedical text and provided services to the target domains. For example, the MedLee system was used to support clinical applications (Friedman et al., 2004), whereas BioNLP shared tasks in 2011 and 2013 focused on extraction of information about pathways and development of biomedical event extraction systems (Miwa, M. et al 2013; Björne J. et al. 2015). Recent developments in and availability of large pre-trained language models (BioBERT, ClinicalBERT, SciBERT, etc.) provide us not only with a chance to advance the research and applications towards language and context understanding, but also to start understanding how the models perform the tasks, as evidenced by the work presented in the next sections.

Looking back: The test of time award

This meeting marking two decades of research in and around the ACL community provides an opportunity to reflect on how we got to where we are. So, following up on a suggestion from Tim Miller, we solicited nominations for a new BioNLP Workshop award: recognition of papers in our field that have "stood the test of time."

As the nominations came in, we quickly realized that the request was underspecified. In what publication venues could a nominee have appeared? Should it be limited to the BioNLP Workshop, or would any venue qualify? How *much* time? Would authors be allowed to nominate their own papers? And what would it mean to have "stood the test," exactly? Number of citations? Actual usage of a system, a technique, a resource, an idea? Would a once-heavily-cited paper that is not cited much any more qualify? And did we need to normalize for the length of time since publication? And who should do the selection from amongst the nominees? What if a paper by a member of the organizing committee was nominated? The organizing committee? An external panel? Open vote of the entire community?

We began with the assumption that we might be able to induce the answers to those questions from the nominations themselves. To facilitate that, we asked nominators to consider writing up a note—of the length of their choice—describing why they felt that their nominee rated recognition. We explicitly allowed self-nomination. And then we waited.

As it turned out, the set of nominations did not answer our questions. Some nominators expressed a well-argued opinion that only publications from the BioNLP Workshop should qualify, but submissions came in from a number of venues. The time spans since their publications varied widely. They covered systems, techniques, resources, and—thank goodness—ideas. Papers were submitted by non-authors, papers were submitted by their own authors, and we had submissions that were co-authored by the organizers of the workshop.

The only thing we did *not* have was a voting mechanism. We considered counting the number of nominations per paper, but several papers were nominated twice; the only one that was nominated three times had two of its nominations from its own authors—not forbidden, but it made the number-of-nominations criterion seem unreasonable; and in any case, elementary power calculations soon convinced us that the total number of nominations was not sufficient to differentiate between one vote, two votes, or three. In the end, we contemplated the set of nominations, saw perfectly good reasons to accept that they had *all*, in one or more ways, "stood the test of time." Consequently, this year we are awarding the BioNLP Test of Time Award to multiple papers—in fact, to all of the papers that were nominated. You will find them listed in Table 1, which accords to all co-authors concerned the right to add "2022 BioNLP Test Of Time Award Recipient" to their CVs. Although we resolved essentially none of the issues that we had identified, this was a tremendously fun exercise, and we look forward to excellent suggestions from the community as to how to answer the questions that we raise above, as well as how to do this next year in a more principled way without quite so glaring an appearance of conflict of interest.

Looking forward: Overview of the work in this volume

BioNLP 2022 received 59 valid submissions, of which 11 were accepted as oral presentations and 32 as posters.

The scope and the depth of the work in this volume reflects the growing rigor and maturity of biomedical language processing. True to the historical inclusiveness of the workshop, the processed text includes scientific publications, clinical notes, and other forms of formal and informal communications, primarily in English, but also in Bangla (Sazzed et al.), Spanish-Catalan (Amin et al.), Spanish (Carrino et al.) and Romanian (Mitrofan et al.)

Advances in literature processing are reflected in the work that presents end-to-end document level relation extraction that leverages coreference resolution and entity extraction (Giorgi et al.); linking citing sentences in a publication to the cited sentences in referenced sources (Roy et al.); and extracting design and evidence from the descriptions of Clinical Trials (Witte et al.)

Aronson, Alan R. Effective mapping of biomedical text to the UMLS Metathesaurus:
the MetaMap program.
Proc. AMIA Symposium, p. 17, 2001.

Björne, Jari, Juho Heimonen, Filip Ginter, Antti Airola, Tapio Pahikkala, and Tapio Salakoski.
Extracting complex biological events with rich graph-based feature sets.
Proc. BioNLP 2009 Workshop Companion Volume (Shared Task), pp. 10-18. 2009.

Chapman WW, Bridewell W, Hanbury P, Cooper GF, Buchanan BG.
A simple algorithm for identifying negated findings and diseases in discharge summaries.
J. Biomed Inform. 2001 Oct;34(5):301-10.

Kim, J-D., Tomoko Ohta, Yuka Tateisi, and Jun'ichi Tsujii.
GENIA corpus–a semantically annotated corpus for bio-textmining.
Bioinformatics 19, no. suppl_1 (2003): i180-i182.

Leaman, Robert, Laura Wojtulewicz, Ryan Sullivan, Annie Skariah, Jian Yang,
and Graciela Gonzalez.
Towards internet-age pharmacovigilance: extracting adverse drug reactions
from user posts to health-related social networks.
Proc. Biomedical Natural Language Processing, pp. 117-125. 2010.

Morgan, Alexander A., Zhiyong Lu, Xinglong Wang, Aaron M. Cohen, Juliane Fluck, Patrick Ruch,
Anna Divoli, Katrin Fundel, Robert Leaman, Jörg Hakenberg, Chengjie Sun, Heng-hui Liu,
Rafael Torres, Michael Krauthammer, William W Lau, Hongfang Liu, Chun-Nan Hsu,
Martijn Schuemie, K Bretonnel Cohen, and Lynette Hirschman.
Overview of BioCreative II gene normalization.
Genome Biology 9, no. 2 (2008): 1-19.

Oronoz M, Gojenola K, Pérez A, de Ilarraza AD, Casillas A.
On the creation of a clinical gold standard corpus in Spanish: Mining adverse drug reactions.
J. Biomed Inform. 2015 Aug;56:318-32.

Pestian, John P., Michael Sorter, Brian Connolly, Kevin Bretonnel Cohen, Cheryl McCullumsmith,
Jeffry T. Gee, Louis-Philippe Morency, Stefan Scherer, Lesley Rohlfs, and STM Research Group.
A machine learning approach to identifying the thought markers of suicidal subjects:
a prospective multicenter trial.
Suicide and Life-Threatening Behavior 47, no. 1 (2017): 112-121.

Sarker, Abeed, Rachel Ginn, Azadeh Nikfarjam, Karen O'Connor, Karen Smith, Swetha Jayaraman,
Tejaswi Upadhaya, and Graciela Gonzalez.
Utilizing social media data for pharmacovigilance: a review.
J. Biomed. Inf. 54 (2015): 202-212.

Tsuruoka, Yoshimasa, Yuka Tateishi, Jin-Dong Kim, Tomoko Ohta, John McNaught,
Sophia Ananiadou, and Jun'ichi Tsujii.
Developing a robust part-of-speech tagger for biomedical text.
Panhellenic Conference on Informatics, pp. 382-392. Springer, Berlin, Heidelberg, 2005.

Wu S, Miller T, Masanz J, Coarr M, Halgrim S, Carrell D, Clark C.
Negation's not solved: generalizability versus optimizability in clinical natural language processing.
PLoS One. 2014 Nov 13;9(11):e112774.

Table 1: The 2022 BioNLP Test Of Time Awardees, in alphabetical order.

The biomedical domain and particularly clinical language processing suffers from a dearth of resources. The community is clearly addressing the need for annotated data by creating new datasets, data augmentation, and exploring approaches to reducing the need for data. We see many efforts in zero-, few-shot training, data augmentation and distant supervision: for causal precedence among chemical interactions (Liang et al.); information extraction (events, named entity, and relation extraction) (Papanikolaou et al., Wang et al., Khandelwal et al., Iinuma et al., Trieu et al., Dhrangadhariya et al., Watanabe et al., Sarrouti et al., Phan et al., Kim et al.); term normalization (Zeng et al.), summarization (Soleimani et al), and cross-lingual transfer (Amin et al.).

The new datasets introduced at BioNLP 2022 include the Medical Video Question Answering Shared Task data (Guota et al.) ; biomedical named-entity annotated corpus for Bangla (Sazzed et al.); ICD coding (Huang et al.); and curation of antibiotic-resistant genes (Chandak et al.)

We are happy to see many efforts on model understanding and analysis. This volume includes work on explaining model decisions on health-related online materials (Boissonnet et al.); explanations of medical coding predictions (Wood-Doughty et al.); entity memorization and recall in pretrained large LMs with positional prompting (Abaho et al.); inter-annotator agreement and its relation to model performance (Richie et al.); and a self-supervised pre-training approach for understanding genetic information (Cahyawijaya et al.).

We notice increased interest in complex tasks of language generation, summarization and question answering. Language generation was studied both in general (Yuan et al.) and for the specific tasks of dialogue generation (Naseem et al., Ngai et al.) and radiology report generation (Yan et al., Tang et al.), Work on summarization includes extractive/abstractive summarization of documents of varying length (Bishop et al.), aspect-based scientific document summarization (Soleimani et al.) and summarization as an approach to calculate seizure frequencies and dates of last seizure (Xie et al.). Question answering was explored on its own (Pappas et al.) and as a tool for risk prediction (Liang et al.), event extraction (Wang et al.), and explaining quality assessment of online materials (Boissonnet et al.)

Clinical language processing shows stable interest in ICD coding (Michalopoulos et al., Falis et al., Wood-Doughty et al.), risk score prediction (Lianf et al.) and the impact of de-identification (Vakili et al.)

The **Medical Video Question Answering Shared Task** co-located with BioNLP 2022 is described in the overview (Gupta et al.) that includes 8 technical reports submitted by the participating teams, in addition to the two papers presented as posters in the workshop(Li et al., Kusa et al.)

Last, but most certainly not least in this era of rampant mental health concerns, approaches to supporting mental health were studied in the works on analysis of speech disfluencies towards automated dementia detection (Farzana et al.) and dialogue generation for psychotherapeutic counselling (Das et al.)

Acknowledging the community

As always, we are deeply grateful to the authors of the submitted papers and to the reviewers (listed elsewhere in this volume) who produced three thorough and thoughtful reviews for each paper in a fairly short review period.

The quality of submitted work continues growing and the Organizers are truly grateful to our amazing Program Committee that helped us determine which work is ready to be presented and which will benefit from additional experiments and analyses suggested by the reviewers.

Finally, we thank everyone who nominated papers for the Test of Time Award—especially for their well-reasoned and insightful discussions of *why* they chose those papers.

As in years past, we are looking forward to a productive workshop, and we hope that new collaborations and research will evolve, continuing contributions of our community to public health and well-being.

References

Aronson AR. Effective mapping of biomedical text to the UMLS Metathesaurus: the MetaMap program. Proc AMIA Symp 2001:17–21.

Björne, Jari, Juho Heimonen, Filip Ginter, Antti Airola, Tapio Pahikkala, and Tapio Salakoski. Extracting complex biological events with rich graph-based feature sets. Proc. BioNLP 2009 Workshop Companion Volume (Shared Task), pp. 10-18. 2009.

Björne J, Salakoski T. TEES 2.2: Biomedical Event Extraction for Diverse Corpora. BMC Bioinformatics. 2015;16 Suppl 16(Suppl 16):S4.

Chapman WW, Bridewell W, Hanbury P, Cooper GF, Buchanan BG. A simple algorithm for identifying negated findings and diseases in discharge summaries. J Biomed Inform. 2001 Oct;34(5):301-10. doi: 10.1006/jbin.2001.1029. PMID: 12123149.

Friedman C, Shagina L, Lussier Y, Hripcsak G. Automated encoding of clinical documents based on natural language processing. J Am Med Inform Assoc. 2004 Sep-Oct;11(5):392-402.

Kim, J-D., Tomoko Ohta, Yuka Tateisi, and Jun'ichi Tsujii. GENIA corpus–a semantically annotated corpus for bio-textmining. Bioinformatics 19, no. suppl_1 (2003): i180-i182.

Leaman, Robert, Laura Wojtulewicz, Ryan Sullivan, Annie Skariah, Jian Yang, and Graciela Gonzalez. Towards internet-age pharmacovigilance: extracting adverse drug reactions from user posts to health-related social networks. Proc. 2010 workshop on biomedical natural language processing, pp. 117-125. 2010.

Morgan, Alexander A., Zhiyong Lu, Xinglong Wang, Aaron M. Cohen, Juliane Fluck, Patrick Ruch, Anna Divoli, Katrin Fundel, Robert Leaman, Jörg Hakenberg, Chengjie Sun, Heng-hui Liu, Rafael Torres, Michael Krauthammer, William W Lau, Hongfang Liu, Chun-Nan Hsu, Martijn Schuemie, K Bretonnel Cohen, and Lynette Hirschman. Overview of BioCreative II gene normalization. Genome Biology 9, no. 2 (2008): 1-19.

Oronoz M, Gojenola K, Pérez A, de Ilarraza AD, Casillas A. On the creation of a clinical gold standard corpus in Spanish: Mining adverse drug reactions. J Biomed Inform. 2015 Aug;56:318-32. doi: 10.1016/j.jbi.2015.06.016. Epub 2015 Jun 30. PMID: 26141794.

Pestian, John P., Michael Sorter, Brian Connolly, Kevin Bretonnel Cohen, Cheryl McCullumsmith, Jeffry T. Gee, Louis-Philippe Morency, Stefan Scherer, Lesley Rohlfs, and STM Research Group. A machine learning approach to identifying the thought markers of suicidal subjects: a prospective multicenter trial. Suicide and Life-Threatening Behavior 47, no. 1 (2017): 112-121.

Pyysalo, S., Ohta, T., Rak, R. et al. Overview of the Cancer Genetics and Pathway Curation tasks of BioNLP Shared Task 2013. BMC Bioinformatics 16, S2 (2015).

Makoto Miwa, Sampo Pyysalo, Tomoko Ohta and Sophia Ananiadou (2013). Wide coverage biomedical event extraction using multiple partially overlapping corpora. BMC Bioinformatics, 14:175

Sarker, Abeed, Rachel Ginn, Azadeh Nikfarjam, Karen O'Connor, Karen Smith, Swetha Jayaraman,

Tejaswi Upadhaya, and Graciela Gonzalez. Utilizing social media data for pharmacovigilance: a review. Journal of Biomedical Informatics 54 (2015): 202-212.

Tsuruoka, Yoshimasa, Yuka Tateishi, Jin-Dong Kim, Tomoko Ohta, John McNaught, Sophia Ananiadou, and Jun'ichi Tsujii. Developing a robust part-of-speech tagger for biomedical text. In Panhellenic Conference on Informatics, pp. 382-392. Springer, Berlin, Heidelberg, 2005.

Wu S, Miller T, Masanz J, Coarr M, Halgrim S, Carrell D, Clark C. Negation's not solved: generalizability versus optimizability in clinical natural language processing. PLoS One. 2014 Nov 13;9(11):e112774. doi: 10.1371/journal.pone.0112774. PMID: 25393544; PMCID: PMC4231086.

Organizing Committee

Organizers:

Dina Demner-Fushman, US National Library of Medicine
Kevin Bretonnel Cohen, University of Colorado School of Medicine, USA
Sophia Ananiadou, National Centre for Text Mining and University of Manchester, UK
Junichi Tsujii, National Institute of Advanced Industrial Science and Technology, Japan

Program Committee:

Sophia Ananiadou, National Centre for Text Mining and University of Manchester, UK
Saadullah Amin, Saarland University, Germany
Emilia Apostolova, Anthem, Inc.
Eiji Aramaki, University of Tokyo, Japan
Timothy Baldwin, University of Melbourne, Australia
Spandana Balumuri, National Institute of Technology Karnataka, India
Steven Bethard, University of Arizona, USA
Robert Bossy, Inrae, Université Paris Saclay, France
Berry de Bruijn, National Research Council Canada
Leonardo Campillos Llanos, Universidad Autonoma de Madrid, Spain
Fenia Christopoulou, National Centre for Text Mining and University of Manchester, UK
Kevin Bretonnel Cohen, University of Colorado School of Medicine, USA
Brian Connolly, Ohio, USA
Mike Conway, University of Utah, USA
Manirupa Das, Amazon Search, Seattle, WA, USA
Surabhi Datta, The University of Texas Health Science Center at Houston, USA
Dina Demner-Fushman, US National Library of Medicine
Dmitriy Dligach, Loyola University Chicago, USA
Kathleen C. Fraser, National Research Council Canada
Travis Goodwin, US National Library of Medicine
Natalia Grabar, CNRS, France
Cyril Grouin, LIMSI - CNRS, France
Tudor Groza, The Garvan Institute of Medical Research, Australia
Deepak Gupta, US National Library of Medicine
Sam Henry, Christopher Newport University, USA
William Hogan, UCSD, USA
Kexin Huang, Stanford University, USA
Brian Hur, University of Melbourne, Australia
Richard Jackson, AstraZeneca
Antonio Jimeno Yepes, IBM, Melbourne Area, Australia
Sarvnaz Karimi, CSIRO, Australia
Nazmul Kazi, Montana State University, USA
Won Gyu Kim, US National Library of Medicine
Ari Klein, University of Pennsylvania, USA
Roman Klinger, University of Stuttgart, Germany
Andre Lamurias, University of Lisbon, Portugal
Majid Latifi, National College of Ireland
Alberto Lavelli, FBK-ICT, Italy
Robert Leaman, US National Library of Medicine
Lung-Hao Lee, National Central University, Taiwan
Ulf Leser, Humboldt-Universität zu Berlin, Germany

Diwakar Mahajan, IBM Thomas J. Watson Research Center, USA
Mark-Christoph Müller, Heidelberg Institute for Theoretical Studies, Germany
Claire Nedellec, INRA, France
Guenter Neumann, DFKI, Saarland, Germany
Aurelie Neveol, LIMSI - CNRS, France
Mariana Neves, German Federal Institute for Risk Assessment, Germany
Yifan Peng, US National Library of Medicine
Francisco J. Ribadas-Pena, University of Vigo, Spain
Anthony Rios, The University of Texas at San Antonio, USA
Angus Roberts, The University of Sheffield, UK
Kirk Roberts, The University of Texas Health Science Center at Houston, USA
Roland Roller, DFKI GmbH, Berlin, Germany
Mourad Sarrouti, Sumitovant Biopharma, Inc., USA
Mario Sänger, Humboldt-Universität zu Berlin, Germany
Diana Sousa, University of Lisbon, Portugal
Michael Spranger, Sony, Tokyo, Japan
Peng Su, University of Delaware, USA
Madhumita Sushil, University of California, San Francisco, USA
Karin Verspoor, The University of Melbourne, Australia
Roger Wattenhofer, ETH Zurich, Switzerland
Leon Weber, Humboldt Universität Berlin, Germany
Nathan M. White, James Cook University, Australia
Davy Weissenbacher, University of Pennsylvania, USA
W John Wilbur, US National Library of Medicine
Amelie Wührl, University of Stuttgart, Germany
Dongfang Xu, Harvard University, USA
Shweta Yadav, University of Illinois Chicago, USA
Jingqing Zhang, Imperial College London, UK
Ayah Zirikly, Johns Hopkins University, USA
Pierre Zweigenbaum, LIMSI - CNRS, France

Additional Reviewers:

Lifang Li, King's College London, UK
Ling Luo, National Library of Medicine, USA
Faith Wavinya Mutinda, University of Tokyo, Japan
Peter Belcak, ETH Zurich, Switzerland
Po-ting Lai, National Library of Medicine, USA
Beni Egressy, ETH Zurich, Switzerland
Zhao Meng, ETH Zurich, Switzerland

Table of Contents

Workshop Program

May 26, 2022

9:00–9:10 **Opening Remarks**

9:10–10:30 **Session 1: Question Answering, Discourse Structure and Clinical Applications**

9:10–9:30 *Explainable Assessment of Healthcare Articles with QA*
Alodie Boissonnet, Marzieh Saeidi, Vassilis Plachouras and Andreas Vlachos

9:30–9:50 *A sequence-to-sequence approach for document-level relation extraction*
John Giorgi, Gary Bader and Bo Wang

9:50–10:10 *Position-based Prompting for Health Outcome Generation*
Micheal Abaho, Danushka Bollegala, Paula Williamson and Susanna Dodd

10:10–10:30 *How You Say It Matters: Measuring the Impact of Verbal Disfluency Tags on Automated Dementia Detection*
Shahla Farzana, Ashwin Deshpande and Natalie Parde

10:30–11:00 **Coffee Break**

11:00–12:30 **Poster Session 1**

Zero-Shot Aspect-Based Scientific Document Summarization using Self-Supervised Pre-training
Amir Soleimani, Vassilina Nikoulina, Benoit Favre and Salah Ait Mokhtar

Data Augmentation for Biomedical Factoid Question Answering
Dimitris Pappas, Prodromos Malakasiotis and Ion Androutsopoulos

Slot Filling for Biomedical Information Extraction
Yannis Papanikolaou, Marlene Staib, Justin Joshua Grace and Francine Bennett

Automatic Biomedical Term Clustering by Learning Fine-grained Term Representations
Sihang Zeng, Zheng Yuan and Sheng Yu

May 26, 2022 (continued)

May 26, 2022 (continued)

VPAI_Lab at MedVidQA 2022: A Two-Stage Cross-modal Fusion Method for Medical Instructional Video Classification
Bin Li, Yixuan Weng, Fei Xia, Bin Sun and Shutao Li

12:30–14:00 Lunch Break

14:00–15:00 Session 2: Summarization and text mining

GenCompareSum: a hybrid unsupervised summarization method using salience
Jennifer Bishop, Qianqian Xie and Sophia Ananiadou

BioCite: A Deep Learning-based Citation Linkage Framework for Biomedical Research Articles
Sudipta Singha Roy and Robert E. Mercer

Low Resource Causal Event Detection from Biomedical Literature
Zhengzhong Liang, Enrique Noriega-Atala, Clayton Morrison and Mihai Surdeanu

15:00–15:30 Coffee Break

15:30–17:00 Poster Session 2

Overview of the MedVidQA 2022 Shared Task on Medical Video Question-Answering
Deepak Gupta and Dina Demner-Fushman

Inter-annotator agreement is not the ceiling of machine learning performance: Evidence from a comprehensive set of simulations
Russell Richie, Sachin Grover and Fuchiang (Rich) Tsui

Conversational Bots for Psychotherapy: A Study of Generative Transformer Models Using Domain-specific Dialogues
Avisha Das, Salih Selek, Alia R. Warner, Xu Zuo, Yan Hu, Vipina Kuttichi Keloth, Jianfu Li, W. Jim Zheng and Hua Xu

BEEDS: Large-Scale Biomedical Event Extraction using Distant Supervision and Question Answering
Xing David Wang, Ulf Leser and Leon Weber

May 26, 2022 (continued)

May 26, 2022 (continued)

Model Distillation for Faithful Explanations of Medical Code Predictions
Zach Wood-Doughty, Isabel Cachola and Mark Dredze

Towards Generalizable Methods for Automating Risk Score Calculation
Jennifer J Liang, Eric Lehman, Ananya Iyengar, Diwakar Mahajan, Preethi Raghavan, Cindy Y. Chang and Peter Szolovits

DoSSIER at MedVidQA 2022: Text-based Approaches to Medical Video Answer Localization Problem
Wojciech Kusa, Georgios Peikos, Óscar Espitia, Allan Hanbury and Gabriella Pasi

17:00–17:10 Closing Remarks

Explainable Assessment of Healthcare Articles with QA

Alodie Boissonnet[1], **Marzieh Saeidi**[2], **Vassilis Plachouras**[2], **Andreas Vlachos**[1,2]
[1]Department of Computer Science, University of Cambridge
[2]Facebook AI, London
{avmb2,av308}@cam.ac.uk, {marzieh,vplachouras,avlachos}@fb.com

Abstract

The healthcare domain suffers from the spread of poor quality articles on the Internet. While manual efforts exist to check the quality of online healthcare articles, they are not sufficient to assess all those in circulation. Such quality assessment can be automated as a text classification task, however, explanations for the labels are necessary for the users to trust the model predictions. While current explainable systems tackle explanation generation as summarization, we propose a new approach based on question answering (QA) that allows us to generate explanations for multiple criteria using a single model. We show that this QA-based approach is competitive with the current state-of-the-art, and complements summarization-based models for explainable quality assessment. We also introduce a human evaluation protocol more appropriate than automatic metrics for the evaluation of explanation generation models.

1 Introduction

The Internet has become an important source of medical advice. According to Rutten et al. (2019), in 2017, 74.4% of the US population first looked for health-related information on the internet, while only 13.3% of the population first asked a physician or healthcare provider. However, poor quality reporting, including misinformation, cherry-picking, exaggerations, etc., is often present online and can be a severe threat to public health. Recent events, such as the Covid-19 pandemic, demonstrate the necessity of developing quality assessment systems for healthcare reports to limit these harms. Fortunately, websites such as HealthNewsReview[1] critically analyze medical articles to identify poor quality reporting and improve the public discourse about healthcare. However, the manual review of medical news is a time-consuming task that would benefit from automated systems to scale up to the volumes needed in today's media ecosystem.

Story #1511
Criterion 1: Does the article adequately discuss the costs of the intervention?
Answer: Not Satisfactory
Explanation: There was no discussion of cost as there was in the competing AP story.

Criterion 2: Does the article adequately quantify the benefits of the treatment/test/product/procedure?
Answer: Satisfactory
Explanation: The story adequately quantified the benefits seen in the study that led to FDA approval.

Criterion 3: ...

Table 1: Example of an article evaluated by the HealthNewsReview website. Each article is evaluated according to ten criteria (two shown) and explanations are given to support the answers.

Assessing the quality of news articles has been the focus of numerous studies that tackle it as a text classification task (Louis and Nenkova, 2013; Chakraborty et al., 2016; Kryscinski et al., 2020). However, explanations for the predictions only recently started receiving attention, despite being necessary to convince the readers of such assessments. For instance, Dai et al. (2020) have built on the evaluation work conducted by the HealthNewsReview website (see Table 1) to automate article quality assessment in healthcare, but have only focused on articles classification, without providing explanations. Likewise, Wright and Augenstein (2021) have also studied exaggeration detection in healthcare as classification, but without explanations.

Beyond quality assessment, previous works have formulated textual explanation generation for classification as summarization (Atanasova et al., 2020; Kotonya and Toni, 2020). However such approaches suffer from a number of shortcomings when applied to the assessment of an article based on multiple criteria. As these approaches always

[1]https://www.healthnewsreview.org

Proceedings of the BioNLP 2022 workshop, Dublin, Ireland, pages 1–9
May 26, 2022.

output a single summary for a given input text, separate models must be trained to generate explanations for each classification label and evaluation criterion (e.g. reliability of sources, lack of information, etc.), as in the example given in Table 1. This considerably reduces the number of available training instances per model, because gold explanations of only one criterion at a time can be used for training, and it also requires developing and maintaining a model per criterion. Summarization-based models are also not appropriate to return an explanation for a label that is justified by the lack of information in the text (see criterion 1 in Table 1).

In this work, we develop an explainable quality assessment system for health news reports, and we evaluate it on the *FakeHealth* corpus (Dai et al., 2020). It differs from previous work as its explanation generation model is based on question answering (QA), which takes into consideration the definition of each evaluation criterion in the form of a question (see Table 1). This approach addresses the limitations of summarisation-based systems: it benefits from a larger training dataset consisting of instances from all criteria and labels at once, can better generate explanations regarding the absence of information, and requires training and maintaining a single model for all criteria.

We compare our approach for explanation generation against a summarization-based system inspired from Kotonya and Toni (2020). Our results show that both approaches are complementary and perform better in different cases. More specifically, summarization-based systems are more appropriate when relevant information is explicitly given in articles, while QA-based systems perform better when relevant information is missing.

Finally, evaluating generated explanations is not an easy task as we should consider both the structure and the sense of texts. Previous works used automatic metrics for the evaluation of explanations, which are known to be insufficient for abstractive text generation. Mani (2002) precisely insisted that assessing the readability and quality of a generated text requires human annotators as no automatic metric can achieve good performance on this task. Likewise, Kryscinski et al. (2019) have recently highlighted that automatic evaluation protocols, usually relying on ROUGE scores, correlate weakly with human judgement and fail to evaluate critical features, such as factual consistency. For this reason, we propose a new human evaluation protocol to assess the fluency, consistency, and factual correctness of the explanations, and we show that automatic metrics are not appropriate for this task.

2 Methodology

Our system starts with classifying articles according to ten evaluation criteria, then generates explanations using QA, taking into account the predicted classification labels. The purpose of the text classification step is to determine whether an article is satisfactory with respect to different evaluation criteria. We consider different options from the literature: logistic regression for its simplicity, BERT-based classification which is commonplace but truncates texts to 512 tokens, and a Longformer-based encoder model (Beltagy et al., 2020), which is able to deal with long input texts like those of our study. Both BERT and Longformer-based classifiers are pre-trained for a large classification task on a biomedical dataset, *PubMed*[2], then fine-tuned on the *FakeHealth* dataset. In line with Beltagy et al. (2020)'s recommendation, we use a classification objective for Longformer classifier, that places a global attention mask on a `[CLS]` token. This token aggregates the representation of the whole text at the beginning of the input text as shown in Table 6 in Appendix C.1, that gives an example of the encoding of input texts and shows the global attention mask of our model. Readers should refer to Beltagy et al. (2020) for further details about attention masks of Longformer models.

The second stage of the pipeline generates abstractive explanations for the previously predicted classes. As the QA approach takes into account the classes and the questions posed by criteria, we only need to train a single model, handling all criteria and classes. Following Soni and Roberts (2020), we have chosen to work with a Longformer-based encoder-decoder that we first train on the open-domain dataset *SQuAD v2.0* (Rajpurkar et al., 2018), and then fine-tune on *FakeHealth*. Because gold explanations in the *FakeHealth* dataset are abstractive, our model learns to write complete explanations despite the pre-training step on *SQuAD* whose explanations are spans of phrases. Even though we always use the same ten questions (shown in Table 9 in Appendix C.2) for fine-tuning

[2] `https://deepai.org/dataset/pubmed`

and evaluation, this approach differs from query-focused summarization because of its ability to generate explanations for information missing from the article which a summarization system cannot handle. We use the QA objective introduced by Beltagy et al. (2020) for Longformer that places a global attention mask on all question tokens (see Table 6 in Appendix C.1), and we feed our model with the article, the criterion, and the class prediction. During training, we use the gold classes of articles to generate explanations, as generating post-hoc explanations for incorrectly predicted labels would not be meaningful.

Following recent previous work on explainable fact-checking in healthcare by Kotonya and Toni (2020), we implement a baseline for the explanation generation task, based on summarization. Because such a system does not take into account the criteria definitions in its input, it cannot combine all criteria together as it would always produce the same explanation for all criteria. Therefore, this approach requires training independent models for each class within a criterion, which results in 30 models (10 criteria × 3 classes) in the case of the *FakeHealth* dataset. We use here a summarization objective for the Longformer model, that applies a global attention mask to the very first token of input texts (see Table 6 in Appendix C.1 and Beltagy et al. (2020)).

3 Human evaluation of explanations

Unlike previous works that assess generated text with automatic metrics, we design a human evaluation protocol that assesses four aspects of explanations: their fluency, consistency, factual correctness, and whether they are indicative of the label that they are supposed to explain. An explanation is considered fluent if it sounds natural, and consistent if it does not contradict itself, include repetitions, or information that is not mentioned in the article. The factual correctness criterion looks for incorrect facts, contradictions with respect to the article, or hallucinations. Finally, generated explanations should allow a human judge to infer correctly the label they are meant to explain.

We conducted two pilot studies in order to assess the quality of our guidelines. As reported in Table 2, Pilot 1 brought to light the ambiguity of the initial version of the guidelines, while Pilot 2 reached higher inter-annotator agreement scores. The new version of the guidelines is more detailed than the first one and provides some examples of what is expected. For instance, instead of asking if an explanation is fluent, the new guidelines specify that explanations should be rated as fluent if they sound natural and their syntactic structure is correct. Thus, the sentence "it's sunny but it's sunny" should not be considered as fluent, while "it's sunny but it's not sunny" should be considered fluent despite the contradiction, which is judged negatively under consistency.

	Fluency	**Factual correctness**	**Guessed class**
Pilot 1	-0.12	0.29	0.76
Pilot 2	0.46	0.49	0.58

Table 2: Inter-annotator agreement scores (averages of Cohen Kappa scores) of the two pilot studies.

Criterion	**Not S.**	**S.**	**Not A.**
1	1431	495	370
2	1505	768	**23**
3	1413	717	**166**
4	1445	848	**3**
5	**286**	1921	**89**
6	1135	1147	**14**
7	1120	1063	**113**
8	538	1457	301
9	672	1543	**81**
10	391	1771	**134**

Table 3: Distribution of articles in each class per criterion. These numbers combine both the *HealthRelease* and *HealthStory* datasets.

The final guidelines used for the evaluation in Section 5 are fully detailed in Appendix B. In Table 2, the consistency criterion is missing as it was added after Pilot 2.

4 Data

We evaluate our QA and summarization-based models on the *FakeHealth* corpus of health news articles, released by Dai et al. (2020). Each article in the dataset was evaluated by at least two experts, according to ten criteria that assess diverse aspects such as "the overclaiming, missing of information, reliability of sources and conflict of interests" (Dai et al., 2020). Dai et al. (2020) found zero to a minor positive correlation between the criteria, which justifies the relevance of all of them. These criteria are reported in Table 9 in Appendix C.2.

For each criterion, articles are annotated with one of three labels, *Not Satisfactory*, *Satisfactory*, and *Not Applicable*, and a textual explanation that justifies the assigned label, as shown in Table 1. The label distribution across criteria is highly unbalanced, *Not Applicable* instances being the rarest. For example, criteria 2, 4, and 6 have at least 65 times more *Not Satisfactory* instances than *Not Applicable* ones (see Table 3).

	Longformer	BERT	LogReg	From gen. expl.
Criterion 1	**0.67**	0.63	0.59	0.61
Criterion 2	**0.43**	0.42	0.40	0.30
Criterion 3	0.52	**0.55**	0.46	0.45
Criterion 4	0.40	0.42	0.36	**0.61**
Criterion 5	0.35	0.30	**0.37**	0.33
Criterion 6	0.42	0.39	0.37	**0.60**
Criterion 7	0.35	0.37	0.36	**0.40**
Criterion 8	0.57	**0.59**	0.49	0.46
Criterion 9	**0.40**	0.37	0.37	0.34
Criterion 10	**0.45**	**0.45**	0.36	0.24
Mean	**0.46**	0.44	0.41	0.43

Table 4: Macro F_1-scores of our different classifiers for each criterion. The last row *Mean* gives the average performance of each model across criteria. The column *From gen. expl.* corresponds to the classification task conducted from generated explanations, as described in Section 5.3.

5 Results

5.1 Quality assessment per criterion

We compare Longformer-based, BERT-based, and Logistic Regression models for the quality of the classification task via their macro F_1-scores for each criterion. Table 4 shows that our Longformer-based models perform the best due to their ability to encode longer texts. The Logistic Regression models also achieves great performance despite its simplicity, but this must be qualified as classes are highly unbalanced and Logistic Regression mostly predicts the dominant class. An analysis broken down by criterion also highlights that all models perform unevenly across criteria. This suggests that some criteria are harder to handle, notably, those requiring external knowledge or subjective judgment (e.g. criterion 5 asking whether articles commit disease-mongering).

We also tried to build a single Longformer-based model handling all classes at once using a QA-based approach that treats criteria as questions and predicts labels, but it performed poorly. We suspect that we have poor results because we perform a classification task with a QA-based model.

5.2 Explanation generation

Table 5 reports the overall performance of both summarization and QA-based approaches for the explanation generation task only. These results show that the QA-based approach performs better than, or as well as, the baseline system. Both approaches achieve similar performance in terms of consistency and factual correctness, but the QA approach produces explanations that are more fluent and that indicate the correct label more often. Table 7 in Appendix C.2 provides some examples of the generated explanations. In these tables, gold explanations correspond to the explanations written by health expert in the *FakeHealth* dataset.

An analysis per class (see Table 5) reveals that the summarization approach performs better for the *Satisfactory* class, while the QA approach performs better for the *Not Satisfactory* and *Not Applicable* classes. This can be explained by the fact that *Satisfactory* articles include the relevant information to the criteria and require models to reuse this information to generate explanations, thus resembling summarization. On the other hand, for the *Not Satisfactory* class, models need to point out missing information and this is naturally harder for a summarization model, but easier for a QA-based one that can generate text about missing information. Finally, the *Not Applicable* class suffers mainly from having very few instances for training (see Table 3). With a single model, the QA approach is able to overcome this issue and generate better explanations.

To achieve the best performance, the previous results suggest combining both systems and using the summarization-based system for *Satisfactory* instances, and the QA-based system for all others. With this combination, 81% of explanations are fluent, 76% consistent, 57% factually correct, and 85% indicate correct labels. The pretty low factual correctness of explanations can be explained by the severeness of guidelines that ask annotators to rate an explanation as factually incorrect as soon as at least one detail is incorrect, regardless of the correctness of all other details.

5.3 Predicting classes from generated explanations

To further test our methodology, we run an experiment in which we first generate explanations,

	Fluency		Consistency		Factual correctness		Correct class		Count
	Sum.	QA	Sum.	QA	Sum.	QA	Sum.	QA	
All classes	74.5	**80**	72.5	72.5	52.5	52	85	**86**	-
Not S.	73.2	**83.5**	67	**73.2**	42.3	**48.5**	87.6	**89.7**	97
S.	**79.6**	76.3	**80.6**	73.1	**63.4**	53.8	**86**	82.8	93
Not A.	40	**80**	50	**60**	50	**70**	50	**80**	10

Table 5: Results of the evaluation of the summarization and QA-based systems per class (as percentages).

and then classify articles from the predicted explanations. We use the same approach as before, i.e. a Longformer-based model with a QA objective fine-tuned on *FakeHealth* articles for explanation generation, and a Longformer-based classifier fine-tuned on predicted explanations. Results are reported in Table 4 and show that classifying articles before generating explanations, achieves better performance. This finding is not surprising as the explanation generation model is influenced by dominant classes and ignores minority classes. Wrong explanations propagate then to the classification task and are responsible for incorrect labels. However, the classification model built from generated explanations performs very well for criteria 4 and 6. Yet, these results should be considered with caution, as classes for these criteria are highly unbalanced (with respectively 3 and 14 instances in the *Not Applicable* class) and the model predicts most of the time the majority class. This ablation study corroborates the recommendations of Kotonya and Toni (2020) and Mani (2002).

5.4 Automatic v. human evaluation

Finally, we investigate the correlation between human judgement and automatic metrics used in previous works (Ermakova et al., 2019), including ROUGE (Lin, 2004) and BLEU (Papineni et al., 2002) scores. Table 8 in Appendix B.3 reports the correlation coefficients between all metrics. Using Kendall's Tau, we find that all these correlations are very low, at most 0.11 with ROUGE scores and 0.07 with the BLEU score. This finding was expected as most of the automatic metrics focus on word overlap, which makes it difficult to check the grammatical and syntactic correctness of explanations, as well as their factual consistency. This conclusion echoes Kryscinski et al. (2019)'s work on automatic evaluation protocols.

6 Conclusion and discussion

In this work, we propose a new QA-based approach to generate explanations for quality assessment systems. This approach allows us to build a single model, able to generate explanations for different criteria and classes, by taking into account the questions related to criteria. We have shown that the QA-based system is competitive with the summarization-based one, and that they are complementary. Notably, the QA-based approach is more appropriate when the relevant information is not explicitly given in articles or for small classes. As for the classification task, Longformer-based models perform best thanks to their ability to deal with long input texts. Finally, we have highlighted that automatic metrics, such as ROUGE, correlate very weakly with human judgment when it comes to evaluating explanation generation models. This paper could serve as a starting point to explore the use of QA models for explainable article assessment.

Acknowledgements

Andreas Vlachos's work at the University of Cambridge is supported by by the ERC grant AVeriTeC (GA 865958) and the EU H2020 grant MONITIO (GA 965576).

References

Pepa Atanasova, Jakob Grue Simonsen, Christina Lioma, and Isabelle Augenstein. 2020. Generating fact checking explanations.

Iz Beltagy, Matthew E. Peters, and Arman Cohan. 2020. Longformer: The long-document transformer. *arXiv:2004.05150*.

Abhijnan Chakraborty, Bhargavi Paranjape, Sourya Kakarla, and Niloy Ganguly. 2016. Stop clickbait: Detecting and preventing clickbaits in online news media. In *2016 IEEE/ACM International Conference on Advances in Social Networks Analysis and Mining (ASONAM)*, pages 9–16.

Enyan Dai, Yiwei Sun, and Suhang Wang. 2020. Ginger cannot cure cancer: Battling fake health news with a comprehensive data repository. *CoRR*, abs/2002.00837.

Liana Ermakova, Jean Valère Cossu, and Josiane Mothe. 2019. A survey on evaluation of summarization methods. *Information Processing Management*, 56(5):1794–1814.

Neema Kotonya and Francesca Toni. 2020. Explainable automated fact-checking for public health claims.

Wojciech Kryscinski, Nitish Shirish Keskar, Bryan McCann, Caiming Xiong, and Richard Socher. 2019. Neural text summarization: A critical evaluation. In *Proceedings of the 2019 Conference on Empirical Methods in Natural Language Processing and the 9th International Joint Conference on Natural Language Processing (EMNLP-IJCNLP)*, pages 540–551, Hong Kong, China. Association for Computational Linguistics.

Wojciech Kryscinski, Bryan McCann, Caiming Xiong, and Richard Socher. 2020. Evaluating the factual consistency of abstractive text summarization. In *Proceedings of the 2020 Conference on Empirical Methods in Natural Language Processing (EMNLP)*, pages 9332–9346, Online. Association for Computational Linguistics.

Chin-Yew Lin. 2004. ROUGE: A package for automatic evaluation of summaries. In *Text Summarization Branches Out*, pages 74–81, Barcelona, Spain. Association for Computational Linguistics.

Annie Louis and Ani Nenkova. 2013. What Makes Writing Great? First Experiments on Article Quality Prediction in the Science Journalism Domain. *Transactions of the Association for Computational Linguistics*, 1:341–352.

Inderjeet Mani. 2002. Summarization evaluation: An overview.

Kishore Papineni, Salim Roukos, Todd Ward, and Wei-Jing Zhu. 2002. Bleu: a method for automatic evaluation of machine translation. In *Proceedings of the 40th Annual Meeting of the Association for Computational Linguistics*, pages 311–318, Philadelphia, Pennsylvania, USA. Association for Computational Linguistics.

Pranav Rajpurkar, Robin Jia, and Percy Liang. 2018. Know what you don't know: Unanswerable questions for SQuAD. In *Proceedings of the 56th Annual Meeting of the Association for Computational Linguistics (Volume 2: Short Papers)*, pages 784–789, Melbourne, Australia. Association for Computational Linguistics.

Lila J. Finney Rutten, Kelly D. Blake, Alexandra J. Greenberg-Worisek, Summer V. Allen, Richard P. Moser, and Bradford W. Hesse. 2019. Online health information seeking among us adults: Measuring progress toward a healthy people 2020 objective. *Public Health Reports*, 134(6):617–625. PMID: 31513756.

Sarvesh Soni and Kirk Roberts. 2020. Evaluation of dataset selection for pre-training and fine-tuning transformer language models for clinical question answering. In *Proceedings of the 12th Language Resources and Evaluation Conference*, pages 5532–5538, Marseille, France. European Language Resources Association.

Dustin Wright and Isabelle Augenstein. 2021. Semi-supervised exaggeration detection of health science press releases.

A Ethical concerns

The ethical concerns of this work are two-fold. First, readers must be aware that such a deep learning model is prone to make mistakes, as evidenced by the results of the experiments we did (see Section 5). Outputs should be treated as an indication or recommendation, rather than the ground truth.

Secondly, our QA-based approach needs to train a single model, by comparison with the summarization-based one that requires 30 models. Having a single model reduces the pressure on computing resources and consequently, on the environment. It also makes the model easier to maintain.

B Human evaluation

B.1 Definition of the evaluation guidelines

To design our human evaluation protocol, we conducted two pilot studies with the same two annotators. To begin with, the first study gathered three annotators who evaluated all explanations generated for the same six articles (three releases and three stories, which results in 60 explanations in total) with the baseline system for explanation generation. They were asked to determine if explanations were written in fluent English, consistent, factually correct, and which classes were suggested by explanations. This evaluation task combined both intrinsic and extrinsic methods to have a complete overview of models' performance, and we assessed to what extent annotators agreed on the evaluation task by looking at inter-annotator agreement scores computed with the Cohen Kappa score. It resulted in a high disagreement among annotators (see Table 2): annotators 1 and 2 even seemed to disagree on the fluency criterion. An in-depth exploration of their annotations revealed that they never agreed when one of them judged that an explanation was not fluent. These low inter-annotator agreement scores seem therefore to be caused by unclear guidelines.

For this reason, more detailed guidelines about the fluency and factual correctness of explanations were defined, and another pilot study was intended to validate them. It gathered two of the three previous annotators, who evaluated all explanations generated for the same five articles (two releases and three stories) with whether the baseline or the QA-based system. We reduced the number of articles to evaluate as evaluation tasks are time-consuming and five articles, resulting in 50 explanations, are enough to validate guidelines. This second evaluation task achieved a much higher inter-annotator agreement reported in Table 2 and confirmed the new evaluation guidelines. However, the agreement score for the guessed classes slightly decreased between the first and second evaluation task. An analysis of annotations highlighted that some criteria could be ambiguous. For example, criterion 5 wonders if articles commit disease-mongering, and if they do, they should be rated as *Not Satisfactory* because it implies that they are less reliable. Consequently, a detailed description of each criterion, extracted from HealthNewsReview's website, has been given to annotators for the last evaluation task to raise all ambiguities.

B.2 Final guidelines

Based on the outcome of the pilot studies, annotators were given the following guidelines:

- Fluency: Is the generated explanation written in fluent English? An explanation should be considered non-fluent if it does not sound natural or its structure is not correct (e.g. paragraphs title). Words case (uppercase or lowercase) should not be taken into account. For example, "it's sunny but it's sunny" should be considered as non-fluent, but "it's sunny but it's not sunny" should be considered as fluent. Likewise, "intro: it's sunny, results: it's sunny, conclusion: it's sunny" should be considered as non-fluent (inappropriate structure).

- Consistency: Is the generated explanation consistent? An explanation should be considered inconsistent if it includes contradiction, repetition, extra information. For example, "it's sunny but it's sunny" should be considered as consistent, but "it's sunny but it's not sunny" should be considered as non-consistent.

- Factual correctness: Are the details (numbers, names, facts, etc.) included in the generated explanation correct? Explanations that contain incorrect facts, contradictions, or hallucinations should be evaluated as not satisfactory; but whether or not the factual details are related to the question should not be taken into consideration.

- Suggested class: According to the generated explanation, how would you classify the article? (*Not Satisfactory*, *Satisfactory*, *Not Applicable*, *Can't tell*) A *Can't tell* class has been added if generated explanations do not help classify articles. A description of what was expected for each criterion was given to annotators to raise all ambiguities. It was taken from the HealthNewsReview website from which explanations had been extracted. The inferred classes are considered correct if it matches the gold classes of articles.

The consistency criterion has been added after the two pilot studies, so we have not evaluated the inter-annotator agreement for it. However, the corresponding guidelines have been defined and detailed similarly to the other evaluation criteria to raise any ambiguity for annotators.

For the real evaluation task, annotators have evaluated ten different articles each. They were the same annotators as for pilot studies, so their inter-annotator agreement was high and we were able to evaluate more articles with great confidence in annotations.

B.3 Correlation with automatic metrics

Table 8 reports the correlation scores between human judgement and automatic metrics used in previous works (Ermakova et al., 2019), including ROUGE (Lin, 2004) and BLEU (Papineni et al., 2002) scores. Using Kendall's Tau, we find that all these correlations are very low, at most 0.11 with ROUGE scores and 0.07 with the BLEU score.

C Model

C.1 Model's Attention

For the Longformer model, Beltagy et al. (2020) defines different global attention masks according to the task to complete. For classification, the `[CLS]` token of input texts receives global attention. For a QA task, the global attention mask is applied to all question tokens, while it is applied to the very first token of input texts for a summarization task. Table 6 illustrates these different attention masks.

C.2 Example of models' outputs

Question-Answering objective
<s> Does the story adequately discuss the costs of the intervention? </s> Satisfactory </s> Triumph for Drug to Straighten Clenched Fingers </s> About one in 20 men is estimated to have Peyronie's, but figures are not precise because people with the condition tend not to discuss it publicly. ... </s>

Summarization objective
<s> Triumph for Drug to Straighten Clenched Fingers </s> About one in 20 men is estimated to have Peyronie's, but figures are not precise because people with the condition tend not to discuss it publicly. ... </s>

Classification objective
[CLS] Triumph for Drug to Straighten Clenched Fingers </s> About one in 20 men is estimated to have Peyronie's, but figures are not precise because people with the condition tend not to discuss it publicly. ... </s>

Table 6: Example of the inputs and global attention masks of the Longformer model for the different training objectives for HealthStory 1613, Criterion 1. The global attention is put on the tokens highlighted in grey.

HealthRelease #528, Criterion 5, Satisfactory
Gold explanation
There is no disease mongering in this release.

Generated explanation – QA-based system
There is no disease mongering in this news release. Obesity is a serious health issue, and the release provides some context about the prevalence of obesity in the US.
Rated as fluent, consistent, factually incorrect, and suggesting a *Satisfactory* label

Generated explanation – baseline system
There is no disease mongering in the news release. The release provides some context about the prevalence and severity of obesity and its consequences.
Rated as fluent, consistent, factually incorrect, and suggesting a *Satisfactory* label

HealthStory #1619, Criterion 1, Satisfactory
Gold explanation
The story notes that a 5.29-ounce tube of the product retails for $39.99 at drugstores.

Generated explanation – QA-based system
The story notes that the cream is available at Sephora.com for $69.99. The story could have done a better job of comparing the cost of the cream to other products on the market.
Rated as fluent, consistent, factually incorrect, and suggesting a *Satisfactory* label

Generated explanation – baseline system
The story states that the cream "takes the redness out of new stretch marks." That's good enough for the story, but it would have been better if the story would have mentioned the cost.
Rated as fluent, consistent, factually correct, and suggesting a *Not Satisfactory* label

Table 7: Examples of explanations generated with our QA-based and summarization-based baseline systems.

	Fluency	**Consistency**	**Factual correctness**	**Correct class**	**Can't tell class**
ROUGE-1 Precision	0.09	0.04	0.03	0.08	-0.07
ROUGE-1 Recall	-0.02	-0.08	-0.05	-0.04	-0.00
ROUGE-1 F1	0.01	-0.05	-0.01	0.00	-0.04
ROUGE-2 Precision	0.08	0.05	0.04	0.09	**-0.11**
ROUGE-2 Recall	0.04	-0.02	-0.01	0.04	-0.09
ROUGE-2 F1	0.06	0.01	0.01	0.07	**-0.11**
ROUGE-L Precision	0.10	0.08	0.05	0.09	-0.09
ROUGE-L Recall	0.01	-0.04	-0.03	-0.01	-0.03
ROUGE-L F1	0.06	0.03	0.02	0.06	-0.08
BLEU	-0.01	-0.07	-0.04	-0.01	-0.03
Length ratio	0.09	0.08	0.05	0.08	-0.06
Cosine similarity	0.08	-0.01	0.03	0.06	-0.05
Euclidean distance	-0.04	0.01	-0.04	-0.02	0.03

Table 8: Correlation between human and automatic evaluation metrics (Kendall Tau correlation coefficient).

Criterion	**Question**
Criterion 1	Does it adequately discuss the costs of the intervention?
Criterion 2	Does it adequately quantify the benefits of the treatment/test/product/procedure?
Criterion 3	Does it adequately explain/quantify the harms of the intervention?
Criterion 4	Does it seem to grasp the quality of the evidence?
Criterion 5	Does it commit disease-mongering?
Criterion 6	Does the story use independent sources and identify conflicts of interest? / Does the news release identify funding sources & disclose conflicts of interest?
Criterion 7	Does it compare the new approach with existing alternatives?
Criterion 8	Does it establish the availability of the treatment/test/product/procedure?
Criterion 9	Does it establish the true novelty of the approach?
Criterion 10	Does the story appear to rely solely or largely on a news release? / Does the news release include unjustifiable, sensational language, including in the quotes of researchers?

Table 9: Datasets' criteria.

A sequence-to-sequence approach for document-level relation extraction

John Giorgi[1,4,5,✉] **Gary D. Bader**[1,2,4,6,7,†] **Bo Wang**[1,3,5,8,†]
[1]Department of Computer Science, University of Toronto
[2]Department of Molecular Genetics, University of Toronto
[3]Department of Laboratory Medicine and Pathobiology, University of Toronto
[4]Terrence Donnelly Centre for Cellular & Biomolecular Research
[5]Vector Institute for Artificial Intelligence
[6]The Lunenfeld-Tanenbaum Research Institute, Sinai Health System
[7]Princess Margaret Cancer Centre, University Health Network
[8]Peter Munk Cardiac Center, University Health Network
[✉]Corresponding author [†]Equal contribution
{john.giorgi, gary.bader}@mail.utoronto.ca
bowang@vectorinstitute.ai

Abstract

Motivated by the fact that many relations cross the sentence boundary, there has been increasing interest in document-level relation extraction (DocRE). DocRE requires integrating information within and across sentences, capturing complex interactions between mentions of entities. Most existing methods are pipeline-based, requiring entities as input. However, jointly learning to extract entities and relations can improve performance and be more efficient due to shared parameters and training steps. In this paper, we develop a sequence-to-sequence approach, seq2rel, that can learn the subtasks of DocRE (entity extraction, coreference resolution and relation extraction) end-to-end, replacing a pipeline of task-specific components. Using a simple strategy we call entity hinting, we compare our approach to existing pipeline-based methods on several popular biomedical datasets, in some cases exceeding their performance. We also report the first end-to-end results on these datasets for future comparison. Finally, we demonstrate that, under our model, an end-to-end approach outperforms a pipeline-based approach. Our code, data and trained models are available at `https://github.com/johngiorgi/seq2rel`. An online demo is available at `https://share.streamlit.io/johngiorgi/seq2rel/main/demo.py`.

1 Introduction

PubMed, the largest repository of biomedical literature, contains over 30 million publications and is adding more than two papers per minute. Accurate, automated text mining and natural language processing (NLP) methods are needed to maximize discovery and extract structured information from this massive volume of text. An important step in this process is relation extraction (RE), the task of identifying groups of entities within some text that participate in a semantic relationship. In the domain of biomedicine, relations of interest include chemical-induced disease, protein-protein interactions, and gene-disease associations.

Many methods have been proposed for RE, ranging from rule-based to machine learning-based (Zhou et al., 2014; Liu et al., 2016). Most of this work has focused on *intra*-sentence binary RE, where pairs of entities within a sentence are classified as belonging to a particular relation (or none). These methods often ignore commonly occurring complexities like nested or discontinuous entities, coreferent mentions (words or phrases in the text that refer to the same entity), inter-sentence and n-ary relations (see Figure 1 for examples). The decision not to model these phenomena is a strong assumption. In GENIA (Kim et al., 2003), a corpus of PubMed articles labelled with around 100,000 biomedical entities, ~17% of all entities are nested within another entity. Discontinuous entities are particularly common in clinical text, where ~10% of mentions in popular benchmark corpora are discontinuous (Wang et al., 2021). In the CDR corpus (Li et al., 2016b), which comprises 1500 PubMed articles annotated for chemical-induced disease relations, ~30% of all relations are inter-sentence. Some relations, like drug-gene-mutation interactions, are difficult to model with binary RE (Zhou et al., 2014).

In response to some of these shortcomings, there has been a growing interest in *document*-level RE (DocRE). DocRE aims to model *inter*-sentence re-

Proceedings of the BioNLP 2022 workshop, Dublin, Ireland, pages 10–25
May 26, 2022. ©2022 Association for Computational Linguistics

Figure 1: Examples of complexities in entity and relation extraction and the proposed linearization schema to model them. CID: chemical-induced disease. GDA: gene-disease association. DGM: drug-gene-mutation.

Complexities	Example	Comment
Discontinuous mentions	Induction by paracetamol of bladder and liver tumours.	Discontinuous mention of bladder tumours.
	paracetamol @DRUG@ bladder tumours @DISEASE@ @CID@ paracetamol @DRUG@ liver tumours @DISEASE@ @CID@	
Coreferent mentions	Proto-oncogene HER2 (also known as erbB-2 or neu) plays an important role in the carcinogenesis and the prognosis of breast cancer.	Two coreferent mentions of HER2.
	her2 ; erbb-2 ; neu @GENE@ breast cancer @DISEASE@ @GDA@	
n-ary, inter-sentence	The deletion mutation on exon-19 of EGFR gene was present in 16 patients, while the L858E point mutation on exon-21 was noted in 10. All patients were treated with gefitinib and showed a partial response.	Ternary DGM relationship crosses a sentence boundary.
	gefitinib @DRUG@ egfr @GENE@ l858e @MUTATION@ @DGM@	

lations between coreferent mentions of entities in a document. A popular approach involves graph-based methods, which have the advantage of naturally modelling inter-sentence relations (Peng et al., 2017; Song et al., 2018; Christopoulou et al., 2019; Nan et al., 2020; Minh Tran et al., 2020). However, like all pipeline-based approaches, these methods assume that the entities within the text are known. As previous work has demonstrated, and as we show in §5.2, jointly learning to extract entities and relations can improve performance (Miwa and Sasaki, 2014; Miwa and Bansal, 2016; Gupta et al., 2016; Li et al., 2016a, 2017; Nguyen and Verspoor, 2019a; Yu et al., 2020) and may be more efficient due to shared parameters and training steps. Existing end-to-end methods typically combine task-specific components for entity detection, coreference resolution, and relation extraction that are trained jointly. Most approaches are restricted to intra-sentence RE (Bekoulis et al., 2018; Luan et al., 2018; Nguyen and Verspoor, 2019b; Wadden et al., 2019; Giorgi et al., 2019) and have only recently been extended to DocRE (Eberts and Ulges, 2021). However, they still focus on binary relations. Ideally, DocRE methods would be capable of modelling the complexities mentioned above without strictly requiring entities to be known.

A less popular end-to-end approach is to frame RE as a *generative* task with sequence-to-sequence (seq2seq) learning (Sutskever et al., 2014). This framing simplifies RE by removing the need for task-specific components and explicit negative training examples, i.e. pairs of entities that *do not* express a relation. If the information to extract is appropriately linearized to a string, seq2seq methods are flexible enough to model all complexities discussed thus far. However, existing work stops short, focusing on intra-sentence binary relations (Zeng et al., 2018; Zhang et al., 2020; Nayak and Ng, 2020; Zeng et al., 2020). In this paper, we extend work on seq2seq methods for RE to the document level, with several important contributions:

- We propose a novel linearization schema that can handle complexities overlooked by previous seq2seq approaches, like coreferent mentions and n-ary relations (§3.1).
- Using this linearization schema, we demonstrate that a seq2seq approach is able to learn the subtasks of DocRE (entity extraction, coreference resolution and relation extraction) jointly, and report the first end-to-end results on several popular biomedical datasets (§5.1).
- We devise a simple strategy, referred to as "entity hinting" (§3.3), to compare our model to existing pipeline-based approaches, in some cases exceeding their performance (§5.1).

2 Task definition: document-level relation extraction

Given a source document of S tokens, a model must extract all tuples corresponding to a relation, R, expressed between the entities, E in the document, $(E_1, ..., E_n, R)$ where n is the number of participating entities, or *arity*, of the relation. Each entity E_i is represented as the set of its coreferent mentions $\{e^i_j\}$ in the document, which are often expressed as aliases, abbreviations or acronyms. All entities appearing in a tuple have at least one mention in the document. The mentions that express a given relation are not necessarily contained within

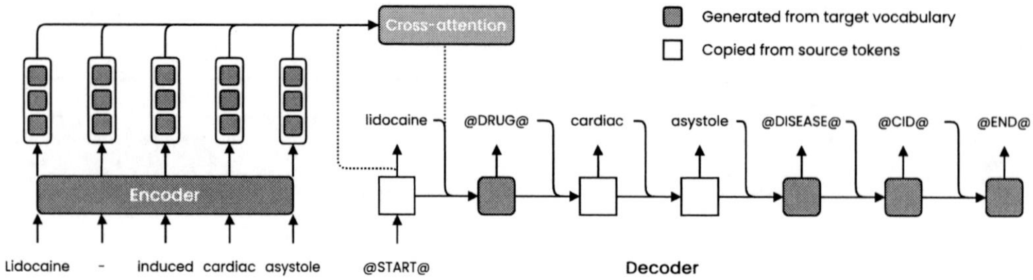

Figure 2: A sequence-to-sequence model for document-level relation extraction. Special tokens are generated by the decoder. Entity mentions are copied from the input via a copy mechanism (not shown). Decoding is initiated by a `@START@` token and terminated when the model generates the `@END@` token. Attention connections shown only for the second timestep to reduce clutter. `CID`: chemical-induced disease.

the same sentence. Commonly, E is assumed to be known and provided as input to a model. We will refer to these methods as "pipeline-based". In this paper, we are primarily concerned with the situation where E is *not* given and must be predicted by a model, which we will refer to as "end-to-end".

3 Our approach: seq2rel

3.1 Linearization

To use seq2seq learning for RE, the information to be extracted must be linearized to a string. This linearization should be expressive enough to model the complexities of entity and relation extraction without being overly verbose. We propose the following schema, illustrated with an example:

X: Variants in the estrogen receptor alpha (ESR1) gene and its mRNA contribute to risk for schizophrenia.
Y: `estrogen receptor alpha ; ESR1 @GENE@ schizophrenia @DISEASE@ @GDA@`

The input text X, expresses a gene-disease association (GDA) between ESR1 and schizophrenia. In the corresponding target string Y, each relation begins with its constituent entities. A semicolon separates coreferent mentions (;), and entities are terminated with a special token denoting their type (e.g. `@GENE@`). Similarly, relations are terminated with a special token denoting their type (e.g. `@GDA@`). Two or more entities can be included before the special relation token to support n-ary extraction. Entities can be ordered if they serve specific roles as head or tail of a relation. For each document, multiple relations can be included in the target string. Entities may be nested or discontinuous in the input text. In Figure 1, we provide examples of how this schema can be used to model various complexities, like coreferent entity mentions and n-ary relations.

3.2 Model

The model follows a canonical seq2seq setup. An encoder maps each token in the input to a contextual embedding. An autoregressive decoder generates an output, token-by-token, attending to the outputs of the encoder at each timestep (Figure 2). Decoding proceeds until a special "end-of-sequence" token (`@END@`) is generated, or a maximum number of tokens have been generated. Formally, X is the *source* sequence of length S, which is some text we would like to extract relations from. Y is the corresponding *target* sequence of length T, a linearization of the relations contained in the source. We model the conditional probability

$$p(Y|X) = \prod_{t=1}^{T} p(y_t|X, y_{<t}) \tag{1}$$

During training, we optimize over the model parameters θ the sequence cross-entropy loss

$$\ell(\theta) = -\sum_{t=1}^{T} \log p(y_t|X, y_{<t}; \theta) \tag{2}$$

maximizing the log-likelihood of the training data.[1]

The main problems with this setup for RE are: 1) The model might "hallucinate" by generating entity mentions that do not appear in the source text. 2) It may generate a target string that does not follow the linearization schema and therefore cannot

[1] See §4.3 for details about the encoder and decoder.

be parsed. 3) The loss function is permutation-sensitive, enforcing an unnecessary decoding order. To address 1) we use two modifications: a restricted target vocabulary (§3.2.1) and a copy mechanism (§3.2.2). To address 2) we experiment with several constraints applied during decoding (§3.2.3). Finally, to address 3) we sort relations according to their order of appearance in the source text (§3.2.4).

3.2.1 Restricted target vocabulary

To prevent the model from "hallucinating" (generating entity mentions that do not appear in the source text), the target vocabulary is restricted to the set of special tokens needed to model entities and relations (e.g. ; and `@DRUG@`). All other tokens must be copied from the input using a copy mechanism (see §3.2.2). The embeddings of these special tokens are initialized randomly and learned jointly with the rest of the model's parameters.

3.2.2 Copy mechanism

To enable copying of input tokens during decoding, we use a copying mechanism (Gu et al., 2016a). The mechanism works by effectively extending the target vocabulary with the tokens in the source sequence X, allowing the model to "copy" these tokens into the output sequence, Y. Our use of the copy mechanism is similar to previous seq2seq-based approaches for RE (Zeng et al., 2018, 2020).

3.2.3 Constrained decoding

We experimented with several constraints applied to the decoder during test time to reduce the likelihood of generating syntactically invalid target strings (strings that do not follow the linearization schema). These constraints are applied by setting the predicted probabilities of invalid tokens to a tiny value at each timestep. The full set of constraints is depicted in Appendix A. In practice, we found that a trained model rarely generates invalid target strings, so these constraints have little effect on final performance (see §5.3). We elected not to apply them in the rest of our experiments.

3.2.4 Sorting relations

The relations to extract from a given document are inherently unordered. However, the sequence cross-entropy loss (Equation 2) is permutation-sensitive with respect to the predicted tokens. During training, this enforces an unnecessary decoding order and may make the model prone to overfit frequent token combinations in the training set (Vinyals et al., 2016; Yang et al., 2019). To partially mitigate this, we sort relations within the target strings according to their order of appearance in the source text, providing the model with a consistent decoding order. The position of a relation is determined by the first occurring mention of its head entity. The position of a mention is determined by the sum of its start and end character offsets. In the case of ties, we then sort by the first mention of its tail entity (and so on for n-ary relations).

3.3 Entity hinting

Although the proposed model can jointly extract entities and relations from unannotated text, most existing DocRE methods provide the entities as input. Therefore, to more fairly compare to existing methods, we also provide entities as input, using a simple strategy that we will refer to as "entity hinting". This involves prepending entities to the source text as they appear in the target string. Taking the example from §3.1, entity hints would be added as follows:

X: `estrogen receptor alpha ; ESR1 @GENE@ schizophrenia @DISEASE@` `@SEP@` Variants in the estrogen receptor alpha (ESR1) gene and its mRNA contribute to risk for schizophrenia.

where the special `@SEP@` token demarcates the end of the entity hint.[2] We experimented with the common approach of inserting marker tokens before and after each entity mention (Zhou and Chen, 2021) but found this to perform worse. Our approach adds fewer extra tokens to the source text and provides a location for the copy mechanism to focus, i.e. tokens left of `@SEP@`. In our experiments, we use entity hinting when comparing to methods that provide ground truth entity annotations as input (§5.1.1). In §5.2, we use entity hinting to compare pipeline-based and end-to-end approaches.

4 Experimental setup

4.1 Datasets

We evaluate our approach on several biomedical, DocRE datasets. We also include one non-biomedical dataset, DocRED. In Appendix B, we list relevant details about their annotations.

[2]Some pretrained models have their own separator token which can be used in place of `@SEP@`, e.g. BERT uses `[SEP]`.

CDR (Li et al., 2016b) The BioCreative V CDR task corpus is manually annotated for chemicals, diseases and chemical-induced disease (CID) relations. It contains the titles and abstracts of 1500 PubMed articles and is split into equally sized train, validation and test sets. Given the relatively small size of the training set, we follow Christopoulou et al. (2019) and others by first tuning the model on the validation set and then training on the combination of the train and validation sets before evaluating on the test set. Similar to prior work, we filter negative relations with disease entities that are hypernyms of a corresponding true relations disease entity within the same abstract (see Appendix C).

GDA (Wu et al., 2019) The gene-disease association corpus contains 30,192 titles and abstracts from PubMed articles that have been automatically labelled for genes, diseases and gene-disease associations via distant supervision. The test set is comprised of 1000 of these examples. Following Christopoulou et al. (2019) and others, we hold out a random 20% of the remaining abstracts as a validation set and use the rest for training.

DGM (Jia et al., 2019) The drug-gene-mutation corpus contains 4606 PubMed articles that have been automatically labelled for drugs, genes, mutations and ternary drug-gene-mutation relationships via distant supervision. The dataset is available in three variants: sentence, paragraph, and document-length text. We train and evaluate our model on the paragraph-length inputs. Since the test set does not contain relation annotations on the paragraph level, we report results on the validation set. We hold out a random 20% of training examples to form a new validation set for tuning.

DocRED (Yao et al., 2019) DocRED includes over 5000 human-annotated documents from Wikipedia. There are six entity and 96 relation types, with ~40% of relations crossing the sentence boundary. We use the same split as previous end-to-end methods (Eberts and Ulges, 2021), which has 3,008 documents in the training set, 300 in the validation set and 700 in the test set[3].

4.2 Evaluation

We evaluate our model using the micro F1-score by extracting relation tuples from the decoder's output (see Appendix D). Similar to prior work, we use a "strict" criteria. A predicted relation is considered correct if the relation type and its entities match a ground truth relation. An entity is considered correct if the entity type and its mentions match a ground truth entity. However, since the aim of DocRE is to extract relations at the *entity*-level (as opposed to the *mention*-level), we also report performance using a relaxed criterion (denoted "relaxed"), where predicted entities are considered correct if more than 50% of their mentions match a ground truth entity (see Appendix E).

Existing methods that evaluate on CDR, GDA and DGM use the ground truth entity annotations as input. This makes it difficult to directly compare with our end-to-end approach, which takes only the raw text as input. To make the comparison fairer, we use entity hinting (§3.3) so that our model has access to the ground truth entity annotations. We also report the performance of our method in the end-to-end setting on these corpora to facilitate future comparison. To compare to existing end-to-end approaches, we use DocRED.

4.3 Implementation, training and hyperparameters

Implementation We implemented our model in PyTorch (Paszke et al., 2017) using AllenNLP (Gardner et al., 2018). As encoder, we use a pretrained transformer, implemented in the Transformers library (Wolf et al., 2020), which is fine-tuned during training. When training and evaluating on biomedical corpora, we use PubMedBERT (Gu et al., 2020), and $\text{BERT}_{\text{BASE}}$ (Devlin et al., 2019) otherwise. In both cases, we use the default hyperparameters of the pretrained model. As decoder, we use a single-layer LSTM (Hochreiter and Schmidhuber, 1997) with randomly initialized weights. We use multi-head attention (Vaswani et al., 2017) as the cross-attention mechanism between encoder and decoder. Select hyperparameters were tuned on the validation sets, see Appendix F for details.

Training All parameters are trained jointly using the AdamW optimizer (Loshchilov and Hutter, 2019). Before training, we re-initialize the top L layers of the pretrained transformer encoder, which has been shown to improve performance and stability during fine-tuning (Zhang et al., 2021b). During training, the learning rate is linearly increased for the first 10% of training steps and linearly decayed to zero afterward. Gradients are scaled to a vector norm of 1.0 before backpropagating. During each forward propagation, the hidden state of the LSTM

[3] `https://github.com/lavis-nlp/jerex`

Table 1: Comparison to existing pipeline-based methods. Performance reported as micro-precision, recall and F1-scores (%) on the CDR and GDA test sets. Results below the horizontal line are not comparable to existing methods. Bold: best scores.

	CDR			GDA		
Method	P	R	F1	P	R	F1
Christopoulou et al. (2019)	62.1	65.2	63.6	–	–	81.5
Nan et al. (2020)	–	–	64.8	–	–	82.2
Minh Tran et al. (2020)	–	–	66.1	–	–	82.8
Lai and Lu (2021)	64.9	67.1	66.0	–	–	–
Xu et al. (2021)	–	–	68.7	–	–	83.7
Zhou et al. (2021)	–	–	**69.4**	–	–	83.9
seq2rel (entity hinting)	68.2	66.2	67.2	84.4	85.3	**84.9**
seq2rel (entity hinting, relaxed)	68.2	66.2	67.2	84.5	85.4	85.0
seq2rel (end-to-end)	43.5	37.5	40.2	55.0	55.4	55.2
seq2rel (end-to-end, relaxed)	56.6	48.8	52.4	70.3	70.8	70.5

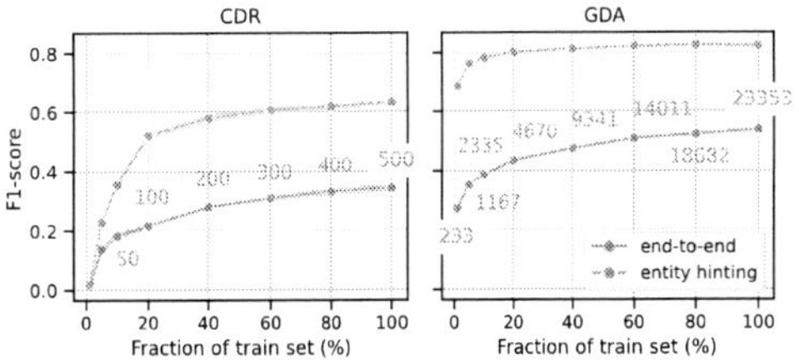

Figure 3: Effect of training set size on performance. Performance reported as the median micro F1-score obtained over five runs with different random seeds on the CDR and GDA validation sets, with and without entity hinting. Error bands correspond to the standard deviation over the five runs. The absolute number of training examples are displayed for each corpus. Some labels are excluded to reduce clutter.

decoder is initialized with the mean of token embeddings output by the encoder. The decoder is regularized by applying dropout (Srivastava et al., 2014) with probability 0.1 to its inputs, and DropConnect (Wan et al., 2013) with probability 0.5 to the hidden-to-hidden weights. As is common, we use teacher forcing, feeding previous ground truth inputs to the decoder when predicting the next token in the sequence. During test time, we generate the output using beam search (Graves, 2012). Beams are ranked by mean token log probability after applying a length penalty.[4] Models were trained and evaluated on a single NVIDIA Tesla V100.[5]

5 Results

5.1 Comparison to existing methods

In the following sections, we compare our model to existing DocRE methods on several benchmark corpora. We compare to existing pipeline-based methods (§5.1.1), including n-ary methods (§5.1.2), and end-to-end methods (§5.1.3). Details about these methods are provided in Appendix G.

5.1.1 Existing pipeline-based methods

In Table 1, we use entity hinting to compare our method to existing pipeline-based methods on CDR and GDA. We also report end-to-end performance, which is not comparable to existing pipeline-based methods but will facilitate future comparisons.

The large performance improvement when using entity hinting (+27-29%) confirms that the model exploits the entity annotations. The fact that relaxed entity matching makes a large difference in the end-to-end setting (+12-15%) suggests that a significant portion of the model's mistakes occur during coreference resolution. Although our method is designed for end-to-end RE, we find that it outperforms existing pipeline-based methods when using entity hinting on GDA. Our method is competitive with existing methods when using entity hinting on the CDR corpus but ultimately underperforms state-of-the-art results. Given that GDA is 46X larger, we speculated that our method might be underperforming in the low-data regime. To determine if this is a contributing factor, we artificially reduce the size of the CDR and GDA training sets and plot the performance as a curve (Figure 3). In all cases besides GDA with entity hinting, performance increases monotonically with dataset size. There is no obvious plateau on CDR even when using all 500 training examples. Together, these results suggest that our seq2seq based approach can outperform existing pipeline-based methods when there are sufficient training examples but underperforms relative to existing methods in the low-data regime.

5.1.2 n-ary relation extraction

In Table 2 we compare to existing n-ary methods on the DGM corpus. With entity hinting, our method significantly outperforms the existing method. The difference in encoders partially explains this large performance gap. Where Jia et al. (2019) use a BiLSTM that is trained from scratch, we use PubMedBERT, a much larger model that has been pretrained on abstracts and full-text ar-

[4]https://docs.allennlp.org/main/api/nn/beam_search/#lengthnormalizedsequencelogprobabilityscorer

[5]https://www.nvidia.com/en-us/data-center/v100/

Table 2: Comparison to existing n-ary methods. Performance reported as micro-precision, recall and F1-scores (%) on the DGM validation set. Results below the horizontal line are not comparable to existing methods. Bold: best scores. † Jia et al. 2019 do not report results on the validation set, so we re-run their paragraph-level model.

Method	P	R	F1
Jia et al. (2019) †	62.9	76.2	68.9
seq2rel (entity hinting)	**84.0**	**84.8**	**84.4**
seq2rel (entity hinting, relaxed)	84.1	84.9	84.5
seq2rel (end-to-end)	68.9	65.9	67.4
seq2rel (end-to-end, relaxed)	78.3	74.9	76.6

ticles from PubMedCentral.[6] However, this does not completely account for the improvement in performance, as recent work that has replaced the BiLSTM encoder of (Jia et al., 2019) with PubMedBERT found that it improves performance by approximately 2-4% on the task of drug-gene-mutation prediction (Zhang et al., 2021a).[7] Our results on the DGM corpus suggest that our linearization schema effectively models n-ary relations without requiring changes to the model architecture or training procedure.

5.1.3 End-to-end methods

In Table 3 we compare to an existing end-to-end approach on DocRED, JEREX (Eberts and Ulges, 2021). To make the comparison fair, we use the same pretrained encoder ($\text{BERT}_{\text{BASE}}$). We find that although our model is arguably simpler (JEREX contains four task-specific sub-components, each with its own loss) it only slightly underperforms JEREX, mainly due to recall. We speculate that one reason for this is a large number of relations per document, which leads to longer target strings and, therefore, more decoding steps. The median length of the target strings in DocRED, using our linearization, is 110, whereas the next largest is 19 in GDA. Improving the decoder's ability to process long sequences, e.g. switching the LSTM for a transformer or modifying the linearization schema to produce shorter target strings, may improve recall and close the gap with existing methods.

5.2 Pipeline vs. End-to-end

In §5.1.1 and §5.1.2, we provide gold-standard entity annotations from each corpus as input to our model via entity hinting (referred to as "gold" hints from here on, see §3.3). This allowed us to compare to existing methods that also provide these annotations as input. However, gold-standard entity annotations are (almost) never available in real-world settings, such as large-scale extraction on PubMed. In this setting, there are two strategies: pipeline-based, where independent systems perform entity and relation extraction, and end-to-end, where a single model performs both tasks. To compare these approaches under our model, we perform evaluations where a named entity recognition (NER) system is used to determine entity hints (referred to as "silver" hints from here on) and when no entity hints are provided (end-to-end).[8] However, this alone does not create a true pipeline, as our model can recover from both false negatives and false positives in the NER step. To mimic error propagation in the pipeline setting, we filter any entity mention predicted by our model that was *not* predicted by the NER system. In Table 4, we

Table 3: Comparison to existing end-to-end methods. Performance reported as micro-precision, recall and F1-scores (%) on the DocRED test set. Results below the horizontal line are not comparable to existing methods. Bold: best scores.

Method	P	R	F1
JEREX (Eberts and Ulges, 2021)	42.8	**38.2**	**40.4**
seq2rel (end-to-end)	**44.0**	33.8	38.2
seq2rel (end-to-end, relaxed)	53.7	41.3	46.7

Table 4: Comparison of pipeline-based and end-to-end approaches. Gold hints use gold-standard entity annotations to insert entity hints in the source text. Silver hints use the entity annotations provided by PubTator. Pipeline is identical to silver entity hints, except that we filter out entity mentions predicted by our model that PubTator does not predict. The end-to-end model only has access to the unannotated source text as input. Performance reported as micro-precision, recall and F1-scores (%) on the CDR test set, with strict and relaxed entity matching criteria. Bold: best scores.

	Strict			Relaxed		
	P	R	F1	P	R	F1
Gold hints	68.2	66.2	67.2	68.2	66.2	67.2
Silver hints	42.4	37.3	39.7	53.0	46.7	49.7
Pipeline	**45.0**	16.9	24.6	**62.5**	23.5	34.1
End-to-end	43.5	**37.5**	**40.2**	56.6	**48.8**	**52.4**

[6] https://www.ncbi.nlm.nih.gov/pmc/

[7] The authors have not released code at the time of writing, so we were unable to evaluate this model on the DGM validation set in order to compare with our method directly.

[8] Specifically, we use PubTator (Wei et al., 2013). PubTator provides up-to-date entity annotations for PubMed using state-of-the-art machine learning systems.

Table 5: Ablation study results. Performance reported as the micro-precision, recall and F1-scores (%) on the CDR and DocRED validation sets. Δ: difference to the complete models F1-score. Bold: best scores.

	CDR				DocRED			
	P	R	F1	Δ	P	R	F1	Δ
seq2rel (end-to-end)	**41.0**	35.1	37.8	–	46.9	**36.1**	**40.8**	–
- pretraining	9.4	6.9	8.0	-29.8	18.5	7.7	10.8	-30.0
- fine-tuning	24.3	20.5	22.2	-15.6	42.4	15.5	22.7	-18.1
- vocab restriction	39.6	32.2	35.5	-2.3	45.2	35.5	39.7	-1.1
- sorting relations	36.1	29.2	32.3	-5.6	**52.9**	17.4	26.2	-14.7
+ constrained decoding	40.8	**35.6**	**38.0**	+0.2	46.8	35.9	40.6	-0.2

present the results of all four settings (gold and silver entity hints, pipeline and end-to-end) on CDR.

We find that using gold entity hints significantly outperforms all other settings. This is expected, as the gold-standard entity annotations are high-quality labels produced by domain experts. Using silver hints significantly drops performance, likely due to a combination of false positive and false negatives from the NER step. In the pipeline setting, where there is no recovery from false negatives, performance falls by another 15%. The end-to-end setting significantly outperforms the pipeline setting (due to a large boost in recall) and performs comparably to using silver hints. Together, our results suggest that performance reported using gold-standard entity annotations may be overly optimistic and corroborates previous work demonstrating the benefits of jointly learning entity and relation extraction (Miwa and Sasaki, 2014; Miwa and Bansal, 2016; Gupta et al., 2016; Li et al., 2016a, 2017; Nguyen and Verspoor, 2019a; Yu et al., 2020).

5.3 Ablation

In Table 5, we present the results of an ablation study. We perform the analysis twice, once on the biomedical corpus CDR and once on the general domain corpus DocRED. Unsurprisingly, we find that fine-tuning a pretrained encoder greatly impacts performance. Training the same encoder from scratch (- pretraining) reduces performance by $\sim$30%. Using the pretrained weights without fine-tuning (- fine-tuning) drops performance by 15.6-18.1%. Restricting the target vocabulary (- vocab restriction, see §3.2.1) has a small positive impact, boosting performance by 1.1%-2.3%. Deliberately ordering the relations within each target string (- sorting relations, see §3.2.4) has a large positive impact, boosting performance by 5.6%-14.7%. This effect is larger on DocRED, likely because it has more relations per document on average than CDR, so ordering becomes more important. Finally, adding constraints to the decoding process (+ constrained decoding) has little impact on performance, suggesting that a trained model rarely generates invalid target strings (see §3.2.3).

6 Discussion

6.1 Related work

Seq2seq learning for RE has been explored in prior work. CopyRE (Zeng et al., 2018) uses an encoder-decoder architecture with a copy mechanism, similar to our approach, but is restricted to intra-sentence relations. Additionally, because CopyRE's decoding proceeds for exactly three timesteps per relation, the model is limited to generating binary relations between single token entities. The ability to decode multi-token entities was addressed in follow-up work, CopyMTL (Zeng et al., 2020). A similar approach was published concurrently but was again limited to intra-sentence binary relations (Nayak and Ng, 2020). Most recently, GenerativeRE (Cao and Ananiadou, 2021) proposed a novel copy mechanism to improve performance on multi-token entities. None of these approaches deal with the complexities of DocRE, where many relations cross the sentence boundary, and coreference resolution is critical.[9]

More generally, our paper is related to a recently proposed "text-to-text" framework (Raffel et al., 2020). In this framework, a task is formulated so that the inputs and outputs are both text strings, enabling the use of the same model, loss function and even hyperparameters across many seq2seq, classification and regression tasks. This framework has recently been applied to biomedical literature to perform named entity recognition, relation extraction (binary, intra-sentence), natural language inference, and question answering (Phan et al., 2021). Our work can be seen as an attempt to formulate the task of DocRE within this framework.

6.2 Limitations and future work

Permutation-sensitive loss Our approach adopts the sequence cross-entropy loss (Equation 2), which is sensitive to the order of predicted tokens, enforcing an unnecessary decoding order on the inherently unordered relations. To partially mitigate this problem, we order relations within the

[9]Concurrent to our work, REBEL (Huguet Cabot and Navigli, 2021) also extends seq2seq methods to document-level RE, achieving strong performance on DocRED. However, the method was not evaluated on n-ary relations.

target string according to order of appearance in the source text, providing the model with a consistent decoding order that can be learned (see §3.2.4, §5.3). Previous work has addressed this issue with various strategies, including reinforcement learning (Zeng et al., 2019), unordered-multi-tree decoders (Zhang et al., 2020), and non-autoregressive decoders (Sui et al., 2020). However, these works are limited to binary intra-sentence relation extraction, and their suitability for DocRE has not been explored. A promising future direction would be to modify our approach such that the arbitrary order of relations is not enforced during training.

Input length restriction Due to the pretrained encoder's input size limit (512 tokens), our experiments are conducted on paragraph-length text. Our model could be extended to full documents by swapping its encoder with any of the recently proposed "efficient transformers" (Tay et al., 2021). Future work could evaluate such a model's ability to extract relations from full scientific papers.

Pretraining the decoder In our model, the encoder is pretrained, while the decoder is trained from scratch. Several recent works, such as T5 (Raffel et al., 2020) and BART (Lewis et al., 2020), have proposed pretraining strategies for entire encoder-decoder architectures, which can be fine-tuned on downstream tasks. An interesting future direction would be to fine-tune such a model on DocRE using our linearization schema.

7 Conclusion

In this paper, we extend generative, seq2seq methods for relation extraction to the document level. We propose a novel linearization schema that can handle complexities overlooked by previous seq2seq approaches, like coreferent mentions and n-ary relations. We compare our approach to existing pipeline-based and end-to-end methods on several benchmark corpora, in some cases exceeding their performance. In future work, we hope to extend our method to full scientific papers and develop strategies to improve performance in the low-data regime and in cases where there are many relations per document.

Acknowledgements

This research was enabled in part by support provided by Compute Ontario (www.computeontario.ca), Compute Canada (www.computecanada.ca) and the CIFAR AI Chairs Program and partially funded by the US National Institutes of Health (NIH) [U41 HG006623, U41 HG003751).

References

Takuya Akiba, Shotaro Sano, Toshihiko Yanase, Takeru Ohta, and Masanori Koyama. 2019. Optuna: A next-generation hyperparameter optimization framework. In *Proceedings of the 25th ACM SIGKDD International Conference on Knowledge Discovery & Data Mining, KDD 2019, Anchorage, AK, USA, August 4-8, 2019*, pages 2623–2631. ACM.

Giannis Bekoulis, Johannes Deleu, Thomas Demeester, and Chris Develder. 2018. Joint entity recognition and relation extraction as a multi-head selection problem. *Expert Systems with Applications*, 114:34–45.

James Bergstra, Rémi Bardenet, Yoshua Bengio, and Balázs Kégl. 2011. Algorithms for hyper-parameter optimization. In *Advances in Neural Information Processing Systems 24: 25th Annual Conference on Neural Information Processing Systems 2011. Proceedings of a meeting held 12-14 December 2011, Granada, Spain*, pages 2546–2554.

Jiarun Cao and Sophia Ananiadou. 2021. GenerativeRE: Incorporating a novel copy mechanism and pretrained model for joint entity and relation extraction. In *Findings of the Association for Computational Linguistics: EMNLP 2021*, pages 2119–2126, Punta Cana, Dominican Republic. Association for Computational Linguistics.

Fenia Christopoulou, Makoto Miwa, and Sophia Ananiadou. 2019. Connecting the dots: Document-level neural relation extraction with edge-oriented graphs. In *Proceedings of the 2019 Conference on Empirical Methods in Natural Language Processing and the 9th International Joint Conference on Natural Language Processing (EMNLP-IJCNLP)*, pages 4925–4936, Hong Kong, China. Association for Computational Linguistics.

Jacob Devlin, Ming-Wei Chang, Kenton Lee, and Kristina Toutanova. 2019. BERT: Pre-training of deep bidirectional transformers for language understanding. In *Proceedings of the 2019 Conference of the North American Chapter of the Association for Computational Linguistics: Human Language Technologies, Volume 1 (Long and Short Papers)*, pages 4171–4186, Minneapolis, Minnesota. Association for Computational Linguistics.

Markus Eberts and Adrian Ulges. 2021. An end-to-end model for entity-level relation extraction using multi-instance learning. In *Proceedings of the 16th Conference of the European Chapter of the Association for Computational Linguistics: Main Volume*, pages 3650–3660, Online. Association for Computational Linguistics.

Matt Gardner, Joel Grus, Mark Neumann, Oyvind Tafjord, Pradeep Dasigi, Nelson F. Liu, Matthew Peters, Michael Schmitz, and Luke Zettlemoyer. 2018. AllenNLP: A deep semantic natural language processing platform. In *Proceedings of Workshop for NLP Open Source Software (NLP-OSS)*, pages 1–6, Melbourne, Australia. Association for Computational Linguistics.

John Giorgi, Xindi Wang, Nicola Sahar, Won Young Shin, Gary D Bader, and Bo Wang. 2019. End-to-end named entity recognition and relation extraction using pre-trained language models. *ArXiv preprint*, abs/1912.13415.

Alex Graves. 2012. Sequence transduction with recurrent neural networks. *ArXiv preprint*, abs/1211.3711.

Jiatao Gu, Zhengdong Lu, Hang Li, and Victor O.K. Li. 2016a. Incorporating copying mechanism in sequence-to-sequence learning. In *Proceedings of the 54th Annual Meeting of the Association for Computational Linguistics (Volume 1: Long Papers)*, pages 1631–1640, Berlin, Germany. Association for Computational Linguistics.

Jinghang Gu, Longhua Qian, and Guodong Zhou. 2016b. Chemical-induced disease relation extraction with various linguistic features. *Database: The Journal of Biological Databases and Curation*, 2016.

Jinghang Gu, Fuqing Sun, Longhua Qian, and Guodong Zhou. 2017. Chemical-induced disease relation extraction via convolutional neural network. *Database: The Journal of Biological Databases and Curation*, 2017.

Yu Gu, Robert Tinn, Hao Cheng, Michael Lucas, Naoto Usuyama, Xiaodong Liu, Tristan Naumann, Jianfeng Gao, and Hoifung Poon. 2020. Domain-specific language model pretraining for biomedical natural language processing. *ArXiv preprint*, abs/2007.15779.

Pankaj Gupta, Hinrich Schütze, and Bernt Andrassy. 2016. Table filling multi-task recurrent neural network for joint entity and relation extraction. In *Proceedings of COLING 2016, the 26th International Conference on Computational Linguistics: Technical Papers*, pages 2537–2547, Osaka, Japan. The COLING 2016 Organizing Committee.

Sepp Hochreiter and Jürgen Schmidhuber. 1997. Long short-term memory. *Neural Computation*, 9:1735–1780.

Pere-Lluís Huguet Cabot and Roberto Navigli. 2021. REBEL: Relation extraction by end-to-end language generation. In *Findings of the Association for Computational Linguistics: EMNLP 2021*, pages 2370–2381, Punta Cana, Dominican Republic. Association for Computational Linguistics.

Sarthak Jain, Madeleine van Zuylen, Hannaneh Hajishirzi, and Iz Beltagy. 2020. SciREX: A challenge dataset for document-level information extraction. In *Proceedings of the 58th Annual Meeting of the Association for Computational Linguistics*, pages 7506–7516, Online. Association for Computational Linguistics.

Robin Jia, Cliff Wong, and Hoifung Poon. 2019. Document-level n-ary relation extraction with multiscale representation learning. In *Proceedings of the 2019 Conference of the North American Chapter of the Association for Computational Linguistics: Human Language Technologies, Volume 1 (Long and Short Papers)*, pages 3693–3704, Minneapolis, Minnesota. Association for Computational Linguistics.

Jin-Dong Kim, T. Ohta, Yuka Tateisi, and Jun'ichi Tsujii. 2003. Genia corpus - a semantically annotated corpus for bio-textmining. *Bioinformatics*, 19 Suppl 1:i180–2.

Po-Ting Lai and Zhiyong Lu. 2021. Bert-gt: Cross-sentence n-ary relation extraction with bert and graph transformer. *Bioinformatics*.

Mike Lewis, Yinhan Liu, Naman Goyal, Marjan Ghazvininejad, Abdelrahman Mohamed, Omer Levy, Veselin Stoyanov, and Luke Zettlemoyer. 2020. BART: Denoising sequence-to-sequence pre-training for natural language generation, translation, and comprehension. In *Proceedings of the 58th Annual Meeting of the Association for Computational Linguistics*, pages 7871–7880, Online. Association for Computational Linguistics.

Fei Li, Meishan Zhang, Guohong Fu, and Donghong Ji. 2017. A neural joint model for entity and relation extraction from biomedical text. *BMC bioinformatics*, 18(1):198.

Fei Li, Yue Zhang, Meishan Zhang, and Donghong Ji. 2016a. Joint models for extracting adverse drug events from biomedical text. In *Proceedings of the Twenty-Fifth International Joint Conference on Artificial Intelligence, IJCAI 2016, New York, NY, USA, 9-15 July 2016*, pages 2838–2844. IJCAI/AAAI Press.

Jiao Li, Yueping Sun, Robin J. Johnson, Daniela Sciaky, Chih-Hsuan Wei, Robert Leaman, Allan Peter Davis, Carolyn J. Mattingly, Thomas C. Wiegers, and Zhiyong Lu. 2016b. Biocreative v cdr task corpus: a resource for chemical disease relation extraction. *ArXiv preprint*, abs/d.

Feifan Liu, Jinying Chen, Abhyuday Jagannatha, and Hong Yu. 2016. Learning for biomedical information extraction: Methodological review of recent advances. *ArXiv preprint*, abs/1606.07993.

Ilya Loshchilov and Frank Hutter. 2019. Decoupled weight decay regularization. In *7th International Conference on Learning Representations, ICLR 2019, New Orleans, LA, USA, May 6-9, 2019*. OpenReview.net.

Yi Luan, Luheng He, Mari Ostendorf, and Hannaneh Hajishirzi. 2018. Multi-task identification of entities, relations, and coreference for scientific knowledge graph construction. In *Proceedings of the 2018 Conference on Empirical Methods in Natural Language Processing*, pages 3219–3232, Brussels, Belgium. Association for Computational Linguistics.

Hieu Minh Tran, Minh Trung Nguyen, and Thien Huu Nguyen. 2020. The dots have their values: Exploiting the node-edge connections in graph-based neural models for document-level relation extraction. In *Findings of the Association for Computational Linguistics: EMNLP 2020*, pages 4561–4567, Online. Association for Computational Linguistics.

Makoto Miwa and Mohit Bansal. 2016. End-to-end relation extraction using LSTMs on sequences and tree structures. In *Proceedings of the 54th Annual Meeting of the Association for Computational Linguistics (Volume 1: Long Papers)*, pages 1105–1116, Berlin, Germany. Association for Computational Linguistics.

Makoto Miwa and Yutaka Sasaki. 2014. Modeling joint entity and relation extraction with table representation. In *Proceedings of the 2014 Conference on Empirical Methods in Natural Language Processing (EMNLP)*, pages 1858–1869, Doha, Qatar. Association for Computational Linguistics.

Guoshun Nan, Zhijiang Guo, Ivan Sekulic, and Wei Lu. 2020. Reasoning with latent structure refinement for document-level relation extraction. In *Proceedings of the 58th Annual Meeting of the Association for Computational Linguistics*, pages 1546–1557, Online. Association for Computational Linguistics.

Tapas Nayak and Hwee Tou Ng. 2020. Effective modeling of encoder-decoder architecture for joint entity and relation extraction. In *The Thirty-Fourth AAAI Conference on Artificial Intelligence, AAAI 2020, The Thirty-Second Innovative Applications of Artificial Intelligence Conference, IAAI 2020, The Tenth AAAI Symposium on Educational Advances in Artificial Intelligence, EAAI 2020, New York, NY, USA, February 7-12, 2020*, pages 8528–8535. AAAI Press.

Dat Quoc Nguyen and Karin Verspoor. 2019a. End-to-end neural relation extraction using deep biaffine attention. In *Advances in Information Retrieval*, pages 729–738, Cham. Springer, Springer International Publishing.

Dat Quoc Nguyen and Karin Verspoor. 2019b. End-to-end neural relation extraction using deep biaffine attention. In *European Conference on Information Retrieval*, pages 729–738. Springer.

Adam Paszke, Sam Gross, Soumith Chintala, Gregory Chanan, Edward Yang, Zachary DeVito, Zeming Lin, Alban Desmaison, Luca Antiga, and Adam Lerer. 2017. Automatic differentiation in PyTorch. In *NIPS Autodiff Workshop*.

Nanyun Peng, Hoifung Poon, Chris Quirk, Kristina Toutanova, and Wen-tau Yih. 2017. Cross-sentence n-ary relation extraction with graph LSTMs. *Transactions of the Association for Computational Linguistics*, 5:101–115.

Long N Phan, James T Anibal, Hieu Tran, Shaurya Chanana, Erol Bahadroglu, Alec Peltekian, and Grégoire Altan-Bonnet. 2021. Scifive: a text-to-text transformer model for biomedical literature. *ArXiv preprint*, abs/2106.03598.

Colin Raffel, Noam Shazeer, Adam Roberts, Katherine Lee, Sharan Narang, Michael Matena, Yanqi Zhou, Wei Li, and Peter J Liu. 2020. Exploring the limits of transfer learning with a unified text-to-text transformer. *Journal of Machine Learning Research*, 21:1–67.

Linfeng Song, Yue Zhang, Zhiguo Wang, and Daniel Gildea. 2018. N-ary relation extraction using graph-state LSTM. In *Proceedings of the 2018 Conference on Empirical Methods in Natural Language Processing*, pages 2226–2235, Brussels, Belgium. Association for Computational Linguistics.

Nitish Srivastava, Geoffrey E. Hinton, Alex Krizhevsky, Ilya Sutskever, and Ruslan Salakhutdinov. 2014. Dropout: a simple way to prevent neural networks from overfitting. *J. Mach. Learn. Res.*, 15:1929–1958.

Dianbo Sui, Yubo Chen, Kang Liu, Jun Zhao, Xiangrong Zeng, and Shengping Liu. 2020. Joint entity and relation extraction with set prediction networks. *ArXiv preprint*, abs/2011.01675.

Ilya Sutskever, Oriol Vinyals, and Quoc V. Le. 2014. Sequence to sequence learning with neural networks. In *Advances in Neural Information Processing Systems 27: Annual Conference on Neural Information Processing Systems 2014, December 8-13 2014, Montreal, Quebec, Canada*, pages 3104–3112.

Yi Tay, Mostafa Dehghani, Samira Abnar, Yikang Shen, Dara Bahri, Philip Pham, Jinfeng Rao, Liu Yang, Sebastian Ruder, and Donald Metzler. 2021. Long range arena : A benchmark for efficient transformers. In *9th International Conference on Learning Representations, ICLR 2021, Virtual Event, Austria, May 3-7, 2021*. OpenReview.net.

Ashish Vaswani, Noam Shazeer, Niki Parmar, Jakob Uszkoreit, Llion Jones, Aidan N. Gomez, Lukasz Kaiser, and Illia Polosukhin. 2017. Attention is all you need. In *Advances in Neural Information Processing Systems 30: Annual Conference on Neural Information Processing Systems 2017, December 4-9, 2017, Long Beach, CA, USA*, pages 5998–6008.

Patrick Verga, Emma Strubell, and Andrew McCallum. 2018. Simultaneously self-attending to all mentions for full-abstract biological relation extraction. In *Proceedings of the 2018 Conference of the North American Chapter of the Association for Computational Linguistics: Human Language Technologies, Volume 1 (Long Papers)*, pages 872–884, New Orleans, Louisiana. Association for Computational Linguistics.

Oriol Vinyals, Samy Bengio, and Manjunath Kudlur. 2016. Order matters: Sequence to sequence for sets. In *4th International Conference on Learning Representations, ICLR 2016, San Juan, Puerto Rico, May 2-4, 2016, Conference Track Proceedings*.

David Wadden, Ulme Wennberg, Yi Luan, and Hannaneh Hajishirzi. 2019. Entity, relation, and event extraction with contextualized span representations. In *Proceedings of the 2019 Conference on Empirical Methods in Natural Language Processing and the 9th International Joint Conference on Natural Language Processing (EMNLP-IJCNLP)*, pages 5784–5789, Hong Kong, China. Association for Computational Linguistics.

Li Wan, Matthew D. Zeiler, Sixin Zhang, Yann LeCun, and Rob Fergus. 2013. Regularization of neural networks using dropconnect. In *Proceedings of the 30th International Conference on Machine Learning, ICML 2013, Atlanta, GA, USA, 16-21 June 2013*, volume 28 of *JMLR Workshop and Conference Proceedings*, pages 1058–1066. JMLR.org.

Yucheng Wang, Bowen Yu, Hongsong Zhu, Tingwen Liu, Nan Yu, and Limin Sun. 2021. Discontinuous named entity recognition as maximal clique discovery. In *Proceedings of the 59th Annual Meeting of the Association for Computational Linguistics and the 11th International Joint Conference on Natural Language Processing (Volume 1: Long Papers)*, pages 764–774, Online. Association for Computational Linguistics.

Chih-Hsuan Wei, Hung-Yu Kao, and Zhiyong Lu. 2013. Pubtator: a web-based text mining tool for assisting biocuration. *Nucleic acids research*, 41(W1):W518–W522.

Thomas Wolf, Lysandre Debut, Victor Sanh, Julien Chaumond, Clement Delangue, Anthony Moi, Pierric Cistac, Tim Rault, Remi Louf, Morgan Funtowicz, Joe Davison, Sam Shleifer, Patrick von Platen, Clara Ma, Yacine Jernite, Julien Plu, Canwen Xu, Teven Le Scao, Sylvain Gugger, Mariama Drame, Quentin Lhoest, and Alexander Rush. 2020. Transformers: State-of-the-art natural language processing. In *Proceedings of the 2020 Conference on Empirical Methods in Natural Language Processing: System Demonstrations*, pages 38–45, Online. Association for Computational Linguistics.

Ye Wu, Ruibang Luo, Henry C. M. Leung, Hing-Fung Ting, and Tak-Wah Lam. 2019. Renet: A deep learning approach for extracting gene-disease associations from literature. In *Research in Computational Molecular Biology*, pages 272–284, Cham. Springer International Publishing.

Benfeng Xu, Quan Wang, Yajuan Lyu, Yong Zhu, and Zhendong Mao. 2021. Entity structure within and throughout: Modeling mention dependencies for document-level relation extraction. In *AAAI*.

Pengcheng Yang, Fuli Luo, Shuming Ma, Junyang Lin, and Xu Sun. 2019. A deep reinforced sequence-to-set model for multi-label classification. In *Proceedings of the 57th Annual Meeting of the Association for Computational Linguistics*, pages 5252–5258, Florence, Italy. Association for Computational Linguistics.

Yuan Yao, Deming Ye, Peng Li, Xu Han, Yankai Lin, Zhenghao Liu, Zhiyuan Liu, Lixin Huang, Jie Zhou, and Maosong Sun. 2019. DocRED: A large-scale document-level relation extraction dataset. In *Proceedings of the 57th Annual Meeting of the Association for Computational Linguistics*, pages 764–777, Florence, Italy. Association for Computational Linguistics.

Bowen Yu, Zhenyu Zhang, Xiaobo Shu, Tingwen Liu, Yubin Wang, Bin Wang, and Sujian Li. 2020. Joint extraction of entities and relations based on a novel decomposition strategy. In *ECAI 2020*, pages 2282–2289. IOS Press.

Daojian Zeng, Haoran Zhang, and Qianying Liu. 2020. Copymtl: Copy mechanism for joint extraction of entities and relations with multi-task learning. In *The Thirty-Fourth AAAI Conference on Artificial Intelligence, AAAI 2020, The Thirty-Second Innovative Applications of Artificial Intelligence Conference, IAAI 2020, The Tenth AAAI Symposium on Educational Advances in Artificial Intelligence, EAAI 2020, New York, NY, USA, February 7-12, 2020*, pages 9507–9514. AAAI Press.

Xiangrong Zeng, Shizhu He, Daojian Zeng, Kang Liu, Shengping Liu, and Jun Zhao. 2019. Learning the extraction order of multiple relational facts in a sentence with reinforcement learning. In *Proceedings of the 2019 Conference on Empirical Methods in Natural Language Processing and the 9th International Joint Conference on Natural Language Processing (EMNLP-IJCNLP)*, pages 367–377, Hong Kong, China. Association for Computational Linguistics.

Xiangrong Zeng, Daojian Zeng, Shizhu He, Kang Liu, and Jun Zhao. 2018. Extracting relational facts by an end-to-end neural model with copy mechanism. In *Proceedings of the 56th Annual Meeting of the Association for Computational Linguistics (Volume 1: Long Papers)*, pages 506–514, Melbourne, Australia. Association for Computational Linguistics.

Ranran Haoran Zhang, Qianying Liu, Aysa Xuemo Fan, Heng Ji, Daojian Zeng, Fei Cheng, Daisuke Kawahara, and Sadao Kurohashi. 2020. Minimize exposure bias of Seq2Seq models in joint entity and relation extraction. In *Findings of the Association for Computational Linguistics: EMNLP 2020*, pages 236–246, Online. Association for Computational Linguistics.

Sheng Zhang, Cliff Wong, Naoto Usuyama, Sarthak Jain, Tristan Naumann, and Hoifung Poon. 2021a. Modular self-supervision for document-level relation extraction. In *Proceedings of the 2021 Conference* on Empirical Methods in Natural Language Processing, pages 5291–5302, Online and Punta Cana, Dominican Republic. Association for Computational Linguistics.

Tianyi Zhang, Felix Wu, Arzoo Katiyar, Kilian Q. Weinberger, and Yoav Artzi. 2021b. Revisiting few-sample BERT fine-tuning. In *9th International Conference on Learning Representations, ICLR 2021, Virtual Event, Austria, May 3-7, 2021*. OpenReview.net.

Deyu Zhou, Dayou Zhong, and Yulan He. 2014. Biomedical relation extraction: from binary to complex. *Computational and mathematical methods in medicine*, 2014.

Wenxuan Zhou and Muhao Chen. 2021. An improved baseline for sentence-level relation extraction. *ArXiv*, abs/2102.01373.

Wenxuan Zhou, Kevin Huang, Tengyu Ma, and Jing Huang. 2021. Document-level relation extraction with adaptive thresholding and localized context pooling. In *AAAI*.

A Constrained decoding

In Figure 4, we illustrate the rules used to constrain decoding. At each timestep t, given the prediction of the previous timestep $t-1$, the predicted class probabilities of tokens that would generate a syntactically invalid target string are set to a tiny value. In practice, we found that a model rarely generates invalid target strings, so these constraints have little effect on final performance (see §3.2.3 and §5.3).

B Details about dataset annotations

In Table 6, we list which complexities (e.g. nested & discontinuous mentions, n-ary relations) are contained within each dataset used in our evaluations. We also report the fraction of relations in the test set that are inter-sentence. We consider a relation intra-sentence if *any* sentence in the document contains *at least one* mention of each entity in the relation, and inter-sentence otherwise. This produces an estimate that matches previously reported numbers for CDR ($\sim$30%). In Yao et al. (2019), the fraction of inter-sentence relations in DocRED is reported as $\sim$40.7%. We can reproduce this value if we consider relations intra-sentence when *all* mentions of an entity exist within a single sentence and inter-sentence otherwise.

C Hypernym filtering

The CDR dataset is annotated for chemical-induced disease (CID) relationships between the most specific chemical and disease mentions in an abstract. Take the following example from the corpus:

Carbamazepine-induced cardiac dysfunction [...] A patient with sinus bradycardia and atrioventricular block, induced by carbamazepine, prompted an extensive literature review of all previously reported cases.

In this example (PMID: 1728915), only (*carbamazepine, bradycardia*) and (*carbamazepine, atrioventricular block*) are labelled as true relations. The relation (*carbamazepine, cardiac dysfunction*), although true, is not labelled as *cardiac dysfunction* is a hypernym of both *bradycardia* and *atrioventricular block*. This can harm evaluation performance, as the prediction (*carbamazepine, cardiac dysfunction*) will be considered a false positive. Therefore, we follow previous work (Gu et al., 2016b, 2017; Verga et al., 2018; Christopoulou et al., 2019; Zhou et al., 2021) by filtering negative relations like these, with disease entities that are hypernyms of a corresponding true relations disease entity within the same abstract, according to the hierarchy in the MeSH vocabulary.[10]

D Parsing the models output

At test time, our model autoregressively generates an output, token-by-token, using beam search decoding (see §3.2). In order to extract the predicted relations from this output, we apply the following steps. First, predicted token ids are converted to a string. We use the `decode()`[11] method of the HuggingFace Transformers tokenizer (Wolf et al., 2020) to do this. For example, after calling `decode()` on the predicted token ids, this string might look like:

```
monoamine oxidase b ; maob @GENE@ parkinson's
disease ; pd @DISEASE@ @GDA@
```

We then use regular expressions to extract any relations from this string that match our linearization schema (see §3.1), which produces a dictionary of nested lists, keyed by relation class:

```
{
  "GDA": [
    [
```

[10] https://meshb.nlm.nih.gov

[11] https://huggingface.co/docs/transformers/main_classes/tokenizer#transformers.PreTrainedTokenizerBase.decode

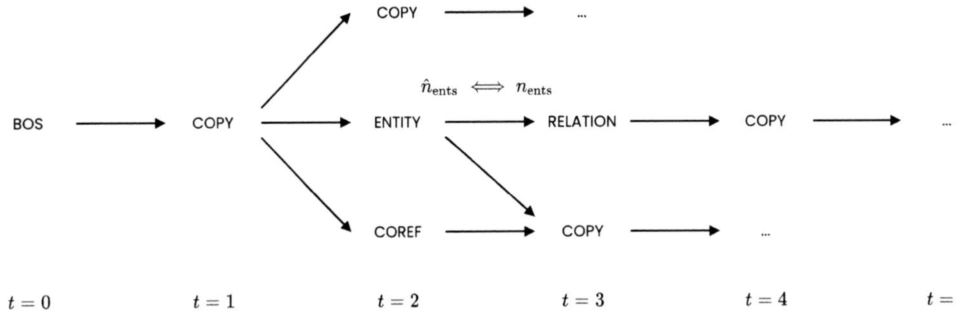

Figure 4: A diagram depicting syntactically valid predictions during decoding at each timestep t. The log probabilities of all other possible predictions are set to a tiny value to prevent the model from producing a syntactically invalid target string. `BOS` is the special beginning-of-sequence token, `COPY` denotes any token copied from the source text, and `COREF` is the special token used to separate coreferent mentions (i.e. ;). `ENTITY` is any special entity token (e.g. `@GENE@`) and `RELATION` any special relation token (e.g. `@GDA@` for gene-disease association). $\hat{n}_{\text{ents}}$ denotes the number of entities predicted by the current timestep and n_{ents} the expected arity of the relation. The special end-of-sequence token (not shown) is always considered valid and its log probability is never modified.

Table 6: Evaluation datasets used in this paper with details about their annotations. Inter-sentence relations (%) are the fraction of relations in the test set that cross sentence boundaries. We consider a relation intra-sentence if any sentence in the document contains at least one mention of each entity in the relation, and inter-sentence otherwise. *This differs from the estimate in Yao et al. (2019), see Appendix B.

Corpus	Nested Mentions?	Discontinuous Mentions?	Coreferent mentions?	n-ary relations?	Inter-sentence relations (%)
CDR (Li et al., 2016b)	✓	✓	✓	✗	29.8
GDA (Wu et al., 2019)	✓	✗	✓	✗	15.6
DGM (Jia et al., 2019)	✗	✗	✓	✓	63.5
DocRED (Yao et al., 2019)	✗	✗	✓	✗	12.5*

```
            [["monoamine oxidase b", "maob"], "GENE"],
            [["parkinson's disease", "pd"], "DISEASE"]
        ]
    ]
}
```

Finally, we apply some normalization steps to the entity mentions. Namely, we strip leading and trailing white space characters, sort entity mentions lexicographically (as their order is not important), and remove duplicate mentions. Similarly, we remove duplicate relations. These steps are applied to both target and model output strings. The F1-score can then be computed by tallying true positives, false positives and false negatives.

E Relaxed entity matching

The aim of DocRE is to extract relations at the *entity*-level. However, it is common to evaluate these methods with a "strict" matching criteria, where a predicted entity $\mathcal{P}$ is considered correct if and only if all its *mentions* exactly match a corresponding gold entities mentions, i.e. $\mathcal{P} = \mathcal{G}$. This penalizes model predictions that miss even a single coreferent mention, but are otherwise correct. A relaxed criteria, proposed in prior work (Jain et al., 2020) considers $\mathcal{P}$ to match $\mathcal{G}$ if more than 50% of $\mathcal{P}$'s mentions belong to $\mathcal{G}$, that is

$$\frac{|\mathcal{P} \cap \mathcal{G}|}{|\mathcal{P}|} > 0.5$$

In this paper, alongside the strict criteria, we report performance using this relaxed entity matching strategy, denoted "relaxed".

F Hyperparameters

In Table 7, we list the hyperparameter values used during evaluation on each corpus, with and without entity hinting. Select hyperparameters were tuned using Optuna (Akiba et al., 2019). The tuning process selects the best hyperparameters according to the validation set micro F1-score using the TPE (Tree-structured Parzen Estimator) algorithm (Bergstra et al., 2011).[12] During tuning, we use greedy decoding (i.e. beam size of one). Once opti-

[12] `https://optuna.readthedocs.io/en/stable/reference/generated/optuna.samplers.TPESampler.html`

Table 7: Hyperparameter values used for each corpus. Hyperparameters values when using entity hinting, if they differ from the values used without entity hinting, are shown in parentheses. Tuned indicates whether or not the hyperparameters were tuned on the validation sets.

Hyperparameter	Tuned?	CDR	GDA	DGM	DocRED
Batch size	✓	4	4	4	4
Training epochs	✓	130 (70)	30 (25)	30 (45)	50
Encoder learning rate	✗	2e-5	2e-5	2e-5	2e-5
Encoder weight decay	✗	0.01	0.01	0.01	0.01
Encoder re-initialized top L layers	✓	1	1 (2)	1	1
Decoder learning rate	✓	1.21e-4 (1.13e-4)	5e-4 (4e-4)	8e-4 (1.5e-5)	7.8e-5
Decoder input dropout	✗	0.1	0.1	0.1	0.1
Decoder hidden-to-hidden weights dropout	✗	0.5	0.5	0.5	0.5
Target embedding size	✗	256	256	256	256
No. heads in multi-head cross-attention	✗	6	6	6	6
Beam size	✓	3 (2)	4 (1)	3 (2)	8
Length penalty	✓	1.4 (0.2)	0.8 (1.0)	0.2 (0.8)	1.4
Max decoding steps	✗	128	96	96	400

mal hyperparameters are found, we tune the beam size (bs) and length penalty (α) using a grid search over the values bs $= \{2...10\}$, with a step size of 1, and $\alpha = \{0.2...2.0\}$, with a step size of 0.2.

G Baselines

This section contains detailed descriptions of all methods we compare to in this paper.

G.1 Pipeline-based methods

These methods are pipeline-based, assuming the entities are provided as input. Many of them construct a document-level graph using dependency parsing, heuristics, or structured attention and then update node and edge representations using propagation.

- Christopoulou et al. (2019) propose EoG, an edge-orientated graph neural model. The nodes of the graph are constructed from mentions, entities, and sentences. Edges between nodes are initially constructed using heuristics. An iterative algorithm is then used to generate edges between nodes in the graph. Finally, a classification layer takes the representation of entity-to-entity edges as input to determine whether those entities express a relation or not. We compare to EoG in the pipeline-based setting on the CDR and GDA corpora.

- Nan et al. (2020) propose LSR (Latent Structure Refinement). A "node constructor" encodes each sentence of an input document and outputs contextual representations. Representations that correspond to mentions and tokens on the shortest dependency path in a sentence are extracted as nodes. A "dynamic reasoner" is then applied to induce a document-level graph based on the extracted nodes. The classifier uses the final representations of nodes for relation classification. We compare to LSR in the pipeline-based setting on the CDR and GDA corpora.

- Lai and Lu (2021) propose BERT-GT, which combines BERT with a graph transformer. Both BERT and the graph transformer accept the document text as input, but the graph transformer requires the neighbouring positions for each token, and the self-attention mechanism is replaced with a neighbour–attention mechanism. The hidden states of the two transformers are aggregated before classification. We compare to BERT-GT in the pipeline-based setting on the CDR and GDA corpora.

- Minh Tran et al. (2020) propose EoGANE (EoG model Augmented with Node Representations), which extends the edge-orientated model proposed by Christopoulou et al. (2019) to include explicit node representations which are used during relation classification. We compare to EoGANE in the pipeline-based setting on the CDR and GDA corpora.

- SSAN (Xu et al., 2021) propose SSAN (Structured Self-Attention Network), which inherits the architecture of the transformer encoder (Vaswani et al., 2017) but adds a novel structured self-attention mechanism to model the coreference and co-occurrence dependencies between an entities mentions. We compare to SSAN in the pipeline-based setting on the CDR and GDA corpora.

- Zhou et al. (2021) propose ALTOP (Adaptive Thresholding and Localized cOntext Pooling), which extends BERT with two modifications. Adaptive thresholding, which learns an optimal threshold to apply to the relation classifier. Localized context pooling, which uses the pre-trained self-attention layers of BERT to create an entity embedding from its mentions and their context. We compare to ALTOP in the pipeline-based setting on the CDR and GDA corpora.

G.2 n-ary relation extraction

These methods are explicitly designed for the extraction of n-ary relations, where $n > 2$.

- Jia et al. (2019) propose a multiscale neural architecture, which combines representations learned over text spans of varying scales and for various sub-relations. We compare to Jia et al. (2019) in the pipeline-based setting on the n-ary DGM corpus.

G.3 End-to-end methods

These methods are capable of performing the sub-tasks of DocRE in an end-to-end fashion with only the document text as input.

- Eberts and Ulges (2021) propose JEREX, which extends BERT with four task-specific components that use BERTs outputs to perform entity mention localization, coreference resolution, entity classification, and relation classification. They present two versions of their relation classifier, denoted "global relation classifier" (GRC) and "multi-instance relation classifier" (MRC). We compare to JEREX-MRC in the end-to-end setting on the DocRED corpus.

Position-based Prompting for Health Outcome Generation

Micheal Abaho[1] **Danushka Bollegala**[1,2*] **Paula R Williamson**[1] **Susanna Dodd**[1]
[1]University of Liverpool, United Kingdom
[2]Amazon
{m.abaho,danushka,prw,shinds}@liverpool.ac.uk

Abstract

Probing factual knowledge in Pre-trained Language Models (PLMs) using prompts has indirectly implied that language models (LMs) can be treated as knowledge bases. To this end, this phenomena has been effective, especially when these LMs are fine-tuned towards not just data, but also to the style or linguistic pattern of the prompts themselves. We observe that, satisfying a particular linguistic pattern in prompts is an unsustainable, time-consuming constraint in the probing task, especially because, they are often manually designed and the range of possible prompt template patterns can vary depending on the prompting task. To alleviate this constraint, we propose using a position-attention mechanism to capture positional information of each word in a prompt relative to the mask to be filled, hence avoiding the need to re-construct prompts when the prompts' linguistic pattern changes. Using our approach, we demonstrate the ability of eliciting answers (in a case study on health outcome generation) to not only common prompt templates like Cloze and Prefix, but also rare ones too, such as Postfix and Mixed patterns whose masks are respectively at the start and in multiple random places of the prompt. More so, using various biomedical PLMs, our approach consistently outperforms a baseline in which the default PLMs representation is used to predict masked tokens.

1 Introduction

Language models (LMs) as knowledge bases (KBs) (LM-as-KB) is a rapidly growing phenomenon attracting a lot of attention in the Natural Language Processing (NLP) community (Petroni et al., 2019; Brown et al., 2020; Shin et al., 2020; Schick and Schütze, 2020b). LM-as-KB implies the usage of LMs as an alternative or at least a proxy for explicit KBs. To achieve LM-as-KB, researchers adopt prompt-based learning (PBL) in which LMs learn to probabilistically predict missing information once given fill-in-the-blank prompt inputs (Liu et al., 2021) such as "Eiffel tower is located in ___". PBL has generally been a success, for example, in a systematic survey of prompting methods, Liu et al. (2021) indicate that *"pre-train, prompt and predict"* is a new paradigm replacing *"pre-train and fine-tune"* paradigm in NLP. Because of this success, the rationale that LMs contain factual retrievable knowledge (LM-as-KB) is ostensibly justified and therefore continually explored.

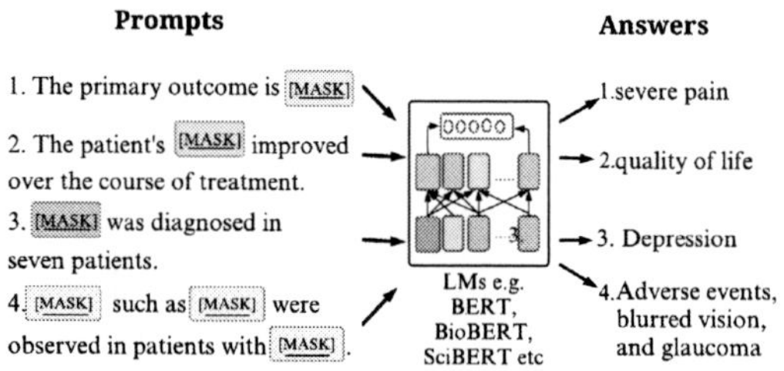

Figure 1: Prompt query variants used for probing evidence (in form of health outcomes) from PLMs, including common styles like Prefix (1) and Cloze (2) style, as well as rare styles Postfix (3) and Mixed (4) styles with [MASK] token/s at the beginning and in multiple positions in the prompt.

The prompt sequences often used in PBL have a masked token or span (denoted by [MASK] in the remainder of the paper) that positionally appears either in the middle (Cloze-style) (Petroni et al., 2019; Schick and Schütze, 2020b; Cui et al., 2021) or at the very end of the sequence (Prefix style) (Qin and Eisner, 2021; Shin et al., 2020). Moreover, we learn that the majority of the PBL tasks probe relational knowledge possessed by pre-trained language models (PLMs) (Jiang et al., 2020b; Petroni et al., 2019; Davison et al., 2019), which implies that the prompt inputs used in querying the PLMs have to contain relational information (such as

*Danushka Bollegala holds concurrent appointments as a Professor at University of Liverpool and as an Amazon Scholar. This paper describes work performed at the University of Liverpool and is not associated with Amazon.

Proceedings of the BioNLP 2022 workshop, Dublin, Ireland, pages 26–36
May 26, 2022.

"*subject-relation-object*" triples). Furthermore, we observe that, a fair amount of time in several PBL tasks is spent reconstructing prompt inputs through manually designing templates (Petroni et al., 2019; Davison et al., 2019) or corrupting prompt inputs through deletion (Lewis et al., 2019), replacement (Raffel et al., 2019) or permutation (Heinzerling and Inui, 2020).

As discussed above, we notice that, the syntactic and semantic structure of prompt inputs is a constraint encountered in PBL, notwithstanding the multitude of constraints that could arise given that PBL is inherently a text generation task (Liu et al., 2021). This constraint will usually require researchers to laboriously prepare supervised data with prompts whose linguistic patterns suit the objective of the prompting task, For instance, (Davison et al., 2019; Jiang et al., 2020a; Heinzerling and Inui, 2020), use templates that reformulate prompts to contain relational information connecting a particular text span to the to-be filled information. However, template-based prompt reformulation has two main challenges. First, it presents a risk of corrupting the grammar of the prompts unwittingly (Davison et al., 2019). Second, the search space of the candidate prompts is too large (Gao et al., 2020) and is practically impossible to create templates that can enumerate all possible linguistic patterns that prompt queries can be tailored to. For example, prompt template patterns with missing information at the beginning and or with multiple missing information in a sequence are yet to be explored in prior works.

To address the above-mentioned challenges, we propose a strategy we denote position-based prompting (PBP), which is less concerned about the linguistic pattern or shape the prompt takes on, but rather focuses on the words (that the prompts are composed of) and their positions relative to the [MASK]. PBP is focused on shifting the emphasis on subject-relation-object triples to the masked positions as well as the interaction of all the other words with the [MASK]s position. PBP is built to automatically adjust from one prompt template to another, which essentially eliminates the need to prepare hand crafted prompts in the event that an LM is to be probed for rare knowledge. In its architecture, PBP enhances contextualised word representations with position-aware representations to solve fill-in-the-blank tasks. In our approach, we fine-tune PLM parameters along with position-oriented parameters to generate position-based contextualised word representations.

To test our approach, we investigate how well biomedical LMs store and recall information relevant to biomedical entities, with a specific interest in health outcomes, which are defined as measurements or observations used to capture and assess the effect of treatments (Williamson et al., 2017). In addition to the Prefix and Cloze styles, we incorporate two rare prompt style patterns that we denote Postfix and Mixed, where the former contains the [MASK] token/s at the beginning of the prompt sequence and the latter has multiple [MASK] token/s in various positions (Figure 1). Our approach obtains mean scores (across several biomedical LMs) in Exact Match (EM) and Partial Match (PM) metrics that are an improvement (2.4% across both metrics) over those obtained using the vanilla PLM representations, reporting a significant improvement of 6.49% in F1 on the EBM-NLP (Nye et al., 2018) dataset. As later defined in section 4.1, EM measures the percentage of predictions of all [MASK] tokens (or spans) that match the ground truth, whereas PM measures the percentage of correctly predicted [MASK] tokens.

2 Entity memorisation and recalling

Large-scale LMs with billions of parameters have already shown to recall facts that were observed in the training data (Heinzerling and Inui, 2020; Jiang et al., 2020a). However, the ground truth for these LMs to achieve this is already laid with systematically handcrafting rules to follow in creating the prompt input sequences they receive at the training stage. For instance, the majority of the prompts created in PBL tasks embed knowledge in form of triples {*subject,relation,object*} such that LMs could correctly predict *object* entities when prompted with a sequence containing a *subject* and *relation* or otherwise predict *subject* entities when prompted with a sequence containing an *object* and a *relation* (Sung et al., 2021; Jiang et al., 2020a; Qin and Eisner, 2021). Whichever the case, models often predict answers as shown in (1).

$$\hat{y}_i = \underset{y_i}{\text{argmax}}\ p([\text{MASK}] = y_i | x_{prompt}) \quad (1)$$

where i is the position of masked token within a prompt x_{prompt}.

In this work, we however do not assume any prior knowledge contained in a prompt, but rather simply locate outcome entities in the sentences extracted from Randomised Clinical Trial (RCT) abstracts and mask them, an approach we refer to as *custom masking*.

3 Method

In addition to formally defining the task we undertake, this section discusses the data used as well as the different stages of our proposed PBP strategy.

3.1 Task

Let us consider an input prompt sequence s with one or more outcomes masked such that $s = x_1, \ldots [M]_i \ldots [M]_j \ldots x_n$, where $[M]$ is a masked token sequence, $[M] = \{x_i\}_{i \geq 1}^{i+|M|}$, $i \in [1, n]$ and $|M|$ is the length of the masked sequence. We consider four different prompt query variants shown in Figure 1: **Prefix prompts** contain $[M]$ at the end of the prompt, **Cloze prompts** contains $[M]$ in the middle of the prompt, **Postfix prompts** contain $[M]$ at the start of the prompt, and **Mixed prompts** where there are several masked sequences distributed across the prompt. The questions we then pose are: (a) *can we determine how knowledgeable biomedical PLMs are of stored facts such as health outcomes?*, and (b) *If queried with any of the above variants, would these PLMs correctly fill in $[M]$s with the correct outcomes?*

3.2 Datasets

Different from previous PBL works, we neither create custom templates nor do we reformulate prompts to follow an ideal linguistic pattern. We use plain raw sentences (that mention health outcomes) extracted from RCT PubMed abstracts, which are contained in the revised version of EBM-NLP (Abaho et al., 2019) and EBM-COMET (Abaho et al., 2021b) datasets. Both of these datasets support evidence based medicine (EBM) tasks such as extraction of health outcomes from clinical trials (Beltagy et al., 2019; Abaho et al., 2021a).

We do not eliminate any of the abstract sentences that do not mention outcomes, because we aim to familiarise the PLM (at fine-tuning) with text or context in RCT abstracts which generally report about outcomes during clinical trial studies (Williamson et al., 2017). We refer to these sentences as *no_blank sequences* and use them alongside the prompt query variants introduced earlier. To our advantage, several sentence segments have no outcome annotations in both the EBM-NLP and EBM-COMET datasets.

3.3 Masked Language model and Prompt engineering

We extract a hidden state h_i for each token in an input prompt s using a domain-specific PLM,

$$\boldsymbol{h_i} = \mathrm{PLM}_\theta(x_i) \tag{2}$$

where $\boldsymbol{h_i}$ is a hidden state for the word x at position i. The matrix of hidden states for the entire input prompt is represented as $\mathbf{H} \in \mathbb{R}^{n \times k}$, where n is number of words in s and k is the hidden state size.

We define a function $f_{\textbf{prompt}}$ that concatenates the h_i in (2) to a randomly initialised d dimensional vector, which we denote as z_t corresponding to one of the four prompt query variants or the additional *no_blank sequences* (introduced in §3.2), where $t \in$ [*prefix, cloze, postfix, mixed, no_blank*]. The function ensures that if an input s is a Prefix prompt, the corresponding vector z_{prefix} is concatenated to each h_i generated from s as shown in (3). This is done to enable knowledge transfer from one prompt query to another. For example, Mixed prompts are by construction a combination of Prefix, Postfix, and Cloze, hence they should benefit from information sharing via a common vector space.

$$f_{\textbf{prompt}}(h_i) = [z_t; h_i] \tag{3}$$

$z_t \in \mathbb{R}^{d_t}$, where z_t is a query type embedding of size d_t.

3.4 Position based conditioning (PBC)

To enrich the token representations, we propose a position-based attention mechanism to steer the model's focus on relevant information in the input prompt. We define a sequence of position ids for each input prompt, where all masked positions take on an id of 0 and all the other tokens take id's relative to the masked position id. For example given a Cloze prompt with m tokens, we assign a mask at position i an id 0, and resulting sequence of position ids is $p = [1-i, 2-i, \ldots, -1, 0, 1, \ldots, (m-1)-i, m-i]$. We compute an attention vector $A^{(s)}$, given by (4), for an input prompt s that allows each token to interact with every other token and retain knowledge of the relative position of the masked tokens in the input sequence.

$$\boldsymbol{A^{(s)}} = \mathrm{softmax}(\mathbf{V}^\top \tanh(\mathbf{W}\mathbf{H}^\top + \mathbf{U}\mathbf{P}_s^\top)) \tag{4}$$

Here, $\boldsymbol{A}^{(s)} \in \mathbb{R}^{n\times 1}$, $\mathbf{V} \in \mathbb{R}^{k_a\times 1}$, k_a is size of attention layer, $\mathbf{W} \in \mathbb{R}^{k_a\times k}$, $\mathbf{P}_s \in \mathbb{R}^{n\times k_p}$ and $\mathbf{U} \in \mathbb{R}^{k_a\times k_p}$. $\mathbf{P}_s$ is a matrix of position embeddings of size k_p extracted for each position p_n in the input prompt s. These embeddings are extracted from a trainable matrix $\mathbf{P} \in \mathbb{R}^{2n\times k_p}$ of randomly initialised vectors of size k_p for all possible positions $2n$ where n is the maximum sequence length, $|\{p_n\}_{-n}^{n-1}| = 2n$. The position based representation of each token is then computed with respect to the type of prompt. For the Prefix, Postfix and Cloze prompts, we obtain a prompt representation M^s given by (5).

$$\mathbf{M}^{(s)} = \boldsymbol{A}^{(s)}\mathbf{H} \tag{5}$$

Here, $\mathbf{M}^{(s)} \in \mathbb{R}^{n\times k}$. For the Mixed prompts in which we have multiple masked positions within the input sequence, we avoid biasing the attention mechanism towards masks at a specific position and thereby considering as many position id sequences as there are masked positions in the input prompt. For example, given a sequence with 3 masked positions, $s = [M], x_2, x_3, [M], x_5, x_6, [M]$, we obtain 3 position id sequences, i.e. the combined position id sequences is,

$$P^{(s)} = \bigcup_i P_i,$$

where each P_i is obtained with respect to the current mask position i. For the example above, we have $P^{(s)}$ = {[0,1,2,3,4,5,6], [-3,-2,-1,0,1,2,3,], [-6,-5,-4,-3,-2,-1,0]}, where the first position id sequence is obtained by treating the $[M]$ at position 1, as mask at i, the second is obtained by treating the $[M]$ at position 4 as mask at i and finally the third by treating $[M]$ at the last position as mask at i. Attention vectors are computed for each position id sequence (P_i) and subsequently used to obtain the prompt representation $\mathbf{M}^s_{P_i}$. We compute the final representation of a Mixed prompt as the mean pool across these different representations,

$$\mathbf{M}^{(s)} = \sum_i^{|P^{(s)}|} \mathbf{M}^s_{P_i} \tag{6}$$

3.5 Prompt fine-tuning

The predicted probability of each vocabulary token is estimated via (7).

$$y = \text{softmax}(f(W_v \mathbf{M}^{(s)^\top}) \tag{7}$$

Therein, $W_v \in \mathbb{R}^{v^*\times k}$, v^* is the vocabulary size and f is a non-linear activation function. We use a BERT-based loss in predicting the masked tokens in each input given by (8).

$$\boldsymbol{L}_{PLM} = -\sum_{s\in\mathcal{T}}\sum_i^n \log P(y_i|s) \tag{8}$$

where $\mathcal{T}$ is the set of training example prompts. Some of the prompt query variants (Postfix and Prefix) are rare in the datasets, and some other prompt sequences are quite lengthy. This poses a challenge particularly when using small PLMs (with few parameters) to recall factual information. In order to mitigate model forgetfulness in such examples, we introduce an auxiliary task that computes a text classification loss as a cross entropy loss given by (9).

$$\boldsymbol{L}_{TC} = -\sum_{s\in\mathcal{T}}\sum_{i\in n} \log P(y_i|y_{<i}, s) \tag{9}$$

The overall training loss is defined as the weighted combination of the two losses as given in (10).

$$\boldsymbol{L} = \boldsymbol{L}_{PLM} + \lambda\boldsymbol{L}_{TC} \tag{10}$$

Similar to (Chronopoulou et al., 2019) and (Schick and Schütze, 2020a), we introduce a weighting parameter $\lambda(>0)$ to adapt the auxiliary losses to the main mask prediction task[1].

3.6 Prediction

Similar to BERT (Devlin et al., 2018), we consider generating outputs in parallel, initially treating the default representations provided by the model in (2) as a baseline and therefore use them to predict tokens in masked positions. We then use position-aware representation obtained using the attention mechanism in §3.4 to predict the mask tokens, calling these results Position-based conditioning (PBC). Lastly, we endeavour to retain the contextual knowledge presented by the PLMs as much as we possibly can by computing an average of the Baseline and PBC representations and term these Contextual PBC.

4 Experiments

In our experiments, we use several PLMs that are pre-trained on clinical texts such as PubMed

[1]Our implementation is publicly available `https://github.com/MichealAbaho/outcome_generation.git`

Dataset-	EBM-COMET						EBM-NLP					
Method-	Baseline		PBC		Contextual PBC		Baseline		PBC		Contextual PBC	
Metric-	EM	PM	EM	PM	EM	PM	EM	PM	EM	PM	EM	PM
BERT	43.12	47.55	43.04	49.84	44.32	55.94	37.40	45.55	41.10	47.00	47.31	51.06
BioBERT	50.71	58.01	50.55	58.61	53.34	59.65	51.15	55.62	51.19	53.80	52.15	54.50
SciBERT	61.17	67.48	62.34	69.85	63.00	70.95	57.12	62.25	57.18	63.75	59.44	63.91
Biomed_RoBERTA	44.01	59.67	44.32	59.73	44.32	62.86	40.45	51.72	47.21	49.81	49.17	55.00
UmlsBERT	31.05	34.61	30.47	35.77	31.88	36.46	28.66	33.15	30.02	38.51	39.16	40.15
Mean score	46.01	53.46	46.14	54.76	47.37	57.17	42.96	49.66	45.34	50.57	49.45	52.92

Table 1: Table reports EM and PM accuracies of the various biomedical Pre-trained Language Models for the outcome recalling experiments. Mean score in a particular column is the average across all results in that column.

abstracts, which often report outcomes such as BioBERT (Lee et al., 2020), SciBERT (Beltagy et al., 2019) and Biomed_RoBERTA (Gururangan et al., 2020). Additionally, we include UmlsBERT because it augments BERT's pre-training input with semantic type embeddings aligned to clinical knowledge (semantic types) in the Unified Medical Language System (UMLS) Metathesaurus (Michalopoulos et al., 2020). We also use BERT (Devlin et al., 2018) as a vanilla PLM that has not been pre-trained specifically on clinical texts.

4.1 Training and Evaluation

Unlike previous works where a particular relation within a prompt e.g. born-in, lives-in etc. might appear multiple times within the train set, in our case, prompts are not semantically related in any way (i.e. their is no relation knowledge that can be transferred over from one prompt to another). Because of the nature of our prompts, we believe it might be harder for the model to memorise them, we therefore opt to train the models until the perplexity on the training data reaches 1 or until the accuracy on the validation data saturates. We examine the model's generalisation ability to transfer knowledge to unseen prompts in few-shot and zero-shot settings. For the few-shot setting, we design experiments where we measure a model's accuracy in generating outcomes (as answers), which it encountered in a small number of prompts during training. The contexts in these evaluation prompts are not encountered during training. For example, consider an evaluation prompt – *"The patient's overall [MASK] improved according to the HRQOL questionnaire"*, the model would not have encountered the context surrounding the *"[MASK]"*. For the zero-shot evaluation, the model would have neither encountered the prompt nor the target outcomes during training. To simulate both the zero- and few-shot settings, we randomly split the datasets into train (80%) and test (20%) splits, and use the latter for the generalisation evaluation task shown in Table 3. We tune all hyperparameters using the validation data, and obtain optimal values as follows: learning rate - 5e-5, batch size - 8, query type embedding size - 50, position embedding size - 300 and an attention layer size - 200. Further details on tuning bounds are provided in the Appendix.

Metrics: We define two different metrics for evaluating the proposed PBP strategy: Exact Match (EM) and Partial Match (PM). EM counts a prediction as 1 only if it matches completely with the correct answer, whereas PM uses the fraction of the overlapping tokens between the predicted and correct answers. Both EM and PM are averaged over all test instances to compute aggregated evaluation metrics, and we report their percentages in the paper.

5 Results

In this section, we evaluate how well the model generates health outcomes when queried to answer a given prompt. For example, *"After patients were given sorafenib, they reported [MASK]"*, the model should correctly generate the outcome *Fatigue* for the *[MASK]*.

5.1 Outcome memorisation and retrieval

Table 1 shows the performance of the proposed PBC method in the outcome generation task. As observed, PBC consistently outperforms the baseline across most of the clinically informed BERT LMs (for both datasets), particularly for the PM results. More interestingly, we notice that Contextual PBC further improves the performance (both in EM and PM), indicating the importance of preserving the contexts in the position-based representations.

	#	Average prompt length	EM	PM
Postfix	65	18.5	48.43	58.51
Prefix	53	9.1	69.23	77.24
Cloze	630	24.2	50.08	60.49
Mixed	2594	38.8	43.68	45.46

Table 2: Exact Match (EM) and Partial Match (PM) accuracies for Outcome memorisation/recalling for the different prompt types using the EBM-COMET dataset.

Comparing the different LMs, we found that, SciBERT performs best followed by Biomed_RoBERTa and BioBERT. Since all tested models follow the original BERT's architecture, we hypothesize that, the nature of corpora used in pre-training the best performing models was responsible for the performance, i.e. unlike UMLSbert and BERT, all the other models are pre-trained on text that includes PubMed abstracts, which often report outcomes. Additionally, we observe that PM results were generally better than EM results, which we attribute to the fact that PM is less strict compared to EM because it rewards the model for correctly generating a few of the tokens in the masked positions. Overall, the results suggest that PBC can be used to effectively retrieve facts such as health outcomes (biomedical entities) by simply augmenting contextual word representations with position-aware representations.

5.1.1 Prompt query variants

In Table 2, we notice that the accuracy with which a model correctly answers Prefix prompts is significantly higher than that of the other prompts. We attribute this performance to the short length of these spans such as the one shown in Table 4 and the average number of tokens to decode per prompt. We also notice that the model struggles to correctly answer Mixed prompts compared to other types of prompts. We attribute this to the fact that, Mixed prompts are generally very long sequences (38.8 tokens on average) and contain multiple masked positions to be predicted.

5.2 Few- and Zero-shot Evaluations

To evaluate the model's generalisability, we fine-tune the model towards a small amount of target outcomes, and then measure the transferability of this knowledge by requiring the model to accurately generate these outcomes in prompts with

	Cloze	Mix	Postfix	Prefix
#	174	613	13	12

Table 3: Number of prompts per prompt type used in evaluation of the few- and zero-shot settings.

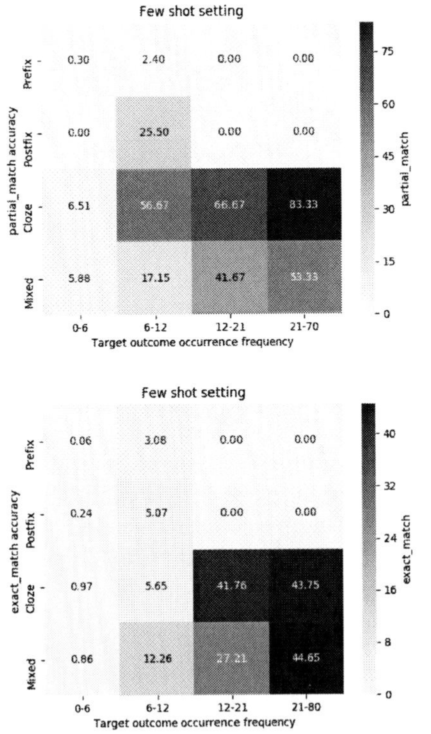

Figure 2: Visualizing the Partial Match and Exact match accuracies when the best model (SciBERT+Contextual PBC+EBM-COMET) is trained with only a certain number of target outcomes.

completely different contexts. Test set prompts in Table 3 are carefully chosen using regular expression matching such that the contexts surrounding the missing outcomes are different from that of similar outcomes observed during training. For example, the model could have been trained on the outcome "adverse events" in five different prompts, and then at evaluation, the model is required to generate the same outcome, however using prompts that are different from those encountered during training. By *different* here we mean that the context (e.g. {ctxt} surrounding masks [M] in Table 4) in the prompt changes during this evaluation. Figure 2 plots shows results of model evaluation on prompts (Table 3). As observed in the plots, the model struggles to generate outcomes it hardly encountered during training (i.e. outcomes appearing

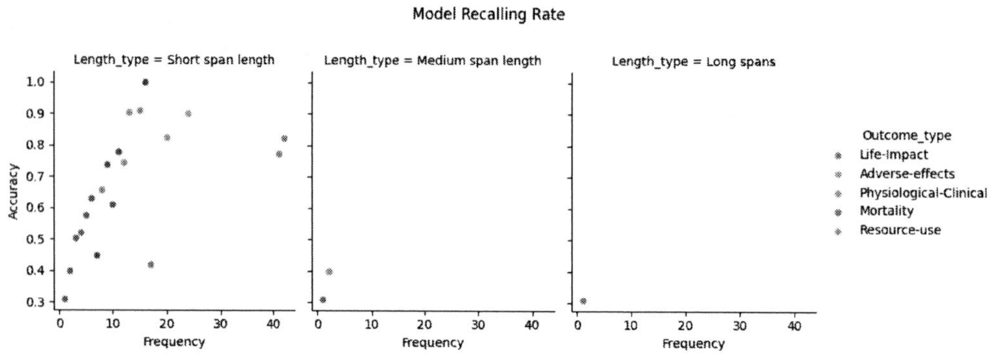

Figure 3: Analysis of the accuracy (PM) with which best model (SciBERT+Contextual PBC+EBM-COMET) recalls different types of factual information (outcome types) with varying span lengths and occurrence frequency (in the dataset).

in 0-6 prompts or 6-12 prompts). This is mostly evident in generating outcomes for Prefix and Postfix prompts, which is because there were not just few evaluated prompts of this types, but there were also few (53 and 65 respectively as shown in Table 2) in the train set. However, we see a trend of performance improvement when the frequency of target outcomes encountered during training increases, particularly for the Mixed and Cloze prompt.

6 Analysis

6.1 Impact of Length and Frequency of Outcomes

We partition the entire set of outcomes in EBM-COMET into 3 different groups based on lengths. Dividing the length of the longest outcome (22) by 3, we get approximately 7 which we use to create 3 groups i.e. 1) "short span length" to represent outcomes that are $\leq$ 7 tokens long, 2) "medium span length" to represent outcomes of 7 $>$ and $\leq$ 14 tokens, and finally 3) "long spans" to represent outcomes of $\geq$ 14 tokens long. Figure 3 shows how well the best model (SciBERT+Contextual PBC+EBM-COMET) performs when recalling outcomes of varying lengths and frequencies. Following prior work on EBM NLP, we endeavour to show the model's outcome recall rate by outcome type, which can be informative in terms of the complexity of modelling these outcomes. We firstly notice the skewed distribution of outcome lengths with short spans dominant in the training sample. Unsurprisingly, we observe a trend of a performance increase as the frequency increases across the left hand plot with short outcomes, implying that the model struggles to recall infrequent outcomes despite their size but easily recalls the more frequent ones.

6.2 Random masking Vs custom masking

Figure 4 shows results of an ablation test in which we replace our custom masking approach with random masking. The key difference between the two is, while custom masking involves masking (or hiding) the outcomes in the prompts, random masking arbitrary masks 15% of the prompts tokens. As shown in the figure, the number of epochs required to reach a perplexity of 1.0 on the train data for the two masking approaches is almost incomparable, with custom masking quickly achieving this in approximately 7 epochs and random masking failing to achieve this, even after 20 epochs. The earliest random masking achieves 1.0 perplexity is 80 epochs for SciBERT, however we only visualise 20 epochs because of space. Besides this, the insight suggests that, custom masking would significantly reduce GPU run-time or otherwise minimise overwhelming computational resources with massive datasets.

6.3 Error Analysis

We analyse the outcomes generated by the best model (SciBERT+Contextual PBC+EBM-COMET) during the few shot evaluation and notice that whilst the model generates correct outcomes for some prompts, it makes various kinds of mistakes. Table 4 includes a fair sample of the most commonly discovered mistakes. **Incomplete outcomes**, such in the Postfix where instead of "Quality of life", the model generates "Life". **Outcomes with irrelevant information**, such as Prefix case where the models generates more than what's ex-

Query Variant	Prompt	Correct	Generated outcomes
Cloze {ctxt} [M] {ctxt}	Self-reported life-time medical diagnosis of **[M]** or use of antidepressants was considered as outcome.	- Depression	- Depression
Postfix [M] {ctxt}	**[M]** was assessed by questionnaires EORTC QLQ-C30, and EORTC QLQ-BR23 at baseline, and at three, six, and nine months.	- Quality of life	- Life
Prefix {ctxt} [M]	Two CMZ patients and one morphine patient showed complete **[M]**.	- pain	- unwanted pain
Mixed {ctxt} [M] {ctxt} [M] {ctxt}	Further additional benefits are better **[M]** and shorter **[M]** compared with standard GVHD prophlaxis without ATLG.	- quality of life (QOL) - immunosuppressive treatment	- immunosuppressive treatment
	The incidence of postoperative **[M]**, **[M]**, **[M]** and **[M]** was similar between the groups	- nausea, - vomiting, - drowsiness, -headache	- anxiety, - depression

Table 4: Example prompts from our test set and their predicted or generated outcomes for the outcome generation task. The Query variant column indicates the type of prompt as well as the prompt structure where {ctxt} implies context which might appear before, after or either ends of a masked sequence span.

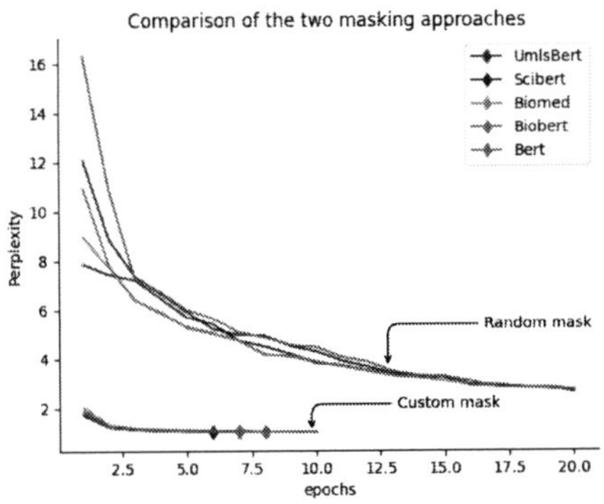

Figure 4: Achieving a target perplexity of 1.0 on the train dataset takes no fewer than 20 epochs with generic random masking of 15% of the input prompt tokens (Devlin et al., 2018) compared to masking target factual information i.e. outcome spans themselves. Hitting target perplexity is shown using a diamond.

pected, "unwanted pain" instead of "pain". Finally, **wrong outcomes**, where the model generates completely unexpected outcomes such as the case in the Mixed prompts.

7 Related work

Interrogating PLMs with fill-in-the-blank prompts to determine their knowledge and awareness of factual information is a trending paradigm in NLP. Despite the emergence of subtle techniques such as automating prompt structuring (Shin et al., 2020; Gao et al., 2020), selectively updating parameters of LMs and prompts (also known as continuous prompting) (Li and Liang, 2021; Qin and Eisner, 2021), or even not tuning at all (Brown et al., 2020), several works including these still heavily rely on handcrafted prompts to use in probing LMs. Our efforts are motivated by the fact that we need not worry about the nature of the prompt, but rather can leverage on information local to the prompt such as word positions to probe the LMs. We attempt to enhance a word's contextualised representation with position based representations to capture the word's position relative to the mask to be filled. Previously some works have used similar position-aware attention over LSTMs for relation extraction, sequence labelling and slot filling tasks in different datasets (Wei et al., 2021; Zhang et al., 2017). To the best of our knowledge, we are the first to use an extra position-attention layer above transformer models such as BERT to solve the fill-in-the-blank prompting task.

8 Conclusion

This paper assesses the possibility of ignoring the constraint of aligning prompts to specific linguistic patterns in prompting tasks that aim to store knowledge in LMs that could later be retrieved or transferred for fact generation tasks. In experiments using clinical domain datasets (supporting EBM tasks), we show that the position-based attention implemented over contextualised LMs can improve the ability of PLMs to recall facts such as outcomes (biomedical entities) encountered during training. We further observe our proposed model is able to generalise across unseen prompts, performing considerably well for Cloze and Mixed (extremely rare in PBL tasks) prompts. With the obtained experimental results, despite not aligning our prompts to commonly followed linguistic patterns, we can positively answer the question posed in §3.1 by claiming that PLMs are knowledgeable of stored facts.

References

Michael Abaho, Danushka Bollegala, Paula Williamson, and Susanna Dodd. 2021a. Detect and classify–joint span detection and classification for health outcomes. *arXiv preprint arXiv:2104.07789*.

Micheal Abaho, Danushka Bollegala, Paula Williamson, and Susanna Dodd. 2019. Correcting crowdsourced annotations to improve detection of outcome types in evidence based medicine. In *CEUR Workshop Proceedings*, volume 2429, pages 1–5.

Micheal Abaho, Danushka Bollegala, Paula Williamson, and Susanna Dodd. 2021b. Assessment of contextualised representations in detecting outcome phrases in clinical trials. *European Journal of Biomedical Informatics*, 17(9).

Iz Beltagy, Kyle Lo, and Arman Cohan. 2019. Scibert: A pretrained language model for scientific text. *arXiv preprint arXiv:1903.10676*.

Austin J Brockmeier, Meizhi Ju, Piotr Przybyła, and Sophia Ananiadou. 2019. Improving reference prioritisation with pico recognition. *BMC medical informatics and decision making*, 19(1):1–14.

Tom B Brown, Benjamin Mann, Nick Ryder, Melanie Subbiah, Jared Kaplan, Prafulla Dhariwal, Arvind Neelakantan, Pranav Shyam, Girish Sastry, Amanda Askell, et al. 2020. Language models are few-shot learners. *arXiv preprint arXiv:2005.14165*.

Alexandra Chronopoulou, Christos Baziotis, and Alexandros Potamianos. 2019. An embarrassingly simple approach for transfer learning from pretrained language models. *arXiv preprint arXiv:1902.10547*.

Leyang Cui, Yu Wu, Jian Liu, Sen Yang, and Yue Zhang. 2021. Template-based named entity recognition using bart. *arXiv preprint arXiv:2106.01760*.

Joe Davison, Joshua Feldman, and Alexander M Rush. 2019. Commonsense knowledge mining from pretrained models. In *Proceedings of the 2019 Conference on Empirical Methods in Natural Language Processing and the 9th International Joint Conference on Natural Language Processing (EMNLP-IJCNLP)*, pages 1173–1178.

Jacob Devlin, Ming-Wei Chang, Kenton Lee, and Kristina Toutanova. 2018. Bert: Pre-training of deep bidirectional transformers for language understanding. *arXiv preprint arXiv:1810.04805*.

Susanna Dodd, Mike Clarke, Lorne Becker, Chris Mavergames, Rebecca Fish, and Paula R. Williamson. 2018. A taxonomy has been developed for outcomes in medical research to help improve knowledge discovery. *Journal of Clinical Epidemiology*, 96:84–92.

Tianyu Gao, Adam Fisch, and Danqi Chen. 2020. Making pre-trained language models better few-shot learners. *arXiv preprint arXiv:2012.15723*.

Suchin Gururangan, Ana Marasović, Swabha Swayamdipta, Kyle Lo, Iz Beltagy, Doug Downey, and Noah A. Smith. 2020. Don't stop pretraining: Adapt language models to domains and tasks. In *Proceedings of the 58th Annual Meeting of the Association for Computational Linguistics*, pages 8342–8360, Online. Association for Computational Linguistics.

Benjamin Heinzerling and Kentaro Inui. 2020. Language models as knowledge bases: On entity representations, storage capacity, and paraphrased queries. *arXiv preprint arXiv:2008.09036*.

Xiaoli Huang, Jimmy Lin, and Dina Demner-Fushman. 2006. Evaluation of PICO as a knowledge representation for clinical questions. *AMIA ... Annual Symposium proceedings. AMIA Symposium*, pages 359–63.

Zhengbao Jiang, Antonios Anastasopoulos, Jun Araki, Haibo Ding, and Graham Neubig. 2020a. X-factr: Multilingual factual knowledge retrieval from pretrained language models. *arXiv preprint arXiv:2010.06189*.

Zhengbao Jiang, Frank F Xu, Jun Araki, and Graham Neubig. 2020b. How can we know what language models know? *Transactions of the Association for Computational Linguistics*, 8:423–438.

Jinhyuk Lee, Wonjin Yoon, Sungdong Kim, Donghyeon Kim, Sunkyu Kim, Chan Ho So, and Jaewoo Kang. 2020. Biobert: a pre-trained biomedical language representation model for biomedical text mining. *Bioinformatics*, 36(4):1234–1240.

Mike Lewis, Yinhan Liu, Naman Goyal, Marjan Ghazvininejad, Abdelrahman Mohamed, Omer Levy, Ves Stoyanov, and Luke Zettlemoyer. 2019. Bart: Denoising sequence-to-sequence pre-training for natural language generation, translation, and comprehension. *arXiv preprint arXiv:1910.13461*.

Xiang Lisa Li and Percy Liang. 2021. Prefix-tuning: Optimizing continuous prompts for generation. *arXiv preprint arXiv:2101.00190*.

Pengfei Liu, Weizhe Yuan, Jinlan Fu, Zhengbao Jiang, Hiroaki Hayashi, and Graham Neubig. 2021. Pre-train, prompt, and predict: A systematic survey of prompting methods in natural language processing. *arXiv preprint arXiv:2107.13586*.

George Michalopoulos, Yuanxin Wang, Hussam Kaka, Helen Chen, and Alex Wong. 2020. Umlsbert: Clinical domain knowledge augmentation of contextual embeddings using the unified medical language system metathesaurus. *arXiv preprint arXiv:2010.10391*.

Benjamin Nye, Junyi Jessy Li, Roma Patel, Yinfei Yang, Iain J. Marshall, Ani Nenkova, and Byron C. Wallace. 2018. A Corpus with Multi-Level Annotations of Patients, Interventions and Outcomes to Support Language Processing for Medical Literature. In *Proc. of the 56th Annual Meeting of the Association for Computational Linguistics*, pages 197–207.

Fabio Petroni, Tim Rocktäschel, Sebastian Riedel, Patrick Lewis, Anton Bakhtin, Yuxiang Wu, and Alexander Miller. 2019. Language models as knowledge bases? In *Proceedings of the 2019 Conference on Empirical Methods in Natural Language Processing and the 9th International Joint Conference on Natural Language Processing (EMNLP-IJCNLP)*, pages 2463–2473, Hong Kong, China. Association for Computational Linguistics.

Guanghui Qin and Jason Eisner. 2021. Learning how to ask: Querying lms with mixtures of soft prompts. *arXiv preprint arXiv:2104.06599*.

Colin Raffel, Noam Shazeer, Adam Roberts, Katherine Lee, Sharan Narang, Michael Matena, Yanqi Zhou, Wei Li, and Peter J Liu. 2019. Exploring the limits of transfer learning with a unified text-to-text transformer. *arXiv preprint arXiv:1910.10683*.

Timo Schick and Hinrich Schütze. 2020a. Exploiting cloze questions for few shot text classification and natural language inference. *arXiv preprint arXiv:2001.07676*.

Timo Schick and Hinrich Schütze. 2020b. It's not just size that matters: Small language models are also few-shot learners. *arXiv preprint arXiv:2009.07118*.

Taylor Shin, Yasaman Razeghi, Robert L Logan IV, Eric Wallace, and Sameer Singh. 2020. Autoprompt: Eliciting knowledge from language models with automatically generated prompts. *arXiv preprint arXiv:2010.15980*.

Mujeen Sung, Jinhyuk Lee, Sean Yi, Minji Jeon, Sungdong Kim, and Jaewoo Kang. 2021. Can language models be biomedical knowledge bases? *arXiv preprint arXiv:2109.07154*.

Wei Wei, Zanbo Wang, Xianling Mao, Guangyou Zhou, Pan Zhou, and Sheng Jiang. 2021. Position-aware self-attention based neural sequence labeling. *Pattern Recognition*, 110:107636.

Paula R. Williamson, Douglas G. Altman, Heather Bagley, Karen L. Barnes, Jane M. Blazeby, Sara T. Brookes, Mike Clarke, Elizabeth Gargon, Sarah Gorst, Nicola Harman, Jamie J. Kirkham, Angus McNair, Cecilia A.C. Prinsen, Jochen Schmitt, Caroline B. Terwee, and Bridget Young. 2017. The COMET Handbook: Version 1.0.

Yuhao Zhang, Victor Zhong, Danqi Chen, Gabor Angeli, and Christopher D Manning. 2017. Position-aware attention and supervised data improve slot filling. In *Proceedings of the 2017 Conference on Empirical Methods in Natural Language Processing*, pages 35–45.

Appendices

A Hyperparameters and Run time

Using BioBERT in the Position based conditioning framework, we perform a grid search through multiple combinations of hyperparameters included in Table Table 5 below. The model is tuned on 20% of EBM-COMET dataset (as a dev set), we obtain the best Partial Match (PM) and Exact Match (EM) accuracies. Table Table 5 shows the range of values (including the lower and upper bound) for which the model is tuned to obtain optimal configurations. Using a shared TITAN RTX 24GB GPU, the baseline model runs for approximately 40 minutes per epoch.

Parameter	Tuned-range	Optimal
Train Batch size	[8,16,32]	16,32
Eval Batch size	[8,16,32]	8
Query type embedding size	[50,100,150]	50
Position embedding size	[100,200,300]	300
Attention layer size	[100,200,300]	200
Optimizer	[Adam, SGD]	Adam
Learning rate	[5e-5, 1e-4, 5e-3, 1e-3]	5e-5

Table 5: Parameter settings for the Position-based conditioning model

B Datasets

B.1 EBM-NLP

EBM-NLP corpus (Nye et al., 2018) is a crowd sourced dataset in which ca.5,000 clinical trial abstracts were annotated with elements in the health literature searching PICO framework (Huang et al., 2006). PICO stands for Participants, Interventions, Comparators and Outcomes. The dataset has supported clinicalNLP research tasks (Beltagy et al., 2019; Brockmeier et al., 2019). The corpus has two versions, (1) the "**starting spans**" in which text spans are annotated with the literal "PIO" labels (I and C merged into I) and (2) the "**hierarchical labels**" in which the annotated outcome "PIO" spans were annotated with more specific labels aligned to the concepts codified by the Medical Subject Headings (MeSH) [2], for instance the Outcomes (O) spans are annotated with more granular (specific) labels which include Physical, Pain, Mental, Mortality and Adverse effects. For the clinical recognition task we attempt, we use the hierarchical version of the dataset. The dataset has however

[2] https://www.nlm.nih.gov/mesh

been discovered to have flawed outcome annotations (Abaho et al., 2019) such as (1) statistical metrics and measurement tools annotated as part of clinical outcomes e.g.“*mean arterial blood pressure*” instead of “*arterial blood-pressure*”,“*Quality of life Questionnaire*” instead of “*Quality of life*” and (2) Multiple outcomes annotated as a single outcome “*Systolic and Diastolic blood- pressure*” instead of “*Systolic blood-pressure*” and “*Diastolic blood-pressure*”.

B.2 EBM-COMET

A biomedical corpus containing 300 PubMed “Randomised controlled Trial” abstracts manually annotated with outcome classifications drawn from the taxonomy proposed by (Dodd et al., 2018). The abstracts were annotated by two experts with extensive experience in annotating outcomes in systematic reviews of clinical trials (Abaho et al., 2021b). Dodd et al. (2018)’s taxonomy hierarchically categorised 38 outcome domains into 5 outcome core areas and applied this classification system to 299 published core outcome sets (COS) in the Core Outcomes Measures in Effectiveness (COMET) database.

C Layer probing

Initially, the hidden state we used (Equation (2)) extracted from the last layer for each of the Biomedical PLMs for all experiments. We however explore an option of extracting a weighted average of representation across all layers (Equation (12)) as a hidden state and study the performance of the models once this hidden state is introduced in the Position based conditioning framework to obtain position-aware representations.

$$h_i^l = \text{PLM}_\theta(x_i) \tag{11}$$

$$h_i = \text{MeanPool}(h_i^1, .., h_i^l, .., h_i^{l_N}) \tag{12}$$

where h_i^l is a hidden state extracted from the lth layer for word x at position i.

We only repeat training experiments using the Contextual PBC setup (subsection 3.6) however this time round using a mean pooled embedding across all layers as the hidden state. We notice that, aggregating a tokens representation by mean pooling across all layers of the transformer-based models does improve the performance in the outcome recalling experiments for both datasets.

Dataset	EBM-COMET			
Method	Contextual PBC (last layer)		Contextual PBC (Mean pool)	
Metric	EM	PM	EM	PM
BERT	43.32	55.94	45.80	57.19
BioBERT	53.34	59.65	53.58	61.22
SciBERT	63.00	70.95	63.15	72.67
Biomed_Roberta	44.32	62.86	45.00	63.17
UmlsBERT	31.88	36.46	33.10	39.21
Mean score	47.37	57.17	48.13	58.70

Table 6: Table reports EM and PM accuracies of the various biomedical Pre-trained Language Models for the outcome recalling experiments using the EBM-COMET and Contextual PBC. Mean score in a particular column is the average across all results in that column.

Dataset	EBM-NLP			
Method	Contextual PBC (last layer)		Contextual PBC (Mean pool)	
Metric	EM	PM	EM	PM
BERT	47.31	51.06	47.45	53.41
BioBERT	52.15	54.50	54.80	55.15
SciBERT	59,44	63.91	60.08	66.93
Biomed_Roberta	49.17	55.00	49.19	56.33
UmlsBERT	39.16	40.15	41.12	42.41
Mean score	49.45	52.92	50.53	54.85

Table 7: Table reports EM and PM accuracies of the various biomedical Pre-trained Language Models for the outcome recalling experiments using the EBM-NLP and Contextual PBC. Mean score in a particular column is the average across all results in that column.

How You Say It Matters: Measuring the Impact of Verbal Disfluency Tags on Automated Dementia Detection

Shahla Farzana, Ashwin Deshpande, and **Natalie Parde**
Natural Language Processing Laboratory
Department of Computer Science
University of Illinois at Chicago
{sfarza3, adeshp27, parde}@uic.edu

Abstract

Automatic speech recognition (ASR) systems usually incorporate postprocessing mechanisms to remove disfluencies, facilitating the generation of clear, fluent transcripts that are conducive to many downstream NLP tasks. However, verbal disfluencies have proved to be predictive of dementia status, although little is known about how various types of verbal disfluencies, nor automatically detected disfluencies, affect predictive performance. We experiment with an off-the-shelf disfluency annotator to tag disfluencies in speech transcripts for a well-known cognitive health assessment task. We evaluate the performance of this model on detecting repetitions and corrections or retracing, and measure the influence of gold-annotated versus automatically detected verbal disfluencies on dementia detection through a series of experiments. We find that removing both gold and automatically-detected disfluencies negatively impacts dementia detection performance, degrading classification accuracy by 5.6% and 3% respectively.

1 Introduction

As populations grow older worldwide, the number of people with Alzheimer's disease (AD) and related dementia is also on the rise (Alzheimer's Association, 2018). Significant changes to speech and language use caused by dementia occur early in disease progression (Bucks et al., 2000). Interesting case studies have demonstrated how diachronic analysis of patients' language use may reveal signs of dementia, using writing samples from British novelists Iris Murdoch, who ultimately perished with Alzheimer's, and Agatha Christie, who was suspected of it (Le et al., 2011). Numerous studies have also sought to automatically detect early signs of the disease and model its progression using speech and writing samples (Becker et al., 1994; Herd et al., 2014; Yancheva et al., 2015; Masrani, 2018; Di Palo and Parde, 2019; Zhu et al., 2019; Fraser et al., 2019; Eyre et al., 2020; Farzana and Parde, 2020; Sarawgi et al., 2020).

Although some studies have pointed to disfluency patterns as an important predictor of AD status (Lopez-de Ipina et al., 2017; Mueller et al., 2018), research in this area has been limited by several factors. Disfluency detection is a challenging and resource-intensive task in itself (Wang et al., 2017; Jamshid Lou and Johnson, 2017; Zayats and Ostendorf, 2019), and may lie out of scope for many interdisciplinary researchers already straddling boundaries between NLP and clinical practice (Valizadeh and Parde, 2022; Kaelin et al., 2021). Rich manual disfluency annotations are present in some datasets common in automated dementia detection (Becker et al., 1994), but off-the-shelf ASR systems do not typically transcribe disfluencies. Moreover, inconsistencies between automatically generated and gold standard transcripts may pose significant challenges for modeling dementia in real-world applications (Balagopalan et al., 2020b), for which ASR will be a necessary component of any speech-based pipeline.

We address these limitations, by investigating the impacts of automatically derived disfluencies on modeling cognitive decline. Our key contributions are as follows:

1. We experiment with an off-the-shelf disfluency detection model to automatically assign word- and phrase-level disfluency tags to samples from the most popular dementia detection dataset, focusing on repetitions and retraces.

2. We measure the influence of these disfluency types on the downstream task of dementia detection by systematically ablating gold-labelled and automatically tagged disfluencies from manual transcripts.

3. We compare AD classification performance on manually and automatically generated transcripts, and compare the removal of gold and

Proceedings of the BioNLP 2022 workshop, Dublin, Ireland, pages 37–48
May 26, 2022. ©2022 Association for Computational Linguistics

automatically detected disfluencies from manual transcripts, to investigate the influence verbal disfluencies have on dementia detection.

This analysis[1] not only paves the way for the discovery of approaches to automated dementia detection that are more suitable for realistic scenarios, but also enhances our understanding of the individual contributions of different disfluency types to this task. We report on related studies and provide relevant background for automatic disfluency detection in §2. We describe our datasets and task setup in §3, and detail our methods in §4. We report the results of our experiments in §5, and further analyze our findings in §6 before concluding in §7.

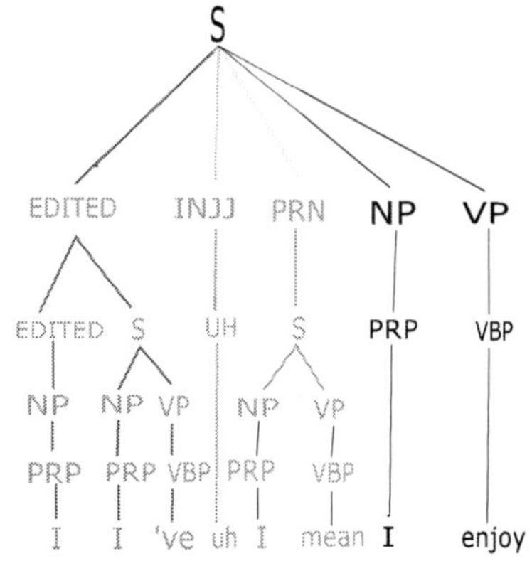

Figure 1: An example of gold labelled parse tree (Jamshid Lou et al., 2019).

2 Related Work

2.1 Studies of Disfluency in the Context of Cognitive Decline

Disfluency, defined as any interruption in the normal flow of speech, is prevalent in spoken language. Verbal disfluency comprises several major subcategories: *false starts*, *repetitions*, *filled pauses* (e.g., “uh,” “um,” etc.), and *sentence corrections* (Shriberg, 1994). Although verbal and nonverbal (*unfilled pauses*) disfluencies are common in spontaneous speech, there is a fine line between normal and abnormal disfluencies. This boundary can be exploited to facilitate modeling cognitive decline.

Studies have found that verbal fluency is an effective indicator of cognitive decline, as fluency declines rapidly for subjects suffering from early stage Mild Cognitive Impairment (MCI) relative to healthy controls (Mueller et al., 2018). Researchers have previously leveraged both acoustic and transcript-based fluency features to automatically detect MCI (Lopez-de Ipina et al., 2017; Mueller et al., 2018). Another study revealed that anomic aphasic subjects tend to produce more disfluent speech than non-aphasic subjects during word retrieval tasks, when examining disfluencies or “stutterings” including part-word repetitions, vocal segregate repetitions, and prolongations (Brown and Cullinan, 1981).

Transcript-based normalized verbal disfluency features (e.g, *filled pause count*, *retracing count*, and *repetition count*) have proved to be discriminative in predicting outcomes from cognitive screening tests such as the Mini Mental State Examination (MMSE) and AD classification, as have concatenations of automatically detected verbal disfluency segments (e.g., *repair onset*, *edit term*, and *fluent words*) with word vectors (Farzana and Parde, 2020; Rohanian et al., 2020, 2021). Automatically extracted non-verbal disfluency features from both transcripts and speech (e.g., *silent pauses*, *speed of articulation*, and *pronounciation*) have also shown performance boosts in AD classification (Yuan et al., 2020; Qiao et al., 2021).

2.2 Automatic Disfluency Detection

Disfluency detection is a key challenge in parsing transcribed speech. Disfluencies are defined structurally with three main components (Shriberg, 1994): the *reparandum*, the *interregnum* and the *repair*. The *reparandum* is replaced by the *repair* segment and the *interregnum* is an optional part of the structure consisting of filled pauses (e.g., “uh”) and discourse connectives (e.g., “I mean”). We present an example disfluency with all three components present below:

$$\overbrace{I\,I\text{'}ve}^{reparandum}\ \overbrace{uh I mean}^{interregnum}\ \overbrace{I enjoy}^{repair} \tag{1}$$

Disfluencies are further categorized into repetition, correction/retracing, and false start (Jamshid Lou and Johnson, 2020a), following established typology of speech repairs (Shriberg, 1994). Repetition contains identical *reparandum* and *repair* segments, whereas the *reparandum* and *repair* differ in correction/retracing. The latter is much harder to detect automatically.

Disfluency detection on pre-segmented utter-

[1]https://github.com/AshwinDeshpande96/Measuring_the_Impact_of_Verbal_Disfluency_Tags_on_Automated_Dementia_Detection

ances from the Switchboard treebank corpus (Godfrey and Holliman, 1993; Marcus et al., 1999) has been the focus of many prior works (Johnson and Charniak, 2004; Charniak and Johnson, 2001; Qian and Liu, 2013; Honnibal and Johnson, 2014). In the Switchboard corpus, reparanda, filled pauses, and discourse connectives are marked by EDITED, INTJ, and PRN labels respectively (illustrated in Figure 1). Conventional syntactic parsers often fail to capture the unconventional relation between reparandum and repair, where repair uses similar words to the reparandum in the same order, functioning as a "rough copy" rather than providing additional information (Johnson and Charniak, 2004; Charniak and Johnson, 2001). Because of the difficulty of addressing disfluency within the task of syntactic parsing, systems have instead been developed to detect and remove disfluency prior to parsing (Charniak and Johnson, 2001; Kahn et al., 2005; Lease and Johnson, 2006). Nonetheless, transition-based dependency parsers designed with special mechanisms to handle disfluencies have proven useful for detecting and removing disfluent words and their dependencies from sentences (Honnibal and Johnson, 2014; Rasooli and Tetreault, 2013; Yoshikawa et al., 2016; Tran et al., 2018). Moreover, encoder-decoder constituency parsing models using lexical and prosodic cues (Tran et al., 2018) have resulted in small performance gains both in parsing and disfluency detection. Augmenting parsing models with location-aware attention mechanisms has also been especially effective for disfluency detection (Tran et al., 2018).

Specialized disfluency detection models frame the problem as a sequence labelling task where each word in the input is labelled as disfluent or not. Neural models (CNNs and LSTMs) have been employed for this (Zayats et al., 2016; Jamshid Lou et al., 2018; Wang et al., 2016) but until recently have not performed very well. A recent state-of-the-art semi-supervised approach introduced a self-attentive model (Wang et al., 2018) that jointly performs syntactic parsing and disfluency detection.

The incremental approach for disfluency detection has been explored on both unsegmented and pre-segmented utterances from manual and automated transcripts using LSTM with different decoding schemes (Hough and Schlangen, 2015, 2017) leveraging joint and multitask settings. Another recent approach introduced the incremental processing of words to a Transformer model (BERT (Devlin et al., 2019)) to detect speech disfluency (Rohanian and Hough, 2021). However, these incremental approaches perform poorly on detecting *reparanda* of longer lengths.

3 Data and Task Setup

We used the ADReSS Challenge corpus for our experiments (Luz et al., 2020). The ADReSS Challenge corpus, developed as part of a shared task for INTERSPEECH 2020, is a benchmark dataset of spontaneous speech in the domain of AD classification and MMSE score prediction. It has been acoustically preprocessed, and is balanced in terms of age and gender. The data consists of audio recordings and manual transcriptions of spoken picture descriptions elicited from participants through the Cookie Theft task from the Boston Diagnostic Aphasia Exam (Roth, 2011). The corpus is a subset of the Pitt corpus,[2] which is itself a subset of the DementiaBank dataset (Becker et al., 1994).

In the Cookie Theft task, an investigator and a participant (in this case, an older adult) carry on a conversation in which the investigator asks the participant to describe what is depicted in an eventful image containing, among other subjects, a boy stealing a cookie from a cookie jar.[3] There is no specific time limit for the conversation, allowing participants to talk as long as they want. In the Pitt corpus and by extension the ADReSS Challenge corpus, these conversations were recorded and manually transcribed using the CHAT transcription protocol (MacWhinney, 2000). Participants were labelled as HC (healthy control with no cognitive decline) or AD (declined cognitively) based on their prior diagnostic test results.

We report the transcript-level mean utterance count and standard deviation (SD) for data collected from AD and HC participants in Table 1, showing that the lengths of conversations across groups were fairly balanced (HC $= 13.79 \pm 5.21$ utterances; AD $= 13.93 \pm 9.54$ utterances). We also report the mean MMSE score and SD for each speaker category, showing a significant difference in cognitive health between groups (HC $= 29.11 \pm 0.98$ MMSE; AD $= 17.06 \pm 5.46$ MMSE). To assess significance, we applied the Mann–Whitney U test (as the normality assump-

[2] https://dementia.talkbank.org/access/English/Pitt.html

[3] We refer interested readers to Karlekar et al. (2018), Mueller et al. (2018), or some others cited in this paper for a copy of the original image.

	AD	HC	Test Statistics
Utterance Count	13.93 (SD=9.54)	13.79 (SD=5.21)	U=135.0 p=0.25
MMSE Score	17.06 (SD=5.46)	29.11 (SD=0.98)	U=47.5 p=0.00

Table 1: Mean utterance count and MMSE score for the AD and HC groups, with standard deviations in parentheses. Statistical significance (p) for differences between groups is reported along with the Mann-Whitney U test statistic.

Ref.: and UM THAT 'S UH that 's about all i can see

Aligned: *** ** **** ** not **** ** ***** all i can see

Figure 2: The reference (*Ref.*) and aligned ASR output for a sample utterance from the ADReSS Challenge corpus. The reference transcript is human-transcribed speech with gold disfluent words (red, capitalized) and fluent words (black). *Aligned* refers to the desired alignment of ASR output with the reference text for making meaningful FER and DER evaluations (Jamshid Lou and Johnson, 2020a).

tion was violated) across the two speaker groups, and we also report the test statistic (U) and significance value (p) for each group in Table 1.

3.1 ASR Setup

We used the phone call enhanced model ($16khz$) of the Google Cloud-based Speech Recognizer to automatically transcribe the audio files in the ADReSS Challenge corpus to facilitate our comparisons of manually and automatically generated transcriptions. Manually segmented utterances were fed to the speech recognizer for transcription. The overall word error rate (WER) for the automatically generated transcripts was 69.47%. To evaluate more fine-grained performance of the speech recognizer, we estimated the fluent and disfluent error rates (FER and DER). We provide the equations for computing both below, where d_f, s_f, i_f, and n_f refer to the number of deleted, substituted, inserted, and total fluent words, respectively, and d_d, s_d, i_d, and n_d refer to the number of deleted, substituted, inserted, and total disfluent words, respectively (Jamshid Lou and Johnson, 2020a):

$$\mathrm{FER} = \frac{d_f + s_f + i_f}{n_f} \tag{2}$$

Group	FER	DER
AD	53.30%	77.60%
HC	47.30%	80.70%
Overall	50.20%	78.80%

Table 2: Rates of ASR error on the ADReSS Challenge dataset, both at the class level (AD and HC) and overall. For DER calcualtion, we consider all the disfluencies in Table 6 as well as the Filled pauses (e.g. *uh, um*)

	Repetition	Retracing
DER	76.50%	61.10%

Table 3: DER of broad disfluency categories (repetition and retracing, as defined in Table 6).

$$\mathrm{DER} = \frac{d_d + s_d + i_d}{n_d} \tag{3}$$

To calculate DER,[4] we considered *word repetition*, *multiple repetition*, *phrase repetition*, *word retracing*, and *phrase retracing*, with additional details regarding each disfluency type provided in Table 4. We show an alignment between gold and automatically generated transcriptions for an example utterance from the ADReSS Challenge corpus in Figure 2. Computing FER for this example would set $d_f = 4$, $s_f = 0$, $i_f = 0$, and $n_f = 8$, resulting in FER=0.5. Computing DER for the same sample would set $d_d = 3$, $s_d = 1$, $i_f = 0$, and $n_f = 4$, resulting in DER=1.0. We report FER and DER across the ADReSS Challenge corpus for AD, HC, and all participants in Table 2 and the breakdown of DER for broad disfluency types (repetition, encompassing *word repetition*, *multiple repetition*, and *phrase repetition*, and retracing, encompassing *word retracing* and *phrase retracing*) in Table 3.

3.2 Disfluency Annotator Setup

We leverage the self-attentive neural parsing model (Jamshid Lou and Johnson, 2020b) to automatically detect disfluencies in the ASR-generated transcripts. The model is trained to jointly parse and detect disfluency using contextualized word embeddings (BERT (Devinney et al., 2020) or ELMO (Peters et al., 2018)) and currently produces state-of-the-art performance with a parsing accuracy of

[4] Although the original DER formulation counts the number of copies, we replace this with the number of deletions since we expect the ASR to transcribe disfluent as well as fluent words.

93.9% and a disfluency detection F_1-score of 0.924 on the Switchboard development set (Jamshid Lou and Johnson, 2020b) in the joint task. We use the pretrained version of the disfluency detector and parser.[5] This version is self-trained on the Switchboard gold parse trees (Marcus et al., 1999) and Fisher Corpus Part 1 (Cieri et al., 2004) and Part 2 (Cieri et al., 2005) silver parse trees, using *BERT-base-uncased* word representations.

4 Methods

4.1 Verbal Disfluency Types

We consider several disfluency types in this investigation: *word repetition*, *phrase repetition*, *word retracing*, and *phrase retracing*. We limit our scope to these disfluency types for two primary reasons: (1) these verbal disfluency types are annotated in our corpus of interest, and (2) automatic detection of these types is challenging. We provide examples of each of these in Table 4.[6] Word and phrase repetition indicate repeated utterance of the same word or phrase in such a way that is disfluent with the natural flow of speech, whereas word and phrase retracing indicate verbal "backtracking" to correct a previously uttered word or phrase. In Table 5, we report the frequencies of these disfluency types across speaker groups.

Disfluency Type	**Example**
Word Repetition	**the [/]** the cabinet door has just swung open
Multiple Repetition	there's nothing going on outside there's just **bushes [x 3]**.
Phrase Repetition	⟨**what are**⟩ [/] what are the instructions ?
Word Retracing	and there are **dishes [/ /]** &uh &uh two cups and a saucer on the sink
Phrase Retracing	and outside the window there's a ⟨**walk with a**⟩ [/ /] &c curved walk with a garden .

Table 4: Example of different types of disfluencies from transcripts annotated using the CHAT protocol (MacWhinney, 2000). Disfluencies are bold-faced followed by disfluency markers. Angle brackets indicate phrase-level disfluencies, whereas [x n] indicates that the word before the marker is repeated n times.

4.2 Automatic Disfluency Annotation

We leveraged the self-attentive neural disfluency annotator described in §3.2, trained on the Penn Treebank-3 SWBD corpus (Marcus et al., 1999) and the Fisher I and II corpora (Cieri et al., 2004, 2005) using a semi-supervised approach (Jamshid Lou and Johnson, 2020b). This multi-task learning setup enables the model to predict both parse trees and disfluency tags for utterances. The disfluency annotator adds word-level annotations to disfluent words, or those acting as *EDITED*, *INTJ*, or *PRN* nodes (illustrated in Figure 1).

We preprocessed both the reference and ASR-generated transcripts by removing punctuation and (for the reference transcripts) existing disfluency tags. We then fed the disfluency annotator one utterance per line, in turn producing both a parse tree and a disfluency-tagged version of the utterance as output. Figure 3 shows an example utterance with: (1) the actual text and disfluency tags from the ADReSS Challenge corpus, considering the disfluency types referred in Table 4; (2) the gold disfluency tags formatted as the expected output from the automatic disfluency annotator; and (3) the predicted word-level disfluency tags from the automatic disfluency annotator. Phrase repetition accuracy for the utterance in Figure 3 would be 100% as both the words in the repeated phrase (highlighted in red) are predicted correctly, whereas phrase retracing accuracy would be 0%, as no words in the retraced phrase (highlighted in blue) are predicted as disfluent.

Table 6 illustrates the performance of the automatic disfluency annotator at predicting different disfluency types for the ADReSS Challenge training set, providing evidence that retracing/correction (especially at the phrase level) is harder to predict than repetition. The annotator often fails to detect cases of *multiple repetition* (accuracy=11.11%, making it lowest among all disfluency types in Table 6), likely because it was intermixed with word-level repetition in the training data.

4.3 Disfluency Removal

We implement two methods for removing disfluencies from transcribed speech, described further in

[5] `https://github.com/pariajm/english-fisher-annotations`

[6] Although *multiple repetition* is coded distinctly from single *word repetition* under the CHAT transcription protocol, we consider both as members of the *word repetition* category.

Disfluency	AD	HC
Word Repetition	96	29
Phrase Repetition	27	17
Word Retracing	48	35
Phrase Retracing	67	46
Total	238	127
Disfluency-Tagged	317	176

Table 5: Frequencies of disfluency types across AD and HC participants, where *Total* refers to the sum of all of our disfluency types of interest (rows 1–4), and *Disfluency-Tagged* refers to the sum of all disfluencies reported (including those not in the focus of this investigation).

Disfluency Type	Accuracy
Word Repetition	72.65%
Phrase Repetition	73.61%
Word Retracing	50.00%
Phrase Retracing	42.64%

Table 6: Percentages of disfluent words in the manually-transcribed ADReSS Challenge training set tagged with different disfluency labels (considering *multiple repetition* as a subset of word repetition) by the Fisher annotator.

Actual text: *his sister <has her> [/] has her <hand up> [//] finger up to her mouth like she's saying.*

Gold tag: *his _ sister _ has E her E has _ her _ hand E up E finger _ up _ to _ her _ mouth _ like _ she _ 's _ saying _*

Predicted tag: *his _ sister _ has E her E has _ her _ hand _ up _ finger _ up _ to _ her _ mouth _ like _ she _ 's _ saying _*

Figure 3: Example utterance annotated by automatic disfluency annotator. **Actual text** represents the gold label annotated utterance from the ADReSS Challenge training set. **Gold tag** represents the expected word level annotation given the gold labels, whereas **Predicted tag** shows the predicted disfluency annotations (fluent words are followed by _ tags and disfluent words are followed by E tags) by the disfluency tagger. Repetition is highlighted in red and retracing in blue.

§4.3.1 and §4.3.2.

4.3.1 Gold Disfluency Removal

We removed gold labelled disfluencies from the manually created reference transcripts. We did this by removing different CHAT transcription tags corresponding to repetition and retracing behaviors. Thus, the text in Figure 3 was converted to:

- **Repetition Removal:** *his sister has her hand up finger up to her mouth like she's saying.*
- **Retracing Removal:** *his sister has her has her finger up to her mouth like she's saying.*

4.3.2 Fisher Disfluency Removal

We removed disfluencies predicted by the Fisher tagger (described in §4.2) from the automatically transcribed speech. To remove words of a particular disfluency type, we matched the relevant segment of text with the predicted tag (see Figure 3) and removed the words tagged as *E* (representing *errors*, or disfluencies). For instance, to remove retracing, the blue segments of actual text and predicted tags in Figure 3 are matched, and since none of the words are predicted as *E*, none are removed. Thus, after the removal of disfluencies according to the Fisher tagger, the text in Figure 3 was converted to:

- **Repetition Removal:** *his sister has her hand up finger up to her mouth like she's saying.*
- **Retracing Removal:** *his sister has her has her hand up finger up to her mouth like she's saying.*

4.4 Classification Setup

4.4.1 Input and Output

The ADReSS Challenge training corpus included data from N=108 participants. The input for a given data point was a sequence of words from the processed transcript, and the output was the class of the speaker: 0 for HC, or 1 for AD. Transcripts were preprocessed to remove disfluency markers, punctuation, and digits. When *multiple repetition* markers followed a word in any utterance, the word was added the specified number of times, and the marker was then removed.

4.4.2 Model

We used Bert-for-Sequence-Classification[7] to implement our model, experimenting with *bert-base-*

[7] https://github.com/huggingface/transformers

uncased as our base model and using the following hyperparameters: learning rate = 2e-5, batch size = 4, epochs = 8, max input length of 256 (a length sufficient to cover most cases). The standard default tokenizer was used. Two special tokens, [CLS] and [SEP], were added to the beginning and the end of each transcript utterance. We chose these model and parameter settings since they attained promising performance in previously published work (Yuan et al., 2020) with leave-one-out cross-validation on the ADReSS Challenge dataset.

5 Experiments

5.1 Experimental Setup

To evaluate the impact of disfluency presence and type on classifying AD status, we performed experiments considering the following conditions:

- **ALLTEXT:** The baseline condition using the original manually-created transcripts, complete with gold disfluencies, preprocessed as defined in §4.4.
- **ASR:** Transcripts are generated using ASR (explained in §3.1), and the ASR-generated transcripts are fed to the model.
- **-REP.:** Repetitions (both word- and phrase-level) are removed from ALLTEXT transcripts using either the gold or Fisher disfluency removal method.
- **-RET.:** Incidents of retracing (both word- and phrase-level) are removed from ALLTEXT using either the gold or Fisher disfluency removal method.
- **-DISF.:** Transcripts are processed so that all cases of word- or phrase-level repetition or retracing are removed. When using the Fisher disfluency removal method, this includes all disfluency-tagged words.

We report accuracy, precision, recall, and F_1 for each condition. When performing development experiments, we observed large performance differences across folds. Such brittleness has also been reported previously (Yuan et al., 2020), and may be attributed to the use of a large model (BERT) for classification on a small dataset. To address this, we perform three runs, each using different random seeds, of five-fold cross-validation and report averages and standard deviations across runs.

5.2 Results

We report our evaluation results in Table 7. As expected, we observe the highest performance in the baseline condition (ALLTEXT), which is comparable to the results in previous literature (Balagopalan et al., 2020a). The ASR condition exhibits the worst performance, with accuracy, F_1 for AD, and F_1 for HC decreasing 17.7%, 14%, and 33% respectively relative to the baseline. This underscores one of our primary motivations in conducting this work—namely, that ASR has a high error rate in real-world settings and particularly in this task environment, and moreover that its mistagging (or in some cases, purposeful removal) of disfluency has a deleterious impact on dementia detection performance. We observe from Tables 2 and 3 that DER is much higher than FER for ASR output. ASR tends to delete or replace repetitive words, increasing overall word error rate and leading to poor performance in the AD detection task. Prior work has clearly suggested that disfluencies are important indicators of cognitive health status (Lopez-de Ipina et al., 2017; Mueller et al., 2018).

Furthermore, performance clearly degrades relative to the baseline when gold disfluencies are removed (-REP.$_G$, -RET.$_G$, and -DISF.$_G$). Although retracing removal caused a slightly higher decrease in accuracy than repetition removal, there is no significant difference in performance between the -REP.$_G$ and -RET.$_G$ conditions across metrics. Accuracy and F_1 decrease 5.6% and 6% (for both AD and HC) compared to the baseline when all gold disfluencies are removed from the transcripts.

Removal of Fisher disfluencies also leads to performance degradation across all metrics. Since the Fisher disfluency annotations are more limited than the gold disfluency labels, performance in this condition (-REP.$_F$, -RET.$_F$, and -DISF.$_F$) degrades less than is observed with gold disfluency removal. Accuracy, F_1 for AD, and F_1 for HC decrease 3%, 4%, and 2% respectively compared to the baseline when all Fisher-predicted disfluencies are removed.

5.3 Distinctive Effects of Disfluency Removal

To further investigate why disfluency removal influences classification performance, we experiment with measures of syntactic complexity, context-free grammar rules, and measures of vocabulary richness[8] to identify linguistic features having mod-

[8]https://github.com/vmasrani/dementia_classifier

	Accuracy	Precision		Recall		F1	
		AD	HC	AD	HC	AD	HC
ALLTEXT	0.843±.015	0.88±.017	0.82±.020	0.80±.028	0.89±.019	0.84±.016	0.85±.013
ASR	0.670±.037	0.69±.062	0.54±.032	0.72±.060	0.52±.121	0.70±.023	0.52±.065
-REP.$_G$	0.797±.034	0.81±.044	0.79±.034	0.78±.049	0.80±.053	0.80±.021	0.79±.036
-RET.$_G$	0.787±.024	0.77±.043	0.80±.012	0.81±.015	0.76±.060	0.80±.017	0.78±.035
-DISF.$_G$	0.787±.020	0.78±.025	0.77±.028	0.76±.040	0.81±.030	0.78±.026	0.79±.020
-REP.$_F$	0.827±.015	0.86±.021	0.80±.010	0.78±.011	0.88±.021	0.82±.014	0.84±.014
-RET.$_F$	0.820±.010	0.86±.013	0.79±.021	0.78±.032	0.87±.019	0.82±.013	0.83±.006
-DISF.$_F$	0.813±.006	0.85±.018	0.78±.004	0.76±.000	0.87±.018	0.80±.008	0.83±.010

Table 7: Five-fold cross-validation results, averaged across three runs with different random seeds on the ADReSS Challenge training set. The subscript G refers to gold disfluency removal and F refers to Fisher disfluency removal.

erate to high correlation with disfluency (as measured by normalised disfluency count, repetition count, and retracing count). We find that disfluency count (considering all disfluencies in Table 4) has significant, high negative Spearman correlation ($r = -0.55$, $p < 0.001$) with type token ratio (TTR). This indicates that verbal disfluencies are highly negatively correlated with vocabulary richness, which is in turn an important feature of AD detection (Masrani, 2018). Some context-free grammar rules (INTJ, INTJ_to_UH, VP_to_VBG, VP_to_AUX) and syntactic complexity features (constituency parse tree height), also key features for AD detection (Masrani, 2018), exhibit moderate correlation with disfluency frequency. Such results show that vocabulary richness and the syntactic structure of language are vulnerable to the deletion of disfluencies, which may in turn lead to classification performance degradation.

6 Discussion

From our corpus analyses, we find that members of the AD group exhibit more verbal disfluency (Table 2), with increased rates of repetition and correction relative to the HC group. This is in line with our expectations, since disfluencies and speech errors are correlated with cognitive functions such as cognitive load, arousal, and working memory (Arciuli et al., 2010; Daneman, 1991); with increased impairment of these functions, hesitations and disfluencies increase. Previous studies have also reported that verbal disfluency frequency can be an important predictor of fine-grained cognitive status of older adults (Farzana et al., 2020). Our evaluation provides evidence that removing both gold-labelled and Fisher-annotated verbal disfluencies leads to changes in AD detection performance, opening intriguing questions for follow-up work that may further tease apart the nature of these contributions.

We speculate that some of these findings may transfer to other conditions as well. For example, studies have also reported that filled pauses are less frequently uttered by children with autism spectrum disorder than typically developed children (Gorman et al., 2016; Irvine et al., 2016). It is possible that incorporating richer disfluency information in speech-based systems for autism detection and monitoring may improve performance similarly to that seen with AD detection.

7 Conclusion

Verbal disfluencies are an important indicator of AD, and current ASR systems fail to capture and label word- and phrase-level disfluencies adequately. Doing so is necessary to generate useful transcripts with minimal human intervention, such that they can be leveraged for successful AD detection. Our future work will focus on training an end-to-end ASR system on disfluent speech so that it can generate richer disfluency annotated transcripts, which will pave the way for building end-to-end speech-based dementia detection systems.

8 Acknowledgements

This work was supported in part by a startup grant from the University of Illinois at Chicago. We thank the anonymous reviewers for their helpful comments.

References

Alzheimer's Association. 2018. 2018 alzheimer's disease facts and figures. *Alzheimer's & Dementia*, 14(3):367–429.

Joanne Arciuli, David Mallard, and Gina Villar. 2010. "Um, i can tell you're lying": Linguistic markers of deception versus truth-telling in speech. *Applied Psycholinguistics*, 31(3):397–411.

Aparna Balagopalan, Benjamin Eyre, Frank Rudzicz, and Jekaterina Novikova. 2020a. To bert or not to bert: Comparing speech and language-based approaches for alzheimer's disease detection. In *INTERSPEECH*.

Aparna Balagopalan, Ksenia Shkaruta, and Jekaterina Novikova. 2020b. Impact of ASR on Alzheimer's disease detection: All errors are equal, but deletions are more equal than others. In *Proceedings of the Sixth Workshop on Noisy User-generated Text (W-NUT 2020)*, pages 159–164, Online. Association for Computational Linguistics.

James T. Becker, François Boiler, Oscar L. Lopez, Judith Saxton, and Karen L. McGonigle. 1994. The Natural History of Alzheimer's Disease: Description of Study Cohort and Accuracy of Diagnosis. *Archives of Neurology*, 51(6):585–594.

Catherine S. Brown and Walter L. Cullinan. 1981. Word-retrieval difficulty and disfluent speech in adult anomic speakers. *Journal of Speech, Language, and Hearing Research*, 24(3):358–365.

R. S. Bucks, S. Singh, J. M. Cuerden, and G. K. Wilcock. 2000. Analysis of spontaneous, conversational speech in dementia of alzheimer type: Evaluation of an objective technique for analysing lexical performance. *Aphasiology*, 14(1):71–91.

Eugene Charniak and Mark Johnson. 2001. Edit detection and parsing for transcribed speech. In *Second Meeting of the North American Chapter of the Association for Computational Linguistics*.

Christopher Cieri, David Graff, Owen Kimball, Dave Miller, and Kevin Walker. 2004. Fisher english training speech part 1 transcripts ldc2004s13.

Christopher Cieri, David Graff, Owen Kimball, Dave Miller, and Kevin Walker. 2005. Fisher english training speech part 2 transcripts ldc2005s13.

M. Daneman. 1991. Working memory as a predictor of verbal fluency. *Journal of Psycholinguistic Research*, 20:445–464.

Hannah Devinney, Jenny Björklund, and Henrik Björklund. 2020. Semi-supervised topic modeling for gender bias discovery in English and Swedish. In *Proceedings of the Second Workshop on Gender Bias in Natural Language Processing*, pages 79–92, Barcelona, Spain (Online). Association for Computational Linguistics.

Jacob Devlin, Ming-Wei Chang, Kenton Lee, and Kristina Toutanova. 2019. BERT: Pre-training of deep bidirectional transformers for language understanding. In *Proceedings of the 2019 Conference of the North American Chapter of the Association for Computational Linguistics: Human Language Technologies, Volume 1 (Long and Short Papers)*, pages 4171–4186, Minneapolis, Minnesota. Association for Computational Linguistics.

Flavio Di Palo and Natalie Parde. 2019. Enriching neural models with targeted features for dementia detection. In *Proceedings of the 57th Annual Meeting of the Association for Computational Linguistics: Student Research Workshop*, pages 302–308, Florence, Italy. Association for Computational Linguistics.

Ben Eyre, Aparna Balagopalan, and Jekaterina Novikova. 2020. Fantastic features and where to find them: Detecting cognitive impairment with a subsequence classification guided approach. In *Proceedings of the Sixth Workshop on Noisy User-generated Text (W-NUT 2020)*, pages 193–199, Online. Association for Computational Linguistics.

Shahla Farzana and Natalie Parde. 2020. Exploring MMSE Score Prediction Using Verbal and Non-Verbal Cues. In *Proceedings of Interspeech 2020*, pages 2207–2211.

Shahla Farzana, Mina Valizadeh, and Natalie Parde. 2020. Modeling dialogue in conversational cognitive health screening interviews. In *Proceedings of the 12th Language Resources and Evaluation Conference*, pages 1167–1177, Marseille, France. European Language Resources Association.

Kathleen C. Fraser, Nicklas Linz, Bai Li, Kristina Lundholm Fors, Frank Rudzicz, Alexandra König, Jan Alexandersson, Philippe Robert, and Dimitrios Kokkinakis. 2019. Multilingual prediction of Alzheimer's disease through domain adaptation and concept-based language modelling. In *Proceedings of the 2019 Conference of the North American Chapter of the Association for Computational Linguistics: Human Language Technologies, Volume 1 (Long and Short Papers)*, pages 3659–3670, Minneapolis, Minnesota. Association for Computational Linguistics.

John Godfrey and Edward Holliman. 1993. Switchboard-1 release 2 ldc97s62.

Kyle Gorman, L. Olson, A. Hill, R. Lunsford, P. Heeman, and J. van Santen. 2016. Uh and um in children with autism spectrum disorders or language impairment. *Autism Research*, 9.

Pamela Herd, Deborah Carr, and Carol Roan. 2014. Cohort profile: Wisconsin longitudinal study (wls). *International journal of epidemiology*, 43(1):34–41.

Matthew Honnibal and Mark Johnson. 2014. Joint incremental disfluency detection and dependency parsing. *Transactions of the Association for Computational Linguistics*, 2:131–142.

Julian Hough and David Schlangen. 2015. Recurrent neural networks for incremental disfluency detection. In *INTERSPEECH 2015, 16th Annual Conference of the International Speech Communication Association, Dresden, Germany, September 6-10, 2015*, pages 849–853. ISCA.

Julian Hough and David Schlangen. 2017. Joint, incremental disfluency detection and utterance segmentation from speech. In *Proceedings of the 15th Conference of the European Chapter of the Association for Computational Linguistics: Volume 1, Long Papers*, pages 326–336, Valencia, Spain. Association for Computational Linguistics.

Christina Irvine, Inge-Marie Eigsti, and Deborah Fein. 2016. Uh, um, and autism: Filler disfluencies as pragmatic markers in adolescents with optimal outcomes from autism spectrum disorder. *Journal of Autism and Developmental Disorders*, 46.

Paria Jamshid Lou, Peter Anderson, and Mark Johnson. 2018. Disfluency detection using auto-correlational neural networks. In *Proceedings of the 2018 Conference on Empirical Methods in Natural Language Processing*, pages 4610–4619, Brussels, Belgium. Association for Computational Linguistics.

Paria Jamshid Lou and Mark Johnson. 2017. Disfluency detection using a noisy channel model and a deep neural language model. In *Proceedings of the 55th Annual Meeting of the Association for Computational Linguistics (Volume 2: Short Papers)*, pages 547–553, Vancouver, Canada. Association for Computational Linguistics.

Paria Jamshid Lou and Mark Johnson. 2020a. End-to-end speech recognition and disfluency removal. In *Findings of the Association for Computational Linguistics: EMNLP 2020*, pages 2051–2061, Online. Association for Computational Linguistics.

Paria Jamshid Lou and Mark Johnson. 2020b. Improving disfluency detection by self-training a self-attentive model. In *Proceedings of the 58th Annual Meeting of the Association for Computational Linguistics*, pages 3754–3763, Online. Association for Computational Linguistics.

Paria Jamshid Lou, Yufei Wang, and Mark Johnson. 2019. Neural constituency parsing of speech transcripts. In *Proceedings of the 2019 Conference of the North American Chapter of the Association for Computational Linguistics: Human Language Technologies, Volume 1 (Long and Short Papers)*, pages 2756–2765, Minneapolis, Minnesota. Association for Computational Linguistics.

Mark Johnson and Eugene Charniak. 2004. A TAG-based noisy-channel model of speech repairs. In *Proceedings of the 42nd Annual Meeting of the Association for Computational Linguistics (ACL-04)*, pages 33–39, Barcelona, Spain.

Vera C Kaelin, Mina Valizadeh, Zurisadai Salgado, Natalie Parde, and Mary A Khetani. 2021. Artificial intelligence in rehabilitation targeting the participation of children and youth with disabilities: Scoping review. *J Med Internet Res*, 23(11):e25745.

Jeremy G. Kahn, Matthew Lease, Eugene Charniak, Mark Johnson, and Mari Ostendorf. 2005. Effective use of prosody in parsing conversational speech. In *Proceedings of Human Language Technology Conference and Conference on Empirical Methods in Natural Language Processing*, pages 233–240, Vancouver, British Columbia, Canada. Association for Computational Linguistics.

Sweta Karlekar, Tong Niu, and Mohit Bansal. 2018. Detecting linguistic characteristics of Alzheimer's dementia by interpreting neural models. In *Proceedings of the 2018 Conference of the North American Chapter of the Association for Computational Linguistics: Human Language Technologies, Volume 2 (Short Papers)*, pages 701–707, New Orleans, Louisiana. Association for Computational Linguistics.

Xuan Le, Ian Lancashire, Graeme Hirst, and Regina Jokel. 2011. Longitudinal detection of dementia through lexical and syntactic changes in writing: a case study of three British novelists. *Literary and Linguistic Computing*, 26(4):435–461.

Matthew Lease and Mark Johnson. 2006. Early deletion of fillers in processing conversational speech. In *Proceedings of the Human Language Technology Conference of the NAACL, Companion Volume: Short Papers*, pages 73–76, New York City, USA. Association for Computational Linguistics.

K. Lopez-de Ipina, U. Martinez-de Lizarduy, P. M. Calvo, B. Beitia, J. Garcia-Melero, M. Ecay-Torres, A. Estanga, and M. Faundez-Zanuy. 2017. Analysis of disfluencies for automatic detection of mild cognitive impartment: a deep learning approach. In *2017 International Conference and Workshop on Bioinspired Intelligence (IWOBI)*, pages 1–4.

Saturnino Luz, Fasih Haider, Sofia de la Fuente, Davida Fromm, and Brian MacWhinney. 2020. Alzheimer's dementia recognition through spontaneous speech: The ADReSS Challenge.

Brian MacWhinney. 2000. *The CHILDES Project: Tools for analyzing talk. transcription format and programs*, volume 1. Psychology Press.

Mitchell P Marcus, Beatrice Santorini, Mary Ann Marcinkiewicz, and Ann Taylor. 1999. Treebank-3 ldc99t42.

Vaden Masrani. 2018. *Detecting dementia from written and spoken language*. Ph.D. thesis, University of British Columbia.

Kimberly D. Mueller, Rebecca L. Koscik, Bruce P. Hermann, Sterling C. Johnson, and Lyn S. Turkstra.

2018. Declines in connected language are associated with very early mild cognitive impairment: Results from the wisconsin registry for alzheimer's prevention. *Frontiers in Aging Neuroscience*, 9.

Matthew E. Peters, Mark Neumann, Mohit Iyyer, Matt Gardner, Christopher Clark, Kenton Lee, and Luke Zettlemoyer. 2018. Deep contextualized word representations. In *Proc. of NAACL*.

Xian Qian and Yang Liu. 2013. Disfluency detection using multi-step stacked learning. In *Proceedings of the 2013 Conference of the North American Chapter of the Association for Computational Linguistics: Human Language Technologies*, pages 820–825, Atlanta, Georgia. Association for Computational Linguistics.

Yu Qiao, Xuefeng Yin, Daniel Wiechmann, and Elma Kerz. 2021. Alzheimer's Disease Detection from Spontaneous Speech Through Combining Linguistic Complexity and (Dis)Fluency Features with Pretrained Language Models. In *Proc. Interspeech 2021*, pages 3805–3809.

Mohammad Sadegh Rasooli and Joel Tetreault. 2013. Joint parsing and disfluency detection in linear time. In *Proceedings of the 2013 Conference on Empirical Methods in Natural Language Processing*, pages 124–129, Seattle, Washington, USA. Association for Computational Linguistics.

Morteza Rohanian and Julian Hough. 2021. Best of both worlds: Making high accuracy non-incremental transformer-based disfluency detection incremental. In *Proceedings of the 59th Annual Meeting of the Association for Computational Linguistics and the 11th International Joint Conference on Natural Language Processing (Volume 1: Long Papers)*, pages 3693–3703, Online. Association for Computational Linguistics.

Morteza Rohanian, Julian Hough, and Matthew Purver. 2020. Multi-modal fusion with gating using audio, lexical and disfluency features for alzheimer's dementia recognition from spontaneous speech. *Interspeech 2020*.

Morteza Rohanian, Julian Hough, and Matthew Purver. 2021. Alzheimer's Dementia Recognition Using Acoustic, Lexical, Disfluency and Speech Pause Features Robust to Noisy Inputs. In *Proc. Interspeech 2021*, pages 3820–3824.

Carole Roth. 2011. Boston diagnostic aphasia examination. In Jeffrey S. Kreutzer, John DeLuca, and Bruce Caplan, editors, *Encyclopedia of Clinical Neuropsychology*, pages 428–430. Springer New York, New York, NY.

Utkarsh Sarawgi, Wazeer Zulfikar, Nouran Soliman, and Pattie Maes. 2020. Multimodal inductive transfer learning for detection of alzheimer's dementia and its severity. In *Proceedings of Interspeech 2020*, pages 2212–2216.

Elizabeth Ellen Shriberg. 1994. Preliminaries to a theory of speech disfluencies. Technical report, University of California, Berkeley.

Trang Tran, Shubham Toshniwal, Mohit Bansal, Kevin Gimpel, Karen Livescu, and Mari Ostendorf. 2018. Parsing speech: a neural approach to integrating lexical and acoustic-prosodic information. In *Proceedings of the 2018 Conference of the North American Chapter of the Association for Computational Linguistics: Human Language Technologies, Volume 1 (Long Papers)*, pages 69–81, New Orleans, Louisiana. Association for Computational Linguistics.

Mina Valizadeh and Natalie Parde. 2022. The AI doctor is in: A survey of task-oriented dialogue systems for healthcare applications. In *Proceedings of the 2022 Conference of the Association for Computational Linguistics*, Dublin, Ireland. Association for Computational Linguistics.

Chunqi Wang, Ji Zhang, and Haiqing Chen. 2018. Semi-autoregressive neural machine translation. In *Proceedings of the 2018 Conference on Empirical Methods in Natural Language Processing*, pages 479–488, Brussels, Belgium. Association for Computational Linguistics.

Shaolei Wang, Wanxiang Che, and Ting Liu. 2016. A neural attention model for disfluency detection. In *Proceedings of COLING 2016, the 26th International Conference on Computational Linguistics: Technical Papers*, pages 278–287, Osaka, Japan. The COLING 2016 Organizing Committee.

Shaolei Wang, Wanxiang Che, Yue Zhang, Meishan Zhang, and Ting Liu. 2017. Transition-based disfluency detection using LSTMs. In *Proceedings of the 2017 Conference on Empirical Methods in Natural Language Processing*, pages 2785–2794, Copenhagen, Denmark. Association for Computational Linguistics.

Maria Yancheva, Kathleen Fraser, and Frank Rudzicz. 2015. Using linguistic features longitudinally to predict clinical scores for Alzheimer's disease and related dementias. In *Proceedings of SLPAT 2015: 6th Workshop on Speech and Language Processing for Assistive Technologies*, pages 134–139, Dresden, Germany. Association for Computational Linguistics.

Masashi Yoshikawa, Hiroyuki Shindo, and Yuji Matsumoto. 2016. Joint transition-based dependency parsing and disfluency detection for automatic speech recognition texts. In *Proceedings of the 2016 Conference on Empirical Methods in Natural Language Processing*, pages 1036–1041, Austin, Texas. Association for Computational Linguistics.

Jiahong Yuan, Yuchen Bian, Xingyu Cai, Jiaji Huang, Zheng Ye, and Kenneth Church. 2020. Disfluencies and Fine-Tuning Pre-Trained Language Models for Detection of Alzheimer's Disease. In *Proc. Interspeech 2020*, pages 2162–2166.

Vicky Zayats and Mari Ostendorf. 2019. Giving attention to the unexpected: Using prosody innovations in disfluency detection. In *Proceedings of the 2019 Conference of the North American Chapter of the Association for Computational Linguistics: Human Language Technologies, Volume 1 (Long and Short Papers)*, pages 86–95, Minneapolis, Minnesota. Association for Computational Linguistics.

Vicky Zayats, Mari Ostendorf, and Hannaneh Hajishirzi. 2016. Disfluency detection using a bidirectional LSTM. In *Proceedings of the 17th Annual Conference of the International Speech Communication Association*, pages 2523–2527. ISCA.

Zining Zhu, Jekaterina Novikova, and Frank Rudzicz. 2019. Detecting cognitive impairments by agreeing on interpretations of linguistic features. In *Proceedings of the 2019 Conference of the North American Chapter of the Association for Computational Linguistics: Human Language Technologies, Volume 1 (Long and Short Papers)*, pages 1431–1441, Minneapolis, Minnesota. Association for Computational Linguistics.

Zero-Shot Aspect-Based Scientific Document Summarization using Self-Supervised Pre-training

Amir Soleimani*
University of Amsterdam
Amsterdam, The Netherlands
a.soleimani@uva.nl

Vassilina Nikoulina
NAVER LABS Europe
Meylan, France
vassilina.nikoulina@naverlabs.com

Benoit Favre
Aix Marseille Univ, Université de Toulon
CNRS, LIS, Marseille, France
benoit.favre@lis-lab.fr

Salah Ait-Mokhtar
NAVER LABS Europe
Meylan, France
salah.ait-mokhtar@naverlabs.com

Abstract

We study the zero-shot setting for the aspect-based scientific document summarization task. Summarizing scientific documents with respect to an aspect can remarkably improve document assistance systems and readers experience. However, existing large-scale datasets contain a limited variety of aspects, causing summarization models to over-fit to a small set of aspects and a specific domain. We establish baseline results in zero-shot performance (over unseen aspects and the presence of domain shift), paraphrasing, leave-one-out, and limited supervised samples experimental setups. We propose a self-supervised pre-training approach to enhance the zero-shot performance. We leverage the PubMed structured abstracts to create a biomedical aspect-based summarization dataset. Experimental results on the PubMed and FacetSum aspect-based datasets show promising performance when the model is pre-trained using unlabelled in-domain data.[1]

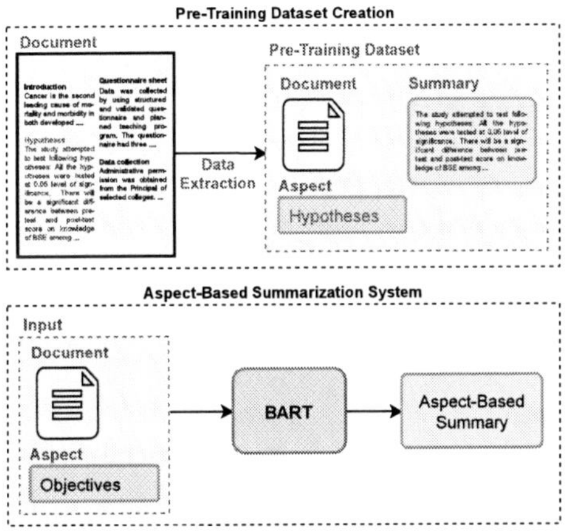

Figure 1: Overview of our approach to create self-supervised pre-training datasets from unlabelled scientific documents. The aspect-based summarization model is pre-trained on unlabelled documents, the section headings as aspects, and the following paragraphs corresponding to the aspects as aspect-based summaries.

1 Introduction

Scientific document summarization aims to summarize research papers, and it is usually considered as generating paper abstracts (Cohan et al., 2018). Compared to the news summarization datasets like CNN/Daily Mail (Hermann et al., 2015) and XSUM (Narayan et al., 2018), scientific papers are significantly longer, follow a standard structure, and contain more technical terms and complex concepts (Yu et al., 2020). Recently, there have been remarkable improvements in the area of scientific document summarization due to the availability of large-scale datasets such as arXiv, PubMed (Cohan et al., 2018), and SUMPUBMED (Gupta et al., 2021) and pre-trained sequence to sequence models such as BART (Lewis et al., 2020) and PEGASUS (Zhang et al., 2020). However, little research has been conducted on aspect-based scientific document summarization.

Aspect-based summarization is the task of summarizing a document given a specific point of interest. Aspect-based scientific document summarization has several advantages for readers to explore articles quickly and facilitates document assistance systems. Collecting a large-scale dataset for this task is extremely costly. Meng et al. (2021) introduce FacetSum, an aspect-based document summarization dataset from mainly management, marketing, and education domains. They employ

*Work done while interning at NAVER LABS Europe.
[1] github.com/asoleimanib/ZeroShotAspectBased

Proceedings of the BioNLP 2022 workshop, Dublin, Ireland, pages 49–62
May 26, 2022. ©2022 Association for Computational Linguistics

structured abstracts from the Emerald database[2] to create summaries from four perspectives (*purpose*, *method*, *findings*, *value*). However, readers may be interested in new aspects beyond proposed annotations or new domains, particularly biomedical area.

Summarization heavily relies on sequence-to-sequence models that require numerous training data. While scientific summarization problem can benefit from large amount of articles with their summaries available (Cohan et al., 2018), the data for aspect-based summarization of scientific papers is scarce. Moreover, most existing methods for aspect-based summarization rely on pre-defined aspects. Adding new aspects would require gathering new data and retraining the whole system.

In this work, we are interested in zero-shot aspect-based summarization of scientific literature. Large pre-trained models such as BERT (Devlin et al., 2019) and BART have demonstrated the high potential of knowledge transfer from self-supervised tasks to downstream tasks. Continuing the BART pre-training task (e.g., token masking and deletion) with domain-related or target datasets can improve the final performance on low-resource domains. However, this process, specifically using domain-related datasets, is substantially time-consuming (Yu et al., 2021). Also, training a summarization model using a second summarization dataset on the same task enhances the performance (Yu et al., 2021). Such approaches only cover limited aspects. We believe a good aspect-based summarization system should establish semantic similarity between aspect and document content. We leverage the *semantic representations* emerging during LM pre-training to allow the model to establish this semantic connection between the aspect and the summary. We also propose an additional pre-training procedure to reinforce this connection. The contributions of this work are the following:

- We establish baselines for aspect-based summarization using two datasets from different domains, biomedical and management, and analyse the zero-shot capabilities of those models on unseen aspects.
- For zero-shot capabilities, we study the effect of domain shift and unseen aspects on aspect-based summarization performance.
- We propose self-supervised pre-training to boost the zero-shot capability of the model and demonstrate its effectiveness.
- Finally, we analyse how different models behave as the amount of supervision decreases.

[2] www.emerald.com

2 Related Work

Abstractive Summarization. Early research on abstractive summarization mainly focused on paraphrasing-based compression methods (Filippova, 2010; Berg-Kirkpatrick et al., 2011). Later motivated by the success of neural attention mechanism (Bahdanau et al., 2014), attention-based sequence-to-sequence models have been developed for abstractive summarization (Rush et al., 2015; Nallapati et al., 2016). Adopting pre-trained transformer models by self-supervised objectives has led to significant improvements in NLP (Devlin et al., 2019). In particular, BART and PEGASUS extend such idea to text generation and have the state of the art performance on abstractive summarization.

Scientific Document Summarization. Scientific documents have complex structures. Extractive summarization under-performs abstractive summarization in scientific documents because information is distributed across documents (Cohan et al., 2018). Different approaches have been proposed to improve models on scientific data, such as a hierarchical encoder with a decoder attending to discourse-level information (Cohan et al., 2018) or summarizing sections separately (Gidiotis and Tsoumakas, 2019). Two-step pipelines is another approach (Gidiotis and Tsoumakas, 2020) to summarize scientific documents. BART is also used in this task (Meng et al., 2021). It can handle long sequences using a hierarchical attention model (Rohde et al., 2021) or simply by extending its positional embedding (Meng et al., 2021). Extended BART might enhance the performance for summaries requiring information spread mostly at the end of papers. However, as BART is not pre-trained on long texts, the extended model would underperform efficient transformers (e.g., Longformer (Beltagy et al., 2020)). We performed some initial experiments by extending BART beyond its default input length and found no significant improvement on average scores (Appendix B). Moreover, our initial experiments exposed similar zero-shot trends across different BART versions. Therefore for computational reasons in follow up experiments, we stick to the standard BART model.

	# Samples (Aspect, Document)				
PubMed	Train: 139.4K / Validation: 7.9K / Test: 8.1K				
	Average Length (# Words)				
	Documents: 3.5K				
	Summaries:				
	Intro.	Objectives	Methods	Results	Conc.
	53	38	76	94	40
	# Samples (Aspect, Document)				
FacetSum	Train: 182.4K/ Validation: 23.7K / Test: 23.7K				
	Average Length (# Words)				
	Documents: 6.6K				
	Summaries:				
		Objectives	Methods	Results	Value
		53	49	66	46

Table 1: Statistics of the PubMed and FacetSum aspect-based scientific summarization datasets.

Aspect-based Summarization. Prior to scientific documents, aspect-based summarization has been primary studied on reviews to summarize opinions (Titov and McDonald, 2008; Lu et al., 2009; Yang et al., 2018; Angelidis and Lapata, 2018), arguments (Wang and Ling, 2016), and news articles (Frermann and Klementiev, 2019; Krishna and Srinivasan, 2018). PMC-SA (Gidiotis and Tsoumakas, 2019) leverages structured scientific abstracts for structured summarization over three sections. In particular, FacetSum, an aspect-based scientific document summarization, has been collected using the structured outline of papers from the Emerald database.

Training separated models per aspects (Hayashi et al., 2020) is not preferable in the zero-shot setting. To integrate aspects and input sequences representations, an attention mechanism over aspects is used for RNNs (Yang et al., 2018), pointer-generator networks (Krishna and Srinivasan, 2018; Fremann and Klementiev, 2019), and Transformer (Xie et al., 2020). Concatenating aspects with documents is a straightforward method result in promising performance using BART (Meng et al., 2021; Tan et al., 2020; Su et al., 2021). We follow this direction and study to what extent models are robust to new aspects and domain shift.

Aspect-based summarization can be seen as a special case of query-based summarization. However, in the query-based literature (Ishigaki et al., 2020; Xu and Lapata, 2021) and datasets (Baumel et al., 2016; Nema et al., 2017) queries are more diverse and mostly long phrases or questions.

Zero-Shot Summarization Hua and Wang (2017) combine in-domain and out-of-domain datasets to improve abstractive summarization on small data. While Magooda and Litman (2020) propose a template-based data synthesis method to improve the small data abstractive summarization. Coavoux et al. (2019) study an unsupervised aspect-based abstractive summarization approach but it is difficult to extend it to predefined aspects. Recently, AdaptSum (Yu et al., 2021) leverages the idea of extra pre-training on BART. They compare intermediate training by a second summarization dataset with continuing BART pre-training using two pre-training approaches: a time-consuming domain-adaptive pre-training (using a corpus related to target) and task-adaptive pre-training (using unlabelled target data). They show intermediate training surpasses continuing the BART pre-training. Similar to our idea of using task-specific self-supervised pre-training, self-supervised generic summaries extracted from the first sentences of Wikipedia documents (Fabbri et al., 2021) and news articles (Zhu et al., 2021) are used to pre-train summarization models for social media, patent document, and news summarization tasks. Duan et al. (2019) also investigate cross-lingual abstractive summarization using a back-translation approach. Zero-shot multi-document summarization has been also studied using pre-trained models (Goodwin et al., 2020). To the best of our knowledge, our paper is the first study investigating zero-shot aspect-based summarization.

3 Methods

In this section, we first present how we formulate the aspect-based summarization problem relying on BART pre-trained model. Then, we propose a method to use unlabelled data for an additional self-supervised pre-training step to improve the zero-shot performance.

3.1 Aspect-Based Summarization

Given an aspect phrase $A = \{A_1, A_2, ..., A_K\}$ containing K words, and a document $D = \{W_1, W_2, ..., W_N\}$ containing N words, the aspect-based summarization task aims to summarize D into summary $S = \{S_1, S_2, ..., S_M\}$ with respect to aspect A using an autoregressive summarization model $S_{t+1} = Model(S_t, X = \{D, A\})$ for $t = \{0, ..., M-1\}$. We use BART, a pre-trained model combining bidirectional and autoregressive transformers, to encode documents and aspects together and generate aspect-based summaries. To combine aspects and documents as input X, we concatenate A to the beginning of D with the following format:

$$X = <s> \{A_1, ..., A_K\} </s> \{W_1, ..., W_N\}$$

where $<s>$ and $</s>$ are the beginning of sentence, and separation tokens, respectively. Finally, we train the model with cross-entropy loss function similar to a generic summarization task.

3.2 Self-Supervised Training

A model can extend its prediction to unseen aspects only if it can make a semantic connection between the aspect and the document content. When only a limited amount of aspects is available, there is a risk that the model treats those as "special tokens" and does not exploit their semantic meaning. Therefore, to make such connection stronger, the model needs more diverse samples. In order to extend it, we propose self-supervised pre-training on (sub-)sections headings from the articles. We assume headings are phrases conveying the central topic of sections and are good alternatives for aspects.

We propose extracting self-supervised samples from the PubMed and FacetSum training sets. Figure 1 explains our extraction method. We use the (sub-)sections headings as aspects. We assign sentences in the corresponding (sub-)sections as aspect-based summaries and truncate the sentences up to 300 characters. We pre-train BART with the extracted dataset using the same cross-entropy loss function used for the final summarization task. While our pre-trained model can theoretically copy text from input to output, it is impossible to copy sentences for most aspects as they are not in the model input range. We experimented with excluding targets from inputs and found no significant difference in the final performance (Table 10 Appendix C).

We assume training a model to generate sentences conditioned on an aspect (heading) helps the model to understand the concept of aspect and learn representations better for diverse aspects. In other words, instead of directly training on labelled aspect-based summarization, we train the model indirectly using a self-supervised approach and later fine-tune it on real summarization samples.

4 Datasets

For our experiments, we consider FacetSum, an aspect-based summarization benchmark built on Emerald articles. In addition, we process PubMed

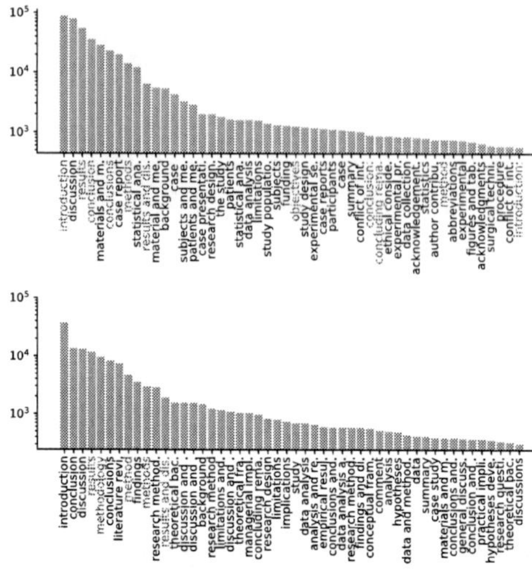

Figure 2: Histogram of 50 most frequent aspects in the self-supervised samples (top: PubMed⋆, bottom: FacetSum⋆). PubMed⋆ has [150K,1.4K,214,33] unique aspects with frequency of higher than [1,10,100,1000] (FacetSum⋆:[96K,841,120,21]). Aspects removed from the NoOverlap datasets are highlighted in red.

and convert into a large aspect-based scientific document summarization dataset. We scraped the PubMed website to collect the structured abstracts corresponding to the papers in the PubMed summarization dataset. We match papers to their web-page using their article ID. We use BeautifulSoup library[3] and leverage the HTML structure of abstracts on their web-page to extract five aspects: *introduction*, *objectives*, *methods*, *results*, and *conclusion*. We manually checked the aspects and their summary and set rules to convert different spellings and typos (e.g., *intro→introduction*, *method→methods*) into the five standard aspects. For papers text and sections, we stick to the PubMed dataset. Table 1 shows the datasets statistics. We slightly change the aspects in FacetSum to make it similar to our dataset and make domain shift study possible (*purpose→objectives*, *method→methods*, *findings→results*).

For self-supervised pre-training we create two self-supervised datasets: *PubMed⋆* and *FacetSum⋆*, from PubMed and FacetSum aspect-based summarization datasets as described in section 3.2. PubMed⋆ and FacetSum⋆ contain 658K and 279K samples and 150K and 96K unique aspects, respectively. Additional dataset PubMed⋆-NoOverlap and

[3]www.crummy.com/software/BeautifulSoup/bs4/doc/

	Model	R-1	R-2	R-L
PubMed Generic	Discourse (Cohan et al., 2018)	38.93	15.37	35.21
	PEGASUS (Zhang et al., 2020)	39.98	15.15	25.23
	BART	**45.04**	**18.45**	**40.62**
PubMed	Greedy Extractive (Oracle)	56.61	39.23	47.58
	BART	**39.03**	**18.47**	**34.10**
	BART-Independent†	38.91	18.21	33.89
	BART Shuffle Aspects	24.21	6.18	19.86
FacetSum Generic	BART (Meng et al., 2021)	45.49	18.10	42.74
	BART-Facet (Meng et al., 2021)	49.29	19.60	45.76
	BART	**49.98**	**19.89**	**46.68**
FacetSum	Greedy Extractive (Oracle)	51.87	32.09	41.55
	BART (Meng et al., 2021)	23.27	10.31	20.29
	BART-Facet (Meng et al., 2021)	**37.97**	15.17	**32.08**
	BART	36.97	**15.50**	31.48
	BART-Independent†	36.77	15.26	31.23
	BART Shuffle Aspects	28.18	6.94	22.71

Table 2: Baselines and the state of the art performance on PubMed and FacetSum generic and aspect-based summarization evaluation sets. Results for the models with † are averaged over all aspects. Results by Meng et al. (2021) are based on BART extended to 10K tokens.

FacetSum*-NoOverlap are the variants in which we exclude aspects that overlap with the main aspects (shown by red in Figure 2). We only exclude aspects containing the main aspects but not semantically equivalent words. These datasets would allow assessing to what extent the model can perform semantic connection with new aspects.

5 Experiments and Results

In this section, we first explain model hyper-parameters. Then, we assess models' ability to make a semantic connection between aspects and summaries in different experimental setups and understand to what extend pre-training helps.

We rely on BART base available through HuggingFace's Transformers library (Wolf et al., 2019). It is trained for each dataset we tackle. Fine-tuning is done on 1 GPU (NVIDIA V100), with a batch size of 64 (8 gradient accumulation steps). We train the model for 10 epochs (2 epochs for self-supervised pre-training) with a learning rate of $3e-4$ and 500 warm-up steps and set the maximum input length to 1024, the BART official length (see Appendix A for a full list of hyper-parameters).

5.1 Baselines Experiments

System performance is evaluated with the ROUGE metric (Lin and Hovy, 2003), the default evaluation metric in the field in absence of universally acceptable semantic and factuality metrics. Table 2 reports R-1, R-2 and R-L scores, measuring the N-gram overlap between the reference and generated summaries for different baseline models. The first part of the table reports the results on generic summarization (summarizing into full abstracts) for a sanity check and compare the ROUGE scores between *off-the-shelf* BART model, as well as the BART model fine-tuned on PubMed or FacetSum.[4] For aspect-based summarization we consider following baselines:

- *Greedy extractive*: an extractive summarization oracle using the greedy extractive (Nallapati et al., 2017) method. We calculate ROUGE-N between every sentence in a document and the reference aspect-based summaries to find top sentences with the highest scores. The best set of sentences in terms of ROUGE-N scores is selected per document, and then scores are aggregated for all samples. The same score chooses sentences for each ROUGE-N score oracle.
- *BART*: BART model fine-tuned on the aspect-based summarization task containing all the available aspects. This is used as a fully supervised baseline for zero-shot experiments.
- *BART-Independent*: BART model trained on each aspect independently; we report an average performance across all the aspects. This baseline is not applicable in zero-shot settings and is reported for comparing baselines.
- *BART Shuffle Aspects*: We evaluate the BART aspect-based summaries generated from a wrong aspect (input document is the same but aspects' summaries are replaced randomly, e.g., *objectives→methods*). This baseline serves as a lower-bound performance.

Table 2 shows the baseline results of the generic and aspect-based summarization models. As expected, *greedy extractive* establishes a maximum oracle extractive summarization performance. BART slightly surpasses *BART-Ind*, showing that training all aspects together results in a better performance. Also, independent training is not applicable in the zero-shot setups. *BART-Shuffle* performs significantly worse than the other models.

[4]We use BART with a length of 1024. We experimented with longer BART models (extending positional embedding to 2,048 and 4,096 tokens) and PEGASUS. We did not see a significant gain in the overall performance of longer BART except the improvement on summaries requiring information from the end of papers (e.g., conclusion). Thus we continued all the experiments with the standard BART (Appendix B).

Model	Introduction	Objectives	Methods	Results	Conclusion
Greedy-Ext.	55.54/38.51/47.09	57.86/37.94/49.65	57.86/37.94/49.65	56.59/40.00/46.09	61.08/44.88/53.81
BART	40.66/**22.12**/36.18	**51.45/31.79/46.09**	40.78/19.08/35.84	**34.73/12.91/30.69**	**34.03/14.11/28.17**
BART-Ind.	**40.76**/22.03/**36.22**	51.11/31.09/45.44	**41.01/19.26/35.99**	34.16/12.40/30.10	33.95/13.76/28.13
BART-Shuf.	26.14/07.14/21.63	27.94/08.51/22.04	24.07/06.14/19.86	20.16/04.08/17.08	24.67/05.78/19.79

Table 3: Baseline and SOTA performance on the PubMed aspect-based summarization dataset (R-1/R-2/R-L).

Model	Objectives	Methods	Results	Value
Greedy-Ext.	54.94/34.27/44.54	49.27/29.82/39.18	53.25/34.35/42.49	50.18/29.97/40.33
BART (Meng et al., 2021)	46.74/27.09/41.21	23.66/07.92/20.53	16.39/04.63/14.33	06.30/01.62/05.07
BART-Facet (Meng et al., 2021)	48.65/27.72/42.55	**33.49**/11.01/**28.07**	**34.46/10.49/28.98**	**35.27/11.44/28.70**
BART	**48.83/29.10/43.46**	32.79/**11.71**/27.64	32.67/10.21/27.43	33.58/10.98/27.38
BART-Ind.	48.77/28.92/43.31	32.59/11.61/27.39	32.26/09.80/26.96	33.47/10.73/27.26
BART-Shuf.	32.52/09.75/26.34	25.86/05.71/20.96	25.76/05.61/20.83	28.48/06.63/22.79

Table 4: Baseline and SOTA performance on the FacetSum aspect-based summarization dataset (R-1/R-2/R-L).

PubMed					FacetSum				
Pre-Train	**Train**	**R-1**	**R-2**	**R-L**	**Pre-Train**	**Train**	**R-1**	**R-2**	**R-L**
Fully Supervised BART Baseline									
-	PubMed	39.03	18.47	34.10	-	FacetSum	36.97	15.50	31.48
Lower-bound BART Shuffle Aspect Baseline									
-	PubMed	24.21	6.18	19.86	-	FacetSum	28.18	6.94	22.71
Domain Shift: Out-Of-Domain Labelled Data & Unlabelled									
-	FacetSum	28.89	10.20	24.52	-	PubMed	31.03	10.04	25.75
PubMed*	FacetSum	**31.31**	**11.53**	**26.79**	FacetSum*	PubMed	**31.67**	**10.34**	**26.25**
PubMed* (No Overlap)	FacetSum	30.37	10.68	25.69	FacetSum* (No Overlap)	PubMed	31.17	10.10	25.90
FacetSum*	FacetSum	28.92	10.12	24.46	PubMed*	PubMed	30.48	9.48	25.29
Only Unlabelled Data									
PubMed*	-	**30.76**	**11.64**	**26.16**	FacetSum*	-	**28.18**	**7.60**	**23.54**
PubMed* (No Overlap)	-	29.70	10.93	25.20	FacetSum* (No Overlap)	-	26.90	6.67	22.45
FacetSum*	-	28.68	9.79	24.30	PubMed*	-	27.24	7.01	22.34

Table 5: Performance on PubMed and FacetSum when out-of-domain training data is available (domain shift) or only unlabelled data is available. PubMed* and FacetSum* are the self-supervised datasets for pre-training.

It indicates that the aspects belonging to a specific paper still demand significantly different summaries. Such a model primarily generates generic summaries rather than aspect-related summaries.

Tables 3 and 4 report the performance in terms of different aspects. In both datasets, *objective* reaches the best ROUGE scores while the performance drops for *results*, *conclusion*, and *value*. A similar phenomenon has been observed by Meng et al. (2021) and can possibly happen due to fact that information needed for summarizing *results*, *conclusion*, and *value* are mostly spread at the end of papers while information about *objectives* is skewed toward the beginning of the papers. The performance drop could be also because we truncate documents into a maximum length (1024 tokens) required by default BART architecture.

5.2 Domain Shift and Unlabelled Experiments

We define different experimental setups concerning the dataset used for pre-training and training. To be zero-shot, a model cannot be trained on in-domain labelled dataset. However, it can be pre-trained on the same unlabelled in-domain dataset (PubMed* or FacetSum*) in a self-supervised approach. This is a real-life case when there are numerous unlabelled but no labelled samples. As shown in Table 5, our proposed in-domain pre-training alleviates the domain shift problem. The best performance on both datasets is when the models trained on an out-of-domain dataset (PubMed or FacetSum) is pre-trained on the unlabelled in-domain dataset (PubMed* or FacetSum*). It gets closer to the fully supervised baseline performance and outperforms the lower-bound. In addition, experiments with only unlabelled data show that our proposed pre-training achieves comparable results with cases where out-of-domain labelled data is available. Interestingly, the models pre-trained on PubMed* performs better on PubMed than the model fine-tuned only on FacetSum*. This does not hold for the same case on the FacetSum experiment. We hypothesize that it might be due to the significantly larger size of PubMed* (658K) compared to FacetSum* (279K).

			PubMed			FacetSum		
Pre-Train	**Train**	**Test**	**R-1**	**R-2**	**R-L**	**R-1**	**R-2**	**R-L**
✗	All - Introduction	Introduction	30.88	11.65	25.66	-	-	-
✓	All - Introduction	Introduction	**40.07**	**21.22**	**35.5**	-	-	-
✓✓	All - Introduction	Introduction	38.76	20.29	33.86	-	-	-
✗	All - Objectives	Objectives	28.97	8.97	22.99	29.08	8.33	23.87
✓	All - Objectives	Objectives	**34.28**	**14.26**	**28.06**	**36.28**	**12.92**	**29.74**
✓✓	All - Objectives	Objectives	30.69	10.60	24.84	29.15	8.28	23.77
✗	All - Methods	Methods	25.68	7.03	21.10	27.32	6.59	22.16
✓	All - Methods	Methods	27.28	7.70	22.23	**28.13**	**6.84**	**22.79**
✓✓	All - Methods	Methods	**27.41**	**7.89**	**22.8**	28.07	6.59	22.63
✗	All - Results	Results	21.28	4.68	17.92	23.82	5.25	19.47
✓	All - Results	Results	**22.86**	**5.05**	**19.51**	23.07	4.80	18.90
✓✓	All - Results	Results	21.12	4.67	17.79	**24.22**	**5.28**	**19.83**
✗	All - Conclusion	Conclusion	27.92	7.36	21.86	-	-	-
✓	All - Conclusion	Conclusion	**31.23**	**9.17**	**24.73**	-	-	-
✓✓	All - Conclusion	Conclusion	30.03	8.13	23.49	-	-	-
✗	All - Value	Value	-	-	-	30.41	7.86	24.22
✓	All - Value	Value	-	-	-	**31.45**	**7.92**	**25.05**
✓✓	All - Value	Value	-	-	-	29.25	7.41	23.52

Table 6: Leave-one-out experiment on PubMed and FacetSum. The models are trained on all aspects except the one which the model is tested on. Considering in-domain training, this table shows unseen aspect performance. ✗: no pre-training except the BART official pre-training. ✓: model is pre-trained on PubMed* or FacetSum* (in-domain). ✓✓: model is pre-trained on PubMed* (No Overlap) or FacetSum* (No Overlap) (in-domain).

		PubMed			FacetSum		
Pre-Train	**Paraphrased Aspect**	**R-1**	**R-2**	**R-L**	**R-1**	**R-2**	**R-L**
✗	Introduction (baseline)	40.66	22.12	36.18	-	-	-
✗	Introduction -> Background ▼	27.98	9.34	23.62	-	-	-
✓	Introduction -> Background	**41.47**	**22.48**	**36.79**	-	-	-
✗	Introduction -> Context ▼	30.37	11.92	25.95	-	-	-
✓	Introduction -> Context	**40.28**	**21.58**	**35.64**	-	-	-
✗	Objectives (baseline)	51.45	31.79	46.09	48.83	29.10	43.46
✗	Objectives -> Objective	**51.37**	**31.66**	**46.03**	**48.91**	**29.17**	**43.52**
✓	Objectives -> Objective	51.10	31.39	45.60	48.51	28.81	43.14
✗	Objectives -> Purpose ▼	36.03	15.93	29.84	46.70	26.11	41.11
✓	Objectives -> Purpose	**49.77**	**29.92**	**44.09**	**48.28**	**28.46**	**42.88**
✗	Objectives -> Aims ▼	28.89	9.29	23.02	30.95	9.64	25.34
✓	Objectives -> Aims	**42.67**	**22.99**	**36.72**	**45.19**	**24.82**	**39.55**
✗	Methods (baseline)	40.78	19.08	35.84	32.79	11.71	27.64
✗	Methods -> Method	40.67	18.75	35.75	**32.94**	11.82	**27.73**
✓	Methods -> Method	**41.13**	**19.24**	**36.07**	32.85	**11.88**	27.69
✗	Methods -> Materials and Methods	**40.84**	**19.16**	**35.82**	**32.98**	11.75	**27.82**
✓	Methods -> Materials and Methods	40.58	19.05	35.58	32.77	**11.80**	27.69
✗	Methods -> Research Design ▼	34.82	14.23	29.74	32.68	11.34	27.41
✓	Methods -> Research Design	**38.22**	**17.18**	**33.12**	**32.84**	**11.81**	**27.62**
✗	Methods -> Methodology	**40.88**	19.13	**35.90**	**32.92**	**11.82**	**27.81**
✓	Methods -> Methodology	40.82	**19.24**	35.75	32.77	**11.82**	27.62
✗	Results (baseline)	34.73	12.91	30.69	32.67	10.21	27.43
✗	Results -> Result	**34.42**	**12.73**	**30.30**	**32.46**	**10.05**	27.21
✓	Results -> Result	34.12	12.53	30.00	**32.46**	9.98	**27.22**
✗	Results -> Discussion ▼	**23.57**	**7.09**	**20.09**	26.12	5.90	21.25
✓	Results -> Discussion	19.80	4.18	16.65	**29.06**	**7.82**	**23.93**
✗	Results -> Finding ▼	24.85	6.01	21.37	26.63	6.40	21.81
✓	Results -> Finding	**29.11**	**9.24**	**25.29**	**32.46**	**10.01**	**27.20**
✗	Conclusion (baseline)	34.03	14.11	28.17	-	-	-
✗	Conclusion -> Conclusions	**33.97**	**14.13**	**28.16**	-	-	-
✓	Conclusion -> Conclusions	33.94	13.92	28.04	-	-	-
✗	Value (baseline)	-	-	-	33.58	10.98	27.38
✗	Value -> Values ▼	-	-	-	32.24	10.59	26.98
✓	Value -> Values	-	-	-	**33.46**	**10.99**	**27.35**

Table 7: Paraphrasing experiment on PubMed and FacetSum. In each section, we evaluate the model trained on all original aspects on a new paraphrased aspect, e.g., *introduction→background* reports the case when *introduction* summaries are assigned to *background*. Considering in-domain training, this table shows unseen aspect performance. Significant drop in no pre-train cases are shown by ▼.

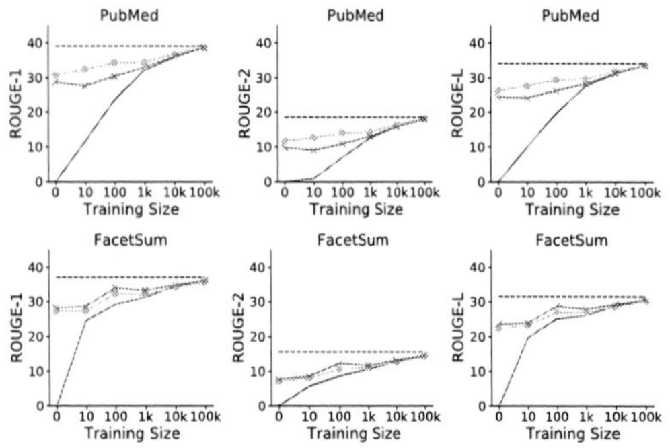

Figure 3: Aspect-based summarization performance with limited supervised examples. Pre-training with in-domain and out-of-domain datasets significantly improves the low-resource training sample performance. Top: evaluation done on PubMed dataset, Bottom: evaluation is done on FacetSum dataset. (—— *BART*, ···•··· *BART + pre-trained on PubMed**, –×– *BART + pre-trained on FacetSum**, - - - *BART fine-tuned on all samples*)

It is also promising that pre-trained models with no aspect overlap with the target aspect perform quite well. Such cases simulate the entirely unseen aspects in real scenarios.

5.3 Unseen Aspect Experiments

Leave-One-Out Experiments. This section studies leave-one-out experiments, aiming to investigate performance on unseen aspects within the same domain. We fine-tune BART for aspect-based summarization on all aspects except one that is left out for evaluation. We repeat the experiments for all the aspects available within our dataset. Table 6 reports the results for this experiment for both PubMed and FacetSum datasets. We compare baseline model (✗) and models enriched with self-supervised pre-training step as described in the section 3.2. The self-supervised pre-training can be done either on all the section headings (✓) or only on those non-overlapping with aspects of interest (✓✓). First, we note that zero-shot performance without self-supervised pre-training performs significantly worse compared to fully supervised models although it is still above random lower bound BART-Shuffle model (cf. tables 3 and 4). The pre-training step allows to significantly improve this performance for most of the aspects. As shown, non-overlapping pre-training (✓✓) also performs better than without pre-training cases except *results* and *value*. *introduction* and *objective* aspects experience the most improvement. As discussed previously (section 5.1) this could be due to the fact that information required to summarize these aspects are skewed toward the beginning of papers (Meng et al., 2021), and therefore is always within the input range of BART.

Paraphrasing Experiments. We study another zero-shot experiment where aspect word is paraphrased for evaluation. This experiment aims to understand to what extent a model can exploit the semantic meaning of aspects to generate good summaries. Table 7 reports results comparing models with and without pre-training. As in the previous experiment, the model without pre-training may significantly drop when replacing the original aspect with its alternative, specially when it does not share common sub-words. However, it still performs better than the random lower bound model meaning that it relies on the semantics of the aspect to some extent (cf. tables 3 and 4). The pre-training step makes the models suffering from a significant drop (▼) more robust to aspects paraphrasing while it does not significantly decline the performance in other cases. This is probably because the model has been exposed to a much richer and more diverse set of aspects during pre-training, and therefore learned to exploit aspect semantics better.

5.4 Few-Shot Experiments

Our final experiment aims at evaluating the summarization performance with limited supervised examples. For this, we train BART on the first 10, 100, 1K, 10K, and 100K training samples from each dataset. We repeat the experiments with the BART models pre-trained on the PubMed* and FacetSum* self-supervised datasets. Figure 3 plots the learning curves behaviour of different models as the amount of supervision grows. We see that models with self-supervised pre-training consistently surpass the baseline model. This superiority is much more significant in the few-shot cases, but the differences fade as more training samples is available and models become fully supervised. As expected, the models pre-trained on in-domain datasets perform better than the out-domain pre-trained models.

6 Conclusion

In this paper, we studied the problem of zero-shot aspect-based summarization of scientific documents. We established various experimental setups to investigate the effect of additional pre-training and intermediate training on the zero-shot performance with respect to domain shift from biomedical to management and unseen aspects. We proposed a self-supervised approach to pre-train the model using unlabelled target datasets. Results indicate that additional pre-training on the target dataset followed by intermediate training results in the best zero-shot performance.

We established leave-one-out and paraphrasing experimental setups to simulate the practical case of facing unseen aspects and showed the promising effect of additional self-supervised pre-training. Our proposed pre-training step improves the performance in the few-shot settings.

Investigating the effect of pre-training in terms of semantics and factuality evaluation scores can be done in the future.

References

Stefanos Angelidis and Mirella Lapata. 2018. Summarizing opinions: Aspect extraction meets sentiment prediction and they are both weakly supervised. In *Proceedings of the 2018 Conference on Empirical Methods in Natural Language Processing*, pages 3675–3686, Brussels, Belgium. Association for Computational Linguistics.

Dzmitry Bahdanau, Kyunghyun Cho, and Yoshua Bengio. 2014. Neural machine translation by jointly learning to align and translate. *arXiv preprint arXiv:1409.0473*.

Tal Baumel, Raphael Cohen, and Michael Elhadad. 2016. Topic concentration in query focused summarization datasets. In *Thirtieth AAAI Conference on Artificial Intelligence*.

Iz Beltagy, Matthew E Peters, and Arman Cohan. 2020. Longformer: The long-document transformer. *arXiv preprint arXiv:2004.05150*.

Taylor Berg-Kirkpatrick, Dan Gillick, and Dan Klein. 2011. Jointly learning to extract and compress. In *Proceedings of the 49th Annual Meeting of the Association for Computational Linguistics: Human Language Technologies*, pages 481–490, Portland, Oregon, USA. Association for Computational Linguistics.

Maximin Coavoux, Hady Elsahar, and Matthias Gallé. 2019. Unsupervised aspect-based multi-document abstractive summarization. In *Proceedings of the 2nd Workshop on New Frontiers in Summarization*, pages 42–47, Hong Kong, China. Association for Computational Linguistics.

Arman Cohan, Franck Dernoncourt, Doo Soon Kim, Trung Bui, Seokhwan Kim, Walter Chang, and Nazli Goharian. 2018. A discourse-aware attention model for abstractive summarization of long documents. In *Proceedings of the 2018 Conference of the North American Chapter of the Association for Computational Linguistics: Human Language Technologies, Volume 2 (Short Papers)*, pages 615–621, New Orleans, Louisiana. Association for Computational Linguistics.

Jacob Devlin, Ming-Wei Chang, Kenton Lee, and Kristina Toutanova. 2019. BERT: Pre-training of deep bidirectional transformers for language understanding. In *Proceedings of the 2019 Conference of the North American Chapter of the Association for Computational Linguistics: Human Language Technologies, Volume 1 (Long and Short Papers)*, pages 4171–4186, Minneapolis, Minnesota. Association for Computational Linguistics.

Xiangyu Duan, Mingming Yin, Min Zhang, Boxing Chen, and Weihua Luo. 2019. Zero-shot cross-lingual abstractive sentence summarization through teaching generation and attention. In *Proceedings of the 57th Annual Meeting of the Association for Computational Linguistics*, pages 3162–3172, Florence, Italy. Association for Computational Linguistics.

Alexander Fabbri, Simeng Han, Haoyuan Li, Haoran Li, Marjan Ghazvininejad, Shafiq Joty, Dragomir Radev, and Yashar Mehdad. 2021. Improving zero and few-shot abstractive summarization with intermediate fine-tuning and data augmentation. In *Proceedings of the 2021 Conference of the North American Chapter of the Association for Computational Linguistics: Human Language Technologies*, pages 704–717, Online. Association for Computational Linguistics.

Katja Filippova. 2010. Multi-sentence compression: Finding shortest paths in word graphs. In *Proceedings of the 23rd International Conference on Computational Linguistics (Coling 2010)*, pages 322–330, Beijing, China. Coling 2010 Organizing Committee.

Lea Frermann and Alexandre Klementiev. 2019. Inducing document structure for aspect-based summarization. In *Proceedings of the 57th Annual Meeting of the Association for Computational Linguistics*, pages 6263–6273, Florence, Italy. Association for Computational Linguistics.

Alexios Gidiotis and Grigorios Tsoumakas. 2019. Structured summarization of academic publications. *arXiv preprint arXiv:1905.07695*.

Alexios Gidiotis and Grigorios Tsoumakas. 2020. A divide-and-conquer approach to the summarization of long documents. *IEEE/ACM Transactions on Audio, Speech, and Language Processing*, 28:3029–3040.

Travis Goodwin, Max Savery, and Dina Demner-Fushman. 2020. Flight of the pegasus? comparing transformers on few-shot and zero-shot multi-document abstractive summarization. In *Proceedings of the 28th International Conference on Computational Linguistics*, pages 5640–5646.

Vivek Gupta, Prerna Bharti, Pegah Nokhiz, and Harish Karnick. 2021. SumPubMed: Summarization dataset of PubMed scientific articles. In *Proceedings of the 59th Annual Meeting of the Association for Computational Linguistics and the 11th International Joint Conference on Natural Language Processing: Student Research Workshop*, pages 292–303, Online. Association for Computational Linguistics.

Hiroaki Hayashi, Prashant Budania, Peng Wang, Chris Ackerson, Raj Neervannan, and Graham Neubig. 2020. WikiAsp: A dataset for multi-domain aspect-based summarization. *arXiv preprint arXiv:2011.07832*.

Karl Moritz Hermann, Tomas Kocisky, Edward Grefenstette, Lasse Espeholt, Will Kay, Mustafa Suleyman, and Phil Blunsom. 2015. Teaching machines to read and comprehend. *Advances in neural information processing systems*, 28:1693–1701.

Xinyu Hua and Lu Wang. 2017. A pilot study of domain adaptation effect for neural abstractive summarization. In *Proceedings of the Workshop on New Frontiers in Summarization*, pages 100–106, Copenhagen, Denmark. Association for Computational Linguistics.

Tatsuya Ishigaki, Hen-Hsen Huang, Hiroya Takamura, Hsin-Hsi Chen, and Manabu Okumura. 2020. Neural query-biased abstractive summarization using copying mechanism. *Advances in Information Retrieval*, 12036:174.

Kundan Krishna and Balaji Vasan Srinivasan. 2018. Generating topic-oriented summaries using neural attention. In *Proceedings of the 2018 Conference of the North American Chapter of the Association for Computational Linguistics: Human Language Technologies, Volume 1 (Long Papers)*, pages 1697–1705, New Orleans, Louisiana. Association for Computational Linguistics.

Mike Lewis, Yinhan Liu, Naman Goyal, Marjan Ghazvininejad, Abdelrahman Mohamed, Omer Levy, Veselin Stoyanov, and Luke Zettlemoyer. 2020. BART: Denoising sequence-to-sequence pre-training for natural language generation, translation, and comprehension. In *Proceedings of the 58th Annual Meeting of the Association for Computational Linguistics*, pages 7871–7880, Online. Association for Computational Linguistics.

Chin-Yew Lin and Eduard Hovy. 2003. Automatic evaluation of summaries using n-gram co-occurrence statistics. In *Proceedings of the 2003 Human Language Technology Conference of the North American Chapter of the Association for Computational Linguistics*, pages 150–157.

Yue Lu, ChengXiang Zhai, and Neel Sundaresan. 2009. Rated aspect summarization of short comments. In *Proceedings of the 18th international conference on World wide web*, pages 131–140.

Ahmed Magooda and Diane Litman. 2020. Abstractive summarization for low resource data using domain transfer and data synthesis. In *The Thirty-Third International Flairs Conference*.

Rui Meng, Khushboo Thaker, Lei Zhang, Yue Dong, Xingdi Yuan, Tong Wang, and Daqing He. 2021. Bringing structure into summaries: a faceted summarization dataset for long scientific documents. *arXiv preprint arXiv:2106.00130*.

Ramesh Nallapati, Feifei Zhai, and Bowen Zhou. 2017. SummaRuNNer: A recurrent neural network based sequence model for extractive summarization of documents. In *Thirty-First AAAI Conference on Artificial Intelligence*.

Ramesh Nallapati, Bowen Zhou, Cicero dos Santos, Çağlar Gůlçehre, and Bing Xiang. 2016. Abstractive text summarization using sequence-to-sequence RNNs and beyond. In *Proceedings of The 20th SIGNLL Conference on Computational Natural Language Learning*, pages 280–290, Berlin, Germany. Association for Computational Linguistics.

Shashi Narayan, Shay B. Cohen, and Mirella Lapata. 2018. Don't give me the details, just the summary! topic-aware convolutional neural networks for extreme summarization. In *Proceedings of the 2018 Conference on Empirical Methods in Natural Language Processing*, pages 1797–1807, Brussels, Belgium. Association for Computational Linguistics.

Preksha Nema, Mitesh M. Khapra, Anirban Laha, and Balaraman Ravindran. 2017. Diversity driven attention model for query-based abstractive summarization. In *Proceedings of the 55th Annual Meeting of the Association for Computational Linguistics (Volume 1: Long Papers)*, pages 1063–1072, Vancouver, Canada. Association for Computational Linguistics.

Tobias Rohde, Xiaoxia Wu, and Yinhan Liu. 2021. Hierarchical learning for generation with long source sequences. *arXiv preprint arXiv:2104.07545*.

Alexander M. Rush, Sumit Chopra, and Jason Weston. 2015. A neural attention model for abstractive sentence summarization. In *Proceedings of the 2015 Conference on Empirical Methods in Natural Language Processing*, pages 379–389, Lisbon, Portugal. Association for Computational Linguistics.

Ivan Sekulić, Amir Soleimani, Mohammad Aliannejadi, and Fabio Crestani. 2020. Longformer for ms marco document re-ranking task. *arXiv preprint arXiv:2009.09392*.

Dan Su, Tiezheng Yu, and Pascale Fung. 2021. Improve query focused abstractive summarization by incorporating answer relevance. *arXiv preprint arXiv:2105.12969*.

Bowen Tan, Lianhui Qin, Eric Xing, and Zhiting Hu. 2020. Summarizing text on any aspects: A knowledge-informed weakly-supervised approach. In *Proceedings of the 2020 Conference on Empirical*

Methods in Natural Language Processing (EMNLP), pages 6301–6309, Online. Association for Computational Linguistics.

Ivan Titov and Ryan McDonald. 2008. A joint model of text and aspect ratings for sentiment summarization. In *Proceedings of ACL-08: HLT*, pages 308–316, Columbus, Ohio. Association for Computational Linguistics.

Lu Wang and Wang Ling. 2016. Neural network-based abstract generation for opinions and arguments. In *Proceedings of the 2016 Conference of the North American Chapter of the Association for Computational Linguistics: Human Language Technologies*, pages 47–57.

Thomas Wolf, Lysandre Debut, Victor Sanh, Julien Chaumond, Clement Delangue, Anthony Moi, Pierric Cistac, Tim Rault, Rémi Louf, Morgan Funtowicz, et al. 2019. Huggingface's transformers: State-of-the-art natural language processing. *arXiv preprint arXiv:1910.03771*.

Yujia Xie, Tianyi Zhou, Yi Mao, and Weizhu Chen. 2020. Conditional self-attention for query-based summarization. *arXiv preprint arXiv:2002.07338*.

Yumo Xu and Mirella Lapata. 2021. Generating query focused summaries from query-free resources. In *Proceedings of the 59th Annual Meeting of the Association for Computational Linguistics and the 11th International Joint Conference on Natural Language Processing (Volume 1: Long Papers)*, pages 6096–6109, Online. Association for Computational Linguistics.

Min Yang, Qiang Qu, Ying Shen, Qiao Liu, Wei Zhao, and Jia Zhu. 2018. Aspect and sentiment aware abstractive review summarization. In *Proceedings of the 27th international conference on computational linguistics*, pages 1110–1120.

Tiezheng Yu, Zihan Liu, and Pascale Fung. 2021. AdaptSum: Towards low-resource domain adaptation for abstractive summarization. In *Proceedings of the 2021 Conference of the North American Chapter of the Association for Computational Linguistics: Human Language Technologies*, pages 5892–5904, Online. Association for Computational Linguistics.

Tiezheng Yu, Dan Su, Wenliang Dai, and Pascale Fung. 2020. Dimsum@ laysumm 20: Bart-based approach for scientific document summarization. *arXiv preprint arXiv:2010.09252*.

Jingqing Zhang, Yao Zhao, Mohammad Saleh, and Peter Liu. 2020. PEGASUS: Pre-training with extracted gap-sentences for abstractive summarization. In *International Conference on Machine Learning*, pages 11328–11339. PMLR.

Chenguang Zhu, Ziyi Yang, Robert Gmyr, Michael Zeng, and Xuedong Huang. 2021. Leveraging lead bias for zero-shot abstractive news summarization. In *Proceedings of the 44th International ACM SIGIR Conference on Research and Development in Information Retrieval*, pages 1462–1471.

A Training Hyper-parameters

BART fine-tuning is done on 1 GPU with 32GB memory (NVIDIA V100) with a batch size of 64. We use a gradient accumulation step of 8 and have 8 training samples per GPU per step. We train the model for 10 epochs (2 epochs for self-supervised pre-training). We use a learning rate of $3e-4$ and 500 warm-up steps. The maximum source length is set to 1024, and the maximum target length is set to 256. We set weight decay to 0.01, maximum gradient norm to 0.1, learning scheduler type to polynomial, label smoothing factor to 0.1, and dropout to 0.1, length penalty to 1.0, and the number of beams to 4.

B BART with Extended Input Length

BART has been pre-trained with a standard maximum input length of 1024 (Lewis et al., 2020). We can simply extend its positional embedding. However, as it has not been pre-trained with extended positional embedding, it would under-perform efficient transformers such as Longformer which is pre-trained on long inputs (Beltagy et al., 2020; Sekulić et al., 2020). In addition, the computational complexity of BART increases quadratically with input length; therefore, extended BART is substantially expensive to be trained. Table 8 and 9 compare the performance of standard BART with BART 2048 and BART 4096. While the extended models enhance the performance for *method*, *results*, *conclusion*, and *value*, which require information spread mostly at the end of papers, the overall improvement is not significant considering extra complexity and excessive training time. The BART-Facet model (Meng et al., 2021), which is an extended BART to 10,000 tokens, confirms the same trend.

C Masked Self-Supervised Pre-training

This section compares our default pre-trained approach with a masked version where we exclude target texts from inputs during the pre-training step. Our goal is to see the performance change when we remove the slight chance of copying sentences from input to output in the default setup. Note, it is impossible to copy sentences for most aspects as they are not in the model input range. Table 10 indicates that the difference between the two cases is insignificant.

D Summarization Examples

This section provides a number of summaries using different experimental setups. Table 11 presents generated summaries in fully-supervised, zero-shot, leave-one-out, and paraphrasing setups. It is not trivial to interpret these examples; however, some simple patterns can be observed. In the absence of in-domain supervised training, summaries are far from perfect, but pre-training can improve summaries when there is domain-shift or unseen aspect. Also, simple paraphrasing (e.g., *conclusion→conclusions*) cannot change the summary significantly unlike when there is no common sub-words between the two aspects (e.g., *objectives→purpose,aims*).

Model	Introduction	Objectives	Methods	Results	Conclusion
BART 1024	**40.66/22.12/36.18**	51.45/31.79/46.09	40.78/19.08/35.84	34.73/12.91/30.69	34.03/14.11/28.17
BART 2048	39.92/21.27/35.33	**52.05/32.30/46.52**	40.01/**20.29/36.89**	38.88/17.28/34.51	36.01/16.39/30.27
BART 4096	39.28/21.53/34.86	**52.05**/32.17/46.39	**44.44**/20.04/36.32	**39.33/18.87/35.13**	**41.13/23.25/36.12**

Table 8: Comparing BART with the standard maximum length of 1024 and the extended BART models on the PubMed aspect-based summarization dataset.

Model	Objectives	Methods	Results	Value
BART 1024	48.83/29.10/43.46	32.79/11.71/27.64	32.67/10.21/27.43	33.58/10.98/27.38
BART 2048	49.82/30.22/44.34	34.64/13.48/29.22	34.16/11.41/28.70	34.19/11.72/27.95
BART 4096	**49.96/30.63/44.58**	**35.20/13.97/29.68**	34.18/**12.04/29.27**	33.95/**11.76**/27.86
BART-Facet 10000 (Meng et al., 2021)	48.65/27.72/42.55	33.49/11.01/28.07	**34.46**/10.49/28.98	**35.27**/11.44/**28.70**

Table 9: Comparing BART with the standard maximum length of 1024 and the extended BART models on the FacetSum aspect-based summarization dataset.

PubMed					FacetSum				
Pre-Train	**Train**	**R-1**	**R-2**	**R-L**	**Pre-Train**	**Train**	**R-1**	**R-2**	**R-L**
Domain Shift: Out-Of-Domain Labelled Data & Unlabelled									
PubMed*	FacetSum	31.31	**11.53**	26.79	FacetSum*	PubMed	**31.67**	**10.34**	**26.25**
PubMed*$_{Masked}$	FacetSum	**31.44**	11.52	**26.83**	FacetSum*$_{Masked}$	PubMed	31.27	10.18	25.96
FacetSum*	FacetSum	**28.92**	**10.12**	**24.46**	PubMed*	PubMed	30.48	9.48	25.29
FacetSum*$_{Masked}$	FacetSum	28.23	9.87	23.75	PubMed*$_{Masked}$	PubMed	**31.21**	**9.91**	**25.87**
Only Unlabelled Data									
PubMed*	-	**30.76**	11.64	**26.16**	FacetSum*	-	28.18	7.60	23.54
PubMed*$_{Masked}$	-	30.73	**11.79**	26.15	FacetSum*$_{Masked}$	-	**28.30**	**7.91**	**23.71**
FacetSum*	-	**28.68**	**9.79**	**24.30**	PubMed*	-	27.24	7.01	22.34
FacetSum*$_{Masked}$	-	28.49	9.63	24.12	PubMed*$_{Masked}$	-	**27.90**	**7.50**	**23.06**

Table 10: Comparing normal self-supervised pre-training using PubMed* and FacetSum* with their masked version. In masked datasets, the target text is masked during training.

Aspect: Objectives
Reference: To evaluate the efficacy and safety of outpatient management of severe ovarian hyperstimulation syndrome (OHSS) requiring placement of a pigtail catheter.
Fully Supervised (Training: PubMed): To evaluate the efficacy and safety of pigtail catheter drainage in the management of severe/critical OHSS in patients who underwent in vitro fertilization and embryo transfer at our centre. (50/36/43)
Zero-Shot (Training: FacetSum): The purpose of this paper is to evaluate the efficacy and safety of pigtail catheter drainage in the management of severe ohss in patients who underwent in vitro fertilization and embryo transfer at the centre between 1999 and 2001. (41/31/36)
Zero-Shot (Training: FacetSum, PubMed*): The purpose of this paper is to assess the efficacy of pigtail catheter drainage in the management of severe/critical ohss in patients who underwent in vitro fertilization. (44/17/32)
Leave-One-Out: The mean age of the patients was 22.5 years (range: 12-40 years). The mean duration of pregnancy was 23.5 months. The mean number of oocytes retrieved per day was 2.5 (range, 1-4). The average number of embryos retrieved per patient was 3 (range = 1-6). The catheter was successfully placed in all patients. No complications were observed during the percutaneous drainage of ascites. (11/0/11)
Leave-One-Out (Training: PubMed*): Patients who underwent in vitro fertilization and embryo transfer at our centre between 1999 and 2001 on both inpatient and outpatient clinics were included in this study. The catheter was placed temporarily for a few days and drainage was completed through one procedure. (15/0/9)
Paraphrasing (Purpose): Percutaneous placement of a pigtail catheter for drainage of ascites caused by severe/critical OHSS was effective and safe (45/21/25)
Paraphrasing (Purpose) (Training: PubMed*): To evaluate the efficacy and safety of percutaneous placement of a pigtail catheter for drainage of ascites caused by severe/critical ovarian syndrome (72/47/54)
Paraphrasing (Aims): Percutaneous placement of a pigtail catheter for drainage of ascites caused by severe/critical OHSS was effective in reducing the incidence of severe OHSS in both inpatient and outpatient populations. (47/20/19)
Paraphrasing (Aims) (Training: PubMed*): We evaluated the efficacy and safety of percutaneous placement of a pigtail catheter for drainage of ascites caused by severe/critical OHSS. (61/43/51)
Aspect: Conclusion
Reference: Nursing interventions based on family needs of hospitalized patients in the ICU increase their satisfaction. Attention to family nursing should be planned especially in the ICUs.
Fully Supervised (Training: PubMed): According to the findings of this study, nursing interventions based on family needs can increase the satisfaction of the families of the hospitalized patients in the ICU. (64/35/45)
Zero-Shot (Training: FacetSum): The family needs of the patients hospitalized in intensive care units of al - zahra (sa. (39/10/24)
Zero-Shot (Training: FacetSum, PubMed*): The study showed that the family needs of the patients in the icu could increase their satisfaction and thus a reduction in stress. (48/29/44)
Leave-One-Out: The aim of this study was to determine the effectiveness of nursing interventions based on family needs on family satisfaction level of hospitalized patients in the neurosurgery intensive care unit of Al-Zahra Hospital, Isfahan, Iran. (48/30/35)
Leave-One-Out (Training: PubMed*): The aim of this study was to determine the effectiveness of nursing interventions based on family needs on family satisfaction level of hospitalized patients in the neurosurgery intensive care unit of Al-Zahra Hospital, Isfahan, Iran. (48/30/35)
Paraphrasing (Conclusions): According to the findings of this study, nursing interventions based on family needs can increase the satisfaction of the families of the hospitalized patients in the intensive care unit. (58/30/40)
Paraphrasing (Conclusions) (Training: PubMed*): The results of this study showed that nursing interventions based on family needs had a positive effect on the family satisfaction level of the hospitalized patients in the intensive care unit. (49/29/38)

Table 11: Examples of summaries using different experimental setups. ROUGE scores are shown at the end of summaries (R1/R2/RL).

Data Augmentation for Biomedical Factoid Question Answering

Dimitris Pappas[1,2], Prodromos Malakasiotis[1,3], and Ion Androutsopoulos[1]

[1]Department of Informatics, Athens University of Economics and Business, Greece
[1]pappasd@aueb.gr, rulller@aueb.gr, ion@aueb.gr
[2]Institute for Language and Speech Processing, Research Center 'Athena', Greece
[2]dpappas@athenarc.gr
[3]Institute of Informatics and Telecommunications, NCSR 'Demokritos', Greece

Abstract

We study the effect of seven data augmentation (DA) methods in factoid question answering, focusing on the biomedical domain, where obtaining training instances is particularly difficult. We experiment with data from the BIOASQ challenge, which we augment with training instances obtained from an artificial biomedical machine reading comprehension dataset, or via back-translation, information retrieval, word substitution based on WORD2VEC embeddings or masked language modeling, question generation, or extending the given passage with additional context. We show that DA can lead to very significant performance gains, even when using large pre-trained Transformers, contributing to a broader discussion of if/when DA benefits large pre-trained models. One of the simplest DA methods, WORD2VEC-based word substitution, performed best and is recommended. We release our artificial training instances and code.

1 Introduction

Question Answering (QA) systems aim to answer natural language questions by searching in structured (Fu et al., 2020; Luo et al., 2018; Yadati et al., 2021) or unstructured data, such as free-text documents (Aghaebrahimian, 2018). Here we consider QA of the latter kind. Fully fledged QA systems for document collections retrieve relevant documents, identify relevant passages, extract and aggregate answer spans etc. (Chen et al., 2017a; Pappas and Androutsopoulos, 2021). There are also different types of questions, e.g., *yes/no, factoid, list, how-to.* Thus, creating realistic datasets to train and evaluate complete QA systems for document collections is resource intensive, especially for systems targeting specialized domains. A prime example is the *biomedical domain*, the focus of this work, where obtaining realistic training (and test) instances requires medical expertise, which is costly and difficult to obtain. Consequently, biomedical datasets for full QA systems contain just a few thousand training instances (Tsatsaronis et al., 2015; Möller et al., 2020) or focus on simpler question types only, e.g., yes/no questions (Jin et al., 2019).

A simplified form of QA for textual data is Machine Reading Comprehension (MRC) (Yang et al., 2015; Rajpurkar et al., 2016; Campos et al., 2016; Chen et al., 2017b; Lai et al., 2017; Joshi et al., 2017; Kwiatkowski et al., 2019; Reddy et al., 2019; Jin et al., 2019; Wang et al., 2020), where the system is given a question and a particular (or a few) passage(s) and the answer must be found therein. In effect, MRC focuses on a particular core stage of a full QA pipeline, identifying answer spans, assuming that relevant documents and passages have already been identified. We also focus on this stage, adopting an MRC setting. Large generic (non domain-specific) MRC datasets have been constructed via crowd-annotation (Rajpurkar et al., 2016, 2018; Yang and Choi, 2019; Joshi et al., 2017), but crowd-annotation on that scale is difficult when biomedical expertise is required. An alternative is to *automatically* generate *cloze-style* MRC datasets. The last sentence or title of a random passage is treated as a question, some part (e.g., named entity) of the 'question' is masked, and the system is required to predict it. This approach has been used to generate large *artificial* cloze-style MRC datasets (Hill et al., 2016; Chen et al., 2016; Bajgar et al., 2016), including biomedical ones (Pappas et al., 2018, 2020). These datasets could be used to augment real ones, but have mostly been used as artificial experimental setups only.

When training examples for end-tasks are limited, as in realistic biomedical QA datasets, the currently dominant approach in NLP is to use pre-trained Transformers (Devlin et al., 2019; Liu et al., 2019; Lan et al., 2019; He et al., 2020; Raffel et al., 2020), possibly pre-trained on domain-specific

Proceedings of the BioNLP 2022 workshop, Dublin, Ireland, pages 63–81
May 26, 2022.

corpora (Lee et al., 2019; Beltagy et al., 2019; Chalkidis et al., 2020), and fine-tune (further train) them on the limited examples of the end-tasks. Nevertheless, increasing the number of end-task examples typically improves performance. One way to achieve this is to employ *data augmentation* (DA) (Shorten et al., 2021; Feng et al., 2021), which adds artificial training instances to a training set, in our case the training set of the end task. It is unclear, however, which DA methods improve most (if at all) the performance of pre-trained models per end-task (Longpre et al., 2019, 2020). Consequently, Feng et al. (2021) recommend exploring when DA is effective for large pre-trained models.

In this paper, we thoroughly examine the impact of DA in biomedical QA, focusing on the factoid questions of the BIOASQ challenge (Tsatsaronis et al., 2015) in an MRC setting, i.e., we assume that relevant text passages, called *snippets* in BIOASQ, have already been identified. We first evaluate on BIOASQ three indicative off-the-shelf pre-trained models, DISTILBERT (Sanh et al., 2019), BIOBERT (Lee et al., 2019), ALBERT (Lan et al., 2019), already fine-tuned on SQUAD (Rajpurkar et al., 2016) or SQUAD-V2 (Rajpurkar et al., 2018), and we select ALBERT as our weak baseline. We also fine-tune ALBERT on BIOASQ, on top of its SQUAD fine-tuning, to obtain a stronger baseline. We then obtain additional artificial training instances from an artificial cloze-style MRC dataset, or via back-translation, information retrieval (IR), word substitution based on WORD2VEC or masked language modeling, question generation, or by extending the given passages with additional context. WORD2VEC-based word substitution, one of the simplest DA methods considered, improves test performance from 76.78% precision-recall AUC (for ALBERT fine-tuned on SQUAD and BIOASQ) to 84.99%. Although we focus on biomedical QA, our work should also be of interest in QA for other specialized domains, e.g., legal QA (Kien et al., 2020; Khazaeli et al., 2021; Zhang and Xing, 2021). Our work is the largest, in terms of DA methods considered, experimental study of DA for QA (Section 4).

Our main contributions are: (1) We present the largest (in terms of methods) experimental comparison of DA methods for QA, focusing on biomedical QA, where obtaining training instances is particularly difficult and costly. (2) We show that DA can lead to very large performance gains, even when using pre-trained Transformers fine-tuned on large generic (SQUAD) and/or small domain-specific (BIOASQ) end-task datasets, contributing to a broader discussion of if/when DA benefits pre-trained models. (3) We show that artificial cloze-style MRC datasets are useful for DA. (4) We show that one of the simplest DA methods, WORD2VEC-based word substitution, is also the best and is, therefore, recommended. (5) We make our artificial training examples and code publicly available.[1]

2 Experimental Setup

2.1 BIOASQ Data in a SQUAD setting

We experiment with data from BIOASQ-8 (2021), Phase B, Task B (Tsatsaronis et al., 2015), which contain English questions of biomedical experts. Each question is accompanied by (i) the gold answer (often several alternative phrasings) and (ii) gold relevant passages, called *snippets* (usually a single sentence each) from biomedical articles; the gold snippets contain the gold answer or other relevant information. There are four question types: *yes/no*, *factoid*, *list*, and questions requiring a *summary*. We focus on factoid questions (e.g., "*Which gene is involved in Giant Axonal Neuropathy?*").

We convert the BIOASQ data to triples each containing a question, a single gold snippet, and the span of the gold answer in the snippet, much as in SQUAD (Rajpurkar et al., 2016). If a question has multiple gold snippets, we produce equally many triples, discarding snippets that do not contain the gold answer. This conversion and considering only factoid questions allow us to use pre-trained Transformers already fine-tuned on SQUAD in a similar setting.[2] A disadvantage of the conversion is that our results are not directly comparable to those of BIOASQ. The goal of our work, however, is to study the effect of different DA methods on a modern Transformer-based QA baseline (and we show that fine-tuning it first on SQUAD helps), not to compete against BIOASQ participants, who often use models tailored to the particular competition.

From the 941 factoid questions of the original BIOASQ data, we obtained 3415 question-snippet-answer triples. We split these in training, development, test subsets (2848, 271, 296 triples, resp.), ensuring no question is in more than one subsets.

[1]See `http://nlp.cs.aueb.gr/publications.html` for links to the code and data.

[2]In the original BIOASQ data, multiple snippets may be given for a particular question, the answer may be present in several of them, and identifying any answer span suffices.

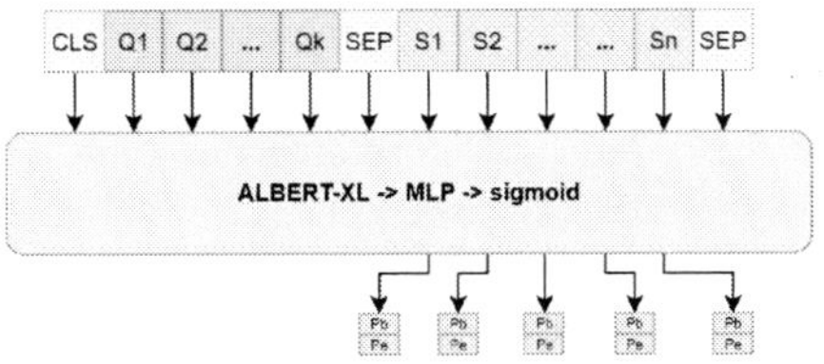

Figure 1: The model used in all of the following experiments. ALBERT-XL is fed with a question and snippet. Its contextualized embeddings are passed through an MLP with sigmoid activations that produces a begin (P_b) and end (P_e) probability per token of the snippet.

2.2 Off-the-shelf Models

As a starting point, we compared the performance of three publicly available pre-trained models that have already been fine-tuned for MRC on SQUAD (Rajpurkar et al., 2016) or SQUAD-V2 (Rajpurkar et al., 2018).[3] At the time of our experiments, ALBERT-based models (Lan et al., 2019) were among the best on the SQUAD leaderboards; here, we use ALBERT fine-tuned on SQUAD-V2. We also considered BIOBERT (Lee et al., 2019), because it is pre-trained on a biomedical corpus; again, we use it fine-tuned on SQUAD-V2. The third model, DISTILBERT (Sanh et al., 2019), was chosen because of its much smaller size, which makes running experiments easier. This model is pretrained on a generic corpus, like the original BERT, and we use it fine-tuned on SQUAD. All three models are used here off-the-self, i.e., they are only evaluated, not trained in any way on BIOASQ data. Throughout this work, we use the development subset of the BIOASQ data to select models and configurations of DA methods, but in this experiment we use the union of the training and development subsets, since no BIOASQ training is involved. ALBERT is the best off-the-shelf model considered (Table 1), hence we use it in all other experiments.[4]

Model	PRAUC (BIOASQ train+dev)
DISTILBERT (SQUAD)	64.27
BIOBERT (SQUAD-V2)	69.22
ALBERT (SQUAD-V2)	**75.05**

Table 1: *Off-the-shelf* pre-trained models, fine-tuned for MRC on SQUAD or SQUAD-V2. We report Precision-Recall AUC (PRAUC, %) on BIOASQ training and development data, since no BIOASQ training is involved.

2.3 Model Architecture Modifications

The results of Table 1 were obtained by feeding the three off-the-shelf models with the concatenation of the question and snippet of each question-snippet-answer BIOASQ triple (training or development), without training of any kind. Following a typical MRC architecture, each model was previously fine-tuned (by others) on SQUAD (or SQUAD-V2) with a shared dense layer on top of each contextualized token embedding (of the snippet only) that the pre-trained model generates. The dense layer produces two logits per token, indicating the model's confidence that the token is the beginning or end of the answer, respectively. Two separate softmax activations operate across all the begin and end logits, respectively, and the answer is the span (of the snippet) whose first and last tokens have the highest sum of begin and end probabilities (and the correct order).[5] The two softmax activations presuppose that there is a single contiguous answer span in each snippet. This is true in SQUAD, but in our BIOASQ data the (single) answer of a triple may consist of multiple non-contiguous spans of the triple's snippet (this happens in 583 out of 2,848 training instances). Hence, in the following experiments, where we further fine-tune ALBERT on BIOASQ or artificial training data, we replace the two softmax activations by two sigmoids that produce the begin and end probability per token of the snippet. Any token whose begin (or end) probability is above a threshold t is treated as the beginning (or end) of an answer span. The PRAUC evaluation measure (discussed below) aggregates results over different t values. We also replace the dense layer on top of the contextualized token embeddings by a Multi-Layer Perceptron (MLP) with a single hidden layer, which performed better on our development data. We use this single typical MRC model architecture (Fig.1) in all the following experiments, since we aim to study the effect of several DA methods, not to propose new MRC model architectures.

2.4 Evaluation Measure

Given a development or test question-snippet-answer triple and a decision threshold t (Section 2.3), we compute precision and recall at the token level, i.e., we measure the ability of the model to identify the tokens of the answer. Precision is the number of correctly identified answer tokens,

[3]We obtained the models from `https://huggingface.co/ktrapeznikov/albert-xlarge-v2-squad-v2`. We use the XL version of ALBERT. The other two models adopt the BERT-BASE architecture; no XL variants were available.

[4]We discuss PRAUC in Sections 2.3 and 2.4.

[5]In SQUAD-V2, additional layers decide if a question is answerable. We do not discuss them to save space.

divided by the number of tokens in the model's answer. Recall is the number of correctly identified answer tokens, divided by the number of tokens in the correct answer. For different thresholds t, we obtain different precision-recall pairs for the same question-snippet-answer triple, which can be plotted as a precision-recall curve. We compute the trapezoidal area under the precision-recall curve (PRAUC), and we then macro-average the PRAUC scores over the test (or development) triples.[6]

2.5 Baselines

We use two baselines that do not involve DA: i) off-the-shelf ALBERT, pre-trained on a generic corpus, already fine-tuned on SQUAD-V2 (last model of Table 1); and ii) same as the first baseline, but further fine-tuned (on top of the fine-tuning on SQUAD-V2) on our BIOASQ training data, with the modified architecture of Section 2.3. Table 2 shows that the second baseline is much stronger. Hence, we report performance gains with DA methods against the second baseline in subsequent sections.[7]

Model	+train ex.	PRAUC (BIOASQ dev)
ALBERT (SQUAD-V2)	0	80.25
+BIOASQ	2,848	89.57

Table 2: Performance of baselines on BIOASQ dev. data. The first one is ALBERT-XL fine-tuned on SQUAD-V2. The second one is also fine-tuned on BIOASQ, with the modified architecture of Fig. 1. We also show the number of domain-specific (BIOASQ) training examples.

3 Data Augmentation Methods

There are two alternatives when using the artificial training instances that DA generates (Yang et al., 2019). In our case, we always start with ALBERT, pre-trained on a generic corpus, and already fine-tuned on SQUAD-V2. In the first alternative, the model is then further fine-tuned on the artificial instances, and is then finally fine-tuned on the end-task data (BIOASQ). In the second alternative, the artificial and the end-task data are mixed, and the model is fine-tuned on the mixed data. In each experiment below, we use the alternative (among the two) that leads to the best development PRAUC.

3.1 Artificial Cloze-style MRC Dataset

For this augmentation method, we use BIOMRC (Pappas et al., 2020), the most recent and largest *artificial* cloze-style biomedical MRC dataset. BIOMRC comes in two versions, LARGE and LITE, with 813k and 100k cloze-style questions, respectively. We use BIOMRC LITE. Each 'question' is the title of a biomedical article, with an entity mentioned in the title hidden. Each question is accompanied by a passage (the abstract of the article), candidate answers (entities mentioned in the abstract), and the gold answer. From each passage we keep only the sentence containing the gold answer as the given snippet, and we generate a question-snippet-answer triple.[8] If more than one sentences of the passage contain the gold answer, we create multiple triples, one for each sentence. We end up with approximately 142k artificial training triples.

ALBERT (SQUAD-V2)	+train ex.	PRAUC (BIOASQ dev)
+BIOASQ	2,848	89.57
+BIOMRC	2,848	78.66
+BIOMRC +BIOASQ	5,696	91.57
+BIOMRC	10,000	91.20
+BIOMRC +BIOASQ	12,848	**93.15**
+BIOMRC	30,000	90.57
+BIOMRC +BIOASQ	32,848	92.19
+BIOMRC	50,000	91.19
+BIOMRC +BIOASQ	52,848	91.51
+BIOMRC	100,000	90.93
+BIOMRC +BIOASQ	102,848	92.39

Table 3: Adding training examples from an *artificial cloze-style* MRC dataset (BIOMRC). The '+train ex.' column shows the number of domain-specific training examples (from BIOMRC and/or BIOASQ) used, on top of examples seen during fine-tuning on SQUAD-V2.

In Table 3, the starting point is the weak baseline of Table 2 (ALBERT fine-tuned on SQUAD-V2). We compare to the strong baseline (the second one of Table 2), which was further fine-tuned on BIOASQ (+BIOASQ). We show results when fine-tuning on BIOMRC (+BIOMRC) instead of BIOASQ, and when fine-tuning on both BIOMRC and BIOASQ (+BIOMRC +BIOASQ), using 10k to 100k randomly sampled BIOMRC examples. Interestingly, fine-tuning on 10k artificial BIOMRC examples leads to better performance (91.20 dev. PRAUC) than fine-tuning on BIOASQ (89.57). The best performance (93.15) is obtained by fine-tuning on both BIOASQ and 10k BIOMRC examples. We attribute this improvement to the resemblance of BIOMRC to BIOASQ data. We see no benefit when adding more than 10k BIOMRC examples, which may be an indication that the useful (for BIOASQ) patterns that the model can learn from BIOMRC are limited.

[6] PRAUC is similar to Mean Average Precision (Manning et al., 2008), but obtains precision-recall points differently.

[7] We also experimented pre-trained ALBERT directly fine-tuned only on BIOASQ, but the performance was much worse.

[8] See the appendix for examples of all the DA methods.

3.2 Back-translation

Back translation (BTR) has been used for data augmentation in several NLP tasks (Feng et al., 2021; Shorten et al., 2021). The training examples are machine-translated from a source to a pivot language and back, obtaining paraphrases. We initially used French as the pivot language, then also Spanish and German, always translating from English to a pivot language and back with Google Translate. For each question-snippet-answer training triple of BIOASQ, we generate two new triples by back-translating either the question or the snippet. If a new triple is identical to the original one, we discard it. We obtained 4,901 new training examples pivoting only to French, and 15,593 when also pivoting to Spanish and German.

ALBERT (SQUAD-V2)	+train ex.	PRAUC (BIOASQ dev)
+BIOASQ	2,848	89.57
+BTR [FR]	2,848	91.84
+BTR [FR] +BIOASQ	5,696	**92.95**
+BTR [FR]	4,901	89.80
+BTR [FR] +BIOASQ	7,749	91.44
+BTR [FR,ES,DE]	2,848	89.80
+BTR [FR,ES,DE] +BIOASQ	5,696	89.99
+BTR [FR,ES,DE]	14,229	92.21
+BTR [FR,ES,DE] +BIOASQ	17,077	92.21

Table 4: Data augmentation via *back-translation* (BTR), using one (FR) or three (FR, ES, DE) pivot languages.

Table 3 shows that adding back-translations to the BIOASQ training data increases development PRAUC from 89.57 to 91.44 (or 92.66) with one (or three) pivot languages. Using back-translations with one pivot (+BTR [FR]) instead of the original BIOASQ data slightly surpasses the strong baseline (89.80 vs. 89.57); and with three pivots, using only back-translations (+BTR [FR,DE,ES]) performs almost the same as adding the original BIOASQ data too (92.52 vs. 92.66). These results show that BTR produces very good training instances and that further benefits may be possible with more pivots. Nevertheless simpler methods (e.g., WORD2VEC-based word substitution, discussed below) offer larger gains with fewer artificial training instances.

3.3 Information Retrieval

Data augmentation based on Information Retrieval (IR) has been found promising in previous open-domain QA work (Yang et al., 2019).[9] Given a question and a gold answer, the question is used as a query to an IR system. Any retrieved document (or passage therein) that includes the gold answer is used to construct a new training example (with the same question and gold answer). We used the open data from the PUBMED Baseline Repository[10] to create a pool of 22.3M biomedical documents. Each document is the concatenation of the title and abstract of a PUBMED article. We indexed all documents with an ElasticSearch retrieval engine[11] and used the 500 top ranked (by BM25) documents per question. From the original 2,848 question-snippet-answer triples, only 289 more were generated, because in most of the retrieved documents no sentence included the entire answer (individual terms of the answer might be scattered in the document). We suspect that the biomedical experts of BIOASQ create questions whose answers cannot be found in large numbers of documents (unlike common questions in open-domain QA datasets), and the few relevant documents (and snippets) of each question have already been included in the BIOASQ training data. Table 5 shows that IR-based augmentation led to very minor gains in our case, because of the very few additional instances.

ALBERT (SQUAD-V2)	+train ex.	PRAUC (BIOASQ dev)
+BIOASQ	2,848	89.57
+IR	289	80.30
+IR +BIOASQ	3,137	**89.80**

Table 5: Data augmentation via *information retrieval* (IR), using PUBMED titles and abstracts as documents.

3.4 Word Substitution

These methods replace words of the original training examples by similar words (e.g., synonyms) from a thesaurus (Jungiewicz and Smywinski-Pohl, 2019; Abdollahi et al., 2020) or words with similar embeddings (Wang and Yang, 2015). More recent work uses large language models, pre-trained to predict masked tokens, which suggest replacements of randomly masked words of the original examples (Kobayashi, 2018; Wu et al., 2019).

3.4.1 WORD2VEC-based Word Substitution

In this case, we use biomedical WORD2VEC (Mikolov et al., 2013; Brokos et al., 2018) embeddings. Given a question-snippet-answer training instance, we consider all the word tokens of the snippet (excluding stop-words). For each token w_i ($i = 1, \ldots, n$) of the snippet, we select the $k_i \leq K$ most similar words w_j ($j = 1, \ldots, k_i$) of

[9] Yang et al. (2019) gained 2.7 to 9.7 F1 percentage points (pp.) in all four datasets they experimented with.

[10] lhncbc.nlm.nih.gov/ii/information/MBR.html

[11] https://www.elastic.co/

the vocabulary, using cosine similarity of word embeddings $(\vec{w}_i, \vec{w}_j)$, that satisfy $\cos(\vec{w}_i, \vec{w}_j) \geq C$. We then produce $(k_1+1)(k_2+1)\ldots(k_n+1)-1$ artificial training instances by replacing each token w_i of the snippet by one of its k_i most similar words (or itself), requiring at least one token of the original snippet to have been replaced. We then randomly sample 10k to 100k of the resulting instances and use them as additional training examples. We set $K = 10$, $C = 0.95$ based on preliminary experiments on development data. Adding 10k of the resulting artificial training examples to the original BIOASQ examples leads to 95.60 development PRAUC, outperforming the strong baseline (89.57) by six percentage points (Table 6). Using only the 10k artificial examples, without any of the original examples, achieves almost identical performance (95.59), which suggests that the generated examples are of high quality. As when using artificial MRC examples (Table 3), adding more than 10k artificial instances provides no further benefit, probably because we end up adding too many minor variants of the same original examples.

ALBERT (SQUAD-V2)	+train ex.	PRAUC (BIOASQ dev)
+ BIOASQ	2,848	89.57
+WORD2VEC	2,848	95.56
+WORD2VEC +BIOASQ	5,696	95.27
+WORD2VEC	10,000	95.59
+WORD2VEC +BIOASQ	12,848	**95.60**
+WORD2VEC	30,000	95.28
+WORD2VEC +BIOASQ	32,848	95.20
+WORD2VEC	50,000	95.16
+WORD2VEC +BIOASQ	52,848	95.13
+WORD2VEC	100,000	95.36
+WORD2VEC +BIOASQ	102,848	95.22

Table 6: Data augmentation with WORD2VEC-*based word substitution*, using biomedical embeddings.

The same DA mechanism could have been applied to questions instead of snippets. In preliminary experiments, we employed an additional pre-trained natural language inference (NLI) model (El Boukkouri et al., 2020) as a *consistency* mechanism to ensure the modified snippets followed from the original ones, but this also greatly reduced the number of artificial training instances we could generate. Performance was better without this mechanism, i.e., generating more artificial instances was better than generating fewer higher quality ones.

3.4.2 Masked LM Word Substitution

Here we use BIOLM (Lewis et al., 2020) and specifically a ROBERTA-LARGE model pre-trained on PUBMED, PMC, and MIMIC-III (Zhu et al., 2018) with a new vocabulary extracted from PUBMED.[12] We use the same process as in WORD2VEC word substitution, but each candidate replacement w_j of an original word w_i of the snippet must now satisfy $p(w_j) \geq P$ (instead of $\cos(\vec{w}_i, \vec{w}_j) \geq C$), where $p(w_j)$ is the probability assigned to w_j by the pre-trained model; we also rank the candidate replacements w_j of each w_i by $p(w_j)$. We set $P = 0.95$, based on preliminary experiments on development data. Table 7 shows that BIOLM-based substitution is almost as good as WORD2VEC-based substitution (94.45 vs. 95.60), but for BIOLM the best performance is obtained with 50k artificial examples (compared to 10k for WORD2VEC). This is probably due to the fact that BIOLM suggests words that fit the particular context of the word being replaced and may, thus, suggest words with very different meanings that can be used in the particular context, adding noisy examples. By contrast, when using WORD2VEC we compare more directly each original word with candidate replacements.[13]

ALBERT (SQUAD-V2)	+train ex.	PRAUC (BIOASQ dev)
+BIOASQ	2,848	89.57
+BIOLM	2,848	91.76
+BIOLM +BIOASQ	5,696	92.37
+BIOLM	10,000	94.06
+BIOLM +BIOASQ	12,848	94.06
+BIOLM	30,000	93.63
+BIOLM +BIOASQ	32,848	93.75
+BIOLM	50,000	93.94
+BIOLM +BIOASQ	52,848	**94.45**
+BIOLM	100,000	93.79
+BIOLM +BIOASQ	102,848	93.84

Table 7: Data augmentation with *word substitution* based on *masked language modeling* using BIOLM.

3.5 Question Generation

Question generation (QG) has been found an effective DA method in open-domain MRC (Alberti et al., 2019; Chan and Fan, 2019; Lopez et al., 2020). The main reported benefit is that it increases the diversity of questions (Qiu and Xiong, 2019; Sultan et al., 2020). Typically QG models are fed with a snippet s, select an answer span a of s, and generate a question q answered by a. We take T5 (Raffel et al., 2020) fine-tuned for QG on a modified version of SQUAD by Lopez et al. (2020)[14] and use it to gen-

[12] We did not use BIOLM as an off-the-shelf QA model (Section 2.2), because it was not available fine-tuned on SQUAD.

[13] WORD2VEC embeddings are not sensitive to the particular context of the snippet and rely exclusively on the (many more) contexts of each word encountered in the pre-training corpus.

[14] The T5 QG model we used is available at https://github.com/patil-suraj/question_generation.

erate alternative questions q' and answer spans a' from the snippets s of the BIOASQ $\langle q, s, a \rangle$ training triples, producing artificial $\langle q', s, a' \rangle$ triples. Multiple artificial triples can be generated from the same original one (the same s), but we require each q' to be answered by a different answer span a' to maximize the diversity of questions. We obtained 3,389 artificial triples from the 2,848 original ones this way. An alternative we explored is to select random snippets s from random PUBMED abstracts, and use the QG model to produce artificial $\langle q', s, a' \rangle$ triples. The alternative approach can generate millions of artificial triples; we generated up to 100k.

ALBERT (SQUAD-V2)	+train ex.	PRAUC (BIOASQ dev)
+BIOASQ	2,848	89.57
+T5@BIOASQ	3,389	84.46
+T5@BIOASQ +BIOASQ	6,237	88.46
+T5@PUBMED	2,848	85.79
+T5@PUBMED +BIOASQ	5,696	89.29
+T5@PUBMED	10,000	87.30
+T5@PUBMED +BIOASQ	12,848	89.34
+T5@PUBMED	30,000	86.65
+T5@PUBMED +BIOASQ	32,848	90.51
+T5@PUBMED	50,000	87.30
+T5@PUBMED +BIOASQ	52,848	**90.69**
+T5@PUBMED	100,000	87.30
+T5@PUBMED +BIOASQ	102,848	90.61

Table 8: Data augmentation via *question generation* using T5. Questions are generated from the training snippets of BIOASQ (T5@BIOASQ) or from random snippets from random PUBMED abstracts (T5@PUBMED).

Table 8 shows that adding to the BIOASQ training data the artificial triples we obtained from BIOASQ (+T5@BIOASQ, BIOASQ) is worse (88.46 vs. 89.57) than our strong baseline (+BIOASQ). Fine-tuning only on the artificial triples (+T5@BIOASQ) is much worse (84.46), i.e., the artificial triples are much less useful, despite being more than the original ones. Adding artificial triples from PUBMED (+T5@PUBMED, BIOASQ) performs only slightly better (90.69) than the strong baseline, when using 50k artificial triples, with no further benefit when using more. A possible explanation for these poor results is the T5was fine-tuned for QG on the open-domain SQUAD dataset. Thus, the generated questions are rather simplistic and not indicative of the specialized questions of BIOASQ. Indeed, most of the generated questions are minor rephrases of the given snippet (e.g., subject replaced by 'what').

3.6 Adding Context

In the original training question-snippet-answer $\langle q, s, a \rangle$ triples, s is usually a single sentence. To help the QA model learn to better distinguish relevant from irrelevant parts of the given snippet, we experimented with an additional DA method, where we find the original article that s comes from and we expand s with the k_1 (and k_2) sentences preceding (and following) it.[15] For each original $\langle q, s, a \rangle$ triple, we create multiple new $\langle q, s', a \rangle$ artificial triples, for different values of $k_1 \geq 0$ and $k_2 \geq 0$, such that $k_1 + k_2 = K$.[16] We experiment with $K = 2$ (three new triples for each original one); then to obtain more artificial examples, we repeat with $K = 4$ (five new triples for each original). To avoid truncation of the input examples, we remove all artificial examples that exceed 500 characters in length. For $K \in \{2, 4\}$, we obtain a development PRAUC score of 94.21 (Table 9), which is surpassed only by the the two embedding-based word substitution methods (Tables 6–7). This DA method was introduced by Yoon et al. (2020), who used it in BIOASQ.[17]

ALBERT (SQUAD-V2)	+train ex.	PRAUC (BIOASQ dev)
+BIOASQ	2,848	89.57
+CONTEXT ($K = 2$)	4,568	93.91
+CONTEXT ($K = 2$) +BIOASQ	7,416	94.05
+CONTEXT ($K \in \{2, 4\}$)	6,428	94.20
+CONTEXT ($K \in \{2, 4\}$) +BIOASQ	9,276	**94.21**

Table 9: Data augmentation by *adding context to the snippet* ($K = 2$ or $K \in \{2, 4\}$ surrounding sentences).

3.7 Final Results

Table 10 shows the performance of all the DA methods considered, on both development and test data. For each DA method, we use the configuration (from Tables 3–9) with the best development score. The test scores are lower than the corresponding development ones, since several hyper-parameters (e.g., K, C in the case of WORD2VEC-based word substitution, number of training epochs) are tuned on the development set. The test set also seems to be harder than the development one, since our weak baseline (ALBERT fine-tuned on SQUAD-V2 with no further training) also performs worse on the test set (77.78 vs. 80.25). Nevertheless, the test scores confirm that WORD2VEC-based word substitution is the best DA method considered, leading to a performance gain of 8.2 percentage points test PRAUC compared to the strong baseline (84.99 vs. 77.78). The ranking of the other DA methods

[15] In BIOASQ, each gold snippet is accompanied by the PUBMED id of the article it was extracted from.

[16] Simply setting $k_1 = k_2$ would risk misguiding the model to always prefer the central sentence. We also experimented with *random* k_1 (or k_2) sentences before (and after) s, but performance was much worse, possibly because the random sentences led to inferior context-aware token embeddings.

[17] Yoon et al. (2020) reported an improvement in BIOASQ's Lenient Accuracy by 2.49 percentage points.

does not change when ranking by test score, instead of development score, with the only exception of adding context to the given passage (+CONTEXT), which is now slightly worse than adding instances from the artificial BIOMRC dataset. Interestingly, all the DA methods, even the weakest IR-based one, improve upon the test score of the strong baseline.

Method	+train ex.	PRAUC (dev)	PRAUC (test)
ALBERT (SQUAD-V2)	0	80.25	77.78
+ BIOASQ	2,848	89.57	76.78
+WORD2VEC +BIOASQ	12,848	**95.60 (+6.03)**	**84.99 (+8.21)**
+BIOLM +BIOASQ	52,848	94.45 (+4.88)	82.76 (+5.98)
+CONTEXT +BIOASQ	9,276	94.21 (+4.64)	81.63 (+4.85)
+BIOMRC +BIOASQ	12,848	93.15 (+3.58)	82.04 (+5.26)
+BTR +BIOASQ	18,441	92.66 (+3.09)	81.27 (+4.49)
+T5@PUBMED +BIOASQ	52,848	90.69 (+1.12)	80.26 (+3.48)
+IR +BIOASQ	3,137	89.80 (+0.23)	78.66 (+1.88)

Table 10: Performance of DA methods on *development* and *test* data, ordered by decreasing development score. For each DA method, we use the configuration (from Tables 3–9) with the best development score.

4 Related Work

DA is a key ingredient of success in deep learning for computer vision (Shorten and Khoshgoftaar, 2019). DA for NLP has been explored less, but is an active research area (Shorten et al., 2021; Feng et al., 2021), with methods ranging from leveraging knowledge graphs (Moussallem et al., 2019) to generating textual data from scratch (Yang et al., 2020; Bayer et al., 2021a). The most common NLP task in DA is text classification (Bayer et al., 2021b). Feng et al. (2021) consider span-based NLP tasks in specialized domains, which includes biomedical MRC, among the most challenging for DA.

Word substitution is a simple and common DA approach in NLP. In thesaurus-based substitution (Jungiewicz and Smywinski-Pohl, 2019; Abdollahi et al., 2020), words are replaced by synonyms or closely related words (e.g., hypernyms). Word embedding substitution (Wang and Yang, 2015) replaces words by others nearby in a pre-trained vector space model (Section 3.4). Alternatively, a random word is removed, inserted (Wei and Zou, 2019a; Miao et al., 2020), or noised with spelling errors (Belinkov and Bisk, 2018). Sentences may also be re-ordered or removed (Shen et al., 2020; Chen et al., 2021). Text generation has also been used in several NLP tasks for adversarial augmentation (Cheng et al., 2020), to paraphrase training examples (Ribeiro et al., 2018; Cai et al., 2020; Xie et al., 2020), or generate new (Anaby-Tavor et al., 2020; Kumar et al., 2020). Back-translation (Sennrich et al., 2016) is also widely used across NLP tasks (Shorten et al., 2021; Feng et al., 2021).

DA work for QA in particular includes back-translation (Du et al., 2019), question generation (Zhang and Bansal, 2019; Alberti et al., 2019; Chan and Fan, 2019; Lopez et al., 2020; Yang et al., 2020), paraphrasing (Dong et al., 2017; Liu et al., 2020), and synonym replacement (Nugraha and Suyanto, 2019), but not in a biomedical setting. The IR-based DA we used (Section 3.3) follows Yang et al. (2019), who experimented in English and Chinese, but not in the biomedical domain. Expanding the passage with surrounding sentences (Section 3.6) follows Yoon et al. (2020), who used this method in BIOASQ. Dhingra et al. (2018) create artificial cloze-style MRC datasets and use them to pre-train neural QA models (not Transformers), which are then fine-tuned on real training examples. By contrast, we use artificial MRC datasets to fine-tune large pre-trained Transformers. All the above studies experimentally compare at most two DA methods; we compare seven. Hence, our work is the largest (in terms of methods considered) experimental study of DA for QA (and possibly NLP).

Longpre et al. (2019) report that back-translation did not improve generalization in (non-biomedical) QA experiments with fine-tuned pre-trained Transformers. Longpre et al. (2020) report that back-translation and Easy Data Augmentation (Wei and Zou, 2019b) are not always effective in text classification when fine-tuning pretrained Transformers, even with small end-task training sets. Consequently, Feng et al. (2021) recommend exploring when DA is effective for large pre-trained models. Our work contributes in this discussion by showing that DA can lead to very significant performance gains, even when using large pre-trained Transformers fine-tuned on large generic (SQUAD) and/or small domain-specific (BIOASQ) end-task datasets.

5 Limitations and Future Work

A limitation of our work is that we consider only DA in the *input space*, i.e., the artificial instances are in textual form, like the original ones, as opposed to, e.g., interpolating feature vectors (Chawla et al., 2002; DeVries and Taylor, 2017; Shorten et al., 2021). We also consider only *offline* augmentation, i.e., the artificial instances are generated once, before training, as opposed to artificial instances generated anew for each training epoch. These two limitations, which are common in DA for NLP, allow generating model-agnostic training instances once and reusing them across training epochs and different models. This greatly reduces computation costs and allows sharing the augmented datasets. Online DA, however, exposes the model to many more synthetic data; and feature space DA may act as layer-specific regularization. One could also exploit ideas from active learning (Ein-Dor et al., 2020; Margatina et al., 2021) to select the most informative, diverse, and representative artificial training instances among those that DA generates. Small subsets of the BIOASQ data could also be used to study the effect of DA in few-shot learning.

References

Mahdi Abdollahi, Xiaoying Gao, Yi Mei, Shameek Ghosh, and Jinyan Li. 2020. Ontology-guided data augmentation for medical document classification. In *Artificial Intelligence in Medicine*, pages 78–88, Cham. Springer International Publishing.

Ahmad Aghaebrahimian. 2018. Linguistically-based deep unstructured question answering. In *Proceedings of the 22nd Conference on Computational Natural Language Learning*, pages 433–443, Brussels, Belgium. Association for Computational Linguistics.

Chris Alberti, Daniel Andor, Emily Pitler, Jacob Devlin, and Michael Collins. 2019. Synthetic QA corpora generation with roundtrip consistency. In *Proceedings of the 57th Annual Meeting of the Association for Computational Linguistics*, pages 6168–6173, Florence, Italy. Association for Computational Linguistics.

Ateret Anaby-Tavor, Boaz Carmeli, Esther Goldbraich, Amir Kantor, George Kour, Segev Shlomov, Naama Tepper, and Naama Zwerdling. 2020. Do not have enough data? deep learning to the rescue! *Proceedings of the AAAI Conference on Artificial Intelligence*, 34(05):7383–7390.

Ondrej Bajgar, Rudolf Kadlec, and Jan Kleindienst. 2016. Embracing data abundance: BookTest Dataset for Reading Comprehension. *CoRR*.

Markus Bayer, Marc-André Kaufhold, Björn Buchhold, Marcel Keller, J. Dallmeyer, and Christian Reuter. 2021a. Data augmentation in natural language processing: A novel text generation approach for long and short text classifiers. *ArXiv*, abs/2103.14453.

Markus Bayer, Marc-André Kaufhold, and Christian Reuter. 2021b. A survey on data augmentation for text classification. *CoRR*, abs/2107.03158.

Yonatan Belinkov and Yonatan Bisk. 2018. Synthetic and natural noise both break neural machine translation. *ArXiv*, abs/1711.02173.

Iz Beltagy, Kyle Lo, and Arman Cohan. 2019. SciBERT: A pretrained language model for scientific text. In *Proceedings of the 2019 Conference on Empirical Methods in Natural Language Processing and the 9th International Joint Conference on Natural Language Processing (EMNLP-IJCNLP)*, pages 3615–3620, Hong Kong, China. Association for Computational Linguistics.

George Brokos, Polyvios Liosis, Ryan McDonald, Dimitris Pappas, and Ion Androutsopoulos. 2018. AUEB at BioASQ 6: Document and Snippet Retrieval. In *Proceedings of the 6th BioASQ Workshop*, pages 30–39, Brussels, Belgium.

Hengyi Cai, Hongshen Chen, Yonghao Song, Cheng Zhang, Xiaofang Zhao, and Dawei Yin. 2020. Data manipulation: Towards effective instance learning for neural dialogue generation via learning to augment and reweight. In *Proceedings of the 58th Annual Meeting of the Association for Computational Linguistics*, pages 6334–6343, Online. Association for Computational Linguistics.

Daniel Fernando Campos, T. Nguyen, M. Rosenberg, Xia Song, Jianfeng Gao, Saurabh Tiwary, Rangan Majumder, L. Deng, and Bhaskar Mitra. 2016. Ms marco: A human generated machine reading comprehension dataset. *ArXiv*, abs/1611.09268.

Ilias Chalkidis, Manos Fergadiotis, Prodromos Malakasiotis, Nikolaos Aletras, and Ion Androutsopoulos. 2020. LEGAL-BERT: The muppets straight out of law school. In *Findings of the Association for Computational Linguistics: EMNLP 2020*, pages 2898–2904, Online. Association for Computational Linguistics.

Ying-Hong Chan and Yao-Chung Fan. 2019. A recurrent BERT-based model for question generation. In *Proceedings of the 2nd Workshop on Machine Reading for Question Answering*, pages 154–162, Hong Kong, China. Association for Computational Linguistics.

Nitesh V. Chawla, Kevin W. Bowyer, Lawrence O. Hall, and W. Philip Kegelmeyer. 2002. Smote: Synthetic minority over-sampling technique. *Journal of Artificial Intelligence Research*, 16(1):321—-357.

Danqi Chen, Jason Bolton, and Christopher D. Manning. 2016. A thorough examination of the CNN/Daily Mail reading comprehension task. In *Proceedings of the 54th Annual Meeting of the Association for Computational Linguistics (Volume 1: Long Papers)*, pages 2358–2367, Berlin, Germany. Association for Computational Linguistics.

Danqi Chen, Adam Fisch, Jason Weston, and Antoine Bordes. 2017a. Reading wikipedia to answer open-domain questions. *CoRR*, abs/1704.00051.

Danqi Chen, Adam Fisch, Jason Weston, and Antoine Bordes. 2017b. Reading Wikipedia to answer open-domain questions. In *Association for Computational Linguistics (ACL)*.

Jiaao Chen, Dinghan Shen, Weizhu Chen, and Diyi Yang. 2021. Hiddencut: Simple data augmentation for natural language understanding with better generalization. *ArXiv*, abs/2106.00149.

Yong Cheng, Lu Jiang, Wolfgang Macherey, and Jacob Eisenstein. 2020. Advaug: Robust adversarial augmentation for neural machine translation. *ArXiv*, abs/2006.11834.

Jacob Devlin, Ming-Wei Chang, Kenton Lee, and Kristina Toutanova. 2019. BERT: Pre-training of deep bidirectional transformers for language understanding. In *Proceedings of the 2019 Conference of the North American Chapter of the Association for Computational Linguistics: Human Language Technologies, Volume 1 (Long and Short Papers)*, pages 4171–4186, Minneapolis, Minnesota.

Terrance DeVries and Graham W. Taylor. 2017. Dataset augmentation in feature space.

Bhuwan Dhingra, Danish Danish, and Dheeraj Rajagopal. 2018. Simple and effective semi-supervised question answering. In *Proceedings of the 2018 Conference of the North American Chapter of the Association for Computational Linguistics: Human Language Technologies, Volume 2 (Short Papers)*, pages 582–587, New Orleans, Louisiana.

Li Dong, Jonathan Mallinson, Siva Reddy, and Mirella Lapata. 2017. Learning to paraphrase for question answering. In *Proceedings of the 2017 Conference on Empirical Methods in Natural Language Processing*, pages 875–886, Copenhagen, Denmark. Association for Computational Linguistics.

Y. Du, W. Guo, and Y. Zhao. 2019. Hierarchical question-aware context learning with augmented data for biomedical question answering. In *2019 IEEE International Conference on Bioinformatics and Biomedicine (BIBM)*, pages 370–375.

Liat Ein-Dor, Alon Halfon, Ariel Gera, Eyal Shnarch, Lena Dankin, Leshem Choshen, Marina Danilevsky, Ranit Aharonov, Yoav Katz, and Noam Slonim. 2020. Active Learning for BERT: An Empirical Study. In *Proceedings of the 2020 Conference on Empirical Methods in Natural Language Processing (EMNLP)*, pages 7949–7962, Online.

Hicham El Boukkouri, Olivier Ferret, Thomas Lavergne, Hiroshi Noji, Pierre Zweigenbaum, and Jun'ichi Tsujii. 2020. CharacterBERT: Reconciling ELMo and BERT for word-level open-vocabulary representations from characters. In *Proceedings of the 28th International Conference on Computational Linguistics*, pages 6903–6915, Barcelona, Spain (Online). International Committee on Computational Linguistics.

Steven Y. Feng, Varun Gangal, Jason Wei, Sarath Chandar, Soroush Vosoughi, Teruko Mitamura, and Eduard Hovy. 2021. A survey of data augmentation approaches for NLP. In *Findings of the Association for Computational Linguistics: ACL-IJCNLP 2021*, pages 968–988, Online.

Bin Fu, Yunqi Qiu, Chengguang Tang, Y. Li, H. Yu, and J. Sun. 2020. A survey on complex question answering over knowledge base: Recent advances and challenges. *ArXiv*, abs/2007.13069.

Pengcheng He, Xiaodong Liu, Jianfeng Gao, and Weizhu Chen. 2020. Deberta: Decoding-enhanced bert with disentangled attention.

Felix Hill, Antoine Bordes, Sumit Chopra, and Jason Weston. 2016. The Goldilocks Principle: Reading Children's Books with Explicit Memory Representations. *CoRR*.

Qiao Jin, Bhuwan Dhingra, Zhengping Liu, William Cohen, and Xinghua Lu. 2019. PubMedQA: A dataset for biomedical research question answering. In *Proceedings of the 2019 Conference on Empirical Methods in Natural Language Processing and the 9th International Joint Conference on Natural Language Processing (EMNLP-IJCNLP)*, pages 2567–2577, Hong Kong, China. Association for Computational Linguistics.

Mandar Joshi, Eunsol Choi, Daniel Weld, and Luke Zettlemoyer. 2017. TriviaQA: A large scale distantly supervised challenge dataset for reading comprehension. In *Proceedings of the 55th Annual Meeting of the Association for Computational Linguistics (Volume 1: Long Papers)*, pages 1601–1611, Vancouver, Canada. Association for Computational Linguistics.

Michał Jungiewicz and Aleksander Smywinski-Pohl. 2019. Towards textual data augmentation for neural networks: synonyms and maximum loss. *Computer Science*, 20.

Soha Khazaeli, Janardhana Punuru, Chad Morris, Sanjay Sharma, Bert Staub, Michael Cole, Sunny Chiu-Webster, and Dhruv Sakalley. 2021. A free format legal question answering system. In *Proceedings of the Natural Legal Language Processing Workshop 2021*, pages 107–113, Punta Cana, Dominican Republic.

Phi Manh Kien, Ha-Thanh Nguyen, Ngo Xuan Bach, Vu Tran, Minh Le Nguyen, and Tu Minh Phuong. 2020. Answering legal questions by learning neural attentive text representation. In *Proceedings of the 28th International Conference on Computational Linguistics*, pages 988–998, Barcelona, Spain (Online). International Committee on Computational Linguistics.

Sosuke Kobayashi. 2018. Contextual augmentation: Data augmentation by words with paradigmatic relations. In *Proceedings of the 2018 Conference of the North American Chapter of the Association for Computational Linguistics: Human Language Technologies, Volume 2 (Short Papers)*, pages 452–457.

Varun Kumar, Ashutosh Choudhary, and Eunah Cho. 2020. Data augmentation using pre-trained transformer models. In *Proceedings of the 2nd Workshop on Life-long Learning for Spoken Language Systems*, pages 18–26, Suzhou, China. Association for Computational Linguistics.

Tom Kwiatkowski, Jennimaria Palomaki, Olivia Redfield, Michael Collins, Ankur Parikh, Chris Alberti, Danielle Epstein, Illia Polosukhin, Jacob Devlin, Kenton Lee, Kristina Toutanova, Llion Jones, Matthew Kelcey, Ming-Wei Chang, Andrew M. Dai, Jakob Uszkoreit, Quoc Le, and Slav Petrov. 2019. Natural questions: A benchmark for question answering research. *Transactions of the Association for Computational Linguistics*, 7:452–466.

Guokun Lai, Qizhe Xie, Hanxiao Liu, Yiming Yang, and Eduard Hovy. 2017. RACE: Large-scale ReAding comprehension dataset from examinations. In *Proceedings of the 2017 Conference on Empirical Methods in Natural Language Processing*, pages 785–794, Copenhagen, Denmark. Association for Computational Linguistics.

Zhenzhong Lan, Mingda Chen, Sebastian Goodman, Kevin Gimpel, Piyush Sharma, and Radu Soricut. 2019. ALBERT: A lite BERT for self-supervised learning of language representations. *CoRR*, abs/1909.11942.

Jinhyuk Lee, Wonjin Yoon, Sungdong Kim, Donghyeon Kim, Sunkyu Kim, Chan Ho So, and Jaewoo Kang. 2019. BioBERT: a pre-trained biomedical language representation model for biomedical text mining. *Bioinformatics*, 36(4):1234–1240.

Patrick Lewis, Myle Ott, Jingfei Du, and Veselin Stoyanov. 2020. Pretrained language models for biomedical and clinical tasks: Understanding and extending the state-of-the-art. In *Proceedings of the 3rd Clinical Natural Language Processing Workshop*, pages 146–157, Online. Association for Computational Linguistics.

Dayiheng Liu, Yeyun Gong, Jie Fu, Yu Yan, Jiusheng Chen, Jiancheng Lv, Nan Duan, and Ming Zhou. 2020. Tell me how to ask again: Question data augmentation with controllable rewriting in continuous space. In *Proceedings of the 2020 Conference on Empirical Methods in Natural Language Processing (EMNLP)*, pages 5798–5810, Online. Association for Computational Linguistics.

Yinhan Liu, Myle Ott, Naman Goyal, Jingfei Du, Mandar Joshi, Danqi Chen, Omer Levy, Mike Lewis, Luke Zettlemoyer, and Veselin Stoyanov. 2019. Roberta: A robustly optimized bert pretraining approach.

Shayne Longpre, Yi Lu, Zhucheng Tu, and Chris DuBois. 2019. An exploration of data augmentation and sampling techniques for domain-agnostic question answering. *CoRR*, abs/1912.02145.

Shayne Longpre, Yu Wang, and Chris DuBois. 2020. How effective is task-agnostic data augmentation for pretrained transformers? In *Findings of the Association for Computational Linguistics: EMNLP 2020*, pages 4401–4411, Online.

Luis Enrico Lopez, Diane Kathryn Cruz, Jan Christian Blaise Cruz, and Charibeth Cheng. 2020. Transformer-based end-to-end question generation. *CoRR*, abs/2005.01107.

Kangqi Luo, Fengli Lin, Xusheng Luo, and Kenny Zhu. 2018. Knowledge base question answering via encoding of complex query graphs. In *Proceedings of the 2018 Conference on Empirical Methods in Natural Language Processing*, pages 2185–2194, Brussels, Belgium. Association for Computational Linguistics.

Christopher D. Manning, Prabhakar Raghavan, and Hinrich Schütze. 2008. *Introduction to Information Retrieval*. Cambridge University Press.

Katerina Margatina, Giorgos Vernikos, Loïc Barrault, and Nikolaos Aletras. 2021. Active learning by acquiring contrastive examples. In *Proceedings of the 2021 Conference on Empirical Methods in Natural Language Processing*, pages 650–663, Online and Punta Cana, Dominican Republic.

Zhengjie Miao, Yuliang Li, Xiaolan Wang, and Wang-Chiew Tan. 2020. Snippext: Semi-supervised opinion mining with augmented data. In *Proceedings of The Web Conference 2020*, page 617–628, New York, NY, USA. Association for Computing Machinery.

Tomás Mikolov, Kai Chen, Greg Corrado, and Jeffrey Dean. 2013. Efficient estimation of word representations in vector space. In *1st International Conference on Learning Representations, ICLR 2013, Scottsdale, Arizona, USA, May 2-4, 2013, Workshop Track Proceedings*.

Timo Möller, Anthony Reina, Raghavan Jayakumar, and Malte Pietsch. 2020. COVID-QA: A question answering dataset for COVID-19. In *Proceedings of the 1st Workshop on NLP for COVID-19 at ACL 2020*, Online. Association for Computational Linguistics.

Diego Moussallem, Mihael Arcan, Axel-Cyrille Ngonga Ngomo, and Paul Buitelaar. 2019. Augmenting neural machine translation with knowledge graphs. *CoRR*, abs/1902.08816.

H. S. Nugraha and S. Suyanto. 2019. Typographic-based data augmentation to improve a question retrieval in short dialogue system. In *2019 International Seminar on Research of Information Technology and Intelligent Systems (ISRITI)*, pages 44–49.

Dimitris Pappas and Ion Androutsopoulos. 2021. A neural model for joint document and snippet ranking in question answering for large document collections. In *Proceedings of the 59th Annual Meeting of the Association for Computational Linguistics and the 11th International Joint Conference on Natural Language Processing (Volume 1: Long Papers)*, pages 3896–3907, Online. Association for Computational Linguistics.

Dimitris Pappas, Ion Androutsopoulos, and Haris Papageorgiou. 2018. BioRead: A new dataset for biomedical reading comprehension. In *Proceedings of the Eleventh International Conference on Language Resources and Evaluation (LREC 2018)*, Miyazaki, Japan. European Language Resources Association (ELRA).

Dimitris Pappas, Petros Stavropoulos, Ion Androutsopoulos, and Ryan McDonald. 2020. BioMRC: A dataset for biomedical machine reading comprehension. In *Proceedings of the 19th SIGBioMed Workshop on Biomedical Language Processing*, pages 140–149, Online. Association for Computational Linguistics.

Jiazuo Qiu and Deyi Xiong. 2019. Generating highly relevant questions. *CoRR*, abs/1910.03401.

Colin Raffel, Noam Shazeer, Adam Roberts, Katherine Lee, Sharan Narang, Michael Matena, Yanqi Zhou, Wei Li, and Peter J. Liu. 2020. Exploring the limits of transfer learning with a unified text-to-text transformer. *Journal of Machine Learning Research*, 21(140):1–67.

Pranav Rajpurkar, Robin Jia, and Percy Liang. 2018. Know what you don't know: Unanswerable questions for SQuAD. In *Proceedings of the 56th Annual Meeting of the Association for Computational Linguistics (Volume 2: Short Papers)*, pages 784–789, Melbourne, Australia. Association for Computational Linguistics.

Pranav Rajpurkar, Jian Zhang, Konstantin Lopyrev, and Percy Liang. 2016. SQuAD: 100,000+ questions for machine comprehension of text. In *Proceedings of the 2016 Conference on Empirical Methods in Natural Language Processing*, pages 2383–2392, Austin, Texas. Association for Computational Linguistics.

Siva Reddy, Danqi Chen, and Christopher D. Manning. 2019. CoQA: A conversational question answering challenge. *Transactions of the Association for Computational Linguistics*, 7:249–266.

Marco Tulio Ribeiro, Sameer Singh, and Carlos Guestrin. 2018. Semantically equivalent adversarial rules for debugging NLP models. In *Proceedings of the 56th Annual Meeting of the Association for Computational Linguistics (Volume 1: Long Papers)*, pages 856–865, Melbourne, Australia. Association for Computational Linguistics.

Victor Sanh, Lysandre Debut, Julien Chaumond, and Thomas Wolf. 2019. Distilbert, a distilled version of BERT: smaller, faster, cheaper and lighter. *CoRR*, abs/1910.01108.

Rico Sennrich, Barry Haddow, and Alexandra Birch. 2016. Improving neural machine translation models with monolingual data. In *Proceedings of the 54th Annual Meeting of the Association for Computational Linguistics (Volume 1: Long Papers)*, pages 86–96, Berlin, Germany.

Dinghan Shen, Ming Zheng, Yelong Shen, Yanru Qu, and Weizhu Chen. 2020. A simple but tough-to-beat data augmentation approach for natural language understanding and generation. *ArXiv*, abs/2009.13818.

Connor Shorten and Taghi M. Khoshgoftaar. 2019. A survey on image data augmentation for deep learning. *Journal of Big Data*, 6(60).

Connor Shorten, Taghi M. Khoshgoftaar, and Borko Furht. 2021. Text data augmentation for deep learning. *Journal of Big Data*, 8(101).

Md Arafat Sultan, Shubham Chandel, Ramón Fernandez Astudillo, and Vittorio Castelli. 2020. On the importance of diversity in question generation for QA. In *Proceedings of the 58th Annual Meeting of the Association for Computational Linguistics*, pages 5651–5656, Online. Association for Computational Linguistics.

G. Tsatsaronis, G. Balikas, P. Malakasiotis, I. Partalas, M. Zschunke, M.R. Alvers, D. Weissenborn, A. Krithara, S. Petridis, D. Polychronopoulos, Y. Almirantis, J. Pavlopoulos, N. Baskiotis, P. Gallinari, T. Artieres, A. Ngonga, N. Heino, E. Gaussier, L. Barrio-Alvers, M. Schroeder, I. Androutsopoulos, and G. Paliouras. 2015. An Overview of the BioASQ Large-Scale Biomedical Semantic Indexing and Question Answering Competition. *BMC Bioinformatics*, 16(138).

Bingning Wang, Ting Yao, Qi Zhang, Jingfang Xu, and Xiaochuan Wang. 2020. Reco: A large scale chinese reading comprehension dataset on opinion. *Proceedings of the AAAI Conference on Artificial Intelligence*, 34(05):9146–9153.

William Yang Wang and Diyi Yang. 2015. That's so annoying!!!: A lexical and frame-semantic embedding based data augmentation approach to automatic categorization of annoying behaviors using #petpeeve tweets. In *Proceedings of the 2015 Conference on Empirical Methods in Natural Language Processing*, pages 2557–2563, Lisbon, Portugal. Association for Computational Linguistics.

Jason Wei and Kai Zou. 2019a. EDA: Easy data augmentation techniques for boosting performance on text classification tasks. In *Proceedings of the 2019 Conference on Empirical Methods in Natural Language Processing and the 9th International Joint Conference on Natural Language Processing (EMNLP-IJCNLP)*, pages 6382–6388, Hong Kong, China. Association for Computational Linguistics.

Jason Wei and Kai Zou. 2019b. EDA: Easy data augmentation techniques for boosting performance on text classification tasks. In *Proceedings of the 2019 Conference on Empirical Methods in Natural Language Processing and the 9th International Joint Conference on Natural Language Processing (EMNLP-IJCNLP)*, pages 6382–6388, Hong Kong, China.

Xing Wu, Shangwen Lv, Liangjun Zang, Jizhong Han, and Songlin Hu. 2019. Conditional bert contextual augmentation. In *International Conference on Computational Science 2019*, pages 84–95, Cham.

Qizhe Xie, Zihang Dai, E. Hovy, Minh-Thang Luong, and Quoc V. Le. 2020. Unsupervised data augmentation for consistency training. *arXiv: Learning*.

Naganand Yadati, Dayanidhi R S, Vaishnavi S, Indira K M, and Srinidhi G. 2021. Knowledge base question answering through recursive hypergraphs. In *Proceedings of the 16th Conference of the European Chapter of the Association for Computational Linguistics: Main Volume*, pages 448–454, Online. Association for Computational Linguistics.

Wei Yang, Yuqing Xie, Luchen Tan, Kun Xiong, M. Li, and Jimmy Lin. 2019. Data augmentation for bert fine-tuning in open-domain question answering. *ArXiv*, abs/1904.06652.

Yi Yang, Wen-tau Yih, and Christopher Meek. 2015. WikiQA: A challenge dataset for open-domain question answering. In *Proceedings of the 2015 Conference on Empirical Methods in Natural Language Processing*, pages 2013–2018, Lisbon, Portugal. Association for Computational Linguistics.

Yiben Yang, Chaitanya Malaviya, Jared Fernandez, Swabha Swayamdipta, Ronan Le Bras, Ji-Ping Wang, Chandra Bhagavatula, Yejin Choi, and Doug Downey. 2020. Generative data augmentation for commonsense reasoning. In *Findings of the Association for Computational Linguistics: EMNLP 2020*, pages 1008–1025, Online. Association for Computational Linguistics.

Zhengzhe Yang and Jinho D. Choi. 2019. FriendsQA: Open-domain question answering on TV show transcripts. In *Proceedings of the 20th Annual SIGdial Meeting on Discourse and Dialogue*, pages 188–197, Stockholm, Sweden. Association for Computational Linguistics.

Wonjin Yoon, Jinhyuk Lee, Donghyeon Kim, Minbyul Jeong, and Jaewoo Kang. 2020. Pre-trained language model for biomedical question answering. In *Machine Learning and Knowledge Discovery in Databases*, pages 727–740, Cham. Springer International Publishing.

Na-Na Zhang and Yinan Xing. 2021. Questions and answers on legal texts based on BERT-BiGRU. *Journal of Physics: Conference Series*, 1828(1):012035.

Shiyue Zhang and Mohit Bansal. 2019. Addressing semantic drift in question generation for semi-supervised question answering. In *Proceedings of the 2019 Conference on Empirical Methods in Natural Language Processing and the 9th International Joint Conference on Natural Language Processing (EMNLP-IJCNLP)*, pages 2495–2509, Hong Kong, China. Association for Computational Linguistics.

Henghui Zhu, Ioannis Ch Paschalidis, and Amir Tahmasebi. 2018. Clinical concept extraction with contextual word embedding. *arXiv preprint arXiv:1810.10566*.

Appendix

A Combining Augmentation Methods

We also tried to combine DA methods. In Table A1, we incrementally add to the training set of the strong baseline (ALBERT fine-tuned on SQUAD-V2, then BIOASQ) artificial training examples obtained from WORD2VEC-based word substitution, then (additionally) training examples obtained by expanding the context of the given passage etc. We started with the artificial examples of the WORD2VEC-based method, which had the best development score, skipped the other (BIOLM-based) word substitution method, then continued with examples from BIOMRC and back-translation, which were the next best in terms of development score. Unfortunately, there was no significant gain, compared to using only the WORD2VEC-based method, which suggests that the DA methods we consider are not complementary. An alternative approach would be to stack DA methods, instead of accumulating their training examples. For example, one could apply the WORD2VEC method to artificial examples produced by increasing the context of the given passages. We leave this for future work.

Method	+train ex.	PRAUC (dev)	PRAUC (test)
ALBERT (SQUAD-V2)	0	80.25	77.78
+BIOASQ	2,848	89.57	76.78
+ WORD2VEC	12,848	**95.60**	84.99
+ CONTEXT	19,276	93.98	83.54
+ BIOMRC	29,276	94.27	85.18
+ BTR	44,869	93.44	83.97

Table A1: Results using a combination of Context Increasing and WORD2VEC data augmentation.

B Examples of Artificial Data

B.1 BIOMRC

Table D3 presents training instances generated from the BIOMRC dataset. Each instance is a triple containing a cloze-style question, a snippet, and the span of the snippet answering a question. This is very similar to the SQUAD setting which we have adopted in our experiments (see Section 2.1).

B.2 Back-translation

Tables D4 and D5 show training instances generated via back-translation of BIOASQ questions and snippets, respectively. The back-translated questions and snippets retain the semantics of the original ones while adding diversity to the training set.

B.3 Information Retrieval

Table D6 contains training instances generated via Information Retrieval. A BIOASQ question is used as a query in a search engine to retrieve PUBMED documents (abstracts and titles). From the retrieved documents all the snippets containing the answer are extracted and used to generate new training triples. Note that although a retrieved snippet may contain the entity that answers the BIOASQ question, it is not always evident that it answers the question, e.g., it may answer another question as is the case in the instance with id 29767248.

B.4 Word Substitution

Tables D7 and D8 presents examples generated via word substitution based on WORD2VEC and BIOLM respectively. Although some substitutions may induce noise, the generated snippets tend to retain the semantics of the original ones and add diversity to the training set.

B.5 Question Generation

Tables D9 and D10 show examples generated via Question Generation using BIOASQ snippets and random snippets from random PUBMED articles respectively. Although, the generated triples introduce diverse answers they contain rather simplistic questions which are not indicative of the specialized questions found in BIOASQ.

B.6 Additional Context

Table D11 contains examples generated by adding context to the original BIOASQ snippets. The additional context provides additional information that helps the model to better distinguish relevant and irrelevant parts of the original snippet.

C Computing Infrastructure

All of our experiments run on a titan-X GPU with 12GB of Memory while all code was compiled for CUDA Version 10.2. The personal computer we used offers 32GB of DDR4-RAM Memory and a 6-core Intel(R) Core(TM) i7-5820K CPU.

D Hyper-parameter tuning

The random seed in all experiments was set to 1. For data augmentation through Information Retrieval (IR), we use an ElasticSearch cluster to retrieve relevant abstracts using BM25 with default parameters.

Due to computational and time restrictions, hyper-parameter tuning was performed with grid-search by training on the original 2,848 BIOASQ examples (Table 2), i.e., without data augmentation, and evaluating on the development data. The 'best' hyper-parameter values were then used in all the augmentation experiments. The hyper-parameter search space (48 settings) and the selected values can be seen in Table D2.

Hyperparameter	choices	best dev. setting
Random Seed	{1}	1
MLP Hidden Size	{50, 100}	100
Total Epochs	{50, 100}	50
Patience	{5}	5
Monitor Score	{AUC, loss}	AUC
Learning Rate	{0.1, 0.01, 2e-5, 5e-5 }	5e-5
Weight Decay	{0.01}	0.01
Warmup Steps	{0}	0
Batch Size	{16, 8}	16

Table D2: Hyper-parameter search space and selected values. We performed a grid-search on a total of 48 different settings. The best choices per hyper-parameter can be seen in the last column.

DA with instances from BIOMRC	
ID	Instance
16061304	**BIOMRC question:** Prognosis of 6644 resected [MASK] in Japan: a Japanese lung cancer registry study. **BIOMRC snippet:** Otherwise, the present TNM staging system seemed to well characterize the stage-specific prognosis in non-small cell lung cancer . **BIOMRC answer:** non-small cell lung cancer
19823942	**BIOMRC question:** Systolic versus diastolic cardiac function variables during [MASK] treatment for breast cancer . **BIOMRC snippet:** epirubicin induces considerable decrease in left ventricular ejection fraction and a high risk of CHF. **BIOMRC answer:** epirubicin
22457372	**BIOMRC question:** Pre-operative education and counselling are associated with [MASK] following carotid endarterectomy: a randomized and open-label study. **BIOMRC snippet:** AIM: To investigate the effect of pre-operative visits and counselling by intensive care unit (intensive care unit) nurses on Patients 's anxiety symptoms following carotid endarterectomy. **BIOMRC answer:** anxiety symptoms

Table D3: Training instances extracted from BIOMRC. Each instance is a triple containing a cloze-style question, a snippet, and the span of the snippet answering the question.

DA via question back-translation	
ID	Instance
8699317	**Pivot language:** French **BIOASQ question:** Which is the gene mutated in type 1 neurofibromatosis? **Back-translated Question:** What is the mutated gene in type 1 neurofibromatosis? **BIOASQ snippet:** An NF1 gene was identified as a gene whose loss of function causes an onset of human disorder, neurofibromatosis type I. **BIOASQ answer:** NF1
11816795	**Pivot language:** Spansih **BIOASQ question:** Which is the primary protein component of Lewy bodies? **Back-translated question:** What is the main protein component of Lewy bodies? **BIOASQ snippet:** The protein alpha-synuclein appears to be an important structural component of Lewy bodies, an observation spurred by the discovery of point mutations in the alpha-synuclein gene linked to rare cases of autosomal dominant PD. **BIOASQ answer:** alpha-synuclein
3056562	**Pivot language:** German **BIOASQ question:** Which type of urinary incontinence is diagnosed with the Q tip test? **Back-translated question:** What type of urinary incontinence does the Q tip test diagnose? **BIOASQ snippet:** Simple clinical tests for support of the urethrovesical junction, such as the Q tip test, are non-specific in patients with stress urinary incontinence. **BIOASQ answer:** stress urinary incontinence

Table D4: Training instances generated via back-translation of BIOASQ questions using French, Spanish, and German as a pivot language. A generated instance contains a back-translated question and the corresponding BIOASQ snippet and answer.

DA via snippet back-translation	
ID	Instance
8699317	**Pivot language:** French **BIOASQ question:** Which is the protein implicated in Spinocerebellar ataxia type 3? **BIOASQ snippet:** Ataxin-3 (AT3) is the protein that triggers the inherited neurodegenerative disorder spinocerebellar ataxia type 3 when its polyglutamine (polyQ) stretch close to the C-terminus exceeds a critical length **Back-translated snippet:** Ataxin-3 (AT3) is the protein that triggers spinocerebellar ataxia type 3 in inherited neurodegenerative disorder when its polyglutamine (polyQ) stretches near the C-terminus exceeds a critical length. **BIOASQ answer:** Ataxin-3
16232326	**Pivot language:** Spanish **BIOASQ question:** Which gene is responsible for the development of Sotos syndrome? **BIOASQ snippet:** Haploinsufficiency of the NSD1 gene has been implicated as the major cause of Sotos syndrome, with a predominance of microdeletions reported in Japanese patients **Back-translated snippet:** NSD1 gene haploinsufficiency has been implicated as the main cause of Sotos syndrome, with a predominance of microdeletions reported in Japanese patients. **BIOASQ answer:** NSD1 gene
11154546	**Pivot language:** German **BIOASQ question:** Abnormality in which vertebral region is important in the Bertolotti's syndrome? **BIOASQ snippet:** Repeated fluoroscopically guided injections implicated a symptomatic L6-S1 facet joint contralateral to an anomalous lumbosacral articulation. **Back-translated snippet:** Repeated fluoroscopic injections implied a symptomatic L6-S1 facet joint contralateral to an abnormal lumbosacral articulation. **BIOASQ answer:** lumbosacral

Table D5: Training instances generated via back-translation of BIOASQ snippets using French, Spanish, and German as a pivot language. A generated instance contains a back-translated snippet and the corresponding BIOASQ question and answer.

DA via Information Retrieval	
ID	Instance
25941473	**BIOASQ question:** Which is the neurodevelopmental disorder associated to mutations in the X- linked gene mecp2? **Retrieved snippet:** Genotype-specific effects of Mecp2 loss-of-function on morphology of Layer V pyramidal neurons in heterozygous female Rett syndrome model mice. **BIOASQ answer:** rett syndrome
28708333	**BIOASQ question:** Which is the molecular target of the immunosuppressant drug Rapamycin? **Retrieved snippet** Conversion from calcineurin inhibitors to mTOR inhibitors as primary immunosuppressive drugs in pediatric heart transplantation. **BIOASQ answer:** mtor
29767248	**BIOASQ question:** What is the target of the drug Olaparib? **Retrieved snippet:** Mechanistically, dual blockade of PI3K and PARP in ARID1A-depleted gastric cancer cells significantly increased apoptosis detected by flow cytometry, and induced DNA damage by immunofluorescent staining. **BIOASQ answer:** parp

Table D6: Training instances generated via IR. A BIOASQ question is used as the query to retrieve PUBMED documents. For each snippet of the retrieved documents that contains the answer, we generate a new training triplet consisting of the BIOASQ question, the snippet and the BIOASQ answer.

DA with word substitution based on WORD2VEC	
ID	Instance
27965160	**BIOASQ question:** Sclerostin regulates what process? **BIOASQ snippet:** Sclerostin is a soluble antagonist of Wnt/b-catenin signaling secreted primarily by osteocytes. Current evidence indicates that sclerostin likely functions as a local/paracrine regulator of bone metabolism rather than as an endocrine hormone. **Snippet after WORD2VEC substitution:** sclerostin is a soluble agonist of wnt-b catenin signaling secreted mainly by osteocytes current evidence suggests that sclerostin likely functions as a localparacrine regulator of bone metabolism rather than as an endocrine hormone **BIOASQ answer:** bone metabolism
22003227	**BIOASQ question:** Which microRNA is the mediator of the obesity phenotype of patients carrying 1p21.3 microdeletions? **BIOASQ snippet:** The study also demonstrated significant enrichment of miR-137 at the synapses of cortical and hippocampal neurons, suggesting a role of miR-137 in regulating local synaptic protein synthesis machinery. CONCLUSIONS: This study showed that dosage effects of MIR137 are associated with 1p21.3 microdeletions and may therefore contribute to the ID phenotype in patients with deletions harbouring this miRNA . **Snippet after WORD2VEC substitution:** the study also demonstrated significant enrichment of mir 137 at the synapses of cortical and hippocampal neurons indicating a implication of mir 137 in regulating local synaptic protein synthesis machinerybr-bconclusionsb this study showed that dosage effects of mir137 are associated with 2q223 microdeletions and might hence contribute to the id phenotype in patients with microinsertions harbouring this micro-rna **BIOASQ answer:** MIR137
21546092	**BIOASQ snippet:** Beck's Medical Lethality Scale (BMLS) was administered to assess the degree of medical injury, and the SAD PERSONS mnemonic scale was used to evaluate suicide risk. **BIOASQ question:** What is evaluated with the SAD PERSONS scale? **Snippet after WORD2VEC substitution:** becks medical lethality scale bmls was administered to evaluate the degree of medical injury and the sad people domain-general scale was utilized to investigate suicide risk **BIOASQ answer:** suicide risk

Table D7: Training instances generated via word substitution based on WORD2VEC. We randomly select at most 10 words of a BIOASQ snippet and substitute each word w_i with its most similar word w_j from the vocabulary of the WORD2VEC model. Highlights of the same color indicate substituted words and the corresponding substitutions.

DA with word substitution based on BIOLM	
ID	Instance
22140526	**BIOASQ question:** Which gene is responsible for red hair? **BIOASQ snippet:** The association signals at the MC1R gene locus from CDH were uniformly more significant than traditional GWA analyses. The CDH test will contribute towards finding rare LOF variants in GWAS and sequencing studies. **BIOASQ snippet after BIOLM substitution:** The association signals at the MC1R 1 identified from CDH were significantly more significant than traditional association analyses. The proposed findings will contribute towards detecting novel risk variants in GWAS and sequencing studies. **BIOASQ answer:** MC1R
26917818	**BIOASQ question:** Dinutuximab is used for treatment of which disease? **BIOASQ snippet:** CONCLUSIONS Dinutuximab is the first anti-GD2 monoclonal antibody approved in combination with GM-CSF, IL-2, and RA for maintenance treatment of pediatric patients with high-risk neuroblastoma who achieve at least a partial response to first-line multiagent, multimodality therapy. **BIOASQ snippet after BIOLM substitution:** CONCLUSIONS Dinutuximab is the first human monoclonal antibody approved in combination with recombinant IL-2, and dexamethasone for maintenance treatment of pediatric patients with high-risk neuroblastoma who achieve at least a partial response to prior multiagent, standard therapy. **BIOASQ answer:** neuroblastoma
27789693	**BIOASQ question:** Which database associates human noncoding SNPs with their three-dimensional interacting genes? **BIOASQ snippet:** 3DSNP: a database for linking human noncoding SNPs to their three-dimensional interacting genes. **BIOASQ snippet after BIOLM substitution:** 3DSNP: a method for linking functional GWAS SNPs to their three-dimensional structural structures **BIOASQ answer:** 3DSNP

Table D8: Training instances generated via word substitution based on BIOLM.We randomly select at most 10 words of a BIOASQ snippet and we substitute each word w_i with the most probable word w_j suggested by BIOLM after masking w_i. Highlights of the same color indicate substituted words and the corresponding substitutions.

DA via Question Generation using BIOASQ snippets	
ID	Instance
21159650	**Generated question:** What enzyme inhibits cullin-RING E3 ubiquitin ligases? **BIOASQ snippet:** MLN4924 is a first-in-class experimental cancer drug that inhibits the NEDD8-activating enzyme, thereby inhibiting cullin-RING E3 ubiquitin ligases and stabilizing many cullin substrates **Generated answer:** NEDD8
17333537	**Generated question:** What type of RNA triggers silencing of inactivation in eutherian mammals? **BIOASQ snippet:** In eutherian mammals X inactivation is regulated by the X-inactive specific transcript (Xist), a cis-acting non-coding RNA that triggers silencing of the chromosome from which it is transcribed **Generated answer:** chromosome
16800744	**Generated question:** What is the human tissue kallikrein family of? **BIOASQ snippet:** The human tissue kallikrein family of serine proteases (hK1-hK15 encoded by the genes KLK1-KLK15) is involved in several cancer-related processes. **Generated answer:** serine proteases

Table D9: Training instances generated using T5. Given a BIOASQ snippet T5selects a span of the snippet and generates a question that can be answered by that span. We select spans different than the ones used in BIOASQ.

DA via Question Generation using random snippets from random PUBMED abstracts	
ID	Instance
26935709	**Generated question:** What can be isolated or in combination with accompanying deformities occurring in the forefoot and/or hindfoot? **PUBMED snippet:** Symptoms can be isolated or in combination with accompanying deformities occurring in the forefoot and/or hindfoot. **Generated answer:** Symptoms
29260288	**Generated question:** What supplementation has been integrated into our practice? **PUBMED snippet:** Vitamin D supplementation has been integrated into our current practice. **Generated answer:** Vitamin D
30706485	**Generated question:** What were connected to a volume-cycled ventilator after sedation, analgesia and endotracheal intubation? **PUBMED snippet:** After sedation, analgesia and endotracheal intubation, pigs were connected to a volume-cycled ventilator. **Generated answer:** pigs

Table D10: Training instances generated using T5. Given a random snippet from a random PUBMED article T5selects a span of the snippet and generates a question that can be answered by that span.

DA by adding context	
ID	Instance
15149039	**BIOASQ question:** Which metabolite activates AtxA? **BIOASQ snippet:** Transcription of the major Bacillus anthracis virulence genes is triggered by CO2, a signal mimicking the host environment. **BIOASQ snippet with additional context:** Transcription of the major Bacillus anthracis virulence genes is triggered by CO2, a signal mimicking the host environment. A 182-kb plasmid, pXO1, carries the anthrax toxin genes and the genes responsible for their regulation of transcription, namely atxA and, pagR, the second gene of the pag operon. AtxA has major effects on the physiology of B. anthracis. It coordinates the transcription activation of the toxin genes with that of the capsule biosynthetic enzyme operon, located on the second virulence plasmid, pXO2. In rich medium, B. anthracis synthesises alternatively two S-layer proteins (Sap and EA1). **Answer:** CO2
16757427	**BIOASQ question:** What tyrosine kinase, involved in a Philadelphia- chromosome positive chronic myelogenous leukemia, is the target of Imatinib (Gleevec)? **BIOASQ snippet:** Imatinib was developed as the first molecularly targeted therapy to specifically inhibit the BCR-ABL kinase in Philadelphia chromosome (Ph)-positive chronic myeloid leukemia (CML). **BIOASQ snippet with additional context:** The second generation of BCR-ABL tyrosine kinase inhibitors. Imatinib was developed as the first molecularly targeted therapy to specifically inhibit the BCR-ABL kinase in Philadelphia chromosome (Ph)-positive chronic myeloid leukemia (CML). Because of the excellent hematologic and cytogenetic responses, imatinib has moved toward first-line treatment for newly diagnosed CML. However, the emergence of resistance to imatinib remains a major problem in the treatment of Ph-positive leukemia. Several mechanisms of imatinib resistance have been identified, including BCR-ABL gene amplification that leads to overexpression of the BCR-ABL protein, point mutations in the BCR-ABL kinase domain that interfere with imatinib binding, and point mutations outside of the kinase domain that allosterically inhibit imatinib binding to BCR-ABL. **Answer:** BCR-ABL

Table D11: Training instances generated by adding context around the original BIOASQ snippet. In the generated snippet the original one is highlighted.

Slot Filling for Biomedical Information Extraction

Yannis Papanikolaou, Marlene Staib, Justin Grace and Francine Bennett
Healx
Cambridge
UK
(yannis.papanikolaou, marlene.staib, justin.grace, francine.bennett)@healx.io

Abstract

Information Extraction (IE) from text refers to the task of extracting structured knowledge from unstructured text. The task typically consists of a series of sub-tasks such as Named Entity Recognition and Relation Extraction. Sourcing entity and relation type specific training data is a major bottleneck in domains with limited resources such as biomedicine. In this work we present a slot filling approach to the task of biomedical IE, effectively replacing the need for entity and relation-specific training data, allowing us to deal with zero-shot settings. We follow the recently proposed paradigm of coupling a Tranformer-based bi-encoder, Dense Passage Retrieval, with a Transformer-based reading comprehension model to extract relations from biomedical text. We assemble a biomedical slot filling dataset for both retrieval and reading comprehension and conduct a series of experiments demonstrating that our approach outperforms a number of simpler baselines. We also evaluate our approach end-to-end for standard as well as zero-shot settings. Our work provides a fresh perspective on how to solve biomedical IE tasks, in the absence of relevant training data. Our code, models and datasets are available at `https://github.com/ypapanik/biomedical-slot-filling`.

1 Introduction

In Information Extraction (IE) we are interested in extracting structured knowledge from unstructured text. This structured knowledge takes most usually the form of directed binary relations between entities, in other words triples of the form *head - relation - tail*, which can then be used to populate a Knowledge Base or a Knowledge Graph with factual information.

The standard approach to perform IE relies on a cascade of Natural Language Processing (NLP) models. First, Named Entity Recognition (NER) is employed to find and extract entities of interest, subsequently Entity Linking (EL) to link the extracted entities to Knowledge Base identifiers and finally Relation Extraction (RE) to identify existing relations between entities.

These individual sub-tasks tasks have attracted a great deal of attention in recent years with methods and datasets fuelling further research (Verga et al., 2018; Zeng et al., 2014, 2015; Lin et al., 2016). IE is largely regarded as a main facilitator of structured data reasoning, such as Knowledge Base Completion.

1.1 Standard Information Extraction vs Slot Filling

A major bottleneck in the above approach is that all modules (NER, EL, RE) need training data specific to the entity or relation types that we are interested in extracting. For instance, a NER model recognizing diseases needs training data annotated with the entity type $disease$ and so forth. The biomedical domain is particularly affected by these limitations, given the vast variety of entity and relation types which are commonly of interest. Additionally, sourcing training data for each sub-task and type is expensive and challenging, requiring subject matter experts. For reference, the UMLS ontology contains 125 semantic (entity) types and 54 relation types.

An alternative approach to standard IE is slot filling. The way IE is conceptualized in slot filling is highly reminiscent of open domain question answering (QA): for a given head-relation query the retriever returns a set of relevant passages, which are then fed to a reader model that then extracts a matching tail entity, the answer. By following such an approach, we can deal with zero-shot settings since, unlike standard IE, we are not seeking to recognize specific entity types or extract specific relation types, but rather do machine reading comprehension, that is, extract answers in response

Proceedings of the BioNLP 2022 workshop, Dublin, Ireland, pages 82–90

to queries. Importantly, this approach extends to relation types that were unseen during training, effectively reducing the need for re-training and re-deployment of a model deployed into production.

Furthermore, standard IE requires processing of every single sentence of the given corpus through its different modules (NER, EL, RE). In contrast, the computational cost of slot filling is much smaller as it performs retrieval and reading comprehension on far fewer queries to extract relations. As an example, Hetionet (Himmelstein et al., 2017) contains around 2.25M relations, but they can be formulated in around 46k distinct queries, of the form *head-relation*[1].

As a final point we summarize below how the two approaches would materialize in a production setting, to make their differences more apparent. We note that standard IE might involve additional tasks, such as coreference resolution (which we do not describe here for simplicity):

Standard IE:

- For each sentence, recognise entities with NER model.
- For each recognised entity, link to an entity identifier from a Knowledge Base, discarding entries that cannot be linked.
- For each sentence that contains more than one recognized entity, extract relations between the entities with a RE model.
- Aggregate relations per sentence, resolving potential conflicts.

Slot filling:

- For each entity in the Knowledge Base and each possible relation type, consider all possible head - relation pairs and construct the relevant queries, in a form *head - relation*[2].
- For each query, retrieve the top k relevant documents with a retriever model.
- For each query-retrieved document pair, perform reading comprehension, extracting zero, one or multiple answers, i.e., relation tails.
- For each answer, link to an entity identifier from a Knowledge Base, discarding entries that cannot be linked.

1.2 Slot Filling: General vs Biomedical Domain

Although similar in most aspects, slot filling in the general domain against slot filling in the biomedical and more broadly the scientific domain differ in a few key ways. The first lies in the link between relations and entities. In the general domain, a specific relation type will often imply a specific entity type as well, whereas this rarely holds in biomedical literature. Consider for example a relation *child-of* in the general domain, where we expect both head and tail of the relation to be entities of type *person*, as opposed to a relation *(up)regulate* in biomedicine where the head might be *gene* or *drug* equivalently. These nuances in the language used render the task of slot filling more challenging in biomedicine.

Another, perhaps more critical aspect relates to retrieval and more specifically how we build and evaluate on a retrieval dataset. In the general domain, a slot filling query, or more broadly a question within the QA framework, will most often have a unique answer[3], whereas this rarely holds when mining the biomedical literature. For instance, consider the examples illustrated in Table 1 coming from two well established general domain benchmarks, Natural Questions (Kwiatkowski et al., 2019) and zsRE (Levy et al., 2017) against two datasets from the biomedical domain, BioASQ (Tsatsaronis et al., 2015) and our slot filling dataset (BioSF).

This difference has a number of implications both for training and evaluation. With respect to training, one of the major successes of neural-based retrieval methods has been attributed to being able to present the model with hard negatives, i.e., examples were a previous version of the retriever (or a simpler statistical retriever) have failed. When, for example, we have a query-answer pair that mentions that Barack Obama's wife is Michelle Obama, and the model returns a passage that does not include the string "Michelle Obama", we can relatively safely consider this a false positive and use that passage as a hard negative. This helps the algorithm correct mistakes and improve. In

[1]In other words, if we were trying to build a KB from biomedical text that would contain these 2.25M relations, we would require to perform around 46k queries on our index to retrieve relevant documents.

[2]With this formulation a head and a tail can be used interchangeably, by just changing the relation type, e.g. a *drug-treats-disease relation* can also be cast to *disease-is treated by-drug* without additional training data.

[3]We are implicitly referring only to factoid queries here which is the case for most open domain QA datasets; queries of list type would have multiple answers in any case.

Dataset	Query	Answer(s)
NQ	when is the next deadpool movie being released	May 18, 2018
NQ	what was the first capital city of australia	Melbourne
zsRE	Elmer George [SEP] spouse	Mari Hulman George
zsRE	Boone River [SEP] mouth of the watercourse	Des Moines River
BioASQ	What are the main indications of lacosamide?	'epilepsy', 'analgesic'
BioASQ	Which metabolite activates AtxA?	'CO2', 'bicarbonate'
BioSF	sildenafil [SEP] regulator	'L765A', 'F786A', 'F820A'
BioSF	Amprenavir [SEP] interacts with	'rifabutin', 'ritonavir'

Table 1: Examples of queries for general domain benchmarks (NQ, zsRE) vs biomedical domain benchmarks (BioASQ, BioSF). Queries in the biomedical domain usually involve multiple valid answers, as opposed to the general domain.

biomedicine on the other hand, if we have an example stating that sildenafil regulates a mutation L765A, we cannot be sure that all alternative strings extracted by the model are true negatives, as there may be other valid answers that we cannot validate due to our Knowledge Base being incomplete. This compromises our ability to build gold standard training data and we are presented with a situation similar to the one encountered in distant supervision, where unlabeled examples are considered as negatives but might be positives in some cases. Practically, this leads to a noisy training set which may reduce model accuracy.

During evaluation of a biomedical retriever, we encounter the same problem, in the sense that we might obtain misleading low performance since unknown correct passages might rank higher than the known correct ones. This leads to an imperfect, i.e., "silver" quality, evaluation regime making it hard to compare approaches and models.

In this work we aim to address the challenges mentioned in the two previous subsections. Specifically,

- We provide a short review of the relevant work in Section 2.
- We contribute a novel formulation of biomedical IE as a slot filling task, to address few-shot or zero-shot settings in Section 3.
- We release a new benchmark for biomedical slot filling, dubbed *BioSF* which we describe in Section 4.
- We train a biomedical dense passage retriever along with a biomedical reading comprehension model for slot filling, using BioSF. We provide the models publicly.
- We present an evaluation of our approach over several baselines on BioSF, which we are able to outperform by a large margin, in Section 5.

2 Related Work

Recent years have witnessed a series of significant advances in the field of QA, primarily owing to the Transformer architecture (Vaswani et al., 2017) and the BERT self-supervised pre-training paradigm (Devlin et al., 2019). These advances, both in terms of methods (Chen et al., 2017; Lin et al., 2019; Guu et al., 2020; Lewis et al., 2020b) and datasets (Kwiatkowski et al., 2019; Yang et al., 2018), motivated researchers to formulate a series of different NLP tasks as open domain QA, including entity linking or relation extraction (Levy et al., 2017; Petroni et al., 2021). In this work we follow this paradigm by formulating biomedical IE as a slot-filling task.

In open domain QA, given a query, a retrieval module first retrieves relevant documents from the knowledge source (such as Wikipedia). A reading comprehension module is then used to extract a span from the relevant documents, the answer. The retrieval step was, up to very recently, dominated by statistical-based approaches, namely BM25 or tf-idf (Chen et al., 2017). ORQA (Lee et al., 2019b) and REALM (Guu et al., 2020) have been the first neural based methods to clearly outperform statistical based retrieval, although they required expensive language model pre-training. Dense Passage Retrieval (DPR) (Karpukhin et al., 2020) has improved upon these methods by employing BERT-based encoders, one for the queries and one for passages. These are jointly optimized during training to classify passages as relevant versus irrelevant. This approach has proved superior to other neural based approaches and has quickly become the preferred method for open domain QA in subsequent work (Lewis et al., 2020b; Izacard and Grave, 2021; Maillard et al., 2021).

Among the subsequent works, Retrieval Augmented Generation (Lewis et al., 2020b) employs an architecture based on DPR and BART (Lewis et al., 2020a) that is optimized end to end during finetuning, to retrieve relevant documents and generate answers to queries. Fusion-in-decoder (Izacard and Grave, 2021) employs DPR or BM25 as retrievers coupled with a T5 language model, to generate answers by attending at multiple passages simultaneously. For simplicity, we are not considering these approaches in this work, leaving their implementation for the biomedical domain for future work.

In an effort to fuel further research on this field, Petroni et al. (2021) introduced KILT, a new benchmark of knowledge intensive tasks, which contains among others two slot filling datasets, *zero-shot RE* which was first presented in (Levy et al., 2017) and *T-REx* introduced by Elsahar et al. (2018). In building our biomedical slot filling dataset we largely follow the conventions and format of KILT, with the intention to ease experimentation.

Finally, Glass et al. (2021) have presented a RAG model specifically finetuned for slot filling on the above datasets, showing significant improvement over the generic alternatives, which were finetuned on Natural Questions (NQ).

3 Biomedical Slot Filling

Formally, let us first define the task of IE. We assume a knowledge source K, consisting of passages p_i. Furthermore, we assume there exists a Knowledge Base that contains a number of entities e_i. Our goal is to extract from K all possible triples of the form $e_a - r_i - e_b$ where $r_i \in R$ and R is the set of possible relation types. For each e_i we assume that it has a specific entity type e_t and that each e_t can be involved in a specific subset of R.

Slot filling further formulates the above task as follows: we first employ a retrieval model M_r that encodes all passages p_i from K. The encoded passages are indexed to allow fast retrieval. At inference, for each e_i of type e_t, we consider all possible relations from R and construct the relevant queries $q_i : e_i - r_i$. Each query is then encoded and the resulting vector is used to query the index, returning the n most similar p_i in terms of the maximum inner product:

$$sim(q_i, p_i) = E_Q(q_i)^T E_P(p_i) \quad (1)$$

where E_Q is the query encoder and E_P is the passage encoder. Subsequently a reader model M_{qa} takes as input the above query and each of the retrieved passages and extracts zero, one or more spans, i.e., answers. Valid answers are considered as those representing an entity e_i.

Here, we adopt as M_r a neural, dense bi-encoder, namely DPR, which uses a different encoder for passages and queries, but any type of retriever can be used such as BM25, where $E_Q = E_P$. We initialize DPR's encoders with the ones presented in (Karpukhin et al., 2020) which were finetuned on the NQ benchmark. We subsequently train DPR on the dataset presented in Section 4, with the following loss function:

$$L(q_i, p_i^+, p_i^-) = -log \frac{e^{sim(q_i, p_i^+)}}{e^{sim(q_i, p_i^+)} + e^{sim(q_i, p_i^-)}} \quad (2)$$

Unlike (Karpukhin et al., 2020), we assume that each training instance is a (q_i, p_i^+, p_i^-) tuple where p_i^+ is a positive, i.e., relevant passage and p_i^- is a negative passage.

Regarding the reader comprehension model M_{qa}, we employ a pretrained BioBERT (Lee et al., 2019a) model and finetune it on the dataset of Section 4. To finetune we follow the standard approach for question answering with BERT where the input is the concatenated query and passage separated by special token *[SEP]* and the outputs are the start and end token positions within the passage. The training objective is the sum of the log-likelihoods of the correct start and end positions. For more details we refer the interested reader to (Devlin et al., 2019).

4 Biomedical Slot Filling Dataset

In order to build a slot filling dataset for biomedicine, we resort to a number of publicly available biomedical NER and RE datasets, summarized in Table 2. Each instance in these datasets contains the relation triple as well as the text where it was found, thus we can easily transform them in a question answering-like format for slot filling. In total, we build two datasets, one to train and evaluate the retriever and one for the reader model respectively.

Specifically for the retriever training, we use negative, i.e., null relation instances, as negatives.

Dataset	relation	relation types	# instances
BioCreative V CDR (Li et al., 2016)	compound-disease	1	15,796
BioCreative VI ChemProt (Krallinger et al., 2017)	compound-protein	9	15,568
DDIExtraction 2013 (Segura Bedmar et al., 2013)	drug-drug	1	32,018

Table 2: Public datasets used to build our biomedical slot filling dataset, BioSF. The relation types for the drug-drug interactions dataset have been merged into one relation dubbed *interacts with*.

Additionally, we have used BM25 to add hard negatives to our dataset, exactly as (Karpukhin et al., 2020; Glass et al., 2021) have done previously. Although, as mentioned above, these negatives might entail some noise, similarly to when following a distant supervision approach we expect the noise to cancel out overall. Both datasets with their training, development and testing splits are released with our code. In the following, we refer to our dataset as *BioSF*.

5 Experiments

In this Section we present the experiments that we conducted, followed by a discussion on their implications. We are interested in evaluating our biomedical DPR retriever, our biomedical slot filling reader and finally the end to end slot filling approach.

5.1 Retrieval

First, we are interested to understand the performance of our approach against different baselines. To that end, we employ BM25 as well as two already finetuned DPR retrievers from (Karpukhin et al., 2020; Glass et al., 2021). BM25 is a well established algorithm for retrieval, outperforming until very recently more sophisticated neural-based approaches. It is also particularly efficient and does not require any training, which makes it a very attractive option for real-world production settings. Nevertheless, it is a statistical, pattern matching based approach lacking the ability to learn semantics or context.

Regarding the general domain DPR models, since they are currently state of the art in the relevant general domain tasks, we seek to see if they can be used successfully for the biomedical domain. Our model is trained on far less data, which is nevertheless domain and task specific, therefore it is crucial to understand which approach fares better.

5.1.1 Experimental Setup

We employ a PubMed dump from April 2020 as our knowledge sourse, filtering to documents that have an abstract and splitting abstracts to roughly 100-token length passages. We also use a smaller subset of one million passages, in order to be able to search for optimal hyper-parameters and allow easy replication of results. In that subset, we randomly sample passages and add the gold passages from BioSF so as to make sure that a perfect retrieval algorithm would be able to retrieve all correct passages and find the answer. We highlight that this is an easier version of the real-world task, where the retriever needs to search among around 29 million passages.

For BM25, we employ the anserini package (Yang et al., 2017), and build a Lucene index on the pre-processed passages, whereas we used the off the shelf Huggingface models () for the general domain DPR retrievers.

For our retriever, we train DPR on the BioSF dataset, for 40 epochs keeping the best model in terms of the validation loss. We use a learning rate of $3e-5$, an Adam optimizer with default options and a training batch size of 32 examples. Subsequently, we encode the passages with the trained passage encoder. Encoding the full 29 million passages takes around 96 GPU hours on a V100. We then build a flat FAISS (Johnson et al., 2019) index for the encoded passages.

5.1.2 Results

Initially, we conduct experiments on the smaller dataset that we described above of one million passages. As we noted in Section 1.2 evaluating retrieval for slot filling or more broadly for QA in the biomedical domain is significantly different than in the general domain since in biomedicine a query has in most cases multiple answers as opposed to the general domain. Table 3 illustrates the results for this first series of experiments.

As we can see the DPR models that have been finetuned on the general domain perform rather

Retriever	hits@1	hits@10	hits@100	index size(Gb)
BM25	21.4	36.1	60.6	**1.1**
DPR-NQ (Karpukhin et al., 2020)	5.5	17.2	37.6	2.9
DPR-multitask (Maillard et al., 2021)	4.2	14.3	33.8	2.9
DPR-zsRE (Glass et al., 2021)	7.6	19.6	37.2	2.9
Bio-DPR(ours)	**31.0**	**55.1**	**72.5**	2.9

Table 3: Evaluation results for retrieval experiments on the BioSF development set using as content one million passages from PubMed. Values in bold show statistically significant results in terms of z-test at p-value of 0.05, whereas for our model we show the average across five different DPR training runs.

Retriever	hits@1	hits@10	hits@100	index size
BM25	11.0	30.3	56.1	**29.4**
DPR-NQ	5.2	17.9	38.9	90.0
DPR-zsRE	2.3	10.2	26.4	90.0
Bio-DPR(ours)	**11.5**	**33.2**	**59.1**	90.0

Table 4: Evaluation results for retrieval experiments on the BioSF development set on full PubMed. Values in bold show statistically significant results for a z-test at p-value of 0.05.

poorly compared to the much lighter and computationally efficient BM25. Nevertheless, our model Bio-DPR, is substantially better than BM25 in all cases, achieving up to 19 points of improvement (in the case of hits@10). These results, are aligned to the results previously presented for the general domain where BM25 has been outperformed by DPR. Nevertheless, in-domain training data seems critical for DPR to perform well for slot filling, a finding also shared in (Maillard et al., 2021).

The same findings apply for the full PubMed knowledge source, as illustrated in Table 4, although the improvement of our model over BM25 is much smaller but still significant.

5.2 Slot Filling Reader

For the reader, we finetune a BioBERT-base and a BioBERT-large model on the BioSF training set. We further include two baselines, one trained on the BioASQ 8 QA dataset and one trained in the zero-shot RE (zsRE) dataset from (Levy et al., 2017). We employ these two baselines to test whether in-domain data from a different task (BioASQ) or general domain data for the same task (zsRE) can be helpful in learning an accurate model.

For all models, we train up to ten epochs, keeping the best performing model on the development set, using a learning rate of $3e-5$, a batch size of 32 and the Adam optimizer with default parameters. Table 5 presents the results. We observe that the baselines perform rather poorly compared to the models trained with in-domain slot filling data - a finding that highlights the importance of building an in-domain dataset for slot-filling.

5.3 End to End Evaluation

Having evaluated both components of our approach, we now turn our attention to the end to end setting, which simulates better a real world scenario. In this setting, we are given a head entity and a relation and we want to correctly extract the tail entity. To evaluate our approach in such a setting, we first use the triples included in the BioSF test set. This dataset contains 3,171 queries with 2.35 answers, i.e. tails, per query on average.

Additionally, we would like to understand how our approach performs in the zero-shot setting, i.e., for entities and relations that our model has not seen during training. To this end, we employ Hetionet (Himmelstein et al., 2017), a network of biomedical knowledge assembled from 29 biomedical Knowledge Bases, containing 24 distinct relation types. We keep nine relation types that our models have not previously seen, e.g., "expresses", "localizes", "treats" and randomly sample 500 queries, with 9.3 answers, i.e. tails, per query on average. We note that this dataset differs substantially to the previous one, in the sense that a query might have far more valid answers. For example, some queries have more than 100 valid answers.

In both cases, we first retrieve the top-100 passages for each query, from the full PubMed knowl-

Model	Data	Exact Match(dev/test)	F1(dev/test)
BioBERT-base	BioASQ	13.10/13.44	17.95/18.64
"	zsRE	16.59/15.77	22.51/22.98
"	BioSF	52.30/54.67	58.82/59.98
BioBERT-large	BioSF	**54.80/55.65**	**60.92/61.55**

Table 5: Evaluation results for the reader experiments on the BioSF development and testing sets. We report the averages across five runs for each model, results in bold show a statistically significant improvement for a z-test at p-value of 0.05.

Setting	Dataset	end-to-end micro-recall
Standard	BioSF test set	24.38
Zero-shot	Hetionet	18.66

Table 6: End to end evaluation of our approach on a standard as well as a zero-shot setting.

edge source, using our bio-DPR model and subsequently we pass all query-passage pairs through our reader model. We evaluate with micro-recall since, as we discussed previously, there might be multiple valid answers not contained in our KB and we aim to examine what percentage of the KB triples we can extract from text. We note again that this is not a perfect evaluation as, besides the issue mentioned above, there might also be triples in Hetionet that do not appear in any sentence in the literature. Table 6 illustrates our results. The recall is substantially low, a finding that is somewhat expected due to the imperfect nature of our evaluation setting, as well the challenging nature of the task, especially in the zero-shot setting. Nevertheless, we consider that these two additional datasets, will enable further research and improved approaches. Overall, the above experiments should be regarded as a stepping stone towards a novel paradigm for biomedical IE, overcoming the shortcomings of the current standard approach.

6 Conclusions and Future Work

In this work we formulated the task of biomedical Information Extraction as a slot filling problem. This approach aims to forgo the need for entity and relation type specific training data, which is scarce and costly to annotate in the biomedical domain. Additionally, this formulation allows to deal with the addition of new relation types, without needing to re-train the relevant models.

Additionally, we have introduced a new biomedical slot filling benchmark and used it to train a biomedical DPR model, a dual BERT-based encoder for retrieval, as well as a biomedical slot filling reader based on BioBERT. In a series of experiments our approach outperforms significantly a number of general domain baselines as well as the simpler BM25 retriever. Furthermore, our results illustrate the importance of in-domain, task-specific training data, in line with findings from recent works (Glass et al., 2021; Maillard et al., 2021).

In future work, we aim to focus on sequence to sequence variants of this work such as the work in (Izacard and Grave, 2021), as well as to conduct a thorough comparison of a standard biomedical IE system against our slot filling approach.

References

Danqi Chen, Adam Fisch, Jason Weston, and Antoine Bordes. 2017. Reading wikipedia to answer open-domain questions. In *Proceedings of the 55th Annual Meeting of the Association for Computational Linguistics (Volume 1: Long Papers)*, pages 1870–1879.

Jacob Devlin, Ming-Wei Chang, Kenton Lee, and Kristina Toutanova. 2019. BERT: pre-training of deep bidirectional transformers for language understanding. In *Proceedings of the 2019 Conference of the North American Chapter of the Association for Computational Linguistics: Human Language Technologies, NAACL-HLT 2019, Minneapolis, MN, USA, June 2-7, 2019, Volume 1 (Long and Short Papers)*, pages 4171–4186. Association for Computational Linguistics.

Hady Elsahar, Pavlos Vougiouklis, Arslen Remaci, Christophe Gravier, Jonathon Hare, Frederique Laforest, and Elena Simperl. 2018. T-REx: A large scale alignment of natural language with knowledge base triples. In *Proceedings of the Eleventh International Conference on Language Resources and Eval-*

uation (LREC 2018), Miyazaki, Japan. European Language Resources Association (ELRA).

Michael Glass, Gaetano Rossiello, Md Faisal Mahbub Chowdhury, and Alfio Gliozzo. 2021. Robust retrieval augmented generation for zero-shot slot filling. In *Proceedings of the 2021 Conference on Empirical Methods in Natural Language Processing*, pages 1939–1949, Online and Punta Cana, Dominican Republic. Association for Computational Linguistics.

Kelvin Guu, Kenton Lee, Zora Tung, Panupong Pasupat, and Ming-Wei Chang. 2020. Realm: Retrieval-augmented language model pre-training. *arXiv preprint arXiv:2002.08909*.

Daniel Scott Himmelstein, Antoine Lizee, Christine Hessler, Leo Brueggeman, Sabrina L Chen, Dexter Hadley, Ari Green, Pouya Khankhanian, and Sergio E Baranzini. 2017. Systematic integration of biomedical knowledge prioritizes drugs for repurposing. *Elife*, 6:e26726.

Gautier Izacard and Édouard Grave. 2021. Leveraging passage retrieval with generative models for open domain question answering. In *Proceedings of the 16th Conference of the European Chapter of the Association for Computational Linguistics: Main Volume*, pages 874–880.

Jeff Johnson, Matthijs Douze, and Hervé Jégou. 2019. Billion-scale similarity search with gpus. *IEEE Transactions on Big Data*.

Vladimir Karpukhin, Barlas Oguz, Sewon Min, Patrick Lewis, Ledell Wu, Sergey Edunov, Danqi Chen, and Wen-tau Yih. 2020. Dense passage retrieval for open-domain question answering. In *Proceedings of the 2020 Conference on Empirical Methods in Natural Language Processing (EMNLP)*, pages 6769–6781.

Martin Krallinger, Obdulia Rabal, Saber A Akhondi, et al. 2017. Overview of the biocreative vi chemical-protein interaction track. In *Proceedings of the sixth BioCreative challenge evaluation workshop*.

Tom Kwiatkowski, Jennimaria Palomaki, Olivia Redfield, Michael Collins, Ankur Parikh, Chris Alberti, Danielle Epstein, Illia Polosukhin, Jacob Devlin, Kenton Lee, et al. 2019. Natural questions: a benchmark for question answering research. *Transactions of the Association for Computational Linguistics*, 7:453–466.

Jinhyuk Lee, Wonjin Yoon, Sungdong Kim, Donghyeon Kim, Sunkyu Kim, Chan Ho So, and Jaewoo Kang. 2019a. Biobert: pre-trained biomedical language representation model for biomedical text mining. *arXiv preprint arXiv:1901.08746*.

Kenton Lee, Ming-Wei Chang, and Kristina Toutanova. 2019b. Latent retrieval for weakly supervised open domain question answering. In *Proceedings of the 57th Annual Meeting of the Association for Computational Linguistics*, pages 6086–6096.

Omer Levy, Minjoon Seo, Eunsol Choi, and Luke Zettlemoyer. 2017. Zero-shot relation extraction via reading comprehension. In *Proceedings of the 21st Conference on Computational Natural Language Learning (CoNLL 2017)*, pages 333–342.

Mike Lewis, Yinhan Liu, Naman Goyal, Marjan Ghazvininejad, Abdelrahman Mohamed, Omer Levy, Veselin Stoyanov, and Luke Zettlemoyer. 2020a. Bart: Denoising sequence-to-sequence pre-training for natural language generation, translation, and comprehension. In *Proceedings of the 58th Annual Meeting of the Association for Computational Linguistics*, pages 7871–7880.

Patrick Lewis, Ethan Perez, Aleksandra Piktus, Fabio Petroni, Vladimir Karpukhin, Naman Goyal, Heinrich Küttler, Mike Lewis, Wen-tau Yih, Tim Rocktäschel, et al. 2020b. Retrieval-augmented generation for knowledge-intensive nlp tasks. *arXiv preprint arXiv:2005.11401*.

Jiao Li, Yueping Sun, Robin J Johnson, Daniela Sciaky, Chih-Hsuan Wei, Robert Leaman, Allan Peter Davis, Carolyn J Mattingly, Thomas C Wiegers, and Zhiyong Lu. 2016. Biocreative v cdr task corpus: a resource for chemical disease relation extraction. *Database*, 2016.

Chen Lin, Timothy Miller, Dmitriy Dligach, Steven Bethard, and Guergana Savova. 2019. A bert-based universal model for both within-and cross-sentence clinical temporal relation extraction. In *Proceedings of the 2nd Clinical Natural Language Processing Workshop*, pages 65–71.

Yankai Lin, Shiqi Shen, Zhiyuan Liu, Huanbo Luan, and Maosong Sun. 2016. Neural relation extraction with selective attention over instances. In *Proceedings of the 54th Annual Meeting of the Association for Computational Linguistics (Volume 1: Long Papers)*, volume 1, pages 2124–2133.

Jean Maillard, Vladimir Karpukhin, Fabio Petroni, Wen-tau Yih, Barlas Oğuz, Veselin Stoyanov, and Gargi Ghosh. 2021. Multi-task retrieval for knowledge-intensive tasks. *arXiv preprint arXiv:2101.00117*.

Fabio Petroni, Aleksandra Piktus, Angela Fan, Patrick Lewis, Majid Yazdani, Nicola De Cao, James Thorne, Yacine Jernite, Vladimir Karpukhin, Jean Maillard, et al. 2021. Kilt: a benchmark for knowledge intensive language tasks. In *Proceedings of the 2021 Conference of the North American Chapter of the Association for Computational Linguistics: Human Language Technologies*, pages 2523–2544.

Isabel Segura Bedmar, Paloma Martínez, and María Herrero Zazo. 2013. Semeval-2013 task 9: Extraction of drug-drug interactions from biomedical texts (ddiextraction 2013). Association for Computational Linguistics.

George Tsatsaronis, Georgios Balikas, Prodromos Malakasiotis, Ioannis Partalas, Matthias Zschunke, Michael R Alvers, Dirk Weissenborn, Anastasia Krithara, Sergios Petridis, Dimitris Polychronopoulos, et al. 2015. An overview of the bioasq large-scale biomedical semantic indexing and question answering competition. *BMC bioinformatics*, 16(1):1–28.

Ashish Vaswani, Noam Shazeer, Niki Parmar, Jakob Uszkoreit, Llion Jones, Aidan N Gomez, Łukasz Kaiser, and Illia Polosukhin. 2017. Attention is all you need. In *Advances in neural information processing systems*, pages 5998–6008.

Patrick Verga, Emma Strubell, and Andrew McCallum. 2018. Simultaneously self-attending to all mentions for full-abstract biological relation extraction. In *Proceedings of the 2018 Conference of the North American Chapter of the Association for Computational Linguistics: Human Language Technologies, Volume 1 (Long Papers)*, volume 1, pages 872–884.

Peilin Yang, Hui Fang, and Jimmy Lin. 2017. Anserini: Enabling the use of lucene for information retrieval research. New York, NY, USA. Association for Computing Machinery.

Zhilin Yang, Peng Qi, Saizheng Zhang, Yoshua Bengio, William Cohen, Ruslan Salakhutdinov, and Christopher D Manning. 2018. Hotpotqa: A dataset for diverse, explainable multi-hop question answering. In *Proceedings of the 2018 Conference on Empirical Methods in Natural Language Processing*, pages 2369–2380.

Daojian Zeng, Kang Liu, Yubo Chen, and Jun Zhao. 2015. Distant supervision for relation extraction via piecewise convolutional neural networks. In *Proceedings of the 2015 Conference on Empirical Methods in Natural Language Processing*, pages 1753–1762.

Daojian Zeng, Kang Liu, Siwei Lai, Guangyou Zhou, and Jun Zhao. 2014. Relation classification via convolutional deep neural network. In *Proceedings of COLING 2014, the 25th International Conference on Computational Linguistics: Technical Papers*, pages 2335–2344.

Automatic Biomedical Term Clustering by Learning Fine-grained Term Representations

Sihang Zeng * **Zheng Yuan** * **Sheng Yu** †
Tsinghua University
{zengsh19,yuanz17}@mails.tsinghua.edu.cn
syu@tsinghua.edu.cn

Abstract

Term clustering is important in biomedical knowledge graph construction. Using similarities between terms embedding is helpful for term clustering. State-of-the-art term embeddings leverage pretrained language models to encode terms, and use synonyms and relation knowledge from knowledge graphs to guide contrastive learning. These embeddings provide close embeddings for terms belonging to the same concept. However, from our probing experiments, these embeddings are not sensitive to minor textual differences which leads to failure for biomedical term clustering. To alleviate this problem, we adjust the sampling strategy in pretraining term embeddings by providing dynamic hard positive and negative samples during contrastive learning to learn fine-grained representations which result in better biomedical term clustering. We name our proposed method as CODER++[1], and it has been applied in clustering biomedical concepts in the newly released Biomedical Knowledge Graph named BIOS[2].

1 Introduction

A critical step for building a biomedical knowledge graph is clustering synonyms terms into concepts (Nicholson and Greene, 2020; Yu et al., 2022). After mining terms from the biomedical corpus or electronic medical records, these terms may belong to an existing concept dictionary or newly discovered concepts. It is hard for humans to link terms to an existing concept dictionary since the volume of the concept dictionary is huge. Furthermore, it is almost impossible for humans to determine if one term is a new concept.

Embedding-based entity linking methods encode terms into a dense space and use similarities among terms for entity linking (Liu et al., 2021; Yuan et al., 2022). Terms that belong to newly discovered concepts should have low similarities to all concepts in the dictionary. Embedding-based entity linking methods can also assist humans in term clustering by providing candidates. However, we find that existing state-of-the-art biomedical term embedding models SapBERT (Liu et al., 2021) and CODER (Yuan et al., 2022) are not sensitive to fine-grained differences (i.e. They provide high similarities for non-synonymous and textually similar term pairs). These term pairs are common, especially in diseases (e.g. *Type 1 Diabetes* v.s. *Type 2 Diabetes*) and chemicals (e.g. *xyloglucan endotransglycosylase* v.s. *xyloglucan endoglucanase*). We suggest the reason comes from the pretraining sampling strategy of SapBERT and CODER. They sample Concept Unique Identifiers (CUIs) from UMLS (Bodenreider, 2004) randomly in the mini-batch. This produces hard positive pairs (i.e. textually different terms with the same CUIs) and easy negative pairs (i.e. textually different terms with different CUIs). Supervised contrastive learning is applied to cluster embeddings under the same CUIs and to keep away embeddings for different CUIs. For benchmarking entity linking tasks, the ability to determine positive pairs is important. For term clustering, it further requests to determine negative pairs. Hard negative pairs are absent in pretraining SapBERT and CODER which lead to unsatisfactory performances in term clustering.

In this paper, we propose a probing experiment to evaluate term clustering on UMLS automatically. This experiment shows SapBERT and CODER have insufficient ability in term clustering. For better term clustering, we propose a dynamic sampling strategy that provides both hard positive and negative pairs to learn fine-grained terms embeddings named CODER++. CODER++ not only reserves the ability to normalize terms but also can distinguish different concepts with similar term names.

* Contributed equally.
† Corresponded author.
[1] Our codes and model will be released at https://github.com/GanjinZero/CODER.
[2] https://bios.idea.edu.cn/

Proceedings of the BioNLP 2022 workshop, Dublin, Ireland, pages 91–96
May 26, 2022.

CODER++ shows decent ability on biomedical entity linking and a significant improvement on biomedical term clustering evaluation.

2 Related Work

Automatic term clustering has long been discussed. Traditional methods use statistical approaches to define similar terms and perform clustering. Lin (1998) defines term similarity based on distributions and Lewis and Croft (1989) forms clusters based on co-occurrence in semantically coherent documents. Kok and Domingos (2008) uses Markov logic for unsupervised concept clustering.

Recent researches focus on deep learning approaches, where biomedical term embeddings can be used for term clustering. Nguyen et al. (2015) identifies biomedical synonyms using word embeddings. SapBERT (Liu et al., 2021; Nguyen et al., 2021) and CODER (Yuan et al., 2022) learn synonyms knowledge from UMLS to provide close embeddings for synonyms. In this work, we improve these embeddings by providing dynamic hard negative samples.

3 Term Clustering Evaluation

We introduce the term clustering evaluation on UMLS as the probing experiment, in which we find that both CODER and SapBERT show poor clustering performance. Through the case study, we find the reason is that both models fail to distinguish between fine-grained biomedical terms, which suggests a refinement is needed to support biomedical term clustering.

3.1 Embedding-based Term Clustering

We use term embeddings including CODER and SapBERT to perform clustering on UMLS terms. We first generate embedding e for each term t in UMLS. The similarity between term t_i and t_j is measured by cosine similarity $S_{ij} = cosine(e_i, e_j)$. If $S_{ij} > \theta$, where θ is a hyperparameter, t_i and t_j are predicted to be clustered. In practice, calculating similarities between all pairs is time-consuming. Instead, for each term t_i, we use the Faiss index (Johnson et al., 2019) to only save terms with top-m similarities with t_i, denoted by $\mathcal{M}_i$. Only when $t_j \in \mathcal{M}_i$ and also $S_{ij} > \theta$, t_j is predicted to be clustered with t_i (i.e. t_i and t_j are synonyms). For convenience, we denote $\mathcal{M} = \bigcup_i \mathcal{M}_i$.

3.2 Large-scale Clustering Evaluation

For evaluation, terms under the same CUI i in UMLS are regarded as ground truth clustering, denoted by $\mathcal{C}_i$. We denote $\mathcal{C} = \bigcup_i \mathcal{C}_i$. Suppose there are n terms, then we have $\binom{n}{2}$ term pairs. For each pair (t_i, t_j), if they are under the same CUI and also predicted to be clustered, then (t_i, t_j) is regarded as true positive (TP). False positive (FP), false negative (FN), and true negative (TN) are defined similarly. Recall, precision and F_1 score can be computed based on TP, FP, FN, and TN. Precision suggests how well a model differentiates between negative term pairs. Recall suggests how well a model clusters terms with similar meanings.

As n is large in practice (over 10M terms in a biomedical terminology like UMLS), it is impossible to enumerate all term pairs to directly count TP, FP, FN, and TN. Nguyen et al. (2021) downsamples negative pairs for evaluation, but this may ignore some hard negative pairs. We propose an efficient algorithm for large-scale clustering evaluation, which reduces the time complexity from $\mathcal{O}(n^2)$ to $\mathcal{O}(n)$ when ground truth cluster $\mathcal{C}_i$ is bounded. The algorithm splits the searching space into two parts, traversing through the Faiss index $\mathcal{M}$ and traversing through the ground truth cluster space $\mathcal{C}$. When traversing through $\mathcal{M}$, we first get pairs with predicted labels to be true, then count how many pairs in $\mathcal{C}$ to obtain TP and FP. When traversing through $\mathcal{C}$, we first get pairs with ground truth label to be true, then count pairs in $\mathcal{M}$ to obtain FN. TN is computed by subtracting TP, FP, and FN from $\binom{n}{2}$ instead of counting which saves time significantly. To speed up the searching process, we also store $\mathcal{C}$ and $\mathcal{M}$ in prefix trees.

3.3 Probing Results

The results of term clustering evaluation in UMLS 2020 AA for CODER and SapBERT are shown in Table 2. We search for the best threshold θ_0 according to the F_1 score. F_1 scores are both low for SapBERT and CODER, which indicates that both models could not differentiate terms well and tend to cluster different terms together. These F_1 scores are much lower than reported in (Nguyen et al., 2021) (0.65 for SapBERT), the reason is they downsample negative pairs in evaluation which underestimates FN. The performance gap between SapBERT and CODER comes from their different sampling strategies.

Term 1	Term 2	Similarity			Same CUI
		CODER	SapBERT	CODER++	
julibroside j2	julibroside c1	0.918	0.918	**0.339**	F
orange colored urine	pink urine	0.738	0.783	**0.451**	F
type 2 diabetes 1	type 1 diabetes	0.908	0.911	**0.502**	F
sb 212047	sb 216754	0.819	0.767	**0.356**	F
early onset	late onset	0.831	0.807	**0.416**	F
ginsenoside rh	ginsenoside rg	0.908	0.979	**0.420**	F
protein phosphatase 1 delta	protein phosphatase 2c delta	0.910	0.832	**0.616**	F
type ii endometrial carcinoma	endometrial cancer stage ii	0.845	0.846	**0.420**	F
headache	cephalgia	0.798	0.741	0.776	T
fhx allergies	fh: allergy	0.879	0.881	0.819	T
herpesvirus murid 004	murine herpesvirus 068	0.634	0.823	0.674	T
tex2	tex2 gene	0.890	0.995	0.921	T
eppin 1 protein, human	eppin protein, human	0.991	0.941	0.834	T
chmp2b gene	chromatin modifying protein 2b	0.743	0.797	0.724	T

Table 1: Similarities of different models between representative term pairs with the same CUI or different CUI. Term pairs with the same CUI are considered positive. Compared with CODER and SapBERT, CODER++ has relatively lower similarities on negative term pairs and moderately higher similarities on positive term pairs.

Model	θ_0	P	R	F_1
SapBERT	0.94	0.302	0.268	0.284
CODER	0.86	0.071	0.401	0.121

Table 2: Results for CODER and SapBERT on term clustering evaluation in UMLS 2020 AA.

3.4 Case Study

We sample term pairs to check why CODER and SapBERT fail on term clustering evaluation. Similarities of representative false positive term pairs for both CODER and SapBERT are shown in the upper part of Table 1. We can observe that CODER and SapBERT embeddings can't distinguish terms with number differences, body part differences, and devices differences. CODER and SapBERT provide similarities for these false positive term pairs as high as true positive term pairs shown in the lower part of Table 1. Hence they tend to cluster terms with highly similar strings but different meanings.

4 Approach

We introduce CODER++ to address the above-mentioned problem. The idea is simple, providing hard negative pairs to reduce false positive term pairs. We focus on how to construct mini-batches to learn fine-grained term representations.

4.1 Term Encoding

CODER++ embeds a term s to a dense representation e with a pretrained language model. We tokenize s into sub-words, and use the representation of [CLS] token for term representation.

4.2 Dynamic Sampling

Positive Sampling For each term t, we sample k terms $p_1, ..., p_k$ with same CUI from UMLS. This adds positive pairs for training. The term p_i can be textually different from t which is considered a hard positive sample.

Possibly Hard Negative Sampling We take terms $n_1, ..., n_m$ with top-m similarities with term t as possibly hard negative samples. It is expensive to find terms with top-m similarities on the fly, and we use the Faiss index instead. For each epoch, we update the Faiss index using the present CODER++. Selected terms can have the same CUI or different CUIs with term t. A not well-trained model has more different CUIs terms as hard negative samples. The model is required to distinguish these fine-grained terms. When the training is progressed, more selected terms will have the same CUI with the term t.

Overall Sample Strategy We first sample terms $\{t_i\}_i$ randomly from the whole term set. For each term t_i, we sample k positive terms $p_{i_1}, ..., p_{i_k}$ and m possibly hard negative terms $n_{i_1}, ..., n_{i_m}$. All these terms $\{t_i, p_{i_1}, ..., p_{i_k}, n_{i_1}, ..., n_{i_m}\}_i$ construct a mini-batch, and we use the CUIs of these terms to guide supervised contrastive learning. An example of mini-batch is visualized in Figure 1. We follow Liu et al. (2021); Yuan et al. (2022) to optimize the model using the Multi-Similarity loss (MS-loss) (Wang et al., 2019) to guide terms with same CUIs

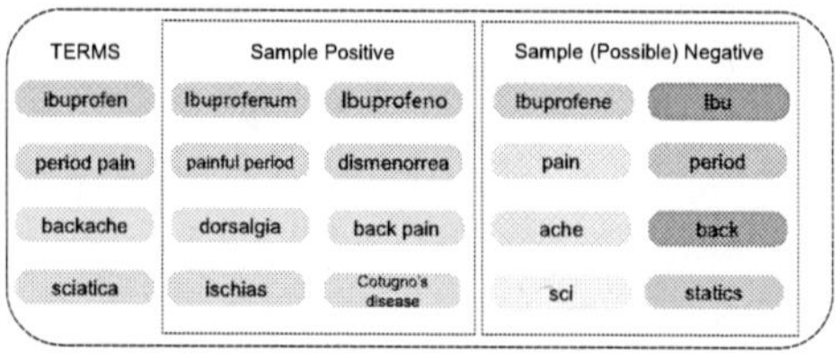

Figure 1: Construction of a mini-batch in CODER++.

similar and terms with different CUIs dissimilar.

$$\mathcal{N}_i := \{j|1 \leq j \leq m, c_i \neq c_j, S_{ij} > \min_{c_k=c_i} S_{ik} - \epsilon\}$$

$$\mathcal{P}_i := \{j|1 \leq j \leq m, c_i = c_j, S_{ij} < \max_{c_k \neq c_i} S_{ik} + \epsilon\}$$

$$\mathcal{L} = \frac{1}{m}\sum_{i=1}^{m}\Big(\frac{\log(1+\sum_{j\in\mathcal{P}_i}\exp(-\alpha(S_{ij}-\lambda)))}{\alpha} + \frac{\log(1+\sum_{j\in\mathcal{N}_i}\exp(\beta(S_{ij}-\lambda)))}{\beta}\Big),$$

where c_i is the CUI of ith term, and ϵ, α, β are hyperparameters.

5 Experiments

5.1 Pre-training

We train CODER++ initialized by CODER with 1,200K training steps [3]. We update the Faiss index every 60K steps. For each mini-batch, we set $k = m = 30$. Training costs 9 days on 8 NVIDIA A100 40GB GPUs. Each GPU samples 16 terms $\{t_i\}$ from UMLS 2020 AA at one time with 8 gradient accumulation steps which indicates a total of $16 \times (1 + 30 + 30) \times 8 \times 8 = 62,464$ terms for each parameter update step. The maximal term length is set to 32. We use AdamW (Loshchilov and Hutter, 2017) as the optimizer with a linear warm-up in the first 10000 steps to a peak of 4e-5 learning rate and a linear decay. The setting of hyperparameters ϵ, α, β in MS-loss is following (Yuan et al., 2022).

5.2 Term Clustering Evaluation

We evaluate CODER++ based on Section 3.2. The result is shown in Figure 2 and Table 3. Table 3 shows that CODER++ greatly outperforms CODER and SapBERT, obtaining 0.732, 0.576, and 0.644 for precision, recall, and F_1 scores respectively. We can see from Figure 2 that CODER++ has a comparable spread in recall with both CODER and SapBERT, which indicates CODER++ reserves the ability of clustering terms

[3] SapBERT can also be used as the initial checkpoint.

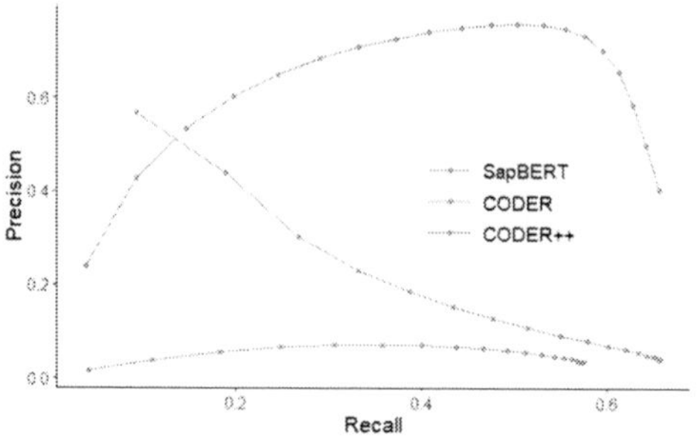

Figure 2: UMLS term clustering evaluation for CODER, SapBERT, and CODER++ under different thresholds.

Model	θ_0	P	R	F_1
SapBERT	0.94	0.302	0.268	0.284
CODER	0.86	0.071	0.401	0.121
CODER++	0.70	**0.732**	**0.576**	**0.644**

Table 3: Results for CODER, SapBERT, and CODER++ on term clustering evaluation in UMLS 2020 AA.

with similar meanings, while achieving much better precision for most thresholds, which indicates a significant improvement in distinguishing terms with different meanings.

5.3 Case Study

We compute similarities for the same term pairs as in Section 3.4 using CODER++, and the results are shown in the upper part of Table 1. It suggests that CODER++ has relatively low similarities on negative term pairs and reduces the FP rate. To check if CODER++ maintains high similarities for positive term pairs, we sample some positive terms pairs and compute the similarities, which are shown in the lower part of Table 1. We observe that CODER++ has moderately high similarities for positive term pairs, which suggests CODER++ reserves the ability to normalize terms with similar meanings.

In conclusion, our dynamic sampling strategy significantly decreases similarities in negative term pairs, while mildly decreasing similarities in positive pairs. The results indicate the efficacy of our dynamic sampling strategy in pretraining.

5.4 Zero-shot Term Normalization

We evaluate CODER++ with zero-shot term normalization on BC5CDR (Li et al., 2016), results are shown in Table 4. CODER++ achieves better performance than CODER and comparable

Model	BC5CDR-d	BC5CDR-c
SapBERT	93.5, 96.0	96.5, 98.2
CODER	92.2, 94.7	95.1, 97.2
CODER++	92.2, 94.9	96.5, 97.9

Table 4: Acc@1 and Acc@5 on BC5CDR for CODER, SapBERT, and CODER++.

Setting	θ_0	P	R	F_1
CODER	0.88	0.273	0.310	0.290
(a)	0.76	0.482	0.289	0.361
(b)	0.74	0.667	0.517	0.583
(c)	0.68	0.830	0.659	**0.735**

Table 5: Ablation study on sampling strategies with $\mathcal{D}_s$ term clustering.

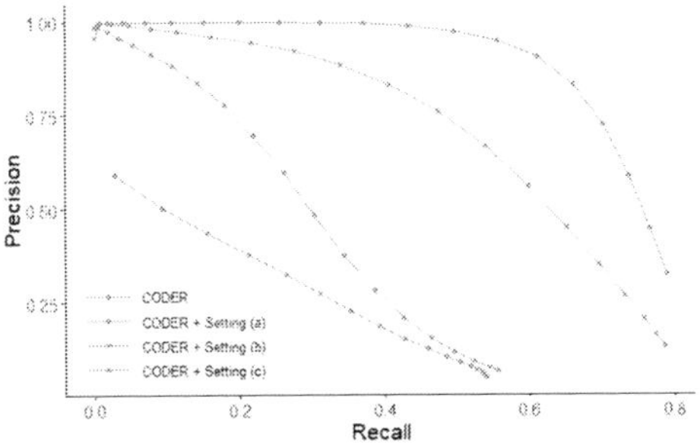

Figure 3: Ablation study on sampling strategies with $\mathcal{D}_s$ term clustering under different thresholds.

performance with SapBERT, which shows that CODER++ generalizes well and reserves the ability to normalize terms with different names.

5.5 Ablation Study

Here we conduct ablation studies on sampling strategies. Ablation studies are based on a sampled subset of UMLS, which consists of 500K terms (denote as $\mathcal{D}_s$). We train models with different settings on $\mathcal{D}_s$ respectively, then use each model to perform clustering evaluation on it:
Setting (a): $k = 1, m = 30$, do not update Faiss.
Setting (b): $k = m = 30$, do not update Faiss.
Setting (c): $k = m = 30$, update Faiss index every epoch (i.e. proposed CODER++).

Figure 3 displays results for thresholds ranging from 0.6 to 0.98, and Table 5 lists the best performances among those thresholds of each model. Setting (a) has much higher precision than the original CODER in all thresholds, which indicates hard negative samples do improve the ability to differentiate negative term pairs. Setting (b) has higher precision and recall than setting (a), especially recall, which indicates simultaneously using positive and negative samples reserves the ability of clustering similar terms while achieving a better capability of differentiating terms. Setting (c) has higher precision than setting (b), which indicates dynamic negative samples greatly enhance the ability to differentiate negative term pairs. The negative sampling under setting (b) is static, the model can easily overfit these samples; while setting (c) will provide new hard negative samples based on the current model. The result is quite intuitive since dynamic negative samples improve precision and recall simultaneously along with all thresholds. In conclusion, dynamic negative sampling with balanced positive sampling is the setting that performs best and we use it for training CODER++.

6 Conclusions

We propose CODER++, a fine-grained biomedical term representation, which benefits from our dynamic sampling strategy that provides hard positive and negative pairs. We propose an automatic large-scale clustering evaluation algorithm. Through a combination of automatic evaluation and the case study, we find CODER++ greatly outperforms CODER and SapBERT on UMLS term clustering and has a much better ability to distinguish different concepts with similar term names. The effectiveness of our dynamic sampling strategy is also proved through an ablation study. Our work can be used for automatic term clustering or recommend candidate similar terms for experts and crowdsourcing participants in human term clustering. Our work also suggests that biomedical term embedding models such as CODER can be further pretrained by focusing on specific information.

Acknowledgments

This work was supported by the National Natural Science Foundation of China (Grant No. 12171270), the Natural Science Foundation of Beijing Municipality (Grant No. Z190024), and the International Digitial Economy Academy.

References

Olivier Bodenreider. 2004. The unified medical language system (umls): integrating biomedical terminology. *Nucleic acids research*, 32(suppl_1):D267–D270.

Jeff Johnson, Matthijs Douze, and Hervé Jégou. 2019.

Billion-scale similarity search with GPUs. *IEEE Transactions on Big Data*, 7(3):535–547.

Stanley Kok and Pedro Domingos. 2008. Extracting semantic networks from text via relational clustering. In *Joint European Conference on Machine Learning and Knowledge Discovery in Databases*, pages 624–639. Springer.

David D Lewis and W Bruce Croft. 1989. Term clustering of syntactic phrases. In *Proceedings of the 13th annual international ACM SIGIR conference on Research and development in information retrieval*, pages 385–404.

Jiao Li, Yueping Sun, Robin J. Johnson, Daniela Sciaky, Chih-Hsuan Wei, Robert Leaman, Allan Peter Davis, Carolyn J. Mattingly, Thomas C. Wiegers, and Zhiyong Lu. 2016. BioCreative V CDR task corpus: a resource for chemical disease relation extraction. *Database*, 2016. Baw068.

Dekang Lin. 1998. Automatic retrieval and clustering of similar words. In *Proceedings of the 36th Annual Meeting of the Association for Computational Linguistics and 17th International Conference on Computational Linguistics - Volume 2*, ACL '98/COLING '98, page 768–774, USA. Association for Computational Linguistics.

Fangyu Liu, Ehsan Shareghi, Zaiqiao Meng, Marco Basaldella, and Nigel Collier. 2021. Self-alignment pretraining for biomedical entity representations. In *Proceedings of the 2021 Conference of the North American Chapter of the Association for Computational Linguistics: Human Language Technologies*, pages 4228–4238.

Ilya Loshchilov and Frank Hutter. 2017. Fixing weight decay regularization in adam. *ArXiv*, abs/1711.05101.

Nhung TH Nguyen, Makoto Miwa, Yoshimasa Tsuruoka, and Satoshi Tojo. 2015. Identifying synonymy between relational phrases using word embeddings. *Journal of Biomedical Informatics*, 56:94–102.

Vinh Nguyen, Hong Yung Yip, and Olivier Bodenreider. 2021. Biomedical vocabulary alignment at scale in the umls metathesaurus. In *Proceedings of the Web Conference 2021*, WWW '21, page 2672–2683, New York, NY, USA. Association for Computing Machinery.

David N Nicholson and Casey S Greene. 2020. Constructing knowledge graphs and their biomedical applications. *Computational and structural biotechnology journal*, 18:1414–1428.

Xun Wang, Xintong Han, Weilin Huang, Dengke Dong, and Matthew R Scott. 2019. Multi-similarity loss with general pair weighting for deep metric learning. In *Proceedings of the IEEE Conference on Computer Vision and Pattern Recognition*, pages 5022–5030.

Sheng Yu, Zheng Yuan, Jun Xia, Shengxuan Luo, Huaiyuan Ying, Sihang Zeng, Jingyi Ren, Hongyi Yuan, Zhengyun Zhao, Yucong Lin, Keming Lu, Jing Wang, Yutao Xie, and Heung-Yeung Shum. 2022. Bios: An algorithmically generated biomedical knowledge graph.

Zheng Yuan, Zhengyun Zhao, Haixia Sun, Jiao Li, Fei Wang, and Sheng Yu. 2022. Coder: Knowledge-infused cross-lingual medical term embedding for term normalization. *Journal of Biomedical Informatics*, page 103983.

BioBART: Pretraining and Evaluation of A Biomedical Generative Language Model

Hongyi Yuan[1] * **Zheng Yuan**[1] * **Ruyi Gan**[2] **Jiaxing Zhang**[2] **Yutao Xie**[2] **Sheng Yu**[1] †
[1]Tsinghua University [2]International Digital Economy Academy
{yuanhy20,yuanz17}@mails.tsinghua.edu.cn
{gaunruyi,zhangjiaxing,xieyutao}@idea.edu.cn
syu@tsinghua.edu.cn

Abstract

Pretrained language models have served as important backbones for natural language processing. Recently, in-domain pretraining has been shown to benefit various domain-specific downstream tasks. In the biomedical domain, natural language generation (NLG) tasks are of critical importance, while understudied. Approaching natural language understanding (NLU) tasks as NLG achieves satisfying performance in the general domain through constrained language generation or language prompting.We emphasize the lack of in-domain generative language models and the unsystematic generative downstream benchmarks in the biomedical domain, hindering the development of the research community. In this work, we introduce the generative language model BioBART that adapts BART to the biomedical domain. We collate various biomedical language generation tasks including dialogue, summarization, entity linking, and named entity recognition. BioBART pretrained on PubMed abstracts has enhanced performance compared to BART and set strong baselines on several tasks. Furthermore, we conduct ablation studies on the pretraining tasks for BioBART and find that sentence permutation has negative effects on downstream tasks.

1 Introduction

Since the advent of ELMo (Peters et al., 2018) and BERT (Devlin et al., 2019), the new pretrain-then-finetune paradigm has brought great performance improvement and dominated the methodology research of the natural language processing (NLP) field. Previous research has illustrated that pretraining language models on the domain-specific corpora can improve the model performance on domain-specific tasks further (Gururangan et al., 2020). With the large-scale publicly accessible corpora from PubMed, researchers have already proposed biomedical domain pretrained language models such as BioBERT (Lee et al., 2020) and PubMedBERT (Gu et al., 2022) to aid the later research.

Natural language generation (NLG) tasks such as dialogue system (Chao et al., 2017) and question answering (Jin et al., 2022) are of critical importance for the biomedical artificial intelligence research, and there is also a trend to approach natural language understanding as NLG tasks in the general domain (Sun et al., 2021; Yan et al., 2021). For example, an entity retrieval task can be solved by constrained natural language generation (Cao et al., 2021). However, there exist two gaps in the research of the biomedical NLG. On the one hand, the architectures of the biomedical pretrained language models are almost all encoder-only transformers. Such architecture is incapable of generating natural languages auto-regressively. A decoder is necessary for language generation (Liu and Lapata, 2019). On the other hand, there are very few in-domain generative language models for biomedicine (Phan et al., 2021). Models pretrained on biomedical corpora may further enhance the performance of current biomedical NLG methods.

To bridge the gaps mentioned above, we propose a biomedical auto-regressive generative language model, BioBART, pretrained on the biomedical corpora. In our work, we adopt BART (Bidirectional and Auto-Regressive Transformers), a generative pretrained language model which achieves state-of-the-art results on different NLG tasks in the general domain (Lewis et al., 2020a). We continuously pretrain BART on PubMed abstracts to achieve biomedical domain adaption only using the text-infilling task. We also collate and evaluate BioBART on the existing biomedical NLG tasks. The in-domain BioBART outperforms BART model and sets strong baselines for several NLG tasks.

The main contributions of our work are summa-

* Contributed equally.
† Corresponded author.

Proceedings of the BioNLP 2022 workshop, Dublin, Ireland, pages 97–109
May 26, 2022.

rized as follows[1]:

1. In aid of the research concerning the biomedical NLG tasks, we collate existing biomedical NLG tasks along with corresponding data and experimental settings. The archived biomedical tasks will be released.

2. We further analyze the influence of the pretraining task of sentence permutation in BART, and we find it brings degradation on the biomedical NLG tasks.

3. We evaluate our BioBART models on various NLG tasks and demonstrate the superb performance over BART. We will release the codes and weights to help reproduce our results.

2 Related Work

2.1 Auto-regressive Language Model

Most of the prestigious language models such as BERT, RoBERTa (Liu et al., 2019) are auto-encoding transformers. The encoder-only architecture prevents the direct implementation of the seq2seq language generation. Several generative auto-regressive language models are proposed to mitigate the problem. The serial GPT models (Radford and Narasimhan, 2018; Radford et al., 2019; Brown et al., 2020) adopt the decoder-only transformer architecture which is a left-to-right language model. They pretrain the models by auto-regressively predicting the upcoming word of sentences. UniLM1 (Dong et al., 2019) and UniLM2 (Bao et al., 2020) implement attention masks to the transformer encoder to achieve unidirectional language modeling. They pretrain their models with a mixture of masked language modeling and auto-regressive language generation. T5 (Raffel et al., 2020) and BART (Lewis et al., 2020a) apply the full transformer architecture, the encoder is used for input sequence encoding and the decoder is used for language generation. T5 and BART are both pretrained by denoising the corrupted corpora. Such models achieve many state-of-the-art results on various NLG tasks and some NLU tasks.

2.2 Biomedical Domain Pretraining

Existing work has shown that pretraining the language models on the domain-specific corpora can bring better model transferability on the corresponding downstream tasks (Gururangan et al., 2020). There are endeavors to adapt language models to the specific domain. BioBERT (Lee et al., 2020) pretrained BERT model using biomedical corpora from PubMed abstracts and PubMed Central (PMC) full-text articles. BlueBERT (Peng et al., 2020) and clinicalBERT (Alsentzer et al., 2019) add electronic medical record (EMR) corpora from MIMIC-III (Johnson et al., 2016) to the pretraining data. Instead of continuous training from the general BERT checkpoint, SciBERT (Beltagy et al., 2019) and PubMedBERT (Gu et al., 2022) are trained from scratch using scientific papers from Semantic Scholar (Ammar et al., 2018) and PubMed articles respectively. (Shin et al., 2020) releases BioMegatron, a larger-size BERT-style language model pretrained on PubMed abstracts, PMC and MIMIC-III. The aforementioned work all use the model architecture of BERT. Other researchers are exploring different language models.

BioELMo (Jin et al., 2019) is pretrained on biomedical corpora based on stacked bidirectional LSTM language model ELMo (Peters et al., 2018). BioELECTRA (Kanakarajan et al., 2021) applies an adversarial training scheme consisting of a discriminator and a generator. They use PubMed abstracts and PMC articles as in-domain pretraining corpora. BioMed-RoBERTa (Gururangan et al., 2020) is initialized from RoBERTa (Liu et al., 2019), with additional training on the scientific papers from Semantic Scholar. Bio-lm (Lewis et al., 2020b) is pretrained on data from PubMed, PMC, and MIMIC-III based on the RoBERTa model. KeBioLM (Yuan et al., 2021) uses Entity as Experts (Févry et al., 2020) model to inject biomedical entity knowledge into the language model, starting from the weights of PubMedBERT. Coder (Yuan et al., 2022b) and SapBERT (Liu et al., 2021) take advantage of the synonyms resource from biomedical knowledge base UMLS (Bodenreider, 2004) and enhance the model with entity knowledge by contrastive pretraining.

Due to the nature of model architecture, encoder-only language models have limited performance on the NLG tasks, such as summarization and question answering. In recent research, SciFive (Phan et al., 2021) is proposed for biomedical NLP tasks. SciFive is pretrained on PubMed abstracts and PMC articles based on T5 architecture. While T5 is avail-

[1]Our codes and pretrained checkpoints can be found at `https://github.com/GanjinZero/BioBART`.

able for NLG tasks, SciFive is focused on evaluating NLU tasks. Compared to SciFive, we choose to use BART as our model backbone and evaluate more on NLG tasks to leverage the power of decoders.

2.3 Biomedical Natural Language Generation

In the biomedical domain, most of the NLP tasks are natural language understanding (NLU) tasks. There are well-archived benchmarks for the evaluation of biomedical NLU, such as BLUE (Gu et al., 2022) and CBLUE (Zhang et al., 2021). NLG tasks are relatively less studied. (Ju et al., 2020) collects the patients and doctors' dialogues and forms a benchmark for Covid-19 related dialogue system. (Ben Abacha et al., 2021) is an annual biomedical NLP competition containing NLG tasks such as medical question (or answer) summarization and figure captions.

Moreover, with the success of GPT-3, there is a novel trend that unifies all the NLP tasks as NLG tasks (McCann et al., 2018; Brown et al., 2020). The traditional NLU tasks can be approached by constrained language generation. Much attention is paid on the NLG methods recently. In the biomedical domain, entities are of primary concern. GENRE (Cao et al., 2021), Yuan et al. (2022a) and BARTNER (Yan et al., 2021) reach the new state-of-the-art by auto-regressive language model on entity linking and named entity recognition tasks. Such methods can be adapted to the biomedical domain.

3 Biomedical Domain Pretraining

BART is a sequence-to-sequence model with a bi-directional encoder and a left-to-right auto-regressive decoder. The model architecture is consistent with the Transformers (Vaswani et al., 2017) except for changing the ReLU activation functions to GeLUs (Hendrycks and Gimpel, 2016). BART is pretrained by denoising the corrupted input documents. The work ablates five different types of corruption noise: text masking, text deletion, text infilling, sentence permutation, and document rotation. As a result, the pretraining documents are corrupted in two ways: 1) **Text Infilling**: For each document, a number of token spans are sampled, and each sample span is replaced with a single mask token. 2) **Sentence Permutation**: A document is split into sentences and sentences are shuffled in random orders. The pretraining objective is to minimize the negative log-likelihood of the original documents.

Prior work has shown that continuous-pretrained models can get competitive results compared with those trained from scratch (Gu et al., 2022). In our work, we continuously pretrain BART on the biomedical domain corpora. We revisit the methods to corrupt input texts. BART keeps the sentence permutation noise because of the significant performance gain on the summarization task, although this noise may lead to slight degradation on other tasks. We run further ablation studies on various biomedical NLG tasks. We show that the model pretrained without sentence permutation has better performance. Further details are listed in Section 5.5. Therefore we only implement the text infilling task to corrupt input texts for pretraining BioBART.

4 Generative Downstream Task

In this section, we introduce the generative downstream tasks in the biomedical domain. We will conduct experiments on these tasks to illustrate the performance of the domain-specific BioBART.

4.1 Dialogue System

A medical dialogue system aims to imitate the human doctor to communicate with human patients in a natural way. Based on the BART-style model, the patients' primitive descriptions and dialogue histories are used as inputs to the model, then the model auto-regressively generates the replies as outputs. The task is trained and evaluated in a sequence-to-sequence fashion.

4.2 Abstractive Summarization

Summarization is a classical NLP task. It is important for healthcare to concisely summarize knowledge-rich biomedical documents. Technically, there are abstractive and extractive approaches to generate better summaries. With the help of large pretrained language models, abstractive summarization methods outperform extractive methods in summary diversity and conciseness (Zhang et al., 2020a; Dou et al., 2021). The abstractive summarization is naturally an NLG task. We follow the BART (Lewis et al., 2020a) work and evaluate our BioBART on the biomedical summarization tasks in the same fashion. The input documents are encoded by the model encoder and the summaries are generated by the decoder auto-regressively.

4.3 Entity Linking

Entity linking is a task that maps entity mentions in texts to its standard entity concepts. Traditional entity linking methods use language models to encode entity concepts from knowledge bases(e.g. UMLS) and mentions into the same dense space and disambiguate mentions by vector similarity. The large memory footprint requirements and difficult model training hinder the development of such methods. Cao et al. (2021) proposes GENRE which uses generative language models to disambiguate entity mentions by auto-regressively generating the standard concept names conditioned on the inputs. (Yuan et al., 2022a) achieves state-of-the-art entity linking performance on various biomedical entity linking datasets by generative methods. We include this leading-edge method to show the superior performance of BioBART.

4.4 Named Entity Recognition

Named entity recognition (NER) is a critical task in the biomedical NLP community which extracts biomedical-related entities from texts. Nested and discontinuous entities widely exist in biomedical papers and EMR due to the multi-granularity semantic meanings and complex syntax structures (Yuan et al., 2020). Well-used sequential labelling framework in NER (Lample et al., 2016) is not directly fitted for nested and discontinuous NER (Finkel and Manning, 2009). Yan et al. (2021) propose BARTNER to model nested and discontinuous NER into seq2seq task by inputting sentences and outputting entities with their entity types one by one. The generative approach of BARTNER achieves state-of-the-art performance on nested and discontinuous NER datasets, and we will use it to evaluate our proposed BioBART can further enhance the performance.

5 Experiments

5.1 Pretraining

Pretraining Corpora There are two main sources of biomedical corpora: PubMed abstracts, PMC articles. In the prior work (Gu et al., 2022), training on both corpora surprisingly leads to a slight degradation in performance compared to solely training on PubMed abstracts. Therefore, we only use PubMed abstracts as the pretraining corpora. The corpora contain about 41 GB of biomedical research paper abstracts on PubMed.

Pretraining Setup We continuously pretrain both large and base versions of BART for 120k steps with a batch size of 2560. We use the same vocabulary as BART to tokenize the texts. Although the input length limitation of BART is 1024, the tokenized PubMed abstracts rarely exceed 512. Therefore, for the sake of training efficiency, we truncate all the input texts to 512 maximum length. We mask 30% of the input tokens and the masked span length is determined by sampling from a Poisson distribution ($\lambda = 3$) as used in BART. We use a learning rate scheduler of 0.02 warm-up ratio and linear decay. The learning rate is set to 1e-4. We train the base version of BioBART on 2 DGX with 16 40GB A100 GPUs for about 100 hours and the large version of BioBART on the same devices for 168 hours with the help of the open-resource framework DeepSpeed (Rajbhandari et al., 2020).

5.2 Dataset for Downstream Task

5.2.1 Dialogue System

CovidDialog (Ju et al., 2020) Concerning the widespread Coronavirus disease 2019 (COVID-19) pandemic, the CovidDialog dataset is proposed to facilitate the development of dialogue system providing COVID-related consultations to people. The dataset is collected from online healthcare forums. It contains 603 consultations about COVID-19 and other related pneumonia, having 1232 utterances in total. Each consultation starts with a description related to patients' medical conditions, then followed the conversation between a doctor and a patient.

5.2.2 Abstractive Summarization

iCliniq, HealthCareMagic Both datasets are extracted from MedDialog (Zeng et al., 2020) dataset, collected from the online healthcare platform. iCliniq contains 31,062 samples and HealthCareMagic contains 226,405 samples. Each sample is comprised of a summary and corresponding dialogues between a patient and a doctor. HealthCareMagic's summaries are more abstractive and are written in a formal style, unlike iCliniq's patient-written summaries. We follow the previous work (Mrini et al., 2021) for training, developing, and testing data separations of both datasets.

MeQSum (Ben Abacha and Demner-Fushman, 2019) The dataset is created for better medical question summarization because the original patients' questions are verbose, causing difficulty for the question-answering system. The dataset contains

Task	Dataset	Train	Dev	Test	Dataset	Train	Dev	Test	Metric
Dialogue	CovidDialog	490	63	61					Rouge,BERTscore, BLEU
Summarization	MeQSum	500	-	500	MEDIQA-ANS	38,166	174	552	Rouge, BERTscore
	iCliniq	24,851	3,105	3,108	MEDIQA-QS	1,000	50	100	
	HealthCareMagic	181,122	22,641	22,642	MEDIQA-MAS	1,104	50	80	
Entity Linking	MedMentions	122,241	40,884	40,157	NCBI	5,784	787	960	Recall@1,@5
	BC5CDR	9,285	9,515	9,654	COMETA	13,489	2,176	4,350	
	AskAPatients	16,826	1,663	1,712					
NER	ShARe13	5,146	669	5,333	ShARe14	10,380	771	7,922	Entity-level F1 score
	CADEC	4,430	898	990	GENIA	50,509	-	5,506	

Table 1: The statistics of the datasets for biomedical generative tasks. The counts for NER are entity counts.

	Covid19-Dialogue				
Model	Rouge-1	Rouge-2	Rouge-L	BLEU	BERTscore
BART BASE	27.24	12.31	25.66	10.36	0.852
BioBART BASE	28.14	12.77	26.32	11.40	0.849
BART LARGE	**29.02**	12.08	26.93	10.96	**0.852**
BioBART LARGE	28.81	**13.79**	**26.96**	**12.05**	0.850
State-of-the-art	-	-	-	7.60	-
Source	-	-	-	(Zhou et al., 2021)	-

Table 2: The main results on Dialogue System task.

1000 patients' health questions selected from a collection distributed by the U.S. National Library of Medicine (Kilicoglu et al., 2018). Each question is annotated with a question summarization by medical experts.

MEDIQA-ANS (Savery et al., 2020) When feeling discomfort, people may turn to the internet for the answers to their medical questions. The raw searching result may be obscure for even medical experts. The dataset is proposed to emphasize the need for a medical answer summarization system in aid of better understanding biomedical materials. It consists of 156 health questions, corresponding answers to these questions, and expert-created summaries (both abstractive and extractive) of these answers. Following the paper, we use BioASQ (Tsatsaronis et al., 2015) to construct training data, MedInfo (Abacha et al., 2019) for validation, and the whole MEDIQA-ANS dataset for testing.

MEDIQA-QS, MEDIQA-MAS Both datasets are derived from the MEDIQA 2021 Tasks (Ben Abacha et al., 2021). MEDIQA-QS dataset aims to incentivize the development of new summarization approaches that address specifically the challenges of long and complex health questions. The dataset provides the validation and test sets, and MeQSum dataset is used as the training set. MEDIQA-MAS aims to prompt research that simultaneously aggregates and summarize the different relevant answers to a medical question. This dataset provides the validation and test sets, and MEDIQA-ANS dataset comprises the training set.

5.2.3 Entity Linking

MedMentions (Mohan and Li, 2019) MedMentions is a large-scale biomedical entity recognition dataset. The commonly used St21pv subset contains 4,392 PubMed abstracts, and over 350,000 mentions are linked to concepts of 21 selected semantic types in UMLS (Bodenreider, 2004).

BC5CDR (Li et al., 2016) BC5CDR is a benchmark for biomedical entity linking. 1500 PubMed article abstracts are annotated with 4409 chemicals, 5818 diseases entities, and 3116 chemical-disease interactions. MeSH ontology, a subset of UMLS is used to annotate entities. We follow most recent work (Angell et al., 2021; Varma et al., 2021) for data pre-processing.

NCBI (Doğan et al., 2014) The dataset is built from 793 PubMed abstracts. It consists of 6892 annotated disease mentions of 790 unique disease concepts. The annotators label all the mentions to concepts in MEDIC ontology (Davis et al., 2012). MEDIC is a medical dictionary that merges the diseases concepts, synonyms, and definitions in MeSH and OMIM and is composed of 9700 unique diseases. We follow BioSyn (Sung et al., 2020) to process data and construct dataset splits.

COMETA (Basaldella et al., 2020) COMETA is derived from the online publicly available and

Model	iCliniq Rouge-1/2/L	iCliniq BERTscore	HealthCareMagic Rouge-1/2/L	HealthCareMagic BERTscore	MEDIQA-QS Rouge-1/2/L	MEDIQA-QS BERTscore
BART BASE	61.43/48.68/**59.71**	**0.941**	46.81/26.19/44.34	0.918	28.82/10.99/26.99	0.896
BioBART BASE	61.07/48.47/59.42	**0.941**	46.67/26.03/44.11	0.918	30.12/11.28/27.44	0.898
BART LARGE	59.87/47.01/58.12	0.938	**47.24/26.54/44.68**	**0.919**	29.97/10.64/28.41	0.901
BioBART LARGE	60.32/47.98/58.69	0.940	46.54/26.14/44.23	**0.919**	31.97/12.39/29.70	**0.903**
State-of-the-art	**62.3/48.7**/58.5	-	46.9/24.8/43.2	-	**35.14/16.08/31.31**	-
Source	(Mrini et al., 2021)		(Mrini et al., 2021)		(Ben Abacha et al., 2021)	

Model	MEDIQA-MAS Rouge-1/2/L	MEDIQA-MAS BERTscore	MEDIQA-ANS(Pages) Rouge-1/2/L	MEDIQA-ANS(Pages) BERTscore	MeQSum Rouge-1/2/L	MeQSum BERTscore
BART BASE	31.63/9.98/27.85	0.859	19.10/6.77/16.90	0.851	52.93/35.79/50.46	0.927
BioBART BASE	**32.90**/11.28/**29.26**	**0.861**	18.97/7.46/16.77	0.850	53.75/36.50/51.27	0.929
BART LARGE	29.32/9.00/26.14	0.857	21.52/9.31/19.15	0.853	53.68/36.80/51.05	0.928
BioBART LARGE	30.60/10.37/27.04	**0.861**	21.58/**9.34/19.18**	**0.857**	**55.61/38.11/53.15**	**0.933**
State-of-the-art	32.15/**16.21**/19.10	-	**23.07**/ 5.41/15.35	-	54.5/37.9/50.2	-
Source	(Ben Abacha et al., 2021)		(Laskar et al., 2021)		(Mrini et al., 2021)	

Table 3: The main results on Summarization tasks.

Model	MedMentions Recall@1/@5	BC5CDR Recall@1/@5	NCBI Recall@1/@5	COMETA Recall@1/@5	AAP Recall@1/@5
BART BASE	69.77/84.59	91.56/94.89	88.54/95.31	78.34/87.40	86.37/94.29
BioBART BASE	71.15/**86.22**	93.01/95.59	89.27/95.31	79.63/88.64	87.51/94.92
BART LARGE	71.49/84.95	92.48/95.26	90.21/95.52	80.70/88.65	88.79/**96.59**
BioBART LARGE	71.78/85.42	**93.26/95.74**	89.90/**95.63**	**81.77/88.87**	**89.40**/95.76
State-of-the-art	**74.6**/ -	91.9/ -	**92.4**/ -	80.1/ -	89.0/ -
Source	(Varma et al., 2021)	(Varma et al., 2021)	(Lai et al., 2021)	(Lai et al., 2021)	(Liu et al., 2021)

Table 4: The main results on Entity Linking tasks.

Model	ShARe13 F1	ShARe14 F1	CADEC F1	GENIA F1
BART BASE	76.63	77.87	68.37	78.06
BioBART BASE	78.78	79.17	68.39	78.43
BART LARGE	79.69	80.34	70.64	78.93
BioBART LARGE	80.75	80.41	70.53	79.93
State-of-the-art	**82.52**	**81.75**	**73.21**	**81.39**
Source		(Li et al., 2021)		

Table 5: The main result on NER tasks.

anonymous health discussion on Reddit. It consists of 20k English biomedical entity mentions expert-annotated with concepts from SNOMED CT. We use the "stratified (general)" split and follow the training and evaluation procedures of SapBert (Liu et al., 2021) and ResCNN (Lai et al., 2021).

AskAPatient (Limsopatham and Collier, 2016) It contains 8,662 phrases from social media. Each phrase can be mapped to one of the 1,036 medical concepts from SNOMED-CT and AMT (the Australian Medicines Terminology). The samples in AskAPatient do not include contextual information. We follow Sung et al. (2020) and Limsopatham and Collier (2016) for data pre-processing and apply the 10-fold evaluation protocol.

5.2.4 Named Entity Recognition

ShARe13, ShARe14, CADEC These three datasets annotate discontinuous adverse drug events entities. The main difference is the annotated data of ShARe tasks (Pradhan et al., 2013; Mowery et al., 2014) comes from MIMIC-II, and CADEC (Karimi et al., 2015) comes from social media. There is only one entity type for these datasets. We follow Yan et al. (2021) for dataset preprocess.

GENIA (Kim et al., 2003) GENIA annotates 2000 MEDLINE abstracts with biological entities. Entities can be nested with others. We follow (Lin et al., 2019) to combine fine-grained entity types into 5 coarse-grained entity types and to construct dataset splits.

All the aforementioned datasets are in English. The statistical overview of the aforementioned datasets is listed in Table 1.

5.3 Fine-tuning details

Dialogue We use BioBART as the dialogue system model. The dialogue history is fed into the encoder and the decoder generates the response auto-regressively. We apply the negative log-likelihood function as the training objective with respect to the reference dialogue response. We fine-tune the model with learning rate 5e-5 for the base version and 1e-5 for the large version for 20 epochs. We run evaluations on the validation set at the end of each epoch and use the checkpoint with the best validation performance for testing. During inference, we use beam search of size 5 to sample responses from the model's outputs. We use Rouge-1/2/L (Lin, 2004), BLEU (Papineni et al., 2002) and BERTscore (Zhang et al., 2020b) as our evaluation metrics. RoBERTa-large (Liu et al., 2019) is used as scorer in BERTscore.

Summarization Similarly, for summarization, the encoder takes the documents as input, and the decoder generates the corresponding summarizations. We minimize the log-likelihood objective to fine-tune the model and apply beam search for inference. Across different summarization datasets, the beam size is set to 5 and we use no length penalty. We fine-tune the model with learning rate 5e-5 for the base version and 1e-5 for the large version for 6 epochs. We run evaluations on the validation set at the end of each epoch and use the checkpoint with the best validation performance for testing. We apply the commonly used Rouge-1/2/L and BERTscore for evaluation metrics. The large version of RoBERTa is used as the scorer in BERTscore.

Entity Linking We follow the method and experimental settings in Yuan et al. (2022a) to implement the generative model for biomedical entity linking tasks. Knowledge-base guided pre-training in Yuan et al. (2022a) has not been applied. The documents with the positions of mentions marked are fed into the encoder and the decoder outputs the corresponding synonyms in the knowledge base directly. We use the top1 and top5 recall (Recall@1 and Recall@5) as the evaluation metrics.

NER We use BARTNER (Yan et al., 2021) as our model. The target type for BARTNER is *word* (i.e. output first BPE of each word in entities). We use the parameters selected by Yan et al. (2021) for all pretrained models and fine-tune for 30 epochs. Entity-level F1 is used as the metric.

5.4 Main Result

In this section, we present the base and large version of BioBART on various generation tasks. We compare our in-domain BioBART with BART to illustrate the effectiveness of domain adaption. We also compare with the existing state-of-the-art results on each dataset to shed light on the superior performance of BioBART. The experimental results are shown in Table 2-5. The best and the second-best scores are highlighted with bold numbers and underlines respectively.

Dialogue We evaluate biomedical dialogue response generation on CovidDialog. For both base and large version, BioBART shows improvement on the automatic metric Rouge. The large BioBART outperforms BART by 1.71 on Rouge-2 and 0.03 on Rouge-L . Our evaluations surpasses the current state-of-the-art on BLEU score by 4.45.

Summarization We present broad experimental results on biomedical summarization datasets. From Table 3, BioBART has competitive or even superior performance on the task. Except for iCliniq and HealthCareMagic, we see consistent improvement on different datasets for both sizes of BioBART. For MeQSum, BioBART large exceeds BART large for 1.93/1.31/2.1 on Rouge-1/2/L and even outperforms the current state-of-the-art. The possible reason that biomedical in-domain pretraining fails on iCliniq and HealthCareMagic is that both datasets are built upon a clinical corpus. There still exists a domain-shifting problem for BioBART pretrained on biomedical scientific articles from PubMed.

On dialogue and summarization tasks, there are minor changes in BERTscore for different models. This is possible because the metric is calculated by other pretranined language models. The implemented RoBERTa may suffer from biomedical domain-shifting and cannot quantify the model performance accurately.

Entity Linking The results on biomedical entity linking tasks are shown in Table 4. For all the tasks, models finetuned based on BioBART have better performance. On AAP, BC5CDR, and COMETA, our results outperform the current discriminative state-of-the-art methods by 0.4, 1.67, and 1.36 points of Recall@1 respectively.

NER The performance improvement of BioBART on ShARe13, ShARe14, and GENIA is significant, while the increase on CADEC is mediocre. For the large models, BioBART improves entity-level F1 scores for 1.06 and 1 on ShARe13 and GENIA datasets. There are promising results for generative biomedical NER methods, while the gap

	CovidDialogue		MeQSum		MEDIQA-MAS	
	Rouge-2/L	BLEU	Rouge-2/L	BERTscore	Rouge-2/L	BERTscore
BART BASE	**12.31**/25.66	10.36	35.79/50.46	0.927	9.98/**27.85**	0.859
w/ TI & SP	10.90/25.46	10.46	34.93/49.28	0.926	9.04/27.43	0.859
w/ TI	11.81/**25.79**	**12.79**	**37.14/51.71**	**0.929**	**10.66**/27.65	**0.862**
	MedMentions		COMETA		ShARe13	CADEC
	Recall@1	Recall@5	Recall@1	Recall@5	F1	F1
BART BASE	69.77	84.59	78.34	87.40	76.63	68.37
w/ TI & SP	70.76	**85.69**	79.31	87.93	77.01	67.98
w/ TI	**70.82**	85.68	**79.43**	**88.28**	**78.03**	**69.22**

Table 6: Ablation study for pretraining tasks. TI is short for Text Infilling and SP is short for Sentence Permutation.

Context(Document)	**BART Response(Summary)**	**BioBART Response(Summary)**
Description: I have type 1 diabetes that I try hard to control, but i struggle with. Are my chances high for poor outcomes with the virus? I'm in such a panic about it.	In brief: diabetes. Your risk of complications with covid-19 is very low. If you are well controlled your risk should not be elevated. If your diabetes is uncontrolled you should not have a particularly high risk.	Diabetes and Covid-19. If you are a type 1 diabetic, you are at high risk for poor outcomes with the virus. Your risk is very high. You should be aware of your risk factors and monitor your glycemic control closely.
SUBJECT: treatment for dry mouth caused by necessary medicine MESSAGE: My provider can't help (I asked.) I am intolerant of all the "sugar alcohols" such as maltilol, ...	What are the treatments for dry mouth caused by sugar alcohols?	What are the treatments for dry mouth?
Are there any new treatments for orthostatic tremor? ... Deep brain stimulation is sometimes done when drugs cannot control a severe, disabling tremor. Sometimes essential tremors or tremors due to Parkinson disease or another disorder require such treatment. ... There is no cure for most forms of tremor, but there are treatments to help manage symptoms. ...	There is no cure for tremor, but there are treatments to help manage the symptoms.	There is no cure for tremor, but there are treatments to help manage the symptoms. The most common ... help relieve symptoms. Deep brain stimulation is sometimes done when drugs cannot control a severe, disabling tremor. Sometimes essential tremors or tremors due to Parkinson disease or another disorder require such treatment.

Table 7: Example dialogue and summaries from the fine-tuned BART and BioBART model excepts of showing the references. The key information and differences are highlighted with colors.

with the current state-of-the-art NER method (Li et al., 2021) is still salient.

5.5 Ablation Study on Pretraining Task

In this section, we test on pretraining with or without the sentence permutation task. We pretrain BART base following the same pretraining settings except for reducing the training step to 40k for efficiency. We fine-tuned the pretrained models on the downstream tasks. The ablation results are shown in Table 6.

From the result, it is illustrated that the model pretrained on isolated text infilling task performs the best. The sentence permutation task downgrades the model's performance even for generative summarization and dialogue system tasks.

5.6 Generated example

Here we demonstrate BioBART's performance qualitatively. In Table 7, we present three generative examples on CovidDialog, MeQSum, and MEDIQA-ANS respectively. In the first example, we can see that BART generates an erroneous instruction of the influence of diabetes. BioBART injected with domain knowledge can correctly give the response. In the second, BART misunderstands the document where sugar alcohol is not the cause of dry mouth. BioBART generates an accurate and concise summary. In the final example, the MEDIQA-ANS document is rather long and BART fails to extract complete information (colored in red). From the examples, we can conclude that BioBART has improvements on biomedical common sense and documents understanding.

6 Conclusions

In this work, we pretrain the biomedical domain generative language model BioBART. We also collect various publicly available benchmarks for biomedical generative tasks to prompt future research. Our experimental results show that continuous pretraining on PubMed abstracts helps the model with domain adaption. BioBART shows great improvements on different benchmarks and achieves competitive or superior results over the current state-of-the-art methods. We also release our pretraining and fine-tuning codes to facilitate future research for reproducibility.

We will explore pretraining generative language models 1) on in-domain vocabularies and from scratch, 2) and with clinical corpora such as EMRs in MIMIC-III (Johnson et al., 2016) or PMC-Patients (Zhao et al., 2022) in the future studies.

Acknowledgements

We appreciate three anonymous reviewers for helpful comments. This work was supported by the National Natural Science Foundation of China (Grant No. 12171270), and the Natural Science Foundation of Beijing Municipality (Grant No. Z190024).

References

Asma Ben Abacha, Yassine Mrabet, Mark E. Sharp, Travis R. Goodwin, Sonya E. Shooshan, and Dina Demner-Fushman. 2019. Bridging the gap between consumers' medication questions and trusted answers. *Studies in health technology and informatics*, 264:25–29.

Emily Alsentzer, John Murphy, William Boag, Wei-Hung Weng, Di Jindi, Tristan Naumann, and Matthew McDermott. 2019. Publicly available clinical bert embeddings. In *Proceedings of the 2nd Clinical Natural Language Processing Workshop*, pages 72–78.

Waleed Ammar, Dirk Groeneveld, Chandra Bhagavatula, Iz Beltagy, Miles Crawford, Doug Downey, Jason Dunkelberger, Ahmed Elgohary, Sergey Feldman, Vu A. Ha, Rodney Michael Kinney, Sebastian Kohlmeier, Kyle Lo, Tyler C. Murray, Hsu-Han Ooi, Matthew E. Peters, Joanna L. Power, Sam Skjonsberg, Lucy Lu Wang, Christopher Wilhelm, Zheng Yuan, Madeleine van Zuylen, and Oren Etzioni. 2018. Construction of the literature graph in semantic scholar. In *NAACL*.

Rico Angell, Nicholas Monath, Sunil Mohan, Nishant Yadav, and Andrew McCallum. 2021. Clustering-based inference for biomedical entity linking. In *Proceedings of the 2021 Conference of the North American Chapter of the Association for Computational Linguistics: Human Language Technologies*, pages 2598–2608, Online. Association for Computational Linguistics.

Hangbo Bao, Li Dong, Furu Wei, Wenhui Wang, Nan Yang, Xiaodong Liu, Yu Wang, Jianfeng Gao, Songhao Piao, Ming Zhou, et al. 2020. Unilmv2: Pseudo-masked language models for unified language model pre-training. In *International Conference on Machine Learning*, pages 642–652. PMLR.

Marco Basaldella, Fangyu Liu, Ehsan Shareghi, and Nigel Collier. 2020. COMETA: A corpus for medical entity linking in the social media. In *Proceedings of the 2020 Conference on Empirical Methods in Natural Language Processing (EMNLP)*, pages 3122–3137, Online. Association for Computational Linguistics.

Iz Beltagy, Kyle Lo, and Arman Cohan. 2019. Scibert: A pretrained language model for scientific text. In *EMNLP*.

Asma Ben Abacha and Dina Demner-Fushman. 2019. On the summarization of consumer health questions. In *Proceedings of the 57th Annual Meeting of the Association for Computational Linguistics, ACL 2019, Florence, Italy, July 28th - August 2*.

Asma Ben Abacha, Yassine Mrabet, Yuhao Zhang, Chaitanya Shivade, Curtis Langlotz, and Dina Demner-Fushman. 2021. Overview of the MEDIQA 2021 shared task on summarization in the medical domain. In *Proceedings of the 20th Workshop on Biomedical Language Processing*, pages 74–85, Online. Association for Computational Linguistics.

Olivier Bodenreider. 2004. The unified medical language system (umls): integrating biomedical terminology. *Nucleic acids research*, 32 Database issue:D267–70.

Tom B. Brown, Benjamin Mann, Nick Ryder, Melanie Subbiah, Jared Kaplan, Prafulla Dhariwal, Arvind Neelakantan, Pranav Shyam, Girish Sastry, Amanda Askell, Sandhini Agarwal, Ariel Herbert-Voss, Gretchen Krueger, T. J. Henighan, Rewon Child, Aditya Ramesh, Daniel M. Ziegler, Jeff Wu, Clemens Winter, Christopher Hesse, Mark Chen, Eric Sigler, Mateusz Litwin, Scott Gray, Benjamin Chess, Jack Clark, Christopher Berner, Sam McCandlish, Alec Radford, Ilya Sutskever, and Dario Amodei. 2020. Language models are few-shot learners. *ArXiv*, abs/2005.14165.

Nicola De Cao, Gautier Izacard, Sebastian Riedel, and Fabio Petroni. 2021. Autoregressive entity retrieval. In *International Conference on Learning Representations*.

Hsiao-Tuan Chao, Lucy Liu, and Hugo J Bellen. 2017. Building dialogues between clinical and biomedical research through cross-species collaborations. In *Seminars in cell & developmental biology*, volume 70, pages 49–57. Elsevier.

Allan Peter Davis, Thomas C Wiegers, Michael C Rosenstein, and Carolyn J Mattingly. 2012. Medic: a practical disease vocabulary used at the comparative toxicogenomics database. *Database*, 2012.

Jacob Devlin, Ming-Wei Chang, Kenton Lee, and Kristina Toutanova. 2019. BERT: Pre-training of deep bidirectional transformers for language understanding. In *Proceedings of the 2019 Conference of the North American Chapter of the Association for*

Computational Linguistics: Human Language Technologies, Volume 1 (Long and Short Papers), pages 4171–4186, Minneapolis, Minnesota. Association for Computational Linguistics.

Rezarta Islamaj Doğan, Robert Leaman, and Zhiyong Lu. 2014. Ncbi disease corpus: a resource for disease name recognition and concept normalization. *Journal of biomedical informatics*, 47:1–10.

Li Dong, Nan Yang, Wenhui Wang, Furu Wei, Xiaodong Liu, Yu Wang, Jianfeng Gao, Ming Zhou, and Hsiao-Wuen Hon. 2019. Unified language model pre-training for natural language understanding and generation. In *Advances in Neural Information Processing Systems*, volume 32. Curran Associates, Inc.

Zi-Yi Dou, Pengfei Liu, Hiroaki Hayashi, Zhengbao Jiang, and Graham Neubig. 2021. GSum: A general framework for guided neural abstractive summarization. In *Proceedings of the 2021 Conference of the North American Chapter of the Association for Computational Linguistics: Human Language Technologies*, pages 4830–4842, Online. Association for Computational Linguistics.

Thibault Févry, Livio Baldini Soares, Nicholas FitzGerald, Eunsol Choi, and Tom Kwiatkowski. 2020. Entities as experts: Sparse memory access with entity supervision. In *Proceedings of the 2020 Conference on Empirical Methods in Natural Language Processing (EMNLP)*, pages 4937–4951, Online. Association for Computational Linguistics.

Jenny Rose Finkel and Christopher D Manning. 2009. Nested named entity recognition. In *Proceedings of the 2009 conference on empirical methods in natural language processing*, pages 141–150.

Yuxian Gu, Robert Tinn, Hao Cheng, Michael R. Lucas, Naoto Usuyama, Xiaodong Liu, Tristan Naumann, Jianfeng Gao, and Hoifung Poon. 2022. Domain-specific language model pretraining for biomedical natural language processing. *ACM Transactions on Computing for Healthcare (HEALTH)*, 3:1 – 23.

Suchin Gururangan, Ana Marasović, Swabha Swayamdipta, Kyle Lo, Iz Beltagy, Doug Downey, and Noah A. Smith. 2020. Don't stop pretraining: Adapt language models to domains and tasks. In *Proceedings of the 58th Annual Meeting of the Association for Computational Linguistics*, pages 8342–8360, Online. Association for Computational Linguistics.

Dan Hendrycks and Kevin Gimpel. 2016. Gaussian error linear units (gelus). *arXiv preprint arXiv:1606.08415*.

Qiao Jin, Bhuwan Dhingra, William Cohen, and Xinghua Lu. 2019. Probing biomedical embeddings from language models. In *Proceedings of the 3rd Workshop on Evaluating Vector Space Representations for NLP*, pages 82–89.

Qiao Jin, Zheng Yuan, Guangzhi Xiong, Qianlan Yu, Huaiyuan Ying, Chuanqi Tan, Mosha Chen, Songfang Huang, Xiaozhong Liu, and Sheng Yu. 2022. Biomedical question answering: A survey of approaches and challenges. *ACM Comput. Surv.*, 55(2).

Alistair E. W. Johnson, Tom J. Pollard, Lu Shen, Li wei H. Lehman, Mengling Feng, Mohammad Mahdi Ghassemi, Benjamin Moody, Peter Szolovits, Leo Anthony Celi, and Roger G. Mark. 2016. Mimic-iii, a freely accessible critical care database. *Scientific Data*, 3.

Zeqian Ju, Subrato Chakravorty, Xuehai He, Shu Chen, Xingyi Yang, and Pengtao Xie. 2020. Coviddialog: Medical dialogue datasets about covid-19. *https://github.com/UCSD-AI4H/COVID-Dialogue*.

Kamal Raj Kanakarajan, Bhuvana Kundumani, and Malaikannan Sankarasubbu. 2021. Bioelectra:pretrained biomedical text encoder using discriminators. In *BIONLP*.

Sarvnaz Karimi, Alejandro Metke-Jimenez, Madonna Kemp, and Chen Wang. 2015. Cadec: A corpus of adverse drug event annotations. *Journal of biomedical informatics*, 55:73–81.

Halil Kilicoglu, Asma Ben Abacha, Yassine Mrabet, Sonya E. Shooshan, Laritza M. Rodriguez, Kate Masterton, and Dina Demner-Fushman. 2018. Semantic annotation of consumer health questions. *BMC Bioinformatics*, 19.

J-D Kim, Tomoko Ohta, Yuka Tateisi, and Jun'ichi Tsujii. 2003. Genia corpus—a semantically annotated corpus for bio-textmining. *Bioinformatics*, 19(suppl_1):i180–i182.

Tuan Lai, Heng Ji, and ChengXiang Zhai. 2021. BERT might be overkill: A tiny but effective biomedical entity linker based on residual convolutional neural networks. In *Findings of the Association for Computational Linguistics: EMNLP 2021*, pages 1631–1639, Punta Cana, Dominican Republic. Association for Computational Linguistics.

Guillaume Lample, Miguel Ballesteros, Sandeep Subramanian, Kazuya Kawakami, and Chris Dyer. 2016. Neural architectures for named entity recognition. In *Proceedings of the 2016 Conference of the North American Chapter of the Association for Computational Linguistics: Human Language Technologies*, pages 260–270, San Diego, California. Association for Computational Linguistics.

Md Tahmid Rahman Laskar, Enamul Hoque, and Jimmy Xiangji Huang. 2021. Domain adaptation with pre-trained transformers for query focused abstractive text summarization. *arXiv preprint arXiv:2112.11670*.

Jinhyuk Lee, Wonjin Yoon, Sungdong Kim, Donghyeon Kim, Sunkyu Kim, Chan Ho So, and Jaewoo Kang. 2020. Biobert: a pre-trained biomedical language representation model for biomedical text mining. *Bioinformatics*, 36:1234 – 1240.

Mike Lewis, Yinhan Liu, Naman Goyal, Marjan Ghazvininejad, Abdelrahman Mohamed, Omer Levy, Veselin Stoyanov, and Luke Zettlemoyer. 2020a. BART: Denoising sequence-to-sequence pre-training for natural language generation, translation, and comprehension. In *Proceedings of the 58th Annual Meeting of the Association for Computational Linguistics*, pages 7871–7880, Online. Association for Computational Linguistics.

Patrick Lewis, Myle Ott, Jingfei Du, and Veselin Stoyanov. 2020b. Pretrained language models for biomedical and clinical tasks: Understanding and extending the state-of-the-art. In *Proceedings of the 3rd Clinical Natural Language Processing Workshop*, pages 146–157, Online. Association for Computational Linguistics.

Jiao Li, Yueping Sun, Robin J Johnson, Daniela Sciaky, Chih-Hsuan Wei, Robert Leaman, Allan Peter Davis, Carolyn J Mattingly, Thomas C Wiegers, and Zhiyong Lu. 2016. Biocreative v cdr task corpus: a resource for chemical disease relation extraction. *Database*, 2016.

Jingye Li, Hao Fei, Jiang Liu, Shengqiong Wu, Meishan Zhang, Chong Teng, Donghong Ji, and Fei Li. 2021. Unified named entity recognition as word-word relation classification. *arXiv preprint arXiv:2112.10070*.

Nut Limsopatham and Nigel Collier. 2016. Normalising medical concepts in social media texts by learning semantic representation. In *Proceedings of the 54th Annual Meeting of the Association for Computational Linguistics (Volume 1: Long Papers)*, pages 1014–1023, Berlin, Germany. Association for Computational Linguistics.

Chin-Yew Lin. 2004. ROUGE: A package for automatic evaluation of summaries. In *Text Summarization Branches Out*, pages 74–81, Barcelona, Spain. Association for Computational Linguistics.

Hongyu Lin, Yaojie Lu, Xianpei Han, and Le Sun. 2019. Sequence-to-nuggets: Nested entity mention detection via anchor-region networks. In *Proceedings of the 57th Annual Meeting of the Association for Computational Linguistics*, pages 5182–5192, Florence, Italy. Association for Computational Linguistics.

Fangyu Liu, Ehsan Shareghi, Zaiqiao Meng, Marco Basaldella, and Nigel Collier. 2021. Self-alignment pretraining for biomedical entity representations. In *Proceedings of the 2021 Conference of the North American Chapter of the Association for Computational Linguistics: Human Language Technologies*, pages 4228–4238.

Yang Liu and Mirella Lapata. 2019. Text summarization with pretrained encoders. In *Proceedings of the 2019 Conference on Empirical Methods in Natural Language Processing and the 9th International Joint Conference on Natural Language Processing (EMNLP-IJCNLP)*, pages 3730–3740, Hong Kong, China. Association for Computational Linguistics.

Yinhan Liu, Myle Ott, Naman Goyal, Jingfei Du, Mandar Joshi, Danqi Chen, Omer Levy, Mike Lewis, Luke Zettlemoyer, and Veselin Stoyanov. 2019. Roberta: A robustly optimized bert pretraining approach. *ArXiv*, abs/1907.11692.

Bryan McCann, Nitish Shirish Keskar, Caiming Xiong, and Richard Socher. 2018. The natural language decathlon: Multitask learning as question answering. *arXiv preprint arXiv:1806.08730*.

Sunil Mohan and Donghui Li. 2019. Medmentions: A large biomedical corpus annotated with {umls} concepts. In *Automated Knowledge Base Construction (AKBC)*.

Danielle L Mowery, Sumithra Velupillai, Brett R South, Lee Christensen, David Martinez, Liadh Kelly, Lorraine Goeuriot, Noemie Elhadad, Sameer Pradhan, Guergana Savova, et al. 2014. Task 2: Share/clef ehealth evaluation lab 2014. In *Proceedings of CLEF 2014*.

Khalil Mrini, Franck Dernoncourt, Seunghyun Yoon, Trung Bui, Walter Chang, Emilia Farcas, and Ndapa Nakashole. 2021. A gradually soft multi-task and data-augmented approach to medical question understanding. In *Proceedings of the 59th Annual Meeting of the Association for Computational Linguistics and the 11th International Joint Conference on Natural Language Processing (Volume 1: Long Papers)*, pages 1505–1515, Online. Association for Computational Linguistics.

Kishore Papineni, Salim Roukos, Todd Ward, and Wei-Jing Zhu. 2002. Bleu: A method for automatic evaluation of machine translation. In *Proceedings of the 40th Annual Meeting on Association for Computational Linguistics*, ACL '02, page 311–318, USA. Association for Computational Linguistics.

Yifan Peng, Qingyu Chen, and Zhiyong Lu. 2020. An empirical study of multi-task learning on bert for biomedical text mining. In *BIONLP*.

Matthew E. Peters, Mark Neumann, Mohit Iyyer, Matt Gardner, Christopher Clark, Kenton Lee, and Luke Zettlemoyer. 2018. Deep contextualized word representations. In *NAACL*.

Long Phan, James T. Anibal, Hieu Tran, Shaurya Chanana, Erol Bahadroglu, Alec Peltekian, and Grégoire Altan-Bonnet. 2021. Scifive: a text-to-text transformer model for biomedical literature. *ArXiv*, abs/2106.03598.

Sameer Pradhan, Noemie Elhadad, Brett R South, David Martinez, Lee M Christensen, Amy Vogel, Hanna Suominen, Wendy W Chapman, and Guergana K Savova. 2013. Task 1: Share/clef ehealth evaluation lab 2013. In *CLEF (Working Notes)*, pages 212–31.

Alec Radford and Karthik Narasimhan. 2018. Improving language understanding by generative pre-training.

Alec Radford, Jeff Wu, Rewon Child, David Luan, Dario Amodei, and Ilya Sutskever. 2019. Language models are unsupervised multitask learners.

Colin Raffel, Noam Shazeer, Adam Roberts, Katherine Lee, Sharan Narang, Michael Matena, Yanqi Zhou, Wei Li, and Peter J. Liu. 2020. Exploring the limits of transfer learning with a unified text-to-text transformer. *Journal of Machine Learning Research*, 21(140):1–67.

Samyam Rajbhandari, Jeff Rasley, Olatunji Ruwase, and Yuxiong He. 2020. Zero: Memory optimizations toward training trillion parameter models. In *Proceedings of the International Conference for High Performance Computing, Networking, Storage and Analysis*, SC '20. IEEE Press.

Max E. Savery, Asma Ben Abacha, Soumya Gayen, and Dina Demner-Fushman. 2020. Question-driven summarization of answers to consumer health questions. *Scientific Data*, 7.

Hoo-Chang Shin, Yang Zhang, Evelina Bakhturina, Raul Puri, Mostofa Patwary, Mohammad Shoeybi, and Raghav Mani. 2020. BioMegatron: Larger biomedical domain language model. In *Proceedings of the 2020 Conference on Empirical Methods in Natural Language Processing (EMNLP)*, pages 4700–4706, Online. Association for Computational Linguistics.

Tianxiang Sun, Xiangyang Liu, Xipeng Qiu, and Xuanjing Huang. 2021. Paradigm shift in natural language processing. *arXiv preprint arXiv:2109.12575*.

Mujeen Sung, Hwisang Jeon, Jinhyuk Lee, and Jaewoo Kang. 2020. Biomedical entity representations with synonym marginalization. In *ACL*.

George Tsatsaronis, Georgios Balikas, Prodromos Malakasiotis, Ioannis Partalas, Matthias Zschunke, Michael R. Alvers, Dirk Weissenborn, Anastasia Krithara, Sergios Petridis, Dimitris Polychronopoulos, Yannis Almirantis, John Pavlopoulos, Nicolas Baskiotis, Patrick Gallinari, Thierry Artières, Axel-Cyrille Ngonga Ngomo, Norman Heino, Éric Gaussier, Liliana Barrio-Alvers, Michael Schroeder, Ion Androutsopoulos, and Georgios Paliouras. 2015. An overview of the bioasq large-scale biomedical semantic indexing and question answering competition. *BMC Bioinformatics*, 16.

Maya Varma, Laurel Orr, Sen Wu, Megan Leszczynski, Xiao Ling, and Christopher Ré. 2021. Cross-domain data integration for named entity disambiguation in biomedical text. In *Findings of the Association for Computational Linguistics: EMNLP 2021*, pages 4566–4575, Punta Cana, Dominican Republic. Association for Computational Linguistics.

Ashish Vaswani, Noam Shazeer, Niki Parmar, Jakob Uszkoreit, Llion Jones, Aidan N Gomez, Ł ukasz Kaiser, and Illia Polosukhin. 2017. Attention is all you need. In *Advances in Neural Information Processing Systems*, volume 30. Curran Associates, Inc.

Hang Yan, Tao Gui, Junqi Dai, Qipeng Guo, Zheng Zhang, and Xipeng Qiu. 2021. A unified generative framework for various NER subtasks. In *Proceedings of the 59th Annual Meeting of the Association for Computational Linguistics and the 11th International Joint Conference on Natural Language Processing (Volume 1: Long Papers)*, pages 5808–5822, Online. Association for Computational Linguistics.

Hongyi Yuan, Zheng Yuan, and Sheng Yu. 2022a. Generative biomedical entity disambiguation via knowledge base-guided pre-training and synonyms-aware fine-tuning. In *Proceedings of the 2022 Conference of the North American Chapter of the Association for Computational Linguistics: Human Language Technologies*.

Zheng Yuan, Yijia Liu, Chuanqi Tan, Songfang Huang, and Fei Huang. 2021. Improving biomedical pretrained language models with knowledge. In *BIONLP*.

Zheng Yuan, Yuanhao Liu, Qiuyang Yin, Boyao Li, Xiaobin Feng, Guoming Zhang, and Sheng Yu. 2020. Unsupervised multi-granular chinese word segmentation and term discovery via graph partition. *Journal of Biomedical Informatics*, 110:103542.

Zheng Yuan, Zhengyun Zhao, Haixia Sun, Jiao Li, Fei Wang, and Sheng Yu. 2022b. Coder: Knowledge-infused cross-lingual medical term embedding for term normalization. *Journal of Biomedical Informatics*, page 103983.

Guangtao Zeng, Wenmian Yang, Zeqian Ju, Yue Yang, Sicheng Wang, Ruisi Zhang, Meng Zhou, Jiaqi Zeng, Xiangyu Dong, Ruoyu Zhang, Hongchao Fang, Penghui Zhu, Shu Chen, and Pengtao Xie. 2020. MedDialog: Large-scale medical dialogue datasets. In *Proceedings of the 2020 Conference on Empirical Methods in Natural Language Processing (EMNLP)*, pages 9241–9250, Online. Association for Computational Linguistics.

Jingqing Zhang, Yao Zhao, Mohammad Saleh, and Peter Liu. 2020a. Pegasus: Pre-training with extracted gap-sentences for abstractive summarization. In *International Conference on Machine Learning*, pages 11328–11339. PMLR.

Ningyu Zhang, Mosha Chen, Zhen Bi, Xiaozhuan Liang, Lei Li, Xin Shang, Kangping Yin, Chuanqi Tan, Jian Xu, Fei Huang, et al. 2021. Cblue: A chinese biomedical language understanding evaluation benchmark. *arXiv preprint arXiv:2106.08087*.

Tianyi Zhang, Varsha Kishore, Felix Wu, Kilian Q. Weinberger, and Yoav Artzi. 2020b. Bertscore: Evaluating text generation with bert. In *International Conference on Learning Representations*.

Zhengyun Zhao, Qiao Jin, and Sheng Yu. 2022. Pmc-patients: A large-scale dataset of patient notes and relations extracted from case reports in pubmed central. *arXiv preprint arXiv:2202.13876*.

Meng Zhou, Zechen Li, Bowen Tan, Guangtao Zeng, Wenmian Yang, Xuehai He, Zeqian Ju, Subrato Chakravorty, Shu Chen, Xingyi Yang, Yichen Zhang, Qingyang Wu, Zhou Yu, Kun Xu, Eric Xing, and Pengtao Xie. 2021. On the generation of medical dialogs for COVID-19. In *Proceedings of the 59th Annual Meeting of the Association for Computational Linguistics and the 11th International Joint Conference on Natural Language Processing (Volume 2: Short Papers)*, pages 886–896, Online. Association for Computational Linguistics.

Incorporating Medical Knowledge to Transformer-based Language Models for Medical Dialogue Generation

Usman Naseem[1], Ajay Bandi[2], Shaina Raza[3], Junaid Rashid[4], Bharathi Raja Chakravarthi[5]
[1]School of Computer Science, University of Sydney, Australia
[2]School of Computer Science and Information Systems, Northwest Missouri State University Maryville, USA
[3]The Dalla Lana School of Public Health, University of Toronto Toronto, Canada
[4]Department of Computer Science and Engineering, Kongju National University, South Korea
[5] Data Science Institute, National University of Ireland Galway, Ireland

Abstract

Medical dialogue systems have the potential to assist doctors in expanding access to medical care, improving the quality of patient experiences, and lowering medical expenses. The computational methods are still in their early stages and are not ready for widespread application despite their great potential. Existing transformer-based language models have shown promising results but lack domain-specific knowledge. However, to diagnose like doctors, an automatic medical diagnosis necessitates more stringent requirements for the rationality of the dialogue in the context of relevant knowledge. In this study, we propose a new method that addresses the challenges of medical dialogue generation by incorporating medical knowledge into transformer-based language models. We present a method that leverages an external medical knowledge graph and injects triples as domain knowledge into the utterances. Automatic and human evaluation on a publicly available dataset demonstrates that incorporating medical knowledge outperforms several state-of-the-art baseline methods.

1 Introduction

Medical dialogue systems, which have gained increasing attention, aim to communicate with patients to enquire about diseases beyond their self-reported and make an automatic diagnosis (Wei et al., 2018; Xu et al., 2019; Lin et al., 2019). It has the potential to substantially automate the diagnostic process while also lowering the cost of gathering information from patients (Kao et al., 2018). In addition, preliminary diagnosis findings that are generated by a medical dialogue system may help doctors make a diagnosis more quickly. Because of these advantages, researchers work on addressing sub-problems in a medical dialogue system, such as natural language understanding (Lin et al., 2019; Shi et al., 2020).

However, the dialogue system for medical diagnosis, on the other hand, has specific requirements for dialogue reasoning in the context of medical knowledge. The diagnosis elicited by the dialogue system should be associated with the underlying medical condition and coherent with medical knowledge. In the absence of medical knowledge, traditional generative dialogue models frequently use neural sequence modelling (Sutskever et al., 2014; Vaswani et al., 2017) and cannot be directly applied to the medical dialogue scenario.

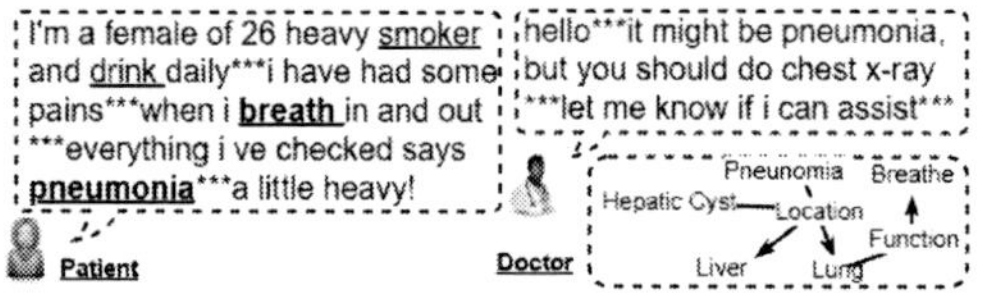

Figure 1: An example of medical dialogue between a patient (left) and a doctor (right).

Recently, transformer-based language models (LMs) (Devlin et al., 2019; Radford et al., 2019; Song et al., 2019) are fine-tuned for medical dialogue tasks. Zeng et al. (2020) collected a MedDialog dataset and fine-tuned various transformer-based LMs which includes a vanilla transformer (Vaswani et al., 2017), GPT (Radford et al., 2019) and BERT-GPT (Wu et al., 2020; Lewis et al., 2020) for medical dialogue generation task. Yang et al. (2020), in another study, presented a CovidDialog dataset and then train dialogue generation models based on Transformer, GPT-based model, and BART (Lewis et al., 2020) and BERT-GPT for medical dialogue generation tasks. These LMs are trained on huge corpus but may not provide a good representation of specific domains (Müller et al., 2020) and need an adequate amount of task-specific data (Dou et al., 2019) in order to establish correlations between diseases and symptoms (see Figure 1). Instead of using publicly available models, we can pre-train a model that emphasizes domain-specificity. On the other hand, pre-training is time-intensive and computationally costly, making it unavailable for most users.

Furthermore, while it is possible to inject

Proceedings of the BioNLP 2022 workshop, Dublin, Ireland, pages 110–115

domain-specific knowledge into LMs during pre-training, this method of acquiring knowledge can be expensive and inefficient. For instance, pre-training data must contain many occurrences of the words "Panadol" and "headache" occurring together for the model to learn that "Panadol" can treat headaches. What other options do we have to make the model an expert in its field besides this one? The knowledge graph (KG), also known as an ontology, was a good solution in the early stages of research. SNOMED-CT (Bodenreider, 2008), in the medical field, and HowNet (Dong et al., 2010), in the field of Chinese conception, are two examples of KGs developed as knowledge was distilled into a structured form. If KG can be incorporated into the LM, it will provide domain knowledge to the computational method, enhancing its effectiveness on domain-specific tasks while significantly lowering the expense of pre-training. To address the limitations mentioned above, this article describes a method for incorporating domain-specific external knowledge into transformer-based LMs for medical dialogue generation tasks. Our contributions are as follows:

- We presented a new method that incorporates medical knowledge to transformer-based language models;
- The proposed method first injects knowledge from a medical knowledge graph into an utterance. Next, the embedding layer transforms the utterance tree into an embedding that is fed to the masked self-attention of a transformer, followed by the decoder to generate the response.
- To evaluate the performance of the proposed method, we performed both automatic and human evaluations. Our results demonstrated that incorporating medical knowledge improves the performance compared to several state-of-the-art baselines on the MedDialog dataset.

2 Methodology

Problem Definition: Given a dialogue, we process a patient-doctor dialogue as a set of pairs $\{(s_i, t_i)\}$, where source s_i is the dialogue from a patient and target t_i is a doctor's response. A dialogue generation model generates *t* from *s*.

Overview of Architecture: As illustrated in Figure 2, the proposed method contains four modules, i.e., knowledge layer, embedding layer, masked transformer encoder, where we extend self-attention to mask-self attention, and transformer decoder. Our knowledge layer injects relevant triples into an input utterance (i.e., conversation) from a KG, converting it to a knowledge-rich utterance tree. Simultaneously, the utterance tree is fed into the embedding layer for token-level representation. The representation from an embedding layer is fed to the masked transformer encoder and decoder to generate a response. We will describe each of these modules in detail in the following discussion.

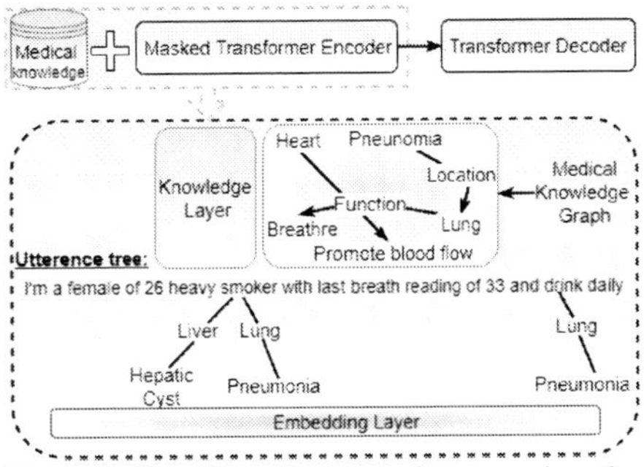

Figure 2: Overall architecture of proposed method

2.1 Knowledge layer

The knowledge layer incorporates domain-specific (medical) knowledge into utterances and transforms them into utterance trees. The knowledge layer generates an utterance tree given an input utterance (s) and a KG. This method involves two stages: query of medical knowledge, referred to as K-Query, and injection of knowledge, referred to as K-Inject. K-Query extracts all entity names from the utterance *s* and queries their correlating triples from knowledge *k*. K-Query can be expressed as follows:

$$E = K\_Query(s, KG), \tag{1}$$

Where *E* is a set of associating triples. K-Inject then injects the queried *E* into the utterance *s* by combining the triples in *E* to their corresponding positions, resulting in an utterance tree *t*. An utterance tree can have different branches; however, its depth is limited, indicating that entity names in triples will not iteratively derive branches. The formulation for K-Inject is as follows:

$$t = K\_Inject(s, E) \tag{2}$$

Knowledge graph: To generate knowledge, we use the medical knowledge graph released by Liu et al. (2021), which is centered on organs and related disorders. A set of 52.6K triplets (head, relation, tail) containing medical information was retrieved. The head and tail represent entities such as organs or diseases. In contrast, the relation indicates the relationship between entities, such as function and treatment. In this study, we employed the English language vocabulary, which has 2,603 triples in total.

2.2 Embedding layer

The embedding layer aims to transform the utterance tree into embedded representations that can be forwarded to the transformer's encoder and then decoder to generate the dialogue. Our embedding layer consists of token, position, and segment embedding layers. However, it differs in that the proposed method's embedding layer receives an utterance tree rather than a token sequence as input. Below, we discuss a method adopted to transform an utterance tree into a sequence that retains its structural information.

Token embedding: In our study, the token embedding, including the vocabulary used, is consistent with the original transformer-based LM (see section 3.3). Each token in the expression tree is transformed into a H dimensional embedding vector by a trainable lookup table. Token embeddings made using the proposed method differ from those made using the original LMs. The utterance tree tokens must first be rearranged before embedding can occur. After incorporating tokens in the branch, we reverse the order of the tokens in the following nodes. Even though this process is simple, it makes the utterance hard to read and loses important structural information that can be solved using soft-position.

Soft-position embedding: Without position embedding, encoders within a transformer will behave similarly to a bag-of-words (BoWs) method, leading to a loss of structural information (i.e., the order of tokens). The position embedding contains all of the structural information in the encoder's input sentence, allowing us to reconstruct the unreadable rearranged utterance. As an alternative to using the transformer encoder's self-attention score for words that appear to be connected but are not, we used masked self-attention (see section 2.3).

Segment embedding: Like the transformer encoder, the proposed method uses segmentation embedding to detect utterances when multiple utterances are included. For instance, when two utterances are fed, `[SEP]` is used to incorporate them. A sequence of segment tags is used to denote the combined utterance.

2.3 Transformer Encoder with Masked-Self Attention

We present a mask-self-attention to avoid false semantic changes, which is a self-attention extension. Mask-self-attention is defined as follow:

$$Q^{i+1}, K^{i+1}, V^{i+1} = h^i W_q, h^i W_k, h^i W_v \quad (3)$$

$$S^{i+1} = softmax(\frac{Q^{i+1} K^{i+1}}{\sqrt{d_k}}) \quad (4)$$

$$h^{i+1} = S^{i+1} V^{i+1} \quad (5)$$

where Wq, Wk, and Wv are model parameters that can be trained. The hidden state of the $i - th$ mask-self-attention blocks is hi. The scaling factor is dk. This process improves the representation but does not affect the original utterance's meaning.

2.4 Transformer Decoder

The knowledge enriched representation from the transformer encoder is fed to the decoder of an original LM to generate a response. The working process of the decoder layers is similar to that of the vanilla transformer decoder layers.

3 Experiments

3.1 Datasets

In this study, we used the English version of MedDialog (Zeng et al., 2020) dataset. Table 1 presents statistics of the MedDialog dataset.

Table 1: Dataset Statistics

Dataset	MedDialog-EN
# dialogues	257,332
# utterances	514,664
# tokens	44,527,872
#diseases	172
Avg. # of utterances	2
Max # of utterances	2
Min # of utterances	2
Avg. # of tokens	87
Max # of tokens	3,672
Min # tokens	1

3.2 Experimental Settings

We used five different LMs, and all configuration and pre-training settings are consistent with the original LMs used (see section3.3). Adam (Kingma and Ba, 2014) optimizer is used to train our model at 1e-6 initial learning rate. We used a batch size

Table 2: Results: Automatic ($BLEU_2$, $BLEU_4$, $METEOR$, $NIST-4$) and Human (5-point scale) evaluation

	Automated Evaluation for MedDialog-EN				Human Evaluation
Model	$BLEU_2$	$BLEU_4$	METEOR	NIST-4	Avg. Score
BERT-GPT (Wu et al., 2020)	5.72	4.82	0.28	0.42	3.70
BERT-GPT+Knowledge (Ours)	9.38	6.07	17.62	0.61	4.00
Performance Increase	3.66↑	1.25↑	17.34↑	0.19↑	0.30↑
Transformer (Vaswani et al., 2017)	2.13	2.28	11.57	0.03	2.70
Transformer+Knowledge (Ours)	2.48	2.46	12.32	0.31	3.00
Performance Increase	0.35↑	0.18↑	0.75↑	0.28↑	0.30↑
mT5 (Xue et al., 2020)	2.59	0.84	0.20	0.41	2.70
mT5+Knowledge (Ours)	7.32	3.63	1.11	0.94	3.00
Performance Increase	4.73↑	2.79↑	0.91↑	0.53↑	0.80↑
BART (Lewis et al., 2020)	15.92	9.72	0.70	2.03	3.90
BART+Knowledge (Ours)	17.25	11.07	1.73	2.07	4.15
Performance Increase	1.33↑	1.35↑	1.03↑	0.04↑	0.250↑
T5 (Raffel et al., 2019)	7.05	1.79	0.95	1.05	3.50
T5+Knowledge (Ours)	15.20	8.96	1.73	1.78	4.00
Performance Increase	8.15↑	7.17↑	0.78↑	0.73↑	0.50↑

of 64 for 50 epochs. We used grid-search optimization to derive the optimal parameters. We divided all datasets into training, validation, and test sets, with an 80:10:10 ratio for all experiments. The number of heads in multi-head attention is set to 12. The trained models were evaluated using automatic metrics such as NIST-4 (Doddington, 2002), $BLEU_2$, $BLEU_4$ (Papineni et al., 2002), and METEOR (Lavie and Agarwal, 2007).

3.3 Baselines

We compared our results with state-of-the-art LMs that are used in previous studies for medical dialogue generation tasks. To be precise, we used BERT-GPT (Wu et al., 2020), Transformer (Vaswani et al., 2017), mT5 (Xue et al., 2020), BART (Lewis et al., 2020), and T5 (Raffel et al., 2019) to compare the performance.

3.4 Results

Automated Evaluation: Table 2 demonstrates the automatic evaluation results achieved by different LMs, with and without knowledge. The results show that adding medical knowledge to LMs improves the performance across all evaluation metrics. For the MedDialogue-EN, we observed an increase in $BLEU_2$ score ranging from 0.35% to 8.15%, for $BLEU_4$, the improvement range is 0.18% to 7.17%, For METEOR, the increase is from 0.91% to 17.34%, and finally, for NIST-4, the increase in performance is in the range of 0.04% to 0.73%. From the results in Table 2, we can conclude that adding medical knowledge to LMs is beneficial and increases the performance of medical dialogue generation tasks.

Human Evaluation: We randomly selected 100 dialog examples for human evaluation. Five medical doctors were asked to rate the generated responses independently on a scale of 1 to 5. The greater the score, the better. The final results are obtained by averaging the ratings provided by various experts. From the human evaluation scores (right column) in Table 2, we deduce that incorporating medical knowledge into LMs generates a more accurate, clinically informative, and human-like response.

4 Conclusion

We present a method for enabling LMs with KGs to achieve domain knowledge like doctors. The proposed method transforms an utterance into a knowledge-enriched utterance tree by injecting medical knowledge from KG. The embedding layer converts the utterance tree into an embedding fed to the masked self-attention of a transformer, followed by the decoder to generate the response using medical dialogue history. Experimental results demonstrated that our method outperforms state-of-the-art LMs trained on general data. Further, through human evaluation, we conclude that generated responses are informative and doctor-like. In future, we aim to expand this work to other tasks and datasets.

References

Olivier Bodenreider. 2008. Biomedical ontologies in action: role in knowledge management, data integration and decision support. *Yearbook of medical informatics*, 17(01):67–79.

Jacob Devlin, Ming-Wei Chang, Kenton Lee, and Kristina Toutanova. 2019. BERT: Pre-training of deep bidirectional transformers for language understanding. In *Proceedings of the 2019 Conference of the North American Chapter of the Association for Computational Linguistics: Human Language Technologies, Volume 1 (Long and Short Papers)*, pages 4171–4186, Minneapolis, Minnesota. Association for Computational Linguistics.

George Doddington. 2002. Automatic evaluation of machine translation quality using n-gram co-occurrence statistics. In *Proceedings of the second international conference on Human Language Technology Research*, pages 138–145.

Zhendong Dong, Qiang Dong, and Changling Hao. 2010. HowNet and its computation of meaning. In *Coling 2010: Demonstrations*, pages 53–56, Beijing, China. Coling 2010 Organizing Committee.

Zi-Yi Dou, Keyi Yu, and Antonios Anastasopoulos. 2019. Investigating meta-learning algorithms for low-resource natural language understanding tasks. *arXiv preprint arXiv:1908.10423*.

Hao-Cheng Kao, Kai-Fu Tang, and Edward Chang. 2018. Context-aware symptom checking for disease diagnosis using hierarchical reinforcement learning. In *Proceedings of the AAAI Conference on Artificial Intelligence*, volume 32.

Diederik Kingma and Jimmy Ba. 2014. Adam: A method for stochastic optimization. *arXiv preprint arXiv:1412.6980*.

Alon Lavie and Abhaya Agarwal. 2007. Meteor: An automatic metric for mt evaluation with high levels of correlation with human judgments. In *Proceedings of the second workshop on statistical machine translation*, pages 228–231.

Mike Lewis, Yinhan Liu, Naman Goyal, Marjan Ghazvininejad, Abdelrahman Mohamed, Omer Levy, Veselin Stoyanov, and Luke Zettlemoyer. 2020. Bart: Denoising sequence-to-sequence pre-training for natural language generation, translation, and comprehension. In *Proceedings of the 58th Annual Meeting of the Association for Computational Linguistics*, pages 7871–7880.

Xinzhu Lin, Xiahui He, Qin Chen, Huaixiao Tou, Zhongyu Wei, and Ting Chen. 2019. Enhancing dialogue symptom diagnosis with global attention and symptom graph. In *Proceedings of the 2019 Conference on Empirical Methods in Natural Language Processing and the 9th International Joint Conference on Natural Language Processing (EMNLP-IJCNLP)*, pages 5033–5042.

Bo Liu, Li-Ming Zhan, Li Xu, Lin Ma, Yan Yang, and Xiao-Ming Wu. 2021. Slake: A semantically-labeled knowledge-enhanced dataset for medical visual question answering. In *2021 IEEE 18th International Symposium on Biomedical Imaging (ISBI)*, pages 1650–1654. IEEE.

Martin Müller, Marcel Salathé, and Per E Kummervold. 2020. Covid-twitter-bert: A natural language processing model to analyse covid-19 content on twitter. *arXiv preprint arXiv:2005.07503*.

Kishore Papineni, Salim Roukos, Todd Ward, and Wei-Jing Zhu. 2002. Bleu: a method for automatic evaluation of machine translation. In *Proceedings of the 40th annual meeting of the Association for Computational Linguistics*, pages 311–318.

Alec Radford, Jeffrey Wu, Rewon Child, David Luan, Dario Amodei, Ilya Sutskever, et al. 2019. Language models are unsupervised multitask learners. *OpenAI blog*, 1(8):9.

Colin Raffel, Noam Shazeer, Adam Roberts, Katherine Lee, Sharan Narang, Michael Matena, Yanqi Zhou, Wei Li, and Peter J Liu. 2019. Exploring the limits of transfer learning with a unified text-to-text transformer. *arXiv preprint arXiv:1910.10683*.

Xiaoming Shi, Haifeng Hu, Wanxiang Che, Zhongqian Sun, Ting Liu, and Junzhou Huang. 2020. Understanding medical conversations with scattered keyword attention and weak supervision from responses. In *Proceedings of the AAAI Conference on Artificial Intelligence*, volume 34, pages 8838–8845.

Kaitao Song, Xu Tan, Tao Qin, Jianfeng Lu, and Tie-Yan Liu. 2019. Mass: Masked sequence to sequence pre-training for language generation. *arXiv preprint arXiv:1905.02450*.

Ilya Sutskever, Oriol Vinyals, and Quoc V Le. 2014. Sequence to sequence learning with neural networks. *Advances in neural information processing systems*, 27.

Ashish Vaswani, Noam Shazeer, Niki Parmar, Jakob Uszkoreit, Llion Jones, Aidan N Gomez, Łukasz Kaiser, and Illia Polosukhin. 2017. Attention is all you need. *Advances in neural information processing systems*, 30.

Zhongyu Wei, Qianlong Liu, Baolin Peng, Huaixiao Tou, Ting Chen, Xuan-Jing Huang, Kam-Fai Wong, and Xiang Dai. 2018. Task-oriented dialogue system for automatic diagnosis. In *Proceedings of the 56th Annual Meeting of the Association for Computational Linguistics (Volume 2: Short Papers)*, pages 201–207.

Qingyang Wu, Lei Li, Hao Zhou, Ying Zeng, and Zhou Yu. 2020. Importance-aware learning for neural headline editing. In *Proceedings of the AAAI Conference on Artificial Intelligence*, volume 34, pages 9282–9289.

Lin Xu, Qixian Zhou, Ke Gong, Xiaodan Liang, Jianheng Tang, and Liang Lin. 2019. End-to-end knowledge-routed relational dialogue system for automatic diagnosis. In *Proceedings of the AAAI Conference on Artificial Intelligence*, volume 33, pages 7346–7353.

Linting Xue, Noah Constant, Adam Roberts, Mihir Kale, Rami Al-Rfou, Aditya Siddhant, Aditya Barua, and Colin Raffel. 2020. mt5: A massively multilingual pre-trained text-to-text transformer. *arXiv preprint arXiv:2010.11934*.

Wenmian Yang, Guangtao Zeng, Bowen Tan, Zeqian Ju, Subrato Chakravorty, Xuehai He, Shu Chen, Xingyi Yang, Qingyang Wu, Zhou Yu, et al. 2020. On the generation of medical dialogues for covid-19. *arXiv preprint arXiv:2005.05442*.

Guangtao Zeng, Wenmian Yang, Zeqian Ju, Yue Yang, Sicheng Wang, Ruisi Zhang, Meng Zhou, Jiaqi Zeng, Xiangyu Dong, Ruoyu Zhang, et al. 2020. Meddialog: Large-scale medical dialogue datasets. In *Proceedings of the 2020 Conference on Empirical Methods in Natural Language Processing (EMNLP)*, pages 9241–9250.

Memory-aligned Knowledge Graph for Clinically Accurate Radiology Image Report Generation

Sixing Yan
Department of Computer Science,
Hong Kong Baptist University,
Hong Kong SAR, China
cssxyan@comp.hkbu.edu.hk

Abstract

Automatic generating the clinically accurate radiology report from X-ray images is important but challenging. The identification of multi-grained abnormal regions in image and corresponding abnormalities is difficult for data-driven neural models. In this work, we introduce a Memory-aligned Knowledge Graph (MaKG) of clinical abnormalities to better learn the visual patterns of abnormalities and their relationships by integrating it into a deep model architecture for the report generation. We carry out extensive experiments and show that the proposed MaKG deep model can improve the clinical accuracy of the generated reports.

1 Introduction

Medical images are complex and hard to understand without specialized expertise. Given that the volume of radiology images is large, automatically generating the reports by the computer-aided system can alleviate the radiologists from the time-consuming reporting task. Recently, many deep learning models are studied in the automated radiology report generation (Han et al., 2018; Xie et al., 2019; Yang et al., 2021; Chen et al., 2020).

The deep encoder-decoder architecture has been commonly adopted in the report generation, where visual features were extracted from the input medical images using a convolutional neural network and fed to a recurrent neural network to generate the report. Different from image captioning which inputs one image and output one sentence, the report has much longer length while the correctness of medical entities generated in the report is the core requirement. More than the requirement of detecting abnormalities accurately like classification, the report is expected to provide the support details of present abnormalities. Thus, generating accurate report with readable and logical descriptions by natural language generation model is the key challenge in the report generation task.

Generating correct reports is impossible if the pathology of abnormal regions and corresponding abnormalities cannot be identified at first. Most existing studies (Liu et al., 2021a; Chen et al., 2020, 2021; You et al., 2021) proposed the attention and memory mechanism to enhance the identification of abnormal regions. However, different status of the same abnormality may have their specifics and the correlations of these visual patterns are ignored. In addition, identifying the actual abnormalities from abnormal regions is also challenging since the complex and rare abnormalities are hard to determined without professional knowledge.

To incorporate the prior medical knowledge, several research (Li et al., 2019; Zhang et al., 2020; Liu et al., 2021b) applied medical knowledge graph of certain abnormalities in the report generation aiming to learn the abnormality relationships. The corresponding representations, i.e., graph embedding, are computed by graph neural network given the input images. However, such representations are affected by the inner-connections of abnormalities for each input where the general characteristic of abnormalities are missing. For example, the representations of "*Effusion*", computed as graph embedding, are different when "*Effusion*" appears with or without "*Atelectasis*". But the general characteristic of "*Effusion*" over all relevant observations, e.g., density or shapes, are only determined by itself independently. This general but independent characteristic is still missed to model by existing approaches which limits the effectiveness the knowledge graph.

To alleviate the above challenges, in this work, we propose to learn the memory-aligned graph model, aiming to enhances the pathology identification and prior medical knowledge incorporation. The memory features of possible abnormal regions are first aligned by the input visual feature in an alternative manner, and concatenated with a universal memory embedding before feeding to

Proceedings of the BioNLP 2022 workshop, Dublin, Ireland, pages 116–122
May 26, 2022. ©2022 Association for Computational Linguistics

the graph attention network to compute the graph embedding. The graph embedding are later learned by the classification and fine-tuned in the report generation. We evaluate the proposed approach using two publicly accessible datasets. The evaluation results show the effectiveness of utilizing memory-aligned knowledge graph in generating the clinically accurate radiology report.

2 The Proposed Method

2.1 Problem Formulation

Given the radiology image with extracted visual features as V, the model aims to generate a radiology report $R = \{y_1, y_2, ...\}$. We introduce a **M**emory-**a**ligned **K**nowledge **G**raph (MaKG) to explore multi-grained features of the abnormalities and their relationships. The multi-grained memory features $\hat{M}$ are first aligned from the memory slots M by V, and concatenated with a meshed memory embedding E to learn the abnormality graph embedding G for generating radiology report R. This process can be formulated as,

$$\{V, M\} \rightarrow \hat{M}; \{\hat{M}, E\} \rightarrow G; G \rightarrow R. \quad (1)$$

Implementation. Following (Chen et al., 2020, 2021; Liu et al., 2021b), we adopt a memory slots $M \in \mathbb{R}^{\mathrm{M} \times \mathrm{D}}$ to record the information of abnormal regions which would indicate the potential abnormalities. The memory slots are initialized as plain learnable vectors and updated together with other modules. The M stands for the total number of the knowledge corresponding to the abnormality identification. We also adopt a $E \in \mathbb{R}^{\mathrm{N} \times \mathrm{D}}$ embedding to model the universal features of each abnormality. The N is equal to the number of the abnormalities. We follow (Zhang et al., 2020) to construct and initialize the abnormality knowledge graph $\mathcal{G} = (\mathcal{V}, \mathcal{E}); |\mathcal{V}| = \mathrm{N}$ which is a universal structure in the training. The nodes $\mathcal{V}$ cover the common chest abnormalities and grouped by their organ or body part appearances as edges $\mathcal{E}$. The graph embedding $G \in \mathbb{R}^{\mathrm{N} \times \mathrm{D}}$ is computed by the graph attention network. A overview of this framework is shown in Fig. 1.

2.2 Memory-aligned Graph Embedding

To learn the visual patterns of possible abnormal regions, we apply Multi-Head Attention (MHA) (Vaswani et al., 2017) to query the responding memory features from the memory slots M. The MHA computes the associated weighted between different features which allows the abnormality-related memory features to be distilled from original M. To align different level of the alignment, we can perform the alignment attention alternatively as,

$$\begin{aligned} V'_{i+1} &= \mathrm{MHA}(M_i, V_i); \\ M'_{i+1} &= \mathrm{MHA}(V'_{i+1}, M_i), \end{aligned} \quad (2)$$

where $V_0 = V$, $M_0 = M$, V'_i and M'_i denote i-th step aligned visual and memory features, respectively. As observed, the patterns of abnormal regions should be learned in different fine-grained ways due to their variable shapes and sizes. Thus, we follow (You et al., 2021) to repeat the alignment K times and obtain multi-grained memory features $\{M'_i\} = \{M'_1, M'_2, ..., M'_{\mathrm{K}}\}$. We then aggregate the multi-grained memory features as $\hat{M} = \mathrm{MHA}(M'_*, M'_*)$, where $M'_* = \bigoplus_{i=1}^{\mathrm{K}} M'_i$ and $\hat{M} \in \mathbb{R}^{\mathrm{M} \times \mathrm{D}}$.

To model the prior knowledge on the global characteristic of each abnormality which may not depend on the current input V, we add an meshed memory embedding $E \in \mathbb{R}^{\mathrm{N} \times \mathrm{D}}$ of which each row represent one particular abnormality. We compute the graph embedding $G \in \mathbb{R}^{\mathrm{N} \times \mathrm{D}}$ using graph attentional layer $\mathrm{GAT}(\cdot)$ (Veličković et al., 2017) as,

$$G = \mathrm{GAT}(\mathrm{FFN}(MW^{\mathrm{G}} \oplus E)) \quad (3)$$

where $\mathrm{FFN}(x) = \mathrm{ReLU}(xW_1^{\mathrm{ff}} + b_1^{\mathrm{ff}})W_2^{\mathrm{ff}} + b_2^{\mathrm{ff}}$, $W_1^{\mathrm{ff}}, W_2^{\mathrm{ff}} \in \mathbb{R}^{\mathrm{D} \times \mathrm{D}}$ and $W^{\mathrm{G}} \in \mathbb{R}^{\mathrm{M} \times \mathrm{N}}$ are learnable parameters, $b_1^{\mathrm{ff}}, b_2^{\mathrm{ff}}$ are learnable bias vectors. We learn G by adding a fully-connected layer with *Sigmoid* activation for each node and serving it as a binary classifier. Each node embedding is used to predict the existence probability of corresponding abnormality, and the classifier is trained using weighted binary cross entropy loss. The details can be found in (Zhang et al., 2020).

2.3 Report Generation by Graph embedding

For each decoding step t, the hidden stats h_t is encoded from the input word features x_t by the standard encoder from Transformer,

$$x_t = w_t + e_t; h_t = \mathrm{MHA}(x_t, x_{1:t}), \quad (4)$$

where w_t and e_t are the word embedding and positional embedding, respectively. A L layers Transformer decoder is employed to generate the proper report by the attending MaKG embeddings G as,

$$\begin{aligned} h'_t &= \mathrm{MHA}(h_t, G); \\ y'_t \sim p_t &= \mathrm{Softmax}(h'_t W + b). \end{aligned} \quad (5)$$

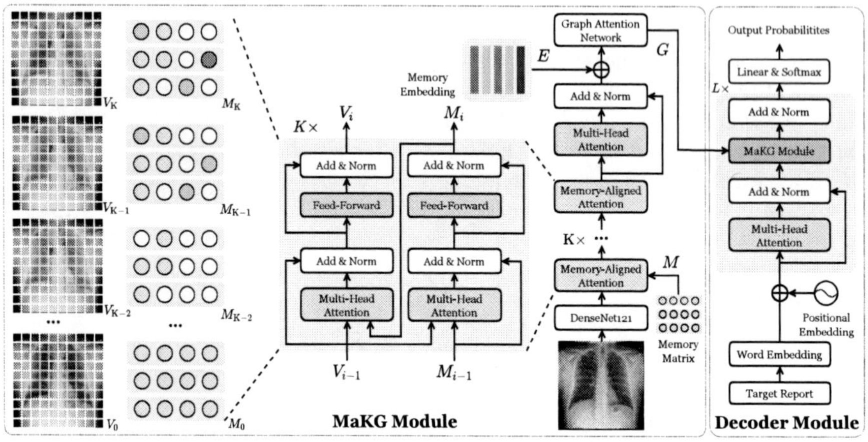

Figure 1: The MaKG-based deep model architecture.

Both encoder and decoder are trained by minimizing the cross-entropy loss $L_{gen}(\theta) = -\sum_{t=1}^{T}\log(p_t|p_{1:t-1})$.

3 Experiments

3.1 Datasets, Metrics and Settings

We use two publicly available datasets IU X-Ray (Demner-Fushman et al., 2016) and MIMIC CXR (Johnson et al., 2019) for evaluating the model performances. For the IU X-Ray dataset, we collect 2,848 reports and 5,696 images containing both frontal and lateral chest X-rays. We partitioned the data into train/validate/test set by 7:1:2 for cross validation. For MIMIC CXR dataset, we follow original split set with train/validate/test size as 222,705 / 1,807 / 3,269 and report the average scores of three different runs.

For report quality, we adopt the language generation metrics including BLEU (Papineni et al., 2002), METEOR (Banerjee and Lavie, 2005), ROUGE (Lin, 2004) and CIDEr (Vedantam et al., 2015). To measure the clinical accuracy, we adopt the Clinical Efficacy (CE) (Chen et al., 2020) and Clinical Metrics (CM) (Miura et al., 2021) for common and critical observation accuracy, and MIRQI (Zhang et al., 2020) to evaluate accuracy of 14[1] observations and their associated attributes. The micro-avg F1-measure scores are reports.

To compare with the proposed model **TRANS.+MAKG**, we employ the basic vanilla Transformer **TRANS.** with three layers, 8 heads and 512 hidden state dimension, and an integration knowledge graph used in (Zhang et al., 2020) denoted as **TRANS.+KG**. We also compare **TRANS.+MAKG** with several report generation models, including **WORDSAT** (Xu et al., 2015), **ADAATTN** (Lu et al., 2017), **SENTSAT** (Krause et al., 2017), **COATTN** (Jing et al., 2018), **SENTKG** (Zhang et al., 2020), **M^{2}TRANS** (Cornia et al., 2020), **R2GEN** (Chen et al., 2020) and **R2GEN-CMN** (Chen et al., 2021).

We adopt DenseNet121 (Huang et al., 2017) to extract the visual features. The dimensions of hidden state and number of heads in MHA are set as 512 and 8. K and M are set as 3 and 20. The model is trained with the learning rate 5e-5 in the end-to-end manner.

3.2 Results on Multi-label Classification

For performance comparisons on image classification, we evaluate the proposed MAKG with the base DENSENET (Huang et al., 2017) and integrating with KG (Zhang et al., 2020) embedded with different graph neural network. Higher or equivalent scores are obtained for most of the classes as shown in Table 1. A possible explanation is that the alignment mechanism of MAKG enhances the learning of the abnormality patterns by distilling the irrelevant regions from the images.

3.3 Results on Report Generation

The main focus of this experiment is to evaluate the effectiveness of applying memory alignment knowledge graph (MaKG) in enhancing the clinical accuracy of the report generation.

[1]14 clincal observations includes: *No finding, Enlarged Cardiomediastinum, Cardiomegaly, Lung lesion, Lung opacity, Edema, Consolidation, Pneumonia, Atelectasis, Pneumothorax, Pleural effusion, Pleural other, Fracture, Support devices*

Class	Integration Module			
	-	KG*	KG	MaKG
Normal/No Finding	0.795	0.807	0.806	**0.821**
Cardiomegaly	0.866	0.913	0.922	**0.930**
Scoliosis	0.664	0.663	0.671	**0.687**
F.B.	0.695	0.671	0.686	**0.727**
Effusion	0.921	0.942	0.950	**0.962**
Thickening	0.733	0.728	0.753	**0.785**
Pneumothorax	0.824	0.843	0.843	**0.889**
H.H	0.860	**0.884**	0.857	0.870
Calcinosis	0.676	0.669	0.669	**0.690**
Emphysema	0.892	0.890	0.902	**0.919**
Pneumonia	0.844	**0.863**	0.835	0.861
Edema	0.897	0.931	0.912	**0.949**
Atelectasis	0.788	0.833	0.823	**0.838**
Cicatrix	0.742	0.734	0.745	**0.774**
Opacity	0.796	0.803	0.806	**0.829**
Lesion	0.597	0.643	0.630	0.647
Airspace Disease	0.830	**0.857**	0.823	0.846
Hypoinflation	0.768	0.775	0.767	**0.791**
Medical Device	0.775	0.805	0.798	**0.825**
Other	0.595	0.596	0.607	**0.653**
Average	0.778	0.792	0.867	**0.879**

Table 1: Performance on multi-label classification (AUC) on IU XRay dataset. The best scores are in bold face and the second best are underlined.

Clinical Accuray Metric As shown in Table 3, TRANS.+MAKG achieves the first and second best performances over all clinical accuracy related metrics, and outperforms TRANS+KG with significantly improvement in MIRQI score which evaluates the accuracy of both abnormalities and their associated attributes. It indicates integrating MaKG is able to enhance the generation of clinically accurate report by providing correct attribute descriptions in the fine-grained level. This observation is important because the correctness of the associated attributes is necessary for the correctness of the abnormality descriptions. The incomplete or incorrect attributes of the same abnormalities would result different or even incorrect follow-up treatments. Noted that TRANS.+MAKG does not obtain the first best score in CE which measures the accuracy of 13 clinical observations and normality observation. However, the best scores of CM and Hits are observed shows that TRANS.+MAKG is able to identify the most critical abnormalities and cover most of the abnormalities that are frequently mentioned in the report repositories.

As observed from Table. 3, no model could detect all evaluated abnormalities for IU XRay dataset. Thus, we further study the detailed results as shown in Table. 2. As observed, there are some abnormalities of which appearance ratio is around 5% in the whole training set which is relatively rare. The failed detection could be caused by different reasons, such too few training data (e.g., "Fracture") or too hard to learning (e.g., "Pneumothorax" which is also very hard for clinicians to determine).

Class (%)	Integration Module		
	-	KG	MaKG
No Finding (31.72%)	**0.603**	0.500	0.456
Enlarged Cardio. (13.3%)	0.000	0.000	**0.034**
Cardiomegaly (15.6%)	0.265	**0.392**	0.341
Lung Lesion (5.2%)	0.000	0.000	**0.054**
Lung Opacity (21.3%)	0.181	0.209	**0.278**
Edema (4.7%)	0.000	0.000	**0.160**
Consolidation (5.2%)	0.000	0.038	**0.073**
Pneumonia (3.0%)	0.000	0.000	0.000
Atelectasis (8.1%)	0.000	0.087	**0.227**
Pneumothorax (6.6%)	0.000	0.000	0.000
Pleural Effusion (10.2%)	0.089	0.172	**0.278**
Pleural Other (1.6%)	0.000	0.000	0.000
Fracture (2.9%)	0.000	0.000	0.000
Support Devices (3.9%)	0.091	0.114	**0.242**

Table 2: Detailed CE evaluation results (F1-measure) of TRANSFORMER and integrating with KG and MAKG in IU XRay dataset, respectively. The best scores are in bold face

Natural Language Generation Metrics As the experimental results show, the higher NLG scores do not always indicate the clinically accurate reports are generated. While the clinical accuracy is a mission-critical requirement for radiology report generation, the generated report is expected to be clinically accurate using relatively readable sentences. The TRANS.+MAKG achieves similar NLG scores which indicates that the integration of MaKG is able to generate more reasonable descriptions of the abnormalities without decreasing the informativeness from TRANS. much. More powerful decoders (e.g., MemroyTrans. (Chen et al., 2020) or AlignTrans. (You et al., 2021)) should be able to enhance the overall performances.

Qualitative Results As shown in Fig. 2, two cases of ground truth and generated reports are visualized. The extracted clinical findings and the associated modifications are also attached. As observed, TRANS.+MAKG is able to detect more correct abnormalities in such cases than TRANS.+KG. It is believed to assistant clinicians to detect the abnormalities which are easy to ignored, thus increases the usability of applying the MaKG in improving the clinical accuracy in the report generation task.

Dataset	Model	NLG Metrics				Clinical Accuracy Metrics			
		B.	**M.**	**R.**	**C.**	**CM**	**CE**	**MIRQI**	**Hits (14)**
IU XRay	WORDSAT (Xu et al., 2015)	0.262	0.383	0.369	0.317	0.094	0.215	0.463	5.6
	ADAATTN (Lu et al., 2017)	0.269	0.379	0.367	0.358	0.240	0.338	0.474	6.6
	SENTSAT (Krause et al., 2017)	0.274	0.372	0.365	0.318	0.106	0.241	0.451	4.8
	COATTN (Jing et al., 2018)	0.256	0.367	0.357	0.307	0.061	0.245	0.438	5.2
	SENTKG (Zhang et al., 2020)	0.271	0.391	0.367	0.304	0.067	0.242	0.490	4.8
	M²TRANS. (Cornia et al., 2020)	0.269	0.299	0.363	0.367	0.104	0.253	0.481	5.6
	R2GEN (Chen et al., 2020)	0.251	0.367	0.342	0.461	0.100	0.322	0.389	9.0
	R2GEN-CMN (Chen et al., 2021)	**0.294**	**0.392**	**0.370**	**0.681**	0.104	0.330	0.462	8.0
	TRANS. (Vaswani et al., 2017)	0.264	0.390	0.357	0.587	0.147	**0.394**	0.486	5.0
	TRANS.+KG	0.265	0.380	0.353	0.593	0.205	0.320	0.504	9.2
	TRANS.+MAKG (ours)	0.265	0.378	0.353	0.523	**0.262**	0.362	**0.515**	**10.8**
MIMIC CXR	WORDSAT (Xu et al., 2015)	0.160	0.284	0.249	0.082	0.354	0.324	0.391	10.0
	ADAATTN (Lu et al., 2017)	0.151	**0.301**	0.248	0.096	0.384	0.366	0.438	12.0
	SENTSAT (Krause et al., 2017)	**0.182**	0.236	0.252	0.073	0.412	0.364	0.411	11.3
	COATTN (Jing et al., 2018)	0.181	0.235	**0.253**	0.070	0.423	0.364	0.418	9.7
	M²TRANS. (Cornia et al., 2020)	0.165	0.299	0.249	0.102	**0.458**	**0.469**	0.518	13.7
	R2GEN (Chen et al., 2020)	0.124	0.158	0.160	**0.170**	0.262	0.296	0.383	13.0
	R2GEN-CMN (Chen et al., 2021)	0.123	0.162	0.163	0.128	0.329	0.356	0.485	10.0
	TRANS. (Vaswani et al., 2017)	0.126	0.160	0.164	0.167	0.286	0.288	0.368	13.0
	TRANS.+KG	0.109	0.280	0.214	0.119	0.406	0.398	0.535	12.0
	TRANS.+MAKG (ours)	0.137	0.284	0.228	0.120	0.455	**0.469**	**0.572**	**14.0**

Table 3: Performance comparison of report generation models. The best scores are in bold face and the second best are underlined."B.", "M." "R." and "C." stand for BLEU, METEOR, ROUGE and CIDEr scores, respectively. The maximum number of "Hits" is 14 which is defined by CheXpert labeling toolkit.

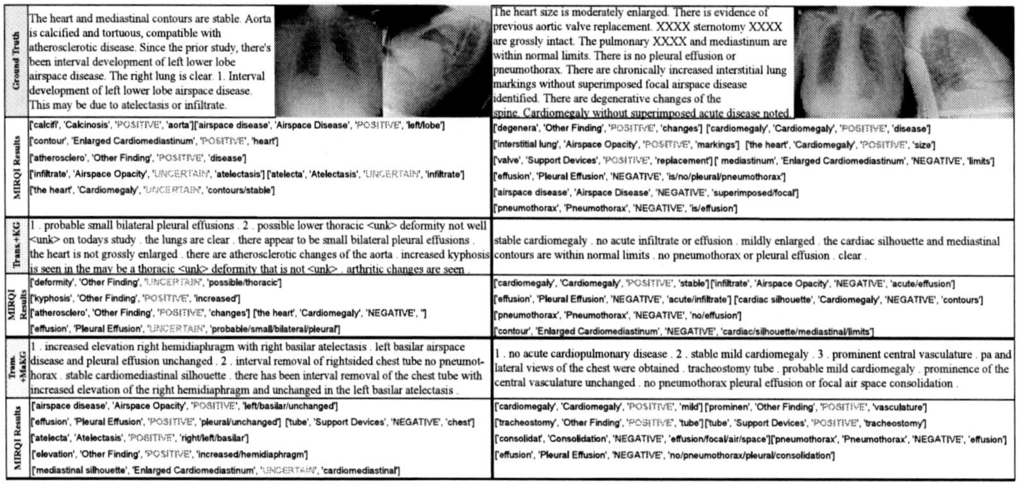

Figure 2: Illustration of reports generated by TRANS.+KG and TRANS.+MAKG. The extracted medical entities by MIRQI evaluation toolkit are attached as ["keyphrase", "category", "negation", "attributes"].

4 Conclusions

In this work, we propose a memory-aligned knowledge graph (MaKG) to enhance the clinically accurate report generation by modeling the relationship between abnormal regions and particular abnormalities. The experiments prove the effectiveness of integrating MaKG with the generation model is able to generate descriptive report with both correct abnormalities and associated attributes. In addition, the proposed MaKG is not limited to the specific knowledge graph structure which give the opportunities on incorporating different professional knowledge for specific medical applications.

References

Satanjeev Banerjee and Alon Lavie. 2005. METEOR: An automatic metric for MT evaluation with improved correlation with human judgments. In *Proceedings of the ACL Workshop on Intrinsic and Ex-*

trinsic Evaluation Measures for Machine Translation and/or Summarization, pages 65–72, Ann Arbor, Michigan. Association for Computational Linguistics.

Zhihong Chen, Yaling Shen, Yan Song, and Xiang Wan. 2021. Cross-modal memory networks for radiology report generation. In *Proceedings of the 59th Annual Meeting of the Association for Computational Linguistics and the 11th International Joint Conference on Natural Language Processing (Volume 1: Long Papers)*, pages 5904–5914.

Zhihong Chen, Yan Song, Tsung-Hui Chang, and Xiang Wan. 2020. Generating radiology reports via memory-driven transformer. In *Proceedings of the Conference on Empirical Methods in Natural Language Processing 2020*, pages 1439–1449.

Marcella Cornia, Matteo Stefanini, Lorenzo Baraldi, and Rita Cucchiara. 2020. Meshed-memory transformer for image captioning. In *Proceedings of the IEEE/CVF Conference on Computer Vision and Pattern Recognition*, pages 10578–10587.

Dina Demner-Fushman, Marc D Kohli, Marc B Rosenman, Sonya E Shooshan, Laritza Rodriguez, Sameer Antani, George R Thoma, and Clement J McDonald. 2016. Preparing a collection of radiology examinations for distribution and retrieval. *Journal of the American Medical Informatics Association*, 23(2):304–310.

Zhongyi Han, Benzheng Wei, Stephanie Leung, Jonathan Chung, and Shuo Li. 2018. Towards automatic report generation in spine radiology using weakly supervised framework. In *Proceedings of the 21th International Conference on Medical Image Computing and Computer-Assisted Intervention*, pages 185–193.

Gao Huang, Zhuang Liu, Laurens Van Der Maaten, and Kilian Q Weinberger. 2017. Densely connected convolutional networks. In *Proceedings of the 30th IEEE Conference on Computer Vision and Pattern Recognition*, pages 4700–4708.

Baoyu Jing, Pengtao Xie, and Eric P. Xing. 2018. On the automatic generation of medical imaging reports. In *Proceedings of the 56th Annual Meeting of the Association for Computational Linguistics*, pages 2577–2586.

Alistair EW Johnson, Tom J Pollard, Nathaniel R Greenbaum, Matthew P Lungren, Chih-ying Deng, Yifan Peng, Zhiyong Lu, Roger G Mark, Seth J Berkowitz, and Steven Horng. 2019. Mimic-cxr-jpg, a large publicly available database of labeled chest radiographs. *arXiv preprint arXiv:1901.07042*.

Jonathan Krause, Justin Johnson, Ranjay Krishna, and Li Fei-Fei. 2017. A hierarchical approach for generating descriptive image paragraphs. In *Proceedings of the IEEE conference on computer vision and pattern recognition*, pages 317–325.

Christy Y Li, Xiaodan Liang, Zhiting Hu, and Eric P. Xing. 2019. Knowledge-driven encode, retrieve, paraphrase for medical image report generation. In *Proceedings of the 34th AAAI Conference on Artificial Intelligence*, pages 6666–6673.

Chin-Yew Lin. 2004. Rouge: A package for automatic evaluation of summaries. In *Text summarization branches out*, pages 74–81.

Fenglin Liu, Xian Wu, Shen Ge, Wei Fan, and Yuexian Zou. 2021a. Exploring and distilling posterior and prior knowledge for radiology report generation. In *Proceedings of the IEEE/CVF Conference on Computer Vision and Pattern Recognition*, pages 13753–13762.

Fenglin Liu, Chenyu You, Xian Wu, Shen Ge, Xu Sun, et al. 2021b. Auto-encoding knowledge graph for unsupervised medical report generation. *Advances in Neural Information Processing Systems*, 34.

Jiasen Lu, Caiming Xiong, Devi Parikh, and Richard Socher. 2017. Knowing when to look: Adaptive attention via a visual sentinel for image captioning. In *Proceedings of the IEEE conference on computer vision and pattern recognition*, pages 375–383.

Yasuhide Miura, Yuhao Zhang, Emily Tsai, Curtis Langlotz, and Dan Jurafsky. 2021. Improving factual completeness and consistency of image-to-text radiology report generation. In *Proceedings of the 2021 Conference of the North American Chapter of the Association for Computational Linguistics: Human Language Technologies*, pages 5288–5304.

Kishore Papineni, Salim Roukos, Todd Ward, and Wei-Jing Zhu. 2002. Bleu: a method for automatic evaluation of machine translation. In *Proceedings of the 40th Annual Meeting of the Association for Computational Linguistics*, pages 311–318.

Ashish Vaswani, Noam Shazeer, Niki Parmar, Jakob Uszkoreit, Llion Jones, Aidan N Gomez, Łukasz Kaiser, and Illia Polosukhin. 2017. Attention is all you need. In *Advances in neural information processing systems*, pages 5998–6008.

Ramakrishna Vedantam, C Lawrence Zitnick, and Devi Parikh. 2015. Cider: Consensus-based image description evaluation. In *Proceedings of the 28th IEEE Conference on Computer Vision and Pattern Recognition*, pages 4566–4575.

Petar Veličković, Guillem Cucurull, Arantxa Casanova, Adriana Romero, Pietro Lio, and Yoshua Bengio. 2017. Graph attention networks. *arXiv preprint arXiv:1710.10903*.

Xiancheng Xie, Yun Xiong, Philip S. Yu, Kangan Li, Suhua Zhang, and Yangyong Zhu. 2019. Attention-based abnormal-aware fusion network for radiology report generation. In *Proceedings of the 24th International Conference on Database Systems for Advanced Applications*, pages 448–452.

Kelvin Xu, Jimmy Ba, Ryan Kiros, Kyunghyun Cho, Aaron Courville, Ruslan Salakhudinov, Rich Zemel, and Yoshua Bengio. 2015. Show, attend and tell: Neural image caption generation with visual attention. In *Proceedings of the 32nd International Conference on Machine Learning*, volume 37 of *Proceedings of Machine Learning Research*, pages 2048–2057.

Xingyi Yang, Muchao Ye, Quanzeng You, and Fenglong Ma. 2021. Writing by memorizing: Hierarchical retrieval-based medical report generation. In *Proceedings of the 59th Annual Meeting of the Association for Computational Linguistics and the 11th International Joint Conference on Natural Language Processing (Volume 1: Long Papers)*, pages 5000–5009. Association for Computational Linguistics.

Di You, Fenglin Liu, Shen Ge, Xiaoxia Xie, Jing Zhang, and Xian Wu. 2021. Aligntransformer: Hierarchical alignment of visual regions and disease tags for medical report generation. In *International Conference on Medical Image Computing and Computer-Assisted Intervention*, pages 72–82. Springer.

Yixiao Zhang, Xiaosong Wang, Ziyue Xu, Qihang Yu, Alan Yuille, and Daguang Xu. 2020. When radiology report generation meets knowledge graph. In *Proceedings of the 34th AAAI Conference on Artificial Intelligence*, pages 12910–12917.

Simple Semantic-based Data Augmentation for Named Entity Recognition in Biomedical Texts

Uyen T.P. Phan[1,2], **Nhung T.H. Nguyen**[3]
[1]Faculty of Information Technology, University of Science, Ho Chi Minh city, Vietnam
[2]Vietnam National University, Ho Chi Minh city, Vietnam
[3]Department of Computer Science, University of Manchester, UK
ptpuyen@fit.hcmus.edu.vn, nhung.nguyen@manchester.ac.uk

Abstract

Data augmentation is important in addressing data sparsity and low resources in NLP. Unlike data augmentation for other tasks such as sentence-level and sentence-pair ones, data augmentation for named entity recognition (NER) requires preserving the semantic of entities. To that end, in this paper we propose a simple semantic-based data augmentation method for biomedical NER. Our method leverages semantic information from pre-trained language models for both entity-level and sentence-level. Experimental results on two datasets: i2b2-2010 (English) and VietBioNER (Vietnamese) showed that the proposed method could improve NER performance.

1 Introduction

In machine learning and especially deep learning approaches, performance of the trained models is often proportional to the size of the training data. Consequently, for a model to achieve acceptable performance, we need a certain amount of labelled data. This would be an issue for low-resource domain and low-resource languages since annotating labelled data is time-consuming and expensive. To address the issue, data augmentation has been proposed to increase the variety of training data without directly collecting or annotating additional data (Feng et al., 2021).

Intuitively, data augmentation for named entity recognition (NER) task is more difficult to perform than for other sentence-level and sentence-pair tasks. Simple operations used to augment a sentence such as token swap, token deletion, and token insertion (Wei and Zou, 2019) may not work well in the case of NER, especially in the biomedical domain. One of the reasons is that a named entity can be composed by multiple tokens and we have to preserve the semantic of entities after applying those operations. For example, consider the following sentence from the i2b2-2010 corpus (Uzuner et al., 2011) with its entities:

She can be given prn $[\text{lasix}]_{Treatment}$ for $[\text{weight gain}]_{Problem}$ or $[\text{shortness of breath}]_{Problem}$.

If we randomly swap the 'lasix' token with 'weight', the sentence is not semantically correct. Similarly, when the 'weight' token is deleted, the remaining 'gain' token is no longer suitable for an entity of Problem. For the insertion operation, if we randomly insert a token into the sentence, the semantic of the sentence will be changed and we will not be able to assign a suitable entity label for it. As a result, it is necessary to have different augmentation methods specified for NER.

There are several model-based data augmentation methods for NER. Chen et al. (2020) proposed Local Additivity-based Data Augmentation (LADA) that can create virtual samples using interpolation technique. Their exeperimental results showed that LADA could help to produce state-of-the-art (SOTA) on two NER benchmarks including CoNLL 2013 (Tjong Kim Sang and De Meulder, 2003) and GermEval 2014 (Benikova et al., 2014). Meanwhile, Nie et al. (2020) took advantages of the rich semantic information in pre-trained word embeddings to create a semantic augmentation module for NER models. They also reported SOTA performance on some social media corpora.

Obviously, model-based methods can help to improve NER performance, but they are often complicated and difficult to implement. In contrast, rule-based methods are simpler and more intepretable than model-based ones, but still effective. Dai and Adel (2020) adjusted simple operations such as replacement and shuffle to preserve the semantic of both entities and sentences. Specifically, they proposed Synonym Replacement (SR) and Mention Replacement (MR). SR replaces a word in a sentence with a word of the same semantics taken from WordNet. MR replaces the whole entity with another random entity in the same entity type based on the training data; the replacement action for each entity is decided based on the binomial distribution.

Proceedings of the BioNLP 2022 workshop, Dublin, Ireland, pages 123–129

As a result, they could improve the NER performance on both MaSciP and i2b2-2010 corpora.

We find two limitations in Dai and Adel (2020)'s approach. Firstly, although the SR operation takes into account the semantic aspect of tokens, it does not consider the semantic at the entity level. Secondly, the MR operation is performed on the entity level randomly, which may cause semantically incorrect sentences. We hypothesise that if we somehow control the semantics in entity and sentence levels in augmentation operations, we could create a meaningful augmented data, hence improving the NER performance. To that end, we propose Semantic Neighbour Replacement (SNR), a simple data augmentation method for biomedical NER that considers the semantic aspects of both entity and sentence levels.

Specifically, at the entity level, unlike MR (Dai and Adel, 2020), we only replace a source entity with a target one if the target entity is in the same entity type and semantically related to the source one. At the sentence level, we only retain sentences that are semantically related to the original sentence. The semantically related entities and sentences are calculated by using pre-trained language models.

We conducted experiments on two biomedical datasets: i2b2-2010 (Uzuner et al., 2011)—an English corpus of clinical records and VietBioNER (Phan et al., 2022)—a Vietnamese corpus of biomedical texts. Experimental results indicate that using SNR, we can improve NER performance on low-resource settings as well as on full training data. In particular, the F1-scores were increased by 0.52% for i2b2-2010 and 1.3% for VietBioNER.

2 Methodology

The core idea of SNR is to replace entities and to control augmented sentences based on semantic similarity. The method can be divided into three consecutive phases: semantic neighbour extraction, entity replacement, and sentence evaluation.

Semantic Neighbour Extraction: Initially, we perform feature extraction for entities using pre-trained language models. An entity embedding is calculated by taking an average of word embeddings in it. Next, we generate sets of semantic neighbors based on cosine similarity. An entity is a semantic neighbor to another entity if both of them belong to the same entity type and have a cosine similarity greater than or equal to a threshold α.

Entity Replacement: During this phase, we generate new sentences by replacing an entity with another random entity in its semantic neighbor set. For each entity type, we just randomly replace one entity of that type in a sentence. As a result, we obtain a set of augmented sentences from original ones.

Sentence Evaluation: Augmented sentences generated in the previous phase are probably semantically incorrect, which may affect the training process. To alleviate the issue, we perform an automatic evaluation to remove augmented sentences that are semantically different from their original sentences. To that end, we firstly represent both original and augmented sentences as vectors by using a pre-trained sentence-level language model. We then use cosine similarity to estimate the semantic similarity between two sentences. If the cosine similarity of an augmented sentence and its original sentence is less than a threshold θ, the augmented sentence will be discarded.

In this paper, the two parameters α and θ will be in ranges of $[0, 1]$. The larger the α, the greater the semantic similarity between entities, but the smaller the number of neighbours. The θ parameter represents the degree of rigour in the automatic evaluation phase. When θ approximates to 1, only sentences that are very close to the meaning of the original sentence are retained. We therefore can keep only a few of the augmented sentences. In contrast, we can keep more sentences as θ approximates to 0. When θ is set to 0, the sentence evaluation phase will be disabled. At this point, we do not discard any augmented sentences from the second phase. We can fine-tune both α and θ to generate suitable augmented data.

3 Experiments

3.1 Datasets

We conduct experiments on the two datasets including i2b2-2010 (English) (Uzuner et al., 2011) and VietBioNER (Phan et al., 2022) (Vietnamese). The i2b2 corpus includes patient records annotated with three named entity categories of Medical Problem, Test, and Treatment. Meanwhile, VietBioNER is constituted by biomedical grey literature specified for tuberculosis. The corpus was annotated with five named entity categories of Organisation, Location, Date and Time, Symptom and Disease, and Diagnostic Procedure. Some statistics of both corpora are reported in Table 1.

	i2b2-2010	VietBioNER
#Sentence	32894	1706
#Sentence in		
Training set	9558	706
Development set	2389	300
Test set	20947	700
Avg. len. of sent.	13	31
#Entity type	3	5
Vocab size	24321	3548

Table 1: The summary statistic of the two datasets.

Following Dai and Adel (2020), to simulate a low-resource setting, we create small, medium and large sets with different numbers of sentences: 50, 150 and 500, respectively. These sentences are randomly selected from the training part of each dataset. It is noted that our small, medium and large splits of the i2b2 dataset are different from those by Dai and Adel (2020). Augmentation methods are only applied on the training set, we use the same development and test sets for all experiments.

3.2 Language Models

For semantic neighbour extraction, we use ClinicalBERT (Alsentzer et al., 2019)—a pre-trained language model on clinical text for the i2b2-2010 dataset and PhoBERT (Nguyen and Nguyen, 2020)—a pre-trained language model on Vietnamese Wikipedia and news for VietBioNER.

In sentence evaluation, we employ Sentence-BERT (SBERT) (Reimers and Gurevych, 2019), a sentence-level language model for sentence embeddings, to represent both original and augmented sentences.

We use all the mentioned models with the initialised weights provided by Hugging Face[1].

Regarding the NER task training, we also fine-tune the aforementioned language models on the two corpora.

3.3 Experiment Settings

To show the effectiveness of the proposed method, we conducted the following experiments:

- Baseline: We only trained NER models on the original training data.
- Baseline combined with augmented data: We trained NER models on the original training set and its augmented data created by the following three methods:
  - Mention Replacement (**MR**): We followed the MR method proposed by Dai and Adel (2020).
  - Entity Replacement (**ER**): We only performed the first two phases of our proposed method. The last phase, Sentence Evaluation, was disabled by setting the parameter θ to 0.
  - Semantic Neighbour Replacement (**SNR**): We performed all three phases of our proposed method.

[1]https://huggingface.co/models, https://huggingface.co/sentence-transformers

		MR	ER	SNR
i2b2-2010	S	17	19	12
	M	67	90	61
	L	242	347	239
	F	4462	7308	4626
VietBioNER	S	21	9	7
	M	76	13	13
	L	256	86	84
	F	347	550	459

Table 2: Number of augmented sentences in each training set. **S**mall, **M**edium, **L**arge, and **F**ull sets contain 50 sentences, 150 sentences, 500 sentences, and the complete training set, respectively.

It is noted that since in this paper we focus on biomedical entities, we only created an augmented data for Symptom_and_Disease and DiagnosticProcedure entities in the case of VietBioNER. We however report the NER performance on all five NE categories.

3.4 Experimental Results

Based on the fine-tuning results on the development sets, we selected $\alpha = 0.8$ for all sets of i2b2-2010; for VietBioNER, $\alpha = 0.65$ for the full set, and $\alpha = 0.85$ for the other sets; and $\theta = 0.9$ for all cases across the corpora. The number of augmented sentences generated in each setting are reported in Table 2. Since SNR discards augmented sentences that are not semantically related to the original ones, it is reasonable that the numbers of augmented sentences by SNR is less than or equal to those by MR and ER.

We trained NER models on a combination of augmented and original sentences, and applied them to the corresponding testing sets. The NER performance in terms of F1-scores on those sets

Method	i2b2-2010				VietBioNER			
	S	M	L	F	S	M	L	F
Baseline	37.13	67.58	75.53	87.21	59.21	70.78	79.48	79.60
+ MR	**39.56**	67.21	76.35	87.54	**60.98**	71.19	79.31	79.00
+ ER (our method)	39.42	68.36	76.33	87.37	59.31	71.94	**79.51**	80.09
+ SNR (our method)	38.75	**69.43**	**76.86**	**87.73**	59.83	**72.14**	79.34	**80.90**

Table 3: NER performance by different augmentation methods in terms of F1-score. Bold numbers indicate the best performance in a specific setting.

		Sentence
i2b2-2010	**Ori**	Her speech was fluent with no [phasic or praxic problems]$_{Problem}$, [dysarthric]$_{Problem}$.
	MR	Her speech was fluent with no [oral lesions]$_{Problem}$, [left coloboma]$_{Problem}$.
	SNR	Her speech was fluent with no [phasic or praxic problems]$_{Problem}$, [slurred speech]$_{Problem}$.
VietBioNER	**Ori**	Tuy nhiên, các xét nghiệm tế bào và vi trùng trong chẩn đoán [lao]$_{Symptom\&Disease}$ có độ nhạy còn thấp. (However, cytology and bacteria tests in the diagnosis of [TB]$_{Symptom\&Disease}$ have low sensitivity.)
	MR	Tuy nhiên, các xét nghiệm tế bào và vi trùng trong chẩn đoán [ho khan]$_{Symptom\&Disease}$ có độ nhạy còn thấp. (However, cytology and bacteria tests in the diagnosis of [dry cough]$_{Symptom\&Disease}$ have low sensitivity.)
	SNR	Tuy nhiên, các xét nghiệm tế bào và vi trùng trong chẩn đoán [bệnh lao]$_{Symptom\&Disease}$ có độ nhạy còn thấp. (However, cytology and bacteria tests in the diagnosis of [TB disease]$_{Symptom\&Disease}$ have low sensitivity.)

Table 4: **Ori**ginal sentences and their augmented sentences with different methods. Blue texts indicates entity replacement.

are reported in Table 3[2]. Generally, we can see that the NER performance was improved when using data augmentation methods on both English and Vietnamese corpora. Detailed results of precision and recall can be found in Appendix A.

Among the four sizes of the data, MR (Dai and Adel, 2020) could obtain the best performance in the small size setting, across the two corpora. This can be explained by the fact that given only 50 sentences in the training, adding more sentences will help the model overcome overfitting. With the medium size sets, MR could improve the performance on VietBioNER but not on i2b2-2010. In contrast, MR could boost F1-scores on the large and full sets on i2b2-2010, but not on VietBioNER.

Regarding SNR, we could have better F1-scores in most settings of medium, large and full sets, on both English and Vietnamese corpora. With the i2b2 English corpus, the proposed methods has an average improvement of 1.23% of F1-scores (SNR) and 0.58% (ER). Meanwhile, that number by MR is 0.26%. For VietBioNER, the average improvement is 0.84%, 0.63%, and -0.12% of F1-scores for SNR, ER, and MR, respectively. It is worth noting that even with a full training set, using SNR to augment the data training could also boost NER performance. In particular, F1-scores were increased by 0.52% for i2b2-2010 and 1.3% for VietBioNER.

Interestingly, while the number of augmented sentences by SNR is lower than those by ER (as shown in Table 2), the NER performance by SNR is better than those by ER in most of the cases across the corpora. This indicates that having augmented sentences semantically related to the original ones in the training data really improves the NER performance, despite the fact that the total number of sentence is not big. For instance, in the case of i2b2-2010, SNR generated about 37% less sentences than ER, but the NER performance by SNR was still better than those by ER.

3.5 Analysis

Although using MR could help improve the NER performance (as illustrated in Table 3), it is inevitable that MR could produce meaningless sentences. We collected such examples and showed them in Table 4. It can be seen that although MR replaced entities in the same type with the original ones, the resulting sentence is meaningless. Meanwhile, SNR controls the semantic at both entity level and sentence level, hence producing a more meaningful sentence close to the original meaning than the one by MR.

Moreover, we observed that most of sentences discarded by the sentence evaluation were semantically incorrect. We report some of discarded sentences in Table 5. It is obvious that the entity re-

[2]We use the IO tagging scheme.

		Sentence
i2b2-2010	Original	He did not sleep at night before and was [extremely fatigued]$_{Problem}$.
	Augmented	He did not sleep at night before and was [some shortness of breath]$_{Problem}$.
VietBioNER	Original	Hình ảnh [X-quang phổi]$_{DiagnosticProcedure}$ chủ yếu là thâm nhiễm 44%... (The [chest X-ray]$_{DiagnosticProcedure}$ image is mainly infiltrative 44%...)
	Augmented	Hình ảnh [chọc dò màng phổi]$_{DiagnosticProcedure}$ chủ yếu là thâm nhiễm 44%... (The [thoracentesis]$_{DiagnosticProcedure}$ image is mainly infiltrative 44%...)

Table 5: Examples of augmented sentences **discarded** by the Sentence Evaluation phase in SNR. Blue texts indicates entity replacement.

placement altered the meaning of those sentences and made them meaningless. As aforementioned, by discarding those sentences, SNR could produce better NER performance, indicating that it is useful to filter augmented sentences based on their semantic relatedness.

4 Conclusion

In this paper, we proposed a semantic-based data augmentation method for the named entity recognition task in the biomedical domain. Our method, namely Semantic Neighbour Replacement (SNR), simply generates more training sentences based on semantics of entity and sentence. Experiments on simulated low-resource settings show that using the proposed method, we can improve F1 score in both English (i2b2-2010) and Vietnamese (VietBioNER) corpora, even on the full training setting. Such results again confirm the importance of semantics in data augmentation. We believe that SNR can be applied to other domains and other languages as long as we have corresponding pretrained language models.

Similar to previous work, our proposed method only augments in-domain data. Therefore, a followup work would be to study cross-domain augmentation method (Chen et al., 2021), in which we can leverage rich-resource data to enrich low-resource ones.

Acknowledgements

We would like to thank the anonymous reviewers for their useful comments. This research was partially funded by the University of Science, VNU-HCM, Vietnam under grant number CNTT2020-04.

References

Emily Alsentzer, John Murphy, William Boag, Wei-Hung Weng, Di Jindi, Tristan Naumann, and Matthew McDermott. 2019. Publicly available clinical BERT embeddings. In *Proceedings of the 2nd Clinical Natural Language Processing Workshop*, pages 72–78, Minneapolis, Minnesota, USA. Association for Computational Linguistics.

Darina Benikova, Chris Biemann, Max Kisselew, and Sebastian Padó. 2014. GermEval 2014 Named Entity Recognition Shared Task: Companion Paper. In *Proceedings of the KONVENS GermEval workshop*, pages 104–112, Hildesheim, Germany.

Jiaao Chen, Zhenghui Wang, Ran Tian, Zichao Yang, and Diyi Yang. 2020. Local additivity based data augmentation for semi-supervised NER. *Proceedings of The 2020 Conference on Empirical Methods in Natural Language Processing*.

Shuguang Chen, Gustavo Aguilar, Leonardo Neves, and Thamar Solorio. 2021. Data augmentation for cross-domain named entity recognition. In *Proceedings of the 2021 Conference on Empirical Methods in Natural Language Processing*, pages 5346–5356, Online and Punta Cana, Dominican Republic. Association for Computational Linguistics.

Xiang Dai and Heike Adel. 2020. An analysis of simple data augmentation for named entity recognition. *CoRR*, abs/2010.11683.

Steven Y. Feng, Varun Gangal, Jason Wei, Sarath Chandar, Soroush Vosoughi, Teruko Mitamura, and Eduard Hovy. 2021. A survey of data augmentation approaches for NLP. In *Findings of the Association for Computational Linguistics: ACL-IJCNLP 2021*, pages 968–988, Online. Association for Computational Linguistics.

Dat Quoc Nguyen and Anh Tuan Nguyen. 2020. PhoBERT: Pre-trained language models for Vietnamese. In *Findings of the Association for Computational Linguistics: EMNLP 2020*, pages 1037–1042.

Yuyang Nie, Yuanhe Tian, Xiang Wan, Yan Song, and Bo Dai. 2020. Named entity recognition for social media texts with semantic augmentation. In *Proceedings of the 2020 Conference on Empirical Methods in Natural Language Processing*.

Uyen Phan, Phuong Nguyen, and Nhung Nguyen. 2022. A Named Entity Recognition Corpus for Vietnamese

Biomedical Texts to Support Tuberculosis Treatment. In *Proceedings of the 13th Language Resources and Evaluation Conference*. European Language Resources Association.

Nils Reimers and Iryna Gurevych. 2019. Sentence-bert: Sentence embeddings using siamese bert-networks. In *Proceedings of the 2019 Conference on Empirical Methods in Natural Language Processing*. Association for Computational Linguistics.

Erik F. Tjong Kim Sang and Fien De Meulder. 2003. Introduction to the CoNLL-2003 shared task: Language-independent named entity recognition. In *Proceedings of the Seventh Conference on Natural Language Learning at HLT-NAACL 2003*, pages 142–147.

Ozlem Uzuner, Brett South, Shuying Shen, and Scott DuVall. 2011. 2010 i2b2/va challenge on concepts, assertions, and relations in clinical text. *Journal of the American Medical Informatics Association : JAMIA*, 18:552–6.

Jason Wei and Kai Zou. 2019. EDA: Easy data augmentation techniques for boosting performance on text classification tasks. In *Proceedings of the 2019 Conference on Empirical Methods in Natural Language Processing and the 9th International Joint Conference on Natural Language Processing (EMNLP-IJCNLP)*, pages 6382–6388, Hong Kong, China. Association for Computational Linguistics.

A Detailed Results

We report the detailed results of precision, recall and F1-scores on i2b2-2010 in Table 6 and VietBioNER in Table 7.

It is expected that NER performances in terms of recall were mostly improved when using the data augmentation methods. Meanwhile, in terms of precision, the increase or decrease of NER performance was dependent on the data augmentation methods as well as the sizes of the training data. Nevertheless, in the case of full training data, using the SNR method, we could improve the NER performance in both recall and precision across corpora.

Method	Small			Medium			Large			Full		
	P	R	F1	P	R	F1	P	R	F1	P	R	F1
Baseline	43.39	32.45	37.13	66.54	68.65	67.58	74.00	77.13	75.53	86.24	88.20	87.21
+ MR	**44.53**	35.59	**39.56**	63.36	71.55	67.21	73.56	79.35	76.35	86.67	88.42	87.54
+ ER	42.44	**36.79**	39.42	**67.24**	69.52	68.36	72.97	**80.02**	76.33	86.47	88.29	87.37
+ SNR	42.49	35.62	38.75	67.11	**71.90**	**69.43**	**74.37**	79.51	**76.86**	**86.92**	**88.55**	**87.73**

Table 6: NER performance on i2b2-2010 by different augmentation methods in terms of **P**recision, **R**ecall and **F1**-score. Bold numbers indicate the best performance in a specific setting.

Method	Small			Medium			Large			Full		
	P	R	F1	P	R	F1	P	R	F1	P	R	F1
Baseline	56.92	61.69	59.21	67.88	73.93	70.78	**77.12**	81.99	79.48	77.49	81.83	79.60
+ MR	**58.91**	**63.19**	**60.98**	67.79	74.96	71.19	76.60	82.23	79.31	76.85	81.28	79.00
+ ER	57.39	61.37	59.31	**69.70**	74.33	71.94	76.50	**82.78**	**79.51**	77.57	**82.78**	80.09
+ SNR	58.87	60.82	59.83	68.92	**75.67**	**72.14**	76.93	81.91	79.34	**79.09**	**82.78**	**80.90**

Table 7: NER performance on VietBioNER by different augmentation methods in terms of **P**recision, **R**ecall and **F1**-score. Bold numbers indicate the best performance in a specific setting.

Auxiliary Learning for Named Entity Recognition with Multiple Auxiliary Training Data

Taiki Watanabe[1,2], **Tomoya Ichikawa**[2], **Akihiro Tamura**[2]
Tomoya Iwakura[1], **Chunpeng Ma**[1], **Tsuneo Kato**[2]
[1] Fujitsu Ltd., [2] Doshisha University
{watanabe-taiki,iwakura.tomoya,ma.chunpeng}@fujitsu.com
ctwg0109@mail4.doshisha.ac.jp
{aktamura,tsukato}@mail.doshisha.ac.jp

Abstract

Named entity recognition (NER) is one of the core technologies for knowledge acquisition from text and has been used for knowledge extraction of chemicals and medicine. As one of the NER improvement approaches, multi-task learning that learns a model from multiple training data has been used. Among multi-task learning, an auxiliary learning method, which uses training data of an auxiliary task for improving its target task, has shown higher NER performance than conventional multi-task learning for improving all the tasks simultaneously. The conventional auxiliary learning method uses only one auxiliary training dataset. We propose *M*ultiple *U*tilization of *N*ER *C*orpora *H*elpful for *A*uxiliary *BLES*sing (**MUNCHABLES**). MUNCHABLES utilizes multiple training datasets as auxiliary training data by the following methods : the first one is to fine-tune the NER model of the target task by sequentially performing auxiliary learning for each auxiliary training dataset, and the other is to use all training datasets in one auxiliary learning. We evaluate MUNCHABLES on eight chemical/biomedical/scientific domain NER tasks, where seven training datasets are used as auxiliary training data. The experiment results show that our proposed methods achieve higher NER performance than conventional multi-task learning methods on average and that NER performance can be improved by using multiple auxiliary training data. Furthermore, the proposed models outperform state-of-the-art models on the datasets.

1 Introduction

Named entity recognition (NER) is a fundamental natural language processing technology for extracting named entity (NE) and technical terms from input texts and has been put to practical use in various situations. For example, NER is used as one of the core technologies for structuring and accumulating information on interrelationships among chemical substances and physical properties of chemical substances, which are reported daily in papers and patents, to develop new materials and products.

NER has been actively studied for a long time, and many NER methods have been proposed. In recent years, neural network (NN)-based methods have become dominant, and a BiLSTM-CRF model (e.g., Huang et al. (2015)), composed of two recurrent neural networks (RNNs) and conditional random fields (CRF), and a Transformer-based model (e.g., Lee et al. (2019)) have achieved high performance in NER.

In addition, it has been reported that the performance of an NER model is improved by multi-task learning, which uses training data of a task different from the target task and simultaneously learns features from multiple NER training datasets (Wang et al., 2019a; Crichton et al., 2017a; Khan et al., 2020; Mehmood et al., 2020; Wang et al., 2019b). Remarkably, Wang et al. (2019a) have shown that, an NER in the biotechnology field (BioNER) with an auxiliary learning method, which is a variant of multi-task learning, achieves higher performance in the target task, compared to a standard multi-task learning method. The auxiliary learning uses a task other than the target task as an auxiliary task for improving the target task performance, in contrast the standard multi-task learning learns models for multiple tasks to improve performance of the multiple tasks.

We propose a new auxiliary learning paradigm that uses multiple NER datasets as auxiliary training data, *M*ultiple *U*tilization of *N*ER *C*orpora *H*elpful for *A*uxiliary *BLES*sing (**MUNCHABLES**), whereas existing auxiliary learning uses only one type of auxiliary training data. Specifically, we propose two types of multi-auxiliary learning: the first one is to fine-tune the NER model of the target task by sequentially performing auxiliary learning for each auxiliary training

Proceedings of the BioNLP 2022 workshop, Dublin, Ireland, pages 130–139
May 26, 2022.

dataset (MUNCHABLES-stack model), and the other is to use all types of training data in single auxiliary learning. As for the latter, we propose two models: one is to concatenate all the multiple auxiliary training datasets and make a batch by randomly selecting data from the auxiliary training dataset (MUNCHABLES-concatenation model), and the other is to change auxiliary training datasets every epoch (MUNCHABLES-iteration model).

We compare the proposed MUNCHABLES models with standard multi-task learning and single auxiliary learning on eight chemical/ biomedical/ scientific domain NER tasks. As for our proposed models, seven training datasets are used as auxiliary training data in each task. The experiment results show that the F1-scores of the proposed models are higher than those of the baselines on average and NER performance can be improved by using multiple auxiliary training datasets. In addition, the proposed models achieve state-of-the-art performance in chemical/biomedical/scientific NER.

2 Existing Multi-Task Learning

This section describes existing multi-task learning methods which use training data of a different task other than the target task. We first outline the NER model used as the base model, and then describe an extension of the NER model to multi-task learning, where multiple tasks are trained simultaneously. In this multi-task learning, the target task and the other tasks are treated equally. Then, we explain an existing auxiliary learning model, which uses training data for a different task from the target task as auxiliary training data.

2.1 Multi-Task Learning Model

In this study, we use the BiLSTM-CRF model proposed by Huang et al. (2015) as our baseline NER model. The BiLSTM-CRF model is a sequence labeling model composed of bi-directional LSTM and CRF.

The BiLSTM-CRM model first computes the intermediate representation of each word in an input sentence using bidirectional LSTM. Let an input sentence be $\mathbf{w} = w_1, w_2, \cdots, w_N$ and the embedding vectors outputted by an embedding layer be $\mathbf{x} = \mathbf{x_1}, \mathbf{x_2}, \cdots, \mathbf{x_N}$. The intermediate representation $\mathbf{e_i}$ of the word w_i is calculated as follows:

$$\overrightarrow{\mathbf{h_i}} = LSTM^{(f)}(\mathbf{x_i}, \overrightarrow{\mathbf{h_{i-1}}}), \tag{1}$$

$$\overleftarrow{\mathbf{h_i}} = LSTM^{(b)}(\mathbf{x_i}, \overleftarrow{\mathbf{h_{i+1}}}), \tag{2}$$

$$\mathbf{h_i} = [\overrightarrow{\mathbf{h_i}}; \overleftarrow{\mathbf{h_i}}], \tag{3}$$

$$\mathbf{e_i} = \mathbf{W^{(e)}} \mathbf{h_i}, \tag{4}$$

where $\rightarrow$ and $\leftarrow$ denote forward and backward directions, respectively, and $LSTM^{(f)}$ and $LSTM^{(b)}$ are forward and backward LSTMs, respectively. ";" denotes the concatenation of vectors. $\mathbf{W^{(e)}} \in R^{k \times d}$ is a weight matrix, d is the dimension of the hidden state vector $\mathbf{h_i}$, and k is the number of labels to be identified.

Then, the intermediate representations $\mathbf{e}$ computed by the bi-directional LSTM are fed to the CRF layer to obtain a label sequence. The score function for the label sequence $\mathbf{y} = (y_1, y_2, \cdots, y_N)$ is defined by using the score matrix $\mathbf{P} = (\mathbf{e_1}, \mathbf{e_2}, \cdots \mathbf{e_N})^T$, which is converted from the intermediate representations $\mathbf{e}$, and the transition score matrix $\mathbf{A}$ as follows:

$$\mathrm{s}(\mathbf{e}, \mathbf{y}) = \sum_{i=0}^{N} A_{y_i, y_{i+1}} + \sum_{i=1}^{N} P_{i, y_i}, \tag{5}$$

where $A_{i,j}$ represents the transition score from the label i to the label j. The output label sequence $\mathbf{y}^*$ is obtained by finding $\mathbf{y}$ that maximizes the score as follows:

$$\mathbf{y}^* = \arg\max_{\tilde{\mathbf{y}} \in \mathbf{Y_w}} \mathrm{s}(\mathbf{e}, \tilde{\mathbf{y}}), \tag{6}$$

where $\mathbf{Y_w}$ is the set of all possible label sequences for the input sentence $\mathbf{w}$.

Using the score function, the output probability of the label sequence $\mathbf{y}$ is defined by the softmax function as follows:

$$p(\mathbf{y}|\mathbf{w}) = \frac{\exp(\mathrm{s}(\mathbf{e}, \mathbf{y}))}{\sum_{\tilde{\mathbf{y}} \in \mathbf{Y_w}} \exp(\mathrm{s}(\mathbf{e}, \tilde{\mathbf{y}}))}. \tag{7}$$

In training, the parameters that minimize the following loss function are obtained:

$$L = - \sum_{(\mathbf{w}, \hat{\mathbf{y}}) \in D} \log(p(\hat{\mathbf{y}}|\mathbf{w})), \tag{8}$$

where D is a training dataset.

Figure 1 shows an overview of the BiLSTM-CRF model extended for multi-task learning. In the model, the word embedding layer and BiLSTM layer are shared by all the training datasets and the weights of these layers are the same on all the tasks. On the other hand, the CRF layer is prepared for each dataset and the weights of the CRF

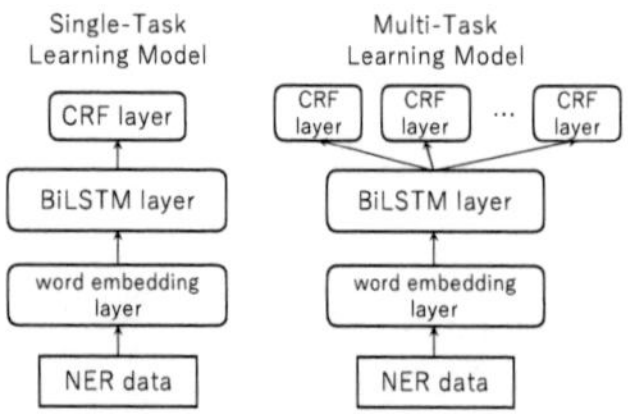

Figure 1: Overview of a multi-task learning model

layer are not shared. The objective function of the multi-task learning model is defined as follows:

$$Loss = \frac{1}{M}\sum_{i=1}^{M} L_i, \tag{9}$$

where L_i $(i = 1, 2, \cdots M)$ is the loss in the CRF layer for each training dataset (see Eq. 8), and M is the number of training datasets.

In the multi-task learning model, training data for the target task and that for the other tasks are treated equally, and thus an NER model common to all the tasks is learned. Larger datasets require more batches during training. In inference, NER is performed by using the CRF layer corresponding to the target task in the learned NER model.

2.2 Auxiliary Learning Model

Wang et al. (2019a) have proposed an auxiliary learning method, which is a multi-task learning method that distinguishes between training data for the target task (main training data) and that for the other task (auxiliary training data), and have improved NER performance for the target task. The auxiliary learning model is trained by using a main batch composed of main training data and an auxiliary batch composed of auxiliary training data. In each iteration, the model parameters are updated by the auxiliary batch first, and then by the main batch. This alternating updates by the main and auxiliary batches are repeated until the loss on the main training data converges.

Algorithm 1 shows the algorithm for the auxiliary learning method. In Algorithm 1, the subscripts denote the target task ($main$) and the auxiliary task (aux). $Epoch$ and $Iteration$ are the number of epochs and the number of iterations for the main task, respectively, and $BatchSize$ is the batch size. The number of iterations for each epoch is the total number of the main training data divided by the batch size (i.e., $Iteration = |D_{main}|/BatchSize$). The $extract$

Algorithm 1 Algorithm of an existing auxiliary learning method

Data: main training dataset D_{main}, auxiliary training dataset D_{aux}

1: **for** $i = 1$ to $EPOCH$ **do**
2:   **for** $j = 1$ to $ITERATION$ **do**
3:     $Batch_{main} = extract(D_{main}, BatchSize)$
4:     $Batch_{aux} = extract(D_{aux}, BatchSize)$
5:     $train(Model, Batch_{aux})$
6:     $train(Model, Batch_{main})$
7:   **end for**
8:   $is\_converge_{main}(Model)$
9: **end for**

Algorithm 2 Algorithm of the MUNCHABLES-concatenation model

Data: main training dataset D_{main}, M auxiliary training datasets $D_{aux}^{(1)}, D_{aux}^{(2)}, \cdots, D_{aux}^{(M)}$

1: $D_{aux} = [D_{aux}^{(1)}; D_{aux}^{(2)}; \cdots; D_{aux}^{(M)}]$
2: **for** $i = 1$ to $EPOCH$ **do**
3:   **for** $j = 1$ to $ITERATION$ **do**
4:     $Batch_{main} = extract(D_{main}, BatchSize)$
5:     $Batch_{aux} = extract(D_{aux}, BatchSize)$
6:     $train(Model, Batch_{aux})$
7:     $train(Model, Batch_{main})$
8:   **end for**
9:   $is\_converge_{main}(Model)$
10: **end for**

---

function in lines 4 and 5 creates a batch by extracting $Batchsize$ data from the training dataset, and the $train$ function in lines 6 and 7 updates the parameters of the NER model $Model$ by using the batch data. The $is\_converge_{main}$ function in line 8 judges whether to stop training or not according to the loss on the target task.

3 MUNCHABLES: Multi-Auxiliary Learning

An existing auxiliary learning method uses only one auxiliary training dataset. In this section, we propose a new auxiliary learning paradigm, multi-auxiliary learning MUNCHABLES, that utilizes multiple training datasets as auxiliary training data. We first propose two MUNCHABLES models that use multiple auxiliary training datasets in single auxiliary learning (MUNCHABLES-concatenation model and MUNCHABLES-iteration model), and then propose a MUNCHABLES model that sequentially fine-tunes a main model by auxiliary learning with each auxiliary training dataset (MUNCHABLES-stack model).

3.1 MUNCHABLES-Concatenation Model

Algorithm 3 Algorithm of the MUNCHABLES-iteration model

Data: main training dataset D_{main}, M auxiliary training datasets $D_{aux}^{(1)}, D_{aux}^{(2)}, \cdots, D_{aux}^{(M)}$

for $i = 1$ to $EPOCH$ **do**
    for $k = 1$ to M **do**
        for $j = 1$ to $ITERATION$ **do**
            $Batch_{main} = extract(D_{main}, BatchSize)$
            $Batch_{aux} = extract(D_{aux}^{(k)}, BatchSize)$
            $train(Model, Batch_{aux})$
            $train(Model, Batch_{main})$
        end for
    end for
    $is\_converge_{main}(Model)$
end for

Algorithm 4 Algorithm of the MUNCHABLES-stack model

Data: main training dataset D_{main}, M auxiliary training datasets $D_{aux}^{(1)}, D_{aux}^{(2)}, \cdots, D_{aux}^{(M)}$

for $k = 1$ to M **do**
    for $i = 1$ to $EPOCH$ **do**
        for $j = 1$ to $ITERATION$ **do**
            $Batch_{main} = extract(D_{main}, BatchSize)$
            $Batch_{aux} = extract(D_{aux}^{(k)}, BatchSize)$
            $train(Model, Batch_{aux})$
            $train(Model, Batch_{main})$
        end for
        $is\_converge_{main}(Model)$
    end for
end for

The MUNCHABLES-concatenation model is a multi-auxiliary learning model that concatenates all the multiple auxiliary training datasets and treats the concatenated training data as one auxiliary training dataset in single auxiliary learning. Algorithm 2 shows the algorithm of the MUNCHABLES-concatenation model. Just like the existing single auxiliary learning model, the MUNCHABLES-concatenation model creates a main batch from the main training data and an auxiliary batch from the concatenated auxiliary training data. Then, the updates of model parameters with the auxiliary batch and with the main batch are repeated alternately until the loss on the main training dataset converges. The difference from the existing single auxiliary learning model is that an auxiliary batch is created from the concatenated data of multiple auxiliary training datasets, and thus an auxiliary batch can contain multiple types of auxiliary training data.

3.2 MUNCHABLES-Iteration Model

The MUNCHABLES-iteration model is a multi-auxiliary learning model which changes training datasets used as an auxiliary training dataset every epoch. Algorithm 3 shows algorithm of the MUNCHABLES-iteration model. The MUNCHABLES-iteration model alternately repeats parameter updates with the main batch created from the main training dataset and those with the auxiliary batch created from an auxiliary training dataset until the loss on the main training dataset converges as well as auxiliary learning models described so far. The difference from the MUNCHABLES-concatenation model is that an auxiliary batch in the MUNCHABLES-iteration model is created from a specific auxiliary training dataset and the source auxiliary training dataset is switched every epoch.

3.3 MUNCHABLES-Stack Model

The MUNCHABLES-stack model is a multi-auxiliary learning model that fine-tunes a main model as many as the number of auxiliary training datasets by sequential auxiliary learning with each auxiliary training dataset. Each auxiliary learning is performed by using a specific auxiliary training dataset as well as the existing single auxiliary learning. When the loss on the main training dataset converges, auxiliary data is switched to a new auxiliary training dataset and subsequently the main model is fine-tuned using the new auxiliary training dataset. Algorithm 4 shows the outline and algorithm of the MUNCHABLES-stack model, respectively. While, in the MUNCHABLES-concatenation model and MUNCHABLES-iteration model, a main model is trained only once (i.e., convergence is only once), in the MUNCHABLES-stack model, a main model is trained as many as the number of auxiliary training datasets.

4 Experiment

4.1 Experiment Settings

We evaluated our proposed models on eight chemical/biomedical/scientific domain NER tasks. Table 1 shows each NER dataset. We compared our three proposed models, the MUNCHABLES-concatenation model (*MUNCH.-Conc*), the MUNCHABLES-iteration model (*MUNCH.-Iter*), and the MUNCHABLES-stack model (*MUNCH.-Stack*), with three baseline models, the single task learning model (*SingleTask*), the standard multi-task learning model (*MultiTask*), and the existing single auxiliary learning model (*SingleAux*), which are described in Section 2. *MultiTask* learns one NER model from all the eight datasets. Our three MUNCHABLES models use all the datasets other than the target task (i.e., seven datasets) as auxiliary training data. *SingleAux* selected an auxiliary training dataset on the development data of the main task. Specifically, *SingleAux* used the model that achieved the best performance (i.e., F1-score) on the development data among seven models each of which is trained by single auxiliary learning with a training dataset for a task other than the target task, for testing. In *MUNCH.-Iter* and *MUNCH.-Stack*, the seven auxiliary training datasets were randomly sorted on condition that auxiliary datasets with the same NE type are not consecutive. We discuss the order of auxiliary training datasets in Section 5.1.

We implemented each NER model by extending the open framework FLAIR (Akbik et al., 2019). For word embeddings, we used Contextual String Embeddings (Akbik et al., 2018) and FastText (Bojanowski et al., 2017) provided by FLAIR, both of which were trained from the PubMed abstracts, a corpus of medical literature. The dimension of the BiLSTM layer was set to 256. We used the SGD optimizer, where a learning rate was adjusted by the following scheduling policy: the learning rate was reduced by a factor of two when the loss per epoch was not less than the minimum loss so far for four consecutive epochs, and training was terminated when the learning rate fell below 1e-4. We used the model at the end of training for testing. In hyperparameter tuning, we tried 0.1 and 0.05 as the initial learning rate and 16 and 32 as the batch size. Four models with these hyperparameter combinations were evaluated on the development data, and the hyperparameter set with the best performance was selected. In testing, we trained an NER model from the training data and the development data, and we reported and compared the performance on the test data. NER performance was evaluated by F1-score.

4.2 Experiment Results

Table 2 shows the experiment results. As can be seen in the table, *SingleAux* outperforms *SingleTask* and *MultiTask* on micro and macro average F1-scores. This suggests that auxiliary learning is more effective than the multi-task learning method where the training data for the target task and the other training data are equally treated. The observation is consistent with previously reported results. Table 2 also shows that *MUNCH.-Iter* and *MUNCH.-Stack* achieve higher performance than *SingleAux* on average and at least one of the MUNCHABLES models is better than *SingleAux* on all the tasks. These results experimentally demonstrate that NER performance can be improved by using multiple auxiliary training datasets in auxiliary learning as in the proposed models, which shows the effectiveness of the proposed auxiliary learning paradigm for NER.

In *MUNCH.-Iter* and *MUNCH.-Conc*, the main model only needs to be trained once, while *MUNCH.-Stack* requires fine-tuning on each auxiliary training dataset individually, so the training time for *MUNCH.-Stack* is longer than the other two MUNCHABLES models. Table 2 shows that *MUNCH.-Stack* achieves the best performance on two out of the eight tasks and its micro and macro average scores are the highest. This indicates the necessity of *MUNCH.-Stack* on some NER tasks even at longer training time.

5 Discussion

5.1 Discussion on the Order of Auxiliary Training Datasets

The performance of *MUNCH.-Iter* and *MUNCH.-Stack* might be affected by the order of auxiliary training datasets ($D_{aux}^{(1)}, D_{aux}^{(2)}, \cdots$ in Algorithms 3 and 4). This section discusses the impact of the order to NER performance.

In the experiments of Section 4, the auxiliary datasets were randomly sorted on condition that auxiliary datasets with the same NE type are not consecutive, in *MUNCH.-Iter* and *MUNCH.-Stack*. However, we conjecture that, in *MUNCH.-Iter* and *MUNCH.-Stack*, the auxiliary training dataset closer to the end of the training of the main model have a larger impact. Based on the conjecture, we sort the auxiliary training datasets in order of the degree of contribution to the performance improvement of the target task. Hereafter, the models are denoted as *MUNCH.-Iter (sort)* and *MUNCH.-Stack (sort)*. Specifically, we first evaluated the performance on the development data of the single auxiliary learning model with each auxiliary training dataset, and then sorted the auxiliary training datasets in ascending order of its single auxiliary learning models' performance and used the sorted training datasets in *MUNCH.-Iter* and *MUNCH.-Stack*.

We describe the order of auxiliary train-

Dataset	Type of NE	# of Sentences			# of Words			# of Annotations		
		Train	Dev	Test	Train	Dev	Test	Train	Dev	Test
NCBI Disease (Doğan et al., 2014)	Disease	5,424	923	940	135,701	23,969	24,497	5,134	787	960
BC5CDR (Li et al., 2016)	Disease	4,560	4,581	4,797	118,170	117,453	124,750	4,182	4,246	4,424
BC5CDR (Li et al., 2016)	Drug/Chem	4,560	4,581	4,797	118,170	117,453	124,750	5,203	5,347	5,385
CHEMDNER (Krallinger et al., 2015)	Drug/Chem	30,682	30,639	26,364	893,685	887,805	767,636	29,478	29,486	25,346
BC2GM (Smith et al., 2008)	Gene/Protein	12,574	2,519	5,038	355,405	71,042	143,465	15,197	3,061	6,325
JNLPBA (Kim et al., 2004)	Gene/Protein	14,690	3,856	3,856	443,653	117,213	114,709	32,178	8,575	6,241
LINNAEUS (Gerner et al., 2010)	Species	11,935	4,078	7,142	281,273	93,877	165,095	2,119	711	1,433
s800 (Pafilis et al., 2013)	Species	5,733	830	1,630	147,291	22,217	42,298	2,557	384	767

Table 1: NER datasets

	NCBI-Disease	BC5CDR Disease	BC5CDR Chem	CHEMD NER	BC2GM	JNLPBA	LINN AEUS	s800	MA. AVG.	MI. AVG.
SingleTask	87.56	86.65	94.12	92.25	83.63	77.31	88.06	75.41	85.62	88.46
MultiTask	87.72	86.12	**94.53**	92.00	83.44	**77.86**	89.06	76.71	85.93	88.44
SingleAux	88.41	86.53	94.27	92.29	83.24	77.71	88.88	76.80	86.02	88.63
MUNCH.-Conc	**89.14**	86.73	94.23	**92.36**	82.57	77.48	**89.46**	76.42	86.05	88.49
MUNCH.-Iter	88.33	86.85	94.52	92.18	82.90	77.78	88.98	**77.20**	86.09	88.52
MUNCH.-Stack	87.69	**86.98**	94.33	92.32	**83.80**	77.62	89.42	76.65	**86.10**	**88.67**

Table 2: Experiment results (F1-score (%)). MA. AVG. and MI. AVG. indicate macro average and micro average, respectively. Each bold font value indicates the best result of each task.

Batch Size		16		32	
Learning Rate		0.05	0.1	0.05	0.1
Single Aux	NCBI-Disease	89.98	90.03	89.69	90.08
	BC5CDR Disease	89.95	90.08	89.87	89.93
	BC5CDR Chem	90.05	90.16	90.00	90.03
	BC2GM	89.95	89.92	89.88	89.91
	JNLPBA	89.90	89.93	89.88	89.89
	LINNAEUS	90.15	90.02	89.90	90.04
	s800	89.88	90.01	89.89	89.94
MUNCH.-Iter (sort)		90.04	90.11	90.03	**90.14**
MUNCH.-Stack (sort)		**90.24**	90.07	89.75	90.14

Table 3: Tuning of hyperparameters and the order of auxiliary training datasets of *MUNCH.-Iter (sort)* and *MUNCH.-Stack (sort)* on the CHEMDNER task (i.e., F1-score (%) on the CHEMDNER development data). In *MUNCH.-Iter (sort)* and *MUNCH.-Stack (sort)*, the best result is shown in bold font value.

Model	*MUNCH.-Iter*		*MUNCH.-Stack*	
Sort	w/o	w/	w/o	w/
NCBI-Disease	88.33	**88.50**	87.69	**87.90**
BC5CDR Disease	**86.85**	**86.85**	**86.98**	86.86
BC5CDR Chem	**94.52**	94.33	94.33	**94.47**
CHEMDNER	92.18	**92.39**	92.32	**92.35**
BC2GM	82.90	**83.59**	83.80	**83.84**
JNLPBA	**77.78**	77.28	**77.62**	77.21
LINNAEUS	**88.98**	88.82	**89.42**	88.87
s800	**77.20**	76.36	**76.65**	76.46
MA. AVG.	**86.09**	86.02	**86.10**	86.00
MI. AVG.	88.52	**88.64**	**88.67**	88.64

Table 4: Impact of the order of auxiliary training datasets. Each bold font value indicates the better result with or without sorting.

ing datasets in *MUNCH.-Iter (sort)* and *MUNCH.-Stack (sort)* for the CHEMDNER task as an example. Table 3 shows the performance on the CHEMDNER development data of *SingleAux* with each auxiliary NER training dataset for all combination of hyperparameters. Note that each model is trained from only training data to evaluate the performance on development data. As for *MUNCH.-Iter (sort)* and *MUNCH.-Stack (sort)*, the auxiliary training datasets are sorted on the basis of the performance of *SingleAux* with the same hyperparameter setting. From the table, for *MUNCH.-Iter (sort)*, the batch size and learning rate were set to 32 and 0.1, respectively, and the order of the auxiliary training datasets was set to "JNLPBA → BC2GM → BC5CDR-Disease → s800 → BC5CDR-Chem → LINNAEUS → NCBI-Disease." For *MUNCH.-Stack (sort)*, the batch size and learning rate were set to 16 and 0.05, respectively, and the order of the auxiliary training datasets was set to "s800 → JNLPBA → BC5CDR-Disease → BC2GM → NCBI-Disease → LINNAEUS → BC5CDR-Chem."

Table 4 shows the performance of *MUNCH.-Iter*,

	NCBI-Disease	BC5CDR Disease	BC5CDR Chem	CHEMDNER	BC2GM	JNLPBA	LINNAEUS	s800
SingleAux (reimpl)	88.41	86.53	94.27	92.29	83.24	77.71	88.88	76.80
BioBERT	**89.36**	86.56	93.44	91.41	84.4	77.59	**89.81**	75.31
HanPaNE	-	-	-	**92.57**	-	-	-	-
SciBERT	88.57	-	-	-	-	77.28	-	-
BioMegatron	87.0	**88.5**	92.5	-	-	-	-	-
SciFive	88.46	87.62	**94.61**	91.56	83.57	77.55	-	76.33
PubMedBERT (PubMed)	87.82	85.62	93.33	-	**84.52**	79.10	-	-
PubMedBERT (+PMC)	88.04	85.76	93.34	-	84.37	**79.16**	-	-
MUNCH.-Conc	89.14	86.73	94.23	92.36	82.57	77.48	89.46	76.42
MUNCH.-Iter	88.33	86.85	94.52	92.18	82.90	77.78	88.98	**77.20**
MUNCH.-Stack	87.69	86.98	94.33	92.32	83.80	77.62	89.42	76.65

Table 5: Comparison with previous results (F-measure (%)). These results are BioBERT (Lee et al., 2019), HanPaNE (Watanabe et al., 2019), SciBERT (Beltagy et al., 2019), BioMegatron (Shin et al., 2020), SciFive (Phan et al., 2021), and PubMedBERT (Gu et al., 2021). Each bold font value indicates the best result of each task.

MUNCH.-Iter (sort), *MUNCH.-Stack*, and *MUNCH.-Stack (sort)* on each test data. Table 4 shows that *MUNCH.-Iter (sort)* obtained a higher micro average than *MUNCH.-Iter* while the macro average of *MUNCH.-Iter (sort)* and the micro and macro averages of *MUNCH.-Stack (sort)* are worse than those of *MUNCH.-Iter* and *MUNCH.-Stack*, respectively. The results indicate that the performance of *MUNCH.-Iter* and *MUNCH.-Stack* are affected by the order of auxiliary training datasets and the performance could be improved by reordering auxiliary training datasets in ascending order of the performance on the development data of *SingleAux* on some NER tasks. We conjecture that sorting order of auxiliary training datasets might be affected by similarity of development data and test data. We will leave its further analysis for future work.

6 Related Work

Previous Methods		MUNCHABLES		*common*
v.s. Method	MA. AVG.	*Iter*	*Stack*	*tasks*
BioBERT	85.99	**86.09**	**86.10**	8
HanPaNE	92.57	92.18	92.32	1
SciBERT	82.93	**83.05**	82.66	2
BioMegatron	89.33	**89.90**	**89.67**	3
SciFive	85.67	**85.68**	85.63	7
PubMedBERT (PubMed)	86.08	86.08	86.08	5
PubMedBERT (+PMC)	86.13	86.08	86.08	5

Table 6: Summary of macro average F-measure (%). The *common tasks* indicates the number of tasks used by both of our MUNCHABLES and previous methods. We compared our MUNCHABLES methods with previous methods in terms of macro average F-measure on the common tasks. The bold font indicates that a MUNCHABLES model is better than the corresponding previous result.

6.1 Comparison with Previous Results

We compared our MUNCHABLES models with previous results[1] including state-of-the-art methods. Table 5 shows the results, and Table 6 shows a summary of the comparison, where we report macro average F-measure on the common tasks used by both of the MUNCHABLES and previous methods. As can be seen in Tables 5 and 6, in general, our MUNCHABLES models obtain competitive or better NER performance than previous results. These results show that our MUNCHABLES models achieve state-of-the-art performance on chemical/biomedical/scientific NER tasks. Another remarkable point is MUNCHABLES can be combined with the previous work. In other words, in order to improve the previous work, we can use MUNCHABLES in the previous work.

v.s. BioBERT BioBERT (Lee et al., 2019) is a BERT-based pre-training model trained with biomedical domain text. We compared BioBERT v1.0 with PubMed + PMC for its pre-training with our MUNCHABLES models. The macro average of BioBERT was 85.99 and those of *MUNCH.-Iter* and *MUNCH.-Stack* are 86.09 and 86.10. Our MUNCHABLES models obtained a higher performance than BioBERT.

v.s. HanPaNE HanPaNE (Watanabe et al., 2019) is a BiLSTM-CRF NER model that jointly learns an LSTM-based chemical compound paraphrase model through multi-task learning. HanPaNE showed 92.57 F-measure on CHEMDNER, which is the state-of-the-art performance on the

[1] If results obtained by different parameters were reported, we listed the results of the model that showed the best macro average F-measure on the NER datasets.

dataset. *MUNCH.-Stack* and *MUNCH.-Iter* are worse than HanPaNE. However, our MUNCHABLES model is compatible with HanPaNE and the models must complement each other. Therefore, we expect higher performance by combining MUNCHABLES with HanPaNE.

v.s. SciBERT SciBERT (Beltagy et al., 2019) is a BERT-based pre-training model trained with scientific domain text. SciBERT was evaluated on NCBI-Diseases and JNLPBA, and the macro average was 82.93. *MUNCH.-Iter* obtained a higher average (i.e., 83.05) than SciBERT.

v.s. BioMegatron BioMegatron (Shin et al., 2020) is a biomedical adaptation of a transformer model called Megatron-LM (Shoeybi et al., 2020). BioMegatron was evaluated on NCBI-Disease, BC5CDR Disease, and BC5CDR Chem. The macro average of BioMegatron with a 50k biomedical domain vocabularies and 345m parameters was 89.33, whereas *MUNCH.-Iter* and *MUNCH.-Stack* showed 89.90 and 89.67, which are higher than BioMegatron.

v.s. SciFive SciFive (Phan et al., 2021) is a domain-specific Text-to-Text Transfer Transformer (T5) (Raffel et al., 2020) model that has been pre-trained on large biomedical corpora. SciFive was evaluated on seven tasks out of the eight tasks except for LINNAEUS. The macro average F-measure of SciFive with PMC pre-training data was 85.67, whereas *MUNCH.-Iter* and *MUNCH.-Stack* were 85.68 and 85.63.

v.s. PubMedBERT PubMedBERT (Gu et al., 2021) is a BERT-based model trained with biomedical domain text from scratch. PubMedBERT (PubMed) was trained with only PubMed and PubMedBERT (+PMC) was trained with PubMed and PMC and these two models were evaluated on NCBI-Disease, BC5CDR Disease, BC5CDR Chem, BC2GM, and JNLPBA for NER. The macro average of PubMedBER (PubMed) was 86.08 and that of PubMedBERT (+PMC) was 86.13. *MUNCH.-Iter* and *MUNCH.-Stack* show the comparable accuracy as PubMedBERT (PubMed), however they showed lower accuracy than PubMedBERT (+PMC). We think that this difference was caused by the pretraining data size. The MUNCHABLES models were pretrained only with PubMed. Therefore, further improvement by increasing the amount of pretraining data is expected. Furthermore, the MUNCHABLES can be incorporated into PubMedBERT, therefore, we expect higher performance by enhancing PubMedBERT with MUNCHABLES.

6.2 Multi-Task Learning

Multi-task learning is employed to boost the performance of NLP systems (Liu et al., 2015; Luong et al., 2016; Dong et al., 2015; Hashimoto et al., 2017), including NER (Liu et al., 2018). Multi-task learning of sequence labeling with language models was proposed (Rei, 2017). Aguilar et al. (2018) and Cao et al. (2018) proposed multi-task learning of NER with word segmentation. Peng and Dredze (2017) proposed multi-task learning that leverages the performance of domain adaptation. Clark et al. (2018) proposed multi-task learning of NER with several NLP tasks such as POS tagging and parsing. Crichton et al. (2017b) and Wang et al. (2018) proposed multi-task learning of several tasks of biomedical NLP to increase NER performance. Watanabe et al. (2019) proposed multi-task learning of NER with chemical compound paraphrase.

Sampling methods for multi-task learning have also been proposed. Guo et al. (2019) is a two-stage mulit-task pipeline, where the first stage automatically selects the most useful auxiliary tasks via a Beta-Bernoulli multi-armed bandit with Thompson Sampling and the second stage learns the training mixing ratio of these selected auxiliary tasks. Kung et al. (2021) proposed a sampling method for training samples of auxiliary tasks based on the assumption that the more similar to the target task is, the more benefit is obtained for the target task.

7 Conclusion

This paper proposed a new auxiliary learning paradigm for NER, MUNCHABLES, that utilizes multiple training datasets as auxiliary training data for improving the performance of its target task. The experiments on eight chemical/biomedical/scientific domain NER datasets, showed that our proposed models achieved higher performance on average than conventional multi-task learning methods and an auxiliary learning method using only one auxiliary training dataset. Moreover, our proposed models achieved the state-of-the-art performance on chemical/biomedical/scientific NER tasks.

References

Gustavo Aguilar, Adrian Pastor López Monroy, Fabio González, and Thamar Solorio. 2018. Modeling noisiness to recognize named entities using multitask neural networks on social media. In *Proceedings of the 2018 Conference of the NAACL-HLT, Volum e 1 (Long Papers)*, pages 1401–1412.

Alan Akbik, Tanja Bergmann, Duncan Blythe, Kashif Rasul, Stefan Schweter, and Roland Vollgraf. 2019. Flair: An easy-to-use framework for state-of-the-art nlp. In *Proceedings of the 2019 NAACL-HLT (Demonstrations)*, pages 54–59.

Alan Akbik, Duncan Blythe, and Roland Vollgraf. 2018. Contextual string embeddings for sequence labeling. In *Proceedings of the 27th COLING*, pages 1638–1649.

Iz Beltagy, Kyle Lo, and Arman Cohan. 2019. SciBERT: A pretrained language model for scientific text. In *Proceedings of the 2019 EMNLP-IJCNLP*, pages 3615–3620, Hong Kong, China. Association for Computational Linguistics.

Piotr Bojanowski, Edouard Grave, Armand Joulin, and Tomas Mikolov. 2017. Enriching word vectors with subword information. *TACL*, 5:135–146.

Pengfei Cao, Yubo Chen, Kang Liu, Jun Zhao, and Shengping Liu. 2018. Adversarial transfer learning for chinese named entity recognition with self-attention mechanism. In *Proceedings of the 2018 EMNLP*, pages 182–192.

Kevin Clark, Minh-Thang Luong, Christopher D. Manning, and Quoc Le. 2018. Semi-supervised sequence modeling with cross-view training. In *Proceedings of the 2018 EMNLP*, pages 1914–1925. Association for Computational Linguistics.

Gamal Crichton, Sampo Pyysalo, Billy Chiu, and Anna Korhonen. 2017a. A neural network multi-task learning approach to biomedical named entity recognition. *BMC bioinformatics*, 18(1):1–14.

Gamal Crichton, Sampo Pyysalo, Billy Chiu, and Anna Korhonen. 2017b. A neural network multi-task learning approach to biomedical named entity recognition. *BMC Bioinformatics*, 18.

Rezarta Islamaj Doğan, Robert Leaman, and Zhiyong Lu. 2014. Ncbi disease corpus: a resource for disease name recognition and concept normalization. *Journal of biomedical informatics*, 47:1–10.

Daxiang Dong, Hua Wu, Wei He, Dianhai Yu, and Haifeng Wang. 2015. Multi-task learning for multiple language translation. In *Proceedings of the 53rd ACL and the 7th IJCNLP (Volume 1: Long Papers)*, pages 1723–1732.

Martin Gerner, Goran Nenadic, and Casey M Bergman. 2010. Linnaeus: a species name identification system for biomedical literature. *BMC bioinformatics*, 11(1):1–17.

Yu Gu, Robert Tinn, Hao Cheng, Michael Lucas, Naoto Usuyama, Xiaodong Liu, Tristan Naumann, Jianfeng Gao, and Hoifung Poon. 2021. Domain-specific language model pretraining for biomedical natural language processing. *ACM Trans. Comput. Healthcare*, 3(1).

Han Guo, Ramakanth Pasunuru, and Mohit Bansal. 2019. AutoSeM: Automatic task selection and mixing in multi-task learning. In *Proceedings of the 2019 Conference of the North American Chapter of the Association for Computational Linguistics: Human Language Technologies, Volume 1 (Long and Short Papers)*, pages 3520–3531, Minneapolis, Minnesota. Association for Computational Linguistics.

Kazuma Hashimoto, caiming xiong, Yoshimasa Tsuruoka, and Richard Socher. 2017. A joint many-task model: Growing a neural network for multiple nlp tasks. In *Proceedings of the 2017 EMNLP*, pages 1923–1933.

Zhiheng Huang, Wei Xu, and Kai Yu. 2015. Bidirectional lstm-crf models for sequence tagging. *arXiv preprint arXiv:1508.01991*.

Muhammad Raza Khan, Morteza Ziyadi, and Mohamed AbdelHady. 2020. Mt-bioner: Multi-task learning for biomedical named entity recognition using deep bidirectional transformers. *arXiv preprint arXiv:2001.08904*.

Jin-Dong Kim, Tomoko Ohta, Yoshimasa Tsuruoka, Yuka Tateisi, and Nigel Collier. 2004. Introduction to the bio-entity recognition task at jnlpba. In *Proceedings of the NLPBA/BioNLP*, pages 70–75. Citeseer.

Martin Krallinger, Obdulia Rabal, Florian Leitner, Miguel Vazquez, David Salgado, Zhiyong Lu, Robert Leaman, Yanan Lu, Donghong Ji, Daniel M Lowe, et al. 2015. The chemdner corpus of chemicals and drugs and its annotation principles. *Journal of cheminformatics*, 7(1):1–17.

Po-Nien Kung, Sheng-Siang Yin, Yi-Cheng Chen, Tse-Hsuan Yang, and Yun-Nung Chen. 2021. Efficient multi-task auxiliary learning: Selecting auxiliary data by feature similarity. In *Proceedings of the 2021 Conference on Empirical Methods in Natural Language Processing*, pages 416–428, Online and Punta Cana, Dominican Republic. Association for Computational Linguistics.

Jinhyuk Lee, Wonjin Yoon, Sungdong Kim, Donghyeon Kim, Sunkyu Kim, Chan Ho So, and Jaewoo Kang. 2019. BioBERT: a pre-trained biomedical language representation model for biomedical text mining. *Bioinformatics*, 36(4):1234–1240.

Jiao Li, Yueping Sun, Robin J Johnson, Daniela Sciaky, Chih-Hsuan Wei, Robert Leaman, Allan Peter Davis, Carolyn J Mattingly, Thomas C Wiegers, and Zhiyong Lu. 2016. Biocreative v cdr task corpus:

a resource for chemical disease relation extraction. *Database*, 2016.

Liyuan Liu, Jingbo Shang, Xiang Ren, Frank Xu, Huan Gui, Jian Peng, and Jiawei Han. 2018. Empower sequence labeling with task-aware neural language model. In *AAAI*.

Xiaodong Liu, Jianfeng Gao, Xiaodong He, Li Deng, Kevin Duh, and Ye-Yi Wang. 2015. Representation learning using multi-task deep neural networks for semantic classification and information retrieval. In *Proceedings of the 2015 NAACL-HLT*, pages 912–921.

Minh-Thang Luong, Quoc V. Le, Ilya Sutskever, Oriol Vinyals, and Lukasz Kaiser. 2016. Multi-task sequence to sequence learning. In *Proceedings of the 2016 ICLR*.

Tahir Mehmood, Alfonso E Gerevini, Alberto Lavelli, and Ivan Serina. 2020. Combining multi-task learning with transfer learning for biomedical named entity recognition. *Procedia Computer Science*, 176:848–857.

Evangelos Pafilis, Sune P Frankild, Lucia Fanini, Sarah Faulwetter, Christina Pavloudi, Aikaterini Vasileiadou, Christos Arvanitidis, and Lars Juhl Jensen. 2013. The species and organisms resources for fast and accurate identification of taxonomic names in text. *PloS one*, 8(6):e65390.

Nanyun Peng and Mark Dredze. 2017. Multi-task domain adaptation for sequence tagging. In *Proceedings of the 2nd Workshop on RepL4NLP*, pages 91–100.

Long N. Phan, James T. Anibal, Hieu Tran, Shaurya Chanana, Erol Bahadroglu, Alec Peltekian, and Grégoire Altan-Bonnet. 2021. Scifive: a text-to-text transformer model for biomedical literature. *CoRR*, abs/2106.03598.

Colin Raffel, Noam Shazeer, Adam Roberts, Katherine Lee, Sharan Narang, Michael Matena, Yanqi Zhou, Wei Li, and Peter J. Liu. 2020. Exploring the limits of transfer learning with a unified text-to-text transformer. *Journal of Machine Learning Research*, 21(140):1–67.

Marek Rei. 2017. Semi-supervised multitask learning for sequence labeling. In *Proceedings of the 55th ACL (Volume 1: Long Papers)*, pages 2121–2130. Association for Computational Linguistics.

Hoo-Chang Shin, Yang Zhang, Evelina Bakhturina, Raul Puri, Mostofa Patwary, Mohammad Shoeybi, and Raghav Mani. 2020. BioMegatron: Larger biomedical domain language model. In *Proceedings of the 2020 EMNLP*, pages 4700–4706, Online. Association for Computational Linguistics.

Mohammad Shoeybi, Mostofa Patwary, Raul Puri, Patrick LeGresley, Jared Casper, and Bryan Catanzaro. 2020. Megatron-lm: Training multi-billion parameter language models using model parallelism.

Larry Smith, Lorraine K Tanabe, Rie Johnson nee Ando, Cheng-Ju Kuo, I-Fang Chung, Chun-Nan Hsu, Yu-Shi Lin, Roman Klinger, Christoph M Friedrich, Kuzman Ganchev, et al. 2008. Overview of biocreative ii gene mention recognition. *Genome biology*, 9(2):1–19.

Xi Wang, Jiagao Lyu, Li Dong, and Ke Xu. 2019a. Multitask learning for biomedical named entity recognition with cross-sharing structure. *BMC bioinformatics*, 20(1):1–13.

Xuan Wang, Yu Zhang, Xiang Ren, Yuhao Zhang, Marinka Zitnik, Jingbo Shang, Curtis Langlotz, and Jiawei Han. 2018. Cross-type biomedical named entity recognition with deep multi-task learning. *Bioinformatics*, page bty869.

Xuan Wang, Yu Zhang, Xiang Ren, Yuhao Zhang, Marinka Zitnik, Jingbo Shang, Curtis Langlotz, and Jiawei Han. 2019b. Cross-type biomedical named entity recognition with deep multi-task learning. *Bioinformatics*, 35(10):1745–1752.

Taiki Watanabe, Akihiro Tamura, Takashi Ninomiya, Takuya Makino, and Tomoya Iwakura. 2019. Multi-task learning for chemical named entity recognition with chemical compound paraphrasing. In *Proceedings of the 2019 EMNLP-IJCNLP*, pages 6244–6249, Hong Kong, China. Association for Computational Linguistics.

SNP2Vec: Scalable Self-Supervised Pre-Training for Genome-Wide Association Study

Samuel Cahyawijaya[1]*, Tiezheng Yu[1]*, Zihan Liu[1]*, Tiffany T.W. MAK[2,3], Xiaopu Zhou[2,3,4], Nancy Y. Ip[2,3,4], Pascale Fung[1]

[1]Center for Artificial Intelligence Research (CAiRE), Department of Electronic and Computer Engineering, The Hong Kong University of Science and Technology, Hong Kong, China
{scahyawijaya,tyuah,zliucr,pascale}@ust.hk

[2]Division of Life Science, State Key Laboratory of Molecular Neuroscience, Molecular Neuroscience Center, The Hong Kong University of Science and Technology, Clear Water Bay, Kowloon, Hong Kong, China
{tiffanytze,xpzhou,boip}@ust.hk

[3]Hong Kong Center for Neurodegenerative Diseases, Hong Kong Science Park, Hong Kong, China
{tiffanytze,xpzhou,boip}@ust.hk

[4]Guangdong Provincial Key Laboratory of Brain Science, Disease and Drug Development; Shenzhen–Hong Kong Institute of Brain Science, HKUST Shenzhen Research Institute, Shenzhen, China
{xpzhou,boip}@ust.hk

Abstract

Self-supervised pre-training methods have brought remarkable breakthroughs in the understanding of text, image, and speech. Recent developments in genomics has also adopted these pre-training methods for genome understanding. However, they focus only on understanding haploid sequences, which hinders their applicability towards understanding genetic variations, also known as single nucleotide polymorphisms (SNPs), which is crucial for genome-wide association study. In this paper, we introduce SNP2Vec, a scalable self-supervised pre-training approach for understanding SNP. We apply SNP2Vec to perform long-sequence genomics modeling, and we evaluate the effectiveness of our approach on predicting Alzheimer's disease risk in a Chinese cohort. Our approach significantly outperforms existing polygenic risk score methods and all other baselines, including the model that is trained entirely with haploid sequences. We release our code and dataset on `https://github.com/HLTCHKUST/snp2vec`.

1 Introduction

Self-supervised pre-training has become an indispensable step for almost all natural language processing (NLP) tasks (Devlin et al., 2019; Liu et al., 2019; Yang et al., 2019). Pre-trained language models, thanks to the usage of massive text corpora, are effective in handling data scarcity and generalizing to unseen examples (Brown et al., 2020; Cahyawijaya et al., 2021; Wilie et al., 2020; Yu et al., 2021; Liu et al., 2021; Winata et al., 2021). Inspired by the success of pre-trained language models, pre-trained genomic models have been proposed to cope with genomic sequence prediction tasks (Zaheer et al., 2020; Ji et al., 2021). However, these models only focus on modeling the four nucleobases (i.e., A, T, C, and G), while ignoring genomic variations in the pre-training stage. Although they are effective in haploid pattern analysis, such as promoter region and chromatin-profile prediction, they fail to tackle more complex and challenging tasks, such as genome-wide association study (GWAS) (The Wellcome Trust Case Control Consortium, 2007; Corvin et al., 2010; Bush and Moore, 2012), which require an in-depth understanding of long genomic sequences and the genomic variation between a homologous chromosome pair.

To address these shortcomings, we introduce a self-supervised pre-training approach called SNP2Vec, which leverages the single-nucleotide polymorphism (SNP, pronounced 'snip') information gathered from a large-scale SNP database to inject genomic variations in the pre-training stage. SNP2Vec enables the model to learn the semantics of a diploid sequence (genotype) pattern in a diploid cell. We apply SNP2Vec to a linear-attention model, Linformer (Wang et al., 2020), to allow the model to encode long genomic sequences for Alzheimer's disease risk prediction in a Chinese cohort. We compare SNP2Vec with non-pretrained models, as well as an existing strong baseline poly-

*These authors contributed equally.

Proceedings of the BioNLP 2022 workshop, Dublin, Ireland, pages 140–154
May 26, 2022. ©2022 Association for Computational Linguistics

genic risk scoring (PRS) model, to demonstrate the effectiveness of our approach.

Our contributions are summarized as follows:

- We are the first to introduce a scalable self-supervised pre-training approach (SNP2Vec) to learn genomic variations, which is popular for genome-wide association study.
- We demonstrate a method for modeling long diploid sequences with a length of >20,000 base pairs (bps) using an attention-based model within a single forward pass.
- We demonstrate the effectiveness of SNP2Vec, which significantly outperforms all the baselines, including a widely-used polygenic risk scoring (PRS) method, by 5-7% accuracy and AUROC for the Alzheimer's disease prediction task in a Chinese elderly cohort.
- We conduct comprehensive analyses to show the effectiveness of SNP encoding and Byte Pair Encoding (BPE) tokenization compared to the other commonly used methods for genomics modeling.

2 Related Works

2.1 Genome-Wide Association Study

To this day, predicting the risk of hereditary diseases from a given genotype is done through genome-wide Associaction Study (GWAS) by applying a polygenic risk score (PRS). PRS utilizes GWAS data to identify important single nucleotide variations (SNVs) over a certain range from the gene of interest. The SNVs are first filtered according to a statistical measure to reduce the bias towards a certain population and the filtered SNVs are then used to build a classifier, which can be applied to a new genotype to determine the likelihood of getting the disease. This method has been applied by many works and has provided valuable insights for researchers to diseases including heart attack, diabetes, and different types of cancer (Lello et al., 2019). Moreover, PRS model has also been used in research and clinical practice for Alzheimer's disease (Zhou et al., 2021). Nevertheless, all these methods fail to incorporate the patterns of the genomics sequence that determines the actual function. This is likely to lead the model towards non-representative bias, especially when the experimental data is small.

2.2 Statistical Modeling for Genomics

Tokenization in Genomics k-mer (synonymous to n-gram) tokenization is the most commonly used tokenization method in existing genome modelling works (Min et al., 2017; Shen et al., 2018). Gapped k-mer tokenization (Ghandi et al., 2014; Shrikumar et al., 2019) is a more efficient variant of k-mer tokenization by introducing the gap parameter L, which constitutes the stride between each k-mer window. However, the gapped k-mer approach will lead to the loss of some information when L is larger than k. In recent years, subword tokenization approaches (Sennrich et al., 2015; Kudo and Richardson, 2018) have also been explored in genomics (Zaheer et al., 2020).

Machine Learning in Genomics The support vector machine (SVM) is a traditional machine learning approach used to quickly and accurately interpret the nonlinear gapped k-mer (Shrikumar et al., 2019). Hill et al. (2018) leverage a deep recurrent neural network (RNN) to discover complex biological rules to decipher RNA protein-coding potential. Zhuang et al. (2019) incorporate convolutional neural network (CNN) to predict enhancer–promoter interactions with DNA sequence data. Shen et al. (2018) introduce a RNN to predict transcription factor binding sites. They treat each k-mer as a word and pre-train a word representation model though word2vec algorithm (Mikolov et al., 2013). Zaheer et al. (2020) propose BigBird and pre-train it on the human reference genome and improves the performance on downstream tasks.

2.3 Self-Supervised Pre-training

Recently, using self-supervised pre-training models on large scale unlabeled data and then fine-tuning them using a small amount of labeled data has become the norm in machine learning. BERT (Devlin et al., 2019) is a deep bidirectional transformer pre-trained on BooksCorpus (Zhu et al., 2015) (800M words) and English Wikipedia (2500M words) for language understanding. Liu et al. (2019) introduces Roberta, which has a similar architecture as BERT but trained on a much larger corpus (160GB of text) and consequently achieves better performance. In recent years, pre-training generative models (Radford et al., 2019; Raffel et al., 2019; Lewis et al., 2019) has significantly improved the performance of various language generation tasks such as machine translation, question answering, conversational AI, etc.

Self-supervised learning approaches have also been adopted in genomics (Zaheer et al., 2020; Ji et al., 2021) and proteomics (Madani et al., 2020; Elnaggar et al., 2021). These methods pre-train models using large-scale unlabelled datasets such as the human reference genome from the Genome Reference Consortium (GRC) (Church et al., 2011; Schneider et al., 2017) and protein sequence databases such as SWISS-PROT and TrEMBL (Boeckmann et al., 2003). In this paper, we focus on genomics and conduct the human reference genome for pre-training. Genomics data does not have the same structure as human languages; it has no known syntax or grammatical rules and it consists of very long sequences with only a number of differences between each human subject.

3 SNP2Vec

Existing pre-training methods in genomics, such as BigBird (Zaheer et al., 2020) and DNABERT (Ji et al., 2021), are only optimized to understand the pattern of a haploid sequence (haplotype) based on the reference genome. This hinders the model from learning genomic variations, which is essential for understanding traits in humans. In contrast to prior works in genomics pre-training, we develop SNP2Vec to enable pre-training for encoding and understanding patterns of genomic variations in a diploid sequence. Figure 1 depicts the overall structure of the SNP2Vec pre-training method. We elaborate on our SNP2Vec method in 3 subsections: 1) SNP Encoding, i.e., how we encode a diploid sequence as a sequence of SNP tokens; 2) Self-Supervised SNP Dataset, i.e., how we construct a self-supervised dataset using the SNP token; 3) Self-supervised SNP Pre-training, i.e., how we perform self-supervised pre-training for learning the sequence pattern of SNP tokens.

3.1 Preliminaries

What are haploid and diploid sequences? A diploid is a cell or organism that has paired chromosomes, one from each parent [1]. Human cells are mostly diploid, except for the sex cells. In this sense, a diploid sequence (genotype) refers to a pair of homologous sequences (allele) inside the diploid chromosome, while a haploid sequence (haplotype) refers to the DNA sequence from the specific allele of the diploid sequence. The haploid sequence is suitable for understanding the regulatory function of a DNA pattern (Zhou and Troyanskaya, 2015; Ouyang et al., 2008), such as determining a binding site for a certain type of protein, as it provides the representation of the actual nucleotides. A diploid sequence, on the other hand, is more suitable for understanding the phenotype (Levy et al., 2007; Wang et al., 2008) over population since it allows understanding of the genomic variations between two homologous DNA sequences, which tells the dosage information and the gene expression level of a variation. These genomic variations are gathered by comparing them to a genome reference sequence, and they can be categorized based on its dosage, i.e., wild-type (normal), heterozygous, or homozygous, and based on their differences, i.e., substitution, insertion, and deletion. The depiction of haploids and diploids along with their variations is shown in Figure 2.

How do we get the haploid and diploid sequence? As most human cells are predominantly diploid, performing genome sequencing on such homologous chromosome pair will produce a diploid sequence rather than a haploid sequence, because the primer binds to both of the homologous regions from each chromosome (Ye et al., 2012). Extracting haploid sequences from a diploid sequence requires an additional step through an estimation process called phasing (Stephens et al., 2001). Despite their effectiveness, the quality of phasing methods (Browning and Browning, 2007, 2009; Patterson et al., 2014) is not perfect and tends to decrease significantly especially when the gap between the SNPs is large (Choi et al., 2018).

3.2 SNP Encoding

We first extend the nucleotide tokens from 5 token types 'A', 'T', 'C', 'G', and 'N' into 11 tokens by adding 6 insertion-deletion (indel) tokens 'AI', 'TI', 'CI', 'GI', 'NI', and 'DEL', where 'XI' token represents any insertion after the nucleotide 'X', and 'DEL' represents the nucleotide deletion. There can be many different possibility for insertion, e.g., a nucleotide 'T' can be inserted into "TG", "TGGG", or "TAAA"; therefore we aggregate all the insertions into a single token to reduce the sparsity of the indel representaton as indel occurs relatively rarely compared to substitution, with an around 1:5 ratio (Chen et al., 2009). To encode a diploid sequence, we construct all the combinations with replacement ($_nC_r^R$) of the 11 haploid

[1] https://www.genome.gov/genetics-glossary/Diploid

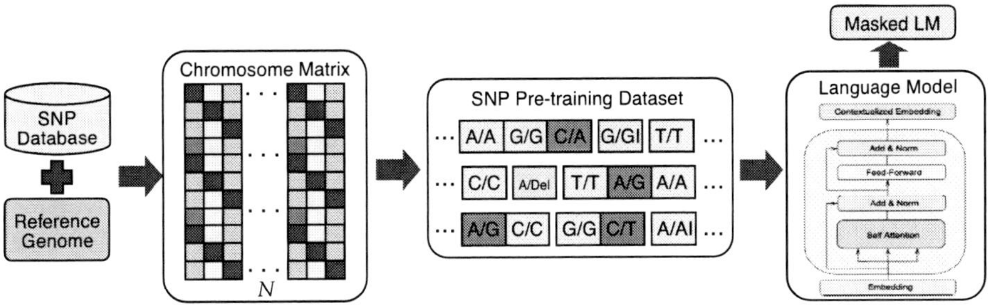

Figure 1: SNP2Vec Pre-training Pipeline. SNP2Vec merges information from reference genome and SNP database to form a chromosome matrix which is then utilized to construst SNP pre-training dataset following the SNP encoding's token format. This pre-training dataset is employed to train a genome language model through the masked language modeling task.

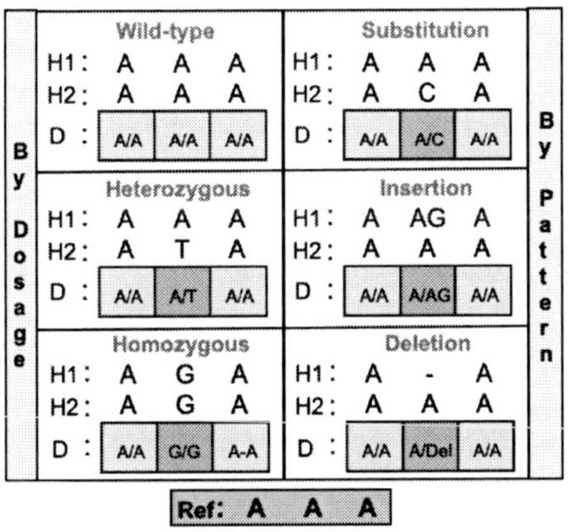

Figure 2: Diploid sequence variations. The box on the top-left shows the wild-type sequence, while others are its variations. **H1** and **H2** denote the haploid sequence for each parent allele. **D** represents the diploid sequence of the two alleles.

and indel tokens with $n = 11$ and $r = 2$, producing a total of 66 types of SNP tokens consisting of wild-type, heterozygous, and homozygous variation tokens. The resulting SNP tokens are represented as 'X_1/X_2'', where 'X_1' and 'X_2' denote aligned nucleotide or indel tokens from the two alleles ordered alphabetically. A depiction of the SNP tokens is shown in Figure 3. To reduce the size and facilitate more straightforward representation for downstream processes such as pre-processing, tokenization, and modeling, we map the SNP tokens into a single character representation. The mapping of the SNP token into a single character representation is shown in Appendix A.

By incorporating the SNP encoding, variant calling information gathered from the DNA sequencing machine can be directly converted into a sequence of SNP tokens, that are then used for the model fine-tuning and inference. However, this is not directly applicable for self-supervised pre-training since DNA sequencing data is hard to obtain and it is unethical to share publicly as it contains very sensitive and personal information of the human subject. In the next section, we discuss in detail how we can construct an inexpensive and reliable pre-training dataset to perform self-supervised pre-training on the SNP tokens by utilizing publicly available genomics data sources.

3.3 Self-Supervised SNP Dataset

Prior self-supervised pre-training approaches in genomics (Zaheer et al., 2020; Ji et al., 2021) only utilize the human reference genome (Church et al., 2011; Schneider et al., 2017) as the unlabelled data for haploid genomics pre-training, the latter does not capture any genomic variations. We extend these haploid modeling techniques into a diploid modeling method, which allows the model to learn patterns of genomic variations by generating unlabelled pre-training data for learning SNP tokens. More specifically, we use the genome sequence from the human reference genome and genome variation from a large-scale SNP database, namely dbSNP (Smigielski et al., 2000), to generate the pre-training data.

Human reference genome The human reference genome is a genome sequence derived from the DNA collected from a number of people (Pollard et al., 2017), which was first released in 2000 and is periodically updated. There are two most commonly used versions of the human reference

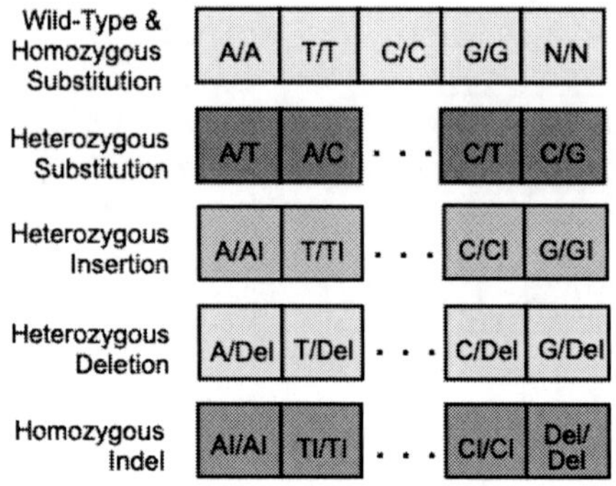

Figure 3: SNP tokens consist of a total of 66 types of token covering all possible variations in a diploid sequence including wild-types, heterozygous variations, and homozygous variations.

genome, namely GRCh37 (Church et al., 2011) [2] and GRCh38 (Schneider et al., 2017) [3]. A human reference genome consists of the genome sequence information for all human chromosomes with ~3B sequence length in total. Most of the positions are mapped and represented as either 'A', 'T', 'C', or 'G', while the others are unmapped and flagged with the unknown ('N') token.

dbSNP dbSNP (Smigielski et al., 2000) [4] is a central public repository of human SNPs. dbSNP covers a broad collection of simple genetic variations with a length of variation ≤50 bps long, which includes single-base nucleotide substitutions, small-scale multi-base deletions, and small-scale multi-base insertions. A single SNP in the dbSNP contains the following information: chromosome number, position in the chromosome, SNP identifier, reference sequence (`REF`), alternative sequence(s) (`ALTS`), probability of the `REF` and `ALTS`, and other metadata. The `REF` is a single-base or multi-base sequence that comes from the human reference genome used for detecting the SNPs. The `ALTS` can consist of one or more alternative variations and each can represent a substitution, a deletion, or an insertion.

Dataset Construction We construct a pre-training dataset consisting of sequences of SNP tokens by combining the sequence information from the human reference genome and the genomic variations from the dbSNP. For each chromosome, we generate an $11 \times N$ matrix, where N is equal to the length of the corresponding chromosome and 11 represents the probability of each nucleotide and indel token. We name this matrix a **chromosome matrix**. We fill the chromosome matrix using all SNPs labelled as `COMMON` in the dbSNP by filling the corresponding matrix position with the `REF` and `ALTS` probability of the corresponding SNP record. Since the SNPs from the dbSNP do not cover all of the genome positions, we fill up all the other gap positions with a probability of 1 to the nucleotide token in the corresponding position on the human reference genome.

For constructing the self-supervised pre-training dataset, we closely follow the setup in the typical NLP pre-training dataset construction pipeline. Specifically, we convert the chromosome matrix into a set of segments S where each segment $s \in S$ comprises of a number of SNP tokens. To construct the sentences S, we sample multiple sequences from different positions of a chromosome. For each position in the sequence, we apply a sampling function F to collect 'X1' and 'X2' (the nucleotide or indel tokens on the corresponding position) and construct the SNP token "X1/X2". The dataset construction method can be applied to all the chromosome pairs except for the sex chromosome, which is always haploid. The details of our dataset construction approach is shown in Algorithm 1.

3.4 Self-Supervised SNP Pre-training

Inspired by BERT (Devlin et al., 2019), SNP2Vec is trained using the masked language modeling (MLM) objective using a transformer-based model (Vaswani et al., 2017). The goal of MLM is to predict the representations of the masked tokens given their neighbouring sequence as the context. As complex genomic tasks, such as disease risk prediction, require the understanding long-genome sequence (>1000 bps), we apply two methods to process long input sequences. First, we apply a transformer variant with a linear-attention mechanism, which enables the model to reduce the computational complexity from $O(N^2)$ to $O(N)$. Second, we apply a BPE tokenization (Sennrich et al., 2015) to encode the sequence of SNP tokens to compress the sequence via aggregation of neighbouring tokens. Unlike k-mer (Min et al., 2017; Ji

[2] `https://www.ncbi.nlm.nih.gov/assembly/GCF_000001405.13/`

[3] `https://www.ncbi.nlm.nih.gov/assembly/GCF_000001405.26/`

[4] `https://www.ncbi.nlm.nih.gov/projects/SNP/snp_summary.cgi`

Algorithm 1 Self-Supervised Pre-training dataset construction for diploid SNP Encoding

Require: C: chromosome matrix
Require: f: SNP sampling function
Require: T: number of iterations
Require: K: start position threshold
Require: L^{inf}: lower bound of segment length
Require: L^{sup}: upper bound of segment length
Initialize $S = \emptyset$
P = sample T positions from range $[0 \dots K]$
for all $p \in P$ **do**
  while $p < |C|$ **do**
    $l \sim \mathcal{U}(L^{inf}, L^{sup})$
    z = segment from p to $p + L$ in C
    s = Sample SNP tokens using f from each position in z
    $S = S \cup s$
    $p = p + l$
  end while
end for

et al., 2021) and gapped k-mer (Ghandi et al., 2014; Shrikumar et al., 2019) tokenizations, BPE tokenization can merge dynamic-length tokens based on their co-occurrences efficiently without losing any information.

4 Experiment Settings

4.1 Dataset

For building the pre-training data, we utilize GRCh37 as the human reference genome and db-SNP version 153 [5] as the SNP database. We utilize a weighted random sampling based on the probability of SNPs on the corresponding position as the sampling function f. For the downstream-task, we construct a dataset of genome sequences for predicting late-onset Alzheimer's disease (LOAD) (Rabinovici, 2019) on a Chinese Cohort from 624 Hong Kong elderly with a minimum age of 65. The subjects are diagnosed with Alzheimer's by a medical professional through the Montreal Cognitive Assessment (MoCA) test (Nasreddine et al., 2005) adjusted for the demographic information. Out of 624 subjects, 384 are labelled as Alzheimer's disease carriers (ADs) and 240 are labelled as non-carriers (NCs). For the genome sequence, we collect sequencing data from the APOE region located in chromosome 19 from each subject, which is known to be highly correlated with Alzheimer's disease in the Chinese cohort (Zhou et al., 2019, 2020). We use BWA-MEM (Li, 2013) assembler to align the sequencing data with the human reference genome.

[5] `https://ftp.ncbi.nih.gov/snp/archive/b153/00readme.txt`

4.2 Training and Evaluation Setting

For our experiment, we build a BPE tokenizer with a vocabulary size of 32,000 tokens. We pre-train a 6-layers linear-attention transformer-based model, Linformer (Wang et al., 2020), using a maximum sequence length of 4,096 tokens, a sequence projection length k of 128 tokens, and a model dimension size of 512. For simplicity, we refer to our pre-trained SNP2Vec model as **Dipformer**. The detail hyperparameters of the BPE tokenizer and the Dipformer model are described in Appendix B. We run MLM pre-training for 200,000 steps with a 15% token replacement rate, where we replace with `[MASK]` 80% of the time, replace with a random token 10% of the time, and keep the token as is 10% of the time. More detail about the pre-training hyperparameter setting is shown in Appendix C.

For the fine-tuning, we apply SNP encoding to the sequencing data, apply BPE tokenization, and add a `[CLS]` token as the prefix of the sequence to gather the sequence representation for predicting the risk of having Alzheimer's disease. We apply fine-tuning for three input sequence length settings, i.e., only APOE gene with 3,611 bps (**APOE only**), APOE with additional 5,000 bps upstream and downstream (**APOE+10k**), and APOE with additional 10,000 bps upstream and downstream (**APOE + 20k**). For each experiment, we apply 10-fold cross validation to ensure the result is significance. We evaluate the model performance using three evaluation metrics: accuracy, area under the ROC curve (AUROC), and area under the precision-recall curve (AUPRC). More detail about the fine-tuning setup is described in Appendix D.

4.3 Baselines

To evaluate the effectiveness of the SNP encoding, we build two different deep learning models using haploid token representation. First, we incorporate DeepSEA (Zhou and Troyanskaya, 2015), a CNN-based model develop for short sequence chromatin profiling tasks (~200-1000 bps), and then we build another Linformer model pre-trained with the human reference genome using haploid tokens, called **Hapformer**. For the haploid token fine-tuning, we generate the haploid sequence from the aligned sequencing data. We generate the variant calling

Models	Acc	AUROC	AUPRC
DeepSEA	0.591	0.579	0.703
PLINK PRS	0.592	0.607	0.705
Hapformer	0.572	0.615	0.715
Dipformer	**0.643**	**0.673**	**0.734**

Table 1: Results of our model and baselines. We refer the pre-trained SNP2Vec model as Dipformer.

data with GATK HaplotypeCaller (McKenna et al., 2010; DePristo et al., 2011) and apply phasing with Beagle (Browning and Browning, 2007, 2009). During fine-tuning, we feed each haploid sequence to the model and fuse the representation using a linear transformation. We also incorporate a logistic regression model from PLINK (Purcell et al., 2007), which is a widely used approach for PRS.

5 Results

The results of our model and baselines are shown in Table 1. We find that Dipformer is able to outperform existing strong baselines, such as DeepSEA and PLINK PRS, by a large margin. This confirms the effectiveness of our SNP2Vec pre-training, and the ability of our Dipformer to capture relevant features for AD prediction. Interestingly, Hapformer, which leverages large amounts of genomic sequences for pre-training, only performs comparably to DeepSEA and PLINK PRS. Moreover, by learning genomic variations in a diploid sequence during the pre-training, Dipformer significantly outperforms Hapformer with an around 5-7% improvement in terms of accuracy and AUROC metrics. This shows that simply using an enormous amount of pre-training data might not necessarily improve the AD prediction, and an effective genomics pre-training approach is essential to guarantee full use of the unlabelled genomics data. More detail on our results is shown in Appendix E.

6 Discussion

6.1 Effect of Different Tokenization Methods in Genomics

In this section, we study different tokenization methods for genome modeling, and explore their effectiveness in terms of capturing genomic patterns and features. We compare BPE tokenization with other common methods, such as k-mer and gapped k-mer (gkm) with various gap parameters. To achive this, we conduct experiments on the chromatin profiling dataset from DeepSEA, which con-

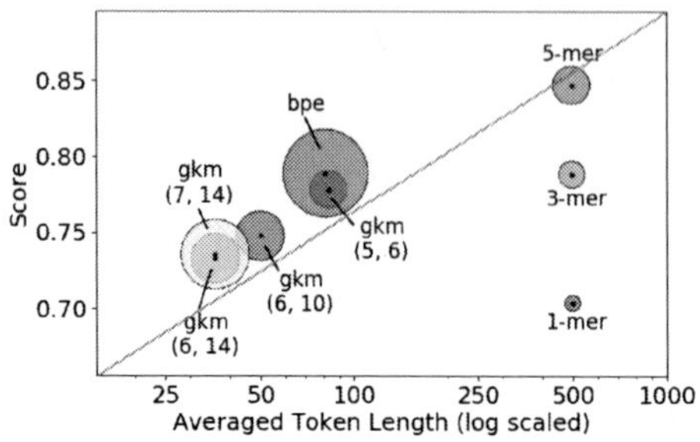

Figure 6: Performance efficiency trade-off of using different tokenization approaches. The score is averaged over the three models (Linear, CNN, and Transformer). The size of the dots represents the vocabulary size of the tokenization method.

sists of 4,863,024 chromatin profiles (4,400,000 training, 8000 validation, and 455,024 test) with 919 labels (690 transcription factor (TF) binding sites, 125 DNase marks, and 104 Histone marks). Three different models are incorporated in this experiment: a linear model with bag-of-word (BoW) representation, a CNN-based model following the DeepSEA architecture, and a transformer model. The models need to predict the TF, DNase, and Histone labels based on the input sequences using various tokenization methods. Hence, for the same model, a more effective tokenization method will lead to a higher prediction accuracy. Additionally, we use the average length of the tokenized sequences to measure the efficiency of different tokenization methods as it determines the input size for the model.

Table 2 provides the effectiveness and averaged token length of different tokenization methods in genome modeling. We find that, on the Linear BoW model, BPE significantly outperforms all other methods except 5-mer. On the CNN model, BPE remarkably surpasses all gapped k-mer methods except for the gkm (5,6). On the Transformer model, BPE performs similarly to 3-mer and gkm (5,6), and significantly outperforms 1-mer and other gkm methods. Moreover, in terms of the averaged score across all three models, BPE performs comparably well to 3-mer, and remarkably outperforms 1-mer and all gkm methods.

Figure 6 illustrates the trade-off between the performance and efficiency of different tokenization methods. We can see that compared to k-mer methods, BPE performs comparably to 3-mer and slightly worse than 5-mer, but it is much more ef-

Tokenization	1-mer	3-mer	5-mer	gkm (5,6)	gkm (6,10)	gkm (6,14)	gkm (7,14)	BPE
Avg. Token Length	500	498	496	83	50	36	36	81.19
BoW Linear	0.499	0.698	0.817	0.771	0.759	0.749	0.753	**0.783***
CNN (DeepSEA)	0.890	0.903	0.898	0.808	0.764	0.749	0.751	**0.811***
Transformer	0.727	0.788	0.825	0.785	0.771	0.761	0.762	**0.789***
Average	0.706	0.796	0.847	0.788	0.765	0.753	0.755	**0.795***

Table 2: Comparison of different tokenization methods in genome modeling (numbers denote the accuracy score), where gkm (k,l) denotes the gapped k-mer tokenization with the gap parameter l constituting the stride between each k-mer window. * denotes that BPE significantly outperforms the underlined baselines with a p-value < 0.01.

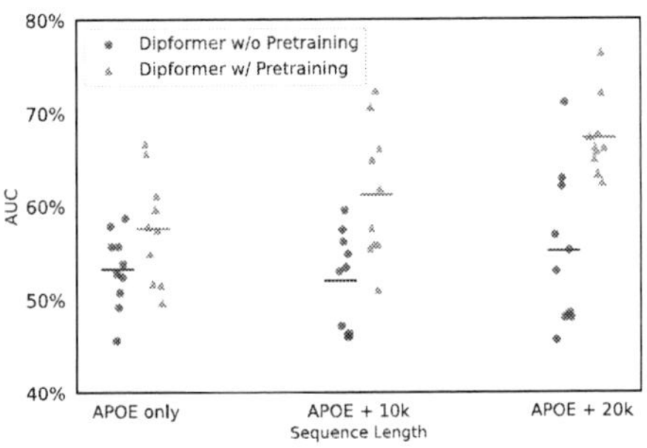

Figure 4: 10-folds AUROC Performance of Dipformer with and without pre-training on the Alzheimer's disease risk prediction over different sequence length input.

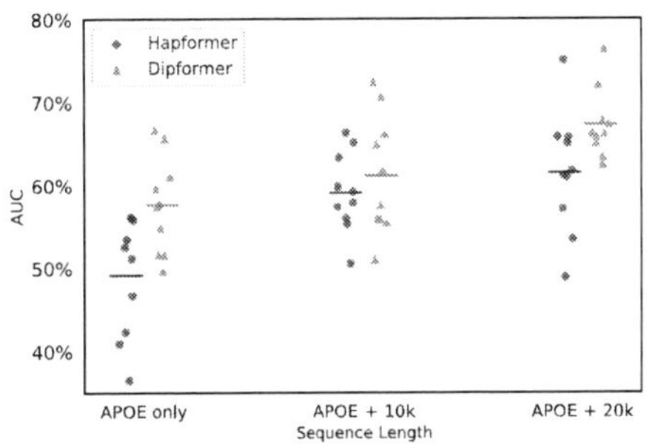

Figure 5: 10-folds AUROC performance of pre-trained Hapformer and Dipformer on the Alzheimer's disease risk prediction over different sequence length input.

ficient due to a much shorter average length. In addition, BPE remarkably outperforms gkm methods with comparable or slightly worse efficiency. Furthermore, from the size of the dots, we can see that BPE has a much larger vocabulary size compared to other methods, which indicates that BPE can potentially capture richer genomics patterns.

6.2 Effect of Pre-training for Disease Risk Prediction

In this section, we focus on exploring the effectiveness of pre-training for disease risk prediction. Figure 4 illustrates the 10-fold AUROC results of our Dipformer model with and without pre-training on Alzheimer's disease risk prediction. The dashes in the figure represent the average AUROC for all 10-fold results. As shown in Figure 4, the average AUROC scores for pre-trained Dipformer significantly outperform the Dipformer without pre-training in all sequence length settings, APOE + 10k, and APOE + 20. Table 3 presents the quantitative results with additional metrics. The accuracy, AUROC, and AUPRC scores of pre-trained Dipformer consistently outperform the non-pre-trained Dipformer in all sequence length settings. By increasing the sequence length, the non-pre-trained Dipformer performs slightly better, while the pre-trained Dipformer improves by a large margin. This shows the importance of pre-training for understanding long-sequence features.

6.3 Effect of the SNP Encoding in Genomics

To study the effect of the SNP encoding, we pre-train and fine-tune a model with the same genomics data but using haploid tokens called Hapformer, as mentioned in the Section 4. Figure 5 shows the 10-fold AUROC results of pre-trained Hapformer and Dipformer on the AD risk prediction over different sequence length inputs. Among all three sequence length settings, Dipformer achieves better average AUROC scores than Hapformer with a p-value of 0.046 for the APOE + 20k setting, which indicates that the improvement of SNP encoding is significant. Meanwhile, the results in Table 3 shows that Dipformer also surpasses Hapformer in all other evaluation metrics. In addition, we also observe that both Hapformer and Dipformer achieve better results when the input sequence is longer. This shows that employing long sequence is essential for handling complex genomics tasks such as disease risk prediction.

Model	Accuracy	AUROC	AUPRC
Without Pre-training			
Dipformer (APOE only)	0.567	0.532	0.667
Dipformer (APOE + 10k)	0.571	0.520	0.608
Dipformer (APOE + 20k)	0.588	0.551	0.668
With Pre-training			
Hapformer (APOE only)	0.524	0.491	0.623
Hapformer (APOE + 10k)	0.565	0.591	0.705
Hapformer (APOE + 20k)	0.572	0.615	0.715
Dipformer (APOE only)	0.611	0.576	0.687
Dipformer (APOE + 10k)	0.574	0.612	0.710
Dipformer (APOE + 20k)	**0.643**	**0.673**	**0.734**

Table 3: Performance of Dipformer and Hapformer on the Alzheimer's disease risk prediction over different lengths of the input sequences.

7 Conclusion

In this paper, we introduce SNP2Vec, a self-supervised pre-training method for understanding genomic variations in a diploid sequence. Unlike prior methods in genomics, SNP2Vec represents each genomics position with a SNP token which allows the model to capture genomic variations which is suitable for understanding complex genomics prediction tasks such as predicting phenotype. By utilizing SNP2Vec, we pre-train a Linformer model called Dipformer and evaluate it for predicting late-onset Alzheimer's disease risk in a Chinese cohort. Experimental results suggest that Dipformer significantly improves the prediction quality by 5-7% Accuracy and AUROC over all other baselines including the widely used polygenic risk score model from PLINK, the haploid-variant of Dipformer, and a CNN-based genomics model called DeepSEA.

8 Future Work

For future works, we expect to focus on model explainability by using multiple analysis methods, such as analyzing the attention behaviour, analyzing the gradient saliency map, etc, to gather and verify insights from the model. Evaluation on larger scale dataset is also necessary to further demostrate the effectiveness of SNP2Vec. Additionally, adoption of SNP2Vec to other hereditary disorders and other complex genomics tasks is also an essential direction for future works.

Acknowledgement

This work has been partially funded by School of Engineering PhD Fellowship Award, the Hong Kong University of Science and Technology and PF20-43679 Hong Kong PhD Fellowship Scheme, Research Grant Council, Hong Kong.

References

Brigitte Boeckmann, Amos Bairoch, Rolf Apweiler, Marie-Claude Blatter, Anne Estreicher, Elisabeth Gasteiger, Maria J. Martin, Karine Michoud, Claire O'Donovan, Isabelle Phan, Sandrine Pilbout, and Michel Schneider. 2003. The SWISS-PROT protein knowledgebase and its supplement TrEMBL in 2003. *Nucleic Acids Research*, 31(1):365–370.

Tom Brown, Benjamin Mann, Nick Ryder, Melanie Subbiah, Jared D Kaplan, Prafulla Dhariwal, Arvind Neelakantan, Pranav Shyam, Girish Sastry, Amanda Askell, Sandhini Agarwal, Ariel Herbert-Voss, Gretchen Krueger, Tom Henighan, Rewon Child, Aditya Ramesh, Daniel Ziegler, Jeffrey Wu, Clemens Winter, Chris Hesse, Mark Chen, Eric Sigler, Mateusz Litwin, Scott Gray, Benjamin Chess, Jack Clark, Christopher Berner, Sam McCandlish, Alec Radford, Ilya Sutskever, and Dario Amodei. 2020. Language models are few-shot learners. In *Advances in Neural Information Processing Systems*, volume 33, pages 1877–1901. Curran Associates, Inc.

Brian L. Browning and Sharon R. Browning. 2009. A unified approach to genotype imputation and haplotype-phase inference for large data sets of trios and unrelated individuals. *The American Journal of Human Genetics*, 84(2):210–223.

Sharon R. Browning and Brian L. Browning. 2007. Rapid and accurate haplotype phasing and missing-data inference for whole-genome association studies by use of localized haplotype clustering. *The American Journal of Human Genetics*, 81(5):1084–1097.

William S Bush and Jason H Moore. 2012. Chapter 11: Genome-wide association studies. *PLoS Comput. Biol.*, 8(12):e1002822.

Samuel Cahyawijaya, Genta Indra Winata, Bryan Wilie, Karissa Vincentio, Xiaohong Li, Adhiguna Kuncoro, Sebastian Ruder, Zhi Yuan Lim, Syafri Bahar, Masayu Khodra, Ayu Purwarianti, and Pascale Fung. 2021. IndoNLG: Benchmark and resources for evaluating Indonesian natural language generation. In *Proceedings of the 2021 Conference on Empirical Methods in Natural Language Processing*, pages 8875–8898, Online and Punta Cana, Dominican Republic. Association for Computational Linguistics.

Jian-Qun Chen, Ying Wu, Haiwang Yang, Joy Bergelson, Martin Kreitman, and Dacheng Tian. 2009. Variation in the ratio of nucleotide substitution and indel rates across genomes in mammals and bacteria. *Molecular biology and evolution*, 26:1523–31.

Yongwook Choi, Agnes P. Chan, Ewen Kirkness, Amalio Telenti, and Nicholas J. Schork. 2018. Comparison of phasing strategies for whole human genomes. *PLOS Genetics*, 14(4):1–26.

Deanna M. Church, Valerie A. Schneider, Tina Graves, Katherine Auger, Fiona Cunningham, Nathan Bouk, Hsiu-Chuan Chen, Richa Agarwala, William M. McLaren, Graham R.S. Ritchie, Derek Albracht, Milinn Kremitzki, Susan Rock, Holland Kotkiewicz, Colin Kremitzki, Aye Wollam, Lee Trani, Lucinda Fulton, Robert Fulton, Lucy Matthews, Siobhan Whitehead, Will Chow, James Torrance, Matthew Dunn, Glenn Harden, Glen Threadgold, Jonathan Wood, Joanna Collins, Paul Heath, Guy Griffiths, Sarah Pelan, Darren Grafham, Evan E. Eichler, George Weinstock, Elaine R. Mardis, Richard K. Wilson, Kerstin Howe, Paul Flicek, and Tim Hubbard. 2011. Modernizing reference genome assemblies. *PLoS Biology*, 9(7):e1001091.

A Corvin, N Craddock, and P F Sullivan. 2010. Genome-wide association studies: a primer. *Psychol. Med.*, 40(7):1063–1077.

Mark A DePristo, Eric Banks, Ryan Poplin, Kiran V Garimella, Jared R Maguire, Christopher Hartl, Anthony A Philippakis, Guillermo del Angel, Manuel A Rivas, Matt Hanna, Aaron McKenna, Tim J Fennell, Andrew M Kernytsky, Andrey Y Sivachenko, Kristian Cibulskis, Stacey B Gabriel, David Altshuler, and Mark J Daly. 2011. A framework for variation discovery and genotyping using next-generation DNA sequencing data. *Nature Genetics*, 43(5):491–498.

Jacob Devlin, Ming-Wei Chang, Kenton Lee, and Kristina Toutanova. 2019. Bert: Pre-training of deep bidirectional transformers for language understanding. In *Proceedings of the 2019 Conference of the North American Chapter of the Association for Computational Linguistics: Human Language Technologies, Volume 1 (Long and Short Papers)*, pages 4171–4186.

A. Elnaggar, M. Heinzinger, C. Dallago, G. Rehawi, Y. Wang, L. Jones, T. Gibbs, T. Feher, C. Angerer, M. Steinegger, D. Bhowmik, and B. Rost. 2021. Prottrans: Towards cracking the language of lifes code through self-supervised deep learning and high performance computing. *IEEE Transactions on Pattern Analysis & Machine Intelligence*, (01):1–1.

Mahmoud Ghandi, Dongwon Lee, Morteza Mohammad-Noori, and Michael A Beer. 2014. Enhanced regulatory sequence prediction using gapped k-mer features. *PLoS computational biology*, 10(7):e1003711.

Steven T Hill, Rachael Kuintzle, Amy Teegarden, Erich Merrill III, Padideh Danaee, and David A Hendrix. 2018. A deep recurrent neural network discovers complex biological rules to decipher rna protein-coding potential. *Nucleic acids research*, 46(16):8105–8113.

Yanrong Ji, Zhihan Zhou, Han Liu, and Ramana V Davuluri. 2021. Dnabert: pre-trained bidirectional encoder representations from transformers model for dna-language in genome. *Bioinformatics*, 37(15):2112–2120.

Taku Kudo and John Richardson. 2018. Sentencepiece: A simple and language independent subword tokenizer and detokenizer for neural text processing. *arXiv preprint arXiv:1808.06226*.

Louis Lello, Timothy G Raben, Soke Yuen Yong, Laurent CAM Tellier, and Stephen DH Hsu. 2019. Genomic prediction of 16 complex disease risks including heart attack, diabetes, breast and prostate cancer. *Scientific reports*, 9(1):1–16.

Samuel Levy, Granger Sutton, Pauline C Ng, Lars Feuk, Aaron L Halpern, Brian P Walenz, Nelson Axelrod, Jiaqi Huang, Ewen F Kirkness, Gennady Denisov, Yuan Lin, Jeffrey R MacDonald, Andy Wing Chun Pang, Mary Shago, Timothy B Stockwell, Alexia Tsiamouri, Vineet Bafna, Vikas Bansal, Saul A Kravitz, Dana A Busam, Karen Y Beeson, Tina C McIntosh, Karin A Remington, Josep F Abril, John Gill, Jon Borman, Yu-Hui Rogers, Marvin E Frazier, Stephen W Scherer, Robert L Strausberg, and J. Craig Venter. 2007. The diploid genome sequence of an individual human. *PLOS Biology*, 5(10):1–32.

Mike Lewis, Yinhan Liu, Naman Goyal, Marjan Ghazvininejad, Abdelrahman Mohamed, Omer Levy, Ves Stoyanov, and Luke Zettlemoyer. 2019. Bart: Denoising sequence-to-sequence pre-training for natural language generation, translation, and comprehension. *arXiv preprint arXiv:1910.13461*.

Heng Li. 2013. Aligning sequence reads, clone sequences and assembly contigs with bwa-mem.

Yinhan Liu, Myle Ott, Naman Goyal, Jingfei Du, Mandar Joshi, Danqi Chen, Omer Levy, Mike Lewis, Luke Zettlemoyer, and Veselin Stoyanov. 2019. Roberta: A robustly optimized bert pretraining approach. *arXiv preprint arXiv:1907.11692*.

Zihan Liu, Yan Xu, Tiezheng Yu, Wenliang Dai, Ziwei Ji, Samuel Cahyawijaya, Andrea Madotto, and Pascale Fung. 2021. Crossner: Evaluating cross-domain named entity recognition. In *Proceedings of the AAAI Conference on Artificial Intelligence*, volume 35, pages 13452–13460.

Ali Madani, Bryan McCann, Nikhil Naik, Nitish Shirish Keskar, Namrata Anand, Raphael R. Eguchi, Po-Ssu Huang, and Richard Socher. 2020. Progen: Language modeling for protein generation.

A. McKenna, M. Hanna, E. Banks, A. Sivachenko, K. Cibulskis, A. Kernytsky, K. Garimella, D. Altshuler, S. Gabriel, M. Daly, and M. A. DePristo. 2010. The genome analysis toolkit: A MapReduce framework for analyzing next-generation DNA sequencing data. *Genome Research*, 20(9):1297–1303.

Tomas Mikolov, Kai Chen, Greg Corrado, and Jeffrey Dean. 2013. Efficient estimation of word representations in vector space. *arXiv preprint arXiv:1301.3781*.

Xu Min, Wanwen Zeng, Ning Chen, Ting Chen, and Rui Jiang. 2017. Chromatin accessibility prediction via convolutional long short-term memory networks with k-mer embedding. *Bioinformatics*, 33(14):i92–i101.

Ziad S. Nasreddine, Natalie A. Phillips, ValÃ©rie BÃ©dirian, Simon Charbonneau, Victor Whitehead, Isabelle Collin, Jeffrey L. Cummings, and Howard Chertkow. 2005. The montreal cognitive assessment, MoCA: A brief screening tool for mild cognitive impairment. *Journal of the American Geriatrics Society*, 53(4):695–699.

Ching Ouyang, David D Smith, and Theodore G Krontiris. 2008. Evolutionary signatures of common human cis-regulatory haplotypes. *PLoS One*, 3(10):e3362.

Murray Patterson, Tobias Marschall, Nadia Pisanti, Leo van Iersel, Leen Stougie, Gunnar W. Klau, and Alexander Schönhuth. 2014. Whatshap: Haplotype assembly for future-generation sequencing reads. In *Research in Computational Molecular Biology*, pages 237–249, Cham. Springer International Publishing.

Thomas D. Pollard, William C. Earnshaw, Jennifer Lippincott-Schwartz, and Graham T. Johnson. 2017. Chapter 7 - chromosome organization. In *Cell Biology (Third Edition)*, third edition edition, pages 107–122. Elsevier.

Shaun Purcell, Benjamin Neale, Kathe Todd-Brown, Lori Thomas, Manuel A R Ferreira, David Bender, Julian Maller, Pamela Sklar, Paul I W de Bakker, Mark J Daly, and Pak C Sham. 2007. PLINK: a tool set for whole-genome association and population-based linkage analyses. *Am. J. Hum. Genet.*, 81(3):559–575.

Gil D. Rabinovici. 2019. Late-onset alzheimer disease. *Continuum (Minneapolis, Minn.)*, 25(1):14–33.

Alec Radford, Jeffrey Wu, Rewon Child, David Luan, Dario Amodei, Ilya Sutskever, et al. 2019. Language models are unsupervised multitask learners. *OpenAI blog*, 1(8):9.

Colin Raffel, Noam Shazeer, Adam Roberts, Katherine Lee, Sharan Narang, Michael Matena, Yanqi Zhou, Wei Li, and Peter J Liu. 2019. Exploring the limits of transfer learning with a unified text-to-text transformer. *arXiv preprint arXiv:1910.10683*.

Valerie A Schneider, Tina Graves-Lindsay, Kerstin Howe, Nathan Bouk, Hsiu-Chuan Chen, Paul A Kitts, Terence D Murphy, Kim D Pruitt, Françoise Thibaud-Nissen, Derek Albracht, Robert S Fulton, Milinn Kremitzki, Vincent Magrini, Chris Markovic, Sean McGrath, Karyn Meltz Steinberg, Kate Auger, William Chow, Joanna Collins, Glenn Harden, Timothy Hubbard, Sarah Pelan, Jared T Simpson, Glen Threadgold, James Torrance, Jonathan M Wood, Laura Clarke, Sergey Koren, Matthew Boitano, Paul Peluso, Heng Li, Chen-Shan Chin, Adam M Phillippy, Richard Durbin, Richard K Wilson, Paul Flicek, Evan E Eichler, and Deanna M Church. 2017. Evaluation of GRCh38 and de novo haploid genome assemblies demonstrates the enduring quality of the reference assembly. *Genome Res.*, 27(5):849–864.

Rico Sennrich, Barry Haddow, and Alexandra Birch. 2015. Neural machine translation of rare words with subword units. *arXiv preprint arXiv:1508.07909*.

Zhen Shen, Wenzheng Bao, and De-Shuang Huang. 2018. Recurrent neural network for predicting transcription factor binding sites. *Scientific reports*, 8(1):1–10.

Avanti Shrikumar, Eva Prakash, and Anshul Kundaje. 2019. Gkmexplain: fast and accurate interpretation of nonlinear gapped k-mer svms. *Bioinformatics*, 35(14):i173–i182.

E M Smigielski, K Sirotkin, M Ward, and S T Sherry. 2000. dbSNP: a database of single nucleotide polymorphisms. *Nucleic Acids Res.*, 28(1):352–355.

Matthew Stephens, Nicholas J. Smith, and Peter Donnelly. 2001. A new statistical method for haplotype reconstruction from population data. *The American Journal of Human Genetics*, 68(4):978–989.

The Wellcome Trust Case Control Consortium. 2007. Genome-wide association study of 14,000 cases of seven common diseases and 3,000 shared controls. *Nature*, 447(7145):661–678.

Ashish Vaswani, Noam Shazeer, Niki Parmar, Jakob Uszkoreit, Llion Jones, Aidan N Gomez, Ł ukasz Kaiser, and Illia Polosukhin. 2017. Attention is all you need. In *Advances in Neural Information Processing Systems*, volume 30. Curran Associates, Inc.

Jun Wang, Wei Wang, Ruiqiang Li, Yingrui Li, Geng Tian, Laurie Goodman, Wei Fan, Junqing Zhang, Jun Li, Juanbin Zhang, Yiran Guo, Binxiao Feng, Heng Li, Yao Lu, Xiaodong Fang, Huiqing Liang, Zhenglin Du, Dong Li, Yiqing Zhao, Yujie Hu, Zhenzhen Yang, Hancheng Zheng, Ines Hellmann, Michael Inouye, John Pool, Xin Yi, Jing Zhao, Jinjie Duan, Yan Zhou, Junjie Qin, Lijia Ma, Guoqing Li, Zhentao Yang, Guojie Zhang, Bin Yang, Chang Yu, Fang Liang, Wenjie Li, Shaochuan Li, Dawei Li, Peixiang Ni, Jue Ruan, Qibin Li, Hongmei Zhu, Dongyuan Liu, Zhike Lu, Ning Li, Guangwu Guo, Jianguo Zhang, Jia Ye, Lin Fang, Qin Hao, Quan Chen, Yu Liang, Yeyang Su, A San, Cuo Ping, Shuang Yang, Fang Chen, Li Li, Ke Zhou, Hongkun Zheng, Yuanyuan Ren, Ling Yang, Yang Gao, Guohua Yang, Zhuo Li, Xiaoli Feng, Karsten Kristiansen, Gane Ka-Shu Wong, Rasmus Nielsen, Richard Durbin, Lars Bolund, Xiuqing Zhang, Songgang Li, Huanming Yang, and Jian Wang. 2008. The diploid genome sequence of an asian individual. *Nature*, 456(7218):60–65.

Sinong Wang, Belinda Z Li, Madian Khabsa, Han Fang, and Hao Ma. 2020. Linformer: Self-attention with linear complexity. *arXiv preprint arXiv:2006.04768*.

Bryan Wilie, Karissa Vincentio, Genta Indra Winata, Samuel Cahyawijaya, X. Li, Zhi Yuan Lim, S. Soleman, R. Mahendra, Pascale Fung, Syafri Bahar, and A. Purwarianti. 2020. Indonlu: Benchmark and resources for evaluating indonesian natural language understanding. In *Proceedings of the 1st Conference of the Asia-Pacific Chapter of the Association for Computational Linguistics and the 10th International Joint Conference on Natural Language Processing*.

Genta Indra Winata, Samuel Cahyawijaya, Zihan Liu, Zhaojiang Lin, Andrea Madotto, and Pascale Fung. 2021. Are multilingual models effective in code-switching? In *Proceedings of the Fifth Workshop on Computational Approaches to Linguistic Code-Switching*, pages 142–153, Online. Association for Computational Linguistics.

Zhilin Yang, Zihang Dai, Yiming Yang, Jaime Carbonell, Russ R Salakhutdinov, and Quoc V Le. 2019. Xlnet: Generalized autoregressive pretraining for language understanding. *Advances in neural information processing systems*, 32.

Jian Ye, George Coulouris, Irena Zaretskaya, Ioana Cutcutache, Steve Rozen, and Thomas L Madden. 2012. Primer-BLAST: a tool to design target-specific primers for polymerase chain reaction. *BMC Bioinformatics*, 13(1):134.

Tiezheng Yu, Zihan Liu, and Pascale Fung. 2021. Adaptsum: Towards low-resource domain adaptation for abstractive summarization. In *Proceedings of the 2021 Conference of the North American Chapter of the Association for Computational Linguistics: Human Language Technologies*, pages 5892–5904.

Manzil Zaheer, Guru Guruganesh, Kumar Avinava Dubey, Joshua Ainslie, Chris Alberti, Santiago Ontanon, Philip Pham, Anirudh Ravula, Qifan Wang, Li Yang, et al. 2020. Big bird: Transformers for longer sequences. *Advances in Neural Information Processing Systems*, 33:17283–17297.

Jian Zhou and Olga G Troyanskaya. 2015. Predicting effects of noncoding variants with deep learning–based sequence model. *Nature Methods*, 12(10):931–934.

Xiaopu Zhou, , Yu Chen, Kin Y. Mok, Timothy C. Y. Kwok, Vincent C. T. Mok, Qihao Guo, Fanny C. Ip, Yuewen Chen, Nandita Mullapudi, Paola Giusti-Rodríguez, Patrick F. Sullivan, John Hardy, Amy K. Y. Fu, Yun Li, and Nancy Y. Ip. 2019. Non-coding variability at the APOE locus contributes to the alzheimer's risk. *Nature Communications*, 10(1).

Xiaopu Zhou, Yu Chen, Fanny C F Ip, Nicole C H Lai, Yolanda Y T Li, Yuanbing Jiang, Huan Zhong, Yuewen Chen, Yulin Zhang, Shuangshuang Ma, Ronnie M N Lo, Kit Cheung, Estella P S Tong, Ho Ko, Maryam Shoai, Kin Y Mok, John Hardy, Vincent C T Mok, Timothy C Y Kwok, Amy K Y Fu, and Nancy Y Ip. 2020. Genetic and polygenic risk score analysis for alzheimer's disease in the chinese population. *Alzheimers Dement. (Amst.)*, 12(1):e12074.

Xiaopu Zhou, Yolanda YT Li, Amy KY Fu, and Nancy Y Ip. 2021. Polygenic score models for alzheimer's disease: From research to clinical applications. *Frontiers in Neuroscience*, 15:346.

Yukun Zhu, Ryan Kiros, Rich Zemel, Ruslan Salakhutdinov, Raquel Urtasun, Antonio Torralba, and Sanja Fidler. 2015. Aligning books and movies: Towards story-like visual explanations by watching movies and reading books. In *Proceedings of the IEEE international conference on computer vision*, pages 19–27.

Zhong Zhuang, Xiaotong Shen, and Wei Pan. 2019. A simple convolutional neural network for prediction of enhancer–promoter interactions with dna sequence data. *Bioinformatics*, 35(17):2899–2906.

A Mapping of SNP Tokens

As our resulting SNP tokens are represented as 'X_1/X_2", to reduce the size and facilitate more straightforward representation for the downstream process in the NLP pipeline, such as pre-processing, tokenization, and modeling, we map all SNP tokens into a single character representation. The mapping of the SNP tokens into a single character representation is shown in Table 4. We use non-alphabetical characters as there are 66 SNP tokens in total, more than the available alphabetical characters, which consists of 52 characters (lower and upper case from 'A' to 'Z') in total. Also note that, all the SNP tokens related to the unkown token 'N' except 'N/N' (such as 'A/N', 'G/NI', 'N/NI', 'NI/NI', etc) are never been used since there is no actual SNP record corresponding to the unknown token 'N'. The combinations of all 'N' and 'NI' tokens are listed on the table only for completion.

B Model Hyperparameters

We develop two Linformer (Wang et al., 2020) models, i.e., Dipformer and Hapformer, which is pre-trained using our proposed SNP tokens and the original nucleotide tokens, respectively. The two models have the same hyperparameter settings resulting in an equal number of parameters. We list all the hyperparameters of our Dipformer and Hapformer models in Table 5.

C Pre-Training Setup

During the pre-training phase, we build the BPE tokenizer with a vocab size of 32,000 for both the SNP tokens and nucleotide tokens datasets. We perform pre-training on both Dipformer and Hapformer models for 200,000 steps using masked language modeling with the cross entropy loss. During the pre-training, we apply a masking strategy similar to BERT (Devlin et al., 2019) with a 15% token replacement rate, where we replace with `[MASK]` 80% of the time, replace with a random token 10% of the time, and keep the token as is 10% of the time. We run the pre-training using 5 units of 2080Ti GPUs and an Intel(R) Xeon(R) Silver 4210 CPU. We use the same hyperparameter settings for pre-training both the Dipformer and Hapformer models. The hyperparameters of our pre-training are shown in Table 6.

D Fine-Tuning Setup

We fine-tune all models on Alzheimer's disease risk prediction on a Chinese cohort consisting of 624 subjects in total, 384 of which are labelled as Alzheimer's disease carriers (ADs) while 240 others are non-carriers (NCs). For predicting Alzheimer's disease, we append a `[CLS]` token as the prefix of the sequence. During the fine-tuning, we take the output of the `[CLS]` token and perform a linear transformation on it to get the disease risk prediction. We evaluate the performance of all models using accuracy, area underthe ROC curve (AUROC), and area under the precision-recall curve (AUPRC). We show all the hyperparameters of the fine-tuning phase in Table 7. We experiment with different learning rate for each model and find that the best setting is achieved when using a learning rate of 1e-4 for models that are not pre-trained (non-pre-trained Dipformer and DeepSEA) and a learning rate of 1e-5 for all pre-trained models (Dipformer and Hapformer).

E Detailed Results

In this section, we show the distribution of the 10-fold results from our experiment in the Alzheimer's disease risk prediction task for all models (Dipformer, Hapformer, DeepSEA, and PLINK) on each evaluation metric. Figure 7 shows the distribution of the best 10-folds accuracy performance on the Alzheimer's disease risk prediction task. Figure 8 shows the distribution of the best 10-folds AUROC performance on the Alzheimer's disease risk prediction task. Figure 9 shows the distribution of the best 10-folds AUPRC performance on the Alzheimer's disease risk prediction task.

Mapping of SNP Tokens											
A/A	A	DEL/A	腌	A/C	嗄	AI/C	爸	C/G	嚓	CI/G	懂
C/C	C	DEL/AI	拔	A/G	阿	AI/CI	比	C/N	拆	CI/GI	答
G/G	G	DEL/C	吃	A/N	呵	AI/G	八	C/T	礤	CI/N	达
N/N	N	DEL/CI	搭	A/T	锕	AI/GI	霸	C/CI	车	CI/NI	第
T/T	T	DEL/G	想	A/AI	吖	AI/N	巴	C/GI	床	CI/T	瘩
AI/AI	B	DEL/GI	香	A/CI	俺	AI/NI	逼	C/NI	穿	CI/TI	沓
CI/CI	D	DEL/N	学	A/GI	安	AI/T	把	C/TI	出	G/GI	高
GI/GI	H	DEL/NI	虚	A/NI	案	AI/TI	笔	GI/N	蝦	G/N	给
NI/NI	O	DEL/T	徐	A/TI	按	N/NI	讷	GI/NI	合	G/NI	股
TI/TI	U	DEL/TI	需	NI/T	喔	N/T	哪	GI/T	虾	G/T	个
DEL/DEL	X	T/TI	拓	NI/TI	侬	N/TI	娜	GI/TI	盒	G/TI	该

Table 4: Mapping of SNP tokens into a single character representation.

Hyperparams	Value
#layers	6
dim	512
k	128
dropout	0.1
num heads	8
dim head	64
num embeddings	32000
single KV head	False
shared KV	False

Table 5: Model Hyperparameters

Hyperparams	Value
batch size	240
optimizer	AdamW
learning rate	1e-4
scheduler $\lambda 1$	1
scheduler $\lambda 2$	0.999991
#steps	200,000
warmup step	1000
loss fn	Cross Entropy
random seed	0

Table 6: Pre-Training Hyperparameters

Hyperparams	Value
batch size	16
optimizer	AdamW
learning rate	[1e-4..1e-6]
scheduler $\lambda 1$	1
scheduler $\lambda 2$	0.999991
#epoch	30
early stopping	3
loss fn	Cross Entropy
random seed	0

Table 7: Fine-Tuning Hyperparameters

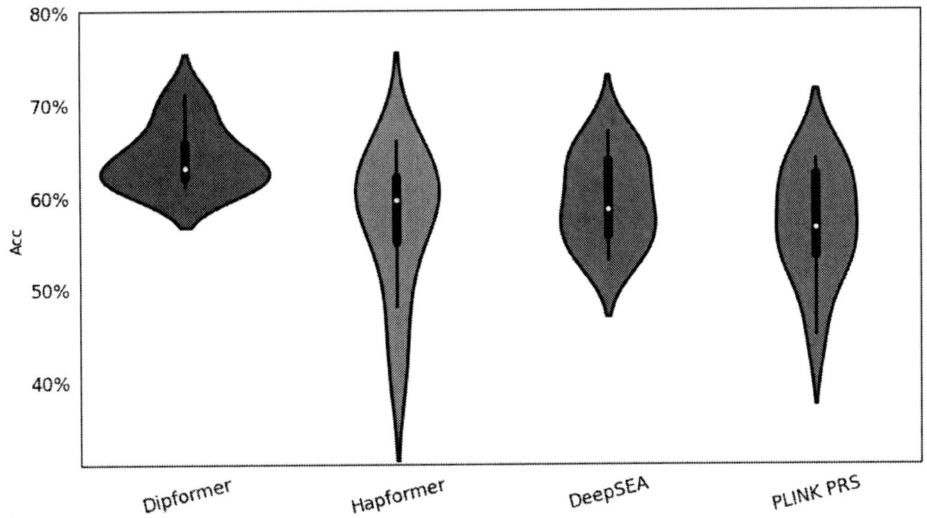

Figure 7: 10-folds accuracy performance of the best Dipformer, Hapformer, DeepSEA, and PLINK models on the Alzheimer's disease risk prediction.

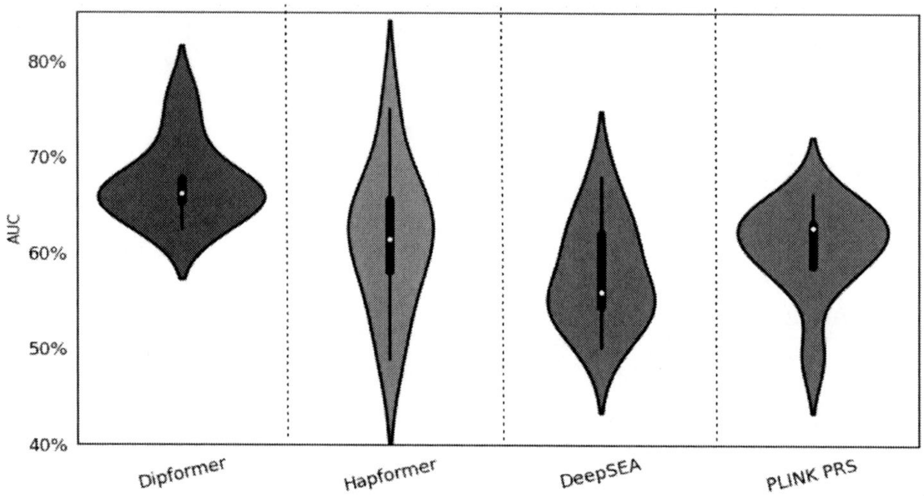

Figure 8: 10-folds AUROC performance of the best Dipformer, Hapformer, DeepSEA, and PLINK models on the Alzheimer's disease risk prediction.

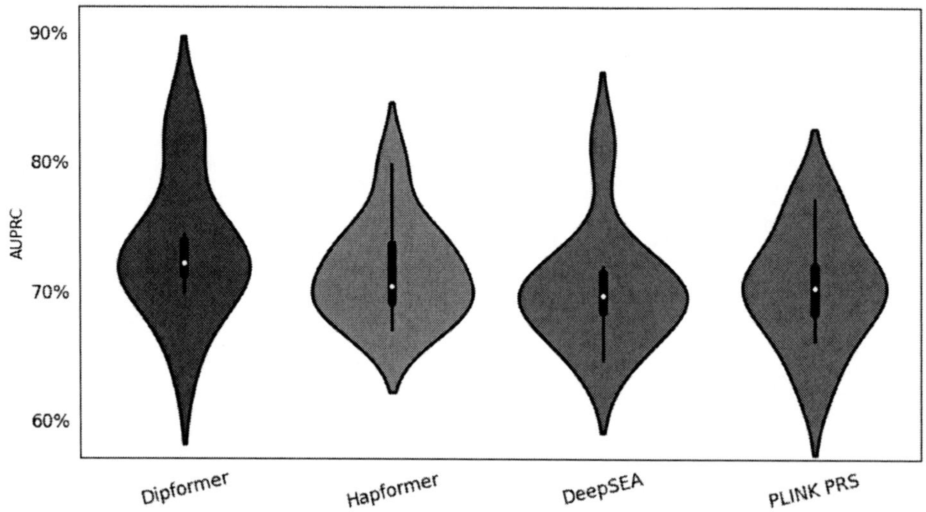

Figure 9: 10-folds AUPRC performance of the best Dipformer, Hapformer, DeepSEA, and PLINK models on the Alzheimer's disease risk prediction.

Biomedical NER using Novel Schema and Distant Supervision

Anshita Khandelwal, Alok Kar, Veera Raghavendra Chikka and **Kamalakar Karlapalem**
Data Science and Analytics Center (DSAC),
Kohli Center for Information Systems (KCIS),
International Institute of Information Technology, Hyderabad (IIIT-H)
{anshita.khandelwal, alok.kar, raghavendra.ch}@alumni.iiit.ac.in,
kamal@iiit.ac.in

Abstract

Biomedical Named Entity Recognition (BMNER) is one of the most important tasks in the field of biomedical text mining. Most work so far on this task has not focused on identification of discontinuous and overlapping entities, even though they are present in significant fractions in real-life biomedical datasets. In this paper, we introduce a novel annotation schema to capture complex entities, and explore the effects of distant supervision on our deep-learning sequence labelling model. For BMNER task, our annotation schema outperforms other BIO-based annotation schemes on the same model. We also achieve higher F1-scores than state-of-the-art models on multiple corpora without fine-tuning embeddings, highlighting the efficacy of neural feature extraction using our model.

1 Introduction

Named entity recognition (NER) consists of identification and classification of named entities in text. Biomedical NER (BMNER) is a crucial problem in healthcare as it is the initial step in solving various tasks, such as relation extraction, semantic role labeling, and clinical decision making (De Bruijn and Martin, 2002)(Hanisch et al., 2003). As compared to NER in other domains, BMNER is a difficult task as labelled data in biomedical domain is less in amount and expensive to obtain, and it requires identification of complex entities that are not common in other domains (Dai, 2018). Recently, deep learning approaches using large unstructured data, such as Bi-LSTM with CRF (Li et al., 2018) and BERT (Symeonidou et al., 2019)(Yu et al., 2019) models have been used to obtain state-of-the-art results on BMNER.

The most common annotation scheme for NER is BIO tagging, where B is for Beginning of entity, I for Inside of entity, and O for Outside of entity. A major assumption of BIO tagging is that an entity is composed of continuous and non-overlapping tokens. As complex entities that defy these assumptions frequently occur in biomedical records, a new scheme is needed to capture them. For this purpose, BIOHD (Tang et al., 2013) was introduced to represent discontinuous entities that may overlap with four new tags : (BH,IH) as shared head tags and (BD,ID) as non-shared non-head tags. However, this scheme fails to capture discontinuous entities that have more than two spans. In this paper, we propose a novel annotation schema BIODT that overcomes this limitation of BIOHD. Our schema includes shared non-head tags and non-shared head tags, and hence captures entities with more than two spans, which BIOHD fails to do.

Distant supervision is a method to generate labelled data from unlabelled data using existing knowledge (Mintz et al., 2009) that is particularly useful to create data for supervised learning algorithms which require large amounts of data. We use this method for BMNER to compensate for the lack of labelled data in the biomedical domain. As we are using an RNN (Recurrent Neural Network) which requires a large amount of training data, distant supervision helps in increasing the amount of annotated records without human effort.

In summary, the main contributions of this paper are as follows:

1. A novel systematic tagging schema to better capture discontinuous entities, that is significantly better(>2%) for prediction of discontinuous entities than BIOHD.
2. A distant supervision approach to biomedical NER, that uses labelled data to generate labels for unlabelled data, without the use of external dictionaries. Our experiments show that distant supervision methods boost the performance of our model, and also outperform state-of-the-art models.

Proceedings of the BioNLP 2022 workshop, Dublin, Ireland, pages 155–160
May 26, 2022.

2 Related Work

Existing solutions for BMNER include traditional NER methods such as dictionary or rule-based approaches, as well as supervised machine learning methods like Markov models (Ponomareva et al., 2007), Conditional Random Fields (CRFs) (Ponomareva et al., 2007)(Sun et al., 2006)(Settles, 2004) and Support Vector Machine (SVM) (Ju et al., 2011)(Kazama et al., 2002). Lately, deep learning approaches using large unstructured data, such as Bi-LSTM with CRF (Li et al., 2018) and BERT (Symeonidou et al., 2019)(Yu et al., 2019) models have been used to obtain state-of-the-art results on BMNER. To deal with scarcity of token-level annotated data required in deep-learning models, some weak-supervision and distant-supervision solutions have been proposed. For the task of BMNER, Mathew et al. (Mathew et al., 2019) introduced a weakly-supervised data augmentation approach for identification of proteins in BioCreative Challenge VI Track 1 dataset(Arighi et al., 2018), using a reference set of entity names from knowledge bases like UniProt (Consortium, 2018) to identify entity mentions on unlabelled data. In 2016, Lee et al. proposed a bagging-based approach using active learning with distant supervision, that uses a semi-automatically constructed dictionary of named entities from Wikipedia (Lee et al., 2016) (Song and Kim, 2015). To the best of our knowledge, no prior work has been done to study the effects of distant supervision on complex entities for NER.

To deal with annotation of complex entities, many methods have been proposed. Annotation schemes like BIOHD (Tang et al., 2013) and BIOHD1234 (Tang et al., 2015) were proposed with four and ten additional tags, respectively, to the commonly used BIO schema. These schemes gave near state-of-the-art results with simple machine learning models. Methods such as representing sentences as hypergraphs (Lu and Roth, 2015) (Muis and Lu, 2016), transition-based models that uses specialized actions and attention mechanisms (Dai et al., 2020), and representing NER task as a structured multi-label classification problem (McDonald et al., 2005) have also been explored. Additionally, a two-stage approach that first detects all continuous parts, then combines them to form discontiguous entities using a classifier (Wang and Lu, 2019) has also been proposed.

3 Annotation Schema

We introduce a new annotation schema called BIODT, which consists of 11 tags: the traditional BIO tags, and 8 additional tags as described below.

1. DB, DI are shared heads of the first term in a discontinuous entity
2. DHB, DHI are shared non-head tags of the subsequent terms in a discontinuous entity
3. TB, TI are non-shared heads of the first term in a discontinuous entity
4. THB, THI are non-shared non-head tags of the subsequent terms in a discontinuous entity

Preference is given to combine shared head tags with shared non-head tags and, similarly, for non-shared tags. For example, in sentence 1 (Figure 1), "aortic root", "descending root" and "dilated" are tagged with shared tags. Similarly, in Sentence 2, "mitral", "leaflet" and "thickened" are tagged with non-shared tags. There are a few cases where shared and non-shared tags can co-occur in a sentence. In Sentence 3 (Figure 1), "ABD" is a shared head tag. If tagged according to BIOHD schema, "tenderness" and "RUQ" would be shared non-head tags, resulting in two entities, "ABD...tenderness" and "ABD...RUQ", which are wrong. In our schema, we tag "tenderness" and "RUQ" with non-shared non-head tags(TH{B,I}), which are combined with the shared head tag(D{B,I}) to form "ABD...tenderness...RUQ". Hence, our schema captures entities that were not captured by the BIOHD schema.

Extracting entities from BIODT tagged sentence:
Discontinuous entities can be obtained from a BIODT tagged sentence using the following simple rules :

- For shared tags :
  1. Each shared non-head tag (DH{B,I}) is joined to each shared head tag (D{B,I}) in the sentence.
  2. If no shared head tag is present, all shared non-head tags in the sentence are combined to form one joined entity.
- For non-shared tags :
  1. All non-shared non-head tags (TH{B,I}) are joined together.
  2. If any non-shared head tag (T{B,I}) is present, then the entity obtained from (1) is joined to each non-shared head tag in the sentence.

3. If no non-shared head tag is present and any shared head tag (D{B,I}) is present, then the entity obtained from (1) is joined to each shared head tag in the sentence.
4. If no head tags(shared/non-shared) are present in the sentence, return entity obtained from (1).

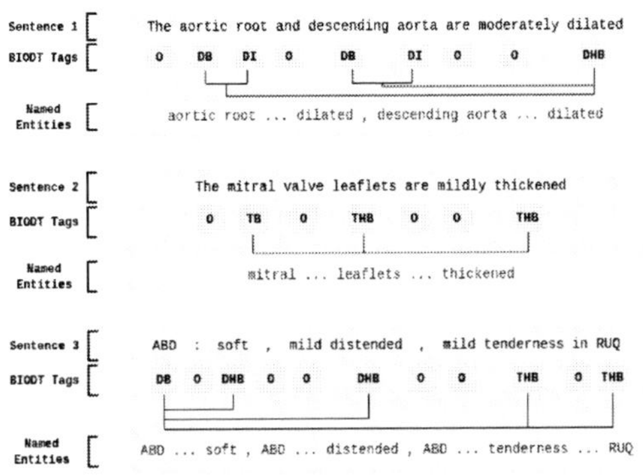

Figure 1: BIODT Schema Examples

4 Approach and Architecture

We use a BiLSTM-CRF network to assign labels for NER, as presented in Figure 2. BiLSTM-CRF is an RNN (Recurrent Neural Network), and is formed by the combination of a BiLSTM (Bidirectional Long-Short Memory) and a CRF (Conditional Network Field). For each sentence, the BiLSTM forms a vector representation for each word, preserving backward and forward context. This vector representation is then used as the input to the CRF, which predicts labels for the words of the sentence.

The labels at the CRF output layer are decoded using the Viterbi algorithm.

4.1 Features and Embeddings

We have used a combination of GloVe word embeddings(Pennington et al., 2014), character embeddings and BERT (Bio+Discharge Summary BERT) embeddings (Alsentzer et al., 2019). Additionally, we have also experimented with part-of-speech(POS) embeddings, case(lower/upper) embeddings, and suffix/prefix embeddings.

4.2 Distant Supervision

We use unlabelled data to generate a larger training set using distant supervision. We trained our baseline model on manually annotated data, then used the model to predict labels on additional unlabelled data to expand our training set.

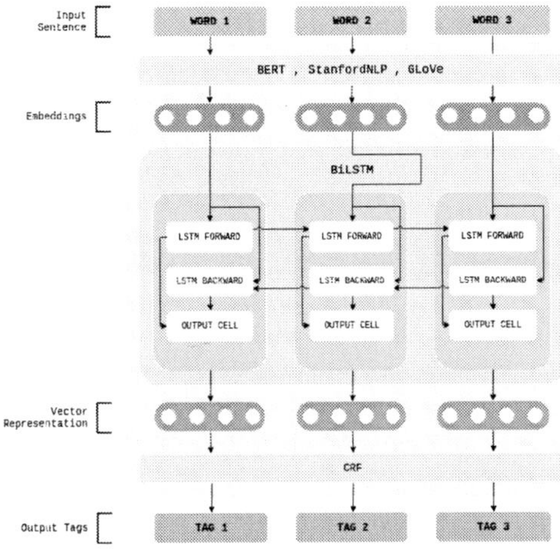

Figure 2: Model Architecture

Our method is 1. train M on D1-train, 2. predict labels on unlabelled dataset D2 and augment this newly labelled dataset to D1-train, 3. train M on D1-train and newly labelled D2, 4. finally, we test M on D1-test.

Here, D1-train and D1-test are train and test partitions of the labelled dataset respectively, D2 is an unlabelled dataset of similar domain, and M is our model.

5 Datasets

We experiment on two datasets from the biomedical domain: ShARe 2013 (Forner et al., 2013) and ShARe 2014 (Cappellato et al., 2014). The datasets contain clinical free-text notes, which include discharge summaries, echo-reports, and radiology reports. An annotated named entity can contain any number of continuous spans, and it maps to a concept in the disorder semantic group of SNOMED-CT (Cornet and de Keizer, 2008).

Since a significant fraction(almost 10%) of the mentions in these datasets are discontinuous(from Table 1), an improvement in discontinuous entity recognition will show noticeable improvement in overall entity recognition.

	ShARe 2013	ShARe 2014
#Records	298	433
#Sentences	18.7k	34.6k
#Total Mentions	11,161	19,131
#Disc. Mentions	1,090	1,710
% Disc. Mentions	9.7	8.9

Table 1: Dataset statistics for ShARe 2013 and ShARe 2014

6 Results and Analysis

For evaluation, we have used scripts provided in ShARe tasks to calculate F-score (F) to evaluate the efficiency of the models in our experimentation.

Our baseline model is a BiLSTM-CRF that uses the features and embeddings mentioned in 4.1, as proposed by Yu et al. (Yu et al., 2019).

We faced a replication crisis while attempting to reproduce the results presented in (Tang et al., 2015) using the proposed BIOHD1234 schema. Hence, we were unable to compare the performance of our schema with that of BIOHD1234.

6.1 Model Evaluation

As can be seen from Table 2, our model outperforms the baseline in both annotation schemes by a small margin. It also gives a better result than the state-of-the-art by 1.6% and 1.1% for both datasets, using BIODT and BIOHD schemes, respectively. Evaluating for discontinuous entities, we find that our model performance is similar to that of the baseline task, with the BIOHD schema slightly underperforming for the ShARe 2013 corpus.

Model	Scheme	ShARe 13	ShARe 14
SSVM (Tang et al., 2013)	BIOHD	75.0	–
SSVM (Tang et al., 2015)	BIOHD1234	78.3	–
Transition-based model (Dai et al., 2020)	HGB	77.7	79.6
Baseline	BIOHD	78.4	79.7
Distant Supervision	BIOHD	78.9	**80.7**
Baseline	BIODT	79.0	80.4
Distant Supervision	BIODT	**79.9**	80.5

Table 2: F1-Scores of other models compared to our model; HGB stands for Hypergraph Based

Dataset	Model	BIOHD	BIODT
ShARe 2013	(Tang et al., 2015)	48.7	-
	Baseline	46.1	51.6
	D. Supervision	45.6	**52.8**
ShARe 2014	Baseline	40.5	44.2
	D. Supervision	41.9	**44.5**

Table 3: F1-Scores with BIOHD and BIODT for discontinuous entities

6.2 Evaluation of Annotation Schema

On entire datasets, BIODT performs similar(within 1%) to BIOHD for all models. The only case where it is not an improvement over BIOHD is when we use our model on ShARe 2014 dataset, where it has a 0.2% less score. As is clear from Table 3, BIODT schema gives a significantly better performance over BIOHD for discontinuous entities(>3%), for all cases.

6.3 Analysis

From Table 2 and Table 3, it can be inferred that while BIODT does not help much for NER in entire datasets, it brings a noticeable improvement compared to BIOHD for discontinuous entities. We believe that for datasets with a higher fraction of discontinuous entities, BIODT will perform better than it has for these experiments.

From Table 2 and Table 3, it is also clear that when used with BIODT schema, distant supervision enhances performance, both for entire datasets and for discontinuous entities.

Limitations of BIODT

Due to the decoding rules of BIODT, some false positives occur even on correctly predicted labels:

```
DB1 DI1 O O DB2 DI2 O O O
DHB1 DHI1 O O O DHB2 DHI2
```

Here, the original entities are :

`(DB1 DI1 DHB1 DHI1) , (DB2 DI2 DHB2 DHI2)`

Now, according to decoding rules, each shared non-head term will combine to each shared head term, hence the entities obtained will be :

1. `DB1 DI1 DHB1 DHI1`
2. `DB1 DI1 DHB2 DHI2`
3. `DB2 DI2 DHB1 DHI1`
4. `DB2 DI2 DHB2 DHI2`

Among these entities, (1) and (4) are correctly decoded, (2) and (3) are not. Even if our model predicts these labels correctly, they will be decoded as false positives. We do not believe that this leads to worse performance of BIODT as compared to BIOHD, as BIOHD faces a similar problem.

7 Conclusion

In this paper, we introduced a novel annotation schema to identify named entities in biomedical data. We have also shown that for the same model, our annotation scheme gives better performance than other BIO-based complex annotation schemes for discontinuous entities. We also explore the distant supervision paradigm to increase our training set for BioNER. Using this, we have achieved state-of-the-art results.

References

Emily Alsentzer, John Murphy, William Boag, Wei-Hung Weng, Di Jindi, Tristan Naumann, and Matthew McDermott. 2019. Publicly available clinical BERT embeddings. In *Proceedings of the 2nd Clinical Natural Language Processing Workshop*, pages 72–78, Minneapolis, Minnesota, USA. Association for Computational Linguistics.

Cecilia Arighi, Lynette Hirschman, Thomas Lemberger, Samuel Bayer, Robin Liechti, Donald Comeau, and Cathy Wu. 2018. Bio-id track overview. In *Proceedings of the BioCreative VI Workshop*, pages 14–19, Bethesda, MD, USA.

Linda Cappellato, Nicola Ferro, Martin Halvey, and Wessel Kraaij, editors. 2014. *Working Notes for CLEF 2014 Conference, Sheffield, UK, September 15-18, 2014*, volume 1180 of *CEUR Workshop Proceedings*. CEUR-WS.org.

The UniProt Consortium. 2018. Uniprot: a worldwide hub of protein knowledge. *Nucleic Acids Research*, 47(D1):D506–D515.

Ronald Cornet and Nicolette de Keizer. 2008. Forty years of snomed: a literature review. *BMC medical informatics and decision making*, 8 Suppl 1:S2.

Xiang Dai. 2018. Recognizing complex entity mentions: A review and future directions. In *Proceedings of ACL 2018, Student Research Workshop*, pages 37–44, Melbourne, Australia. Association for Computational Linguistics.

Xiang Dai, Sarvnaz Karimi, Ben Hachey, and Cecile Paris. 2020. An effective transition-based model for discontinuous NER. In *Proceedings of the 58th Annual Meeting of the Association for Computational Linguistics*, pages 5860–5870, Online. Association for Computational Linguistics.

Berry De Bruijn and Joel Martin. 2002. Getting to the (c) ore of knowledge: mining biomedical literature. *International journal of medical informatics*, 67(1):7–18.

Pamela Forner, Roberto Navigli, Dan Tufis, and Nicola Ferro, editors. 2013. *Working Notes for CLEF 2013 Conference , Valencia, Spain, September 23-26, 2013*, volume 1179 of *CEUR Workshop Proceedings*. CEUR-WS.org.

Daniel Hanisch, Juliane Fluck, Heinz-Theodor Mevissen, and Ralf Zimmer. 2003. Playing biology's name game: Identifying protein names in scientific text. In *Proceedings of the 8th Pacific Symposium on Biocomputing, PSB 2003, Lihue, Hawaii, USA, January 3-7, 2003*, pages 403–414.

Z. Ju, J. Wang, and F. Zhu. 2011. Named entity recognition from biomedical text using svm. In *2011 5th International Conference on Bioinformatics and Biomedical Engineering*, pages 1–4.

Jun'ichi Kazama, Takaki Makino, Yoshihiro Ohta, and Jun'ichi Tsujii. 2002. Tuning support vector machines for biomedical named entity recognition. In *Proceedings of the ACL-02 Workshop on Natural Language Processing in the Biomedical Domain*, pages 1–8, Phildadelphia, Pennsylvania, USA. Association for Computational Linguistics.

Sunghee Lee, Yeongkil Song, Maengsik Choi, and Harksoo Kim. 2016. Bagging-based active learning model for named entity recognition with distant supervision. In *Proceedings of the 2016 International Conference on Big Data and Smart Computing (BigComp)*, BIGCOMP '16, page 321–324, USA. IEEE Computer Society.

Fei Li, Meishan Zhang, Bo Tian, Bo Chen, Guohong Fu, and Donghong Ji. 2018. Recognizing irregular entities in biomedical text via deep neural networks. *Pattern Recognition Letters*, 105(C):105–113.

Wei Lu and Dan Roth. 2015. Joint mention extraction and classification with mention hypergraphs. In *Proceedings of the 2015 Conference on Empirical Methods in Natural Language Processing*, pages 857–867, Lisbon, Portugal. Association for Computational Linguistics.

Joel Mathew, Shobeir Fakhraei, and José Luis Ambite. 2019. Biomedical named entity recognition via reference-set augmented bootstrapping.

Ryan McDonald, Koby Crammer, and Fernando Pereira. 2005. Flexible text segmentation with structured multilabel classification. In *Proceedings of Human Language Technology Conference and Conference on Empirical Methods in Natural Language Processing*, pages 987–994, Vancouver, British Columbia, Canada. Association for Computational Linguistics.

Mike Mintz, Steven Bills, Rion Snow, and Dan Jurafsky. 2009. Distant supervision for relation extraction without labeled data. In *Proceedings of the Joint Conference of the 47th Annual Meeting of the ACL and the 4th International Joint Conference on Natural Language Processing of the AFNLP: Volume 2 - Volume 2*, ACL '09, page 1003–1011, USA. Association for Computational Linguistics.

Aldrian Obaja Muis and Wei Lu. 2016. Learning to recognize discontiguous entities. In *Proceedings of the 2016 Conference on Empirical Methods in Natural Language Processing*, pages 75–84, Austin, Texas. Association for Computational Linguistics.

Jeffrey Pennington, Richard Socher, and Christopher D Manning. 2014. Glove: Global vectors for word representation. *Proceedings of the Empiricial Methods in Natural Language Processing (EMNLP 2014)*, 12:1532–1543.

Natalia Ponomareva, Paolo Rosso, Ferran Pla, and Antonio Molina. 2007. Conditional random fields vs. hidden markov models in a biomedical named entity recognition task.

Burr Settles. 2004. Biomedical named entity recognition using conditional random fields and rich feature sets. JNLPBA '04, page 104–107, USA. Association for Computational Linguistics.

Yeongkil Song and Harksoo Kim. 2015. Semi-automatic construction of a named entity dictionary based on active learning. In *Computer Science and its Applications*, pages 65–70, Berlin, Heidelberg. Springer Berlin Heidelberg.

Chengjie Sun, Yi Guan, Xiaolong Wang, and Lei Lin. 2006. Biomedical named entities recognition using conditional random fields model. In *Fuzzy Systems and Knowledge Discovery*, pages 1279–1288, Berlin, Heidelberg. Springer Berlin Heidelberg.

Anthi Symeonidou, Viachaslau Sazonau, and Paul Groth. 2019. Transfer learning for biomedical named entity recognition with biobert. In *Posters and Demo Track of the 15th International Conference on Semantic Systems. (Poster and Demo Track at SEMANTiCS 2019)*, number 2451 in CEUR Workshop Proceedings, pages 126–130, Aachen.

Buzhou Tang, Qingcai Chen, Xiaolong Wang, Yonghui Wu, Yaoyun Zhang, Min Jiang, Jingqi Wang, and Hua Xu. 2015. Recognizing disjoint clinical concepts in clinical text using machine learning-based methods. *AMIA Annual Symposium Proceedings*, 2015:1184–1193.

Buzhou Tang, Yonghui Wu, Min Jiang, Joshua C Denny, and Hua Xu. 2013. Recognizing and encoding discorder concepts in clinical text using machine learning and vector space model. In *Workshop of ShARe/CLEF eHealth Evaluation Lab*.

Bailin Wang and Wei Lu. 2019. Combining spans into entities: A neural two-stage approach for recognizing discontiguous entities. In *Proceedings of the 2019 Conference on Empirical Methods in Natural Language Processing and the 9th International Joint Conference on Natural Language Processing (EMNLP-IJCNLP)*, pages 6216–6224, Hong Kong, China. Association for Computational Linguistics.

X. Yu, W. Hu, S. Lu, X. Sun, and Z. Yuan. 2019. Biobert based named entity recognition in electronic medical record. In *2019 10th International Conference on Information Technology in Medicine and Education (ITME)*, pages 49–52, Los Alamitos, CA, USA. IEEE Computer Society.

Improving Supervised Drug-Protein Relation Extraction with Distantly Supervised Models

Naoki Iinuma, Makoto Miwa and **Yutaka Sasaki**
Computational Intelligence Laboratory
Toyota Technological Institute
2-12-1 Hisakata, Tempaku-ku, Nagoya, Aichi, 468-8511, Japan
inaoki2628@gmail.com
{makoto-miwa, yutaka.sasaki}@toyota-ti.ac.jp

Abstract

This paper proposes novel drug-protein relation extraction models that indirectly utilize distant supervision data. Concretely, instead of adding distant supervision data to the manually annotated training data, our models incorporate distantly supervised models that are relation extraction models trained with distant supervision data. Distantly supervised learning has been proposed to generate a large amount of pseudo-training data at low cost. However, there is still a problem of low prediction performance due to the inclusion of mislabeled data. Therefore, several methods have been proposed to suppress the effects of noisy cases by utilizing some manually annotated training data. However, their performance is lower than that of supervised learning on manually annotated data because mislabeled data that cannot be fully suppressed becomes noise when training the model. To overcome this issue, our methods indirectly utilize distant supervision data with manually annotated training data. The experimental results on the DrugProt corpus in the BioCreative VII Track 1 showed that our proposed model can consistently improve the supervised models in different settings.

1 Introduction

Drug-protein relations are important for drug discovery, metabolic, and drug response modeling, and their textual evidence is important in the development of evidence-based medicine. However, since drug-protein interactions are reported in the literature and the number of relevant articles is rapidly increasing (Coordinators, 2016), it is difficult for pharmacologists to read every single article to determine the interactions. Therefore, automatic interaction extraction from text has attracted much attention. The related shared tasks (Krallinger et al., 2021, 2017) are being conducted at BioCreative, an international workshop that aims to evaluate text mining and information extraction in the biological domain.

For drug-protein relation extraction, models using deep learning have achieved high performance. A typical deep learning model takes as input a sentence and the drug and protein mentions in the sentence, and predicts the relationship between the drug and the protein as expressed in the sentence. Gu et al. (2022) extracted the relationships using a large neural network model pretrained on a large biomedical literature (PubMedBERT). Deep learning models suffer from the problem of the huge cost of manually annotated training data.

A distantly supervised learning method has been proposed by Mints et al. (2009). The method enables the creation of a large amount of training data at low cost. However, this method still has the problem of producing data with incorrect labels, which become noise during training. Several methods have been proposed to mitigate the effects of such noisy examples. One of the most commonly-used methods is multi-instance learning (Riedel et al., 2010), where the distant supervision data is treated as a bag of instances corresponding to pairs in the database. Zeng et al. (2015) proposed a method to train instances with the representation with the highest prediction probability of the target label in the bag. Ji et al. (2017) proposed a method to weight instances in the bag so that correctly labeled instances will have large weights while noisy cases have small weights. Beltagy et al. (2019a) proposed a method of learning with distant supervision data by utilizing some of manually annotated training data to learn the weights. Although such methods show performance improvement in the distantly supervised training setting, the performance is still lower than that of the methods trained on manually annotated training data.

This study proposes a novel method of using distantly supervised relation extraction models for supervised drug-protein relation extraction. By using the model trained over the easy-to-create distant supervision data, we aim to improve the performance

Proceedings of the BioNLP 2022 workshop, Dublin, Ireland, pages 161–170
May 26, 2022. ©2022 Association for Computational Linguistics

of supervised drug-protein relation extraction while reducing the cost of building additional manually annotated data and the effect of noisy instances in the distant supervision data.

Our contributions are as follows:

1. We generate distant supervision data for drug-protein relation extraction from domain databases. By utilizing four databases, we create distant supervision data of the same scale as that of general domain distant supervision data.

2. We propose to utilizing representations obtained from a distantly supervised model for ordinary supervised training. The performance in extracting relations between drugs and proteins was consistently improved for two models (i.e., PubMedBERT and BioRoBERTa-large (Lewis et al., 2020)) with different parameter sizes.

3. The proposed method showed consistent performance improvement regardless of the data size of the manually annotated training data, indicating that it is effective for utilizing distantly supervised model to improve the extraction performance.

2 Methods

We propose a novel method for extracting drug-protein relations from manually annotated training data. The method uses a model trained on distant supervision data, which we call a *distantly supervised model*. By utilizing the distantly supervised model, we aim to improve the extraction performance while reducing the influence of noisy instances included in the distant supervision data.

In the following sections, we will explain the baseline relation extraction model in Section 2.1, the construction of distant supervision data from databases in Section 2.2, and the methods for utilizing the distantly supervised model in Section 2.3.

2.1 Relation Extraction Model

We describe a supervised relation extraction model that is used as the baseline in this research. The model predicts the relation for a given entity pair from the input sentence.

First, the mentions of target drug and protein in the input sentence are masked with *"DRUG"* and *"PROTEIN"*, respectively. Table 1 shows an example of this preprocessing. The sentence contains three drug mentions (*androstenedione*, *oestrone*, *oestrone*) and one protein mention (*aromatase*), so three drug-protein pairs are created and their mentions are replaced.

Next, the input sentence with the target protein and drug entities is encoded with BERT (Devlin et al., 2019) to generate a feature representation vector $\boldsymbol{h}$ that represents the input sentence. For this vector, we use the representation vector of the [CLS] token since it contains the features of the whole sentence in BERT. Finally, based on the feature representation vector, the model then generates a prediction vector that represents the prediction probability for each relation by using one fully-connected layer and the softmax function. The model predicts the relation that has the maximum prediction probability. The optimizer is Adam (Kingma and Ba, 2015), and the model is trained to minimize the cross-entropy loss.

2.2 Building Distant Supervision Data

An overview of the process of building distant supervision data is shown in Figure 1. In this method, we use a medical literature database PubMed (Coordinators, 2016), a drug database DrugBank (DS et al., 2018), a protein database UniProt (Consortium, 2020), and a chemical substance database Comparative Toxicogenomics Database (CTD) (Davis et al., 2020). From these databases, we extract about 33 million articles, about 500 thousand drug entries, and about 570 thousand protein entries to create distant supervision data. In the following, we explain the process of building distant supervision data using these databases.

First, drug and protein entities are extracted from the medical literature in PubMed, as shown in Figure 1-(i). Sentence segmentation and entity extraction modules in SciSpacy (Neumann et al., 2019), a tool specialized for processing biomedical and scientific literature, are used to analyze the medical literature and extract drug entities and protein entities as named entities in the literature.

Next, we create relational triples as shown in Figure 1-(ii). ID relation triples are extracted from DrugBank. Here, an ID relation triple is a triple of drug ID, relation name, and protein ID. We create relation triples from the ID relation triples by mapping the IDs to their names using drug and protein name dictionaries. The drug name dictionary is created by mapping drug IDs to drug names and its synonyms on the information in DrugBank

Target drug	Target protein	Preprocessed input sentence
androstenedione	*aromatase*	*The **PROTEIN** enzyme, which converts **DRUG** to oestrone, regulates the availability of oestrogen so support the growth of hormone-dependent beast tumours.*
oestrone	*aromatase*	*The **PROTEIN** enzyme, which converts androstenedione to **DRUG**, regulates the availability of oestrogen so support the growth of hormone-dependent beast tumours.*
oestrogen	*aromatase*	*The **PROTEIN** enzyme, which converts androstenedione to oestrone, regulates the availability of **DRUG** so support the growth of hormone-dependent beast tumours.*

Table 1: Examples of preprocessing of drug-protein pairs in the sentence *The aromatase enzyme, which converts androstenedione to oestrone, regulates the availability of oestrogen so support the growth of hormone-dependent beast tumours.* (PMID:15341993)

DrugBank	DrugProt
ligand, binder, binding	DIRECT-REGULATOR
partial agonist	AGONIST-ACTIVATOR
inverse agonist	AGONIST-INHIBITOR
blocker, partial antagonist	ANTAGONIST
inducer, stimulator	INDIRECT -UPREGULATOR
product of	PRODUCT-OF
activator	ACTIVATOR
inhibitor	INHIBITOR
agonist	AGONIST
antagonist	ANTAGONIST
substrate	SUBSTRATE

Table 2: Mapping of relationships

and CTD. Similarly, a protein name dictionary is created from UniProt and CTD.

Then, as shown in Figure 1-(iii), the distant supervision data is created by strict matching the named entities extracted from the PubMed literature with drug and protein names in the relation triples after lowercasing the entities and names.

Finally, as shown in Figure 1-(iv), we map the relation types in DrugBank, which are the original labels of the distant supervision data, to the relation types in the DrugProt task (Krallinger et al., 2021) using a mapping dictionary as shown in Table 2. We manually build the mapping dictionary based on the relation annotation guideline (Rabal et al., 2021) in the DrugProt corpus.

2.3 Relation Extraction Using Distantly Supervised Models

We propose two alternatives to utilize the distantly supervised model. One is the initialization approach that initializes the supervised model with the distantly supervised model (`Initialization`), and the other is the mixture approach that combines representations obtained from a fixed distantly supervised model and representations obtained from a supervised model in training the supervised model (`Mixture`).

2.3.1 Initialization

In the task of natural language processing, pretraining on datasets close to the domain sometimes improves the performance of the model on the target dataset. (Beltagy et al., 2019b) Following this line, for `Initialization`, we perform pretraining using distant supervision data to initialize the model for supervised learning. Specifically, we first train the relation extraction model described in Section 2.1 using the distant supervision data, use the model parameters to initialize another relation extraction model for supervised learning, and then train the relation extraction model using manually annotated training data.

2.3.2 Mixture

For `Mixture`, we pretrain a relation extraction model explained in Section 2.1 using distant supervision data to extract additional features from the input. Similarly, another relation extraction model is pretrained with manually annotated training data[1]. The two pretrained feature extraction models, i.e., BERT, are used to mix the feature representations. In training, the feature extraction model pretrained on the distant supervision data is fixed, while the feature extraction model trained on the manually annotated training data is not fixed and further fine-tuned[2].

Predictions are made by mixing representations obtained from the model pretrained with distant supervision data and representations obtained from the model that is specific to supervised training with manually annotated training data as shown in Figure 2. We propose two mixing methods that use the importance weights of the representations, which mix the representations obtained from dis-

[1]We find this pretraining can improve the performance in our preliminary experiments.

[2]In our preliminary experiments, we tried to fine-tune the feature extraction model pretrained on the distant supervision data, but the performance with the model was lower than one with fixed parameters.

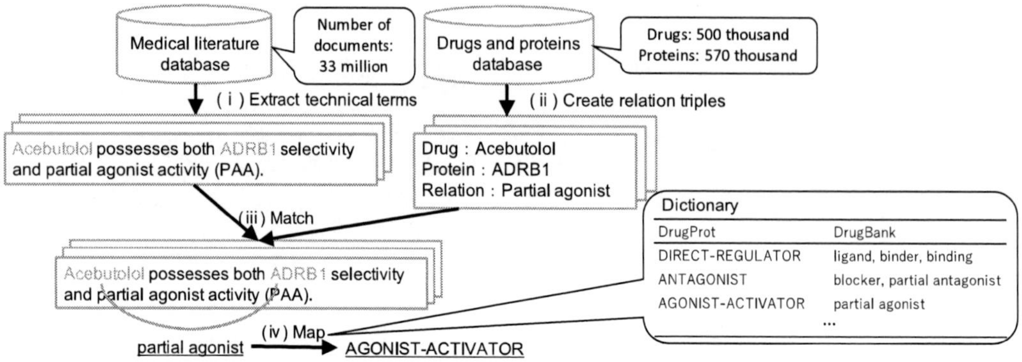

Figure 1: Overview of the creation of distant supervision data

tant supervision data with those obtained from manually annotated training data.

First, as shown in Figure 2-(i), the representations $\boldsymbol{h_{ds}}$ obtained from the fixed BERT model in the fixed pre-trained distantly supervised model are mixed with the representations $\boldsymbol{h_{sv}}$ from the BERT model in another relation extraction model that is pre-trained on the manually annotated training data. Next, as shown in Figure 2-(ii), we mix the representations $\boldsymbol{h_{ds}}$, $\boldsymbol{h_{sv}}$. In mixing the representations, we propose two mixing methods, `Add` and `Concat`, which are defined as follows:

$$\boldsymbol{h}_{\texttt{Add}} = \alpha \boldsymbol{h_{ds}} + \beta \boldsymbol{h_{sv}} \quad (1)$$

$$\boldsymbol{h}_{\texttt{Concat}} = [\alpha \boldsymbol{h_{ds}}; \beta \boldsymbol{h_{sv}}] \quad (2)$$

$[\cdot;\cdot]$ denotes the concatenation of vectors. α and β are the importance weights of each feature, which are scalar-valued parameters that are trained during training. Here, `Add`, as shown in Eq. (1), sums $\boldsymbol{h_{ds}}$ and $\boldsymbol{h_{sv}}$ after multipying the corresponding weight, which indicate the importance, to each representation. `Concat` is mixed by concatenating $\boldsymbol{h_{ds}}$ and $\boldsymbol{h_{sv}}$ after multiplying weights to the parameters, as shown in Eq. (2).

Finally, as shown in Figure 2-(iii), the obtained representations, i.e., $\boldsymbol{h}_{\texttt{Add}}$ or $\boldsymbol{h}_{\texttt{Concat}}$, are used to predict the relation between the drug and the protein with one fully connected layer (FC) and the softmax function. The model is trained on the manually annotated training data to minimize the cross-entropy loss.

3 Experimental Settings

In this section, we explain the settings for the data sets, tasks and hyper-parameter tuning.

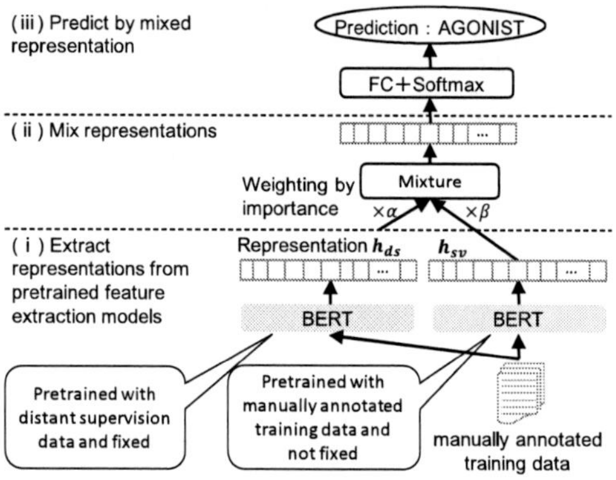

Figure 2: Overview of the `Mixture` of the representations

We used the data set from the BioCreative VII Track 1 - Text mining drug and chemical-protein interactions (DrugProt) (Krallinger et al., 2021) for the evaluation. This data set is composed of documents annotated with drug mentions, protein mentions, and their relations. The DrugProt corpus consists of train, develop, and test. Since the annotations for the test data are not publicly available, this study evaluates the model on the development data. In addition, the distant supervision data built by the method in Section 2.2 were used to train the model. The number of instances per relation in the DrugProt corpus and the distant supervision data are shown in Table 3. We followed the task setting of DrugProt. The task is to classify a given pair of a drug and a protein into 13 relation types or no relation. We evaluated the performance with the F-score on each relation type and the micro-averaged F-score on all relation types. Micro-averaged F-score is also shown for reference. We used the

	DrugProt train	develop	Distant supervision data
ANTAGONIST	972	218	69,234
AGONIST	659	131	89,704
AGONIST -ACTIVATOR	29	10	875
AGONIST -INHIBITOR	13	2	1,107
DIRECT -REGULATOR	2,250	458	18,945
ACTIVATOR	1,429	246	31,745
INHIBITOR	5,392	1,152	173,400
INDIRECT-DOWN-REGULATOR	1,330	332	0
INDIRECT-UP-REGULATOR	1,379	302	11,981
PART-OF	88	625	0
PRODUCT-OF	921	158	1,565
SUBSTRATE	2,003	495	2,311
SUBSTRATE _PRODUCT-OF	25	3	0
Total	17,288	3,765	400,867

Table 3: The number of instances per relation in the DrugProt corpus and the distant supervision data

official evaluation script[3] provided by the task organizers.

We used the Successive Halving Algorithm from the open-source hyper-parameter auto-optimization framework Optuna (Akiba et al., 2019) for hyper-parameter tuning. We chose the dropout rate from the region of [0.0, 0.5], the learning rate of Adam from the region of [1e-6, 1e-4], the weight decay of Adam from the region of [1e-10, 1e-3]. Hyper-parameters are determined by a parameter search to maximize the micro-averaged F-score on the development data of the DrugProt corpus[4].

4 Results

To evaluate the proposed method, we conducted three experiments: evaluation of the performance of extracting drug-protein relations, analysis of prediction results, and comparison of extraction performance on small-scale manually annotated training data. In this section, we describe these three experiments.

4.1 Drug-Protein Relation Extraction

We conducted experiments to compare the extraction performance of the proposed method with a baseline trained only on manually annotated training data. As the baselines, we trained relation extraction models based on PubMedBERT and BioRoBERTa-large, both of which were pretrained in a domain close to the dataset, with manually annotated training data. BioRoBERTa-large is a large-scale pretrained model with a parameter size approximately three times larger than PubMedBERT. The baseline model with BioRoBERTa-large is the same as the model by Yoon et al. (2021) that achieved the high performance of 77.46% on the development data without external knowledge. [5]

The results are shown in Table 4. First, we focus on the performance of the proposed methods when they are applied to the PubMedBERT baseline model. For all the proposed methods, the prediction performance for AGONISTand PRODUCT-OF, which have less manually annotated training data, is greatly improved. This is because the representations obtained from the distantly supervised model can compensate for the lack of manually annotated data. Besides, the performance of AGONIST-ACTIVATORand AGONIST-INHIBITOR, which have particularly less manually annotated training data, was significantly improved by `Initialization`, but not by `Mixture`. This shows that the representations obtained from the distantly supervised model with `Initialization`more directly influenced the performance than those with `Mixture`. In addition, `Add` and `Concat`, which mixed the representations from the distantly supervised model data with the representations specific to the supervised model, improved the micro-averaged F scores by 0.6 and 0.8 points, respectively. This indicates that `Mixture` is a more effective way to use distantly supervised model than `Initialization`.

Next, we discuss the performance of the proposed method for the BioRoBERTa baseline. Overall, the proposed method improves the micro-averaged F-score by 0.5 points. Furthermore, when we compare the F-score of each relation, the performance of all relations except ACTIVATOR, ANTAGONIST, and SUBSTRATE is improved or maintained. From these results with two different BERT models, we show that the proposed

[3]`https://github.com/tonifuc3m/drugprot-evaluation-library`

[4]This setting can cause overfitting to the development data sets, but since this is an official development set, we decided to report the best score to make the scores comparable to other methods in the shared task.

[5]Weger et al. (Weber et al., 2021) showed a slightly better performance with 78.3% on the development data by adding input start and end markers for target entities in the sentences, instead of masking the target entities like us. Since our main focus is not investigating a better baseline model, we leave investigating the representation of target entities for future work.

method can improve the performance regardless of the parameter size of the model.

4.2 Analysis of Prediction Results

We show the confusion matrices between gold labels and predicted labels by the baseline and the proposed method to analyze the prediction tendency of the two methods, and visually check the prediction cases. The confusion matrix is a table that visualizes the differences in two different sets of labels for instances. It has gold labels in the row direction and predicted labels in the column direction, and each element has the number of cases for the pair of gold and predicted labels. For the proposed method, we used a model that employs `Mixture` with `Concat`, which showed the best performance improvement from the baseline in the approach to utilize the distantly supervised model as shown in Section 4.1, based on PubMedBERT. The confusion matrices of the baseline and the proposed method are shown in Figure 3. The left and right confusion matrices are for the baseline and the proposed method, respectively.

First, we focus on the cases of different predictions in relation types. We can see that the number of cases that the proposed method mistakenly predicts INHIBITOR for DIRECT-REGULATOR is reduced from 14 to 2. Some example cases, where the predictions are improved by the proposed method, are shown in Table 5. The reason for the incorrect prediction by the baseline model is that the sentence contains "inhibit", "inhibited", and "inhibition", which are important for predicting the INHIBITOR type. For these cases, the baseline may predict the relations as INHIBITOR even though the sentence indicated DIRECT-REGULATOR between DRUG and PROTEIN entities. The reason why the proposed method was able to correctly predict such cases may be that the proposed method uses representation obtained from distantly supervised models that are trained on large-scale distant supervision data, and thus places more emphasis on the context than on word-level expressions.

Conversely, the number of cases in which the proposed method predicted INHIBITOR for the instances with the gold INDIRECT-DOWNREGULATOR type has increased from 19 to 25. The cases where the baseline made a correct prediction and the proposed method made a wrong prediction are shown in Table 6. The reason why the proposed method made such incorrect predictions in these cases may also be due to the existence of inhibit, inhibited, and inhibition in the sentences, which are important for predicting INHIBITOR, similarly to the baseline's wrong predictions for the cases in Table 5. This is because the sentence contains "inhibit", "inhibited", and "inhibition", which are important for predicting both INHIBITOR and INDIRECT-DOWNREGULATOR. Furthermore, the context of the cases is similar because these types are both related to the cases that drugs inhibit proteins. Therefore, the proposed method is likely to make INHIBITOR predictions based on such keywords for cases that the prediction is difficult with the context, without much consideration on the differences in the actions of drugs on proteins.

Then, we focus on the cases where the miss prediction is made between a relation type and a negative type. We can see that the proposed method reduces the number of cases in which the negative examples are mistakenly predicted as the INHIBITOR type from 204 to 178, the number of cases in which the negative examples are mistakenly predicted as PRODUCT-OF from 70 to 44, and the number of cases in which the negative examples are mistakenly predicted as SUBSTRATE from 142 to 86. The cases, where the baseline incorrectly predicted the negative cases as PRODUCT-OF while the proposed method correctly predicted them, are shown in Table 7. The numbers of improved cases and example cases suggest that the proposed method is more context-sensitive in its prediction than the baseline model.

These results suggest that the proposed method places more emphasis on contextual expression than on word expressions in making predictions compared to the baseline models. However, for cases where it is difficult to make predictions based on context, we found that the proposed method made incorrect predictions.

4.3 Performance Comparison with Small-Scale Manually Annotated Training Data

This section examines the effectiveness of the proposed method in training with small-scale manually annotated training data. We aim to improve the performance of drug-protein interaction extraction while reducing the cost of creating additional manually annotated training data by utilizing distant supervision data that have low creation costs. In

	PubMedBERT					BioRoBERTa		#Manually-annotated instances
	Manual data	Distant data	+ Init	+ Mix (Add)	+ Mix (Concat)	Manual data	+ Mix (Concat)	
INDIRECT -DOWNREGULATOR	76.7	0.0	74.6	77.7	78.7	79.3	79.9	1,330
INDIRECT-UPREGULATOR	73.3	1.9	75.1	73.7	73.6	75.6	76.2	1,379
DIRECT-REGULATOR	65.9	6.1	62.1	66.9	67.7	66.9	69.4	2,250
ACTIVATOR	77.3	5.2	70.6	77.5	76.7	75.7	73.8	1,429
INHIBITOR	84.2	29.4	84.7	84.6	84.3	85.1	86.1	5,392
AGONIST	75.5	6.7	79.7	78.2	77.0	76.1	77.2	659
AGONIST-ACTIVATOR	0.0	0.0	46.2	0.0	0.0	0.0	0.0	29
AGONIST-INHIBITOR	0.0	0.0	80.0	0.0	0.0	0.0	0.0	13
ANTAGONIST	90.6	26.0	89.6	92.2	91.8	91.7	90.2	972
PRODUCT-OF	59.0	10.6	63.7	62.9	62.5	61.2	62.0	921
SUBSTRATE	69.5	13.1	69.1	68.4	69.9	72.7	71.8	2,003
SUBSTRATE_PRODUCT-OF	0.0	0.0	0.0	0.0	0.0	0.0	0.0	25
PART-OF	71.7	0.0	70.6	72.2	71.7	72.8	74.4	886
Macro-averaged F-score	57.2	7.6	66.6	58.0	58.0	58.2	58.5	—
Micro-averaged F-score	76.2	16.6	75.6	76.8	77.0	77.5	78.0	—

Table 4: Relation extraction performance on the development data set. +Init, +Mix (Add), and +Mix (Concat) denote `Initialization`, `Add` of `Mixture`, and `Concat` of `Mixture`, respectively

DRUG inhibit (125) i - PROTEIN binding to recombinant rat eta receptors.
N - (diphenylmethyl) - 2 - phenyl - 4 - quinazolinamine (DRUG), n - (2, 2 - diphenylethyl) - 2 - phenyl - 4 - quinazolinamine (sori - 20040), and n - (3, 3 - diphenylpropyl) - 2 - phenyl - 4 - quinazolinamine (sori - 20041) partially inhibited [(125) i] 3beta - (4'- iodophenyl) tropan - 2beta - carboxylic acid methyl ester (rti - 55) binding, slowed the dissociation rate of [(125) i] rti - 55 from the PROTEIN, and partially inhibited [(3) h] dopamine uptake.
DRUG (parent compound), has moderate affinity for the PROTEIN (competitive inhibition).

Table 5: Improved cases with wrong predictions by the baseline model. The baseline model mistakenly predicted INHIBITOR for DIRECT-REGULATOR for the DRUG and PROTEIN pairs.

Section 4.1, we trained models using all manually annotated training data and confirmed that the proposed method can improve the performance of the baseline models. To verify the effectiveness of the proposed method in training with a small amount of manually annotated training data, we trained with only a small portion of the manually annotated training data and compared the performance of relation extraction between the PubMedBERT baseline model and the model with the proposed method. We checked the performance of the proposed method on the development data when the model was trained with the small number of cases, we chose the number from [3, 5, 7, 10, 20, 50, 100, 200, 500, 1,000], for each relation in the manually annotated training data. For the proposed method, we used a model that mixes feature representations with `Concat`, which showed the best performance improvement from the baseline with Section 4.1.

The results are shown in Figure 4. As in the case of Section 4.1, we did not obtain a significant performance improvement over the baseline as we saw when training with all manually annotated training data, but the performance consistently improved for all the cases. This indicates that the proposed method can improve performance by using representations obtained from the distantly supervised model, regardless of the number of cases of manually annotated training data.

5 Conclusions

We aimed to improve the performance of drug-protein relation extraction by creating distant supervision data at low cost and utilizing the model pre-trained on the data while reducing the noise contained in the distant supervision data. We proposed two methods of utilizing distant supervision data. Both methods improved the prediction performance from the baseline for relation types with less manually annotated training data. In addition, the method that mixes representations also improved the F-scores for many relation types, some of them have a large amount of manually annotated training data, as well as the micro-averaged F-score, demonstrating the effectiveness of the proposed method. In addition, we showed that the performance im-

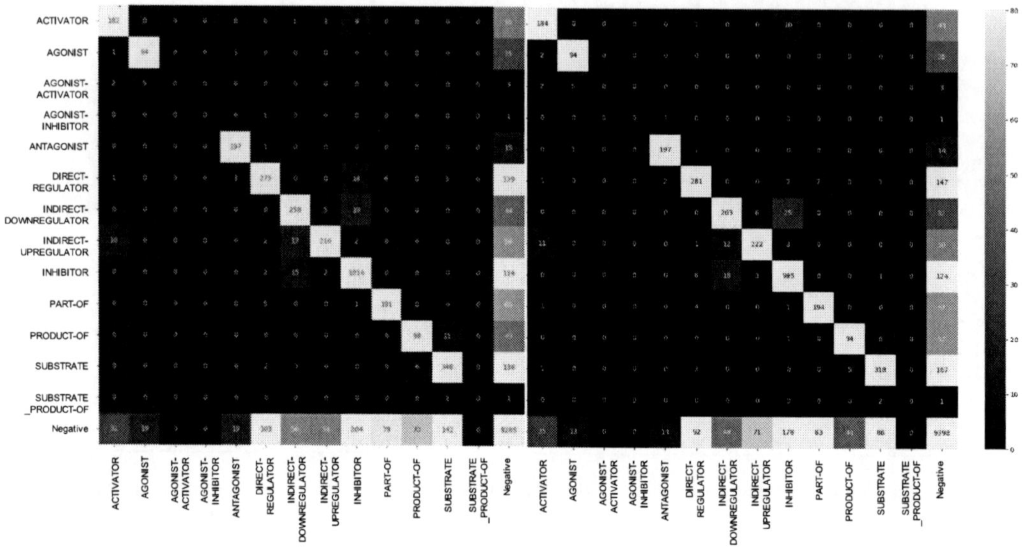

Figure 3: The confusion matrices (left: baseline, right: proposed method)

The upregulation of calpain, PROTEIN and caspase - 3 activity were further inhibited by treatment with DRUG in the presence of ald.
The mechanism of action of DRUG was related to the inhibition of the cleavage of pro - caspase - 1, PROTEIN and pro - il - 18 which in turn suppressed the activation of nlrp3 inflammasome.

Table 6: Deteriorated cases with wrong predictions by the proposed model. The model wrongly predicted INHIBITOR, instead of INDIRECT-DOWNREGULATOR, for the DRUG and PROTEIN pairs.

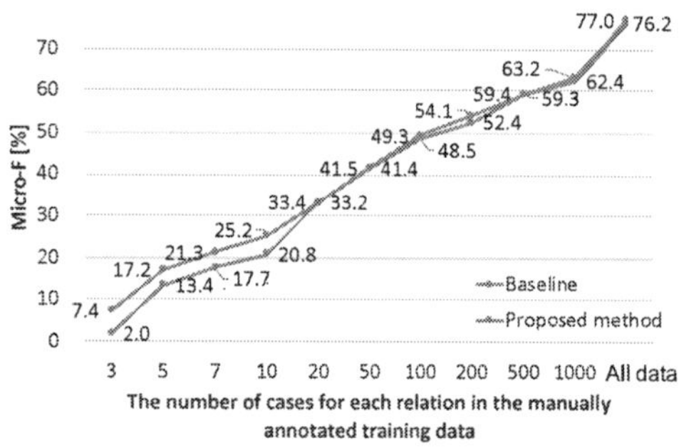

Figure 4: Micro-averaged F-scores for the number of manually annotated training instances for each relation type

provement was independent of the parameter size of the model and the number of cases of manually annotated training data.

To improve the extraction performance, we plan to investigate the `Mixture` method for its way of mixing representations and pretraing.

Acknowledgements

This work was supported by JSPS Grant-in-Aid for Scientific Research JP20K11962.

References

Takuya Akiba, Shotaro Sano, Toshihiko Yanase, Takeru Ohta, and Masanori Koyama. 2019. Optuna: A next-generation hyperparameter optimization framework. In *Proceedings of the 25rd ACM SIGKDD International Conference on Knowledge Discovery and Data Mining*.

Iz Beltagy, Kyle Lo, and Waleed Ammar. 2019a. Combining distant and direct supervision for neural relation extraction. In *Proceedings of the 2019 Conference of the North American Chapter of the Association for Computational Linguistics: Human Language Technologies, Volume 1 (Long and Short Papers)*, pages 1858–1867, Minneapolis, Minnesota. Association for Computational Linguistics.

Iz Beltagy, Kyle Lo, and Arman Cohan. 2019b. Scibert: Pretrained language model for scientific text. In *EMNLP*.

The structures solved after the diffusion of oligosaccharides (either maltotetraose, g4 or maltopentaose, g5) into PROTEIN / glc1p crystals show the formation of DRUG and elongation of the oligosaccharide chain
DRUG biosynthesis in plants : molecular and functional characterization of PROTEIN and three isoforms of folylpolyglutamate synthetase in arabidopsis thaliana.
Knockdown of nadph oxidase, nox5 - s, a variant lacking calcium - binding domains, by nox5 sirna significantly inhibited acid - induced increase in PROTEIN expression, thymidine incorporation, and DRUG production.

Table 7: Improved cases with wrong predictions by the baseline model. The baseline model mistakenly predicted PRODUCT-PRODUCT-OF for the negative DRUG and PROTEIN pairs.

The UniProt Consortium. 2020. UniProt: the universal protein knowledgebase in 2021. *Nucleic Acids Research*, 49(D1):D480–D489.

NCBI Resource Coordinators. 2016. Database resources of the national center for biotechnology information. *Nucleic Acids Res.*

Allan Peter Davis, Cynthia J Grondin, Robin J Johnson, Daniela Sciaky, Jolene Wiegers, Thomas C Wiegers, and Carolyn J Mattingly. 2020. Comparative Toxicogenomics Database (CTD): update 2021. *Nucleic Acids Research*, 49(D1):D1138–D1143.

Jacob Devlin, Ming-Wei Chang, Kenton Lee, and Kristina Toutanova. 2019. Bert: Pre-training of deep bidirectional transformers for language understanding.

Wishart DS, Feunang YD, Guo AC, Lo EJ, Marcu A, Grant JR, Sajed T, Johnson D, Li C, Sayeeda Z, Assempour N, Iynkkaran I, Liu Y, Maciejewski A, Gale N, Wilson A, Chin L, Cummings R, Le D, Pon A, Knox C, and Wilson M. 2018. Drugbank 5.0: a major update to the drugbank database for 2018. *Nucleic Acids Res.*

Yu Gu, Robert Tinn, Hao Cheng, Michael Lucas, Naoto Usuyama, Xiaodong Liu, Tristan Naumann, Jianfeng Gao, and Hoifung Poon. 2022. Domain-specific language model pretraining for biomedical natural language processing. *ACM Transactions on Computing for Healthcare*, 3(1):1–23.

Guoliang Ji, Kang Liu, Shizhu He, and Jun Zhao. 2017. Distant supervision for relation extraction with sentence-level attention and entity descriptions. *Proceedings of the AAAI Conference on Artificial Intelligence*, 31(1).

Diederik Kingma and Jimmy Ba. 2015. Adam: A method for stochastic optimization. In *Proceedings of the 3rd International Conference for Learning Representations*.

Martin Krallinger, Obdulia Rabal, Saber Ahmad Akhondi, Martín Pérez Pérez, Jesus Santamaría, Gael Pérez Rodríguez, Georgios Tsatsaronis, Ander Intxaurrondo, José Antonio Baso López, Umesh K. Nandal, Erin M. van Buel, Anjana Chandrasekhar, Marleen Rodenburg, Astrid Lægreid, Marius A. Doornenbal, Julen Oyarzábal, Anália Lourenço, and Alfonso Valencia. 2017. Overview of the biocreative vi chemical-protein interaction track.

Martin Krallinger, Obdulia Rabal, Antonio Miranda-Escalada, and Alfonso Valencia. 2021. DrugProt corpus: Biocreative VII Track 1 - Text mining drug and chemical-protein interactions.

Patrick Lewis, Myle Ott, Jingfei Du, and Veselin Stoyanov. 2020. Pretrained language models for biomedical and clinical tasks: Understanding and extending the state-of-the-art. In *Proceedings of the 3rd Clinical Natural Language Processing Workshop*, pages 146–157, Online. Association for Computational Linguistics.

Mike Mintz, Steven Bills, Rion Snow, and Daniel Jurafsky. 2009. Distant supervision for relation extraction without labeled data. In *Proceedings of the Joint Conference of the 47th Annual Meeting of the ACL and the 4th International Joint Conference on Natural Language Processing of the AFNLP*, pages 1003–1011, Suntec, Singapore. Association for Computational Linguistics.

Mark Neumann, Daniel King, Iz Beltagy, and Waleed Ammar. 2019. ScispaCy: Fast and Robust Models for Biomedical Natural Language Processing. In *Proceedings of the 18th BioNLP Workshop and Shared Task*, pages 319–327, Florence, Italy. Association for Computational Linguistics.

Obdulia Rabal, Jose Antonio López, Astrid Lagreid, and Martin Krallinger. 2021. DrugProt corpus relation annotation guidelines [ChemProt - Biocreative VI].

Sebastian Riedel, Limin Yao, and Andrew McCallum. 2010. Modeling relations and their mentions without labeled text. In *Machine Learning and Knowledge Discovery in Databases*, pages 148–163, Berlin, Heidelberg. Springer Berlin Heidelberg.

Leon Weber, Mario Sänger, Samuele Garda, Fabio Barth, Christoph Alt, and Ulf Leser. 2021. Humboldt@ drugprot: Chemical-protein relation extraction with pretrained transformers and entity descriptions. In *Proceedings of the BioCreative VII Challenge Evaluation Workshop*.

Wonjin Yoon, Sean Yi, Richard Jackson, Hyunjae Kim, Sunkyu Kim, and Jaewoo Kang. 2021. Using knowledge base to refine data augmentation for biomedical relation extraction. In *Proceedings of the BioCreative VII Challenge Evaluation Workshop*.

Daojian Zeng, Kang Liu, Yubo Chen, and Jun Zhao. 2015. Distant supervision for relation extraction via piecewise convolutional neural networks. In *Proceedings of the 2015 Conference on Empirical Methods in Natural Language Processing*, pages 1753–1762, Lisbon, Portugal. Association for Computational Linguistics.

Named Entity Recognition for Cancer Immunology Research Using Distant Supervision

Hai-Long Trieu [1,3], **Makoto Miwa** [1,2] and **Sophia Ananiadou** [3]
[1]Artificial Intelligence Research Center (AIRC),
National Institute of Advanced Industrial Science and Technology (AIST), Japan
[2]Toyota Technological Institute, Japan
[3]National Centre for Text Mining, University of Manchester, United Kingdom
long.trieu@aist.go.jp, makoto-miwa@toyota-ti.ac.jp,
sophia.ananiadou@manchester.ac.uk

Abstract

Cancer immunology research involves several important cell and protein factors. Extracting the information of such cells and proteins and the interactions between them from text are crucial in text mining for cancer immunology research. However, there are few available datasets for these entities, and the amount of annotated documents is not sufficient compared with other major named entity types. In this work, we introduce our automatically annotated dataset of key named entities, i.e., T-cells, cytokines, and transcription factors, which engages the recent cancer immunotherapy. The entities are annotated based on the UniProtKB knowledge base using dictionary matching. We build a neural named entity recognition (NER) model to be trained on this dataset and evaluate it on a manually-annotated data. Experimental results show that we can achieve a promising NER performance even though our data is automatically annotated. Our dataset also enhances the NER performance when combined with existing data, especially gaining improvement in yet investigated named entities such as cytokines and transcription factors.

1 Introduction

Cancer immunology research has a central focus on T lymphocytes (*T-cells*), which engage the immune system in fighting against cancer (Luckheeram et al., 2012; Waldman et al., 2020; Kim et al., 2021). The development of T-cells can be guided by *cytokines* and *transcription factors* (Hosokawa and Rothenberg, 2018). *Transcription factors* (TF) are nuclear proteins that bind specific gene sequences and involved in decision-making processes during T-cell differentiation (Naito et al., 2011; Xia et al., 2019). Meanwhile, *cytokines* are signaling molecules secreted and sensed by immune and other cell types (Kveler et al., 2018). Extracting *T-cell*, *cytokine*, and *TF* entities and the interactions between them can be crucial for text mining in cancer immunology research.

However, there are few existing datasets containing these entities to train text mining models. At the core of text mining tasks, the named entity recognition (NER) task also lacks such datasets for training NER models to detect these named entities, which may limit the development of text mining systems in this cancer immunology research field. There is an existing T-cell related named entity dataset called TCRE (Czech and Hammerbacher, 2019), but the amount of annotated data is also limited to only 89 documents. Several knowledge bases related to immune system have been proposed such as immuneXpresso (Kveler et al., 2018) and DES-Tcell (AlSaieedi et al., 2021), which contain cell type and cytokine information, but they lack utilizing and evaluating with modern NER models on these named entities.

In this paper, as a step to fill these gaps and promote the development of text mining systems on these named entities in cancer immunology research articles, we present our automatically annotated dataset containing named entities of *T-cell*, *cytokine* and *TF*, which are important for mining and understanding cancer immunology research articles. The entities in the dataset are automatically annotated using dictionary matching based on the UniProtKB (UniProt-Consortium, 2021), a knowledgebase of protein sequences with functional information.[1] From the annotations of cytokine and TF entries in UniProtKB, a dictionary is constructed to annotate cytokine and TF named entities in their referenced PubMed articles. Additionally, we utilized the existing JNLPBA corpus, which contains manually annotated protein named entities, to annotate cytokine and TF entities. We build a NER model based on the span-based model with pre-trained BERT. We trained the NER model on our automatically annotated dataset and evaluated the model on an existing manually annotated T-cell related named entity TCRE dataset (Czech

[1]https://www.uniprot.org/uniprot/

Proceedings of the BioNLP 2022 workshop, Dublin, Ireland, pages 171–177
May 26, 2022. ©2022 Association for Computational Linguistics

Item	cytokine	TF
# UniProtKB entries	1,001	3,418
# Dictionary size	6,859	20,055
# Collected articles	585	1,903

Table 1: UniProtKB entries and annotated data

Data	#Docs.	#Entities		
		CT	CY	TF
KB-T-cell	386	340	744	2,891
Dic-T-cell	761	2,686	1,752	2,686
TCRE	89	1,006	235	114

Table 2: Statistics of the datasets (*Docs*: documents; CT (cell type), CY (cytokine), TF (transcription factor))

and Hammerbacher, 2019). We achieve a promising result that the NER model trained on our automatically annotated data gains a slightly lower performance than a supervised NER model trained on a manually annotated data, although our data is automatically annotated. Furthermore, our data enhances NER performance when combined with the existing manually annotated data.

2 Approach

We present our datasets containing three named entity types: cell_type, cytokine, and transcription factor (TF). The datasets are automatically annotated using dictionary matching with the entries in the UniProtKB in two different ways.

2.1 UniProtKB

Cytokine and TF queries From the UniProtKB, we obtain entries by querying *cytokine*. We filtered the options to keep only *Reviewed* annotations (manually annotated, added by expert biocuration team) and for *Human* organism. Similarly, we conducted for *transcription factor*. They are equivalent to the following queries.

- *cytokine AND reviewed:yes AND organism:"Homo sapiens (Human) [9606]".*
- *transcription factor AND reviewed:yes AND organism:"Homo sapiens (Human) [9606]"*

UniProtKB entries We obtained 1,001 entries for cytokine and 3,418 entries for TF from UniProtKB. Each entry contains protein names, gene names, and referenced PubMed articles, etc.

UniProtKB-dictionary We built a dictionary containing protein and gene names of the cytokine and TF entries in UniProtKB, which we named *UniProtKB-dictionary*.

Collecting PubMed references For each UniProtKB entry, there is a list of referenced PubMed articles. We collect the referenced articles' abstract texts from PubMed for each entry. Since there is a large number of references, we only collect the abstracts that contain a large number ($\geq k$) of cytokine/TF protein and gene names (we set $k = 20$, which we based on several preliminary experiments to remove abstracts containing few annotations). We present the statistics of UniProtKB entries and related annotated data in Table 1.

2.2 Automatically Annotated Datasets

We constructed two automatically annotated datasets using the *UniProtKB-dictionary*. The statistics for automatically annotated datasets are presented in Table 2.

2.2.1 Knowledge-based Annotation (KB-T-cell)

Annotating cytokine and TF From the *UniProtKB dictionary*, we identify the position of each name in the collected articles by strict text matching to annotate cytokine and TF named entities.

Annotating cell_type We found that JNLPBA (Collier and Kim, 2004) is a large manually annotated dataset for NER, which contains named entities of cell_type, protein, etc. Therefore, we utilized the JNLPBA data to train a NER model to predict cell_type named entities in the collected articles. We build a neural-based NER method with span-based and pre-trained BERT model, which we present in §3. These cell_type entities are combined with the cytokine and TF named entities, and we named *KB-T-cell*.

2.2.2 Dictionary-based Re-annotation (Dic-T-cell)

Since the JNLPBA dataset contains protein entities while CT and TF are proteins, we utilized the annotated protein names in the JNLPBA to annotate cytokine and TF entities. Specifically, if an annotated protein name in the JNLPBA is included in the *UniProtKB-dictionary*, we re-annotate it as cytokine or TF, correspondingly. We ignored documents which do not contain any matched CT/TF entity. We named this dataset as *Dic-T-cell*.

3 NER model

We explain the NER model to be trained on the annotated datasets. We build a neural-based NER model using a span-based method (Lee et al., 2017; Luan et al., 2018) and finetuned pre-trained BERT (Devlin et al., 2019). Specifically, each sentence is split into sub-word sequences, which are passed through the BERT layer for contextual representations. Then, for each span (i.e., a sequence of continuous words in a sentence), its representation is calculated by concatenating the representations of the first, last, and averaged sub-words of the span, which follows (Sohrab and Miwa, 2018a; Trieu et al., 2020). Finally, each span representation is passed to classifiers to predict named entity types for each span.

4 Experiments

4.1 Data

We used our datasets *KB-T-cell* and *Dic-T-cell* to train NER models using the NER model introduced in §3 and evaluated NER performance.

TCRE For evaluation data, we employed the TCRE (Czech and Hammerbacher, 2019), an existing manually annotated data which contains 89 documents of cell_type, cytokine, and TF named entities. We utilized this data for training supervised NER models and for evaluation. The original TCRE dataset contains a mixture of both abstract and full-text documents. For the scope of this paper, we aim at utilizing only abstracts from both UniProtKB's references and JNLPBA data. Therefore, we used only the abstract documents and the abstract section of full-text documents from the TCRE data.

The data statistics of the datasets are presented in Table 2.

4.2 Settings

Cross validation We conducted k-fold cross validation evaluation on the TCRE dataset. Since the TCRE data size is quite small, we set $k = 3$ to ensure a reasonable amount of data in the test set. For each fold, we further randomly split the training set into train/development sets so that we can tune hyper-parameters to get the best models on the development set. Finally, all of our reported results are based on the TCRE test set in each fold.

NER training settings Our model was implemented on PyTorch (Paszke et al., 2017). We used the BERT model from the PyTorch Pretrained BERT repository[2] as our BERT layer. We employed the pre-trained SciBERT model (Beltagy et al., 2019) trained on large-scale biomedical texts. The model is trained on multiple GPUs in the AI Bridging Cloud Infrastructure (ABCI)[3]. We train the model with the Adam optimizer (Kingma and Ba, 2015), gradient clipping, dropout, and L2 regularization. The model is trained with early-stopping, and the training mini-batch size is set as 16.

Evaluation settings We compared the following NER models, which mostly differ in the training data settings.

1. **Matching-NER**: we created a baseline using dictionary matching. The dictionary is built from the entity's texts of the JNLPBA training data (for cell_type) and the UniProtKB-dictionary for cytokine and TF.

2. **Supervised-NER**: we used the training set of the TCRE data to train the NER model.

3. **KB-NER**, **Dic-NER**, **KB-Dic-NER**: we train the NER models on our annotated datasets: KB-T-cell, Dic-T-cell, and merged the KB-T-cell and Dic-T-cell, respectively.

4. **Enhanced-KB-NER**, **Enhanced-Dic-NER**, **Enhanced-KB-Dic-NER**: we merge the training set of the TCRE with the KB-T-cell, Dic-T-cell, and merged KB-T-cell and Dic-T-cell, respectively, to train NER models.

The results are reported based on the commonly used micro-averaged precision (P), recall (R), and F-score (F) metrics at entity level.

4.3 Results

We compare the results of different NER models on each data fold in Table 3.

Enhancement Using our automatically annotated dataset, we achieved the best performance with 2-5% point improvements in F-score (Enhanced-KB-NER) in comparison with the Supervised-NER in all of the data folds.

[2]https://github.com/huggingface/pytorch-pretrained-BERT/tree/34cf67fd6c

[3]https://abci.ai/

Model	Fold-1			Fold-2			Fold-3		
	P	R	F	P	R	F	P	R	F
Matching-NER	39.88	66.16	49.76	39.54	68.63	50.17	38.05	69.27	49.12
Supervised-NER	68.67	66.92	67.78	70.92	70.75	70.84	73.36	74.23	73.80
KB-NER	64.55	62.09	63.29	71.34	54.01	61.48	63.85	57.21	60.35
Dic-NER	63.19	61.58	62.37	66.67	60.38	63.37	67.00	64.30	65.62
KB-Dic-NER	65.33	66.16	65.74	71.74	62.26	66.67	67.07	65.48	66.27
Enhanced-KB-NER	**72.98**	**73.54**	**73.26**	**75.12**	**76.89**	**75.99**	**75.71**	**75.89**	**75.80**
Enhanced-Dic-NER	71.11	72.02	71.55	70.14	73.11	71.59	73.23	75.65	74.42
Enhanced-KB-Dic-NER	72.18	73.28	72.73	72.86	72.17	72.51	74.13	75.18	74.65

Table 3: Comparison NER results of the models (the best scores are in bold)

Model	Fold-1			Fold-2			Fold-3		
	CT	CY	TF	CT	CY	TF	CT	CY	TF
Matching-NER	65.18	1.45	15.07	66.42	6.00	18.44	65.96	6.86	5.97
Supervised-NER	71.22	56.64	41.18	76.36	56.36	32.14	76.15	65.45	57.78
KB-NER	69.57	31.46	52.38	73.70	18.95	0.00	70.79	20.95	8.00
Dic-NER	72.81	3.33	0.00	79.50	5.56	3.03	76.00	13.19	0.00
KB-Dic-NER	73.62	22.54	35.29	79.21	8.33	0.00	78.06	18.69	8.00
Enhanced-KB-NER	76.32	62.50	63.77	82.16	60.66	43.48	80.65	68.91	21.74
Enhanced-Dic-NER	77.49	55.32	18.18	81.61	36.51	39.44	79.21	60.34	27.03
Enhanced-KB-Dic-NER	77.55	64.08	30.77	81.33	41.44	37.68	80.06	63.64	15.79

Table 4: Results on each entity type in F-score (%). The underline scores are higher than the Supervised-NER's.

Supervised vs. unsupervised When training NER models on our automatically annotated datasets (KB-NER, Dic-NER, KB-Dic-NER), the performance is lower than the Supervised-NER, which is trained on a time-consuming manually annotated data. The degraded performance is about 5-7% points in F-score, which are acceptable considering that our datasets are automatically annotated. We can further improve the quality of our datasets in future work, such as filtering noisy annotations.

Dictionary matching Since our automatically annotated data is based on the dictionary built from the UniProtKB and JNLPBA, we may raise a question whether using only the dictionary with the same vocabulary is still enough. The results of KB-NER and Dic-NER show that our automatically annotated data can improve from 11-15% in comparison with the Matching-NER.

KB vs Dic Table 3 also shows that the NER models based on the KB-T-cell (KB-NER, Enhanced-KB-NER) obtain higher performance than those based on the Dic-T-cell (Dic-NER, Enhanced-Dic-NER). When combining these two datasets, the performance decreased even though the data size of the Dic-T-cell is mostly double of the KB-T-cell, which indicates that we need to investigate a better combination. Another possible direction can be filtering noisy annotations of the Dic-T-cell.

4.4 Analyses and Discussions

We further investigate the detailed performance on each entity type: cell_type, cytokine, and TF. The results from Table 4 show that the Enhanced-KB-NER achieves improvements on all entity types except for the TF entity type in Fold-3.

Comparing the performance among the entity types between the Supervised-NER and the enhanced models, the CT type performance gains improvement (3-5% points) in most cases. The reason may come from the quality of the CT type in the large manually annotated JNLPBA data. Meanwhile, the improvement of the CY type is 3-6% points, and the improvement of TF is 11-22% points. When training only on our automatically annotated datasets (KB-NER, Dic-NER), we still obtain the higher performance for the CT type. We obtain some reasonable performance in cytokines (lower than the Supervised-NER but much better than the Matching-NER).

Limitation The performance of CY and TF from KB-NER and Dic-NER is low in most cases. There is no correct TF prediction (Dic-NER in Fold-1 and Fold-3, KB-NER in Fold-2). For CY, the performance is also low from Dic-NER (3% to 13% F-score), but it is slightly better in KB-NER (18% to 31% F-score). These results show a challenge to extract CY and TF entities based on only our automatically annotated corpus. This work is our first investigation in utilizing the UniProtKB and the existing JNLPBA corpus for our research goal in extracting T-cell related entities, and we accept this limitation in this first version. It is required to conduct further investigation and improvement especially for these CY and TF types in future work.

Future work We would like to improve the performance of CY and TF. We also plan to conduct the evaluation not only on the TCRE task but other NER tasks such as JNLPBA (Collier and Kim, 2004), NCBI (Doğan et al., 2014), and BC5CDR (Li et al., 2016). Additionally, we intend to extend our corpus for other tasks such as relation and event extraction on these T-cell named entities.

5 Related Work

Distant supervision methods for NER have been investigated in several previous works. (Shang et al., 2018) revised the LSTM-CRF NER model (Lample et al., 2016) and utilized the MeSH database for chemical and disease entities. Some methods are proposed to reduce noisy annotations for Chinese NER(Yang et al., 2018), or general domain OntoNotes (Liang et al., 2020; Meng et al., 2021).

The span-based method has been used to build our NER model in this work. The method was proposed and employed in previous work (Lee et al., 2017; Luan et al., 2018; Sohrab and Miwa, 2018b; Trieu et al., 2020), which have shown the advantages in extracting nested or continuous text sequences and successful in many sequence labeling tasks such as NER or coreference resolution.

Immunotherapy has achieved remarkable advances in recent years and can be important cancer treatment in future (Falzone et al., 2018; Zhang and Chen, 2018; Kruger et al., 2019). However, there are few related work or annotated datasets in text mining on this domain. immuneXpresso (Kveler et al., 2018) is a text mining engine related to mammalian immune system, and NER is evaluated on cells and cytokine using dictionary matching. DES-Tcell (AlSaieedi et al., 2021) is a knowledgebase containing concepts of T-cell and other types of drugs, diseases, genes, etc in PubMed documents. However, it lacks utilizing novel text mining methods in the creation and evaluation the extracted data including NER tasks.

For the datasets used in our work, TCRE is manually annotated by Czech and Hammerbacher (2019) containing cell_type, cytokine, and TF entities, which are closed to our goal, and we used for our evaluation. A limitation of the TCRE is that it contains only 89 documents, which is insufficient to train powerful NER models. Therefore, our annotation method in this work can advance the task in extracting T-cell named entities. JNLPBA (Collier and Kim, 2004) contains manually annotated cell_type and protein entities. Meanwhile, UniProtKB (UniProt-Consortium, 2021) is a large and useful knowledgebase containing protein sequences annotated by experts with corresponding PubMed references. The UniProtKB and JNLPBA are leveraged to build our corpus.

6 Conclusion

We introduce our automatically annotated dataset for NER containing cell_type, cytokine, and TF entities, which are important in cancer immunology research, using a distant supervision method. The dataset is automatically annotated based on the entries in the UniProtKB knowledge base. We built a dictionary of the protein and gene names of cytokines and TF from the UniProtKB annotations. We then collected referenced PubMed articles and annotated these names in the texts using text matching with the dictionary entries. Additionally, we utilized the large manually annotated JNLPBA dataset, which contains cell_type and protein named entities to build our dataset. We trained NER models on our automatically annotated dataset and evaluated them on a manually annotated T-cell corpus. The results show that our automatically annotated dataset helps to improve the NER performance by extracting more named entities of cytokines and TF accurately. For future work, we plan to improve and extend our dataset to extract interactions or events related to these entities for text mining in cancer immunology research.

Acknowledgements

This paper is supported by the Artificial Intelligence Research Center (AIRC, Japan) and BBSRC, Japan Partnering Award, BB/P025684/1.

References

Ahdab AlSaieedi, Adil Salhi, Faroug Tifratene, Arwa Bin Raies, Arnaud Hungler, Mahmut Uludag, Christophe Van Neste, Vladimir B Bajic, Takashi Gojobori, and Magbubah Essack. 2021. Des-tcell is a knowledgebase for exploring immunology-related literature. *Scientific reports*, 11(1):1–11.

Iz Beltagy, Kyle Lo, and Arman Cohan. 2019. SciBERT: A pretrained language model for scientific text. In *Proceedings of the 2019 Conference on Empirical Methods in Natural Language Processing and the 9th International Joint Conference on Natural Language Processing (EMNLP-IJCNLP)*, pages 3606–3611, Hong Kong, China. Association for Computational Linguistics.

Nigel Collier and Jin-Dong Kim. 2004. Introduction to the bio-entity recognition task at JNLPBA. In *Proceedings of the International Joint Workshop on Natural Language Processing in Biomedicine and its Applications (NLPBA/BioNLP)*, pages 73–78, Geneva, Switzerland. COLING.

Eric Czech and Jeff Hammerbacher. 2019. Extracting T cell function and differentiation characteristics from the biomedical literature. *bioRxiv*, page 643767.

Jacob Devlin, Ming-Wei Chang, Kenton Lee, and Kristina Toutanova. 2019. BERT: Pre-training of deep bidirectional transformers for language understanding. In *Proceedings of NAACL*, pages 4171–4186.

Rezarta Islamaj Doğan, Robert Leaman, and Zhiyong Lu. 2014. Ncbi disease corpus: a resource for disease name recognition and concept normalization. *Journal of biomedical informatics*, 47:1–10.

Luca Falzone, Salvatore Salomone, and Massimo Libra. 2018. Evolution of cancer pharmacological treatments at the turn of the third millennium. *Frontiers in pharmacology*, page 1300.

Hiroyuki Hosokawa and Ellen V Rothenberg. 2018. Cytokines, transcription factors, and the initiation of T-cell development. *Cold Spring Harbor perspectives in biology*, 10(5):a028621.

Ji-Hae Kim, Kun-Joo Lee, and Seung-Woo Lee. 2021. Cancer immunotherapy with T-cell targeting cytokines: IL-2 and IL-7. *BMB reports*, 54(1):21.

Diederik P Kingma and Jimmy Lei Ba. 2015. Adam: A method for stochastic optimization. In *Proceedings of the 3rd International Conference on Learning Representations (ICLR2015)*.

Stephan Kruger, Matthias Ilmer, Sebastian Kobold, Bruno L Cadilha, Stefan Endres, Steffen Ormanns, Gesa Schuebbe, Bernhard W Renz, Jan G D'Haese, Hans Schloesser, et al. 2019. Advances in cancer immunotherapy 2019–latest trends. *Journal of Experimental & Clinical Cancer Research*, 38(1):1–11.

Ksenya Kveler, Elina Starosvetsky, Amit Ziv-Kenet, Yuval Kalugny, Yuri Gorelik, Gali Shalev-Malul, Netta Aizenbud-Reshef, Tania Dubovik, Mayan Briller, John Campbell, et al. 2018. Immune-centric network of cytokines and cells in disease context identified by computational mining of PubMed. *Nature biotechnology*, 36(7):651–659.

Guillaume Lample, Miguel Ballesteros, Sandeep Subramanian, Kazuya Kawakami, and Chris Dyer. 2016. Neural architectures for named entity recognition. In *Proceedings of the 2016 Conference of the North American Chapter of the Association for Computational Linguistics: Human Language Technologies*, pages 260–270.

Kenton Lee, Luheng He, Mike Lewis, and Luke Zettlemoyer. 2017. End-to-end neural coreference resolution. In *Proceedings of the 2017 Conference on Empirical Methods in Natural Language Processing*, pages 188–197.

Jiao Li, Yueping Sun, Robin J Johnson, Daniela Sciaky, Chih-Hsuan Wei, Robert Leaman, Allan Peter Davis, Carolyn J Mattingly, Thomas C Wiegers, and Zhiyong Lu. 2016. Biocreative v cdr task corpus: a resource for chemical disease relation extraction. *Database*, 2016.

Chen Liang, Yue Yu, Haoming Jiang, Siawpeng Er, Ruijia Wang, Tuo Zhao, and Chao Zhang. 2020. Bond: Bert-assisted open-domain named entity recognition with distant supervision. In *Proceedings of the 26th ACM SIGKDD International Conference on Knowledge Discovery & Data Mining*, pages 1054–1064.

Yi Luan, Luheng He, Mari Ostendorf, and Hannaneh Hajishirzi. 2018. Multi-task identification of entities, relations, and coreference for scientific knowledge graph construction. In *Proceedings of the 2018 Conference on Empirical Methods in Natural Language Processing*, pages 3219–3232.

Rishi Vishal Luckheeram, Rui Zhou, Asha Devi Verma, and Bing Xia. 2012. CD4+ T cells: differentiation and functions. *Clinical and developmental immunology*, 2012.

Yu Meng, Yunyi Zhang, Jiaxin Huang, Xuan Wang, Yu Zhang, Heng Ji, and Jiawei Han. 2021. Distantly-supervised named entity recognition with noise-robust learning and language model augmented self-training. In *Proceedings of the 2021 Conference on Empirical Methods in Natural Language Processing*, pages 10367–10378.

Taku Naito, Hirokazu Tanaka, Yoshinori Naoe, and Ichiro Taniuchi. 2011. Transcriptional control of T-cell development. *International immunology*, 23(11):661–668.

Adam Paszke, Sam Gross, Soumith Chintala, Gregory Chanan, Edward Yang, Zachary DeVito, Zeming Lin, Alban Desmaison, Luca Antiga, and Adam Lerer. 2017. Automatic differentiation in pytorch. In *NIPS-W*.

Jingbo Shang, Liyuan Liu, Xiaotao Gu, Xiang Ren, Teng Ren, and Jiawei Han. 2018. Learning named entity tagger using domain-specific dictionary. In *Proceedings of the 2018 Conference on Empirical Methods in Natural Language Processing*, pages 2054–2064, Brussels, Belgium. Association for Computational Linguistics.

Mohammad Golam Sohrab and Makoto Miwa. 2018a. Deep exhaustive model for nested named entity recognition. In *Proceedings of the 2018 Conference on Empirical Methods in Natural Language Processing*, pages 2843–2849.

Mohammad Golam Sohrab and Makoto Miwa. 2018b. Deep exhaustive model for nested named entity recognition. In *Proceedings of EMNLP*, pages 2843–2849. ACL.

Hai-Long Trieu, Thy Thy Tran, Khoa NA Duong, Anh Nguyen, Makoto Miwa, and Sophia Ananiadou. 2020. Deepeventmine: end-to-end neural nested event extraction from biomedical texts. *Bioinformatics*, 36(19):4910–4917.

UniProt-Consortium. 2021. UniProt: the universal protein knowledgebase in 2021. *Nucleic acids research*, 49(D1):D480–D489.

Alex D Waldman, Jill M Fritz, and Michael J Lenardo. 2020. A guide to cancer immunotherapy: from T cell basic science to clinical practice. *Nature Reviews Immunology*, 20(11):651–668.

Anliang Xia, Yan Zhang, Jiang Xu, Tailang Yin, and Xiao-Jie Lu. 2019. T cell dysfunction in cancer immunity and immunotherapy. *Frontiers in immunology*, 10:1719.

Yaosheng Yang, Wenliang Chen, Zhenghua Li, Zhengqiu He, and Min Zhang. 2018. Distantly supervised ner with partial annotation learning and reinforcement learning. In *Proceedings of the 27th International Conference on Computational Linguistics*, pages 2159–2169.

Hongming Zhang and Jibei Chen. 2018. Current status and future directions of cancer immunotherapy. *Journal of Cancer*, 9(10):1773.

Intra-Template Entity Compatibility based Slot-Filling for Clinical Trial Information Extraction

Christian Witte
Bielefeld University, Germany
cwitte@
techfak.uni-bielefeld.de

Philipp Cimiano
Bielefeld University, Germany
cimiano@
cit-ec.uni-bielefeld.de

Abstract

We present a deep learning based information extraction system that can extract the design and results of a published abstract describing a Randomized Controlled Trial (RCT). In contrast to other approaches, our system does not regard the PICO elements as flat objects or labels but as structured objects. We thus model the task as the one of filling a set of templates and slots; our two-step approach recognizes relevant slot candidates as a first step and assigns them to a corresponding template as second step, relying on a learned pairwise scoring function that models the compatibility of the different slot values. We evaluate the approach on a dataset of 211 manually annotated abstracts for Type 2 Diabetes and Glaucoma, showing the positive impact of modelling intra-template entity compatibility. As main benefit, our approach yields a structured object for every RCT abstract that supports the aggregation and summarization of clinical trial results across published studies and can facilitate the task of creating a systematic review or meta-analysis.

1 Introduction

The evidence based medicine (EBM) paradigm (Sackett et al., 1996) propagates that individual medical decisions are taken on the basis of the best available clinical evidence. The activity of summarizing the existing body of evidence is a core activity to support EBM and its most prominent instrument is the systematic review. Creating a systematic review involves a high effort, involving on average 67.3 weeks and involving 5 authors per review on average (Borah et al., 2017). Keeping systematic reviews up to date involves an even much higher and continuous effort (Koch, 2006; Beller et al., 2013).

Thus, there is increased interest in partially automatizing the creation of systematic reviews (O'Connor et al., 2019). A significant hindrance for the automation of systematic reviews is that data needs to be extracted by hand from published studies. This problem could be alleviated if publications were machine readable, or could be turned into a structured, machine readable form by information extraction methods (Liu et al., 2016; Wu et al., 2020).

The methods that so far have been applied to the automatic extraction of information from clinical trial publications follow the PICO framework and attempt to extract the Population, Intervention, Comparator and Outcomes from a publication. Most approaches formalize the task as a tagging or classification problem. Some approaches for instance attempt to tag spans in the text and label them with the PICO elements (e.g. (Trenta et al., 2015)). Others classify complete text segments into these classes (Boudin et al., 2010; Jin and Szolovits, 2018).

However, the PICO elements denote structured objects rather than plain tags or classes. An intervention is described by a drug, frequency of administration, administration route, dose, etc. An outcome is described by a certain increase or decrease of a value from a baseline condition, refers to a certain primary or secondary endpoint, and there are outcomes for each arm of a trial that need to be compared to each other. In spite of being structured objects, most previous work treats these elements as flat and unstructured. Treating them as such makes the automatic aggregation and summarization of results challenging if not impossible.

Towards treating information extraction from clinical publications as a problem of predicting structured elements, we model the task as a template extraction task in which each template consists of a number of slots to be extracted. In Table 1 we provide an overview of all the templates we consider in this work and the number and types of slots they have.

Towards extracting these templates and thus a

Proceedings of the BioNLP 2022 workshop, Dublin, Ireland, pages 178–192
May 26, 2022. ©2022 Association for Computational Linguistics

structured representation of a clinical trial and its results, we present a novel deep learning architecture. The architecture first labels spans of text as candidate slot fillers of a particular slot in a first step. In a second step, the filler is assigned to an instance of a template. With this two-step architecture, we can transform each clinical trial abstract into a structured representation that supports downstream aggregation of results.

As there can be multiple interventions, arms and outcomes in a given study, an important challenge is to predict how many instances of each template occur in a given clinical trial publication. We leave this subpart of the problem for future work and assume that the number of interventions, arms and outcomes is known a priori. This assumption is reasonable as this information is typically contained in existing registries for trials such as `https://www.clinicaltrials.gov/`.

When assigning slot fillers to templates, it is important to model the dependencies between the different slots as some values might be compatible while others not. We model this compatibility by a trained function that predicts a compatibility score.

In summary our contributions are as follows:

- We propose a new approach to extracting evidence from clinical trial publications that consists in instantiating a set of pre-defined templates. As a result, the key findings of a clinical trial can be represented in a fully structured and machine-readable form that supports down-stream aggregation. To the best of out knowledge, we present the first template-filling IE approach in the clinical trial domain.
- We present a novel two-step deep learning based architecture that first recognizes slot candidates and then assigns these candidates to instances of templates. At a second step, candidates for slot fillers are assigned to a template instance.
- We show that it is possible to extract fine grained candidates of slot fillers from 37 classes yielding very good results of micro $F_1 = 76.21\%$ on the Glaucoma and $F_1 = 76.49\%$ the Type 2 Diabetes Mellitus (T2DM) dataset (Sanchez-Graillet et al., 2021), respectively.
- We introduce an intra-template entity compatibility optimization procedure for distributing entities to template instance of the same type. We show the impact of including a function for scoring the compatibility of slot assignments, and show that it improves extraction results in terms of F-Measure by 6.34% and 3.95% on the Glaucoma and T2DM dataset, respectively.

2 Related Work

The template extraction and slot filling task we address is related to the field of event extraction (Frisoni et al., 2021) where the goal is to extract so called *event triggers* and the arguments of the events. Our templates can be seen as complex events and our slots as arguments thereof.

Wang et al. (2020) adopt the question answering paradigm to extract events from biomedical texts. They introduce two different types of questions for extracting event triggers and event arguments. However, in their approach the extraction of event arguments also relies on the extraction of event triggers.

Adel et al. (2018) introduce a framework for task-independent template-based information extraction. Their approach first identifies text spans representing slot-fillers as in our approach. However, their system relies on the successful identification of anchor spans representing template instances as they cast the assignment of slot-fillers to template instances as a binary classification between anchor spans and other text spans. The slot filling system proposed by Zhang et al. (2017) is a neural architecture that can exploit the combination of semantic similarity-based attention and position-based attention. The authors address a relation extraction task and develop a large corpus of annotated relations, TACRED (Zhang et al., 2017).

More recent work has framed the task of relation extraction in the biomedical field as a slot filling task as well (Papanikolaou and Bennett, 2021). However, the work is limited to extracting binary relationships (drug-drug, compound-drug and compound-disease).

Early work on extracting information from text describing clinical trials has focused on the classification of sentences into sections of papers describing Randomized Controlled Trials (RCTs), e.g. Methods, Results, etc. (McKnight and Srinivasan, 2003; Hirohata et al., 2008; Chung, 2009). Such systems tackle a very coarse-grained information

Template Name	#Slots	Slots
Arm	7	AdverseEffect, FinalNumPatientsArm, Intervention, NumPatientsLeftArm, NumberPatientsArm, Outcome, RelFinalNumPatientsArm,
ClinicalTrial	15	analysesHealthCondition, AllocationRatio, AnalysisApproach, Arm, CTDesign, CTduration, ConclusionComment, DiffBetweenGroups, EvidQualityIndicator, FinalNumberPatientsCT, NumPatientsLeftCT, NumberPatientsCT, ObjectiveDescription, Population, RelNumPatientsLeftCT
DiffBetweenGroups	8	ConfIntervalDiff, DiffGroupAbsValue, DiffGroupRelValue, Outcome1, Outcome2, PvalueDiff, StandardDevDiff, StandardErrorDiff
Endpoint	4	AggregationMethod, BaselineUnit, EndPointDescription, MeasurementDevice
Intervention	5	Duration, Frequency, Interval, Medication, RelativeFreqTime
Medication	6	ApplicationCondition, DeliveryMethod, DoseDescription, DoseUnit, DoseValue, Drug
Outcome	26	BaselineValue, ChangeValue, ConfIntervalBL, ConfIntervalChangeValue, ConfIntervalNumAffected, ConfIntervalResValue, Endpoint, NumberAffected, ObservedResult, PValueBL, PValueChangeValue, PValueNumAffected, PValueResValue, PercentageAffected, RelativeChangeValue, ResultMeasuredValue, SdDevBL, SdDevChangeValue, SdDevNumAffected, SdDevResValue, SdErrorBL, SdErrorChangeValue, SdErrorNumAffected, SdErrorResValue, SubGroupDescription, TimePoint
Population	7	in AvgAge, Country, Ethnicity, Gender, MaxAge, MinAge, Precondition
Publication	6	describes, Author, Journal, PMID, PublicationYear, Title

Table 1: Template types and corresponding slots

extraction task as they do not extract the actual content or results of a published RCT, but only extract correspondences between content and the standard sections used to describe a clinical trial in a publication. Such a sentence classification task can support the indexation and thus retrieval of information from a published RCT, but does not support the use case we consider, i.e. the aggregation of evidence across published trials.

Beyond the classification of sentences into sections of an article, other authors have considered the classification of sentences into PICO elements, that is classifying a sentence in a published clinical trial with respect to whether it describes the Population, Intervention, Comparator or an Outcome (Demner-Fushman and Lin, 2007; Chung, 2009; Boudin et al., 2010; Jin and Szolovits, 2018). Such approaches are able to extract information at a more detailed granularity, but they still do not support aggregation of evidence across studies as the mere classification of sentences with respect to PICO elements does not provide a semantic structure that can be used to describe the key results of a study.

The work by Trenta et al. (2015) goes one step further in that it tags spans of text in an RCT abstract into the PICO classes, considering the following classes: patient group, intervention, arm, control arm, measured outcome, etc. Trenta et al. (2015) rely on maximum entropy models and use integer linear programming to define constraints on the classified tokens, e.g., such that *Results* can not occur in the *Methods* section. They show that their approach is able to extract evidence tables from RCT abstracts. Yet, the different spans extracted are only indirectly related to each other in the model of Trenta et al. (2015). This gap is addressed by the approach of Nye et al. (2020), which beyond extracting PICO elements (intervention arms, outcome measures, results) also relates the different snippets to each other, yielding a relational structure.

Inspired by the work of Trenta et al. (2015) as well as Nye et al. (2020) we go one step further in extracting a complete structured object from an RCT abstract comprising of nine main template types with overall 85 slots. To our knowledge, this is thus the most fine-grained representation that so far has been considered by an information extraction system in the clinical domain.

3 Model

As already mentioned in the introduction, our proposed model consists of a two-step architecture. The first component, the entity extraction (EE) module, identifies spans of slot filler candidates (SFCs). We assume that we have a set of template types $\mathcal{T} = t_1, ..., t_{|\mathcal{L}|}$ which correspond to the template types depicted in Table 1, where $\mathcal{L}$ denotes the number of template types. We refer to the slot j of template t_i as $s_{i,j}$. The set of all slots is $\mathcal{S} = \bigcup_{i,j}\{s_{i,j}\}$ and the set of slots of template type t is $\mathcal{S}_t = \bigcup_j\{s_{t,j}\}$.

The set of all SFCs extracted within an abstract is denoted by $\mathcal{E}$. Formally speaking, the entity extraction module implements a function f_{EE} that maps each slot filler candidate into a slot type, i.e. $f_{EE} : \mathcal{E} \rightarrow \mathcal{S}$.

The second component, the template assignment (TA) module, maps each slot filler to a particular instance of a template. Hereby, we can have multiple instances of a given template type. For instance, in the general case a clinical study might describe multiple interventions, multiple endpoints and multiple outcomes. We denote the i-th instance of template t by $T_i^{(t)}$. The set of all template instances is thus $\theta = \bigcup\{T_i^{(t)}\}$ and the number of template instances of template type t is denoted by m_t. The second component thus realizes a function $f_{TA} : \mathcal{E} \rightarrow \theta$. We denote the template type to which SFC e_j has been assigned to as y_{e_j}.

Take the following sentence as an example: *Mean 24-h IOP with BTFC was significantly lower than with latanoprost (18.9 vs 21.2 mmHg; p < 0.001).* The first component would recognize the spans *18.9* and *21.2* and map them both to the slot type `ResultMeasuredValue`. Then the TA module assignes these identifies SFCs to template instances of type Outcome, together with other SCFs extracted from other sentences.

Note that both modules fully specify a mapping from entities detected in the clinical trial abstract to fully instantiated templates, where f_{EE} identifies and classifies text spans into slots and f_{EA} identifies the appropriate instance of a template.

We describe both modules in more detail subsequently. In particular, as the assignment of text spans to slots and template instances should not be modelled completely independently, we introduce an additional component that computes an overall score for a given template instance that quantifies the compatibility of the assigned text spans to all of the slots of the template instance. These scores can be regarded as factors as used in factor graph models (Kschischang et al., 2001). In order to reduce the complexity, we model the interaction between different slots in a pairwise fashion, limiting the scope of these factors to two slots.

3.1 Entity Extraction Module

The entity extraction module identifies token spans in the input document which either represent named entities or literals. The extracted token spans are later assigned to slots by the module described in section 3.2. We represent documents $\mathcal{D}$ by a sequence of sentences $(s_1, \ldots, s_{n_S})$ where each sentence s_i in turn is represented by a sequence of tokens $(w_1^{(s_i)}, \ldots, w_{n_{s_i}}^{(s_i)})$, where n_S denotes the number of sentences in document $\mathcal{D}$ and n_{s_i} denotes the number of tokens of sentence s_i. We adopt the Bidirectional Encoder Representations from Transformers (BERT) (Devlin et al., 2019) architecture for computing contextualized token representations within the input document. A BERT layer is a stack of K identical Transformers (Vaswani et al., 2017) which captures pairwise token dependencies via an attention mechanism. Since most BERT implementations limit the length of input sequences by k_{max}, we split the sequence of sentences of the input document into n_C subsequences (chunks) if the number of tokens of the document exceeds this upper bound. We use the special token $[SEP]$ to separate sentences within a given chunk c_i and prepend the special token $[CLS]$ to each chunk which allows for capturing global context information for each chunk. The output for chunk c_i of the K-th Transformer of the BERT layer is a sequence of contextualized vectors $\mathbf{h}_1^{(c_i)}, \ldots, \mathbf{h}_{n_{c_i}}^{(c_i)} \in R^{d_{bert}}$, where the vector $\mathbf{h}_j^{(c_i)}$ represents the j-th token of chunk c_i, d_{bert} denotes the dimension of the BERT model and n_{c_i} denotes the number of tokens in chunk c_i.

Entity extraction is implemented through two dense layers which independently predict which tokens are start and/or end positions of entities which are referenced by a slot. This is achieved by using the set of slots $\mathcal{S}$ as entity types. Then the predicted entity type indirectly specifies the type of the template the entity has to be assigned to since no pair of template types shares the same set of slots. More formally, the two dense layers are given by

$$\hat{\mathbf{y}}_{j,start}^{(c_i)} = softmax(\mathbf{W}_{start}\mathbf{h}_j^{(c_i)} + \mathbf{b}_{start}) \quad (1)$$

$$\hat{\mathbf{y}}_{j,end}^{(c_i)} = softmax(\mathbf{W}_{end}\mathbf{h}_j^{(c_i)} + \mathbf{b}_{end}) \quad (2)$$

where $\mathbf{W}_{start}, \mathbf{W}_{end} \in R^{(|\mathcal{S}|+1)\times d_{bert}}$, $\mathbf{b}_{start}, \mathbf{b}_{end}, \in R^{d_{bert}}$.

The prediction of the slot is performed as follows:

$$\begin{aligned} \hat{y}_{j,start}^{(c_i)} &= \arg\max \hat{\mathbf{y}}_{j,start}^{(c_i)} \\ \hat{y}_{j,end}^{(c_i)} &= \arg\max \hat{\mathbf{y}}_{j,end}^{(c_i)} \end{aligned}$$

At inference time we join the predicted start and end positions by assigning the closest predicted end position p_{end} of type t within the same sentence to each predicted start position p_{start} of type t under the constraint $p_{start} <= p_{end}$.

Finally we compute a vector representation $\mathbf{e}_k$ for each extracted SFC e_k by summing the vectors $\mathbf{h}_j^{c_i}$ of the corresponding start and end tokens of the SFC, followed by a dense layer with a ReLu activation function (Agarap, 2018).

3.2 Template Assignment Module

The TA module described in this section assigns each SFC $e_j \in \mathcal{E}$ extracted by the entity extraction module to a template in θ. As we know the slot y_{e_j} that e_j has been assigned to, the template type t of y_{e_j} determines the subset θ_t of template instances in the set θ that e_j can be assigned to. This reduces the search space considerably and essentially allows us to model the template assignment task as the one of inducing a partition.

Let's assume that SFCs are grouped into $|\mathcal{L}|$ disjoint subsets $\mathcal{E}_1, \ldots, \mathcal{E}_{|\mathcal{L}|}$ according to their type t, that is:

$$\mathcal{E}_t = \{e_j \in \mathcal{E} \mid y_{e_j} \in \mathcal{S}_t\}, \quad t \in \mathcal{L} \tag{3}$$

The task of template assignment can be reduced to the task of partitioning each set $\mathcal{E}_t$ into a partition $\mathcal{P}_t = \{\mathcal{T}_1^{(t)}, \ldots, \mathcal{T}_{m_t}^{(t)}\}$ of $\mathcal{E}_t$ where each set $\mathcal{T}_i^{(t)}$ contains the SFCs assigned to template instance $T_i^{(t)}$.

We call a partition $\mathcal{P}_t$ of the set $\mathcal{E}_t$ valid if each SFC $e_j \in \mathcal{E}_t$ is assigned to exactly one partition $\mathcal{T}_i^{(t)} \in \mathcal{P}_t$ and we denote the set of all valid partitions for the set $\mathcal{E}_t$ as $\mathcal{U}_t$.

We propose a pairwise intra-template entity compatibility optimization objective which measures the joint compatibility of the SFCs within the sets $\mathcal{T}_i^{(t)}$ of a partition. Let $q : \mathcal{E} \times \mathcal{E} \rightarrow [0, 1]$ denote the function which measures the compatibility between two SFCs e_j, e_k, where $q(e_j, e_k) = 1$ means maximal compatibility and $q(e_j, e_k) = 0$ means minimal compatibility. Note that we assume that q is symmetric in its arguments, i.e., $q(e_j, e_k) = q(e_k, e_j)$. Then the mean pairwise entity compatibility score $h(\mathcal{T}_i^{(t)})$ for the set $T_i^{(t)}$ is given by

$$h(\mathcal{T}_i^{(t)}) = \frac{1}{\frac{m_t!}{2(m_t-2)!}} \sum_{e_j, e_k \in \mathcal{T}_i^{(t)}, j<k} q(e_j, e_k) \tag{4}$$

and the compatibility score for partition $\mathcal{P}_t$ is the sum of the mean pairwise compatibility scores of each template set $\mathcal{T}_i^{(t)} \in \mathcal{P}_t$:

$$\sum_{\mathcal{T}_i^{(t)} \in \mathcal{P}_t} h(\mathcal{T}_i^{(t)}) \tag{5}$$

Given these definitions, we seek the partition $\hat{\mathcal{P}}_t \in \mathcal{U}_t$ which maximizes the compatibility score defined by Eq. (5). Hence the optimization problem proposed by our approach is given by

$$\hat{\mathcal{P}}_t = \arg\max_{\mathcal{P}_t \in \mathcal{U}_t} \sum_{\mathcal{T}_i^{(t)} \in \mathcal{P}_t} h(\mathcal{T}_i^{(t)}) \tag{6}$$

for all template types $t \in \mathcal{L}$. For arbitrary large entity sets $\mathcal{E}_t$, the sets $\mathcal{U}_t$ of valid partitions can become very large because of the combinatorial explosion, and hence finding the exact solution of the optimization problem defined by Eq. (6) can become intractable. Therefore we propose an approximate optimization method based on beam search which maintains a set $\mathcal{B}_t^{(z)}$ of n_B candidate solutions in each iteration z which are gradually refined. We define a candidate solution i for template type t as a pair $(\mathcal{E}_t^{(i)}, \mathcal{P}_t^{(i)})$, where $\mathcal{P}_t^{(i)}$ denotes the candidate partition and $\mathcal{E}_t^{(i)} \subseteq \mathcal{E}_t$ denotes the set of entities of that candidate solution which are not yet assigned to any template set $\mathcal{T}_i^{(t)} \in \mathcal{P}_t^{(i)}$. In each iteration z, we compute all successors of all candidate solutions $(\mathcal{E}_t^{(i)}, \mathcal{P}_t^{(i)}) \in \mathcal{B}_t^{(z)}$ by assigning an entity $e_j \in \mathcal{E}_t^{(i)}$ to a template set $\mathcal{T}_i^{(t)} \in \mathcal{P}_t^{(i)}$, which yields a set of new candidate solutions $\tilde{\mathcal{B}}_t^{(z)}$. Next we rank all candidate solutions in $\tilde{\mathcal{B}}_t^{(z)}$ by computing the mean intra-template entity compatibility score defined by Eq (5) for each candidate partition of the respective candidate solutions and keep only the best n_B ones, which yields the new beam $\mathcal{B}_t^{(z+1)}$ for the next iteration. After all entities for template type t have been assigned to a template after Z iterations, the partition $\mathcal{P}_t^{(i)}$ of the best ranked final candidate solution $(\mathcal{E}_t^{(i)}, \mathcal{P}_t^{(i)}) \in \mathcal{B}_t^{(Z)}$ is returned. The initial seed sets $\mathcal{B}_t^{(0)}$ of candidate solutions for each template type t are given by

$$\mathcal{B}_t^{(0)} = \{(\mathcal{E}_t, \{\mathcal{T}_i^{(t)}\}_{i=1}^{m_t})\}, \quad \mathcal{T}_i^{(t)} = \{\} \tag{7}$$

More details of the optimization procedure can be found in algorithm 1.

We implement the pairwise entity compatibility function $q(e_i, e_j)$ through summing the vector representations $\mathbf{e}_i$ and $\mathbf{e}_j$ of the corresponding entities

Data: Set of SFCs $\mathcal{E}$; entity compatibility function g; beam size n_B
Result: Partitions $\mathcal{P}_i, \ldots, \mathcal{P}_{|\mathcal{L}|}$
for $t \in \mathcal{L}$ **do**
  Compute set of SFCs $\mathcal{E}_t$ for template type t by Eq. (3)
  Compute beam seed set $\mathcal{B}_t^{(0)}$ defined by Eq. (7)
  $z \leftarrow 0$
  for $i \in \{1, \ldots, |\mathcal{E}_t|\}$ **do**
    Initialize set of successor candidate solutions $\tilde{\mathcal{B}}_t^{(z)}$ as empty set
    for $(\mathcal{E}_t^{(i)}, \mathcal{P}_t^{(i)}) \in \mathcal{B}_t^{(z)}$ **do**
      for $e_k \in \mathcal{E}_t^{(i)}$ **do**
        for $\mathcal{T}_j^{(t)} \in \mathcal{P}_t^{(i)}$ **do**
          Remove e_k from $\mathcal{E}_t^{(i)}$ which yields the set $\tilde{\mathcal{E}}_t^{(i)}$
          Add e_k to set $\mathcal{T}_j^{(t)}$ which yields $\tilde{\mathcal{T}}_j^{(t)}$
          Replace $\mathcal{T}_j^{(t)}$ in $\mathcal{P}_t^{(i)}$ by $\tilde{\mathcal{T}}_j^{(t)}$ which yields $\tilde{\mathcal{P}}_t^{(i)}$
          Add new candidate solution $(\tilde{\mathcal{E}}_t^{(i)}, \tilde{\mathcal{P}}_t^{(i)})$ to set $\tilde{\mathcal{B}}_t^{(z)}$
        end
      end
    end
    Rank all candidate solutions in $\tilde{\mathcal{B}}_t^{(z)}$ by Eq. (5)
    Keep the best ranked n_B candidate solutions from $\tilde{\mathcal{B}}_t^{(z)}$ which yields new batch $\tilde{\mathcal{B}}_t^{(z)}$
    $z \leftarrow z + 1$
  end
  Get best ranked candidate solution $(\hat{\mathcal{E}}_t^{(i)}, \hat{\mathcal{P}}_t^{(i)})$ from $\mathcal{B}_t^{(|\mathcal{E}_t|)}$
  $\mathcal{P}_t \leftarrow \hat{\mathcal{P}}_t^{(i)}$
end

Algorithm 1: Pseudo-code of our proposed approximate optimization method for maximizing the mean intra-template entity compatibility when assigning the extracted entities to templates

followed by a dense layer with sigmoid activation function. More formally:

$$\hat{q}(e_i, e_j) = \sigma(\mathbf{w}_{comp} \odot (\mathbf{e}_i + \mathbf{e}_j) + b_{comp}) \quad (8)$$

where $\mathbf{w}_{comp} \in R^{d_{bert}}$, $b_{comp} \in R$ and $\odot$ denotes the scalar product of two vectors.

3.3 Model Training

We train the model in end-to-end fashion by jointly minimizing the loss of the EE module and the TA module. The loss L_{EE} of the EE module is given by the cross entropy between the predicted SFC start position $\hat{\mathbf{y}}_{j,start}^{(c_i)}$ and ground truth SFC start position $\mathbf{y}_{j,start}^{(c_i)}$ plus the cross entropy between predicted SFC end positions $\hat{\mathbf{y}}_{j,end}^{(c_i)}$ and the ground truth SFC end positions $\mathbf{y}_{j,end}^{(c_i)}$.

The loss L_{TA} of the TA module is given by the cross entropy between the ground truth compatibility scores $q^*(e_i, e_j)$ and the predicted compatibility scores $\hat{q}(e_i, e_j)$ for all pairs of SFCs (e_i, e_j) in a given training set. If two SFCs e_i are assigned to the same template instance in the gold standard, then $q^*(e_i, e_j) = 1$, otherwise $q^*(e_i, e_j) = 0$, Note that we only consider pairs of slot-filler candidates which are assigned to the same template type.

The complete model is trained by minimizing the loss $L_{EE} + L_{TA}$ with respect to model parameters which are given by the parameters of the BERT encoder, the parameters of the dense layers defined by (1), (2), (8) and the parameters of the layer which is used to compute the vector representation $\mathbf{e}_k$ of the SFCs.

4 Experiments

We conduct experiments on two public datasets (Sanchez-Graillet et al., 2021) which contain RCT abstracts from the Glaucoma and Type 2 Diabetes Mellitus (T2DM) domain, respectively. The corpora of both datasets are annotated at two levels: At the first level, salient entities which describe components of the PICO elements are annotated. The second level comprises template-based annotations of complex PICO elements and their interactions.

4.1 Experimental Setting

In all our experiments, we use a BERT model pretrained on biomedical and life sciences literature abstracts [1]. We use the same train/validation/test split as in (Sanchez-Graillet et al., 2021), Table 2 shows the number of abstracts included in the train, validation and test sets of the respective datasets. All models are trained with the AdamW optimizer (Loshchilov and Hutter, 2017) for 30 epochs with an inital learning rate of $3 * 10^{-5}$ and with a linear warm-up phase over the first 10% of training steps. Further, we use batches of exactly one abstract and set the beam size of the intra-template compatibility optimization algorithm depicted in 1 to 50.

We score a predicted SFC as correct if there is a SFC in the corresponding sentence in the test set with the same label, start and end position. Further, we use the Hungarian algorithm (Kuhn, 1955) for aligning predicted and ground truth templates for

[1] https://tfhub.dev/google/experts/bert/pubmed/2

Table 2: Number of abstracts in the train, validation and test sets

	# Abstracts training set	# Abstracts validation set	# Abstracts test set
Glaucoma	69	17	21
T2DM	68	16	20

each template class, using the pairwise micro F_1 as optimization objective.

As a baseline, we implement a *greedy assignment* approach to assign SFCs to template instances: Given the set $\mathcal{E}_t$ of extracted SFCs for template type t, we repeatedly loop over the template instances T_t^k, randomly pick a SFC from $\mathcal{E}_T$, assign this entity to T_t^k and remove it from $\mathcal{E}_t$. This is repeated until the set $\mathcal{E}_t$ is empty, i.e., all SFCs for template type t have been assigned.

4.2 Results

Extraction of slot filler candidates: Our approach can extract 37 types of slot filler candidates (see Table 1). The results in terms of Precision, Recall and F-Measure for all slot types are given in Table 8 in the Appendix. Overall, the model yields a micro-averaged F-Measure of 0.80 (P=0.80, R=0.73) on the Glaucoma dataset as well as F=0.76 (P=0.80, R=0.73) on the T2DM dataset. Table 3 shows the top 10 slot types with the best extraction results. Similarly, table 4 shows the five slot types with the worst prediction results.

Template extraction: Table 8 in the Appendix shows the prediction results of the SFCs on the Glaucoma and T2DM test sets. The entries "-" indicate that the corresponding slots are not used in the respective data set. Table 5 shows the aggregated results over each template type by averaging the F-values for all slots of the corresponding template. Note that Table 5 only contains template types which could have more than one instance, whereas Table 1 shows all template types. Overall, our proposed model yields a micro F_1 score of 62.27% on the Glaucoma corpus and 64.38% on the T2DM corpus, with a gain of 6,34% in micro-averaged F_1 compared to greedy assignment on the Glaucoma dataset and 3,95% on the T2DM dataset, showing the superiority of our proposed intra-template entity compatibility (ITC) algorithm. For both datasets, the instances of template Arm are extracted best with mean F_1 of 91% and 93% on the Glaucoma and T2DM dataset, respectively. The templates types that have the worst performance are Endpoint for the Glaucoma dataset (mean F=48%)

Table 3: Top 10 slot types for the Glaucoma and T2DM datasets

Slot Name	F_1
Glaucoma	
PMID	1.00
PublicationYear	1.00
RelativeChangeValue	1.00
SdErrorChangeValue	1.00
Title	0.94
SdDevResValue	0.94
NumberPatientsCT	0.93
ChangeValue	0.92
HealthCondition	0.91
NumberPatientsArm	0.91
T2DM	
NumberAffected	1.00
PMID	1.00
PublicationYear	1.00
Journal	0.97
PercentageAffected	0.95
Author	0.94
NumberPatientsArm	0.93
NumberPatientsCT	0.93
ChangeValue	0.90
CTDesign	0.88

Table 4: Slot types with the worst prediction results for the Glaucoma and T2DM datasets

Slot Name	F_1
Glaucoma	
ObservedResult	0.00
Drug	0.27
Precondition	0.28
PointDescription	0.32
ObjectiveDescrip-	0.49
T2DM	
ConfIntervalDiff	0.00
ObservedResult	0.00
SdDevChangeValue	0.25
SdDevBL	0.38
Precondition	0.41

h

Table 5: Aggregated slot-filling results (mean F_1 and overall micro F_1) (ITC=Intra-Template Compatibility)

	Glaucoma		T2DM	
	Greedy Assignment	ITC	Greedy assignment	ITC
DiffBetweenGroups	0.58	**0.64**	0.47	**0.48**
Arm	0.85	**0.91**	0.93	0.93
Intervention	0.53	**0.73**	**0.68**	0.58
Medication	0.89	0.89	0.57	**0.77**
Outcome	0.41	**0.61**	**0.47**	0.44
Endpoint	**0.51**	0.48	0.56	**0.60**
Micro Average	0.56	**0.62**	0.60	**0.64**

and Outcome for the T2DM dataset (mean F=44%).

On the Glaucoma dataset, for four out of six template types, our proposed ITC algorithm yields better performance than the greedy assignment in terms of mean F_1, for one template type (Medication) the performance is equal and for one out of six template types the performance is worse. On the T2DM dataset, for three out of six template types our ITC algorithm performs better than greedy assignment, for one template type (Arm) the performance is equal and and for two out out six template types the performance is worse.

We also conducted a study simulating perfect entity extraction by performing the second step with gold standard SFCs. The results in Table 6 show that results are significantly better with perfect SFC identification, yielding an increase of more than 0.20 points in micro averaged F-Measure for the Glaucoma dataset and more than 0.15 points on the T2DM dataset. This shows the importance of good entity recognition and extraction models.

Table 7 shows the effect of the beam size on the template extraction results. Overall, we see that the beam size has a negligible effect on the results.

Case study: As a case study, we compare the predicted structure to the gold standard structure for one published clinical trial in the test set of the T2DM corpus. We cherry pick the study with the best results in terms of micro-averaged F_1, that is $F_1 = 0.85$. The selected paper is the publication by Shankar et al. (2017). Table 10 contrasts the instances of templates specified in the gold standard vs. the instances of templates extracted by our approach. Overall, the results are very good, clearly showing the potential of our approach and hinting at the fact that the task can be solved to a satisfactory extent. Regarding the *Population* studied in the paper, our method can extract a corresponding condition, but is not able to explicitly extract the countries in which the population was recruited (USA, Australia). With except of the health condition (type 2 diabetes mellitus), all other elements describing the characteristics of the *Clinical Trial* are extracted correctly. Most of the relevant endpoints are extracted correctly, albeit not always the correct units are extracted. Two endpoints are conflated into one: fasting plasma glucose and 2 - h post - meal glucose with the result that one endpoint has a unit (mg/dl) but no endpoint description. The medications for the two arms (sinagliptin vs. placebo) are extracted correctly. The dose value of sinagliptin is mistaken for the dose value of the placebo unfortunately. Most of the outcome values are extracted correctly, but the percentage of patients affected is not extracted. The p values reporting significance of results when comparing the two arms / groups are extracted perfectly.

Table 11 shows the instances of templates specified in the gold standard vs. the instances of templates extracted by our approach for the abstract from the T2DM test set with the worst prediction result in terms of micro $F_1 = 0.57$. The corresponding publication can be found in (Klein et al., 2014). Although our system gets the Publication metadata, the Clinical Trial design, Arms and Medications right to a great extent, it makes a number of important errors in the categories Endpoints and Outcomes.

5 Conclusion

We have presented a twp-step neural architecture based on a transformer model that can induce a structured representation from an abstract describing a Randomized Controlled Trial (RCT). The architecture performs extraction of candidate slot fillers as a first step by identifying spans of 37

Table 6: Aggregated slot-filling results comparing the settings with perfect entity recognition using gold standard entity annotations and entity recognition by our model (mean F_1 and overall micro F_1)

	Glaucoma		T2DM	
	Ground Truth SFCs	Predicted SFCs	Ground Truth SFCs	Predicted SFCs
DiffBetweenGroups	**0.87**	0.64	**0.77**	0.48
Arm	**1.00**	0.91	**1.00**	0.93
Intervention	**0.83**	0.73	**1.00**	0.58
Medication	**0.92**	0.89	**0.93**	0.77
Outcome	**0.69**	0.61	**0.59**	0.44
Endpoint	**0.90**	0.48	**0.82**	0.60
Micro Average	**0.83**	0.62	**0.81**	0.64

Table 7: Effect of the beam size on template extraction results (mean F_1 and overall micro F_1)

	Glaucoma					T2DM				
	10	20	30	40	50	10	20	30	40	50
DiffBetweenGroups	0.64	0.60	0.60	0.60	0.64	0.50	0.48	0.48	0.47	0.48
Arm	0.91	0.91	0.91	0.91	0.91	0.93	0.93	0.93	0.93	0.93
Intervention	0.73	0.73	0.73	0.73	0.72	0.58	0.58	0.58	0.58	0.58
Medication	0.89	0.89	0.89	0.89	0.77	0.77	0.79	0.79	0.77	0.77
Outcome	0.58	0.61	0.57	0.56	0.61	0.42	0.43	0.42	0.43	0.44
Enpoint	0.48	0.48	0.48	0.48	0.48	0.62	0.61	0.60	0.62	0.60
Micro Average	0.62	0.62	0.62	0.62	0.62	0.64	0.64	0.64	0.64	0.64

different classes. At a second step, it assigns the extracted candidate slot fillers into nine main templates. We have shown that our approach can extract candidate slot fillers reliably, yielding micro F-Measures of 76.21% and 76.49% on our Glaucoma and T2DM dataset, respectively. In terms of extraction of templates, our approach yields micro F-measures of 62.27% and 64.38% averaged over all slots on our Glaucoma and T2DM dataset, respectively. The structure of our templates is inspired by the C-TrO ontology (Sanchez-Graillet et al., 2019) and induces the most fine-grained and accurate representation of a published RCT that has been considered so far by any information extraction system. In future work we intend to show that our information extraction approach indeed supports the aggregation of results across clinical trials. Further, we plan to use the intra-template compatibility scores to infer the number of template instance for template types which could have several instances. This can be regarded as an additional layer on top of our proposed optimization algorithm. In addition, we plan to predict links between template instances.

References

Heike Adel, Laura Ana Maria Oberländer, Sean Papay, Sebastian Padó, and Roman Klinger. 2018. Dere: A task and domain-independent slot filling framework for declarative relation extraction. In *Proceedings of the 2018 Conference on Empirical Methods in Natural Language Processing: System Demonstrations*, pages 42–47.

Abien Fred Agarap. 2018. Deep learning using rectified linear units (relu). *arXiv preprint arXiv:1803.08375*.

Elaine M Beller, Joyce Kee-Hsin Chen, Una Li-Hsiang Wang, and Paul P Glasziou. 2013. Are systematic reviews up-to-date at the time of publication? *Systematic Reviews*, 2:36.

Rohit Borah, Andrew W Brown, Patrice L Capers, and Kathryn A Kaiser. 2017. Analysis of the time and workers needed to conduct systematic reviews of medical interventions using data from the prospero registry. *BMJ Open*, 7(2).

Florian Boudin, Jian-Yun Nie, Joan C Bartlett, Roland Grad, Pierre Pluye, and Martin Dawes. 2010. Combining classifiers for robust pico element detection. *BMC medical informatics and decision making*, 10(1):1–6.

Grace Y Chung. 2009. Sentence retrieval for abstracts of randomized controlled trials. *BMC Medical Informatics and Decision Making*, 9:10.

Dina Demner-Fushman and Jimmy Lin. 2007. Answering clinical questions with knowledge-based and statistical techniques. *Computational Linguistics*, 33(1):63–103.

Jacob Devlin, Ming-Wei Chang, Kenton Lee, and Kristina Toutanova. 2019. Bert: Pre-training of deep bidirectional transformers for language understanding. In *Proceedings of NAACL-HLT*, pages 4171–4186.

Giacomo Frisoni, Gianluca Moro, and Antonella Carbonaro. 2021. A survey on event extraction for natural language understanding: Riding the biomedical literature wave. *IEEE Access*, 9:160721–160757.

Kenji Hirohata, Naoaki Okazaki, Sophia Ananiadou, and Mitsuru Ishizuka. 2008. Identifying sections in scientific abstracts using conditional random fields. In *Proceedings of the Third International Joint Conference on Natural Language Processing: Volume-I*.

Di Jin and Peter Szolovits. 2018. Pico element detection in medical text via long short-term memory neural networks. In *Proceedings of the BioNLP 2018 workshop*, pages 67–75.

David J Klein, Tadej Battelino, DJ Chatterjee, Lisbeth V Jacobsen, Paula M Hale, Silva Arslanian, and NN2211-1800 Study Group. 2014. Liraglutide's safety, tolerability, pharmacokinetics, and pharmacodynamics in pediatric type 2 diabetes: a randomized, double-blind, placebo-controlled trial. *Diabetes technology & therapeutics*, 16(10):679–687.

GG Koch. 2006. No improvement–still less than half of the cochrane reviews are up to date. In *XIV Cochrane Colloquium, Dublin, Ireland*.

F.R. Kschischang, B.J. Frey, and H.-A. Loeliger. 2001. Factor graphs and the sum-product algorithm. *IEEE Transactions on Information Theory*, 47(2):498–519.

Harold W Kuhn. 1955. The hungarian method for the assignment problem. *Naval research logistics quarterly*, 2(1-2):83–97.

Feifan Liu, Jinying Chen, Abhyuday Jagannatha, and Hong Yu. 2016. Learning for biomedical information extraction: Methodological review of recent advances. *arXiv preprint arXiv:1606.07993*.

Ilya Loshchilov and Frank Hutter. 2017. Decoupled weight decay regularization. *arXiv preprint arXiv:1711.05101*.

Larry McKnight and Padmini Srinivasan. 2003. Categorization of sentence types in medical abstracts. In *AMIA Annual Symposium Proceedings*, volume 2003, page 440. American Medical Informatics Association.

Benjamin E Nye, Ani Nenkova, Iain J Marshall, and Byron C Wallace. 2020. Trialstreamer: mapping and browsing medical evidence in real-time. In *Proceedings of the conference. Association for Computational Linguistics. North American Chapter. Meeting*, volume 2020, page 63. NIH Public Access.

Annette M O'Connor, Guy Tsafnat, Stephen B Gilbert, Kristina A Thayer, Ian Shemilt, James Thomas, Paul Glasziou, and Mary S Wolfe. 2019. Still moving toward automation of the systematic review process: a summary of discussions at the third meeting of the international collaboration for automation of systematic reviews (icasr). *Systematic reviews*, 8:57.

Yannis Papanikolaou and Francine Bennett. 2021. Slot filling for biomedical information extraction.

David L Sackett, William MC Rosenberg, JA Muir Gray, R Brian Haynes, and W Scott Richardson. 1996. Evidence based medicine. *BMJ*, 313(7050):170.

Olivia Sanchez-Graillet, Philipp Cimiano, Christian Witte, and Basil Ell. 2019. C-TrO: An Ontology for Summarization and Aggregation of the Level of Evidence in Clinical Trials. In *Proc. of the 5th Joint Ontology Workshops (JOWO): Ontologies and Data in the Life Sciences*.

Olivia Sanchez-Graillet, Christian Witte, Frank Grimm, and Philipp Cimiano. 2021. An annotated corpus of clinical trial publications supporting schema-based relational information extraction. *Journal of Biomedical Semantics*. Under review.

R Ravi Shankar, Yuqian Bao, Ping Han, Ji Hu, Jianhua Ma, Yongde Peng, Fan Wu, Lei Xu, Samuel S Engel, and Weiping Jia. 2017. Sitagliptin added to stable insulin therapy with or without metformin in chinese patients with type 2 diabetes. *Journal of diabetes investigation*, 8(3):321–329.

Antonio Trenta, Anthony Hunter, and Sebastian Riedel. 2015. Extraction of evidence tables from abstracts of randomized clinical trials using a maximum entropy classifier and global constraints. *arXiv preprint arXiv:1509.05209*.

Ashish Vaswani, Noam Shazeer, Niki Parmar, Jakob Uszkoreit, Llion Jones, Aidan N Gomez, Łukasz Kaiser, and Illia Polosukhin. 2017. Attention is all you need. In *Advances in neural information processing systems*, pages 5998–6008.

Xing David Wang, Leon Weber, and Ulf Leser. 2020. Biomedical event extraction as multi-turn question answering. In *Proceedings of the 11th International Workshop on Health Text Mining and Information Analysis*, pages 88–96.

Stephen Wu, Kirk Roberts, Surabhi Datta, Jingcheng Du, Zongcheng Ji, Yuqi Si, Sarvesh Soni, Qiong Wang, Qiang Wei, Yang Xiang, et al. 2020. Deep learning in clinical natural language processing: a methodical review. *Journal of the American Medical Informatics Association*, 27(3):457–470.

Yuhao Zhang, Victor Zhong, Danqi Chen, Gabor Angeli, and Christopher D. Manning. 2017. Position-aware attention and supervised data improve slot filling. In *Proceedings of the 2017 Conference on Empirical Methods in Natural Language Processing*, pages 35–45, Copenhagen, Denmark. Association for Computational Linguistics.

A Supplementary Material

Table 8: Results of the slot-filler candidate extraction on the Glaucoma and T2DM test sets

Slot Name	Glaucoma Precision	Glaucoma Recall	Glaucoma F_1	T2DM Precision	T2DM Recall	T2DM F_1
analysesHealthCondition	0.97	0.86	0.91	0.64	0.58	0.61
Author	1.00	1.00	1.00	0.88	1.00	0.94
BaselineUnit	0.62	0.48	0.54	0.81	0.81	0.81
BaselineValue	0.90	0.67	0.77	0.80	0.60	0.69
CTDesign	0.76	0.83	0.79	0.85	0.91	0.88
CTduration	0.84	0.94	0.89	0.80	0.84	0.82
ChangeValue	0.97	0.88	0.92	0.90	0.90	0.90
ConclusionComment	0.85	0.79	0.81	0.83	0.31	0.45
ConfIntervalDiff	-	-	-	0.00	0.00	0.00
Country	0.81	0.89	0.85	0.89	0.44	0.59
DiffGroupAbsValue	0.75	0.67	0.71	0.84	0.70	0.76
DoseUnit	0.61	0.82	0.70	0.84	0.80	0.82
DoseValue	0.72	0.68	0.70	0.87	0.73	0.80
Drug	0.40	0.21	0.27	0.84	0.76	0.80
EndPointDescription	0.32	0.33	0.32	0.68	0.80	0.74
Frequency	0.89	0.71	0.79	0.71	0.57	0.63
Journal	0.76	0.76	0.76	1.00	0.95	0.97
NumberAffected	0.63	1.00	0.77	1.00	1.00	1.00
NumberPatientsArm	0.88	0.94	0.91	1.00	0.87	0.93
NumberPatientsCT	0.93	0.93	0.93	0.93	0.93	0.93
ObjectiveDescription	0.56	0.43	0.49	0.50	0.44	0.47
ObservedResult	0.00	0.00	0.00	0.00	0.00	0.00
PMID	1.00	1.00	1.00	1.00	1.00	1.00
PValueChangeValue	0.50	0.75	0.60	0.83	0.45	0.59
PercentageAffected	0.82	0.95	0.88	0.96	0.94	0.95
Precondition	0.42	0.22	0.29	0.57	0.32	0.41
PublicationYear	1.00	1.00	1.00	1.00	1.00	1.00
PvalueDiff	0.49	0.68	0.57	0.82	0.94	0.88
RelativeChangeValue	1.00	1.00	1.00	-	-	-
RelativeFreqTime	0.44	0.67	0.53	-	-	-
ResultMeasuredValue	0.85	0.79	0.82	0.75	0.95	0.84
SdDevBL	1.00	0.67	0.80	0.60	0.27	0.38
SdDevChangeValue	0.89	0.67	0.76	0.22	0.29	0.25
SdDevResValue	0.87	1.00	0.93	0.41	1.00	0.58
SdErrorChangeValue	1.00	1.00	1.00	-	-	-
TimePoint	0.60	0.71	0.65	0.63	0.57	0.60
Title	1.00	0.88	0.94	0.77	0.77	0.77
micro average:	0.80	0.73	0.76	0.80	0.73	0.77

Table 9: F_1 scores of the assignment of slot-filler candidates to template instances on the Glaucoma and T2DM test sets. The ITC (Intra-Template Compatibility) columns show the results of our proposed method

	Glaucoma		T2DM	
Slot Name	Greedy Assignment	ITC	Greedy Assignment	ITC
	DiffBetweenGroups			
ConfIntervalDiff	-	-	0.00	0.00
PvalueDiff	0.57	0.57	0.83	**0.83**
DiffGroupAbsValue	0.59	**0.71**	0.58	**0.62**
mean	0.58	**0.64**	0.47	**0.48**
	Arm			
NumberPatientsArm	0.85	**0.91**	**0.93**	0.92
mean	0.85	**0.91**	0.93	0.93
	Intervention			
Frequency	0.79	0.79	**0.68**	0.58
RelativeFreqTime	0.27	**0.67**	-	-
mean	0.53	**0.73**	**0.68**	0.58
	Medication			
Drug	**0.37**	0.34	0.73	**0.83**
DoseValue	**0.70**	0.65	0.46	**0.63**
DoseUnit	0.71	**0.79**	0.49	**0.87**
mean	0.89	0.89	0.57	**0.77**
	Outcome			
ResultMeasuredValue	**0.44**	0.41	**0.61**	0.56
TimePoint	**0.55**	0.50	**0.55**	0.35
PValueChangeValue	0.40	**0.42**	0.59	0.59
PercentageAffected	0.65	**0.65**	0.72	**0.89**
SdErrorChangeValue	0.29	**1.00**	-	-
BaselineValue	0.39	**0.69**	0.34	**0.46**
SdDevBL	0.56	**0.80**	**0.25**	0.13
RelativeChangeValue	0.00	**1.00**	-	-
ChangeValue	**0.79**	0.73	**0.73**	**0.71**
SdDevResValue	0.37	**0.74**	0.42	0.42
NumberAffected	0.46	0.46	**0.88**	0.50
SdDevChangeValue	0.48	**0.57**	0.13	**0.25**
ObservedResult	0.00	0.00	0.00	0.00
mean	0.41	**0.61**	**0.47**	0.44
	Endpoint			
EndPointDescription	**0.29**	0.28	0.56	**0.63**
BaselineUnit	**0.73**	0.68	0.55	**0.57**
mean	**0.51**	0.48	0.60	**0.64**
micro average	0.56	**0.62**	0.60	**0.64**

	Gold Standard	Predicted
	Population	
Country Precondition	usa, australia chinese patients with type 2 diabetes mellitus receiving stable insulin therapy alone or in combination with metformin	patients with inadequate glycemic control on insulin (glycated hemoglobin (hba1c) at 7.5% and at 11%)
	Publication	
Author	shankar rr I bao y I han p I hu j I ma j I peng y I wu f I xu l I engel ss I jia w	engel ss I shankar rr I bao y I han p I hu j I ma j I peng y I wu f I xu l I jia w
Journal	j diabetes investig	j diabetes invest ig .
PMID	27740719	27740719
PublicationYear	2017	2017
Tile	sitagliptin added to stable insulin therapy with or without metformin in chinese patients with type 2 diabetes .	sitagliptin added to stable insulin therapy with or without metformin in chinese patients with type 2 diabetes .
	Clinical Trial	
healthCondition Design Duration NumberPatients ObjectiveDescription	type 2 diabetes mellitus randomized 24 weeks 467 we evaluated the tolerability and efficacy of the addition of sitagliptin in chinese patients with type 2 diabetes mellitus receiving stable insulin therapy alone or in combination with metformin.	 randomized 24 weeks 467 we evaluated the tolerability and efficacy of the addition of sitagliptin in chinese patients with type 2 diabetes mellitus receiving stable insulin therapy alone or in combination with metformin .
	Endpoints	
BaselineUnit: EndPointDescription	% hba1c	 hba1c
EndPointDescription	hba1c of < 7.0%	hba1c of < 7.0%
BaselineUnit EndPointDescription	mg / dl 2 - h post - meal glucose	mg / dl fasting plasma glucose, 2 - h post - meal glucose
BaselineUnit EndPointDescription	mg / dl fasting plasma glucose	mg/dl
BaselineUnit EndPointDescription	mg / dl hypoglycemia (symptomatic or asymptomatic)	 hypoglycemia
BaselineUnit: EndPointDescription	 bodyweight	 bodyweight
	Medications	
DoseUnit DoseValue Drug	mg 100 sitagliptin	mg sitagliptin
DoseUnit DoseValue Drug	100 placebo	 placebo
	Outcomes	
ChangeValue	0.7	
ChangeValue	0.3	0.3
PercentageAffected TimePoint	16 week 24	 week 24
PercentageAffected	8	
ChangeValue	26.5	26.5
ChangeValue	14.4	14.4
ChangeValue	10.7	10.7
NumberAffected PercentageAffected	64 27.4	 27.4
NumberAffected PercentageAffected ObservedResult	51 21.9 neither group had a significant change from baseline in bodyweight.	 21.9, 8
	Differences between groups	
PvalueDiff	p < 0.001	p < 0.001
PvalueDiff	p = 0.013	p = 0.013
PvalueDiff	p < 0.001	p < 0.001

Table 10: Predicted and gold standard structures for the abstract of the clinical trial described in Shankar, R Ravi et al. "Sitagliptin added to stable insulin therapy with or without metformin in Chinese patients with type 2 diabetes." Journal of diabetes investigation vol. 8,3 (2017): 321-329. doi:10.1111/jdi.12585; within one template type, horizontal lines separate different instances of the same template type

	Gold Standard	Predicted
	Population	
AvgAge	14 . 8	
Country	ohio	
MaxAge	17	
MinAge	10	
Precondition	youth treated with diet / exercise alone or with metformin and having a hemoglo bin a1c (hba1c) level of 6 . 5 - 11 %; youth	
	Publication	
Author	battelino t; arslanian s; jacobsen lv; chatterjee dj; klein dj; hale pm	lopez x; neufeld n; battelino t; blumer j; arslanian s; bone m; randell t; jacobsen lv; chatterjee dj; hazan l; ferry r; christensen m; tsalikian e; toltzis p; de schepper j; wadwa rp; wintergerst k; klein dj; barrett t; hale pm
Journal	diabetes technol ther .	diabetes technol ther .
PMID	25036533	25036533
PublicationYear	2014	2014
	Clinical Trial	
analysesHealthCondition	type 2 diabetes	type 2 diabetes
CTDesign	randomized | double - blind	randomized
CTduration	5 weeks	5 weeks
	Arms	
NumberPatientsArm	14	14
NumberPatientsArm	7	7
	Endpoints	
EndPointDescription	severe hypoglycemia	hba1c
EndPointDescription	gastrointestinal aes	
EndPointDescription BaselineUnit	hba1c %	 %
EndPointDescription BaselineUnit	body weight kg	 kg
	Medications	
DoseUnit Drug	mg liraglutide	mg placebo | liraglutide
	Outcomes	
ObservedResult hasSdDevBL	no serious adverse events 35 . 6	
ObservedResult TimePoint	were most common at lower liraglutide doses during dose escalation . 5 weeks	
ResultMeasuredValue BaselineValue	12 113 . 2	
ResultMeasuredValue	1 . 7	1 . 7 | 0 . 3
ChangeValue TimePoint	0 . 86 5 weeks	0 . 86
ChangeValue	0 . 04	0 . 50 | 0 . 04
ChangeValue BaselineValue	0 . 50 8 . 1	
ChangeValue	0 . 54	0 . 54
	Differences between groups	
hasPvalueDiff	p = 0 . 9703	p = 0 . 9703
hasPvalueDiff	p = 0 . 0007	p = 0 . 0007

Table 11: Predicted and gold standard structures for the abstract of the clinical trial described in Klein, David J et al. "Liraglutide's safety, tolerability, pharmacokinetics, and pharmacodynamics "pediatric type 2 diabetes: a randomized, double-blind, placebo-controlled trial" Diabetes technology & therapeutics vol. 16,19 (2014): 679-687. doi:10.1089/dia.2013.0366; within one template type, horizontal lines separate different instances of the same template type; "|" separates SFCs

Pre-trained Biomedical Language Models for Clinical NLP in Spanish

Casimiro Pio Carrino, Joan Llop, Marc Pàmies, Asier Gutiérrez-Fandiño
Jordi Armengol-Estapé, Joaquín Silveira-Ocampo, Alfonso Valencia
Aitor Gonzalez-Agirre and **Marta Villegas**
Barcelona Supercomputing Center
marta.villegas@bsc.es

Abstract

This work presents the first large-scale biomedical Spanish language models trained from scratch, using large biomedical corpora consisting of a total of 1.1B tokens and an EHR corpus of 95M tokens. We compared them against general-domain and other domain-specific models for Spanish on three clinical NER tasks. As main results, our models are superior across the NER tasks, rendering them more convenient for clinical NLP applications. Furthermore, our findings indicate that when enough data is available, pre-training from scratch is better than continual pre-training when tested on clinical tasks, raising an exciting research question about which approach is optimal. Our models and fine-tuning scripts are publicly available at HuggingFace and GitHub.

1 Introduction and Background

The success of Transformer-based models in the general domain (Devlin et al., 2019) soon encouraged the development of language models for domain-specific scenarios (Chalkidis et al., 2020; Gutiérrez-Fandiño et al., 2021; Tai et al., 2020; Araci, 2019; Lee and Hsiang, 2019). Specifically, in the biomedical domain, there has been a proliferation of models (Peng et al., 2019; Beltagy et al., 2019; Alsentzer et al., 2019; Gu et al., 2021) since the first BioBERT (Lee et al., 2019) model was published. Unfortunately, there is still a significant lack of biomedical and clinical models in languages other than English, despite the increasing efforts of the NLP community (Névéol et al., 2014; Schneider et al., 2020). Consequently, general-domain pre-trained language models supporting Spanish, such as mBERT (Devlin et al., 2019) and BETO (Cañete et al., 2020), have been often used as a proxy to build domain-specific systems in the absence of genuine alternatives. For instance, Sun and Yang (2019) used mBERT and BioBERT on the PharmaCoNER (Gonzalez-Agirre et al., 2019) dataset, using a fine-tuning strategy aimed to maximize the results.

Very recently, new pre-trained clinical language models for Spanish have been published (López-García et al., 2021) by further pre-training the mBERT, BETO and XLM-RoBERTa (Conneau et al., 2020) models with a corpus of Spanish clinical cases with about 64M tokens. In our work, we go one step further to address the language gap for Spanish and train two Transformer-based language models from scratch. We employed biomedical and clinical corpora (including clinical texts) gathered by ourselves. We evaluated our models with three different Named Entity Recognition (NER) tasks, since NER constitutes a core task in many clinical NLP scenarios. They obtained significant gains over the general-domain models, and matched or outperformed the domain-specific models in all tasks.

2 Corpora

We built two corpora of very different sizes and nature: an Electronic Health Record (EHR) corpus and a biomedical one. The **EHR corpus** contains 95M tokens from more than 514k clinical documents (including discharge reports, clinical course notes and X-ray reports). The **biomedical corpus** includes Spanish data from a variety of sources for a total of 1.1B tokens across 2,5M documents, namely:

- **Medical crawler**:[1] Crawler of more than 3,000 URLs belonging to Spanish biomedical and health domains (Carrino et al., 2021).
- **Clinical cases misc.**: A miscellany of medical content, essentially clinical cases. Note that a clinical case report is different from a scientific publication where medical practitioners share patient cases and it is also different from a clinical note or document.

[1] https://zenodo.org/record/4561970

Proceedings of the BioNLP 2022 workshop, Dublin, Ireland, pages 193–199
May 26, 2022.

- **Scielo**:[2] Scientific publications written in Spanish crawled from the Spanish SciELO server in 2017.
- **BARR2 Background**:[3] Biomedical Abbreviation Recognition and Resolution (BARR2) containing Spanish clinical case study sections from a variety of clinical disciplines.
- **Wikipedia (Life Sciences)**: Wikipedia articles crawled on 04/01/2021 with the Wikipedia API python library[4] starting from the "Ciencias_de_la_vida" category up to a maximum of 5 subcategories. Multiple links to the same article are discarded to avoid repeated content.
- **Patents**: Google Patent in Medical Domain for Spain (Spanish). The accepted codes (Medical Domain) for JSON files of patents are: "A61B", "A61C","A61F", "A61H", "A61K", "A61L","A61M", and "A61P".
- **EMEA**:[5] Spanish-side documents extracted from parallel corpora made out of PDF documents from the European Medicines Agency.
- **Mespen (MedlinePlus)**:[6] Spanish-side articles extracted from a collection of Spanish-English parallel corpus consisting of biomedical scientific literature. The collection of parallel resources are aggregated from the MedlinePlus source.
- **PubMed**: Open-access Spanish abstracts from the PubMed repository crawled in 2017.

For each biomedical resource, we applied a cleaning pipeline with customized operations designed to read data in different formats, split it into sentences, detect the language, remove noisy and ill-formed sentences, deduplicate and eventually output the data with their original document boundaries. Finally, to remove repetitive content, we concatenated the entire corpus and deduplicated it again, obtaining about 1.1B words. These preprocessing steps were applied to all data except the EHR corpus, which was left in its original form. Table 1 shows detailed statistics of each component of the corpus.

Source	**No. tokens**
Medical crawler	903,558,136
Clinical cases misc.	102,855,267
EHRs documents*	95,267,204
Scielo	60,007,289
BARR2 Background	24,516,442
Wikipedia (Life Sciences)	13,890,501
Patents	13,463,387
EMEA	5,377,448
Mespen (MedlinePlus)	4,166,077
PubMed	1,858,966

Table 1: List of individual sources in the training corpora. The number of tokens refers to *white-spaced* tokens on cleaned untokenized text. Documents from the EHR corpus are marked with an asterisk.

3 Models Pre-training

The models presented in this work were pre-trained from scratch employing a RoBERTa (Liu et al., 2019) base model with 12 self-attention layers. Following the original training, we only used Masked Language Modeling (MLM) as the pre-training objective with Subword Masking (SWM), as in (Liu et al., 2019).

We tokenized the training corpus with the Byte-Level BPE algorithm (Radford et al., 2019), employed in the original RoBERTa, and learned a vocabulary of 50,262 tokens.

We run the training for 48 hours on 16 NVIDIA V100 GPUs of 16GB VRAM, using Adam optimizer (Kingma and Ba, 2015) with a peak learning rate of 0.0005, 10,000 warm-up steps and an effective batch size of 2,048 sentences.[7] Other hyper-parameters were left in their default values as in the original RoBERTa training configuration. Training was performed at the document level, preserving document boundaries.[8] We performed a train-validation split based on the number of documents, choosing a total of 2,000 documents for the validation set, corresponding to less than 1% of the entire corpus' documents. We then select the model with the lowest perplexity on the validation set as the best model.

We used the corpora described in the previous section to produce two RoBERTa models: a biomedical language model training

[2] https://zenodo.org/record/2541681

[3] https://temu.bsc.es/BARR2/downloads/background_set.raw_text.tar.bz2

[4] https://github.com/martin-majlis/Wikipedia-API/

[5] http://opus.nlpl.eu/download.php?f=EMEA/v3/moses/en-es.txt.zip

[6] https://zenodo.org/record/3562536

[7] Through gradient accumulation as implemented in Fairseq (Ott et al., 2019)

[8] We believe document-level training may be crucial to promote the modelling of long-range dependencies and push the model towards the comprehension of entire documents.

only with the so-called biomedical resources (`bsc-bio-es`),[9] and a BIO-EHR language model that uses both the biomedical and EHR corpus (`bsc-bio-ehr-es`).[10] We trained the latter model, the biomedical-EHR, to assess if adding a relatively small EHR data to a large-scale corpora has a positive impact on real-world clinical NLP tasks.

4 NER Fine-tuning

We tested and evaluated our models by fine-tuning the NER task, a key component of information extraction tasks in the clinical domain. Indeed, we used it as a testbed to evaluate the effectiveness of our pre-trained models. Following the usual fine-tuning method, employed both for general-domain models (Devlin et al., 2019; Liu et al., 2019) and domain-specific ones (Lee et al., 2019), we added a standard linear layer as a token classification head, and the BIO tagging schema (Sang and Buchholz, 2000) to solve the NER tasks. During fine-tuning, both the pre-trained model and the classification layer's parameters are learned with stochastic gradient descent. We used an Adam (Kingma and Ba, 2015) optimizer and searched for an optimal learning rate out of [8e-6, 1e-5, 2e-5, 3e-5, 5e-5] with linear decay and no warm-up steps. We used a batch size of 32 sequences with a maximum length of 512 tokens and a gradient accumulation of 2 steps, resulting in a total batch size of 64. We trained each configuration using three random seeds. The rest of hyper-parameters were left to the default values of HuggingFace's codebase (Wolf et al., 2019). The complete list of hyper-parameter values is displayed in Appendix B.

We applied this fine-tuning strategy to three different NER datasets. The first two use annotations on curated medical data (clinical cases extracted from medical literature), whereas the last one uses medical records from the ICTUSnet project.[11] More details are given below.

PharmaCoNER (Gonzalez-Agirre et al., 2019) is a track on chemical and drug mention recognition from Spanish medical texts. The authors compiled a manually classified collection of clinical case report sections derived from open access Spanish medical publications, named the Spanish Clinical Case Corpus (SPACCC). The corpus contained a total of 1,000 clinical cases and 396,988 words and was manually annotated, with a total of 7,624 entity mentions, corresponding to four different mention types.[12]

CANTEMIST (Miranda-Escalada et al., 2020) is a shared task focused on named entity recognition of tumor morphology, in Spanish. The CANTEMIST corpus[13] is a collection of 1,301 oncological case reports written in Spanish, with a total of 63,016 sentences and 1,093,501 tokens.

The **ICTUSnet** dataset consists of 1,006 hospital discharge reports of patients admitted for stroke from 18 different Spanish hospitals. It contains more than 79,000 annotations for 51 different variables. The dataset is part of the ICTUSnet project, whose main objective was the development of an information extraction system to support domain experts when identifying relevant information in discharge reports.

Finally, we remark that our main goal is a head-to-head comparison between different models to assess the best model pre-training choice. We were not aiming at maximizing results on the NER tasks and therefore we decided not to use sophisticated classification layers that might improve the performances, such as Conditional Random Field (Lafferty et al., 2001) layers on top of Bidirectional Long Short-Term Memory Recurrent Networks (Panchendrarajan and Amaresan, 2018). We argue that a simpler token classification layer better evaluates the quality of model representations than a task-specific layer. Unlike Sun and Yang (2019), where authors fine-tuned for 200 epochs (obtaining the best results using 100 epochs), we limit the fine-tuning to 20 epochs, and we do not merge the train and development sets in order to improve the results. We consider that fine-tuning for 200 epochs goes against the pre-training/fine-tuning philosophy that states that fine-tuning should be a relatively inexpensive step (Devlin et al., 2019), and also that fine-tuning for less epochs evaluates better the pre-training strategy.

5 Evaluation and Results

Each fine-tuning was executed on 4 NVIDIA V100 GPUs of 16GB VRAM. It took around 0.5, 1 and

[9] https://huggingface.co/PlanTL-GOB-ES/bsc-bio-es

[10] https://huggingface.co/PlanTL-GOB-ES/bsc-bio-ehr-es

[11] https://ictusnet-sudoe.eu/es/

[12] For a detailed description, see https://temu.bsc.es/pharmaconer/

[13] CANTEMIST corpus: https://doi.org/10.5281/zenodo.3878178

		Average of all configurations			Best on development set		
Task	Model	F1	Precision	Recall	F1	LR	Epoch
PharmaCoNER	bsc-bio-es*	$0.8907_{0.01}$	$0.8736_{0.01}$	$0.9085_{0.00}$	$0.8939_{0.01}$	5e-5	15
	bsc-bio-ehr-es*	$\mathbf{0.8913_{0.01}}$	$\mathbf{0.8758_{0.01}}$	$\mathbf{0.9073_{0.01}}$	$\mathbf{0.8954_{0.01}}$	3e-5	10
	XLM-R-Galén	$0.8754_{0.01}$	$0.8591_{0.02}$	$0.8924_{0.01}$	$0.8883_{0.00}$	5e-5	15
	BETO-Galén	$0.8537_{0.02}$	$0.8399_{0.02}$	$0.8680_{0.01}$	$0.8741_{0.01}$	5e-5	20
	mBERT-Galén	$0.8594_{0.01}$	$0.8469_{0.02}$	$0.8722_{0.01}$	$0.8760_{0.00}$	5e-5	15
	mBERT	$0.8671_{0.01}$	$0.8540_{0.02}$	$0.8809_{0.01}$	$0.8729_{0.00}$	3e-5	13
	BioBERT	$0.8545_{0.01}$	$0.8502_{0.01}$	$0.8590_{0.01}$	$0.8533_{0.01}$	2e-5	12
	roberta-base-bne	$0.8474_{0.02}$	$0.8430_{0.02}$	$0.8520_{0.02}$	$0.8680_{0.01}$	5e-5	13
CANTEMIST	bsc-bio-es*	$0.8220_{0.01}$	$0.7939_{0.02}$	$0.8522_{0.01}$	$0.8351_{0.00}$	5e-5	20
	bsc-bio-ehr-es*	$\mathbf{0.8340_{0.01}}$	$\mathbf{0.8141_{0.01}}$	$\mathbf{0.8551_{0.01}}$	$\mathbf{0.8449_{0.00}}$	5e-5	20
	XLM-R-Galén	$0.8078_{0.02}$	$0.7755_{0.02}$	$0.8431_{0.01}$	$0.8259_{0.00}$	5e-5	15
	BETO-Galén	$0.8153_{0.01}$	$0.7933_{0.02}$	$0.8387_{0.01}$	$0.8332_{0.01}$	5e-5	20
	mBERT-Galén	$0.8168_{0.01}$	$0.7919_{0.02}$	$0.8435_{0.01}$	$0.8304_{0.00}$	5e-5	20
	mBERT	$0.8116_{0.01}$	$0.7923_{0.02}$	$0.8319_{0.01}$	$0.8257_{0.00}$	5e-5	16
	BioBERT	$0.8070_{0.01}$	$0.7848_{0.02}$	$0.8306_{0.01}$	$0.8219_{0.00}$	5e-5	20
	roberta-base-bne	$0.7875_{0.03}$	$0.7733_{0.03}$	$0.8023_{0.02}$	$0.8161_{0.00}$	5e-5	15
ICTUSnet	bsc-bio-es*	$0.8727_{0.01}$	$0.8359_{0.01}$	$\mathbf{0.9131_{0.01}}$	$0.8804_{0.00}$	5e-5	19
	bsc-bio-ehr-es*	$\mathbf{0.8756_{0.00}}$	$0.8418_{0.01}$	$0.9122_{0.00}$	$0.8781_{0.00}$	2e-5	18
	XLM-R-Galén	$0.8716_{0.01}$	$0.8375_{0.01}$	$0.9087_{0.01}$	$\mathbf{0.8809_{0.00}}$	5e-5	17
	BETO-Galén	$0.8498_{0.01}$	$0.8226_{0.01}$	$0.8791_{0.01}$	$0.8551_{0.00}$	5e-5	20
	mBERT-Galén	$0.8509_{0.01}$	$0.8219_{0.01}$	$0.8820_{0.01}$	$0.8576_{0.00}$	5e-5	17
	mBERT	$0.8631_{0.01}$	$0.8301_{0.01}$	$0.8989_{0.01}$	$0.8646_{0.01}$	2e-5	20
	BioBERT	$0.8521_{0.00}$	$0.8132_{0.01}$	$0.8950_{0.01}$	$0.8503_{0.00}$	2e-5	16
	roberta-base-bne	$0.8677_{0.01}$	$\mathbf{0.8456_{0.01}}$	$0.8910_{0.01}$	$0.8769_{0.00}$	5e-5	18

Table 2: Fine-tuning results of the models for each dataset on the test set. In bold, the best results for metric and task. Subscript numbers indicate the standard deviations. Our models are marked with an asterisk.

2 hours to complete the PharmaCoNER, CANTEMIST and ICTUSnet tasks, respectively.

We then report the overall best scores on the test set, obtained by using the best model's hyper-parameters on the development set for each dataset (the standard deviation is computed using all the seeds for that configuration). Finally, we also report the models' average scores and standard deviations by computing statistics across all the seeds and the learning rates used for each dataset. The average scores are helpful to indicate which model is more robust to the variation of hyper-parameters, which are the learning rate and initial seed in our case. A higher average score and a smaller standard deviation minimizes the risk of obtaining poor results when performing an extensive hyper-parameter search is not feasible.

We compared our models with a general-domain Spanish model (`roberta-base-bne`) (Gutiérrez-Fandiño et al., 2022), a general-domain multilingual model that supports Spanish (`mBERT`), a domain-specific English model (`BioBERT`), and three domain-specific models based on continual pre-training: `mBERT-Galén` (based on mBERT), `BETO-Galén` (based on BETO, a general-domain Spanish model), and `XLM-R-Galén` (based on XLM-RoBERTa, a general-domain multilingual model supporting Spanish). The results are shown in Table 2. The last two columns report the learning rate and epoch in which the best configuration on the development set was achieved

Our models obtained significantly better performances than the general-domain models, namely `mBERT` and `roberta-base-bne`. Compared to the domain-specific Galén models, our average models' scores surpassed them on the clinical NER tasks. However, when looking at the best on development score on the ICTUSnet dataset, the `XLM-R-Galén` model outperformed our models. We also highlight that our models exhibit smaller standard deviations. This makes them more robust and a good option if not enough computational resources are available to experiment with the different hyper-parameter configurations.

6 Conclusions and Future Work

This work presents the first large-scale biomedical Spanish language models trained from scratch, using a large biomedical corpora for a total of 1.1B tokens and an EHR corpus of 95M tokens. We fine-tuned the models on three clinical NER tasks and compared them with both general-domain and other available Spanish clinical models. The results show the superiority of our models across the NER tasks, making them competitive candidates for clinical NLP applications. Our findings demonstrate the benefits of pre-training from scratch, as seen in Gu et al. (2021). Regarding continual pre-training, the benefits are not clear, especially when continual pre-training is performed with small data, as in the case of the `mBERT-Galén`, `XLM-R-Galén`, and `BETO-Galén` (note that `mBERT` outperforms `mBERT-Galén` in two out of three tasks). Our work raises exciting research questions about which pre-training approach is optimal to tackle challenging clinical NLP tasks. We will devote future efforts to address the previous question in detail by providing new models based on continual pre-training and extending our evaluation setting to a diverse range of tasks.

7 Data Availability

Our work encourages the development of Clinical and Biomedical NLP applications for Spanish. Therefore, we released our pre-trained models and the best on dev set fine-tuned models under the Apache License 2.0 in the HuggingFace models hub under the following links:

Pre-trained models

- bsc-bio-es
- bsc-bio-ehr-es

Fine-tuned models

- bsc-bio-ehr-es-pharmaconer
- bsc-bio-ehr-es-cantemist

Moreover, to guarantee reproducibility, we share the script used to fine-tuned our pre-trained model in the official GitHub repository: `https://github.com/PlanTL-GOB-ES/lm-biomedical-clinical-es`.

Acknowledgements

This work was funded by the Spanish State Secretariat for Digitalization and Artificial Intelligence (SEDIA) within the framework of the Plan-TL.[14]

[14]`https://plantl.mineco.gob.es/Paginas/index.aspx`

References

Emily Alsentzer, John Murphy, William Boag, Wei-Hung Weng, Di Jindi, Tristan Naumann, and Matthew McDermott. 2019. Publicly available clinical BERT embeddings. In *Proceedings of the 2nd Clinical Natural Language Processing Workshop*, pages 72–78, Minneapolis, Minnesota, USA. Association for Computational Linguistics.

Dogu Araci. 2019. Finbert: Financial sentiment analysis with pre-trained language models. *CoRR*, abs/1908.10063.

Iz Beltagy, Kyle Lo, and Arman Cohan. 2019. SciBERT: A pretrained language model for scientific text. In *Proceedings of the 2019 Conference on Empirical Methods in Natural Language Processing and the 9th International Joint Conference on Natural Language Processing (EMNLP-IJCNLP)*, pages 3615–3620, Hong Kong, China. Association for Computational Linguistics.

Casimiro Pio Carrino, Jordi Armengol-Estapé, Ona de Gibert Bonet, Asier Gutiérrez-Fandiño, Aitor Gonzalez-Agirre, Martin Krallinger, and Marta Villegas. 2021. Spanish biomedical crawled corpus: A large, diverse dataset for spanish biomedical language models.

José Cañete, Gabriel Chaperon, Rodrigo Fuentes, Jou-Hui Ho, Hojin Kang, and Jorge Pérez. 2020. Spanish pre-trained bert model and evaluation data. In *PML4DC at ICLR 2020*.

Ilias Chalkidis, Manos Fergadiotis, Prodromos Malakasiotis, Nikolaos Aletras, and Ion Androutsopoulos. 2020. LEGAL-BERT: The muppets straight out of law school. In *Findings of the Association for Computational Linguistics: EMNLP 2020*, pages 2898–2904, Online. Association for Computational Linguistics.

Alexis Conneau, Kartikay Khandelwal, Naman Goyal, Vishrav Chaudhary, Guillaume Wenzek, Francisco Guzmán, Edouard Grave, Myle Ott, Luke Zettlemoyer, and Veselin Stoyanov. 2020. Unsupervised cross-lingual representation learning at scale. In *Proceedings of the 58th Annual Meeting of the Association for Computational Linguistics*, pages 8440–8451, Online. Association for Computational Linguistics.

Jacob Devlin, Ming-Wei Chang, Kenton Lee, and Kristina Toutanova. 2019. BERT: Pre-training of deep bidirectional transformers for language understanding. In *Proceedings of the 2019 Conference of the North American Chapter of the Association for Computational Linguistics: Human Language Technologies, Volume 1 (Long and Short Papers)*, pages 4171–4186, Minneapolis, Minnesota. Association for Computational Linguistics.

Aitor Gonzalez-Agirre, Montserrat Marimon, Ander Intxaurrondo, Obdulia Rabal, Marta Villegas, and Martin Krallinger. 2019. PharmaCoNER: Pharmacological substances, compounds and proteins named entity recognition track. In *Proceedings of The 5th Workshop on BioNLP Open Shared Tasks*, pages 1–10, Hong Kong, China. Association for Computational Linguistics.

Yu Gu, Robert Tinn, Hao Cheng, Michael Lucas, Naoto Usuyama, Xiaodong Liu, Tristan Naumann, Jianfeng Gao, and Hoifung Poon. 2021. Domain-specific language model pretraining for biomedical natural language processing. *ACM Trans. Comput. Healthcare*, 3(1).

Asier Gutiérrez-Fandiño, Jordi Armengol-Estapé, Aitor Gonzalez-Agirre, and Marta Villegas. 2021. Spanish legalese language model and corpora.

Asier Gutiérrez-Fandiño, Jordi Armengol-Estapé, Marc Pàmies, Joan Llop-Palao, Joaquin Silveira-Ocampo, Casimiro Pio Carrino, Carme Armentano-Oller, Carlos Rodriguez-Penagos, Aitor Gonzalez-Agirre, and Marta Villegas. 2022. Maria: Spanish language models. *Procesamiento del Lenguaje Natural*, 68(0):39–60.

Diederik P. Kingma and Jimmy Ba. 2015. Adam: A method for stochastic optimization. In *3rd International Conference on Learning Representations, ICLR 2015, San Diego, CA, USA, May 7-9, 2015, Conference Track Proceedings*.

John D. Lafferty, Andrew McCallum, and Fernando C. N. Pereira. 2001. Conditional random fields: Probabilistic models for segmenting and labeling sequence data. In *Proceedings of the Eighteenth International Conference on Machine Learning*, ICML '01, page 282–289, San Francisco, CA, USA. Morgan Kaufmann Publishers Inc.

Jieh-Sheng Lee and Jieh Hsiang. 2019. Patentbert: Patent classification with fine-tuning a pre-trained BERT model. *CoRR*, abs/1906.02124.

Jinhyuk Lee, Wonjin Yoon, Sungdong Kim, Donghyeon Kim, Sunkyu Kim, Chan Ho So, and Jaewoo Kang. 2019. BioBERT: a pre-trained biomedical language representation model for biomedical text mining. *Bioinformatics*, 36(4):1234–1240.

Yinhan Liu, Myle Ott, Naman Goyal, Jingfei Du, Mandar Joshi, Danqi Chen, Omer Levy, Mike Lewis, Luke Zettlemoyer, and Veselin Stoyanov. 2019. Roberta: A robustly optimized BERT pretraining approach. *CoRR*, abs/1907.11692.

Guillermo López-García, José M Jerez, Nuria Ribelles, Emilio Alba, and Francisco J Veredas. 2021. Transformers for clinical coding in spanish. *IEEE Access*, 9:72387–72397.

A Miranda-Escalada, E Farré, and M Krallinger. 2020. Named entity recognition, concept normalization and clinical coding: Overview of the cantemist track for cancer text mining in spanish, corpus, guidelines, methods and results. In *Proceedings of the Iberian Languages Evaluation Forum (IberLEF 2020), CEUR Workshop Proceedings*.

Aurélie Névéol, H. Dalianis, G. Savova, and Pierre Zweigenbaum. 2014. Clinical natural language processing in languages other than english: opportunities and challenges. *Journal of Biomedical Semantics*, 9.

Myle Ott, Sergey Edunov, Alexei Baevski, Angela Fan, Sam Gross, Nathan Ng, David Grangier, and Michael Auli. 2019. fairseq: A fast, extensible toolkit for sequence modeling. In *Proceedings of the 2019 Conference of the North American Chapter of the Association for Computational Linguistics (Demonstrations)*, pages 48–53, Minneapolis, Minnesota. Association for Computational Linguistics.

Rrubaa Panchendrarajan and Aravindh Amaresan. 2018. Bidirectional LSTM-CRF for named entity recognition. In *Proceedings of the 32nd Pacific Asia Conference on Language, Information and Computation*, Hong Kong. Association for Computational Linguistics.

Yifan Peng, Shankai Yan, and Zhiyong Lu. 2019. Transfer learning in biomedical natural language processing: An evaluation of BERT and ELMo on ten benchmarking datasets. In *Proceedings of the 18th BioNLP Workshop and Shared Task*, pages 58–65, Florence, Italy. Association for Computational Linguistics.

Alec Radford, Jeffrey Wu, Rewon Child, David Luan, Dario Amodei, and Ilya Sutskever. 2019. Language models are unsupervised multitask learners.

Erik F. Tjong Kim Sang and Sabine Buchholz. 2000. Introduction to the conll-2000 shared task: Chunking. *CoRR*, cs.CL/0009008.

Elisa Terumi Rubel Schneider, João Vitor Andrioli de Souza, Julien Knafou, Lucas Emanuel Silva e Oliveira, Jenny Copara, Yohan Bonescki Gumiel, Lucas Ferro Antunes de Oliveira, Emerson Cabrera Paraiso, Douglas Teodoro, and Cláudia Maria Cabral Moro Barra. 2020. BioBERTpt - a Portuguese neural language model for clinical named entity recognition. In *Proceedings of the 3rd Clinical Natural Language Processing Workshop*, pages 65–72, Online. Association for Computational Linguistics.

Cong Sun and Zhihao Yang. 2019. Transfer learning in biomedical named entity recognition: An evaluation of BERT in the PharmaCoNER task. In *Proceedings of The 5th Workshop on BioNLP Open Shared Tasks*, pages 100–104, Hong Kong, China. Association for Computational Linguistics.

Wen Tai, H. T. Kung, Xin Dong, Marcus Comiter, and Chang-Fu Kuo. 2020. exBERT: Extending pre-trained models with domain-specific vocabulary under constrained training resources. In *Findings of the Association for Computational Linguistics: EMNLP 2020*, pages 1433–1439, Online. Association for Computational Linguistics.

Thomas Wolf, Lysandre Debut, Victor Sanh, Julien Chaumond, Clement Delangue, Anthony Moi, Pierric Cistac, Tim Rault, Rémi Louf, Morgan Funtowicz, and Jamie Brew. 2019. Huggingface's transformers: State-of-the-art natural language processing. *CoRR*, abs/1910.03771.

A Pre-training Hyper-parameters

The hyper-parameters used for pre-training our models are shown in Table 3.

Hyper-parameter	Value
Number of Layers	12
Hidden size	768
FNN inner hidden size	3072
Attention Heads	12
Attention Head size	64
Dropout	0.1
Attention Dropout	0.1
Warmup Steps	10k
Peak Learning Rate	5e-4
Batch Size	2,048
Weight Decay	0.01
Max Steps	125k
Learning Rate Decay	Linear
Adam ϵ	1e-6
Adam β_1	0.9
Adam β_2	0.98
Gradient Clipping	0.0

Table 3: Hyper-parameters used for pre-training.

B Fine-tuning Hyper-parameters

The hyper-parameters used for fine-tuning the models on various tasks are shown in Table 4.

Hyper-parameter	Value
Learning Rates	{0.8, 1, 2, 3, 5}e-5
Learning Rate Decay	Linear
Warmup Steps	0
Batch Size	64
Weight Decay	0.0
Max. Training Epochs	20
Adam ϵ	1e-8
Adam β_1	0.9
Adam β_2	0.999
Gradient Clipping	1.0

Table 4: Hyper-parameters used for fine-tuning.

Few-Shot Cross-lingual Transfer for Coarse-grained De-identification of Code-Mixed Clinical Texts

Saadullah Amin[1,2], Noon Pokaratsiri Goldstein[1,2], Morgan K. Wixted[1,2]
Alejandro García-Rudolph[3], Catalina Martínez-Costa[4], Günter Neumann[1,2]
[1]Department of LST, Saarland University, Germany [2]DFKI, SIC, Germany
[3]Department of Research and Innovation, Institut Guttmann, UAB, Spain
[4]Department Informática y Sistemas, Universidad de Murcia, IMIB-Arrixaca, Spain
{saadullah.amin,noon.pokaratsiri,morgan.wixted,guenter.neumann}@dfki.de
agarciar@guttmann.com cmartinezcosta@um.es

Abstract

Despite the advances in digital healthcare systems offering curated structured knowledge, much of the critical information still lies in large volumes of unlabeled and unstructured clinical texts. These texts, which often contain protected health information (PHI), are exposed to information extraction tools for downstream applications, risking patient identification. Existing works in de-identification rely on using large-scale annotated corpora in English, which often are not suitable in real-world multilingual settings. Pre-trained language models (LM) have shown great potential for *cross-lingual transfer* in low-resource settings. In this work, we empirically show the *few-shot cross-lingual transfer* property of LMs for named entity recognition (NER) and apply it to solve a low-resource and real-world challenge of code-mixed (Spanish-Catalan) clinical notes de-identification in the stroke domain. We annotate a gold evaluation dataset to assess few-shot setting performance where we only use a few hundred labeled examples for training. Our model improves the zero-shot F1-score from 73.7% to 91.2% on the gold evaluation set when adapting Multilingual BERT (mBERT) (Devlin et al., 2019) from the MEDDOCAN (Marimon et al., 2019) corpus with our few-shot cross-lingual target corpus. When generalized to an out-of-sample test set, the best model achieves a human-evaluation F1-score of 97.2%.

1 Introduction

With growing interest and innovations in data-driven digital technologies, privacy has become an important legal topic for the technology to be regulations-compliant. In Europe, the General Data Protection Regulation (GDPR) (Regulation, 2016) requires data owners to have a legal basis for processing personally identifiable information (PII), which also includes the explicit consent of the subjects. In cases where explicit consent is not possible, anonymization is often seen as a resorted-to solution. Clinical texts contain rich information about patients, including their gender, age, profession, residence, family, and history, that is useful for record keeping and billing purposes (Johnson et al., 2016; Shickel et al., 2017).

```
Pathological history - Ischemic stroke in the
territory of the left MCA of undetermined
etiology that occurred on [**** DATE ****] -
DM type 2.

COPD, presenting a single hospital admission
on [**** DATE ****] at [**** LOCATION ****].

Pleurisy at [**** AGE ****].

Sialolithiasis 30 years ago.

Work accident [**** DATE ****] with bilateral
calcaneal fracture.

IQx the left one.

Intervened anal fistula [**** DATE ****]
Interval incidents.

General condition: BEG Skin and mucous
membranes: Well hydrated.
```

Figure 1: The process of text de-identification involves the removal of a predefined set of direct identifiers in text (Elliot et al., 2016). For clinical notes, this set is often the PHI categories (or types) defined by the Health Insurance Portability and Accountability Act (HIPAA) (Gunn et al., 2004). The example here shows a de-identified excerpt of a patient note from the Spanish-Catalan stroke dataset used in this study (translated into English here for readability).

In this work, we focus on the task of removing PHI from clinical texts, also called de-identification (Fig. 1). We address a real-world challenge where the target texts are code-mixed (Spanish-Catalan) and domain-constrained (stroke). To avoid high annotation costs, we consider a more realistic setting where we annotate a gold evaluation corpus and a few hundred examples for training. Our approach is motivated by strong performance of pre-trained LMs in few-shot cross-lingual transfer for NER with high sample efficiency (see Fig. 2) in comparison to supervised or unsupervised approaches.

Proceedings of the BioNLP 2022 workshop, Dublin, Ireland, pages 200–211
May 26, 2022.

Our contributions are summarized as follows:

- We empirically show the few-shot cross-lingual transfer property of multi-lingual pre-trained LM, mBERT, for NER.
- We apply this property to solve a low-resource problem of code-mixed and domain-specific clinical note de-identification.
- We annotate a few-shot training corpus and a gold evaluation set, minimizing annotation costs while achieving a significant performance boost without needing a large-scale labeled training corpus.

2 Related Work

GDPR-compliant anonymization requires complete and irreversible removal of any information that may lead to a subject's data being identified (directly or indirectly) from a dataset (Elliot et al., 2016). However, de-identification is limited to removing specific predefined direct identifiers; further replacement of such direct identifiers with pseudonyms is referred to as pseudonymization (Alfalahi et al., 2012). Generally, de-identification can be seen as a subset of anonymization despite interchangeable usage of the terms in the literature (Chevrier et al., 2019). We focus on solving the problem of de-identification in the clinical domain as a sequence labeling task, specifically named entity recognition (NER) (Lample et al., 2016).

2.1 Clinical De-identification

2014 i2b2/UTHealth (Stubbs and Uzuner, 2015), and the 2016 CEGS N-GRID (Stubbs et al., 2017) shared tasks explore the challenges of clinical de-identification on diabetic patient records and psychiatric intake records respectively. Earlier works include machine learning and rule-based approaches (Meystre et al., 2010; Yogarajan et al., 2018), with Liu et al. (2017) and Dernoncourt et al. (2017) being the first to propose neural architectures. Friedrich et al. (2019) propose an adversarial approach to learn privacy-preserving text representations; Yang et al. (2019) use domain-specific embeddings trained on unlabeled corpora. While most works have mainly focused on English, some efforts have been made for Swedish (Velupillai et al., 2009; Alfalahi et al., 2012) and Spanish (with a synthetic dataset at the MEDDOCAN shared task (Marimon et al., 2019)).

As outlined in Lison et al. (2021), a significant challenge in clinical text de-identification is the lack of labeled data. Hartman et al. (2020) show that a small number of manually labeled PHI examples can significantly improve performance. Prior works in few-shot NER consider the problem where a model is trained on one or more source domains and tested on unseen domains with a few labeled examples per class, some of which with entity tags different from those in the source domains (Yang and Katiyar, 2020). Models are trained with prototypical methods, noisy supervised pre-training, or self-labeling (Huang et al., 2020).

We consider a setting where the target and source domains share the *same* entity (PHI) tags, but with a few labeled examples in the target domain (or language). A similar setup has been employed in few-shot question answering (Ram et al., 2021).

3 Problem Statement

We approach the de-identification problem as an NER task. Given an input sentence $\mathbf{x}$ with N words: $\mathbf{x} = [x_i]_{i=1:N}$, we feed it to an encoder $f_\phi : \mathbb{R}^N \to \mathbb{R}^{N\times d}$ to obtain a sequence of hidden representations $\mathbf{h} = [h_i]_{i=1:N}$

$$\mathbf{h} = f_\phi(\mathbf{x}).$$

We feed $\mathbf{h}$ into the NER classifier which is a linear classification layer with the `softmax` activation function to predict the PHI label of $\mathbf{x}$:

$$p_\theta(\mathbf{Y}|\mathbf{x}) = \texttt{softmax}(\mathbf{W}^T\mathbf{h} + \mathbf{b}).$$

$p_\theta(\mathbf{Y}|\mathbf{x}) \in \mathbb{R}^{N\times|\mathcal{P}|}$ is the probability distribution of PHI labels for sentence $\mathbf{x}$ and $\mathcal{P}$ is the PHI label set. $\theta = [\phi, \mathbf{W} \in \mathbb{R}^{d\times|\mathcal{P}|}, \mathbf{b} \in \mathbb{R}^{|\mathcal{P}|}]$ denote the set of learnable parameters and d being the hidden dimension. The model is trained to minimize the per-sample negative log-likelihood:

$$\mathcal{L} = -\frac{1}{N}\sum_{i=1}^{N} \log p_\theta(Y_i = y_i|x_i). \qquad (1)$$

For pre-trained LMs, this setting corresponds to NER fine-tuning (Wu and Dredze, 2019). When we jointly fine-tune on more than one NER dataset, we refer to it as multi-task learning.

Definition 1 (Few-Shot NER). *Given an entity label set $\mathcal{P}$, we define the task of few-shot NER as having access to $K \leq M$ labeled sentences containing each element $p \in \mathcal{P}$ at least once, where K is a small number (e.g., in $[50, 500]$) and M is orders of magnitude larger (e.g., ≥ 1000).*

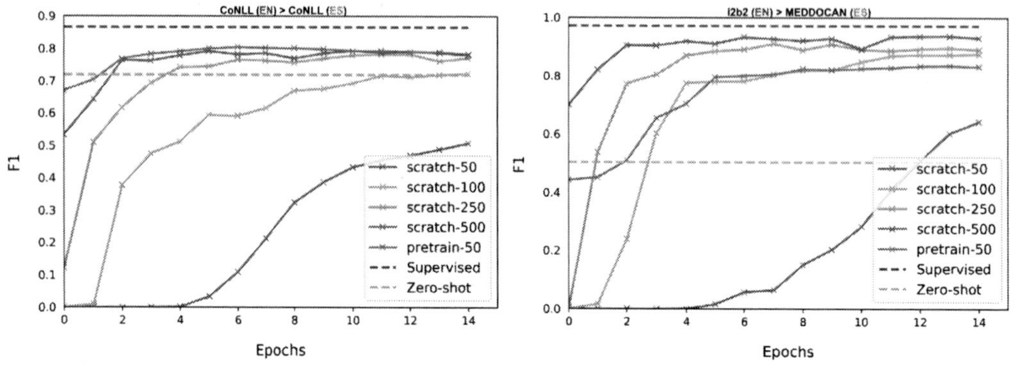

Figure 2: **Few-shot cross-lingual NER transfer of mBERT**. We compare different transfer learning scenarios from English (EN) to Spanish (ES) for two pairs of datasets as preliminary study to investigate the effectiveness of the few-shot cross-lingual transfer in mBERT: CoNLL-2003 to CoNLL-2002 (*left*) and i2b2-2014 to MEDDOCAN (*right*). We use supervised fine-tuning on the full training set of the target language (ES) as the *upper bound* and the zero-shot score of the model (pre-trained only on the source language (EN) training set) as the *lower bound* for target language (ES) performance. We then consider 50, 100, 250, and 500 examples from the target language as few-shot training corpora and train the models for 15 epochs. The models without any pre-training on the source corpus (scratch) eventually outperform the *lower bound* as the number of examples grow; with sufficient epochs, the model with only 50 target-language samples reaches more than 10% gain in de-identification task (*right: scratch-50*). However, we find the cross-lingual transfer-learning strategy to be most sample efficient when pre-trained on *source* language—50-shot performance (*left & right: pretrain-50*) comparable to 500-shot (*left & right: scratch-500*). We apply this strategy to address the real-world challenges of Spanish-Catalan de-identification.

Definition 2 (Few-Shot NER Transfer). *Given an NER dataset in a source domain (or language), we define the task of few-shot cross-domain (or cross-lingual) NER transfer as adapting a model trained on the source domain (or language) to a target domain (or language) with access to a few-shot corpus (Definition 1).*

This setting is different from prior studies in NER transfer including few-shot (Huang et al., 2020), unsupervised (Keung et al., 2020), and semi-supervised NER (Amin and Neumann, 2021).

CoNLL (EN) → CoNLL (ES)	F1
Pires et al. (2019)	73.59
Wu and Dredze (2019)	74.96
Keung et al. (2019)	74.30
Wang et al. (2020)	75.77
Wu et al. (2020b)	77.30
Wu et al. (2020a)	76.75
Chen et al. (2021)	**79.00**
few-50 (or pretrain-50)	78.30

Table 1: Cross-lingual transfer results on CoNLL. *few-50* represents our fine-tuning of EN trained mBERT with 50 random labeled samples from ES.

3.1 Few-Shot Cross-Lingual Transfer

mBERT (Devlin et al., 2019) has been shown to achieve strong performance for zero-shot cross-lingual transfer tasks, including NER (Wu and Dredze, 2019; Pires et al., 2019). Adversarial learning has been applied with limited gains (Keung et al., 2019) in unsupervised approaches to improve zero-shot NER transfer, whereas feature alignments have shown better results (Wang et al., 2020). Meta-learning with minimal resources (Wu et al., 2020b) and word-to-word translation (Wu et al., 2020a) have shown further performance gains. The current state-of-the-art approach (Chen et al., 2021) combines token-level adversarial learning with self-labeled data selection and knowledge distillation.

To investigate the few-shot transferability of mBERT, we consider two pairs of datasets with English as the source language and Spanish as the target language: the CoNLL-2003/CoNLL-2002 (Tjong Kim Sang, 2002a,b) in the *general domain* and i2b2/MEDDOCAN (Stubbs and Uzuner, 2015; Marimon et al., 2019) in the *clinical domain*. We report results of our preliminary study in Fig. 2. We observed that with as few as 50 random labeled training samples from the target language, we obtain substantial gains for both datasets, with near state-of-the-art on CoNLL (Table 1). We refer to this as *few-shot cross-lingual transfer* property of mBERT for NER. Our study highlights that the property holds for different domains (general

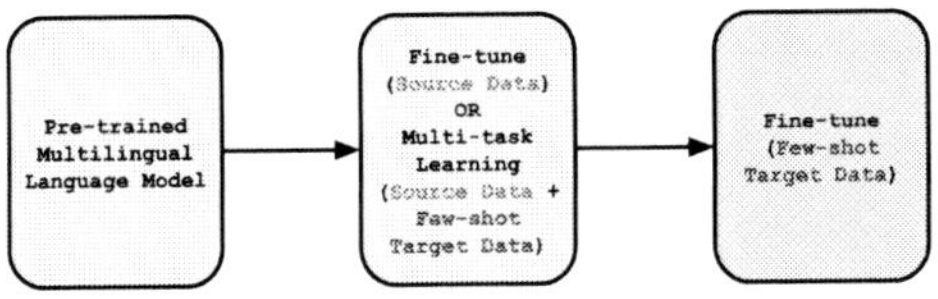

Figure 3: Our *few-shot cross-lingual transfer* strategy for clinical text de-identification.

and clinical), where the latter focuses on the de-identification task. We leave a large-scale study on more datasets with different languages and domains as future work.

Compared to supervised (unsupervised) methods, which use complete labeled (unlabeled) target data, our few-shot approach is *sample-efficient* and alleviates the need of complex pipelines (Wu et al., 2020b,a; Chen et al., 2021) and large-scale annotations. Furthermore, Keung et al. (2020) highlights the spurious effects of using source data as a development set and recommends using target data as a development set for model selection in NER transfer. Our findings and those in Hartman et al. (2020) motivate us to *(a)* propose an optimal *few-shot cross-lingual transfer* strategy (outlined in Fig. 3), *(b)* annotate a target development set, and *(c)* construct an annotated *few-shot target corpus* for effective cross-lingual transfer learning.

4 Data and Annotation

Our dataset consists of stroke patient records collected at Instiut Guttmann.[1] Table 2 summarizes the raw data statistics and Table A.1 in Appendix A describes the topics present in the texts. We set aside 100 randomly sampled notes for out-of-sample generalizability evaluation and consider the remaining notes for our development and few-shot corpora sampling; 396k sentences are tokenized in the process.

Following the protocol in Gao et al. (2021) for constructing manually annotated distantly supervised relation extraction test sets, we train mBERT on the MEDDOCAN corpus, using coarse-grained PHI categories {DATE, AGE, LOCATION, NAME, CONTACT, PROFESSION, ID} with the BIO scheme (Farber et al., 2008), for evaluation and few-shot training data selection. We use the trained model to make predictions on the dataset and observe that the model predicts PHI on only

Patients	Notes	ES	CA	Other
1,500	327,775	42.8%	53.0%	4.2%

Table 2: Raw statistics of the Spanish (ES)-Catalan (CA) data from stroke domain.

50k out of the 396k sentences. A dataset of 5000 sentences ($<2\%$ of raw sentences) is constructed from a mix of randomly sampled 2500 sentences from this 50k and 2500 from the remaining sentences. We split the dataset into two partitions of 2500 sentences for independent annotation by two annotators. The annotation is performed one sentence at a time by applying one of the 7 coarse-grained PHI labels to each token using the T2NER-ANNOTATE toolkit (Amin and Neumann, 2021). Each annotator's confidence level between 1-5 is recorded for the token-level labels for each sample. To record the inter-annotator agreement, we use token-level Cohen's kappa (Cohen, 1960) statistic reaching a value of 0.898. In total, the two annotators agreed on 3924 sentences, resulting in our final evaluation set. To save annotation costs for developing a *few-shot target corpus*, we resolved the disagreements to obtain a 384-sentence few-shot corpus for training (see Appendix B for annotation details).

Our source dataset (the MEDDOCAN corpus) consists of 1000 synthetically generated clinical case studies in Spanish (Marimon et al., 2019). The corpus was selected manually by a practicing physician and augmented with PHI from discharge summaries and clinical records. In contrast, our target corpus focuses on the stroke domain and contains PHI from real-world records. Since the target data is code-mixed between Spanish and Catalan, with the majority (53%) being Catalan, the transfer from Spanish source data (MEDDOCAN) is cross-lingual.[2]

5 Experiments and Results

We conduct our experiments with the T2NER framework (Amin and Neumann, 2021).[3] For the baseline, we consider zero-shot performance on the evaluation set of the mBERT encoder fine-tuned on the MEDDOCAN training set consisting of 16,299 samples. We then fine-tune it on the few-shot tar-

[1] `https://www.guttmann.com/ca/institut-universitari-guttmann-uab`

[2] Although similar, Spanish and Catalan are distinct languages. The domain of MEDDOCAN is missing an explicit mention in Marimon et al. (2019).

[3] `https://github.com/suamin/T2NER`

Transfer Strategy	Precision	Recall	F1
FINE-TUNE (M)	80.1	68.2	73.7
FINE-TUNE (M) → FINE-TUNE (F)	83.5	94.2	88.6
MULTI-TASK (M + F)	86.0	93.3	89.5
MULTI-TASK (M + F) → FINE-TUNE (F)	**87.7**	**95.0**	**91.2**

Table 3: Results on the development set from the code-mixed stroke data. M denotes the MEDDOCAN (Marimon et al., 2019) training set (source) normalized to 7 PHIs (see Appendix C) at sentence level and F denotes our few-shot target corpus. Here multi-task learning refers to the joint fine-tuning on two datasets.

get corpus as outlined in Fig. 3. Following the multi-task learning (Lin et al., 2018) approach in T2NER, we jointly fine-tune mBERT on the MEDDOCAN and few-shot target corpora. Since the few-shot corpus is much smaller, the multi-task learning helps the model transfer. It further acts as a regularization approach by sharing parameters between the datasets. To improve performance on the target data, we further fine-tune with the few-shot target corpus after the first step of fine-tuning to have improved target performance; for the model to be an expert in target (Cao et al., 2020). All the models are trained for 3 epochs with a learning rate of 3e-5 and linear warm-up of 10%. For few-shot fine-tuning only, the model is trained for 25 epochs.

Table 3 shows our results. Fine-tuning the baseline mBERT model with the few-shot target corpus improves the F1-score from 73.7% to 88.6%, a substantial gain of 14.9%, highlighting the effectiveness of *few-shot cross-lingual transfer* with mBERT. The significant increase in recall (26% points) compared to precision (3.4% points) suggests an increase in the model's capacity to recognize domain-specific entities. Multi-task fine-tuning improves the F1-score to 89.5%; further fine-tuning on the few-shot target corpus boosts the best model's performance to 91.2%. Fig. 4 shows per-PHI-label scores on the development set, along with their frequency. The model performs almost perfectly on DATE and AGE, since most DATE and AGE labeled segments are similar between Spanish and Catalan as they are simple numbers (for DATE) and numbers followed by the word (for AGE; '*edad*' in both Spanish and Catalan). There are some differences in time expressions, e.g., day of the week, as the words are distinctly dissimilar. However, structurally there is only a slight difference. Further, the model struggles with the ID class due to low sample size (5 instances in the few-shot corpus), and it is generally challenging to disambiguate between an alphanumeric string

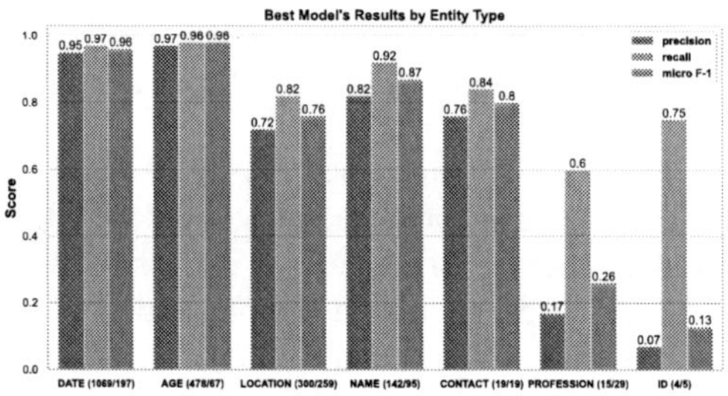

Figure 4: NER metrics on the evaluation set for each entity type with their frequency (dev/few-train).

and a PHI ID, as also noted by the ID class' high recall. Our error analysis reveals high false positives for the PROFESSION label in Catalan, e.g.: '*Coloma de Gramenet*' (a LOCATION) and '*Dialogant*' (being able to communicate) are both labeled as PROFESSION.

To test the model's generalizability, we tokenize the 100 out-of-sample notes into sentences and make predictions with our best model. The resulting annotated sentences are reconstructed into patient notes, which are manually evaluated by two reviewers (one external and one annotator) for occurrences of true and false positives and negatives. The model achieves precision, recall, and F1-scores of 95.1%, 99.3%, and 97.1% respectively on the out-of-sample notes, highlighting the effectiveness of our approach.

6 Conclusion

We address the task of clinical notes de-identification in a low-resource scenario. By investigating the *few-shot cross-lingual transfer* property of mBERT, we propose a strategy that significantly boosts zero-shot performance while keeping the number of required annotated samples low. Our results highlight the effectiveness of the proposed strategy for the task with a potential for future applications in other low-resource scenarios.

Acknowledgments

The authors would like to thank the anonymous reviewers and Josef van Genabith for their helpful feedback. The work was partially funded by the European Union (EU) Horizon 2020 research and innovation programme through the project Precise4Q (777107), the German Federal Ministry of Education and Research (BMBF) through the project CoRA4NLP (01IW20010), the grant RTI2018-099039-J-I00 funded by MCIN/AEI/10.13039/501100011033/ and "FEDER", by the EU. The authors also acknowledge the cluster compute resources provided by the DFKI.

References

Alyaa Alfalahi, Sara Brissman, and Hercules Dalianis. 2012. Pseudonymisation of personal names and other phis in an annotated clinical swedish corpus. In *Third Workshop on Building and Evaluating Resources for Biomedical Text Mining (BioTxtM 2012) Held in Conjunction with LREC*, pages 49–54. Citeseer.

Saadullah Amin and Guenter Neumann. 2021. T2NER: Transformers based transfer learning framework for named entity recognition. In *Proceedings of the 16th Conference of the European Chapter of the Association for Computational Linguistics: System Demonstrations*, pages 212–220, Online. Association for Computational Linguistics.

Yu Cao, Meng Fang, Baosheng Yu, and Joey Tianyi Zhou. 2020. Unsupervised domain adaptation on reading comprehension. In *The Thirty-Fourth AAAI Conference on Artificial Intelligence, AAAI 2020, The Thirty-Second Innovative Applications of Artificial Intelligence Conference, IAAI 2020, The Tenth AAAI Symposium on Educational Advances in Artificial Intelligence, EAAI 2020, New York, NY, USA, February 7-12, 2020*, pages 7480–7487. AAAI Press.

Weile Chen, Huiqiang Jiang, Qianhui Wu, Börje Karlsson, and Yi Guan. 2021. AdvPicker: Effectively Leveraging Unlabeled Data via Adversarial Discriminator for Cross-Lingual NER. In *Proceedings of the 59th Annual Meeting of the Association for Computational Linguistics and the 11th International Joint Conference on Natural Language Processing (Volume 1: Long Papers)*, pages 743–753, Online. Association for Computational Linguistics.

Raphaël Chevrier, Vasiliki Foufi, Christophe Gaudet-Blavignac, Arnaud Robert, Christian Lovis, et al. 2019. Use and understanding of anonymization and de-identification in the biomedical literature: scoping review. *Journal of medical Internet research*, 21(5):e13484.

Jacob Cohen. 1960. A coefficient of agreement for nominal scales. *Educational and psychological measurement*, 20(1):37–46.

Franck Dernoncourt, Ji Young Lee, Ozlem Uzuner, and Peter Szolovits. 2017. De-identification of patient notes with recurrent neural networks. *Journal of the American Medical Informatics Association*, 24(3):596–606.

Jacob Devlin, Ming-Wei Chang, Kenton Lee, and Kristina Toutanova. 2019. BERT: Pre-training of deep bidirectional transformers for language understanding. In *Proceedings of the 2019 Conference of the North American Chapter of the Association for Computational Linguistics: Human Language Technologies, Volume 1 (Long and Short Papers)*, pages 4171–4186, Minneapolis, Minnesota. Association for Computational Linguistics.

Mark Elliot, Elaine Mackey, Kieron O'Hara, and Caroline Tudor. 2016. The anonymisation decision-making framework. ukan.

Benjamin Farber, Dayne Freitag, Nizar Habash, and Owen Rambow. 2008. Improving NER in Arabic using a morphological tagger. In *Proceedings of the Sixth International Conference on Language Resources and Evaluation (LREC'08)*, Marrakech, Morocco. European Language Resources Association (ELRA).

Max Friedrich, Arne Köhn, Gregor Wiedemann, and Chris Biemann. 2019. Adversarial learning of privacy-preserving text representations for de-identification of medical records. In *Proceedings of the 57th Annual Meeting of the Association for Computational Linguistics*, pages 5829–5839, Florence, Italy. Association for Computational Linguistics.

Tianyu Gao, Xu Han, Yuzhuo Bai, Keyue Qiu, Zhiyu Xie, Yankai Lin, Zhiyuan Liu, Peng Li, Maosong Sun, and Jie Zhou. 2021. Manual evaluation matters: Reviewing test protocols of distantly supervised relation extraction. In *Findings of the Association for Computational Linguistics: ACL-IJCNLP 2021*, pages 1306–1318, Online. Association for Computational Linguistics.

Patrick P Gunn, Allen M Fremont, Melissa Bottrell, Lisa R Shugarman, Jolene Galegher, and Tora Bikson. 2004. The health insurance portability and accountability act privacy rule: a practical guide for researchers. *Medical care*, pages 321–327.

Tzvika Hartman, Michael D Howell, Jeff Dean, Shlomo Hoory, Ronit Slyper, Itay Laish, Oren Gilon, Danny Vainstein, Greg Corrado, Katherine Chou, et al. 2020. Customization scenarios for de-identification of clinical notes. *BMC medical informatics and decision making*, 20(1):1–9.

Jiaxin Huang, Chunyuan Li, Krishan Subudhi, Damien Jose, Shobana Balakrishnan, Weizhu Chen, Baolin

Peng, Jianfeng Gao, and Jiawei Han. 2020. Few-shot named entity recognition: A comprehensive study. *ArXiv preprint*, abs/2012.14978.

Alistair EW Johnson, Tom J Pollard, Lu Shen, Li-wei H Lehman, Mengling Feng, Mohammad Ghassemi, Benjamin Moody, Peter Szolovits, Leo Anthony Celi, and Roger G Mark. 2016. Mimic-iii, a freely accessible critical care database. *Scientific data*, 3(1):1–9.

Phillip Keung, Yichao Lu, and Vikas Bhardwaj. 2019. Adversarial learning with contextual embeddings for zero-resource cross-lingual classification and NER. In *Proceedings of the 2019 Conference on Empirical Methods in Natural Language Processing and the 9th International Joint Conference on Natural Language Processing (EMNLP-IJCNLP)*, pages 1355–1360, Hong Kong, China. Association for Computational Linguistics.

Phillip Keung, Yichao Lu, Julian Salazar, and Vikas Bhardwaj. 2020. Don't use English dev: On the zero-shot cross-lingual evaluation of contextual embeddings. In *Proceedings of the 2020 Conference on Empirical Methods in Natural Language Processing (EMNLP)*, pages 549–554, Online. Association for Computational Linguistics.

Guillaume Lample, Miguel Ballesteros, Sandeep Subramanian, Kazuya Kawakami, and Chris Dyer. 2016. Neural architectures for named entity recognition. In *Proceedings of the 2016 Conference of the North American Chapter of the Association for Computational Linguistics: Human Language Technologies*, pages 260–270, San Diego, California. Association for Computational Linguistics.

Ying Lin, Shengqi Yang, Veselin Stoyanov, and Heng Ji. 2018. A multi-lingual multi-task architecture for low-resource sequence labeling. In *Proceedings of the 56th Annual Meeting of the Association for Computational Linguistics (Volume 1: Long Papers)*, pages 799–809, Melbourne, Australia. Association for Computational Linguistics.

Pierre Lison, Ildikó Pilán, David Sanchez, Montserrat Batet, and Lilja Øvrelid. 2021. Anonymisation models for text data: State of the art, challenges and future directions. In *Proceedings of the 59th Annual Meeting of the Association for Computational Linguistics and the 11th International Joint Conference on Natural Language Processing (Volume 1: Long Papers)*, pages 4188–4203, Online. Association for Computational Linguistics.

Zengjian Liu, Buzhou Tang, Xiaolong Wang, and Qingcai Chen. 2017. De-identification of clinical notes via recurrent neural network and conditional random field. *Journal of biomedical informatics*, 75:S34–S42.

Montserrat Marimon, Aitor Gonzalez-Agirre, Ander Intxaurrondo, Heidy Rodriguez, Jose Lopez Martin, Marta Villegas, and Martin Krallinger. 2019. Automatic de-identification of medical texts in spanish: the meddocan track, corpus, guidelines, methods and evaluation of results. In *IberLEF@ SEPLN*, pages 618–638.

Stephane M Meystre, F Jeffrey Friedlin, Brett R South, Shuying Shen, and Matthew H Samore. 2010. Automatic de-identification of textual documents in the electronic health record: a review of recent research. *BMC medical research methodology*, 10(1):1–16.

Telmo Pires, Eva Schlinger, and Dan Garrette. 2019. How multilingual is multilingual BERT? In *Proceedings of the 57th Annual Meeting of the Association for Computational Linguistics*, pages 4996–5001, Florence, Italy. Association for Computational Linguistics.

Ori Ram, Yuval Kirstain, Jonathan Berant, Amir Globerson, and Omer Levy. 2021. Few-shot question answering by pretraining span selection. In *Proceedings of the 59th Annual Meeting of the Association for Computational Linguistics and the 11th International Joint Conference on Natural Language Processing (Volume 1: Long Papers)*, pages 3066–3079, Online. Association for Computational Linguistics.

Protection Regulation. 2016. Regulation (eu) 2016/679 of the european parliament and of the council. *Regulation (eu)*, 679:2016.

Benjamin Shickel, Patrick James Tighe, Azra Bihorac, and Parisa Rashidi. 2017. Deep ehr: a survey of recent advances in deep learning techniques for electronic health record (ehr) analysis. *IEEE journal of biomedical and health informatics*, 22(5):1589–1604.

Amber Stubbs, Michele Filannino, and Özlem Uzuner. 2017. De-identification of psychiatric intake records: Overview of 2016 cegs n-grid shared tasks track 1. *Journal of biomedical informatics*, 75:S4–S18.

Amber Stubbs and Özlem Uzuner. 2015. Annotating longitudinal clinical narratives for de-identification: The 2014 i2b2/uthealth corpus. *Journal of biomedical informatics*, 58:S20–S29.

Erik F. Tjong Kim Sang. 2002a. Introduction to the CoNLL-2002 shared task: Language-independent named entity recognition. In *COLING-02: The 6th Conference on Natural Language Learning 2002 (CoNLL-2002)*.

Erik F. Tjong Kim Sang. 2002b. Introduction to the CoNLL-2002 shared task: Language-independent named entity recognition. In *COLING-02: The 6th Conference on Natural Language Learning 2002 (CoNLL-2002)*.

Sumithra Velupillai, Hercules Dalianis, Martin Hassel, and Gunnar H Nilsson. 2009. Developing a standard for de-identifying electronic patient records written in swedish: precision, recall and f-measure in a manual and computerized annotation trial. *International journal of medical informatics*, 78(12):e19–e26.

Zirui Wang, Jiateng Xie, Ruochen Xu, Yiming Yang, Graham Neubig, and Jaime G. Carbonell. 2020. Cross-lingual alignment vs joint training: A comparative study and A simple unified framework. In *8th International Conference on Learning Representations, ICLR 2020, Addis Ababa, Ethiopia, April 26-30, 2020*. OpenReview.net.

Qianhui Wu, Zijia Lin, Börje F. Karlsson, Biqing Huang, and Jianguang Lou. 2020a. Unitrans : Unifying model transfer and data transfer for cross-lingual named entity recognition with unlabeled data. In *Proceedings of the Twenty-Ninth International Joint Conference on Artificial Intelligence, IJCAI 2020*, pages 3926–3932. ijcai.org.

Qianhui Wu, Zijia Lin, Guoxin Wang, Hui Chen, Börje F Karlsson, Biqing Huang, and Chin-Yew Lin. 2020b. Enhanced meta-learning for cross-lingual named entity recognition with minimal resources. In *Proceedings of the AAAI Conference on Artificial Intelligence*, volume 34, pages 9274–9281.

Shijie Wu and Mark Dredze. 2019. Beto, bentz, becas: The surprising cross-lingual effectiveness of BERT. In *Proceedings of the 2019 Conference on Empirical Methods in Natural Language Processing and the 9th International Joint Conference on Natural Language Processing (EMNLP-IJCNLP)*, pages 833–844, Hong Kong, China. Association for Computational Linguistics.

Xi Yang, Tianchen Lyu, Qian Li, Chih-Yin Lee, Jiang Bian, William R Hogan, and Yonghui Wu. 2019. A study of deep learning methods for de-identification of clinical notes in cross-institute settings. *BMC Medical Informatics and Decision Making*, 19(5):1–9.

Yi Yang and Arzoo Katiyar. 2020. Simple and effective few-shot named entity recognition with structured nearest neighbor learning. In *Proceedings of the 2020 Conference on Empirical Methods in Natural Language Processing (EMNLP)*, pages 6365–6375, Online. Association for Computational Linguistics.

Vithya Yogarajan, Michael Mayo, and Bernhard Pfahringer. 2018. A survey of automatic de-identification of longitudinal clinical narratives. *ArXiv preprint*, abs/1810.06765.

A Additional Dataset Details

In addition to the 7 coarse-grained PHI entities in (Stubbs and Uzuner, 2015), our dataset contains cross-sentence recurring entities about topics that may be of interest in the clinical domain. These topics are grouped by their potential clinical application areas and are summarized in Table A.1.

Topic Areas	**Subcategories**
Diagnostics & Treatments	Ischemic vs Hemorrhagic Affected areas and vessels Comorbidities Medication history Associated lifestyle factors Treatments and interventions
Symptoms & Monitoring	Vital signs Lab results and cultures Pain and comfort Bladder and bowel controls
Long-term Care & Discharge Planning	Mobility Cognitive ability Nutrition Psychosocial factors

Table A.1: Topics and subcategories in the corpus.

The label frequency distribution, as noted in Fig. 4, is consistent with general characteristics of medical notes, which usually highlight notable events such as symptom onsets, procedures, admissions, transfers, and discharges, in addition to the date of each documentation. As a result, they tend to contain a higher frequency for the DATE PHI. In addition, the lower occurrence of the NAME PHI compared to the AGE and LOCATION entities is consistent with how healthcare providers usually communicate patient's information.

Healthcare providers are trained to refer to patients simply by their age, gender, and the appropriate diagnosis to avoid inadvertently sharing HIPAA-sensitive information, e.g., *"a 60-year-old male with ischemic stroke admitted on [DATE] from [LOCATION] (...)"*. The patient's name may be used at the beginning of a medical note; however, subsequent anaphoric references are often accomplished via pronouns, omitting the NAME entity in the process. In addition, as it is applicable to Spanish medical records, nominative pronouns anaphorically referencing a patient may be omitted as they are grammatical in Spanish.

We avoid releasing our dataset due to presence of real PHI information. We will consider replacing the real PHI with synthetic ones, similar to MEDDOCAN, for a GDPR-compliant release.

Description	Observation
Sentences annotated by annotator (A)	4400
Sentences annotated by annotator (B)	4343
Sentences annotated and revised by (A, B)	**4314**
Agreements	3924
Disagreements	390
Token-level Cohen's Kappa score	0.898
DEVELOPMENT CORPUS	**3924**
w entity mentions	1493
w/o entity mentions	2431
FEW-SHOT TARGET CORPUS	**384**
w entity mentions	369
w/o entity mentions	15

Table A.2: Dataset annotation and final statistics.

B Annotation

B.1 Annotator Profile

Two graduate research assistants completed the annotation of the dataset. Both annotators have at a minimum CEFR[4] B2-C1 Spanish (Castilian) proficiency. One annotator also has clinical experience in the cardiovascular and cerebrovascular specialty, including knowledge of Spanish medical terminology. Neither annotator has formal training in Catalan; both have prior experience working with text data in the language in this domain.

B.2 Annotation Guidelines

The annotation process followed criteria for each entity as described in Stubbs and Uzuner (2015). The 7 entities: AGE, CONTACT, DATE, ID, LOCATION, NAME, and PROFESSION represent a larger granularity of the 18 HIPAA-defined PHI (Stubbs and Uzuner, 2015). We examined the training sets of i2b2 (Stubbs and Uzuner, 2015) and MEDDOCAN (Marimon et al., 2019) and adapted the i2b2 annotation guidelines to create our own annotation guidelines. This step was necessary since we only focused on coarse-grained PHI types compared to fine-grained types considered in these two datasets. The adjusted guidelines utilized in this annotation process are summarized in Table A.5.

B.3 Annotation Procedure

Both annotators reviewed and revised their work without discussion or knowledge of the other annotator's work. In cross-revision, the reviewing annotator only made corrections when *labeling*

[4]https://www.coe.int/en/

Annotator	Sample sentence	Explanation
A	"Actualmente reside en XXXX-Xxxx Xxxxxxxx, Treballadora Social." [Currently resides in XXXX-Xxxxx Xxxxxxxx, social worker.]	The underlined words are grouped as a single word token. From the context it's clear that 'XXXX' belongs to 'LOCATION' and 'Xxxxx Xxxxxxxx' are 'NAME' entities.
B	"Lmarxa. [sic]" [March or walks]	Annotator does not have enough context to understand this token to annotate.

Table A.3: Examples of sentences skipped by annotators and rationales.

inconsistencies were due to a lack of medical terminology comprehension. During revision, no changes to the original annotator's confidence level rating were made.

Confidence Level: The criteria for the confidence levels are annotator dependent as summarized in Table A.6 with examples. PHI has been manually modified from the original data to preserve privacy while maintaining exemplary characteristics for each label entity type.

Skipped Sentences: Each annotator followed an independent set of criteria to exclude sentences from annotation, as demonstrated by examples in Table A.3.

B.4 Annotator Disagreements and Resolution

An attempt was made to review the 390 sentences where our annotators disagreed to find a resolution. Main sources of disagreement are due to: (a) annotation criteria discrepancy, (b) ambiguity between related entities, and (c) annotation errors. After further revision to correct identified errors and clarify ambiguous annotation criteria, agreement was reached for 384 sentences while 6 sentences were left un-annotated due to insufficient context. Confidence levels from both annotators were left unchanged. These sentences constitute our *few-shot target corpus* in the pipeline explained in Fig. 3.

Inclusion Criteria Discrepancy: Most disagreements are related to discrepancy in the inclusion of surrounding words such as determiners, punctuation marks, and descriptive phrases. This is prevalent particularly in the LOCATION and PROFESSION entities. One annotator considered denoted sentences with these characteristics a lower confidence level of 4 compared to sentences without determiners or punctuation marks surrounding LOCATION tokens. The resolution step changed the annotations to be more consistent with the annotation guidelines described in Table A.5.

Ambiguity Between Related Entities: Another source of disagreements in the LOCATION PHI stems from abbreviation usage and confusion with the NAME PHI. In instances where the syntax is ambiguous, annotators may not be able to infer correctly that certain unknown abbreviations are place names. Since it is common that places are named after people's names and vice versa, lack of contextual information creates unresolvable ambiguity regarding the NAME and LOCATION entities. DATE and AGE also demonstrate a similar disagreement behavior. In particular, numerical and text expressions involving 'years' may express age or time depending on context.

Annotation Errors: A few disagreements are due to mislabeling or erroneous omissions. There are fewer than 5 instances in the 390 disagreements. Notable errors are associated with mislabeling proper names that resemble valid named entities. For instance, some assessment tools are named after people or place names e.g. Barcelona Test and Boston (Naming) Test.

C MEDDOCAN Normalization

The original MEDDOCAN dataset (Marimon et al., 2019) provides document level de-identification annotations, following 2014 i2b2/UTHealth (Stubbs and Uzuner, 2015), of 1000 clinical notes which are divided into 500, 250 and 250 for training, validation and testing respectively. It contains 29 fine-grained entity types classified into 8 coarse-grained PHI types (Table A.4). Compared to i2b2 (2014), MEDDOCAN has an additional OTHER category which we normalize to O in the BIO scheme, resulting in 7 coarse-grained PHI types considered in this work. We tokenize the 500 training notes resulting in 16,299 sentences. The conversion script is available in T2NER. [5]

[5] https://github.com/suamin/T2NER/blob/master/utils/convert_i2b2style_xml_to_conll.py

PHI	Fine-grained Types
AGE	EDAD_SUJETO_ASISTENCIA
CONTACT	NUMERO_TELEFONO, NUMERO_FAX, CORREO_ELECTRONICO, URL_WEB
DATE	FECHAS
ID	ID_ASEGURAMIENTO, ID_CONTACTO_ASISTENCIAL, NUMERO_BENEF_PLAN_SALUD, IDENTIF_VEHICULOS_NRSERIE_PLACAS, IDENTIF_DISPOSITIVOS_NRSERIE, IDENTIF_BIOMETRICOS, ID_SUJETO_ASISTENCIA, ID_TITULACION_PERSONAL_SANITARIO, ID_EMPLEO_PERSONAL_SANITARIO, OTRO_NUMERO_IDENTIF
LOCATION	HOSPITAL, INSTITUCION, CALLE, TERRITORIO, PAIS, CENTRO_SALUD
NAME	NOMBRE_SUJETO_ASISTENCIA, NOMBRE_PERSONAL_SANITARIO
PROFESSION	PROFESION
OTHER	SEXO_SUJETO_ASISTENCIA, FAMILIARES_SUJETO_ASISTENCIA, OTROS_SUJETO_ASISTENCIA, DIREC_PROT_INTERNET

Table A.4: MEDDOCAN PHI (coarse-grained) and fine-grained types (Marimon et al., 2019).

PHI	Criteria
AGE	Annotate only the numerical part of the expression; include both numerical and word expressions of age (e.g. 36 or thirty-six). Include the words 'years', 'months', and 'days' when they express age. Include expressions that describe an age group e.g. 'adolescent', 'recently born', 'new born'. Include punctuation associated with age, including separate tokens, e.g. in his/her 30's.
CONTACT	All forms of contact information, e.g. pager, phone numbers, e-mail address. Physical or mailing address is annotated as 'Location' Include punctuation and symbols that occur with contact information, e.g. include all tokens in '(123) 456-789'.
DATE	Include days of the week and months. Include punctuation in all formats. Include the word 'year' and 'month' that are part of a date-time expression, e.g. 'the year 2000'. Include prepositions that are part of a date-time expression, e.g. include the word 'of' in '5th of May'.
ID	Include all identification numbers such as Medical Record Number (MRN), Social Security Number (SSN), Document ID, device lot number, etc. Include any alpha-numeric expressions appearing in the beginning of the document or next to a name that's not formatted as a date. When separated by punctuation, annotate all parts of the expression including punctuation, e.g. include all tokens in '12-34-5678'. Exclude the ID descriptive words and associated punctuation, e.g. exclude 'MRN' and ':' in 'MRN: 1234567'.
LOCATION	Include all place names and all parts of an address: street name, city, state, county, province, region, and country. Include punctuation in address. Include Zip/postal codes. Include organization names. Include words that specify location when they appear as part of a 'Location' entity, e.g. include the word 'Center' in 'Social Security Center'.
NAME	Include only the person's names. Include punctuation between first and last names when present Exclude titles and salutations.
PROFESSION	Include all professional titles, e.g. annotate 'MD' in the phrase 'X works as an MD'. Exclude professional titles in name suffixes, e.g. exclude 'MD' in the phrase 'Dr. X Y, MD'. Include professional and occupational descriptions, e.g. annotate 'carpentry in the phrase 'X works in carpentry'. Annotate the entire expression describing a profession, e.g. annotate all tokens in a phrase such as 'worker in a cafeteria'. Exclude workplace names; annotate workplace names as 'Location' instead.

Table A.5: Adjusted annotation guidelines with examples for each PHI type.

Level	Annotator	Criteria	Sample Sentence	Explanation
1	A	Annotator is unable to assign labels due to insufficient contextual information from the given sentence.	"PASe." [PASe.] or [ENTer.]	The token may be an unknown acronym or an oddly typed imperative form of the verb "to enter". Insufficient context.
	B	Annotator is unable to assign labels due to lack of comprehension.	"Allitat, en DDLL." [n/a]	Annotator did not understand this sentence in Catalan sufficiently to annotate.
2	A	Annotator is unsure about the assigned labels due to contextual ambiguity.	""50 años." [50 years]	Without any surrounding context, the years can be 'AGE' or a temporal expression; annotator thinks it's most likely to be AGE, but does not feel confident enough to make a determination.
	B	Annotator is unsure about the assigned labels due to lack of medical knowledge or terminology.	"Uroculltiu [sic] 13.01: + per *A. baumanii* multiR." Urine Culture 13.01: + for MDR *A. baumanii*	Annotator omitted this sentence due to uncertainty about the word *A. baumanii*, whether it could be a NAME or a non-labelled entity.
3	A	Annotator is confident about the labels, but some context may be missing that could change the entity labels.	"712345678)."	This is likely a phone number CONTACT, but may also be an ID entity.
	B	Annotator is confident about the sentence in general, but has some doubt due to presumed lack of specialized knowledge.	"hipoTA [sic] asintomática." [Asymptomatic hypotension.]	Annotator did not specify any label but was unsure whether there was a labelled entity or not.
4	A	Annotator is confident about the labels, but the sentence may have some inconsistencies with the gold standard sentences.	"El marido la vió y llamó a la ambulancia e ingresó en el hospital de Xxxxxxx." [The spouse saw her and called the ambulance and she was admitted to Xxxxxxx hospital.]	Annotator was unsure whether to only annotate 'Xxxxxxx' or 'hospital de Xxxxxxx' or 'el hospital de Xxxxxxx' as LOCATION
	B	Annotator is confident about the labels, but the sentence may have some inconsistencies with the gold standard sample sentences.	"Torna d'Oftalmologia de Xxx Xxxx (Dra. [sic]" [Returns from Xxx Xxxx Ophthalmology (Dra.]	Annotation unsure whether or not to include Ophthalmology as part of 'LOCATION'
5	A	Annotator is confident and there's no ambiguity regarding name entities of the labels. This could mean that the sentences have no entities to be annotated or that all the entities needing annotations are consistent with the gold standard sample sentences.	"Cito a control el próximo 25.12.20 y doy pautas a la esposa." [I make a follow-up appointment for the upcoming date 25.12.20 and I give the prescription to the wife.]	It's clear that '25.12.20' is a DATE PHI.
	B	It is clear to the annotator that the sentence has no entities to be annotated or that the entities are consistent with gold standard annotation. This could be either apparent at first glance or because the sentence has been seen several times before, which increases the annotator's confidence regarding the assigned label(s).	"Cito a control el próximo 25.12.20 y doy pautas a la esposa." [I make a follow-up appointment for the upcoming date 25.12.20 and I give the prescription to the wife.]	It's clear that '25.12.20' is a DATE PHI.

Table A.6: Confidence level criteria and examples as reported by the two annotators. In instances where PHI entities are utilized in the examples, we replaced the characters with generic alphanumeric characters or with fictitious information (while maintaining the same PHI type).

VPAI_Lab at MedVidQA 2022: A Two-Stage Cross-modal Fusion Method for Medical Instructional Video Classification

Bin Li[1]*, Yixuan Weng[2*], Fei Xia[2,3*], Bin Sun[1], Shutao Li[1]
[1] College of Electrical and Information Engineering, Hunan University
[2] National Laboratory of Pattern Recognition, Institute of Automation, CAS
[3] School of Artificial Intelligence, University of Chinese Academy of Sciences
{libincn, sunbin611, shutao_li}@hnu.edu.cn,
wengsyx@gmail.com, xiafei2020@ia.ac.cn

Abstract

This paper introduces the method of VPAI_Lab team's experiments on BioNLP 2022 shared task 1 Medical Video Classification (MedVidCL). Given an input video, the MedVidCL task aims to correctly classify it into one of three following categories: Medical Instructional, Medical Non-instructional, and Non-medical. Inspired by its dataset construction process, we divide the classification process into two stages. The first stage is to classify videos into medical videos and non-medical videos. In the second stage, for those samples classified as medical videos, we further classify them into instructional videos and non-instructional videos. In addition, we also propose the cross-modal fusion method to solve the video classification, such as fusing the text features (question and subtitles) from the pre-training language models and visual features from image frames. Specifically, we use textual information to concatenate and query the visual information for obtaining better feature representation. Extensive experiments show that the proposed method significantly outperforms the official baseline method by 15.4% in the F1 score, which shows its effectiveness. Finally, the official results show that our method ranks the Top-1 on the official unseen test set. All the experimental codes are open-sourced at `https://github.com/Lireanstar/MedVidCL`.

1 Introduction

One of the key goals of artificial intelligence (AI) is to develop a multimodal system that uses natural language queries to facilitate communication with the visual world (i.e., images, videos) (Cukurova et al., 2019). In recent years, the gap between language and visual understanding has narrowed (Guo et al., 2016; Lu et al., 2019) due to the development of pre-trained models (Devlin et al., 2018) and the introduction of large-scale language-vision datasets (Lei et al., 2018, 2020a,b). Improvements have been made in numerous vision-and-language tasks, such as visual classification (Servières et al., 2021), video question answering (Huang et al., 2020) and natural language video localization (Yuan et al., 2019; Chen et al., 2019; Zhang et al., 2020).

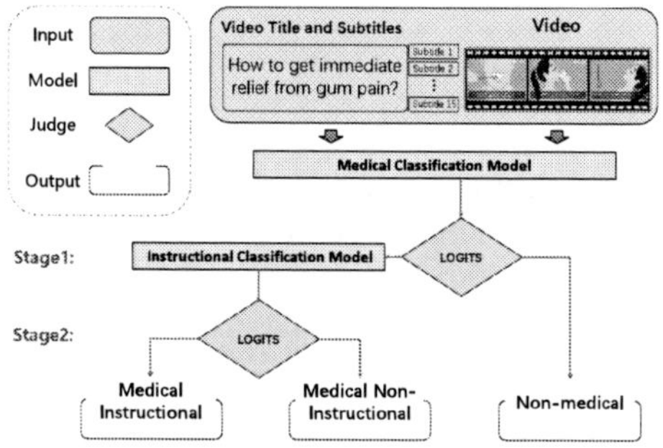

Figure 1: The overview of the proposed two-stage cross-modal fusion method.

The recent proliferation of online videos has changed the way people acquire information and knowledge. More and more people are accustomed to using instructional videos to teach or learn specific tasks. Medical instructional videos are more suitable and conducive to conveying key information through both visual and verbal communication in an effective and efficient manner (Gupta et al., 2022; Gupta and Demner-Fushman, 2022).

To better distinguish medical instructional videos from other videos, MedVidQA proposes Medical Video Classification (MedVidCL) task[1]. Given an input video, the MedVidCL task aims to correctly classify it into one of three following categories: Medical Instructional, Medical Non-instructional, and Non-medical.

Inspired by its dataset construction process (Gupta et al., 2022), we divide the classification process into two stages. As shown in Figure 1, given the question "How to get immediate relief

*These authors contribute equally to this work.

[1]https://codalab.lisn.upsaclay.fr/competitions/1058

Proceedings of the BioNLP 2022 workshop, Dublin, Ireland, pages 212–219
May 26, 2022. ©2022 Association for Computational Linguistics

from gum pain?" and subtitles and videos, we first classify it into medical and non-medical videos by the Medical Classification Model in stage one. If the input is classified as a medical video, in the second stage, we further classify it into medical instructional videos and non-instructional videos through the Instructional Classification Model.

For monomodal (Language) setting, it is feasible to classify the input video with the corresponding subtitle texts since the content of the video is directly related to its subtitles (Mahdisoltani et al., 2018; Perez-Martin et al., 2021). We choose various pre-trained models combined with our designed two-stage method to perform video classification. The experimental results show that pre-trained language models can achieve better semantic understanding.

Moreover, visual information is equally important for the MedVidCL task. To make full use of the information of visual and textual modality, we perform feature extraction on them separately and perform the query concatenation mechanism (Zhang et al., 2020) for better feature representation.

In this paper, we propose a two-stage cross-modal fusion method, by fusing the extracted visual features and textual features from the pre-trained language model. Compared with the official multimodal method, our multimodal method improves by 15.4% in F1, and the results show the effectiveness of our cross-modal method.

2 Proposed Approach

In this section, we will elaborate on the proposed approach for the medical video classification (MedVidCL) track. As the pre-training language method can enhance the performance of semantic representation queried by the text subtitles (Perez-Martin et al., 2022), we design the two-stages cross-modal fusion method, which is described in turn as follows.

2.1 Two-stage modeling for classification

Acquisition of the MedVidCL dataset mainly goes through (1) Extraction of medical and health-related tasks from WikiHow[2]; (2) Identification of relevant health-related tasks; (3) Expert label annotation for medical instructional videos. Therefore, we consider that the overall three-category (non-medical, medical instructional, and non-medical instructional) can be turned into a two-stage taxonomy. We first perform the binary classification of medical-related then process the binary classification of medical instructional-related.

[2]https://www.wikihow.com/Main-Page

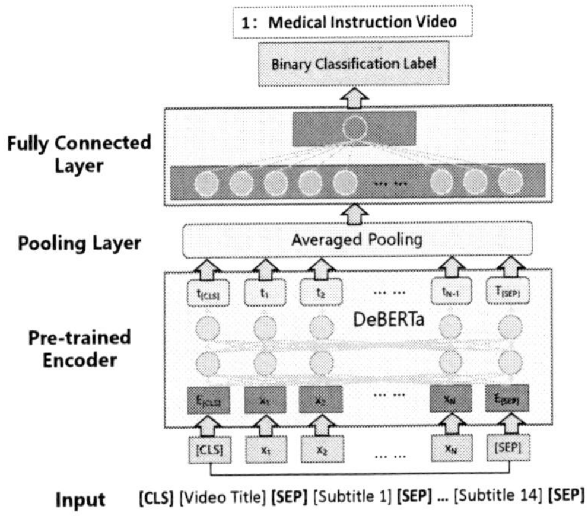

Figure 2: An example of the language-only classification model. Given the video title and its subtitles, it is required to perform the binary classification on the MedVidCL datasets.

2.2 Language-only video classification

Because the content of the video is directly related to its subtitles, it is feasible to use the corresponding subtitle texts to perform the classification of the input video (Miech et al., 2020). As shown in Figure 2, we concatenate the video title with the subtitles which are segmented into text spans for text encoding. Then the tokenized tokens are encoded through the DeBERTa model (He et al., 2020) for learning well-formed representations. An averaged pooling with the fully connected layer is designed to obtain the final features for the binary classification prediction.

2.3 Cross-modal video classification

When people watch videos, they may not always judge the video contents through the subtitle texts. For the non-audio parts, the visual information counts a lot (Gabeur et al., 2020). Therefore, for each subtitle span, we can add the visual feature to predict the video content. As shown in Figure 3, we design the cross-modal video classification model. Specifically, we focus on the feature joint alignment of video frames and subtitle text. The binary classification is performed after mapping the subtitle spans with their corresponding video frame into the same vector space. For the text modality, we input the subtitle texts into the pre-trained model

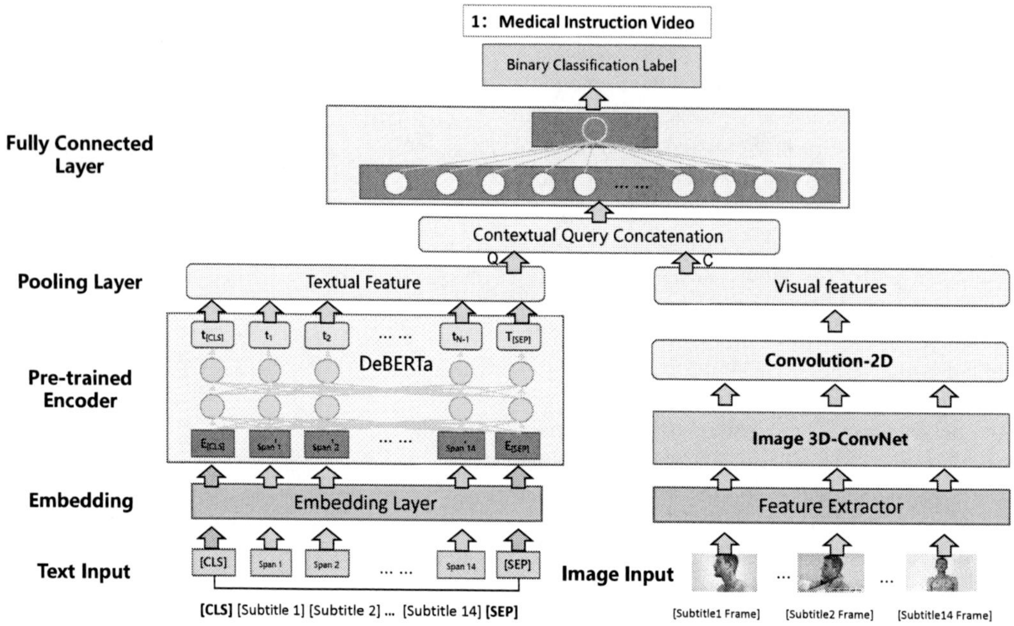

Figure 3: An example of the cross-modal classification model. The cross-modal features encoded separately are sent to the context query concatenation module for joint alignment. The binary classification is performed through the fully connected layer.

Video Category	Train	Validation	Test	Total
Medical Instructional	789	100	600	1489
Medical Non-instructional	2394	100	500	2994
Non-Medical	1034	100	500	1634
Total	4217	300	1600	6117

Table 1: Statistics of the medical video classification task dataset for the experiments.

for obtaining the textual feature. For the visual modality, we extract the raw frames with down-sampling, where 20 frames are derived from each video at a uniform time interval. Then we utilized a 3D ConvNet (I3D) module (Balaguer and Gobbetti, 1995) with the Convolution-2D for obtaining the visual features, which was pre-trained on the Kinetics dataset (Kay et al., 2017). We perform the Context Query Concatenation (Cq_Concat) (Zhang et al., 2020) for joint alignment of the textual features (Q) and the visual features (C) for the final binary classification prediction.

2.4 Late fusion method

Since there is a huge gap between visual features (Zhang et al., 2021) and language features, we design the late fusion method to use the Bagging algorithm (Breiman, 1996) to obtain the results of the above two models. Specifically, we use the logits from the different models for this ensemble method, where these logits are summed together before the softmax. We adopt the softmax to perform the final prediction. The Bagging algorithm is used during the prediction, which can effectively reduce the variance of the final prediction by bridging the prediction bias of different models, enhancing the overall generalization ability of the system.

3 Experimental setup

3.1 Data Description

Recently, with the rapid development of the video field, the informative nature of video has changed the way human beings obtain information (Lin et al., 2019). Medical Video CLassification (MedVidCL) (Gupta et al., 2022) is a data set about medical instructional video classification, which has been validated by human annotators. The medical classification datasets contain a collection of 6,617 videos, and it is required to classify the video into "Medical Instructional", "Medical Non-Instructional", and "Non-Medical" classes.

The statistics of the medical video classification datasets (seen datasets for experiments) are shown in Table 1. To construct the MedVidCL dataset, the

Experimental Items		Medical-related			Instructional-related		Overall
Method	F1	Precision	Recall	F1	Precision	Recall	F1(macro)
SVM	/	/	/	0.802	1.000	0.670	0.874
BERT-Base-Uncased	/	/	/	0.915	0.960	0.875	0.929
RoBERTa-Base	/	/	/	0.934	0.980	0.893	0.947
BigBird-Base	/	/	/	0.942	0.982	0.907	0.957
DeBERTa One-Stage	0.996	0.996	0.996	0.992	0.984	1.000	0.963
DeBERTa Two-Stage	0.980	0.980	0.980	0.936	**1.000**	0.880	0.934
BigBird One-Stage	0.996	0.996	0.996	0.994	0.996	0.990	0.983
BigBird Two-Stage	0.996	0.996	0.996	0.998	0.996	0.990	0.985
Ensemble	**0.996**	**0.996**	**0.996**	**0.998**	0.999	**1.000**	**0.988**

Table 2: Results of the monomodal with language on the seen test set.

Experimental Items		Medical-related			Instructional-related		Overall
Method	F1	Precision	Recall	F1	Precision	Recall	F1(macro)
L + V (I3D) + LSTM	/	/	/	0.726	0.797	0.667	0.757
L + V (ViT) + LSTM	/	/	/	0.773	0.902	0.677	0.814
L + V (I3D) + Transformer	/	/	/	0.727	0.762	0.695	0.748
L + V (ViT) + Transformer	/	/	/	0.791	0.922	0.692	0.824
Ours (One-Stage) + DeBERTa + I3D	0.990	0.988	0.992	0.984	0.969	1.000	0.967
Ours (Two-Stage) + DeBERTa + I3D	**0.998**	**1.000**	**0.996**	0.986	0.973	1.000	0.971
Ours (One-Stage) + BigBird + I3D	0.992	0.992	0.992	0.986	0.981	0.992	0.975
Ours (Two-Stage) + BigBird + I3D	0.992	0.988	0.995	0.973	0.947	1.000	0.977
Ensemble	0.994	0.994	0.994	**0.992**	**0.984**	**1.000**	**0.978**

Table 3: Results of the Multimodal with Language (L) and Vision (V) on the seen test set.

organizers first train the machine learning model based on the data marked by medical experts from HowTo100M and YouTube8M datasets (Abu-El-Haija et al., 2016). After that, the videos with high confidence are selected and sorted out with the machine learning method (Gupta et al., 2022).

3.2 Evaluation metrics

We follow the standard evaluation metrics of answer prediction in MedVidQA. The performance of the system is evaluated through two evaluation indicators (Gupta and Demner-Fushman, 2022). Each experiment was conducted for 10 rounds with different random seeds for eliminating the random bias, and we select the model with the highest F1 score on the valid set and then report its score on the test set. The metrics are introduced as follows.

1. F1 Score on Medical Instructional class.
2. Average macro-level F1 score across all the classes.

The calculation equation of each metric is shown as follows.

$$Precision = \frac{TP}{TP + FP}$$

$$Recall = \frac{TP}{TP + FN}$$

$$F1 = \frac{2 * PrecisionRecall}{Precision + Recall}$$

$$MacroF1 = \frac{\sum_{i=1}^{n} F1_i}{n}$$

3.3 Experimental Details

The Method provided by the organizer[3] provides four baseline methods based on different features. In order to obtain language features, the organizer extracted the subtitle information in the video using the Pytube library[4]. In addition, the organizer uses 3D convolution (I3D) (Carreira and Zisserman, 2017) to extract the visual features of the video in units of every second. And then they train statistical classifiers such SVM (Cortes and Vapnik, 1995) method, Transformer model (Vaswani et al., 2017), VIT model (Dosovitskiy et al., 2021) and LSTM model (Hochreiter and Schmidhuber, 1997). The details of each baseline are introduced below.

[3]Specific implementations can refer to https://github.com/deepaknlp/MedVidQACL/tree/master/MedVidCL

[4]https://github.com/pytube/pytube

Method	Med-Inst Precision	Med-Inst Recall	Med-Inst F1	Macro F1
BigBird Two-Stage (Monomodal)	0.9949	0.9775	0.9861	0.9893
Ours (Two-Stage) + DeBERTa + I3D	0.9948	0.9750	0.9848	0.9884
Ensemble	**0.9974**	**0.9775**	**0.9873**	**0.9901**

Table 4: Submitted official results of the unseen test set.

1. **Monomodal (Language)** They utilize the pre-trained Transformer models from Hugging Face(Wolf et al., 2020) such as BERT-Base-Uncase(Devlin et al., 2019) , RoBERTa-Base(Liu et al., 2019) and BigBird-Base(Zaheer et al., 2020).

2. **Monomodal (Vision)** After extracting features from I3D or VIT, the organizer uses the LSTM network and transformers network to build classifiers.

3. **Multimodal (Language + Vision)** After the text features and visual features are obtained, they are concatenated and then connected to a full connection layer for classification.

Our method use the DeBERTa-large-v3 (He et al., 2021) model. The DeBERTa improves the BERT (Devlin et al., 2019) and RoBERTa (Liu et al., 2019) models using disentangled attention and enhanced mask decoder. It shares the base model with 24 layers and 1024 hidden size. We formulate the original three-classification task (one-stage) into two two-classification tasks (two-stage). And we trained two models separately to support our video classification under the two-stage setting. In the experimental table, we will report and compare the testing effect in one-stage and two-stage settings respectively.

3.4 Implementation details

We train the model based on the Pytorch framework (Paszke et al., 2019) and use the huggingface[5] (Wolf et al., 2020) framework. When training the model, we employ the AdamW optimizer (Loshchilov and Hutter, 2017). The default learning rate is set to 1e-5 with the warm-up (He et al., 2016). Four RTX3090 GPUs with 24G memory are implemented for all experiments.

We use the [SEP] token to concatenate the title and subtitles of the video. Experiments are carried out in maximum lengths 512. When it is necessary to distinguish whether it is a medical video or not at the first stage, we set the "Medical Non-instructional" video and the "Medical Instructional" video as the same category. When turning into the second stage, we exclude "Non-medical" video samples. All the experimental codes are open-source at `https://github.com/Lireanstar/MedVidCL`.

4 Results and discussions

In this section, we introduce the experimental results of the monomodal in language and the multimodal in language-version where the further discussions and official results are also presented.

4.1 Experimental results

The experimental results of the monomodal with language on the seen test set are shown in Table 2, and the multimodal with language and vision can be found in Table 3. For the one-stage setting, we implement the classification for three categories of prediction. For the proposed two-stage setting, we exclude the non-medical category for the first stage classification, then perform the two-categories classification to differentiate the medical instructional and medical non-instructional videos for the final result. Specifically, for the monomodal results, our method outperforms all the baselines in the overall scores. What excites us is that the SVM method achieves the same recall score (1.000) as the deep learning DeBERTa model on medical instructional-related classification, indicating that the subtitle information of the video has strong semantics. As for the multimodal settings, the proposed two-stage cross-modal fusion method outperforms the one-stage cross-modal fusion method, which demonstrates its effectiveness. The proposed method is significantly ahead of the baseline methods. We believe that is because our model can recognize visual features more efficiently combined with the strong pre-trained language model. Moreover, the ensemble method can be a wise choice to enhance the final score compared with other single models. In the end, we find that the proposed method with cross-modal fusion can achieve similar perfor-

[5]https://github.com/huggingface/transformers

mance to the monomodal methods, which demonstrates the superiority of our proposed method.

4.2 Official results

As shown in Table 4, we present the results of official submissions on the unseen test set. Further conclusions can be found that the monomodal modality (language) can indeed effectively identify the semantic information from the video, which outperforms the cross-modal setting. It is in line with the experimental results under the test set. We perform the ensemble method by adding the logits generated from the two single model, and adopt the Softmax for the final prediction. Finally, by adopting an ensemble method, we achieve the Top-1 score in the final official stage.

5 Conclusion

This paper introduces our approach to solving the medical video classification (MedVidCL) task in BioNLP of the ACL2022. Specifically, we propose the two-stage method with cross-modal fusion using the pre-trained language model. We report the performance of our model compared on the test set in monomodal and multimodal settings. The experimental results show that our method obtains the best performance on the seen test set and unseen official test set, which proves that our method is effective. Also, experimental results show that language understanding is better than multimodal video understanding. In the future, we will further study how to design a more efficient structure to jointly learn the representation in visual language for better multimodal video understanding.

Acknowledgement

This work is supported by the National Key Research and Development Project of China (2018YFB1305200) and the National Natural Science Fund of China (62171183, 61801178).

References

Sami Abu-El-Haija, Nisarg Kothari, Joonseok Lee, Paul Natsev, George Toderici, Balakrishnan Varadarajan, and Sudheendra Vijayanarasimhan. 2016. Youtube-8m: A large-scale video classification benchmark. *arXiv preprint arXiv:1609.08675*.

Jean-Francis Balaguer and Enrico Gobbetti. 1995. i3d: a high-speed 3d web browser. In *Proceedings of the first symposium on Virtual reality modeling language*, pages 69–76.

Leo Breiman. 1996. Bagging predictors. *Machine learning*, 24(2):123–140.

Joao Carreira and Andrew Zisserman. 2017. Quo vadis, action recognition? a new model and the kinetics dataset. In *Computer Vision and Pattern Recognition*.

Jingyuan Chen, Lin Ma, Xinpeng Chen, Zequn Jie, and Jiebo Luo. 2019. Localizing natural language in videos. In *Proceedings of the AAAI Conference on Artificial Intelligence*, volume 33, pages 8175–8182.

Corinna Cortes and Vladimir Vapnik. 1995. Support-vector networks. *Machine Learning*, 20:273–297.

Mutlu Cukurova, Carmel Kent, and Rosemary Luckin. 2019. Artificial intelligence and multimodal data in the service of human decision-making: A case study in debate tutoring. *British Journal of Educational Technology*, 50(6):3032–3046.

Jacob Devlin, Ming-Wei Chang, Kenton Lee, and Kristina Toutanova. 2018. Bert: Pre-training of deep bidirectional transformers for language understanding. *arXiv preprint arXiv:1810.04805*.

Jacob Devlin, Ming-Wei Chang, Kenton Lee, and Kristina Toutanova. 2019. BERT: Pre-training of deep bidirectional transformers for language understanding. In *Proceedings of the 2019 Conference of the North American Chapter of the Association for Computational Linguistics: Human Language Technologies, Volume 1 (Long and Short Papers)*, pages 4171–4186, Minneapolis, Minnesota. Association for Computational Linguistics.

Alexey Dosovitskiy, Lucas Beyer, Alexander Kolesnikov, Dirk Weissenborn, Xiaohua Zhai, Thomas Unterthiner, Mostafa Dehghani, Matthias Minderer, Georg Heigold, Sylvain Gelly, Jakob Uszkoreit, and Neil Houlsby. 2021. An image is worth 16x16 words: Transformers for image recognition at scale. *ICLR*.

Valentin Gabeur, Chen Sun, Karteek Alahari, and Cordelia Schmid. 2020. Multi-modal transformer for video retrieval. In *ECCV 2020-European Conference on Computer Vision*, volume 12349, pages 214–229. Springer.

Yanming Guo, Yu Liu, Ard Oerlemans, Songyang Lao, Song Wu, and Michael S Lew. 2016. Deep learning for visual understanding: A review. *Neurocomputing*, 187:27–48.

Deepak Gupta, Kush Attal, and Dina Demner-Fushman. 2022. A dataset for medical instructional video classification and question answering. *arXiv preprint arXiv:2201.12888*.

Deepak Gupta and Dina Demner-Fushman. 2022. Overview of the MedVidQA 2022 Shared Task on Medical Video Question Answering. In *Proceedings of the 21st SIGBioMed Workshop on Biomedical Language Processing, ACL-BioNLP 2022*. Association for Computational Linguistics.

Kaiming He, Xiangyu Zhang, Shaoqing Ren, and Jian Sun. 2016. Deep residual learning for image recognition. In *2016 IEEE Conference on Computer Vision and Pattern Recognition (CVPR)*, pages 770–778.

Pengcheng He, Xiaodong Liu, Jianfeng Gao, and Weizhu Chen. 2020. Deberta: Decoding-enhanced bert with disentangled attention. *arXiv preprint arXiv:2006.03654*.

Pengcheng He, Xiaodong Liu, Jianfeng Gao, and Weizhu Chen. 2021. Deberta: Decoding-enhanced bert with disentangled attention. In *International Conference on Learning Representations*.

Sepp Hochreiter and Jürgen Schmidhuber. 1997. Long Short-Term Memory. *Neural Computation*, 9(8):1735–1780.

Deng Huang, Peihao Chen, Runhao Zeng, Qing Du, Mingkui Tan, and Chuang Gan. 2020. Location-aware graph convolutional networks for video question answering. In *Proceedings of the AAAI Conference on Artificial Intelligence*, volume 34, pages 11021–11028.

Will Kay, Joao Carreira, Karen Simonyan, Brian Zhang, Chloe Hillier, Sudheendra Vijayanarasimhan, Fabio Viola, Tim Green, Trevor Back, Paul Natsev, et al. 2017. The kinetics human action video dataset. *arXiv preprint arXiv:1705.06950*.

Jie Lei, Licheng Yu, Mohit Bansal, and Tamara Berg. 2018. Tvqa: Localized, compositional video question answering. In *Proceedings of the 2018 Conference on Empirical Methods in Natural Language Processing*, pages 1369–1379.

Jie Lei, Licheng Yu, Tamara Berg, and Mohit Bansal. 2020a. Tvqa+: Spatio-temporal grounding for video question answering. In *Proceedings of the 58th Annual Meeting of the Association for Computational Linguistics*, pages 8211–8225.

Jie Lei, Licheng Yu, Tamara L Berg, and Mohit Bansal. 2020b. Tvr: A large-scale dataset for video-subtitle moment retrieval. In *European Conference on Computer Vision*, pages 447–463. Springer.

Ji Lin, Chuang Gan, and Song Han. 2019. Tsm: Temporal shift module for efficient video understanding. In *Proceedings of the IEEE/CVF International Conference on Computer Vision*, pages 7083–7093.

Yinhan Liu, Myle Ott, Naman Goyal, Jingfei Du, Mandar Joshi, Danqi Chen, Omer Levy, Mike Lewis, Luke Zettlemoyer, and Veselin Stoyanov. 2019. Roberta: A robustly optimized bert pretraining approach. *arXiv preprint arXiv:1907.11692*.

Ilya Loshchilov and Frank Hutter. 2017. Fixing weight decay regularization in adam. *ArXiv*, abs/1711.05101.

Jiasen Lu, Dhruv Batra, Devi Parikh, and Stefan Lee. 2019. Vilbert: Pretraining task-agnostic visiolinguistic representations for vision-and-language tasks. *Advances in neural information processing systems*, 32.

Farzaneh Mahdisoltani, Guillaume Berger, Waseem Gharbieh, David Fleet, and Roland Memisevic. 2018. Fine-grained video classification and captioning. *arXiv preprint arXiv:1804.09235*, 5(6).

Antoine Miech, Jean-Baptiste Alayrac, Lucas Smaira, Ivan Laptev, Josef Sivic, and Andrew Zisserman. 2020. End-to-end learning of visual representations from uncurated instructional videos. In *Proceedings of the IEEE/CVF Conference on Computer Vision and Pattern Recognition*, pages 9879–9889.

Antoine Miech, Dimitri Zhukov, Jean-Baptiste Alayrac, Makarand Tapaswi, Ivan Laptev, and Josef Sivic. 2019. Howto100m: Learning a text-video embedding by watching hundred million narrated video clips. In *Proceedings of the IEEE/CVF International Conference on Computer Vision*, pages 2630–2640.

Adam Paszke, Sam Gross, Francisco Massa, Adam Lerer, James Bradbury, Gregory Chanan, Trevor Killeen, Zeming Lin, Natalia Gimelshein, Luca Antiga, Alban Desmaison, Andreas Kopf, Edward Yang, Zachary DeVito, Martin Raison, Alykhan Tejani, Sasank Chilamkurthy, Benoit Steiner, Lu Fang, Junjie Bai, and Soumith Chintala. 2019. Pytorch: An imperative style, high-performance deep learning library. In *Advances in Neural Information Processing Systems*, volume 32. Curran Associates, Inc.

Jesus Perez-Martin, Benjamin Bustos, Silvio Jamil F Guimarães, Ivan Sipiran, Jorge Pérez, and Grethel Coello Said. 2021. Bridging vision and language from the video-to-text perspective: A comprehensive review. *arXiv preprint arXiv:2103.14785*.

Jesus Perez-Martin, Benjamin Bustos, Silvio Jamil F Guimarães, Ivan Sipiran, Jorge Pérez, and Grethel Coello Said. 2022. A comprehensive review of the video-to-text problem. *Artificial Intelligence Review*, pages 1–75.

Myriam Servières, Valérie Renaudin, Alexis Dupuis, and Nicolas Antigny. 2021. Visual and visual-inertial slam: State of the art, classification, and experimental benchmarking. *Journal of Sensors*, 2021.

Ashish Vaswani, Noam Shazeer, Niki Parmar, Jakob Uszkoreit, Llion Jones, Aidan N. Gomez, Łukasz Kaiser, and Illia Polosukhin. 2017. Attention is all you need. In *Proceedings of the 31st International Conference on Neural Information Processing Systems*, NIPS'17, page 6000–6010, Red Hook, NY, USA. Curran Associates Inc.

Thomas Wolf, Lysandre Debut, Victor Sanh, Julien Chaumond, Clement Delangue, Anthony Moi, Pierric Cistac, Tim Rault, Rémi Louf, Morgan Funtowicz, Joe Davison, Sam Shleifer, Patrick von Platen, Clara Ma, Yacine Jernite, Julien Plu, Canwen Xu, Teven Le

Scao, Sylvain Gugger, Mariama Drame, Quentin Lhoest, and Alexander M. Rush. 2020. Transformers: State-of-the-art natural language processing. In *Proceedings of the 2020 Conference on Empirical Methods in Natural Language Processing: System Demonstrations*, pages 38–45, Online. Association for Computational Linguistics.

Yitian Yuan, Tao Mei, and Wenwu Zhu. 2019. To find where you talk: Temporal sentence localization in video with attention based location regression. In *Proceedings of the AAAI Conference on Artificial Intelligence*, volume 33, pages 9159–9166.

Manzil Zaheer, Guru Guruganesh, Avinava Dubey, Joshua Ainslie, Chris Alberti, Santiago Ontañón, Philip Pham, Anirudh Ravula, Qifan Wang, Li Yang, and Amr Ahmed. 2020. Big bird: Transformers for longer sequences. *arXiv: Learning*.

Hao Zhang, Aixin Sun, Wei Jing, Liangli Zhen, Joey Tianyi Zhou, and Rick Siow Mong Goh. 2021. Natural language video localization: A revisit in span-based question answering framework. *IEEE transactions on pattern analysis and machine intelligence*.

Hao Zhang, Aixin Sun, Wei Jing, and Joey Tianyi Zhou. 2020. Span-based localizing network for natural language video localization. In *Proceedings of the 58th Annual Meeting of the Association for Computational Linguistics*, pages 6543–6554.

GenCompareSum: a hybrid unsupervised summarization method using salience

Jennifer A Bishop[1], Qianqian Xie[1], Sophia Ananiadou[1,2]
[1]National Centre for Text Mining, Department of Computer Science, The University of Manchester, Manchester, United Kingdom
[2]Alan Turing Institute, London, United Kingdom
{jennifer.bishop-2@ postgrad.,qianqian.xie@,sophia.ananiadou@}manchester.ac.uk

Abstract

Text summarization (TS) is an important NLP task. Pre-trained Language Models (PLMs) have been used to improve the performance of TS. However, PLMs are limited by their need of labelled training data and by their attention mechanism, which often makes them unsuitable for use on long documents. To this end, we propose a hybrid, unsupervised, abstractive-extractive approach, in which we walk through a document, generating salient textual fragments representing its key points. We then select the most important sentences of the document by choosing the most similar sentences to the generated texts, calculated using BERTScore. We evaluate the efficacy of generating and using salient textual fragments to guide extractive summarization on documents from the biomedical and general scientific domains. We compare the performance between long and short documents using different generative text models, which are finetuned to generate relevant queries or document titles. We show that our hybrid approach out-performs existing unsupervised methods, as well as state-of-the-art supervised methods, despite not needing a vast amount of labelled training data.

1 Introduction

Recent advancements in transformer-based architectures have enabled improvements in natural language processing (NLP) tasks. The use of encoder-decoder models, such as the T5 language model (Raffel and al., 2020) in generative linguistic tasks, such as abstractive summarization (Cachola et al., 2020) and query generation (Nogueira and Lin., 2019; Klein and Nabi, 2019), have been shown to significantly improve performance over existing methods. Bidirectional-encoder transformer architectures, namely BERT-based PLMs (Devlin et al., 2018) have also proven to be powerful for a broad range of NLP tasks, including text summarization (Liu and Lapata, 2019).

Whilst transformers have made great advancements in their ability at capturing semantic knowledge, they have also introduced new limitations. Firstly, they are restricted by the number of tokens that they can process at any one time. Another issue is the computational cost of finetuning the attention mechanisms embedded in transformers. These constraints are challenging for recent text summarization methods, often resulting in analysis being done on a truncated version of a document (Cachola et al., 2020; Liu and Lapata, 2019; Xu et al., 2020; Zhong et al., 2020; Dou et al. 2021; Zhang and Zhao, 2020). Since summarization should be able to succinctly capture the meaning of very long documents in a few sentences, the requirement to truncate a document before summarization is a major disadvantage. As a result, recent works have shifted their attention towards addressing the issue of long document summarization (Xiao and Carenini, 2020; Grail et al., 2021; Rohde et al., 2021; Xiao and Carenini, 2019). However, these are mostly supervised methods, requiring large amounts of labelled training data, which are often unavailable or time-consuming and costly to produce.

We address the challenges of supervised methods by adopting a hybrid unsupervised approach, where the PLMs are required only to act on short sections of the document at any time, meaning that our method can be extended to any document length. Furthermore, by nature of it being an unsupervised approach, it does not require manually labelled training data for the extractive summarization task. To-date, unsupervised methods for text summarization have generally used graph-based methods (Erkan and Radev, 2004; Mihalcea and Tarau, 2004; Liang et al 2021; Zheng and Lapata, 2019; Done et al., 2021), the more recent of these using

Proceedings of the BioNLP 2022 workshop, Dublin, Ireland, pages 220–240
May 26, 2022. ©2022 Association for Computational Linguistics

transformer-based embeddings to calculate weights between the nodes in the graph (Zheng and Lapata, 2019; Done et al., 2021). We differ from these previous approaches as we do not use a graph-based model and instead evaluate the effectiveness of a novel approach – generating and using salient textual fragments to guide the extractive summarization. Moreover, earlier unsupervised, graph-based methods have been criticised in their ability to effectively represent documents which present multiple facts (Liang et al., 2021). Our method addresses this by generating multiple salient texts per document, thus enabling it to represent multiple facts per document.

Text summarization methods are divided into extractive and abstractive groupings. Extractive methods select the most relevant sentences from a document and abstractive methods consider the most relevant pieces of information to produce new textual fragments which convey the core message. Although abstractive summarization has the potential to be more succinct and readable, in its current state it cannot be trusted to be factually consistent (Wallace et al., 2021), making it unsuitable in many practical applications, such as summarization of biomedical articles for use by clinicians. Furthermore, Huang (2020) showed extractive techniques to outperform their abstractive counterparts in human evaluation. See (2017) recognises the advantage of hybrid extractive-abstractive summarization methods and uses a pointer-generator approach, where the model is mostly abstractive, but identifies and copies key facts directly from the source document to try to reduce factual inconsistency. We consider these factors and choose also to opt for a hybrid approach; however, we differ from See (2017) in our use of abstractive models. Specifically, we use transformer-based models for the generation of salient points, but ultimately, we generate an extractive summarization to ensure factual consistency.

Our method, GenCompareSum is a two-step hybrid summarization approach. GenCompareSum first splits a document into sections of several sentences and walks through them, generating salient textual fragments which represent each section. We experiment with different generative models, which are finetuned to predict either queries or document titles, that best represent a section of the document. Our method then uses these generated textual fragments to guide an unsupervised extractive summarization by calculating the BERTScore similarity between each of the generated texts and each of the sentences in the source document. We then select the sentences with the highest scores to form the extractive summarization. We evaluate our approach on short and long versions of data sets from the biomedical and scientific domains. Furthermore, we compare the use of different PLMs for generating salient textual fragments.

Our main contributions are as follows:

1. A novel two-step unsupervised hybrid abstractive-extractive summarization method, which generates salient textual fragments - queries and document titles - which represent sections of a document, and then uses them to guide the extractive summarization step.
2. The fusion of state-of-the-art PLMs with unsupervised approaches, to achieve a summary which harnesses the semantic knowledge of transformer-based models, whilst being extendable to any length document, without requiring a large corpus of training data.
3. Evaluation results demonstrate our hybrid method outperforms both existing unsupervised methods and state-of-the-art supervised methods, both on long and short documents.

2 Methods

We propose GenCompareSum, a hybrid abstractive-extractive model, which makes use of transformer-based architectures but is extendable to any document length, can represent multiple facts, and does not require vast amounts of training data. The method is comprised of two steps: first, using a generative model to produce salient textual fragments, i.e., queries or document titles, which represent key points from across a document, then a comparison between these salient fragments and each sentence, to select the most important sentences from across the document. A representation of our method can be seen in Figure 1. We make our code publicly available[1].

[1] `https://github.com/jbshp/GenCompareSum`

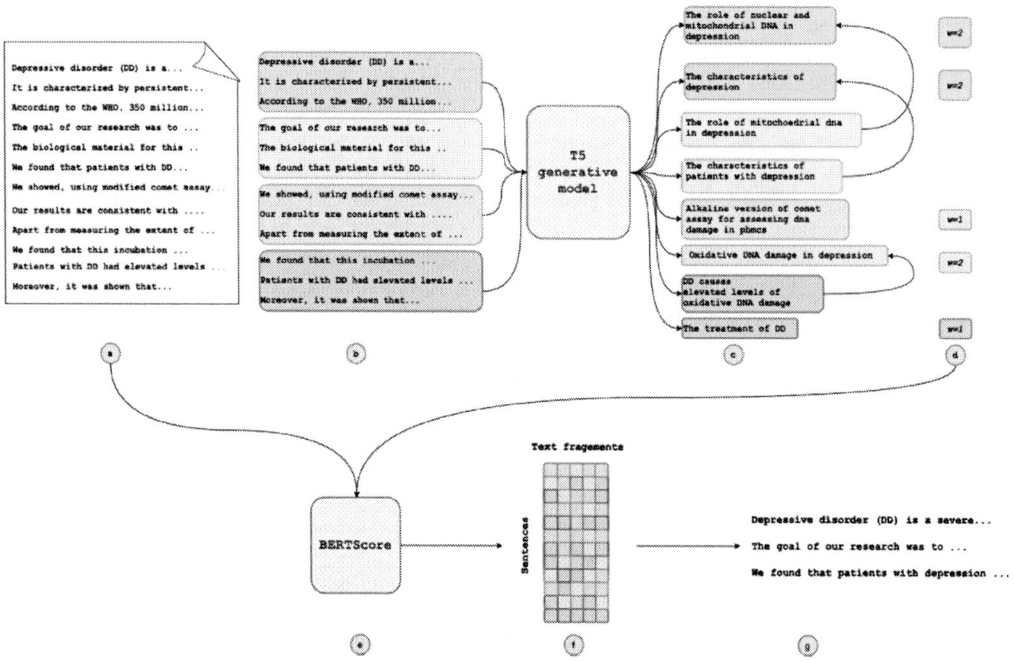

Figure 1: GenCompareSum pipeline. (a) We split the document into sentences. (b) We combine these sentences into sections of several sentences. (c) We feed each section into the generative text model and generate several text fragments per section. (d) We aggregate the questions, removing redundant questions by using n-gram blocking. Where aggregation occurs, we apply a count to represent the number of textual fragments which were combined and use this as a weighting going forwards. The highest weighted textual fragments are then selected to guide the summary. (e) The similarity between each sentence from the source document and each selected textual fragment is calculated using BERTScore. (f) We create a similarity matrix from the scores calculated in the previous step. These are then summed over the textual fragments, weighted by the values calculated in step (d), to give a score per sentence. (g) The highest scoring sentences are selected to form the summary.

2.1 Text splitting

Given a document D, we first split it into sentences s, such that $D = \{s_1, \dots, s_n\}$, using the Stanford CoreNLP software package (Manning et al., 2014). We then combine sentences into document sections of x sentences, i.e., $D = \{p_1, \dots, p_m\}$; $m = ceil\left(\frac{n}{x}\right)$. We chose not to use any pre-defined sections already existing within the documents as we found that the documents were not consistently extracted into their sections well across the different datasets. Splitting the document into a consistent number of sentences per section removes the requirement for high quality text extraction into document sections. The number of sentences x used to form the short text sections was decided via experimentation on the validation data sets.

2.2 Salient text generation

T5 (Raffel and al., 2020) is a sequence-to-sequence model, pre-trained on a cleaned and pre-processed version of the Common Crawl[2] data set – a data set consisting of textual content scraped from the internet. T5-based models have been shown to be high performing sequence-to-sequence models across a range of generative tasks, from question generation (Nogueira and Lin, 2019), to graph-to-text generation (Ribeiro et al., 2021), to generative common-sense reasoning (Yuchen Lin, et al., 2020), to abstractive text summarization (Zhang and Zhao, 2020; Goodwin, 2020). The T5 model uses an encoder-decoder architecture and is pre-trained via an unsupervised task in which 15% of tokens are masked; the masked words can be individual words or a span of words; the target of the training objective is to predict these

[2] https://commoncrawl.org

masked words, given the un-masked tokens and their respective positions. For downstream tasks, the pre-trained T5 model is finetuned using pairs of input and output sequences. A diagram of the T5 architecture and its pre-training and finetuning settings can be seen in Appendix A. We experiment with several T5-based models for the salient text generation task.

We use each section, p, as an input to a generative model to give k salient texts t, which aim to encapsulate the key facts of that section:

$$\{t_1, \ldots, t_k\} = text\_gen(p). \quad (1)$$

In the case where p is longer than 512 tokens, it is truncated. We then aggregate the generated textual fragments from across the document sections to give $T = \{t_1, \ldots, t_{mk}\}$.

For the generation of the textual fragments, we first experiment with a T5-based model finetuned with a query generation task in the general domain. This model, provided textual input, aims to generate queries which ask the most relevant questions of it. We use docTTTTTquery (Nogueira and Lin, 2019), a question generation model trained on the MS-MARCO data set (Bajaj et al., 2018), which is a question-answer data set generated from Bing's[3] search query logs. Surita et al. (2020), showed this pre-trained model to be effective at generating questions for long, biomedical texts.

Second, we follow the approach taken by Nogueira and Lin (2019) and finetune our own model on long-answer - query pairs from the biomedical domain, details of which can be found in Appendix B. We refer to this model as 't5-med-query'.

Last, we experiment using an open-source T5-based model, finetuned on abstract-title pairs from the scientific domain[4]. This approach has shown to be effective at proxying highly abstractive summaries (Cachola et al., 2020). We apply this model to our problem space, generating potential 'titles' for each document section. We refer to this model as 't5-s2orc-title'.

2.3 N-gram blocking

N-gram blocking is a technique which is applied to reduce redundancy and improve coverage in summarization models (Liu and Lapata, 2019). We apply n-gram blocking to the generated textual fragments, resulting in $T^* \subseteq T$, where $T^* = \{t_1, \ldots t_{l, l \leq mk}\}$. Where we have removed generated texts by applying this technique, we keep a count of how many times a similar textual fragment was seen before n-gram blocking. We associate this count with the remaining generated text after n-gram blocking. We refer to these counts as weights, which can be described by $w = \{w_1, \ldots, w_l\}$, such we have one weight associated with each generated textual fragment remaining after n-gram blocking. A visualization of this can be seen in steps c and d of Figure 1. We then take the top $q; q < l$ generated texts after ordering by the weight.

2.4 Text vector comparison

BERT-based comparisons have been shown to outperform traditional sentence comparative metrics like TF-IDF when used in unsupervised summarization tasks (Done et al., 2021). Furthermore, they have been demonstrated to align better with human judgement of text similarity than n-gram matching approaches during evaluation, likely due to their ability to match based on semantic meaning and their penalization of word re-ordering which changes a text's meaning (Zhang et al,. 2020). BERTScore (Zhang,et al., 2020) uses BERT-based token embeddings, calculates the cosine similarity between them and uses greedy matching to match each token in the first text to its most similar token in the second; these scores are averaged across the sentences to give precision, recall and F1 scores which quantify the similarity between two texts.

[3] https://www.bing.com

[4] https://huggingface.co/doc2query/S2ORC-t5-base-v1

Data set	Instances			Input length – Truncated document		Input length – Full document		Target length	
	Train	Val	Tokens	Tokens	Sentences	Tokens	Sentences	Tokens	Sentences
PubMed	117108	6631	3209	525	20	3209	124	208	9
S2ORC	47474	9490	4312	523	19	4312	154	250	9
CORD-19	31505	6299	5240	525	18	5240	206	232	8
ArXiv	202917	6436	6515	528	20	6515	249	279	11

Table 1: Description of the four data sets used in the extractive summarization experiments. For each data set, we give the number of articles in each of the train, validation and test splits, the mean number of tokens and sentences in the input research article, as well as the mean number of tokens and sentences in the gold summary (abstract) of the articles.

We weight the score by w, the count representing the number of textual fragments which were aggregated during n-gram blocking to give:

$$score_i = \sum_1^{Z=q} w_z * BERTScore(s_i, t_z) \quad (2)$$

We then select the sentences with the highest score to form our summary and reorder them back into the sequence that they appear within the original document.

3 Experiments

3.1 Data sets

We evaluate the efficacy of our hybrid summarization model with four publicly available data sets from the biomedical and scientific domains. All four data sets consist of full-article research papers and their corresponding abstracts. In line with previous literature, we use their abstracts as the target summaries. The data sets included in our experiments are CORD-19 (Wang et al., 2020), PubMed and ArXiv (Cohan et al. 2018), and S2ORC (Lo et al., 2020). The CORD-19 data set used is the version released on 2020-06-28, containing 57,037 articles relating to COVID-19. The S2ORC data set is a large corpus of scientific literature across several domains; we select a random subset of 63,709 articles tagged as being from the biological and biomedical domains. The PubMed and ArXiv data sets are from the biomedical and scientific domains respectively.

For the S2ORC and CORD-19 data sets, we split the data set by sampling randomly to create training/validation/test sets using the ratio 75/15/10. For the PubMed and ArXiv data sets, we use the train/validation/test sets given in the resources associated with the original paper.

Since most previous literature using transformer-based models in their methods either evaluates them on short or truncated texts (Cachola et al., 2020; Liu and Lapata, 2019; Xu et al., 2020; Zhong et al., 2020; Dou et al., 2021; Zhang and Zhao, 2020), we also create short data sets for evaluating our models. We create these data sets by truncating documents to the end of the sentence which contains their 512th token. We evaluate our models both on the short and full-text versions of the four data sets described above. Table 1 gives, for each data set, the mean number of tokens and sentences for the documents and their target summaries.

As training is not required for unsupervised models, for these methods only the test data sets are used. To train the supervised method, BERTExtSum (Liu and Lapata, 2019), which we implement for comparison, we use the training data set to train the model and the validation data set to select the best performing epoch for evaluation on the test set.

3.2 Parameter selection

To select the optimal parameters for our models, we take a constant but random sample of 1000 articles from the PubMed validation data set and experiment with different combinations of the parameters, details of which can be found in Appendix C. Different methods for calculating text similarity were also compared, namely, BERTScore, SimCSE (Gao et al., 2021) and Sentence Transformers (Reimers and Gurevych, 2019), with BERTScore shown to be the highest performing against a ROUGE metric for the extractive summarization task, details of this analysis can be found in Appendix D.

3.3 Implementation details

We run all experiments requiring GPUs on NVIDIA Quadro RTX 6000 hardware. We report all our results in terms of ROUGE-1, ROUGE-2 and ROUGE-L scores (Lin, 2004), calculated using pyrouge[5] python package.

Several extractive text summarization methods are compared across the short and full-text versions of the four scientific data sets. For the short-text data sets, we take 6 sentences to generate our predictive summary. We choose to give results on a short-text summary for a fair comparison against supervised methods, which are restricted by the length of document that they can easily summarize. For the full-text articles, the number of sentences that we select for the predictive summary is the same as the average number of sentences in the target summaries for a given data set, shown in Table 1. E.g., for the PubMed data set, we select 9 sentences to summarize the full text article.

3.4 Related work

ORACLE summaries indicate the upper bound for extractive text summarization. We calculate ORACLE summaries by adapting code from Liu and Lapata (2019), which applies greedy sentence selection to maximise ROUGE scores.

As baseline methods for comparison, we implement the LEAD method, taking the first n sentences to form the summary, and the RANDOM method, taking a random sample of n sentences to form the summary.

We also compare our method to unsupervised extractive methods, LexRank (Erkan and Radev, 2004), TextRank (Mihalcea and Tarau, 2004) and SumBasic (Nenkova and Vanderwende, 2005), all of which were implemented using the sumy[6] package. LexRank (Erkan and Radev, 2004) and TextRank (Mihalcea and Tarau, 2004) are both graph-based models, based on Google's PageRank algorithm (Brin and Page, 1998), which assume that the sentences with the highest centrality are the most important and use these to form a summary. SumBasic simply assumes that sentences containing the words which are used with the highest frequency across the whole document will be the most important.

Additionally, we compare our method to BERTExtSum (Liu and Lapata, 2019), a state-of-the-art supervised method using BERT-based transformer models. For evaluation on the short data sets, where the documents are truncated at the end of the sentence containing the 512th token, we use their implementation without modification to train and evaluate the models. For the full-text article, we adapt their code, denoted BERTExtSum*, to cycle through the article in 512 token-length blocks and predict the best sentences to select from across this cycle. However, due to hardware limitations and the computational intensity of the attention calculation, we were still required to truncate the document at 1024 tokens to evaluate this method.

Lastly, we implement GenCompareSum and compare the performance between using different generative text models: docTTTTTquery, t5-med-query, and t5-s2orc-title.

4 Experimental Results

4.1 Automatic evaluation

We report the results of our unsupervised hybrid abstractive-extractive method on the extractive summarization task in Table 2.

For the short documents, our method GenCompareSum (t5-s2orc-title) performs best across three out of four of the data sets, and second-best for the fourth data set. There is no clear 'second-best' model out of the methods compared for the short data sets.

[5] https://github.com/bheinzerling/pyrouge

[6] https://github.com/miso-belica/sumy

Model	**PubMed**			**S2ORC**			**CORD-19**			**ArXiv**		
	R1	R2	RL	R1	R2	RL	R1	R2	RL	R1	R2	RL
Short Document												
ORACLE	47.27	22.85	43.20	49.29	25.42	45.52	43.47	17.75	39.28	47.29	18.49	41.90
RANDOM	34.98	10.82	31.37	34.69	11.06	31.30	31.64	7.91	28.10	34.53	8.88	30.26
LEAD	35.39	12.07	32.28	**40.50**	**16.72**	**37.68**	34.80	10.17	31.67	34.35	8.75	30.61
LexRank	38.48	13.05	34.92	39.44	14.57	36.13	35.65	10.17	32.11	38.98	11.44	34.64
TextRank	38.15	12.99	34.77	40.17	14.84	36.63	36.25	10.61	32.53	37.97	11.58	33.53
SumBasic	36.11	11.06	32.67	35.99	11.99	32.87	33.63	8.82	30.22	37.14	9.83	33.06
BERTExtSum	38.78	14.47	35.43	39.41	16.14	36.38	34.68	10.34	31.42	39.36	11.74	35.09
GenCompareSum (docTTTTTquery)	37.82	13.12	32.41	38.31	14.27	35.17	33.77	9.73	30.66	38.59	11.49	34.50
GenCompareSum (t5-med-query)	38.54	13.67	35.06	38.96	14.78	35.80	36.77	11.24	33.29	38.92	11.59	34.76
GenCompareSum (t5-s2orc-title)	**39.19**	14.35	**35.65**	40.16	15.84	36.91	**36.84**	**11.35**	**33.35**	**39.66**	**12.30**	**35.38**
Long Document												
ORACLE	61.76	36.78	57.61	64.11	39.21	60.16	59.10	32.09	54.63	60.16	32.17	54.97
RANDOM	37.26	11.19	33.66	37.12	10.23	33.73	33.37	7.70	29.98	34.20	8.70	30.64
LEAD	37.23	11.11	33.67	40.50	16.72	37.68	34.61	10.17	31.68	34.70	10.27	31.37
LexRank	41.02	15.83	37.18	42.60	15.84	38.97	39.50	12.65	35.68	33.94	12.09	30.62
TextRank	34.53	12.98	30.99	36.58	13.23	33.10	32.99	10.39	24.47	26.57	9.20	23.74
SumBasic	40.61	12.42	36.54	36.63	10.43	33.68	33.88	8.24	30.86	33.18	7.75	30.29
BERTExtSum*	41.87	16.01	**38.51**	**43.56**	**17.85**	**40.40**	38.95	12.17	35.48	**40.65**	14.01	**36.89**
GenCompareSum (docTTTTTquery)	40.54	14.77	36.83	40.78	14.24	37.43	36.84	11.19	33.51	38.19	12.76	34.55
GenCompareSum (t5-med-query)	41.60	15.67	37.79	41.84	15.10	38.35	39.33	12.31	35.74	37.17	11.97	33.95
GenCompareSum (t5-s2orc-title)	**42.10**	**16.51**	38.25	43.39	16.84	39.82	**41.02**	**13.79**	**37.25**	39.96	**15.15**	36.19

Table 2: Results of the extractive summarization task on the PubMed, ArXiv, s2orc and CORD-19 data sets. The short text version of the data set consists of the articles truncated at the end of the sentence containing the 512th token. We select 6 sentences for the short text summary. For the full-text document prediction, we use select the average number of sentences in the gold summaries of the respective data sets, which are given in Table 1, PubMed: 9, S2ORC: 9, CORD-19: 8, ArXiv: 11. Bold font indicates the top result within a data set, underlined font indicates the second-best result.

Interestingly, for the S2ORC data set, the method outperforming all others is LEAD, i.e., taking the first sentences from the document as the predictive summary. However, in evaluation of the full-text version of the S2ORC data set, it does not hold that LEAD is the best method, and it is seen to be outperformed by several other methods.

For the long document data sets, GenCompareSum (t5-s2orc-title) outperforms all other unsupervised models. A strong unsupervised baseline, LexRank has been shown in prior literature to give competitive performance when compared to supervised approaches (Cohan et al., 2018; Subramanian, Li and Pilault, 2020). In-line with these works, we show LexRank to be the best-performing unsupervised method after our own.

Our method, GenCompareSum (t5-s2orc-title), outperforms LexRank by a large margin – an average ΔR1, ΔR2, ΔR1 of 2.35, 1.47, 2.27 across the four data sets. We also demonstrate a slight performance increase over our implementation of the supervised method BERTExtSum*, which we adapted to run over longer documents. The same calculation across the data sets with BERTExtSum* shows us outperforming $\Delta R1, \Delta R2, \Delta R1$ by 0.36, 0.56, 0.06 across the four data sets. Given that our method is unsupervised, and therefore does not require labelled training data and can be extended to any document length, we believe this is a significant improvement.

Considering the different implementations of GenCompareSum, we can see that, as expected, our results show that finetuning on in-domain data gives notable performance increases. Table 2 shows that the $\Delta R1$ between an out-of-domain query generation model (docTTTTTquery) and a query generation model trained on biomedical data (t5-med-query) were as high as 3 and 2.49 for the short and long articles respectively, for the CORD-19 biomedical data set. However, for the ArXiv data set, which consists of predominantly physical and computer science related research articles, the performance decreased when using the t5-med-query generative model instead of the general domain docTTTTTquery model.

Our best-performing GenCompareSum model, t5-s2orc-title, uses a generative PLM finetuned on document-title pairs from the S2ORC data set to guide the extractive summarization. In many ways, a title can be considered as a highly abstractive summarization (Cachola et al., 2020). A major advantage of this finding is that, although it does require training data to finetune this generative model, document-title pairs are readily available across many domains, thus a model can easily be trained for a specific task without needing extensive manual labelling effort. Furthermore, this model, although finetuned on biomedical and scientific data, is finetuned on a very broad range of documents within these fields. We demonstrate that, despite the broad coverage of fields in its training data, it performs very well when applied to data from a more specific domain, e.g., biomedicine in the PubMed and CORD-19 data sets.

Lastly, we observe that there is big difference in ORACLE scores between the short and full text data sets. Although our models out-perform all other methods evaluated for both short and full text documents, the gap between the best predictive scores in our experiments and the ORACLE upper bound is large for long documents, suggesting that much more research could be done in this space. Furthermore, based on this observation, we also hypothesise that predicting summaries from short documents is a significantly easier task than doing the same for long documents. This is supported by TextRank performing worse on the long documents than on the truncated versions. We believe this is explainable both by the fact that there are much fewer sentences to choose from within a shorter document (we select approximately 32% of all sentences across the data sets for short document predictions and 5% for the full documents), thus less room for error. Furthermore, previous work has shown that often the most important parts of the document are towards the beginning of it (Zheng et al., 2019), implying that there is less 'noise' (i.e., unimportant sentences) to select from a truncated document.

4.2 Qualitative analysis

In Appendix E we provide a randomly sampled PubMed document, the associated generated salient fragments, and the predicted extractive summary given by each of the three GenCompareSum methods. We also provide the gold summary (document abstract) for comparison. In Appendix F, we give the same for a randomly sampled document from the ArXiv data set. In this section, we comment on the difference between the texts generated by the difference T5-based models and hypothesise on how this influences the extractive summary.

The docTTTTTquery model produces questions which are relatively general and imply little biomedical knowledge when provided the PubMed document as input. In this setting, it produces textual fragments such "what is nlrp3". Interestingly, it does manage to produce more complex texts from sections of the ArXiv data set, such as: "what is the contribution of the spiral arm to the resonant structure in the solar neighborhood?".

In comparison, the t5-med-query model, whilst also generating questions, better encapsulates biomedical concepts when given a document from the PubMed dataset, e.g., "what is the role of nuclear and mitochondrial dna damage and repair in people with depression?". However, in line with the ROUGE results given in Section 4.1, it seems to perform less well on out-of-domain (i.e., scientific rather than biomedical) literature, and appears to default to a more general question generation model, generating texts for the ArXiv document such as "what is the effect of a spiral arm?".

The t5-s2orc-title model generates texts which read much more like very short, highly abstractive summaries. E.g., for the PubMed article, it generated the textual fragment: "the role of the nuclear and mitochondrial dna in depression" and for the ArXiv article it generated: "the spiral arm contribution to the resonant structure of the solar neighborhood". Although outperformed by the title-generation model t5-s2orc-title in the automatic evaluation, on analysis of the generated textual fragments, the query generation models do seem to effectively represent the important facts from an article, especially in the biomedical domain. We hypothesise that our use of BERTScore, to calculate the similarity between salient texts and document sentences, favours the title generation model due to it calculating the similarity between words in different texts and not being designed to answer questions. In future work, we would like to experiment further with the combination of the query generation models and extractive question answering approaches for the extractive summarization task.

5 Future work

In this section we suggest future directions for our research. Firstly, we highlight that our method is generalizable and not restricted to T5-based architectures for the generation of salient text fragments. Therefore, we would like to experiment with different models for this step, e.g., BART-based (Lewis, 2019) models, or models trained with different data.

Another interesting direction would be the inclusion of zoning into the method. As mentioned previously, we chose not to use an article's pre-defined sections as they are often not available. However, it would be interesting to predict a classification for a sentence within the text (e.g., 'Results' for a scientific article), and to incorporate this into the model.

We would also like to evaluate our models on other data sets and domains in future research, e.g., clinical notes, and would like to carry out a human evaluation to validate the results, ideally with experts from the same domain as the data being summarized. Human evaluation would allow for aspects such as fluency, factual consistency, and coherence to be assessed, which have been shown not to necessarily align well with ROUGE evaluation in previous works (Kryscinski et al., 2019; Huang et al., 2020).

Lastly, our analysis (details in Appendix D) showed that BERTScore was the best performing method for calculating text similarity for the extractive summarization task, outperforming methods using sentence embeddings. We hypothesise that this could be due to our evaluation metric being ROUGE scoring, which favours methods that produce summaries containing exactly the same words as the gold summary, rather than semantically similar sentences. Experimentation into different evaluation metrics for extractive summarization, including human evaluation, and how they correlate to the performance of our methods when using different models for calculating text similarity, is also an interesting direction for future work.

6 Conclusion

In this work we propose GenCompareSum, a novel two-step unsupervised hybrid abstractive-extractive method for text summarization. We evaluate the efficacy of using PLMs to generate salient textual fragments which represent the key points of a document – experimenting with generation of both queries and document titles - and using them to guide the second step, extractive summarization. We show that that our unsupervised method, which can be extended to any length of document and does not require a corpus of annotated training data, outperforms over both strong supervised and unsupervised baselines on long and short documents. Furthermore, we show that our best-performing model uses title-document pairs for the generative task, which are readily available across many domains without the need for manual labelling effort.

Author contributions

JA Bishop proposed the research themes, developed the code, conducted the experiments, and drafted the manuscript. Q Xie and S Ananiadou supervised all steps of the work and revised the manuscript. All authors approved the final version of the manuscript.

References

Bajaj, Payal, et al. 2018. "MS MARCO: A Human Generated MAchine Reading

COmprehension Dataset." *arXiv:1611.09268.*

Beltagy, Iz, Kyle Lo, and Arman Cohan. 2019. "SciBERT: A pretrained language model for scientific text." *arXiv preprint arXiv:1903.10676.*

Brin, Sergey, and Lawrence Page. 1998. "The anatomy of a large-scale hypertextual web search engine." *Computer networks and ISDN systems* 30 (1-7): 107-117.

Cachola, Isabel, Kyle Lo, Arman Cohan, and Daniel S. Weld. 2020. "TLDR: Extreme Summarization of Scientific Documents." *Findings of the Association for Computational Linguistics: EMNLP 2020.* 4766-4777.

Cohan, Arman, et al. 2018. "A discourse-aware attention model for abstractive summarization of long documents." *NAACL HLT 2018 - 2018 Conference of the North American Chapter of the Association for Computational Linguistics: Human Language Technologies - Proceedings of the Conference.* 615-621.

Cohan, Arman, et al. 2020. "Specter: Document-level representation learning using citation-informed transformers." *arXiv preprint arXiv:2004.07180.*

Devlin, Jacob, Ming-Wei Chang, Kenton Lee, and Kristina Toutanova. 2018. "Bert: Pre-training of deep bidirectional transformers for language understanding." *arXiv preprint arXiv:1810.04805* .

Dou, Zi-Yi, Pengfei Liu, Hiroaki Hayashi, and Zhengbao and Neubig, Graham Jiang. 2021. "GSum: A General Framework for Guided Neural Abstractive Summarization." *Proceedings of the 2021 Conference of the North American Chapter of the Association for Computational Linguistics: Human Language Technologies.* 4830-4842.

Erkan, Gunes, and Dragomir R. Radev. 2004. "Lexrank: Graph-based lexical centrality as salience in text summarization." *Journal of artificial intelligence research* 22: 457-479

Gao, Tianyu, Xingcheng Yao, and Danqi Chen. 2021. "Simcse: Simple contrastive learning of sentence embeddings." *arXiv preprint arXiv:2104.08821.*

Travis Goodwin, Max Savery, and Dina Demner-Fushman. 2020. Towards Zero-Shot Conditional Summarization with Adaptive Multi-Task Fine-Tuning. In *Findings of the Association for Computational Linguistics: EMNLP 2020,* pages 3215–3226, Online. Association for Computational Linguistics.

Grail, Quentin, Julien Perez, and Eric Gaussier. 2021. "Globalizing BERT-based Transformer Architectures for Long Document Summarization." *Proceedings of the 16th Conference of the European Chapter of the Association for Computational Linguistics: Main Volume.*

Huang, Dandan, et al. 2020. "What have we achieved on text summarization?" In *Proceedings of the 2020 Conference on Emperical Methods in Natural Language Processing (EMNLP),* pages 446-469. Association for Computational Linguistics.

Jin, Qiao, et al. 2019. "PubMedQA: A Dataset for Biomedical Research Question Answering." *arXiv preprint arXiv:1909.06146.*

Klein, Tassilo, and Moin Nabi. 2019. "Learning to answer by learning to ask: Getting the best of gpt-2 and BERT worlds." *arXiv preprint arXiv:1911.02365.*

Kryscinski, Wojciech, Nitish Shirish Keskar, Bryan McCann, Caiming Xiong, and Richard Socher. 2019. Neural Text Summarization: A Critical Evaluation. In *Proceedings of the 2019 Conference on Empirical Methods in Natural Language Processing and the 9th International Joint Conference on Natural Language Processing (EMNLP-IJCNLP),* pages 540–551, Hong Kong, China. Association for Computational Linguistics.

Lewis, Mike, Yinhan Liu, Naman Goyal, Marjan Ghazvininejad, Abdelrahman Mohamed, Omer Levy, Ves Stoyanov, and Luke Zettlemoyer. 2019. "Bart: Denoising sequence-to-sequence pre-training for natural language generation, translation, and comprehension." *arXiv preprint arXiv:1910.13461.*

Liang, Xinnian, et al. 2021. "Improving unsupervised extractive summarization with facet-aware modeling." *Findings of the Association for Computational Linguistics: ACL-IJCNLP* .

Lin, Chin-Yew. 2004. "ROUGE: A package for automatic evaluation of summaries." *Text Summarization Branches Out.* pages 74-81. Association for Computational Linguistics.

Liu, Yang, and Mirella Lapata. 2019. "Text summarization with pretrained encoders." In *Preceedings of the 2019 Conference onon Empirical Methods in Natural Language Processing and 9th International Joint Conference on Natural Language Processing (EMNLP-IJCNLP),* pges 3730-3740, Hong Kong, China. Association for Computational Linguistics.

Lo, Kyle, Lucy Lu Wang, Mark Neumann, Rodney Kinney, and Daniel. Weld. 2020. "S2ORC: The Semantic Scholar Open Research Corpus." *Proceedings of the 58th Annual Meeting of the Association for Computational Linguistics.* pages 4969-4983, Online. Association for Computational Linguistics.

Möller, Timo, Anthony Reina, Raghavan Jayakumar, and Malte Pietsch. 2020. "COVID-QA: A Question Answering Dataset for COVID-19" *Proceedings of the 1st Workshop on NLP for COVID-19 at ACL 2020,* Online. Association for Computational Linguistics.

Manning, Christopher D, and et al. 2014. "The Stanford CoreNLP natural language processing toolkit." *Proceedings of 52nd annual meeting of the association for computational linguistics: system demonstrations.*

Mihalcea, Rada, and Paul Tarau. 2004. "Textrank: Bringing order into text." *Proceedings of the 2004 conference on empirical methods in natural language processing.* Pages 404-411. Barcelona, Spain. Association for Computational Linguistics.

Nenkova, Ani, and Lucy Vanderwende. 2005. "The impact of frequency on summarization." *Microsoft Research, Redmond, Washington, Tech. Rep. MSR-TR-2005* 101.

Nentidis, Anastasios et al. 2021. "Overview of BioASQ 2021: The ninth BioASQ challenge on Large-Scale Biomedical Semantic Indexing and Question Answering." *Experimental IR Meets Multilinguality, Multimodality, and Interaction* 239–263.

Nogueira, Rodrigo, and Jimmy Lin. 2019. "From doc2query to docTTTTTquery." *Online preprint.*

Raffel, Colin, et al. 2020. "Exploring the limits of transfer learning with a unified text-to-text transformer." *Journal of Machine Learning Research* 21: 1-67.

Reimers, Nils, and Iryna Gurevych. 2019. "Sentence-bert: Sentence embeddings using siamese bert-networks." *arXiv preprint arXiv:1908.10084.*

Leonardo F. R. Ribeiro, Martin Schmitt, Hinrich Schütze, and Iryna Gurevych. 2021. Investigating Pretrained Language Models for Graph-to-Text Generation. In *Proceedings of the 3rd Workshop on Natural Language Processing for Conversational AI,* pages 211–227, Online. Association for Computational Linguistics.

Rohde, Tobias, Xiaoxia Wu, and Yinhan Liu. 2021. "Hierarchical learning for generation with long source sequences." *arXiv preprint arXiv:2104.07545.*

See, Abigail, Peter J Liu, and Christopher D Manning. 2017. "Get to the point: Summarization with pointer-generator networks." *ACL 2017 - 55th Annual Meeting of the Association for Computational Linguistics, Proceedings of the Conference (Long Papers).* pages 1073-1083, Ancouver, Canada, Association for Computational Linguistics.

Subramanian, Sandeep, Raymond Li, Jonathan Pilault and Chris Pal. 2020. "On Extractive and Abstractive Neural Document Summarization with Transformer Language Models." In *Proceedings ofthe 2020 Conference on Empirical Methods in Natural Language Processing.* pages 9308–9319, Online. Association for Computational Linguistics.

Surita, Gabriela, Rodrigo Nogueira, and Roberto Lotufo. 2020. "Can questions summarize a corpus? Using question generation for characterizing COVID-19 research." *arXiv:2009.092900.*

Wallace, Byron C, Sayantan Saha, Frank Soboczenski, and Iain J Marshall. 2021. "Generating (Factual?) narrative summaries of RCTs: Experiments with neural multi-document summarization." *AMIA Jt Summits Transl Sci Proc.* 605–614.

Wang, Lucy Lu, and et al. 2020. "CORD-19: The covid-19 open research dataset." *arXiv preprint arXiv:2004.10706v4.*

Xiao, Wen, and Giuseppe Carenini. 2019. "Extractive summarization of long documents by combining global and local context." *arXiv preprint arXiv:1909.08089.*

Xiao, Wen and Giuseppe Carenini. 2020. "Systematically Exploring Redundancy Reduction in Summarizing Long Documents." *Proceedings of the 1st Conference of the Asia-Pacific Chapter of the Association for Computational Linguistics and the 10th International Joint Conference on Natural Language Processing.* pages 516-528. Association for Computational Linguistics.

Xu, Shusheng, Xingxing Zhang, Yi Wu, Furu Wei, and Ming Zhou. 2020. "Unsupervised Extractive Summarization by Pre-training Hierarchical Transformers" *Findings of the Association for Computational Linguistics: EMNLP 2020.* pages 1784-1795, Online. Association for Computational Linguistics.

Yuchen Lin, Bill, Wangchunshu Zhou, Ming Shen, Pei Zhou, Chandra Bhagavatula, Yejin Choi, and Xiang Ren. 2020. CommonGen: A Constrained Text Generation Challenge for Generative Commonsense Reasoning. In Findings of the Association for Computational Linguistics: EMNLP 2020, pages 1823–1840, Online. Association for Computational Linguistics.

Zhang, Jingqing, Yao Zhao, Mohammad Saleh, and Peter Liu. 2020. "PEGASUS: Pre-training with Extracted Gap-sentences for Abstractive Summarization" *Proceedings of the 37th International Conference on Machine Learning.* 1328-11339.

Zhang, Tianyi, and et al. 2020. "BERTScore: Evaluating Text Generation with BERT" *International Conference on Learning Representations.*

Zheng, Hoa, and Mirella Lapata. 2019. "Sentence Centrality Revisited for Unsupervised Summarization." In *Procedings of the 57th Annual Meeting of the Association for Computational Liguistcs*. pages 6236-6247, Florence, Italy. Association for Computational Linguistics.

Zhong, M, P Liu, Y Chen, D Wang, X Qiu, and X. Huang. 2020. "Extractive summarization as text matching" *Proceedings of the 58th Annual Meeting of the Association for Computational Linguistics.* Pages 6197-6208, Online. Association for Computational Linguistics.

Zhu, Ming, Aman Ahuja, Da-Cheng Juan, Wei Wei, and Chandan K. Reddy. 2020. "Question answering with long multiple-span answers" *Findings of the Association for Computational Linguistics: EMNLP 2020.* Pages 3840-3849 , Online. Association for Computational Linguistics.

Appendix A. T5 model architecture.

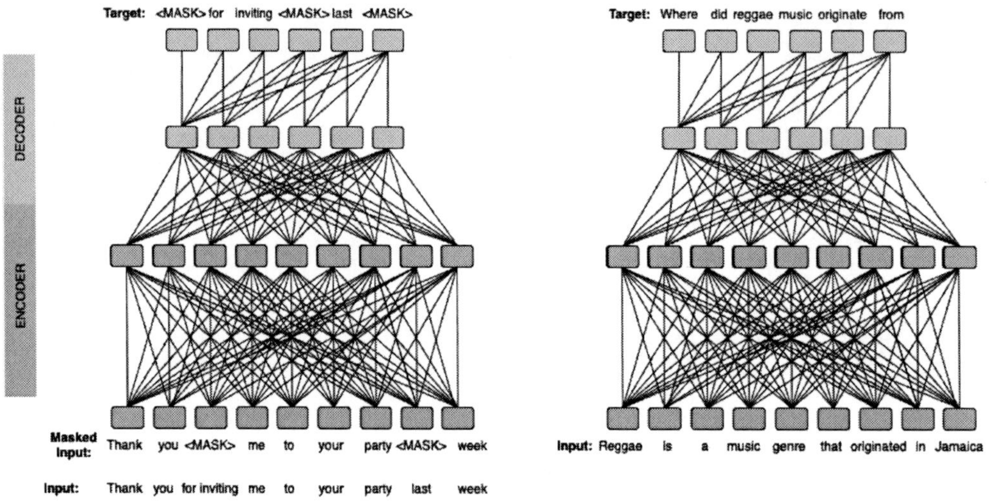

Figure 2: Representations of the two training settings of the T5 encoder-decoder model. The left diagram shows the unsupervised pretraining task, in which a tokenized text containing masked spans is passed to the encoder and the output target of the decoder is the prediction of the masked spans. The right diagram shows the supervised downstream task, where the pre-trained model is finetuned on pairs of tokenized sequences.

Appendix B. t5-med-query training.

To finetune our GenCompareSum (t5-med-query) model, we combine four biomedical data sets to make a large corpus of text-question pairs, where the questions can be answered by the long textual input. From the BioAsq data set (Nentidis et al., 2021), 3,433 'ideal answer'-question pairs were used, 2,720 text-question pairs from COVID-QA (Möller et al., 2020), where the paragraph containing the answer is used as the text input, 61,244 context-question pairs from PubMedQA (Jin et al. 2019), where the 'context' refers to the abstract without its 'conclusion' section, and 27,722 long answer-question pairs from the MASH-QA (Zhu et al. 2020) data set. The t5-base model is loaded and finetuned on this data set for 5 epochs, with a batch size of 8.

Appendix C. Parameter selection.

Parameter Name	Parameter Definition	Parameter range experimented with	Optimal parameter selected
T5 model temperature	Controls randomness of generative text model predictions	0.2-1	0.5
T5 input size (x)	Number of sentences used to form sections to input to T5 text generation model	2-12	4
T5 predictions per input (k)	Number of salient texts generated per section passed to the model	2-6	3
T5 prediction n-gram blocking	Number of consecutive word matches used to determine whether a generated text should be removed due to redundancy when compared to another generated text	No n-gram blocking, n=3, n=4	4
T5 generated texts used for comparison (q)	Number of generated texts used for comparison to the original document sentences	4-12	10
BERTScore embedding model	Base model used in BERTScore package for word-embedding comparison	bert-base-uncased[7], facebook/bart-large-mnli[8], allenai/longformer-large-4096[9], allenai/scibert_scivocab_uncased[10]	bert-base-uncased
BERTScore batch size	Batch size in BERTScore package	64 (used default)	64
Score weighting	Optional multiplication of scores by frequency of question occurrence	True/False	True
Sentence selecton n-gram blocking	Number of consecutive word matches used to determine whether a selected sentence should be removed due to redundancy when compared to another selected sentence	No n-gram blocking, n=3, n=4	4

Table 3: Parameters experimented with, and selected for use, in the GenCompareSum models.

[7] https://huggingface.co/bert-base-uncased

[8] https://huggingface.co/facebook/bart-large-mnli

[9] https://huggingface.co/allenai/longformer-base-4096

[10] https://huggingface.co/allenai/scibert_scivocab_uncased

Appendix D. Analysis of methods for calculating text similarity

In this section we compare different methods for calculating the similarity between the generated salient text fragments and the document sentences. We use our best performing model, GenCompareSum (s2orc-title), and implement different models for the text comparison step. We present results for the extractive summarization task on the PubMed 'Short Document' data set.

We compare BERTScore, a method which uses word embeddings to calculate the similarity between texts, with two other methods to calculate the similarity between texts using sentence embeddings. Sentence Transformers (Reimers and Gurevych, 2019) is trained with a triplet / siamese bert-based architecture and a training objective designed to minimize distances between similar sentences. We implement this method with their python package[11]. We compare both their suggested base model for the general domain 'all-mpnet-base-v2' and a model trained to calculate document-level similarity for scientific documents 'allenai-specter' (Cohan et al., 2020). We also implement SimCSE (Gao et al., 2021), which generates sentence embeddings with a model trained using contrastive learning. For this method, we use the general-domain base model which is suggested to be the best performing in SimCSE's documentation[12]. For the BERTScore method, we experiment with base models from the general domain, namely 'bert-base-uncased'[13], which was used in our implementations to give the results in Table 2 of the main manuscript, and a base model pretrained on data from the scientific domain (Beltagy et al., 2019), 'allenai/scibert_scivocab_cased'.

Table 4 gives the results. We can observe that BERTScore, implemented with a base model from the general domain, outperforms all other methods compared for calculating text similarity on the extractive summarization task when evaluated using ROUGE metrics.

Text similarity method	R1	R2	RL
BERTScore (bert-base-uncased)	39.19	14.35	35.65
BERTScore (allenai/scibert_scivocab_cased)	37.78	13.40	34.45
SentenceTransformer (all-mpnet-base-v2)	39.03	14.20	35.45
SentenceTransformer(allenai-specter)	38.20	13.41	34.67
SimCSE (princeton-nlp/sup-simcse-roberta-large)	38.62	13.73	35.07

Table 4: A comparison of ROUGE-1,-2 and -L results for the PubMed Short Document data set on the extractive summarization task, using different methods for calculating text similarity between generated salient texts and the document's sentences. The method is given in the first column, with the base model used in its implementation given in brackets.

[11] https://github.com/UKPLab/sentence-transformers

[12] https://github.com/princeton-nlp/SimCSE

[13] https://huggingface.co/bert-base-uncased

Appendix E. Example output of our method on a PubMed article.

PubMed Sample Document and Predictions	
PubMed Sample Document	https://www.ncbi.nlm.nih.gov/pmc/articles/PMC4329942/
PubMed Sample Abstract (Target Summary)	depressive disorder (dd) , including recurrent dd (rdd) , is a severe psychological di sease , which affects a large percentage of the world population . although pathogenes is of the disease is not known , a growing body of evidence shows that inflammation t ogether with oxidative stress may contribute to development of dd . since reactive oxy gen species produced during stress may damage dna , we wanted to evaluate the exten t of dna damage and efficiency of dna repair in patients with depression. material / we measured and compared the extent of endogenous dna damage single - and double - st rand breaks , alkali - labile sites , and oxidative damage of the pyrimidines and purine s in peripheral blood mononuclear cells isolated from rdd patients (n = 40) and healt hy controls (n = 46) using comet assay . we also measured dna damage evoked by h ydrogen peroxide and monitored changes in dna damage during repair incubation. we found an increased number dna breaks , alkali - labile sites , and oxidative modificatio n of dna bases in the patients compared to the controls . exposure to hydrogen peroxid e evoked the same increased damage in both groups . examination of the repair kineti cs of both groups revealed that the lesions were more efficiently repaired in the contro ls than in the patients. the first time we showed that patients with depression , compar ed with non - depresses individuals , had more dna breaks , alkali - labile sites , and o xidative dna damage , and that those lesions may be accumulated by impairments of t he dna repair systems . more studies must be conducted to elucidate the role of dna da mage and repair in depression .
Salient Texts - GenCompareSum (docTTTTTquery)	• what is nlrp3 • how long does it take for dna damage to be repaired • what is the oxidative modification of purines • what is the main activator of nlrp3 • what is the damage caused by dna repair • what is the role of mitochondrial dna in depression • what is oxidative damage in dna • does oxog cause depression • what is the dna damage response • what is the oxidative damage of pyrimidines and purines?
Predicted Summary - GenCompareSum (docTTTTTquery)	• since the findings described above are inconsistent , we wanted to determine if the oxidative modification of purines , like 8 - oxog , and pyrimidines are present in a higher degree in patients with depression than in controls . • to achieve these objectives , we measured and compared the extent of endog enous dna damage single - and double - strand breaks , alkali - labile sites , a nd oxidative damage of the pyrimidines and purines in pbmcs isolated from dd patients and healthy controls . • we evaluated the level of basal endogenous dna damage by subjecting pbmcs to comet assay procedure immediately after isolation from blood . • moreover , we estimated the extent of oxidative dna damage by employing m odified comet assay with 2 glycosylases : nth removing oxidized pyrimidines and hogg1 excising oxidized purines . • figure 3 shows mean dna damage changes in pbmcs of the patients with depr ession and the controls without psychiatric disturbances during the repair inc ubation . • the goal of our research was to examine the susceptibility of rdd patients to d na damage induced by oxidative stress by measuring the level of endogenous dna damage , including oxidative dna damage , the amount of dna damage in duced by h2o2 , and efficiency of dna damage repair in the patients as compa red to the controls without psychological disturbances . • apart from measuring the extent of endogenous dna damage , we also estimat ed the amount of dna damage induced by the incubation of pbmcs with h2o2 and efficiency of its repair .

	• additionally , we monitored the repair efficiency of the induced dna damage . • moreover , nlrp3 inflammasome , activation of which was detected in the pat ients pbmcs , was also found to inhibit dna repair after induction of oxidative stress .
Salient Texts - GenCompareSum (t5-med-query)	• what was the purpose of the study? • what is the alkaline version of the comet assay? • what is the effect of pbmcs on basal endogenous dna damage? • what is the incubation time for dna repair? • what is the role of nuclear and mitochondrial dna damage and repair in peopl e with depression? • is it possible to study the susceptibility of rdd patients to dna damage induce d by oxidative stress? • what is recurrent depressive disorder? • what is the association between 8 - oxog and depression in japanese office w orkers? • which is the most versatile nlr? • what enzymes are bifunctional glycosylases?
Predicted Summary - GenCompareSum (t5-med-query)	• moreover , we also wanted to know if the patients have elevated levels of oth er kinds of dna damage , such as strand breaks . • we evaluated the level of basal endogenous dna damage by subjecting pbmcs to comet assay procedure immediately after isolation from blood . • figure 2 shows basal endogenous dna damage and the damage induced after 10 - min incubation with 20 m h2o2 in pbmcs isolated from the patients and controls without psychiatric disturbances . • figure 3 shows mean dna damage changes in pbmcs of the patients with depr ession and the controls without psychiatric disturbances during the repair inc ubation . • figure 5 compares basal endogenous dna damage and the level of this param eter at the end of the repair incubation in pbmcs of the patients and the contr ols measured by the alkaline version of comet assay . • the goal of our research was to examine the susceptibility of rdd patients to d na damage induced by oxidative stress by measuring the level of endogenous dna damage , including oxidative dna damage , the amount of dna damage in duced by h2o2 , and efficiency of dna damage repair in the patients as compa red to the controls without psychological disturbances . • apart from measuring the extent of endogenous dna damage , we also estimat ed the amount of dna damage induced by the incubation of pbmcs with h2o2 and efficiency of its repair . • additionally , we monitored the repair efficiency of the induced dna damage . • there is a need for further studies to define the role of nuclear and mitochond rial dna damage and repair in people with depression , and their implications for clinical outcome .
Salient Texts - GenCompareSum (t5-s2orc-title)	• dna damage in patients with depression. • oxidative dna damage in depression • the oxidative dna damage in patients with renal failure • activation of nlrp3 by oxygen species in pbmc patients. • activation of mitochondrial nlrp3 in patients with pbmcs. • urinary 8-oxog in japanese office workers • the use of the alkaline version of comet assay for assessing dna damage in pb mcs • the role of the nuclear and mitochondrial dna in depression. • the role of the dna repair rate in the repair of pbmcs in patients with squamou s cell carcinoma.
Predicted Summary - GenCompareSum (t5 -s2orc-title)	• in agreement with this , activation of nlrp3 in pbmcs of the patients was acco mpanied by increased lipid peroxidation , which can be attributed to increase d oxidative stress and elevated mitochondrial ros (mtros) production .

	• moreover , we induced oxidative dna damage in those pbmcs by incubating t hem with hydrogen peroxide , measured the kinetics of removing of such da mage , and compared the results between the patients and the controls . • we evaluated the level of basal endogenous dna damage by subjecting pbmcs to comet assay procedure immediately after isolation from blood . • figure 2 shows basal endogenous dna damage and the damage induced after 10 - min incubation with 20 m h2o2 in pbmcs isolated from the patients and controls without psychiatric disturbances . • figure 3 shows mean dna damage changes in pbmcs of the patients with depr ession and the controls without psychiatric disturbances during the repair inc ubation . • it is possible that increased oxidative dna damage occurs only in patients wit h more severe forms of depression , or in later stages of the disease develop ment . • these results indicate that in the patients , oxidative dna damage is less effici ently removed than in the controls . • moreover , nlrp3 inflammasome , activation of which was detected in the pat ients pbmcs , was also found to inhibit dna repair after induction of oxidative stress . • for the first time , we showed that patients with depression had elevated level s of dna breaks , alkali - labile sites , and oxidative dna damage , and that the se lesions may be accumulated by impairments of dna repair pathways .

Appendix F. Example output of our method on an ArXiv article.

ArXiv Sample Document and Predictions	
ArXiv Sample Document	https://arxiv.org/abs/0906.4682
ArXiv Sample Abstract (Target Summary)	we study the phase space available to the local stellar distribution using a galactic pot ential consistent with several recent observational constraints . we find that the induced phase space structure has several observable consequences . the spiral arm contribution to the kinematic structure in the solar neighborhood may be as important as the one produced by the galactic bar . we suggest that some of the stellar kinematic groups in the solar neighborhood , like t he hercules structure and the kinematic branches , can be created by the dynamical res onances of self - gravitating spiral arms and not exclusively by the galactic bar . a structure coincident with the arcturus kinematic group is developed when a hot stell ar disk population is considered , which introduces a new perspective on the interpret ation of its extragalactic origin . a bar - related resonant mechanism can modify this kinematic structure . we show that particles in the dark matter disk - like structure predicted by recent lcd m galaxy formation experiments , with similar kinematics to the thick disk , are affect ed by the same resonances , developing phase space structures or dark kinematic grou ps that are independent of the galaxy assembly history and substructure abundance . we discuss the possibility of using the stellar phase space groups as constraints to no n - axisymmetric models of the milky way structure .
Salient Texts - GenCompareSum (docTTTTTquery)	• what is the role of the bar in the local kinematic structure • what is the effect of the non axisymmetric galactic structure on the solar neig hborhood kinematic distribution? • what is the shape of the solar structure at @xmath27 • what is the structure of the hercules branch • what is the effect of a spiral arm • what is the hercules structure • how does the kinematics of the disk affect the galaxy? • which of the following structure is a contribution to the solar neighborhood k inematics? • what type of spiral arm is used to measure observations made in the solar nei ghborhood • what is the contribution of the spiral arm to the resonant structure in the solar neighborhood?
Predicted Summary - GenCompareSum (docTTTTTquery)	• however , it is unclear whether there is any dependence of the induced local solar neighborhood kinematics on the detailed galactic structure . • in order to study the effect of the non - axisymmetric galactic structure on th e solar neighborhood kinematic distribution , we have performed numerical i ntegrations of test particle orbits on the galactic plane , adopting the initial c onditions discussed in sect . • the induced kinematic distribution at the end of the simulation is studied by c onsidering the particles inside a circle of radius @xmath7 centered at the sol ar position . • therefore we focused on the recently induced kinematic structure in the solar neighborhood . • with these initial conditions , we can study the relatively rapid induced effect s of the non - axisymmetric component on the local kinematics . • we conclude that the contribution of the spiral arms to the solar neighborhoo d kinematics may be comparable to that of the bar . • in our simulations the positions of these kinematic arches are modified when the bar is added to the model . • furthermore , these simulations show the important role of the bar in the dev elopment of the local kinematic structure .

	• the spiral arm contribution to the resonant structure in the solar neighborhoo d may be comparable to that of the galactic bar . • in summary , the imprints of the non - axisymmetric galactic structure on the local stellar kinematics are strong .
Salient Texts - GenCompareSum (t5-med-query)	• what is the effect of dark matter kinematics on the bar - and spiral arm - indu ced phase space structure? • what is the main argument of @xcite? • what is the structure of the hercules? • what is the solar neighborhood? • what is the kinematic distribution of the particles? • what is the relationship between spiral arms and stellar behavior? • what is the galactic potential? • what is the required condition for a thick disk? • what is the difference between ic3 and ic2? • why is the observed velocity field a useful parameter for predicting the beha vior of galaxies?
Predicted Summary - GenCompareSum (t5-med-query)	• however , it is unclear whether there is any dependence of the induced local solar neighborhood kinematics on the detailed galactic structure . • moreover , the initial conditions hardly consider the evolution of the mw . • the induced kinematic distribution at the end of the simulation is studied by c onsidering the particles inside a circle of radius @xmath7 centered at the sol ar position . • therefore we focused on the recently induced kinematic structure in the solar neighborhood . • we conclude that the contribution of the spiral arms to the solar neighborhoo d kinematics may be comparable to that of the bar . • in our simulations the positions of these kinematic arches are modified when the bar is added to the model . • another unexpected aspect of the bar - and spiral arm - induced phase space s tructure is the effect on the local dark matter kinematics . • the spiral arm contribution to the resonant structure in the solar neighborhoo d may be comparable to that of the galactic bar . • the main differences to previous studies are the arm force contrast and force field shape? • in summary , the imprints of the non - axisymmetric galactic structure on the local stellar kinematics are strong .
Salient Texts - GenCompareSum (t5-s2orc-title)	• dark matter kinematics in the solar neighborhood • a note on the arcturus structure in a $xmath26$ plane • dark kinematic groups in the dark disk • the spiral arm contribution to the resonant structure of the solar neighborhoo d • the birth of stars in the disk with small velocity dispersion • the solar neighborhood kinematics and the spiral arms • spiral arms in the mw-type galaxies • the hercules branch of a galactic model using only a bar • theoretical study of the bar and spiral arm perturbations in the xci model • dark matter currents in the galactic dark disk
Predicted Summary - GenCompareSum (t5 -s2orc-title)	• in @xcite we presented a study of the solar neighborhood kinematic groups using a sample of 24,190 stars . • lastly , we investigate effects on the local dark matter kinematics , in particul ar in the disk - like dark matter structure recently predicted by lcdm models . • the induced kinematic distribution at the end of the simulation is studied by c onsidering the particles inside a circle of radius @xmath7 centered at the sol ar position . • therefore we focused on the recently induced kinematic structure in the solar neighborhood .

	• we conclude that the contribution of the spiral arms to the solar neighborhoo d kinematics may be comparable to that of the bar . • another unexpected aspect of the bar - and spiral arm - induced phase space s tructure is the effect on the local dark matter kinematics . • our results show that these models generate dark matter currents inside the g alactic dark disk . • the spiral arm contribution to the resonant structure in the solar neighborhoo d may be comparable to that of the galactic bar . • we show that the galactic non - axisymmetric potential develops dark kinema tic groups in the dark disk predicted in cosmological simulations of galaxy f ormation . • in summary , the imprints of the non - axisymmetric galactic structure on the local stellar kinematics are strong .

BioCite: A Deep Learning-based Citation Linkage Framework for Biomedical Research Articles

Sudipta Singha Roy
University of Western Ontario
ssinghar@uwo.ca

Robert E. Mercer
University of Western Ontario
mercer@csd.uwo.ca

Abstract

Research papers reflect scientific advances. Citations are widely used in research publications to support the new findings and show their benefits, while also regulating the information flow to make the contents clearer for the audience. A citation in a research article refers to the information's source, but not the specific text span from that source article. In biomedical research articles, this task is challenging as the same chemical or biological component can be represented in multiple ways in different papers from various domains. This paper suggests a mechanism for linking citing sentences in a publication with cited sentences in referenced sources. The framework presented here pairs the citing sentence with all of the sentences in the reference text, and then tries to retrieve the semantically equivalent pairs. These semantically related sentences from the reference paper are chosen as the cited statements. This effort involves designing a citation linkage framework utilizing sequential and tree-structured siamese deep learning models. This paper also provides a method to create an automatically generated corpus for such a task.

1 Introduction

Research articles from different domains use varying writing styles and formats. They serve different purposes as well. A research publication may discuss current research trends, a novel discovery, or alternative approaches to solving a problem in a given domain. While writing a research article, the author mentions prior research that was either significant in resolving the same topic or impacted the author's views mentioned in the current research paper. This referencing another document in a research piece is referred to as a *citation* (Houngbo, 2017). This way, citations establish connections between distinct research literature as well as alleviating authors' writing burden by preventing them from having to write the same thing mentioned in another research article again. Simultaneously, it assists readers in acquiring prior knowledge about a subject that may be necessary to comprehend the ideas contained in the ongoing research work.

The idea of citation indexing was first introduced in 1964 where indexes contain the references in a research document. Citation-based bibliometrics are utilized to evaluate the significance of a research work (Garfield, 1972). In response to the growing popularity of citation indexing, a more critical analysis of citing was later suggested. Garzone and Mercer (Garzone and Mercer, 2000) devised a mechanism for determining the objective of a reference in biochemistry and physics research publications. Moreover, citations help to keep track of the logical argumentation across various research articles (Mercer, 2016). Prominent applications of citation incorporate maintaining the trail of scientific research argumentation across different research articles (Palau and Moens, 2009) and summarization of these documents (Radev et al., 2000).

In scientific research publications, a citation refers to the source article from which the cited notion is drawn. However, in experimental biomedical research articles, a citing sentence usually only relates to a small text span of the cited document's contents. This small span of text can be from the method section, result analysis section or any other section of the reference document (Singha Roy et al., 2020). The above-mentioned applications would substantially benefit if such a text span could be extracted from the original document. It would also free up the readers from having to read the full document to locate the cited piece of text.

The citation linkage task is more complicated for biomedical research papers as the same chemical or biological component has various representation formats and the use of these variations is very common in such research articles. For example, the chemical compound carbon dioxide can be represented as CO_2 as well as $O{=}C{=}O$, whereas in some

Proceedings of the BioNLP 2022 workshop, Dublin, Ireland, pages 241–251
May 26, 2022. ©2022 Association for Computational Linguistics

articles the writers write the whole name in plain text (*carbon dioxide*). Similarly, there are multiple representations to indicate the same reactions between various genes, chemicals, and drugs. On top of that, the only human annotated corpus available for the citation linkage task in the biomedical domain is from (Houngbo and Mercer, 2017) which comes with 3857 sentence pairs which are highly imbalanced with only 2% positive samples and 98% negative samples. The size and imbalanced nature of this corpus makes it difficult to train deep learning models on this dataset. To overcome this, we propose an automatically generated corpus for this task containing 74,568 sentence pairs.

This paper has two objectives: first, introducing an automatically generated corpus for the citation linkage task for biomedical research papers and second, providing a framework for this task to retrieve the cited text span from the reference paper given the citing sentence by means of measuring the semantic similarities between the citing sentence and candidate cited sentences from the referenced paper. The cited text span can be a single sentence, part of a sentence or even one or more paragraphs (Houngbo and Mercer, 2017). However, for this task this text span is restricted to a single sentence like Li et al. (2017). Considering the first objective, we introduce an automatically generated corpus containing 74,568 sentence pairs and also an approach to annotate data automatically without any human effort. The quality of the data annotation is evaluated by annotating a portion of the dataset by human experts and then measuring Cohen's κ among the human annotators' decisions and the automatically generated annotation labels. Sentence pairs from this dataset are used only for training the models for the citation linkage task. And for the second aspect, we have investigated multiple sequential and tree-structured neural networks and presented one ensemble architecture, which we call BioCite, that computes the semantic similarity between the citing statement and all of the sentences in the referred document. The performance of the model is tested against the expert annotated dataset from Houngbo and Mercer (2017) which contains citing sentences that refer to methods statements in the cited documents. The outline for the paper is: Section 2 gives a brief description of the citation linkage task and Section 3 mentions and discusses a few prominent works for the citation linkage task. Section 4 discusses the automatically generated corpus creation and the framework design. The performance of the models are reported and analyzed in Section 5. The parameters of the models are also described in this section. The paper ends with a brief summary and possible future directions of this research.

2 Citation Linkage

Citations construct semantic bridges between citing and cited manuscripts. To support the findings, claims and hypotheses, authors cite several resources while preparing manuscripts. They also try to address the results and findings of the other research works. It is also important to mention others' works, in order to demonstrate the authors' significance and progress with their current work.

A citation in any research paper focuses on some specific sections of the referenced article acknowledged as the *citation context*. This citation context often focuses on a specific idea or issue in the referenced manuscript (Houngbo, 2017). The intent of a using citation is to provide the readers with the apposite background information for a better understanding of the concepts introduced in the citing paper. The citation context can reveal information about a cited publication's hypotheses, findings, methodologies, etc. In order to improve the performance or make the method compatible with the domain for which it is intended to be used, an author may adapt or modify the method described in the citing paper to the extent necessary. Aside from that, the author may undertake experiments based on the idea presented in a cited paper to confirm or refute the idea presented in that work. References to the hypotheses and methodologies that were employed in the referenced paper aid the readers to grasp the concepts presented in the current work.

However, citations only provide the source of information which is being referred. The current citation indexing approach does not provide a way to indicate which text span from the cited research manuscript is actually being touched on. It provides no method other than going through the whole referenced article for the reader if he or she wants to grasp the idea properly. On the other hand, research articles that include detailed information on the study's discoveries, as well as relevant background information, are more appealing to readers. This necessity has influenced the work we are presenting in this paper.

The author can cite a paper by paraphrasing the

statements from the cited paper. He or she can also elaborate some statements from the cited paper. For example in the citing statement, "DNA samples are frequently harmed by exposure to excessively acidic environment", Wang et al. (2009) explains that "pH4" is an "excessively acidic environment" when citing "DNA is fairly stable in mildly acidic solutions, although the beta glycosidic link in the purine bases is hydrolyzed at around pH4." (Bonin et al., 2003). Sometimes these citations are the interpretations of the cited statements, e.g., the citing sentence "Different PCR buffer systems and/or Taq polymerases may produce variable results in real time PCR." (Huijsmans et al., 2010) is nothing but an interpretation of the cited sentence "There is a significant disparity between the outcomes obtained using the various DNA polymerase-buffer solutions." (Wolffs et al., 2004). As these examples demonstrate, precise mapping between words and sentences is required to establish a connection between the citing and cited sentences.

This paper provides a citation linkage framework for biomedical research articles along with an automatically generated corpus comprising 74,568 sentence pairs. The framework at first generates sentence pairs with the citing sentence and all the sentences from the referenced paper. Then, the model measures the semantic similarity scores between the sentences in each pair. Based on these similarity scores, it retrieves the actual cited sentences from the referenced manuscript. We have formulated this semantic similarity measurement task as a binary classification task where each sentence pair is predicted with either label 1 or label 0. Sentence pairs predicted with label 1 are selected as the cited sentences given the particular citing sentence.

3 Related Work

The study of citations in scientific research has led to a lot of work. Citation analysis attempts to identify which section (i.e., abstract of the paper, introduction of the problem statement, description of methods, analysis of result, etc.) of the referenced article this sentence refers to (Garfield, 1972; Garzone and Mercer, 2000). However, this form of study cannot pinpoint the citation span.

Another type of work is to determine the citation span. PolyU (Cao et al., 2016) applied RankSVM over chunks of sentences to predict the cited text span. Baruah and Kolla (2018) computed cosine similarity of word embeddings for the citation linkage task. Yeh et al. (2019) applied majority voting to six machine learning classifiers over the lexical, knowledge-based, corpus-based, syntactic and surface features for this task.

The CL-SciSumm Shared Task tries to solve three aspects: find the cited text span given the citation sentence ("citance"), identifying the discourse facet of the cited sentence and summarise the referred article using only the text spans that are quoted many times in the referenced document. However, the later two sub-tasks are out of the scope of this work. Ma et al. (2017) applied different classifiers and voting mechanism over similarity, rule and position-based features to determine the similarity between the citing and cited statements for CL-SciSumm-17. The citation linkage between citing and cited sentence pairs was determined by Li et al. (2017) utilizing inverse document frequency and Jaccard similarity. In their following works, they computed the sentence vectors by concatenating 200 dimensional word vectors (Li et al., 2018) and then applying a convolutional neural network (CNN) over that concatenated vector representation (Li et al., 2019). In both cases, the cited text span is determined by measuring the cosine similarities between the citing and candidate cited statements. Other works, such as AbuRa'ed et al. (2017) have also worked with the CL-SciSumm corpus.

Recently, BERT-based models have been deployed for the citation linkage task and are being used in many experiments. Gidiotis et al. (2020) fine-tuned BERT to determine the referred cited sentences from the cited document. Zerva et al. (2019) applied a CNN over SciBERT-based features (Beltagy et al., 2019) to determine which text span in the cited article is actually being referred. They concatenated the features from the BERT-based model for feature generation. Umapathy et al. (2020) utilized key-phrase similarity using the Rapid Automated Keyword Extraction Algorithm (Rose et al., 2010) and a BERT-based architecture for cited text span identification.

However, only a few citation linkage works are found for biomedical research papers. Citation linkage for biomedical research articles is more challenging due to various representations of the same component. One notable work for this domain is from 2017, where Houngbo and Mercer (2017) used traditional machine learning approach over their own small expert-annotated corpus. And so far, this is the only human annotated corpus for the citation linkage task in the biomedical domain.

4 BioCite: Description of the Framework

The development of the framework involves two major steps: creating a balanced automatically generated training corpus of reasonable size and building a framework for determining the referred statements from the cited document for a particular citing statement.

4.1 Corpus Creation

The only expert-annotated corpus for the biomedical domain to serve the purpose of our work is from Houngbo and Mercer (2017) which comes with only 3857 sentence pairs. For training, the major problem with this dataset is the class imbalance: only 81 positive pairs which is only 2% of the corpus. Eventually, training any model with this corpus would make it biased towards negative outcome. At the same time, manually annotating enough data from biomedical and biochemical research articles for this task is time consuming. So, we have created an automatically generated corpus of 74,568 sentence pairs spanning three biomedical sub-domains: biochemistry, cell biology and chemical biology. We are calling this corpus automatically generated as no human annotation has been used for generating these sentence pairs. For the validation and testing of the models, we have used the validation and testing sets from the Houngbo and Mercer (2017) corpus (800 samples with 20 positive ones for validation and 3057 samples containing 61 positives for test set). The sentence pairs in the training set are annotated with 0 (not semantically similar) or 1 (semantically similar) to make it compatible with the validation and test set.

We collected 28,310 research documents from BioMed Central spanning multiple biomedical subdomains. From these documents, 138 are randomly chosen from the above-mentioned three subdomains and then corresponding citing statements from 2736 papers (manually collected) citing these 138 articles are extracted manually. The citing statements are then paired-up with all of the sentences from the corresponding cited documents, ending-up with 522,398 pairs.

Sentences of each pair are fed individually to the Sent2Vec (Pagliardini et al., 2018) model, which is trained over all of the research documents we

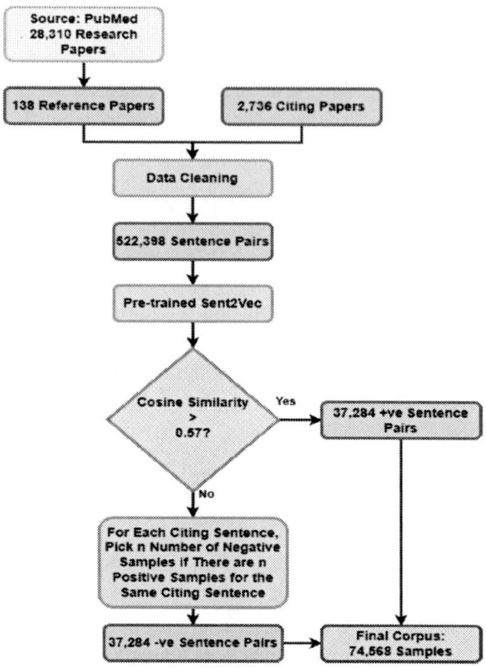

Figure 1: automatically generated corpus build-up: Sentence pair creation and annotation.

accumulated, and the cosine similarity between the paired sentences is measured. Pairs with cosine similarity value greater than a cutoff value 0.57 (selected after testing against the validation set) are labelled 1, 0 otherwise. We experimented with different cut-off values and plotted the results on AUC and ROC curves while testing on the validation set from the expert annotated corpus (Houngbo and Mercer, 2017). From there, we chose the cut-off value for which the best validation accuracy was found. From there As there are many fewer positive samples than negative ones, for each citing statement, negative samples are randomly chosen for each citing sentence to balance the classes. In this automatically generated corpus, for each citing sentence, an equal number of positive and negative samples are preserved. The overall process of this corpus creation is illustrated in Figure 1.

4.2 Semantic Similarity Measurement Module

The aim of building this citation linkage framework is to link the citing sentence to the referenced text span in the referenced biomedical research article. To solve this challenge, we have used a variety of supervised deep learning-based models to estimate the semantic similarity between the citing and cited text span where the text span is limited to a single sentence. The predictions of these models are set to binary class labels: 0 and 1. Here 1 indicates that the candidate cited and the particular citing statement are semantically similar and it can be interpreted as the candidate cited sentence is truly being referenced by the citing sentence and if the prediction value is 0, it represents the candidate cited sentence is not being referred.

The base of the sequential and tree-structured neural network models is InferSent (Conneau et al., 2017): a siamese architecture. This is a supervised sentence representation model which is able to work with sentence pairs and has been used in many cases for semantic relatedness measurement tasks (Ahmed et al., 2019; Reimers and Gurevych, 2019). The overview of the training process of InferSent for the semantic similarity measurement task is portrayed in Figure 2. In InferSent two identical encoder neural network topologies are used with identical parameter settings. The citing sentence (S_{citing}) and the cited sentence (S_{cited}) are encoded by them in parallel. This is followed by generating a feature map that concatenates concatenation, absolute point-wise difference, and point-wise multiplication. This feature map is then loaded into the dense and *softmax* layers in sequence to predict the binary class label. As the encoder models, four sequential and four tree-structured neural networks are used. The functioning principles of these models are first outlined, and then the ensembles of them are discussed. The best encoder model for the BioCite framework is chosen in the end based on the performances of the investigated models.

4.3 Sequential Encoders

As the encoder for the InferSent model, four sequential models are applied. The first one is the Bi-LSTM with a following max-pooling layer. The second encoder model applies inner attention (Liu et al., 2016) over the Bi-LSTM output features for producing the sentence representations. The third encoder model utilizes the hierarchical attention (Yang et al., 2016) in place of inner attention over the Bi-LSTM. This attention mechanism was introduced for document classification where at the first layer it attends on the words for generating sentence representation and in the second layer it attends over the sentences for paragraph or document representation. As our work is limited to single sentences, we have used only the first layer of this attention mechanism. This approach is designed in such a way that it can focus on four different parts of the sentence. Thus it generates four sentence representations, which are concatenated to form the sentence vector. The last sequential encoder we investigated is the hierarchical CNN with four layers of convolution operations, each followed by one max-pooling operation. These four feature maps are concatenated in the end to generate the sentence representation vector.

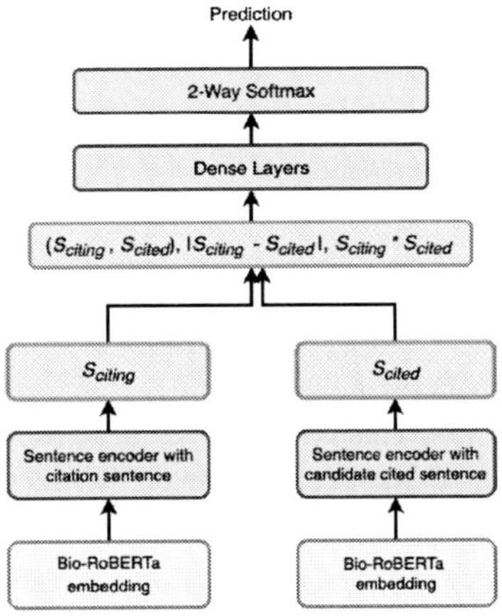

Figure 2: InferSent training for the citation linkage task.

4.4 Tree-Structured Encoders

Sequential neural networks provide reasonable sentence representations. However, they can't preserve structural information and miss semantic compositionality. Tree-structured neural networks, on the other hand, can preserve both semantic and syntactic properties of the text by working with the parse tree. For the tree-structured neural network models we investigated the dependency and constituency tree-transformers with both multi-head and multi-branch attention mechanisms over child nodes' representations (Ahmed et al., 2019). For completeness, we provide details of these tree-transformers that are developed therein.

A constituency tree contains words at leaf nodes only, whereas a dependency tree has a word at each node. So, while traversing a dependency tree, it is required to consider both the child and corresponding parent nodes whereas for constituency tree, only after traversing every sub-tree the non-terminal intermediate nodes can be calculated. So, in both cases, the children nodes are considered. This approach (Ahmed et al., 2019) uses self attention mechanism for attending the child nodes. This attention mechanism uses three matrices: *key*, *value* and *query* like the transformer model (Vaswani et al., 2017) (Equ. 1).

$$\alpha = \texttt{softmax}(\frac{query\ key^T}{\sqrt{d_k}})value \quad (1)$$

Here d_k is the dimension of the *key*, *value* and *query* matrices. For this experiment the dimension of all these matrices are kept the same. n copies of these matrices are generated for n branches of the multi-branch attention mechanism. Here, n is the number of branches to be used. Then scaled dot product is used as in Equ. 2:

$$\beta_i = \alpha_{i\in[1,n]}(query_i\ \omega_i^q, key_i\ \omega_i^k, value_i\ \omega_i^v) \quad (2)$$

where ω_i^q, ω_i^k, ω_i^v are the hyper-parameter weight matrices for *query*, *key*, and *value*, respectively.

Following this scaled dot product operation, a residual connection is employed over these tensors β. A layer-wise batch normalization is used in the following step which is multiplied with a scaling factor τ (Equ. 3). Over every $\tilde{\beta}$, position-wise CNN (PCNN) is then employed (Equ. 4). By applying weighted summation then, the attention encoded semantic sub-spaces' representation are generated (Equ. 5). Here $\gamma \in \mathcal{R}^n$ is a hyper-parameter. In the end, another residual connection is established with `BranchAttn` which is then fed to a non-linearity function tanh and an element-wise summation function `EWS` is done to produce the parent node representation (Equ. 6) (Ahmed et al., 2019).

$$\tilde{\beta}_i = \texttt{LayerNorm}(\beta_i\omega_i^b + \beta_i) \times \tau_i \quad (3)$$

$$\texttt{PCNN}(x) = \texttt{Conv}(\textit{Relu}(\texttt{Conv}(x) + b_1)) + b_2 \quad (4)$$

$$\texttt{BranchAttn} = \sum_{i=1}^{n} \gamma_i \texttt{PCNN}(\tilde{\beta}_i) \quad (5)$$

$$\texttt{ParentNodeRep} = \texttt{EWS}(\tanh((\tilde{\chi} + \chi)\omega + b)) \quad (6)$$

For multi-head attentions, attention matrices *key*, *value* and *query* are projected h times (Vaswani et al., 2017) and it is calculated as follows:

$$\texttt{MultiHead}(query, key, value) = \texttt{Concat}(\texttt{head}_1, ..., \texttt{head}_h)W^O \quad (7)$$

where, for each head,

$$\texttt{head}_i = \alpha(queryW_i^q, keyW_i^k, valueW_i^v) \quad (8)$$

All of the Ws are the hyper-parameter matrices which get updated during training.

4.5 Ensemble Architectures

After investigating the sequential and tree-structured neural network models, we experimented with two ensemble models. The first ensemble architecture utilizes all the models investigated here. After all the models are trained separately, each sentence pair is fed to all the models in parallel. Each model individually predicts the semantic similarity score and in the end, the final similarity value is selected by applying a winner-takes-all approach (Roy et al., 2018) over all the predictions. In the second approach we used only the tree-transformer models. The dependency tree-transformer is able to preserve the word level dependency between different part of the sentence, whereas the constituency tree-transformer can preserve phrase-level information. To benefit from both of these models, we concatenated the feature representations generated from both of the tree-transformers and used it as the vector representation of the sentence. This sentence vector is then fed to a multi-layer perceptron for the similarity score prediction.

5 Experimental Setup and Result Analysis

In this section, the experimental setup and the results of the models investigated for the citation linkage task are discussed. As the human annotated test data is highly imbalanced, apart from F-1 score, Matthews Correlation Coefficient (MCC) and Balanced accuracy (BAcc) are also used to assess the performance of the models.

5.1 Experimental Setup

Sent2Vec was trained with various parameter settings. The cutoff value and the best model are chosen based on the MCC and BAcc over the validation set. The best hyper-parameter settings for Sent2Vec are: 500d sentence embedding, window size 20, learning rate 0.2, negative sampling loss function and sampling threshold 0.0001. For the four sequential models: hierarchical CNN (hCNN), Bi-LSTM with max pooling, hierarchical and inner attentions over Bi-LSTM; the learning rates (LR) were initialized to 0.1. With a drop in validation accuracy, the LR is multiplied by 0.2. The batch size and LR threshold are set to 50 and 0.0001, respectively. For training, stochastic gradient descent is used as the optimizer. For hCNN, 4 layers of convolution are used followed by max-pooling.

Table 1: Statistics of the annotations by the experts and the automatically generated corpus for the 1500 samples

	Annotator Group 1	Annotator Group 2	The Automatically Generated Corpus
Positive samples (in total)	731	709	750
Negative Samples (in total)	769	791	750

Table 2: Analysis of the agreements among the expert annotators and the automatically generated corpus

	Between Annotator Groups 1 and 2	Between Annotator Group 1 and the Automatically Generated Corpus	Between Annotator Group 2 and the Automatically Generated Corpus
Agreed Positive Samples	706	715	701
Agreed Negative Samples	765	750	750
Cohen's κ	0.96	0.95	0.93

Four context vectors are used for both hierarchical and inner attention mechanisms to focus on 4 distinct parts which are concatenated for final sentence representations. For all of the tree-structured transformer models, 6 parallel heads are used with 50d query, value and key matrices where 6 position-wise convolution layers are used for multi-branch attention. Two layers of CNN (first layer: 341 1d kernel and no dropout, second layer: 300 1d kernels, 0.1 dropout) are used in the PCNN layer as the composition function which is the same as Ahmed et al. (2019). For parameter tuning, Adagrad (Duchi et al., 2011) with LR 0.0002 is used in all cases.

5.2 Performance Analysis

We first evaluate the quality of the automatically generated corpus. For analyzing the quality of the data annotation, we randomly picked 750 positive and 750 negative samples (labelled as such in the automatically generated corpus) from the 74,568 citing and candidate cited sentence pairs. These 1500 sentence pairs were provided to two groups of expert annotators. Each group consisted of three people and each person annotated 500 samples. So, each 500 sample chunk was annotated by two individuals, one from each group. Each reviewer also mentioned their confidence level for each sample annotation. We then used Cohen's κ (Cohen, 1960) to compute inter-annotator reliability between the human annotators and the automatically generated corpus. The overall statistics are shown in Table 1. The first group identified 731 positive and 769 negative samples in the 1500 sentence pairs, and the second group identified 709 positive and 791 negative samples. Table 2 shows the annotator groups' decisions agreed for 706 positive and 765 negative samples. The reliability factor κ found here is 0.96. While comparing the annotation provided by the automatically generated corpus against the first and second annotator groups, we see that the annotation decisions match for 715 and 701 positive samples between the automatically generated corpus and groups 1 and 2, respectively. For negative samples, the agreed decisions are 750 samples in both cases. The κ values are 0.95 (between first annotator group and the automatically generated corpus) and 0.93 (between second annotator group and the automatically generated corpus). These values indicate that the automatically generated corpus annotations match the experts' annotations quite well. When interpreting these high κ values, it is important to recall that the data given to the annotators were balanced (50/50 split of positive and negative samples). From Table 2 it is clear that the human annotators have high agreement for both of their positive and negative choices.

Next we provide the citation linkage task outcomes. To compare the performance of the model against the previous models, we evaluated the model with the gold standard human annotated data from Houngbo and Mercer (2017) because the previous models were tested against this gold standard corpus. This corpus focusses on citations of methods used in the citing and cited articles. Houngbo (2017) suggests that in most cases the citation refers to single sentences in the cited articles. As an example, the citing statement "Recently, Chauhan et al. employed SVM to predict the ATP binding residues in ATP binding proteins using amino acid

Table 3: Performance analysis of different architectures for the citation linkage task for biomedical research articles. Models tagged with † are the investigated ones in this work. Here, CT: constituency tree, DT: dependency tree, MB: multi-branch attention, MH: multi-head attention, TP: true-positive, FP: false-positive, TN: true-negative, FN: false-negative.

	Model	TP	FP	TN	FN	F1	MCC	BAcc (in %)
Previous Works	Houngbo	34	995	2001	27	0.06	0.07	61.27
	Li	39	779	2217	22	0.09	0.12	68.97
Sequential Models	Hierarchical CNN †	45	580	2416	16	0.13	0.19	77.21
	Bi-LSTM + Max-pooling †	54	361	2635	7	0.23	0.31	88.24
	Inner attentive Bi-LSTM †	55	372	2624	6	0.23	0.31	88.87
	Hierarchical Attentive Bi-LSTM †	56	355	2641	5	0.24	0.33	89.98
Tree Structured	DT-Transformer (MH) †	57	301	2695	4	0.27	0.36	91.70
	DT-Transformer (MB) †	58	287	2709	3	0.29	0.38	92.75
	CT-Transformer (MH) †	57	315	2681	4	0.26	0.35	91.46
	CT-Transformer (MB) †	57	309	2687	4	0.27	0.36	91.56
Ensemble	Winner-takes-all ensemble †	59	253	2743	2	0.32	0.41	94.14
	BioCite †	**60**	**240**	**2756**	**1**	**0.33**	**0.42**	**95.17**

sequences and their evolutionary profiles" (Firoz et al., 2011) indicates the cited sentence "Our SVM module predicts a score for each residue in protein (in range of -1.0 to 1.0), we define a threshold to discriminate ATP interacting and non-interacting residues" (Chauhan et al., 2009). Another approach for such a task could have been ranking the candidate sentences as was one of the methods done by Houngbo (2017). However, for the final classification step we used `softmax`, which gives a probability to every possible outcome, so this approach could easily be modified to be a ranking approach.

Table 3 reflects multiple performance metrics found for the models used here along with the results from a few prominent works. Among the sequential models, Bi-LSTM with the hierarchical attention mechanism fed with Bio-RoBERTa embeddings performs the best based on the MCC and BAcc (0.33 and 89.98% accordingly). However, it can correctly extract only 56 out of 61 positive samples. The inner attentive Bi-LSTM and simple Bi-LSTM followed by a max-pooling layer captures 54 and 55 positive samples correctly with the same MCC (0.31) and F1 score (0.23). However, the inner attentive Bi-LSTM model earns a slightly higher BAcc (88.87%) as it predicts more negative samples correctly.

The tree-structured models outperform all of the sequential models to extract the cited statements from the referenced documents. The reason for this is the constituency tree-transformer is able to capture phrase level information and the dependency tree-transformer is able to preserve word level dependencies. In biomedical articles, biological components' chemical names may comprise multiple words. The constituency tree-transformer has the capability to work better with such phrase level text. And in a lot of cases, the citing statements are complex in nature. The dependency tree-transformer deals with such cases well. Another important thing to notice here is that tree-transformers with multi-branch attention perform better than the tree-transformers with multi-head attention as multi-branch attention applies multiple heads in each branch and is thus able to obtain more information about each sentence (Fan et al., 2020). Here, both the constituency and dependency tree-transformers with multi-head attention mechanism predict 57 positive samples correctly. Multi-branch attentive dependency tree-transformer predicts 58 positive samples correctly. Constituency tree-transformer with multi-branch attention predicts 57 positive samples correctly. However, it predicts 6 more negative samples correctly attaining a 0.10 percentage point improvement in BAcc.

The two ensemble architectures investigated here improve the performance of the citation linkage task for biomedical research articles. The first approach ensembles all of the investigated individual models with the winner takes all selection process. This approach considers all the outcomes from different models and the outcome with the highest probability is chosen as the final prediction. It successfully predicts 59 positive samples out of 61 with 94.14% BAcc, 0.41 MCC and 0.32 F1 score which are higher compared to any of the standalone models. The second ensemble architecture considers only dependency and constituency tree-transformers with multi-branch attention. There are two reasons behind choosing only these two models for ensemble in this case: firstly, the major intention was to investigate how the model performs if we combine both the word dependency and phrase level information, and secondly, these two models showed better performance among all individual models. This ensemble architecture extracts 60 true positive cited statements given the citing statements for the citation linkage task. It also achieves 95.17% BAcc, 0.42 MCC and 0.33 F1 score. As the best performance is attained by this last ensemble architecture, for the BioCite citation linkage framework, we choose this approach for extracting cited statements from the referenced biomedical research article given the citing statement from the citing paper. Is the computationally more expensive ensemble model justified for predicting only a few more true-positives? We notice that the increase in true-positives is approximately 2%. This increase, especially in a larger corpus, would seem to justify the extra computational cost. However, it should also be noted that the false-positives have decreased by almost 20%. The applications noted in the introduction will benefit substantially by such a decrease in false-positives. This decrease in false-positives further justifies the extra computational cost of the ensemble model.

Now, there remains one more question to be discussed. Which one is actually improving the performance, the automatically generated corpus or the model? From Table 3, it is clear that, the performance of BioCite is better than the other models. To check the effectiveness of the proposed automatically generated corpus, we trained all the models over the human annotated small corpus (Houngbo and Mercer, 2017). In this experiment we found all the investigated models' accuracies were very high (around 98%). However, the BAcc, MCC and the F1 scores were very poor as the models are strongly biased towards the negative outcome. This gives evidence of the effectiveness of training models over our proposed automatically generated corpus. Furthermore, analyzing the outcomes and going through the predictions of the sentence pairs, we found that this model can successfully predict cited sentence given the citing statement when chemical components and reactions are presented in different ways. For example: the cited sentence "DNA is fairly stable in mildly acidic solutions, although the beta glycosidic link in the purine bases is hydrolyzed at around pH4." (Bonin et al., 2003) is predicted successfully for the citing sentence "DNA samples are frequently harmed by exposure to excessively acidic environment.", Wang et al. (2009). It indicates that this model has the ability to resolve "pH4" as an "excessively acidic environment" and "hydrolyzed" with "harmed".

6 Conclusion

Biomedical literature is complex in nature due to having complex biological and chemical component names. Our framework, BioCite, performs well when dealing with the human annotated test set containing research articles accumulated from the biomedical domain and outperforms the previous prominent works. However, there are still a few avenues to investigate. The text span used here is a single sentence. In future, it can be expanded to the paragraph level which would capture the contextual information as well. Graph-based neural networks which perform well when working with paragraphs (Zhang et al., 2020) could be used. Moreover, BERT-based models can be explored as well.

References

Ahmed AbuRa'ed, Luis Chiruzzo, and Horacio Saggion. 2017. What sentence are you referring to and why? Identifying cited sentences in scientific literature. In *Proceedings of the International Conference Recent Advances in Natural Language Processing, RANLP 2017*, pages 9–17.

Mahtab Ahmed, Muhammad Rifayat Samee, and Robert E Mercer. 2019. You only need attention to traverse trees. In *Proceedings of the 57th Annual Meeting of the Association for Computational Linguistics*, pages 316–322.

Gaurav Baruah and Maheedhar Kolla. 2018. Klick labs at cl-scisumm 2018. In *BIRNDL@ SIGIR*.

Iz Beltagy, Kyle Lo, and Arman Cohan. 2019. Scibert: A pretrained language model for scientific text. *arXiv preprint arXiv:1903.10676*.

Serena Bonin, F Petrera, B Niccolini, and Giorgio Stanta. 2003. PCR analysis in archival postmortem tissues. *Molecular Pathology*, 56(3):184–186.

Ziqiang Cao, Wenjie Li, and Dapeng Wu. 2016. Polyu at cl-scisumm 2016. In *Proceedings of the joint workshop on bibliometric-enhanced information retrieval and natural language processing for digital libraries (BIRNDL)*, pages 132–138.

Jagat S Chauhan, Nitish K Mishra, and Gajendra PS Raghava. 2009. Identification of atp binding residues of a protein from its primary sequence. *BMC bioinformatics*, 10(1):1–9.

Jacob Cohen. 1960. A coefficient of agreement for nominal scales. *Educational and psychological measurement*, 20(1):37–46.

Alexis Conneau, Douwe Kiela, Holger Schwenk, Loic Barrault, and Antoine Bordes. 2017. Supervised learning of universal sentence representations from natural language inference data. *arXiv preprint arXiv:1705.02364*.

John Duchi, Elad Hazan, and Yoram Singer. 2011. Adaptive subgradient methods for online learning and stochastic optimization. *Journal of machine learning research*, 12(7).

Yang Fan, Shufang Xie, Yingce Xia, Lijun Wu, Tao Qin, Xiang-Yang Li, and Tie-Yan Liu. 2020. Multi-branch attentive transformer. *arXiv preprint arXiv:2006.10270*.

Ahmad Firoz, Adeel Malik, Karl H Joplin, Zulfiqar Ahmad, Vivekanand Jha, and Shandar Ahmad. 2011. Residue propensities, discrimination and binding site prediction of adenine and guanine phosphates. *BMC biochemistry*, 12(1):1–12.

Eugene Garfield. 1972. Citation analysis as a tool in journal evaluation. *Science*, 178(4060):471–479.

Mark Garzone and Robert E Mercer. 2000. Towards an automated citation classifier. In *Conference of the Canadian Society for Computational Studies of Intelligence*, pages 337–346. Springer.

Alexios Gidiotis, Stefanos Stefanidis, and Grigorios Tsoumakas. 2020. Auth@ clscisumm 20, laysumm 20, longsumm 20. In *Proceedings of the First Workshop on Scholarly Document Processing*, pages 251–260.

Hospice Houngbo and Robert E Mercer. 2017. Investigating citation linkage with machine learning. In *Canadian Conference on Artificial Intelligence*, pages 78–83. Springer.

Kokou Hospice Houngbo. 2017. *Investigating Citation Linkage Between Research Articles*. Ph.D. thesis, The University of Western Ontario.

Cornelis JJ Huijsmans, Jan Damen, Johannes C van der Linden, Paul HM Savelkoul, and Mirjam HA Hermans. 2010. Comparative analysis of four methods to extract DNA from paraffin-embedded tissues: Effect on downstream molecular applications. *BMC Research Notes*, 3(1):239.

Lei Li, Liyuan Mao, Yazhao Zhang, Junqi Chi, Taiwen Huang, Xiaoyue Cong, and Heng Peng. 2018. Computational linguistics literature and citations oriented citation linkage, classification and summarization. *International Journal on Digital Libraries*, 19(2-3):173–190.

Lei Li, Yazhao Zhang, Liyuan Mao, Junqi Chi, Moye Chen, and Zuying Huang. 2017. CIST@CLSciSumm-17: Multiple features based citation linkage, classification and summarization. In *BIRNDL@ SIGIR (2)*, pages 43–54.

Lei Li, Yingqi Zhu, Yang Xie, Zuying Huang, Wei Liu, Xingyuan Li, and Yinan Liu. 2019. CIST@CLSciSumm-19: Automatic scientific paper summarization with citances and facets. *BIRNDL2019*.

Yang Liu, Chengjie Sun, Lei Lin, and Xiaolong Wang. 2016. Learning natural language inference using bidirectional LSTM model and inner-attention. *arXiv preprint arXiv:1605.09090*.

Shutian Ma, Jin Xu, Jie Wang, and Chengzhi Zhang. 2017. Njust @ clscisumm-17 shutian. In *Proceedings of the First Workshop on Scholarly Document Processing*.

Robert Mercer. 2016. Locating and extracting key components of argumentation from scholarly scientific writing. *Dagstuhl Reports*, 6(4):3–15.

Matteo Pagliardini, Prakhar Gupta, and Martin Jaggi. 2018. Unsupervised learning of sentence embeddings using compositional n-gram features. In *Proceedings of the 2018 Conference of the North American Chapter of the Association for Computational Linguistics: Human Language Technologies, Volume 1 (Long Papers)*, pages 528–540.

Raquel Mochales Palau and Marie-Francine Moens. 2009. Argumentation mining: The detection, classification and structure of arguments in text. In *Proceedings of the 12th International Conference on Artificial Intelligence and Law*, pages 98–107. ACM.

Dragomir R. Radev, Hongyan Jing, and Malgorzata Budzikowska. 2000. Centroid-based summarization of multiple documents: Sentence extraction, utility-based evaluation, and user studies. In *NAACL-ANLP 2000 Workshop: Automatic Summarization*.

Nils Reimers and Iryna Gurevych. 2019. Sentence-bert: Sentence embeddings using siamese bert-networks. *arXiv preprint arXiv:1908.10084*.

Stuart Rose, Dave Engel, Nick Cramer, and Wendy Cowley. 2010. Automatic keyword extraction from individual documents. *Text mining: applications and theory*, 1:1–20.

Sudipta Singha Roy, Sk Imran Hossain, MAH Akhand, and Kazuyuki Murase. 2018. A robust system for noisy image classification combining denoising autoencoder and convolutional neural network. *International Journal of Advanced Computer Science and Applications*, 9(1):224–235.

Sudipta Singha Roy, Robert E Mercer, and Felipe Urra. 2020. Investigating citation linkage as a sentence similaritymeasurement task using deep learning. In *33th Canadian Conference on Artificial Intelligence*.

Anjana Umapathy, Karthik Radhakrishnan, Kinjal Jain, and Rahul Singh. 2020. Citeqa@ clscisumm 2020. In *Proceedings of the First Workshop on Scholarly Document Processing*, pages 297–302.

Ashish Vaswani, Noam Shazeer, Niki Parmar, Jakob Uszkoreit, Llion Jones, Aidan N Gomez, Łukasz Kaiser, and Illia Polosukhin. 2017. Attention is all you need. *Advances in Neural Information Processing Systems*, 30.

Yuker Wang, Victoria EH Carlton, George Karlin-Neumann, Ronald Sapolsky, Li Zhang, Martin Moorhead, Zhigang C Wang, Andrea L Richardson, Robert Warren, Axel Walther, et al. 2009. High quality copy number and genotype data from FFPE samples using molecular inversion probe (MIP) microarrays. *BMC Medical Genomics*, 2(1):8.

Petra Wolffs, Halfdan Grage, Oskar Hagberg, and Peter Rådström. 2004. Impact of DNA polymerases and their buffer systems on quantitative real-time PCR. *Journal of Clinical Microbiology*, 42(1):408–411.

Zichao Yang, Diyi Yang, Chris Dyer, Xiaodong He, Alex Smola, and Eduard Hovy. 2016. Hierarchical attention networks for document classification. In *Proceedings of the 2016 Conference of the North American Chapter of the Association for Computational Linguistics: Human Language Technologies*, pages 1480–1489.

Jen-Yuan Yeh, Tien-Yu Hsu, Cheng-Jung Tsai, Pei-Cheng Cheng, and Jung-Yi Lin. 2019. On identifying cited texts for citances and classifying their discourse facets by classification techniques. *Journal of Information Science & Engineering*, 35(1).

Chrysoula Zerva, Minh-Quoc Nghiem, Nhung TH Nguyen, and Sophia Ananiadou. 2019. Nactemuom@ cl-scisumm 2019. In *BIRNDL@ SIGIR*, pages 167–180.

Yufeng Zhang, Xueli Yu, Zeyu Cui, Shu Wu, Zhongzhen Wen, and Liang Wang. 2020. Every document owns its structure: Inductive text classification via graph neural networks. In *Proceedings of the 58th Annual Meeting of the Association for Computational Linguistics*, pages 334–339.

Low Resource Causal Event Detection from Biomedical Literature

Zhengzhong Liang, Enrique Noriega-Atala, Clayton Morrison and Mihai Surdeanu
The University of Arizona, Tucson, AZ
{zhengzhongliang, enoriega, claytonm, msurdeanu}@email.arizona.edu

Abstract

Recognizing causal precedence relations among the chemical interactions in biomedical literature is crucial to understanding the underlying biological mechanisms. However, detecting such causal relation can be hard because: (1) many times, such causal relations among events are not *explicitly* expressed by certain phrases but *implicitly* implied by very diverse expressions in the text, and (2) annotating such causal relation detection datasets requires considerable expert knowledge and effort. In this paper, we propose a strategy to address both challenges by training neural models with in-domain pre-training and knowledge distillation. We show that, by using very limited amount of labeled data, and sufficient amount of unlabeled data, the neural models outperform previous baselines on the causal precedence detection task, and are ten times faster at inference compared to the BERT base model.

1 Introduction

Since 2011, more than one million new articles are added to PubMed every year (Vardakas et al., 2015). The growth rate of newly published articles makes it hard to keep up with the important discoveries just by reading them. Therefore, tremendous efforts have been made to automate knowledge discovery from biomedical papers by extracting the biochemical events described in the literature (Kim et al., 2009, 2012; Nédellec et al., 2013).

In addition to the extraction of the biochemical events, there are existing efforts to detect the causal relationships among them (Mihăilă et al., 2013; Hahn-Powell et al., 2016), i.e., whether the occurrence of one event necessarily leads to the occurrence of another event. Knowing the causal precedence order of the events helps to describe more accurately the underlying mechanisms of biological processes described on the scientific literature. However, annotating such causal event pairs requires significant domain expertise and effort (Hahn-Powell et al., 2016).

In this work, we investigate multiple strategies for improving the detection of causal precedence relations within biochemical events. The contributions of this paper are the following:

(1) We propose and investigate multiple neural architectures for detection of causal precedence among biochemical interactions trained with a few hundred annotated training examples and numerous weakly-supervised training examples.

(2) We analyze the impact of in-domain pre-training and distillation on the performance of the proposed architectures, and conclude that several compact BERT architectures can benefit from in-domain pre-training, and can potentially benefit from further distillation.

(3) Lastly, we study a hybrid methodology that combines neural models with the traditional rule/feature-based methods in a sieve-based framework, and observe that well-trained neural models can largely replace the rule/feature-based methods and do not benefit from the sieve framework.[1]

2 Related Work

The detection of causal precedence among chemical interactions from text is a long-standing problem. Early methods include rule-based approaches (Khoo et al., 2000) and machine learning-based approaches (Girju, 2003; Blanco et al., 2008; Akkasi and Moens, 2021). Other work (Sorgente et al., 2013; Hahn-Powell et al., 2016; Dasgupta et al., 2018) have also explored the combination of rule-based methods, machine learning-based or neural-based methods.

Recently, large pre-trained language models (LLM) have increased the state-of-the-art performance of many natural language processing tasks

[1]The code and data can be found at: https://github.com/clulab/releases/tree/master/acl2022-bionlp-causal

Proceedings of the BioNLP 2022 workshop, Dublin, Ireland, pages 252–263
May 26, 2022. ©2022 Association for Computational Linguistics

(Devlin et al., 2019; Liu et al., 2019b; Raffel et al., 2020). However, such LLM models require enormous computational resources, making it hard to deploy them in many applications. One popular approach to reduce the memory footprint of LLMs is distillation (Sanh et al., 2019; Jiao et al., 2020; Wu et al., 2020; Wang et al., 2020). Distillation trains a relatively small model to imitate the behavior of a larger model, such as a LLM, trading off performance for a significant reduction in the amount of parameters.

Tang et al. (2019b) shows that it is possible to distill BERT-large to a **task-specific** compact LSTM model with approximately $\frac{1}{300}$ of the model's original parameters while maintaining a comparable performance. Wasserblat et al. (2020) and Adhikari et al. (2020) investigated whether the performance of the distilled model depends on the nature of the task and the size of the student model. Turc et al. (2019) found that the general domain pre-training of the compact model is essential and helpful to the distillation on the downstream tasks. In addition, various data augmentation techniques are proposed to improve the distillation process with very limited labeled training data (Mukherjee and Awadallah, 2019; Tang et al., 2019a; Melas-Kyriazi et al., 2019). Finally, several works explored whether cross-task distillation helps the compact models to learn more robust representations (Liu et al., 2019a; Pan et al., 2021). To the best of our knowledge, this work is the first to investigate model distillation specifically for the task of causal precedence detection in the biomedical domain.

3 Dataset

We use of a dataset of causal precedence annotations of biochemical interactions (Hahn-Powell et al., 2016). The dataset contains 858 interaction pairs. Each pair is annotated with one of three classes: *E1 precedes E2*, *E2 precedes E1*, and *no precedence relationship*, with 109, 27 and 722 instances, respectively. Table 1 contains a few examples of the annotations.

Working with this dataset presents multiple challenges. Firstly, it's small, with only total of 858 annotated examples. The scarcity of training data is a challenge for a model with a relatively large number of parameters to pick up training signal from the data. Second, prediction of some examples requires more than the shallow understanding of linguistic knowledge (i.e., understanding the phrases such as "leading to"), and also requires understanding the underlying mechanistic process described in the phrase. For example, in the last row of Table 1, the model needs to understand "FoxO1 can bind to ATG7" and "FoxO1 and ATG7 complex" are referring to the same event, so that there is no precedence between them. Finally, we are aiming at obtaining a compact model that can be efficiently deployed without a GPU and with high processing speed.

4 Approach

4.1 Neural-based Approaches

We propose two neural-based architectures: A BiLSTM (Graves and Schmidhuber, 2005) and a fine-tuned BERT model (Devlin et al., 2019).

Both architectures take as input the text span containing both biochemical interactions (events). The text span is encoded as:

```
[E1 tokens] + [SEP] + ...   +
     [SEP] + [E2 tokens]
```

Where `...` represents the text between both events. If E1 is adjacent to E2 (i.e., there is no text between them), the input sequence becomes:

```
[E1 tokens] + [SEP] + [E2 tokens]
```

How much context to include in the input is a design choice. An alternative design is to include more text in the input sequence, such as the text preceding `E1` and the text following `E2`. However, the model might fail to learn to concentrate on the most essential part for the causal relation detection when the context is too long, especially considering there are very limited labeled data in our task. Therefore we did not include the context preceding `E1` and following `E2` in our current model. We leave the impact of such design choices to future work.

BiLSTM

We use a single layer BiLSTM with input dimension of 100 and hidden dimension $h \in \{200, 700, 750\}$. The output of the BiLSTM model H is a tensor of size $l \times 2h$, where l is the number of tokens in the input and h is the hidden dimension of the BiLSTM. The output vector tensor H is then max-pooled over the sequence, creating a vector H' with size $2h$. The pooled hidden representation H' is then passed to a 2-layer MLP to predict the class of the input sequence.

Text spans with a pair of biochemical interactions	Label	Explanation
IKKalpha then phosphorylates the C-terminal region of p100 *leading to* subsequent processing of the p100 and RelB complex into p52 and RelB and its translocation into the nucleus	E1 precedes E2	The expression "leading to" suggests the precedence relationship.
Given that an oxidant inhibits the catalytic action of Cdc25 on wt Ras that is an enhancement of the wt Ras bound GDP, the oxidant evidently targets the ternary complex.	E2 precedes E1	The expression "enhancement of" indicates the precedence relationship.
We next studied the effect of these growth factors on the tyrosine phosphorylation of Gab1 and its binding to SHP-2. EGF, *but not* IGF or PDGF, led to both increased tyrosine phosphorylation of Gab1 and binding to SHP-2, suggesting a selective effect of EGF on Ras and MAPK activation mediated by Gab1 and SHP-2.	No precedence	The expression "but not" indicates there is no precedence relationship.
FoxO1 can bind to ATG7, which is an important regulator in autophagosome expansion, and the FoxO1 and ATG7 complex may impact autophagy in human colon cancer HCT116 cells or in HeLa cells.	No precedence	The two events are equivalent although the expressions are a little different.

Table 1: Examples of relations in the causal precedence dataset. Each example contains a span of text from either one or two adjacent sentences. The text contains a pair of biochemical interactions. The first interaction (E1) is colored in red and the second (E2) in blue. The boundary of each event is extracted by REACH. The classification problem is to predict whether there is an existence of a causal precedence relations between E1 and E2.

BERT

For BERT, in addition to the common encoding, we prepend a `[CLS]` token to the input sequence. Then, the sequence is passed through BERT, generating a list of embeddings (with size h) of all l input tokens. Then a 2-layer MLP is placed on top of the embedding of the `[CLS]` token to obtain the final prediction result.[2]

We evaluate 4 pre-trained variants of BERT:

BERT-base: The original BERT-base model released by Google. It contains approximately 110M parameters. In the experiment we use the `bert-base-uncased` model provided by the huggingface library.[3]

BioBERT-base: (Lee et al., 2020) This model has the same amount of parameters as BERT-base. It was further pre-trained on PubMed papers. We use the `BioBERT-base-cased V1.1` in our experiments.[4]

BERT-L8H128A2: (Turc et al., 2019) A compact BERT model pre-trained on the same corpus as BERT-base, but with only 8 layers, hidden size of 128 and 2 attention heads. It has 5.5M parameters.

BERT-L4H256A4: (Turc et al., 2019) Similar to BERT-L8H128A2 but with only 4 layers, hidden size of 256 and 4 attention heads. It contains 11M parameters.

4.2 Pre-training

Previous works have shown that both general-domain pre-training (Turc et al., 2019) and in-domain pre-training (Lee et al., 2020) can improve the model's performance on the down-stream tasks. Gururangan et al. (2020) shows that even the pre-training in a non-target but similar-to-target domain can help with the later fine-tuning. In this work we investigate whether in-domain pre-training can help with the compact BiLSTM or BERT classifiers.

Pre-training Corpus

We use REACH, a bio-medical domain information extraction tool (Valenzuela-Escárcega et al., 2018), to extract 10,000 biomedical papers from PMC Open Access.[5] The corpus is composed of papers that contain biochemical events, such as *phosphorylation*, *methylation* and a few others.[6] We cleaned the text (e.g., remove the sentences that are too short, usually citations) and split the sentences using the NLTK toolkit (Bird et al., 2009). The total number of sentences is 1.5M. We use the sentences of 9,000 papers as the training set and the sentences of the remaining 1,000 papers as the eval-

[2] We use the `BertForSequenceClassification` function from the huggingface library.

[3] `https://huggingface.co/transformers/v3.0.2/model_doc/bert.html?highlight=bertforsequenceclassification`

[4] `https://github.com/dmis-lab/biobert`

[5] `https://www.ncbi.nlm.nih.gov/pmc/tools/openftlist/`

[6] For the complete list of the keywords we use to retrieve the papers, please see Appendix A.

uation set. The evaluation set here is solely used to determine when to stop training the language model in the pre-training stage and is not used for the evaluation of the causal relation detection task. For the rest of this work, we will refer to this corpus as PMC-10000.

BiLSTM Pre-training

We investigate the impact of pre-training to a BiLSTM model in two ways. First, we evaluate whether it is helpful to train a skip-gram model (Mikolov et al., 2013) on PMC-10000 to use as input to the LSTM. We also evaluate whether it is helpful to pre-train the model using a language modeling task. We employ a similar protocol to (Mousa and Schuller, 2017): Given an input sequence of tokens $[t_1, t_2, ..., t_l]$, the forward LSTM is taught to predict the tokens $[t_2, t_3, ..., t_l]$ and the backward LSTM is taught to predict the tokens $[t_{l-1}, t_2, ..., t_1]$.

BERT Pre-training

We train the model using the standard Masked Language Modeling (MLM) on PMC-10000 with whole-word masking but without Next Sentence Prediction (NSP) task. The length of each sentence is limited to 50 (after applying the sub-word tokenizer). The mask probability is set to 0.15, as in (Devlin et al., 2019). The model is trained with a batch size of 64 using Adam optimizer with the learning rate of 5e-5 for 12 epochs (for a total of approximately 284K optimization steps).

4.3 Distillation

Although the large language models such as BERT and BioBERT have shown strong performance on various tasks, they consume a lot of computation resources and could have a high inference latency when deployed without a GPU. Such a high inference latency is undesirable when thousands and millions of biomedical publications need to be processed. Therefore we are motivated to develop a compact model that can be deployed with a low inference latency even without a GPU.

However, compact models usually could not reach a comparable performance as large pre-trained language models. Therefore we seek to use knowledge distillation to transfer the knowledge of a large language model into compact neural models.

We first fine-tune BioBERT-base with the causal precedence dataset. For each labeled event pair, the model is trained to predict the precedence relationship using a cross-entropy loss. The fine-tuned model will serve as the teacher during the distillation process. We train several BiLSTM student models and compact BERT (BERT-L8H128A2 and BERT-L4H256A4) student models. Following (Tang et al., 2019b), the loss between the teacher and the student is formulated as the Mean Square Error (MSE) loss between the logits of the teacher $z^{(B)}$ and the logits of the student $z^{(S)}$.

$$L = ||z^{(B)} - z^{(S)}||_2^2$$

A distillation process may suffer from a small labeled training set, and data augmentation techniques are frequently used to obtain numerous unlabeled data (Tang et al., 2019b). Similarly, we use both the labeled data D_l and unlabeled data D_u for distillation. However, we don't use data augmentation to obtain D_u, but generate D_u by processing 88,000 PubMed articles with REACH (Valenzuela-Escárcega et al., 2018) and extract 20,001 unlabeled event pairs.

4.4 Baselines

We consider a rule-based heuristic and a feature-based classifier, both of which are proposed and elaborated in (Hahn-Powell et al., 2016). Here we briefly introduce these two baselines, and more details can be found in (Hahn-Powell et al., 2016).

Rule-based heuristic

The event pair causal precedence relation is predicted using a few hand-written deterministic rules. There are three types of rules: **intra-sentence** rules, **inter-sentence** relations and **verbal-tense**.[7]

Feature-based classifier

Event pairs are transformed into a feature vector representation using hand-crafted rules. The encoded pairs are used to train a SVM. Some of the features include the interaction type (i.e. "phosphorylation", "ubiquitination"), the text between the events, coreference resolution, etc.

5 Results

5.1 The Impact of Pre-training

Table 2 shows the impact of in-domain pre-training (as detailed in section 4). For each row, we run experiments with five different random seeds and

[7] A slightly more detailed description can be found in Appendix B.

Model	Dev P.	Dev R.	Dev. F1	Test P.	Test R.	Test F1
BiLSTM-small-w2v-VO1	0.489 (0.028)	0.481 (0.021)	0.484 (0.009)	0.534 (0.025)	0.325 (0.033)	0.403 (0.027)
BiLSTM-small-w2v-VO2	0.326 (0.028)	0.646 (0.067)	0.430 (0.015)	0.404 (0.055)	0.554 (0.075)	0.459 (0.014)
BiLSTM-large-w2v-VO1	0.447 (0.034)	0.545 (0.039)	0.489 (0.014)	0.528 (0.027)	0.400 (0.041)	0.454 (0.026)
BiLSTM-large-w2v-VO2	0.317 (0.014)	0.683 (0.014)	0.433 (0.015)	0.384 (0.008)	0.586 (0.037)	**0.464** (0.012)
BiLSTM-large-w2v-ID-VO1	0.496 (0.040)	0.628 (0.032)	**0.552** (0.021)	0.471 (0.023)	0.421 (0.060)	0.442 (0.033)
BiLSTM-large-w2v-ID-VO2	0.418 (0.029)	0.715 (0.039)	0.526 (0.014)	0.357 (0.039)	0.523 (0.036)	0.422 (0.027)
BiLSTM-large-WP	0.404 (0.024)	0.609 (0.073)	0.484 (0.032)	0.413 (0.038)	0.426 (0.067)	0.418 (0.050)
BERT-L8H128A2	0.340 (0.022)	0.655 (0.047)	0.446 (0.018)	0.407 (0.040)	0.523 (0.049)	0.456 (0.035)
BERT-L8H128A2-Bio	0.375 (0.015)	0.650 (0.031)	0.475 (0.005)	0.491 (0.043)	0.557 (0.064)	0.518 (0.030)
BERT-L8H128A2-Bio-RV	0.364 (0.020)	0.709 (0.042)	0.481 (0.022)	0.449 (0.031)	0.630 (0.017)	0.524 (0.021)
BERT-L4H256A4	0.351 (0.006)	0.561 (0.031)	0.431 (0.010)	0.499 (0.045)	0.485 (0.023)	0.491 (0.027)
BERT-L4H256A4-Bio	0.408 (0.029)	0.622 (0.051)	0.491 (0.015)	0.554 (0.048)	0.549 (0.041)	**0.548** (0.007)
BERT-L4H256A4-Bio-RV	0.420 (0.036)	0.612 (0.030)	**0.497** (0.026)	0.557 (0.065)	0.525 (0.035)	0.537 (0.026)
BERT	0.420 (0.037)	0.605 (0.053)	0.492 (0.013)	0.537 (0.045)	0.512 (0.057)	0.520 (0.030)
BioBERT	0.437 (0.031)	0.705 (0.055)	**0.537** (0.019)	0.547 (0.079)	0.539 (0.072)	**0.535** (0.023)

Table 2: The impact of in-domain pre-training for the BiLSTM and BERT architectures. w2v and w2v-ID are the general-domain/in-domain Word2Vec embeddings. VO1 and VO2 are the two options to build the LSTM vocabulary. WP is the LSTM pre-trained by the language modeling task using WordPiece tokenizer. RV is the reduced vocabulary for BERT. All of these models are discussed throughout Section 5.1.

report mean and standard deviation of the different metrics. Each experiment is a 5-fold cross validation, using 64% of the dataset for training, 16% for validation and 20% for testing. Each model is trained for 40 epochs. The validation F1 is used for early stopping using a patience counter of 5. We used Adam optimizer (Kingma and Ba, 2015). For the LSTM models, we experimented with different hidden sizes, word embedding options and vocabulary options (explained later in the text), and the learning rate is set to 1e-4. For all BERT-based models the learning rate is set to 2e-5.

All of the models in Table 2 contain less than 12M parameters, with the exception of *BERT* and *BioBERT*, which have approximately 110M parameters. Table 3 shows a detailed presentation of the model's size and inference time. Results show that pre-training the compact BERT models on PMC-10000 boosts the models' performance, obtaining test F1s even slightly higher than the large BioBERT. On the other hand, the pre-training of LSTM models using PMC-10000 does not help.

BERT-based models

Rows *BERT* and *BioBERT* in table 2 show the performance of BERT models fine-tuned for the causal precedence task. BioBERT showed both higher F1 scores on dev and test sets, and a lower discrepancy between the dev and test scores compared with other compact models. Since this is a small dataset, we hypothesize that the in-domain pre-training of BioBERT boosts the performance of the fine-tuned model compared to the open domain BERT.

W2V embeddings

For the LSTM models, the *w2v* embeddings were trained using *Word2Vec* over 1 million PubMed papers as introduced in (Hahn-Powell et al., 2016), whereas *w2v-ID* embeddings were trained using the same method but on the PMC-10000 corpus. Both w2v and w2v-ID were trained with biomedical papers, but w2v-ID's corpus is smaller and focused on narrower topics. Results show that the *w2v-ID* embeddings trained on PMC-10000 largely increase the dev scores of the models, which doesn't transfer to the test scores, suggesting the models are overfitting to the dev examples. We suspect that the reason for this is that the PMC-10000 corpus is too small, and not diverse enough for the *w2v-ID* embeddings to learn general and robust representations.

The vocabulary of LSTM models

We found that the composition of the vocabulary used by the LSTM models can impact their performance. We tried two different vocabularies: `VO1`, which contains any word that appears in the training set; and `VO2`, which contains words that occur at least twice in the training set. To deal with out-of-vocabulary words (OOV), `VO1` uses the `unk` vector as trained by Word2Vec (not fine-tuned on our causal detection dataset) whereas in `VO2` the `unk` vector is further fine-tuned in our causal detection dataset. The trade-off is that fine-tuning the `unk` embedding should yield a more accurate representation for it, but it also reduces the vocabulary size. We found that using `V02` works better than `V01` for *w2v* but for *w2v-ID*.[8] This is likely due to the fact that *w2v-ID* already obtains a fairly accurate `unk` embedding through in-domain pre-training, so that the model benefits more from a larger vocabulary than a fine-tuned `unk` embedding. On the other hand, for *w2v*, the `unk` embedding is not good enough without fine-tuning.

LSTM sizes

Since the size of the model may affect LSTM architecture's performance on some tasks (Adhikari et al., 2020), we investigate the impact of the model size. Results show that for our task the larger LSTM works slightly better than the smaller LSTM but the difference is negligible. Note that the BiLSTM-small is only about 1/8 of BiLSTM-large (size comparison in Table 3).

In-domain BiLSTM language modeling

LSTM-large-WP is trained with the language modeling task introduced in section 4 using the PMC-10000 corpus. However, if we use the regular vocabulary, its size and the embedding layer's size would be large. We reduce both sizes using two strategies: (1) we use the same WordPiece tokenization algorithm that BERT uses; (2) to further reduce the number of embedding vectors, we keep only the top 10,000 tokens by corpus frequency in PMC-10000 and use only 10,000 token pieces. In perspective, the WordPiece tokenization model of *bert-base-uncased* has 30,522 tokens. With this approach the vocabulary size and the number of embeddings is reduced by $\frac{2}{3}$.

Our results show that pre-training the LSTM model using in-domain language modeling task does not help with the fine-tuning of our causal precedence detection task. The pre-trained LSTM has a relatively large gap between the dev F1 (0.484) and test F1 (0.418), and the test F1 is even lower than using the *w2v-ID* embeddings. This is probably because the LSTM is not pre-trained on the general domain corpus (like BERT), therefore it doesn't benefit from any transfer learning signal.

In-domain pre-training of BERT

BERT-L8H128A2 and BERT-L4H256A4 are pre-trained on BookCorpus (Zhu et al., 2015) and English Wikipedia (the same as the regular BERT) but not trained on any in-domain datasets (such as any PubMed articles). Our results show that fine-tuning BERT-L8H128A2 yields similar results to BiLSTM-large. Fine-tuning BERT-L4H256A4 yields better results than the LSTM models, but it has twice the number of parameters than the BiLSTM-large model (comparison in Table 3).

However, if we pre-train them on PMC-10000, corresponding to models BERT-L8H128A2-Bio and BERT-L4H256A4-Bio, both the dev and test F1 scores largely improve (the improvement ranges from 0.03 to 0.06) compared to the equivalent models without in-domain pre-training.

The size of *BERT-L4H256A4-Bio* is much larger than other compact models in the table. This is mostly explained by the size of the embedding layer. We experiment reducing the embedding layer size using the similar approach as with the BiLSTM model: Keep the top 10,000 word pieces by frequency of the *base-base-uncased* tokenizer in PMC-10000 and resize the vocabulary to 10,000. The original pre-trained embeddings are used to initialize the embedding layers of the Reduced Vocab BERT-L4H256A4 (see Appendix C for details). The new models resulting of this procedure are identified by the *-RV* suffix in tables 2, 3 and 4.

Both BERT-L8H128A2-Bio-RV and BERT-L4H256A4-Bio-RV are pre-trained on PMC-10000 before fine-tuned on the causal precedence dataset. Previous work has shown that larger vocabulary sizes could slightly boost the performance of BERT-based models (Conneau et al., 2020). We observed different impacts of vocabulary reduction on BERT-L8H128A2 and BERT-L2H256A4. The test F1 of BERT-L4H256A4 drops from 0.548 to 0.537 whereas that of BERT-L2H128A2 even increases from 0.518 to 0.524. This shows that the impact of the vocabulary size to the BERT's performance is task- and model-dependent. Further, it is possible to gain some improvement by reducing the vocabulary size of BERT.

Model size and inference time

Table 3 shows the number of parameters of the models and their inference times on CPU and GPU. In general, all compact BERT models yield much better inference time than LSTM models on CPU. For example, both BiLSTM-large and BERT-L8H128A2 have approximately 5M parameters, with an inference time on CPU are 0.026s and 0.013s, respectively. This clearly shows the

[8]See the "W2V embedding" section of Section 5.1 for the explanation of *w2v* and *w2v-ID*.

Model	# Param.	# Embd. Param.	CPU Inf. T	GPU Inf. T
BiLSTM-small-VO2	0.66M	0.14M	0.007	0.002
BiLSTM-large-w2v(-ID)-VO2	5.40M	0.14M	0.026	0.006
BiLSTM-large-WP	5.63M	1M	0.031	0.009
BERT-L8H128A2(-Bio)	5.58M	3.91M	0.013	0.007
BERT-L8H128A2-Bio-RV	2.95M	1.28M	0.014	0.007
BERT-L4H256A4(-Bio)	11.17M	7.81M	0.011	0.005
BERT-L4H256A4-Bio-RV	5.92M	2.56M	0.011	0.004
BioBERT	108.31M	22.27M	0.119	0.012

Table 3: Model sizes and inference times. For all models, we show the total number of parameters, the number of parameters in the embedding layers (which can be reduced by reducing the model's vocabulary), the average CPU and GPU inference time (seconds per input sequence). The numbers are averaged across 5 runs of all examples.

transformer architecture of BERT is better suited for parallelization. Furthermore, BERT-L4H256A4 has about twice number of parameters as BERT-L8H256A4, but it has smaller inference time (0.011s vs 0.013s) because of fewer layers.

5.2 The Impact of Distillation

Previous work shows that knowledge distillation from a large model (teacher) to a compact model (student) does not always work and is highly dependent on the nature of task. For example, Wasserblat et al. (2020) found that distillation can be helpful for the tasks that require general lexical semantics. However, the distillation on our dataset is very challenging because: (1) there are only about 580 labeled training samples for the teacher, and (2) after fine-tuning, our teacher can only reach a 0.54 test F1 (BioBERT in Table 2).

We adopt a three-stage pipeline for distillation. (1) The teacher model (BioBERT) is fine-tuned on the labeled training data. (2) The teacher model runs inference on the labeled data (and optionally on the unlabeled data) to get the predictions scores for each example. (3) The student model is trained to reproduce the teacher's score on each training example with the loss function introduced in Section 4. Depending on how many unlabeled data to use, we evaluate 3 distillation settings: labeled, labeled + 2k unlabeled and labeled + 20k unlabeled. The results are shown in Table 4.

The impact of distillation on out-of-domain pre-trained models

Among the models we evaluate, BiLSTM(-small/large)-w2v and BERT-L4H256A4 were not pre-trained using PMC-10000, the in-domain corpus. For BiLSTM(-small/large)-w2v, distillation using only the labeled data is not helpful compared with direct fine-tuning. However, distillation becomes helpful when more unlabeled data are used. With BiLSTM-small-w2v, the testing F1 score increases from 0.452, when only labeled data is used for distillation, to 0.489, when using the labeled and 2k unlabeled examples for distillation. The testing F1 further improves to 0.496 when using labeled and 20k unlabeled examples for distillation. A similar trend is also found for BiLSTM-large-w2v. The trend for BERT-L4H256A4 is slightly different. When we use only labeled data, labeled data plus 2k unlabeled examples, and labeled data plus 20k unlabeled examples for distillation, testing F1 scores are 0.502, 0.499 and 0.516, respectively. Both BiLSTM(-small/large)-w2v and BERT-L4H256A4, using labeled data, plus 20k unlabeled examples for distillation attain better testing F1 scores compared to only using the labeled data for fine-tuning (table 2). It shows that in general, out-of-domain pre-trained models can largely benefit from distillation, especially when there are sufficient unlabeled data.

The impact of distillation on in-domain pre-trained models

We observed a similar pattern when distilling in-domain, pre-trained models. For most cases, the model's testing F1 score increased as more unlabeled data became available for distillation. The testing F1 scores of BiLSTM-large-WP increased from 0.400 to 0.430 and 0.487 when using either only labeled data, labeled data + 2k unlabeled examples, labeled + 20k unlabeled examples, respectively for distillation. Similar trends are also found for BERT-L4H256A4-Bio and BERT-L4H256A4-Bio-RV. The only exception we observed was BiLSTM-large-w2v-ID, whose testing F1 score was 0.441 when using the labeled data for distillation, then peaked at 0.477 when using labeled + 2k unlabeled data, just to decrease to

Model	Dev P.	Dev R.	Dev. F1	Test P.	Test R.	Test. F1
labeled						
BiLSTM-small-w2v-VO2	0.387 (0.014)	0.600 (0.090)	**0.467** (0.024)	0.426 (0.034)	0.491 (0.073)	0.452 (0.037)
BiLSTM-large-w2v-VO2	0.393 (0.026)	0.703 (0.045)	**0.503** (0.012)	0.414 (0.058)	0.508 (0.061)	0.450 (0.022)
BiLSTM-large-w2v-ID-VO1	0.524 (0.051)	0.587 (0.051)	0.550 (0.029)	0.488 (0.046)	0.407 (0.037)	0.441 (0.028)
BiLSTM-large-WP	0.462 (0.047)	0.598 (0.053)	**0.518** (0.036)	0.444 (0.051)	0.373 (0.051)	0.400 (0.024)
BERT-L4H256A4	0.364 (0.015)	0.592 (0.058)	**0.450** (0.024)	0.468 (0.044)	0.550 (0.042)	**0.502** (0.011)
BERT-L4H256A4-Bio	0.389 (0.022)	0.645 (0.048)	0.483 (0.008)	0.514 (0.044)	0.569 (0.040)	0.537 (0.007)
BERT-L4H256A4-Bio-RV	0.417 (0.031)	0.625 (0.057)	**0.498** (0.020)	0.554 (0.066)	0.555 (0.038)	**0.551** (0.035)
labeled + unlabeled 2k						
BiLSTM-small-w2v-VO2	0.467 (0.031)	0.574 (0.085)	**0.510** (0.029)	0.541 (0.049)	0.452 (0.057)	**0.489** (0.037)
BiLSTM-large-w2v-VO2	0.480 (0.031)	0.616 (0.082)	**0.535** (0.033)	0.561 (0.088)	0.438 (0.050)	**0.483** (0.018)
BiLSTM-large-w2v-ID-VO1	0.531 (0.027)	0.650 (0.066)	**0.583** (0.038)	0.535 (0.047)	0.434 (0.033)	**0.477** (0.023)
BiLSTM-large-WP	0.494 (0.027)	0.568 (0.079)	**0.525** (0.034)	0.498 (0.044)	0.384 (0.057)	**0.430** (0.037)
BERT-L4H256A4	0.394 (0.034)	0.591 (0.071)	**0.468** (0.013)	0.520 (0.066)	0.498 (0.076)	**0.499** (0.010)
BERT-L4H256A4-Bio	0.393 (0.024)	0.688 (0.045)	**0.499** (0.015)	0.509 (0.060)	0.584 (0.021)	0.541 (0.024)
BERT-L4H256A4-Bio-RV	0.408 (0.030)	0.678 (0.031)	**0.508** (0.018)	0.532 (0.057)	0.592 (0.030)	**0.558** (0.023)
labeled + unlabeled 20k						
BiLSTM-small-w2v-VO2	0.464 (0.027)	0.580 (0.080)	**0.512** (0.028)	0.539 (0.065)	0.468 (0.054)	**0.496** (0.029)
BiLSTM-large-w2v-VO2	0.458 (0.039)	0.629 (0.043)	**0.527** (0.015)	0.524 (0.082)	0.473 (0.054)	**0.490** (0.027)
BiLSTM-large-w2v-ID-VO1	0.512 (0.057)	0.575 (0.090)	0.537 (0.055)	0.521 (0.040)	0.399 (0.060)	**0.449** (0.039)
BiLSTM-large-WP	0.474 (0.026)	0.572 (0.040)	**0.517** (0.005)	0.563 (0.050)	0.432 (0.024)	**0.487** (0.021)
BERT-L4H256A4	0.382 (0.026)	0.617 (0.037)	**0.470** (0.008)	0.492 (0.075)	0.553 (0.062)	**0.516** (0.044)
BERT-L4H256A4-Bio	0.388 (0.028)	0.666 (0.046)	0.489 (0.014)	0.508 (0.057)	0.602 (0.039)	0.547 (0.014)
BERT-L4H256A4-Bio-RV	0.393 (0.017)	0.647 (0.043)	0.488 (0.007)	0.526 (0.062)	0.602 (0.033)	**0.558** (0.021)

Table 4: The compact model's performance using distillation with different amount of unlabeled data. The improved F1s (compared with the fine-tuned model's F1 in Table 2) are shown in **bold** text. All experiments are run for 5 seeds and 5-fold cross validation. The standard deviation across 5 random seeds is shown in the parenthesis.

Model	Dev P.	Dev R.	Dev. F1	Test P.	Test R.	Test. F1
Rule	0.534	0.272	0.360	0.523	0.170	0.257
SVM	0.361	0.407	0.383	0.395	0.364	0.379
Rule -> SVM	0.367	0.537	**0.436** ↑	0.383	0.445	**0.412** ↑
Rule -> BiLSTM-small-FT	0.325	0.713	**0.445** ↑	0.393	0.622	**0.476** ↑
Rule -> BiLSTM-large-FT	0.314	0.739	**0.441** ↑	0.381	0.659	**0.482** ↑
BiLSTM-small-FT -> SVM	0.308	0.767	**0.439** ↑	0.362	0.649	**0.461** ↑
BiLSTM-large-FT -> SVM	0.299	0.785	0.433 -	0.348	0.662	0.456 ↓
Rule -> BiLSTM-small-DS	0.414	0.618	0.493 ↓	0.482	0.507	0.492 ↓
Rule -> BiLSTM-large-DS	0.406	0.645	0.497 ↓	0.472	0.513	0.488 ↓
Rule -> BERT-L4-Bio-RV-DS	0.357	0.647	0.459 ↓	0.476	0.616	0.535 ↓
Rule -> BERT-L4-Bio-RV-FT	0.376	0.625	0.469 ↓	0.503	0.562	0.529 ↓
BiLSTM-small-DS -> SVM	0.377	0.692	0.487 ↓	0.411	0.593	0.485 ↓
BiLSTM-large-DS -> SVM	0.369	0.717	0.486 ↓	0.409	0.590	0.482 ↓
BERT-L4-Bio-RV-DS -> SVM	0.339	0.731	0.463 ↓	0.417	0.684	0.518 ↓
BERT-L4-Bio-RV-FT -> SVM	0.349	0.708	0.467 ↓	0.423	0.624	0.504 ↓
BiLSTM-s -> BERT-L4-DS	0.376	0.678	0.482 ↓	0.498	0.630	0.552 ↓
BiLSTM-l -> BERT-L4-DS	0.374	0.685	0.482 ↓	0.491	0.635	0.550 ↓
Rule + BiLSTM-l + BERT-L4	0.488	0.609	**0.540** ↑	0.585	0.478	0.521 ↓
SVM + BiLSTM-l + BERT-L4	0.458	0.612	0.523 ↓	0.551	0.495	0.517 ↓

Table 5: Results of the sieve models. "X -> Y" means model X's prediction is firstly used in the sieve then Y. FT means fine-tuning (entries in Table 2) and DS means distillation using **20k unlabeled data and labeled data** (entries in Table 4). The upside and downside arrows besides the scores indicate whether the sieve score outperforms the best individual model in the sieve. In the last four rows, BiLSTM-s is BiLSTM-small-DS, BiLSTM-l is BiLSTM-large-DS and BERT-L4 is BERT-L4H256A4-Bio-RV-DS. For all BiLSTM models we use w2v general embedding with VO2. The last two rows of the table are the ensemble model's performance.

0.449 when using labeled + 20k unlabeled data. This indicates in general the in-domain pre-trained models can still benefit from distillation when there are sufficient unlabeled data.

5.3 Comparison among Rule-based, Feature-based and Neural-based Models

Previous work has explored combining multiple models by using a sieve method (Mirza, 2014; Hahn-Powell et al., 2016). Generally speaking, a sieve method starts by using the model with the highest precision to predict the class of an input. If the prediction is positive, it is returned as the result, otherwise the input is forwarded to the model with the second best precision, and the process is repeated until a model makes a positive prediction or all the models are exhausted. In this work we explore the performance of a sieve method composed of multiple combinations of feature/rule/neural-based models. We rank the models by their decreasing precision in the development set. Table 5 contains the performance of different sieves.

Rules and SVM complement each other

As shown in Table 5, combining the rule-based and feature-based models into a sieve results in an improvement over either of them individually. However, this sieve is not on par with the performance of sieves that contain neural models.

Non-distilled neural models are complemented by non-neural models

Table 5 shows that in three out of four cases, sieves with a rule-based classifier or SVM classifier boost the performance of the LSTM models that are fine-tuned but not distilled. The benefits are more evident for the rule classifier than for the SVM classifier.

Well-trained neural models are not complemented by non-neural models

For models distilled with labeled and 20k unlabeled examples (BiLSTM-small/large-DS, BERT-L4-DS) and the model pre-trained both in the general domain and in the target domain (BERT-L4-FT), neither the rule-based classifier nor the SVM classifier result on increase the performance when sieved. This hints that well-trained neural models could learn to represent the same high-level handcrafted features in the Rule and SVM classifier.

Different neural models are not likely to complement each other in a sieve

As the last two rows of Table 5 show, although BiLSTM and BERT are very different models, they do not tend to complement each other in a sieve.

6 Conclusion

In this work we trained several neural models for causal precedence detection in the biomedical literature. To help with the deployment of neural models on systems without a GPU, we restricted the sizes of our architectures to approximately $\frac{1}{20}$ of the size of a state-of-the-art language model such as BERT. Moreover, to overcome the challenge of scarcity of labeled training data, we used in-domain unlabeled data combined with pre-training and distillation and obtained robust neural models. Finally, we compared our neural models with previous rule-based and feature-based classifiers and found the in-domain pre-trained models can mostly replace them.

Acknowledgements

Thank Gus Hahn-Powell for his help on replicating the data and experiments from his previous work (Hahn-Powell et al., 2016). This work was supported by the Defense Advanced Research Projects Agency (DARPA) under the World Modelers and Automating Scientific Knowledge Extraction (ASKE, HR00111990011) programs. Mihai Surdeanu declares a financial interest in lum.ai. This interest has been properly disclosed to the University of Arizona Institutional Review Committee, and is managed in accordance with its conflict of interest policies.

References

Ashutosh Adhikari, Achyudh Ram, Raphael Tang, William L. Hamilton, and Jimmy Lin. 2020. Exploring the limits of simple learners in knowledge distillation for document classification with DocBERT. In *Proceedings of the 5th Workshop on Representation Learning for NLP*, pages 72–77, Online. Association for Computational Linguistics.

Abbas Akkasi and Mari-Francine Moens. 2021. Causal relationship extraction from biomedical text using deep neural models: A comprehensive survey. *Journal of Biomedical Informatics*, 119:103820.

Steven Bird, Ewan Klein, and Edward Loper. 2009. *Natural language processing with Python: analyzing text with the natural language toolkit*. " O'Reilly Media, Inc.".

Eduardo Blanco, Nuria Castell, and Dan Moldovan. 2008. Causal relation extraction. In *Proceedings of the Sixth International Conference on Language Resources and Evaluation (LREC'08)*.

Alexis Conneau, Kartikay Khandelwal, Naman Goyal, Vishrav Chaudhary, Guillaume Wenzek, Francisco Guzmán, Edouard Grave, Myle Ott, Luke Zettlemoyer, and Veselin Stoyanov. 2020. Unsupervised cross-lingual representation learning at scale. In *Proceedings of the 58th Annual Meeting of the Association for Computational Linguistics*, pages 8440–8451, Online. Association for Computational Linguistics.

Tirthankar Dasgupta, Rupsa Saha, Lipika Dey, and Abir Naskar. 2018. Automatic extraction of causal relations from text using linguistically informed deep neural networks. In *Proceedings of the 19th Annual SIGdial Meeting on Discourse and Dialogue*, pages 306–316, Melbourne, Australia. Association for Computational Linguistics.

Jacob Devlin, Ming-Wei Chang, Kenton Lee, and Kristina Toutanova. 2019. BERT: Pre-training of deep bidirectional transformers for language understanding. In *Proceedings of the 2019 Conference of the North American Chapter of the Association for Computational Linguistics: Human Language Technologies, Volume 1 (Long and Short Papers)*, pages 4171–4186, Minneapolis, Minnesota. Association for Computational Linguistics.

Roxana Girju. 2003. Automatic detection of causal relations for question answering. In *Proceedings of the ACL 2003 workshop on Multilingual summarization and question answering*, pages 76–83.

Alex Graves and Jürgen Schmidhuber. 2005. Framewise phoneme classification with bidirectional lstm and other neural network architectures. *Neural networks*, 18(5-6):602–610.

Suchin Gururangan, Ana Marasović, Swabha Swayamdipta, Kyle Lo, Iz Beltagy, Doug Downey, and Noah A. Smith. 2020. Don't stop pretraining: Adapt language models to domains and tasks. In *Proceedings of the 58th Annual Meeting of the Association for Computational Linguistics*, pages 8342–8360, Online. Association for Computational Linguistics.

Gus Hahn-Powell, Dane Bell, Marco A. Valenzuela-Escárcega, and Mihai Surdeanu. 2016. This before that: Causal precedence in the biomedical domain. In *Proceedings of the 15th Workshop on Biomedical Natural Language Processing*, pages 146–155, Berlin, Germany. Association for Computational Linguistics.

Xiaoqi Jiao, Yichun Yin, Lifeng Shang, Xin Jiang, Xiao Chen, Linlin Li, Fang Wang, and Qun Liu. 2020. TinyBERT: Distilling BERT for natural language understanding. In *Findings of the Association for Computational Linguistics: EMNLP 2020*, pages 4163–4174, Online. Association for Computational Linguistics.

Christopher SG Khoo, Syin Chan, and Yun Niu. 2000. Extracting causal knowledge from a medical database using graphical patterns. In *Proceedings of the 38th annual meeting of the association for computational linguistics*, pages 336–343.

Jin-Dong Kim, Ngan Nguyen, Yue Wang, Jun'ichi Tsujii, Toshihisa Takagi, and Akinori Yonezawa. 2012. The genia event and protein coreference tasks of the bionlp shared task 2011. In *BMC bioinformatics*, volume 13, pages 1–12. BioMed Central.

Jin-Dong Kim, Tomoko Ohta, Sampo Pyysalo, Yoshinobu Kano, and Jun'ichi Tsujii. 2009. Overview of bionlp'09 shared task on event extraction. In *Proceedings of the BioNLP 2009 workshop companion volume for shared task*, pages 1–9.

Diederik P Kingma and Jimmy Ba. 2015. Adam: A method for stochastic optimization. In *ICLR (Poster)*.

Jinhyuk Lee, Wonjin Yoon, Sungdong Kim, Donghyeon Kim, Sunkyu Kim, Chan Ho So, and Jaewoo Kang. 2020. BioBERT: a pre-trained biomedical language representation model for biomedical text mining. *Bioinformatics*, 36(4):1234–1240.

Linqing Liu, Huan Wang, Jimmy Lin, Richard Socher, and Caiming Xiong. 2019a. Mkd: a multi-task knowledge distillation approach for pretrained language models. *arXiv preprint arXiv:1911.03588*.

Yinhan Liu, Myle Ott, Naman Goyal, Jingfei Du, Mandar Joshi, Danqi Chen, Omer Levy, Mike Lewis, Luke Zettlemoyer, and Veselin Stoyanov. 2019b. Roberta: A robustly optimized bert pretraining approach. *arXiv preprint arXiv:1907.11692*.

Luke Melas-Kyriazi, George Han, and Celine Liang. 2019. Generation-distillation for efficient natural language understanding in low-data settings. In *Proceedings of the 2nd Workshop on Deep Learning Approaches for Low-Resource NLP (DeepLo 2019)*, pages 124–131, Hong Kong, China. Association for Computational Linguistics.

Claudiu Mihăilă, Tomoko Ohta, Sampo Pyysalo, and Sophia Ananiadou. 2013. Biocause: Annotating and analysing causality in the biomedical domain. *BMC bioinformatics*, 14(1):1–18.

Tomas Mikolov, Kai Chen, Greg Corrado, and Jeffrey Dean. 2013. Efficient estimation of word representations in vector space. *arXiv preprint arXiv:1301.3781*.

Paramita Mirza. 2014. Extracting temporal and causal relations between events. In *Proceedings of the ACL 2014 Student Research Workshop*, pages 10–17, Baltimore, Maryland, USA. Association for Computational Linguistics.

Amr Mousa and Björn Schuller. 2017. Contextual bidirectional long short-term memory recurrent neural network language models: A generative approach to sentiment analysis. In *Proceedings of the 15th Conference of the European Chapter of the Association for Computational Linguistics: Volume 1, Long Papers*, pages 1023–1032.

Subhabrata Mukherjee and Ahmed Hassan Awadallah. 2019. Distilling transformers into simple neural networks with unlabeled transfer data. *arXiv preprint arXiv:1910.01769*, 1.

Claire Nédellec, Robert Bossy, Jin-Dong Kim, Jung-Jae Kim, Tomoko Ohta, Sampo Pyysalo, and Pierre Zweigenbaum. 2013. Overview of bionlp shared task 2013. In *Proceedings of the BioNLP shared task 2013 workshop*, pages 1–7.

Haojie Pan, Chengyu Wang, Minghui Qiu, Yichang Zhang, Yaliang Li, and Jun Huang. 2021. Meta-KD: A meta knowledge distillation framework for language model compression across domains. In *Proceedings of the 59th Annual Meeting of the Association for Computational Linguistics and the 11th International Joint Conference on Natural Language Processing (Volume 1: Long Papers)*, pages 3026–3036, Online. Association for Computational Linguistics.

Colin Raffel, Noam Shazeer, Adam Roberts, Katherine Lee, Sharan Narang, Michael Matena, Yanqi Zhou, Wei Li, and Peter J Liu. 2020. Exploring the limits of transfer learning with a unified text-to-text transformer. *Journal of Machine Learning Research*, 21:1–67.

Victor Sanh, Lysandre Debut, Julien Chaumond, and Thomas Wolf. 2019. Distilbert, a distilled version of bert: smaller, faster, cheaper and lighter. *arXiv preprint arXiv:1910.01108*.

Antonio Sorgente, Giuseppe Vettigli, and Francesco Mele. 2013. Automatic extraction of cause-effect relations in natural language text. *DART@ AI* IA*, 2013:37–48.

Raphael Tang, Yao Lu, and Jimmy Lin. 2019a. Natural language generation for effective knowledge distillation. In *Proceedings of the 2nd Workshop on Deep Learning Approaches for Low-Resource NLP (DeepLo 2019)*, pages 202–208, Hong Kong, China. Association for Computational Linguistics.

Raphael Tang, Yao Lu, Linqing Liu, Lili Mou, Olga Vechtomova, and Jimmy Lin. 2019b. Distilling task-specific knowledge from bert into simple neural networks. *arXiv preprint arXiv:1903.12136*.

Iulia Turc, Ming-Wei Chang, Kenton Lee, and Kristina Toutanova. 2019. Well-read students learn better: On the importance of pre-training compact models. *arXiv preprint arXiv:1908.08962*.

Marco A Valenzuela-Escárcega, Özgün Babur, Gus Hahn-Powell, Dane Bell, Thomas Hicks, Enrique Noriega-Atala, Xia Wang, Mihai Surdeanu, Emek Demir, and Clayton T Morrison. 2018. Large-scale automated machine reading discovers new cancer-driving mechanisms. *Database*, 2018. Bay098.

Konstantinos Z Vardakas, Grigorios Tsopanakis, Alexandra Poulopoulou, and Matthew E Falagas. 2015. An analysis of factors contributing to pubmed's growth. *Journal of Informetrics*, 9(3):592–617.

Wenhui Wang, Furu Wei, Li Dong, Hangbo Bao, Nan Yang, and Ming Zhou. 2020. Minilm: Deep self-attention distillation for task-agnostic compression of pre-trained transformers. *Advances in Neural Information Processing Systems*, 33:5776–5788.

Moshe Wasserblat, Oren Pereg, and Peter Izsak. 2020. Exploring the boundaries of low-resource BERT distillation. In *Proceedings of SustaiNLP: Workshop on Simple and Efficient Natural Language Processing*, pages 35–40, Online. Association for Computational Linguistics.

Bowen Wu, Huan Zhang, MengYuan Li, Zongsheng Wang, Qihang Feng, Junhong Huang, and Baoxun Wang. 2020. Towards non-task-specific distillation of bert via sentence representation approximation. In *Proceedings of the 1st Conference of the Asia-Pacific Chapter of the Association for Computational Linguistics and the 10th International Joint Conference on Natural Language Processing*, pages 70–79.

Yukun Zhu, Ryan Kiros, Rich Zemel, Ruslan Salakhutdinov, Raquel Urtasun, Antonio Torralba, and Sanja Fidler. 2015. Aligning books and movies: Towards story-like visual explanations by watching movies and reading books. In *Proceedings of the IEEE international conference on computer vision*, pages 19–27.

A Types of Events of PMC-10000

phosphorylation, phosphorylates, ubiquitination, ubiquitinates, hydroxylation, hydroxylates, sumoylation, sumoylates, glycosylation, glycosylates, acetylation, acetylates, farnesylation, farnesylates, ribosylation, ribosylates, methylation, methylates, binding, binds, activation, activates.

B More Description of the Rule-based Classifier

The event pair causal precedence relation is predicted using a few hand-written deterministic rules. There are three types of rules: The first type are **intra-sentence** rules, where the two events are in the same sentence. Patterns of this type operate over the syntactic dependency graph of the sentence. The second type are rules for **inter-sentence** relations, where the two events occur on different sentences and a dependency graph is not available. These kind of rules use the presence of patterns, such as "leads to", "result in" to predict causal precedence. The third kind of rules, also for inter-sentence event pairs, use verbal-tense information. Phrases such as "has been phosphorylated" are used to detect the existence of causal precedence.

C Reducing the Vocabulary of BERT-L4H256A4 and Resizing the Embeddings

The original vocabulary of the *bert-base-uncased* model has a size of 30,522. As discussed in Section 5.1, we count the frequency of the word pieces in PMC-10000 and only maintain the top 10000 most frequent word pieces.

The next step would be to resize the embedding layer of BERT-L4H256A4. Note that the original embeddings of BERT-L4H256A4 are pre-trained in the language modeling task on BookCorpus and English Wikipedia. We don't want to lose such information during the resizing of the embedding layer by initializing the 10000 token embeddings randomly. Instead, the new embedding weights are initialized with the values of the corresponding weights of the original embedding layer (i.e., all the embedding weights in the new embedding layer reuses the pre-trained weights of the old embedding layer).

Overview of the MedVidQA 2022 Shared Task on Medical Video Question-Answering

Deepak Gupta and Dina Demner-Fushman
Lister Hill National Center for Biomedical Communications
National Library of Medicine, National Institutes of Health
Bethesda, MD, USA
{firstname.lastname}@nih.gov

Abstract

In this paper, we present an overview of the MedVidQA 2022 shared task, collocated with the 21st BioNLP workshop at ACL 2022. The shared task addressed two of the challenges faced by medical video question answering: (i) a video classification task that explores new approaches to medical video understanding (labeling), and (ii) a visual answer localization task. Visual answer localization refers to identification of the relevant temporal segments (start and end timestamps) in the video where the answer to the medical question is being shown or illustrated. A total of thirteen teams participated in the shared task challenges, with eleven system descriptions submitted to the workshop. The descriptions present mono-modal and multi-modal approaches developed for medical video classification and visual answer localization. This paper describes the tasks, the datasets, evaluation metrics, and baseline systems for both tasks. Finally, the paper summarizes the techniques and results of the evaluation of the various approaches explored by the participating teams.

1 Introduction

With the increasing interest in using artificial intelligence (AI) to support clinical decision-making, improving patient engagement, patient health and well-being (HHS, 2021), there is a need to explore the efficient algorithms for medical language-video understanding. Further, the recent surge in availability of online educational videos on diverse medical and health-related topics demands the development of effective systems that can understand medical videos to provide the best possible answers to consumers' first aid, medical emergency, and medical educational questions.

Video Question Answering (VQA) is an emerging and challenging task that requires the understanding of video, language, and their interaction to correctly provide the answer to the question. The majority of the existing studies (Lei et al., 2018; Xue et al., 2018; Li et al., 2020a; Chadha et al., 2020) on video question answering are focused on open-domain videos such as movies (Tapaswi et al., 2016), TV shows (Lei et al., 2018, 2020a), and games (Mun et al., 2017). Moreover, the primary objective of the existing VQA studies is to develop a system that can provide natural language answers to the users' questions about the video. Some works, such as Anne Hendricks et al. (2017); Lei et al. (2020b); Wang et al. (2020) focus on natural language frame/video localization, but most of them aim to find the video segment that has semantic understanding equivalent to the natural language query. The existing VQA approaches, however, do not take into account the real-world scenarios, where people interact through natural language questions and expect relevant and concise temporal segments from the videos as answers to their questions. Consider a health-related question "*How can I ease my neck pain?*". The textual answer (*cf.* Fig. 1) to the given health-related question will be hard to understand and act upon without visual assistance. In order to provide a visual answer to the question, the first step is to identify the most relevant medical video that has a series of steps describing the detailed visual answer to the question. The second and most important step is to locate the relevant temporal segment in the video that is suitable to be a visual answer (*cf.* Fig. 1) to the question.

Towards solving these challenges, we introduced the MedVidQA 2022 shared task[1], which aims to explore and develop efficient algorithms for video question answering that remain understudied in the medical domain. In the first task (medical video classification) of the MedVidQA 2022 shared task, participants are asked to develop a system that can categorize the video into medical instructional, medical non-instructional, and non-medical. The

[1] https://medvidqa.github.io/

Proceedings of the BioNLP 2022 workshop, Dublin, Ireland, pages 264–274
May 26, 2022.

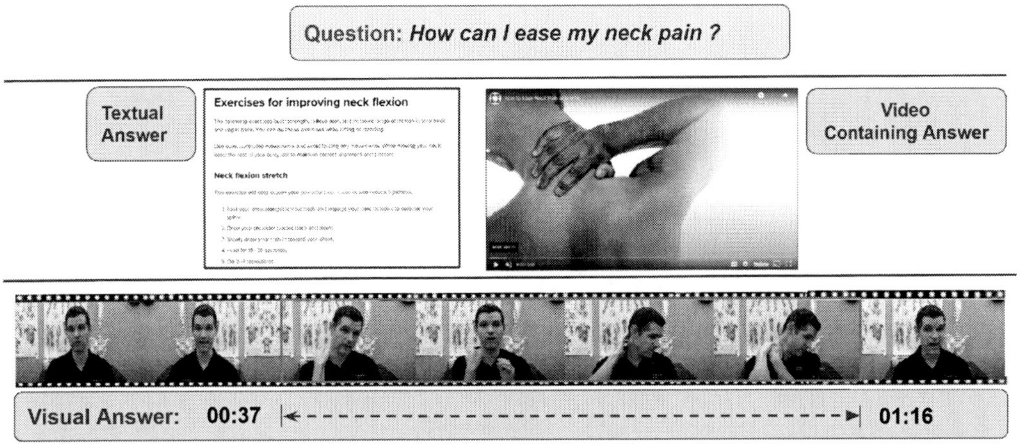

Figure 1: An example of a health-related question, textual answer, video containing the answer, and visual answer (temporal segment) from the video. The textual answer (**center-left**) is retrieved from the web. It contains a series of steps to relieve neck pain by improving neck flexion. The suggested steps in textual answer might be difficult to follow for a consumer who has little or no medical knowledge. The top video (**center-right**) retrieved from the YouTube search contains the answer; however, one has to watch the entire video to find the appropriate temporal segment from the video, which could be served as a visual answer to the question. Unlike the textual and video containing the answer, locating the appropriate temporal segment (**bottom**) which has the visual answer is easy to follow and also eliminates the need to watch the entire video to find the answer.

second task (medical visual answer localization) aims to effectively localize the visual answer to the given medical or health-related question in a given video.

2 MedVidQA 2022 Task Descriptions

Following creation of the dataset for video question answering (Gupta et al., 2022), we consider the following tasks:

2.1 Task 1: Medical Video Classification (MVC)

Given an input video, the task is to categorize the video into one of the following classes:

- **Medical Instructional**: A medical instructional video for non-professionals should clearly demonstrate a medical procedure, providing enough details to reproduce the procedure and achieve the desired results without prior training. The accompanying narrative should be to the point, and should clearly describe the steps in the visual content. A video is medical instructional if a valid medical or health-related question is aligned with it, and it explains/answers the medical question with a demonstration. The demonstration should be a tutorial/educational video where someone (e.g., a doctor or a medical professional) demonstrates a procedure related to the medical question or a how-to video about the medical or health-related question.
- **Medical Non-instructional**: A medical video can be categorized into a medical non-instructional if it discusses medical-related topics without any visual demonstration.
- **Non-medical**: A video can be categorized as non-medical if the video is neither medical instructional nor medical non-instructional.

We have provided the link to the sample videos for each class in Fig. 2.

2.2 Task 2: Medical Visual Answer Localization (MVAL)

Given a medical or health-related question and a video, the task aims to locate the temporal segments (start and end timestamps) in the video where the answer to the medical question is being shown, or the explanation is illustrated in the video. A similar task in the literature is established as natural language frame localization (Anne Hendricks et al., 2017; Miech et al., 2019), where the task is to find the video segment that has equivalent semantics as to the natural language. In contrast, the introduced task seeks to find a video segment with a visual answer to the natural language query. The MVAL task can be considered as finding a series of

(a) **Medical Instructional** (b) **Medical Non-Instructional** (c) **Non-medical**

Figure 2: Sample videos from each category in Medical Video Classification Task

Video Category	Train	Validation	Test
Medical Instructional	789	100	400
Medical Non-instructional	2,394	100	426
Non-medical	1,034	100	382
Total	4,217	300	1,208

Table 1: The dataset statistics for the MVC task. Training and validation datasets statistics are borrowed from `MedVidCL` corpus (Gupta et al., 2022).

Dataset Detail	Train	Validation	Test
Medical instructional videos	800	49	50
Video duration (hours)	86.37	4.54	5.13
Mean video duration (seconds)	388.68	333.89	369.62
Questions and visual answers	2,710	145	153
Minimum question length	5	6	6
Maximum question length	25	21	18
Mean question length	11.67	11.76	11.20
Minimum visual answer length (seconds)	3	10	4
Maximum visual answer length (seconds)	298	267	257
Mean visual answer length (seconds)	62.29	66.81	60.45

Table 2: The dataset statistics for the MVAL task. Training and validation datasets statistics are borrowed from `MedVidQA` corpus (Gupta et al., 2022).

"*medical instructional activity-based frame localization*" where a potential solution first searches for all medical instructional activity for a given medical question and then localizes a particular activity that is aligned to medical or health-related question in an untrimmed medical-instructional video. The sample health-related question and the visual answer are shown in Fig. 1.

3 Data Description

3.1 MVC Dataset

The `MedVidCL`[2] (Gupta et al., 2022) training and validation datasets are provided to train and validate the system for MVC task. A human-assisted two-stage approach was used to construct the `MedVidCL` dataset. In the first stage, human-annotated videos were used to train a machine learning model that predicts the appropriate category for the input video. In the second stage, only high-confidence (classifier probability ≥ 0.8) videos from HowTo100M (Miech et al., 2019) and YouTube8M (Abu-El-Haija et al., 2016) dataset are selected and manually validated. The automatically predicted video category is then updated, if needed. This strategy was used to construct the `MedVidCL` dataset. The videos in the training dataset are taken from YouTube[3]; however, the validation and test dataset contain the videos from HowTo100M and YouTube8M datasets. We have provided the detailed statistics of the datasets used for the MVC task in Table 1.

3.2 MVAL Dataset

The `MedVidQA` datasets are created from the top-4 videos returned by the YouTube search in response to the WikiHow[4] health-related query. The dataset contains 800 medical instructional videos in the training and 50 medical instructional videos in the validation set. `MedVidQA` contains medical-informatics expert-curated instructional questions and timestamps in the video, which serve as the visual answer to the questions. For the test dataset, we followed the dataset creation strategy similar to `MedVidQA` creation. We selected 50 YouTube videos from the search results in response to the diverse set of WikiHow queries. The instructional questions and visual answer timestamps were manually created by watching these 50 videos. We have

[2]https://osf.io/pc594/

[3]https://www.youtube.com/

[4]https://www.wikihow.com/Main-Page

provided the detailed statistics of the dataset used for the MVAL task in Table 2.

4 Evaluation

4.1 Evaluation Metrics

4.1.1 MVC Evaluation

To evaluate the performance of the MVC task, we use the following evaluation metrics:

Medical-Inst Precision: It measures the proportion of Medical Instructional class predictions that are actually correct.

$$\text{Med-Inst Precision} = \frac{TP_{medinst}}{TP_{medinst} + FP_{medinst}} \quad (1)$$

where, $TP_{medinst}$ and $FP_{medinst}$ are the True positive and False positive corresponding to the Medical Instructional class.

Medical-Inst Recall: It measures the proportion of actual Medical Instructional class video that were predicted correctly.

$$\text{Med-Inst Recall} = \frac{TP_{medinst}}{TP_{medinst} + FN_{medinst}} \quad (2)$$

where, $TP_{medinst}$ and $FN_{medinst}$ are the True positive and False negative corresponding to the Medical Instructional class.

Medical-Inst F1-score: It is the harmonic mean between precision $P_{medinst}$ and recall $R_{medinst}$ for the Medical Instructional video category.

$$\text{Med-Inst F1-score} = \frac{2 \times P_{medinst} \times R_{medinst}}{P_{medinst} + R_{medinst}} \quad (3)$$

Macro-averaged F1-score: It is the average harmonic mean between precision and recall, where the precision and recall are calculated per video category.

$$\text{Macro-F1} = \sum_{l \in \mathcal{L}} \frac{2 \times P_l \times R_l}{P_l + R_l} \quad (4)$$

where, P_l and R_l are the precision and recall corresponding to the class $l \in \mathcal{L}$.

Since the goal of the MVC task is to effectively predict Medical Instructional video, we consider **Medical-Inst F1-score** as our primary metric to rank the submission. We used the Scikit-learn (Pedregosa et al., 2011) implementation[5] of the precision, recall and macro-averaged F1-score metrics.

4.1.2 MVAL Evaluation

Following Gupta et al. (2022), we evaluated the performance of the MVAL task using the following metrics:

Mean Intersection over Union (mIoU): For a given question q_i, IoU is computed as the ratio of intersection area over union area between predicted and ground-truth temporal visual answer segments. It ranges from 0 to 1. A larger IoU means the predicted and ground-truth temporal visual answer segments match better, and IoU = 1.0 denotes exact match. The mIoU is defined as the average temporal IoUs for all questions (N) in the test set. Formally,

$$\text{mIoU} = \frac{1}{N} \sum_{i=1}^{i=N} IoU(q_i) \quad (5)$$

Rαn, IoU = μ is another metric used to evaluate the performance of the MVAL system. It denotes the percentage of questions for which, out of the top-n retrieved temporal segments, at least one predicted temporal segment having IoU with ground-truth is larger than μ. We asked the participants to submit only the top-1 temporal segment as the visual answer to the question; therefore, we have $n = 1$. Formally,

$$< R\alpha 1, IoU = \mu > = \frac{1}{N} \sum_{i=1}^{i=N} s(q_i, \mu), \text{ and} \quad (6)$$

$$s(q_i, \mu) = \begin{cases} 1, & \text{if } IoU(q_i) \geq \mu \\ 0, & \text{otherwise} \end{cases} \quad (7)$$

We evaluated the participants' submission by considering $\mu = \{0.3, 0.5, 0.7\}$ and for brevity, we denote the $< R\alpha 1, IoU = \mu >$ metric with IoU=μ. Since the IoU=0.7 is the most restrictive metric amongst all the MVAL metrics, we use IoU=0.7 as our primary metric to rank the participants' submissions. The implementation of the evaluation metric is released here[6].

[5] https://scikit-learn.org/stable/modules/generated/sklearn.metrics.classification_report.html

[6] https://github.com/deepaknlp/MedVidQACL/blob/master/MedVidQA/util/runner_utils_t7.py

4.2 Baseline Systems

4.2.1 MVC Baselines

Monomodal (Language) Baseline: In the first baseline, we consider extracting the English subtitles from the videos using the `pytube`[7]. The extracted subtitles are used to fine-tune the BERT-Base-Uncased (Devlin et al., 2019) pre-trained language model (PLM) to classify the video category.

Monomodal (Vision) Baseline: The monomodal vision-based baseline is built upon the video frames, which are extracted from each video at a uniform time interval. In order to extract the frame features, we considered the pre-trained ViT (Dosovitskiy et al., 2021) model as the feature extractor. The sequence of frame features is passed to the LSTM (Hochreiter and Schmidhuber, 1997) network for video category prediction.

Multimodal Baseline: For the multimodal baseline, we consider utilizing both video subtitles and video frames features (extracted from ViT) to predict the video category. The features are passed to the LSTM network to learn their sequence representation. We then concatenated the language and vision representation and passed the concatenated features to a feed-forward layer to predict the video category.

4.2.2 MVAL Baselines

VSL-BASE: Following Gupta et al. (2022), we consider the VSL-BASE as the first baseline for MVAL task, where the visual answer span is predicted using a multimodal fusion-based technique introduced by Zhang et al. (2020). In the VSL-BASE a Transformer-based encoder is used to encode the question, and video frames features (obtained from I3D (Carreira and Zisserman, 2017)), and thereafter, both features are fused with the help of attention mechanism. The joint feature representation is used to predict the start and end timestamps of the visual answer.

VSL-QGH: This baseline is the extension of the VSL-BASE introduced by Zhang et al. (2020), where the target temporal segment in the video is considered as the foreground and the rest of the video as the background. With the VSL-QGH technique, the network is trained by extending the span of the foreground to cover its preceding and following video frames. We follow the experimental details discussed in Gupta et al. (2022) to obtain the results on the test dataset.

[7]https://pypi.org/project/pytube/

Team Name	Team Affiliations	MVC	MVAL
ALIBABA_DAMO	Alibaba Damo Research	✓	✓
BAIDU AI TEAM	Baidu AI Team	✓	✓
SJTU_YITU	SJU/YITU	✓	✓
TENCENT AI RESEARCH	Tencent AI Research	✓	✓
CMU_HKUST	CMU/HKUST	✓	✓
VPAI_LAB (Li et al., 2022a)	Hunan University/CAS	✓	✗
CHICHEALTH	Chic Health	✓	✓
PAHT	Pingan Health Tech	✓	✓
I AM BERT	No Information Available	✓	✗
LINGJING	Hunan University/CAS	✗	✓
UWASHINGTON	University of Washington	✓	✓
CS	No Information Available	✗	✓
DOSSIER (Kusa et al., 2022)	TU Wien	✗	✓

Table 3: Participating teams and their task participation at MedVidQA 2022 shared task

Team Name	MVC		MVAL	
	Language	Vision	Language	Vision
ALIBABA_DAMO	✓	✗	✓	✓
BAIDU AI TEAM	✓	✗	✓	✗
SJTU_YITU	✓	✗	✓	✗
TENCENT AI RESEARCH	✓	✗	✓	✓
CMU_HKUST	✓	✗	✓	✗
VPAI_LAB	✓	✓	NA	NA
CHICHEALTH	✓	✗	✓	✗
PAHT	NA	NA	✓	✗
I AM BERT	NA	NA	NA	NA
LINGJING	NA	NA	✓	✓
UWASHINGTON	✓	✗	✓	✗
CS	NA	NA	NA	NA
DOSSIER	NA	NA	✓	✓

Table 4: Participating teams and their submissions considering the language (video subtitles) and vision (video frames) to build their approaches for MedVidQA 2022 shared task

5 Participating Teams and Methods

5.1 Participating Teams

We use the CodaLab platform to release the datasets, registration, and submissions of the participating teams. In total, 13 teams from Asia (China), Europe (Germany), and North America (USA) continents participated in the MedVidQA 2022 shared task and submitted 30 and 43 individual runs for the MVC and MVAL task, respectively. We have provided (*cf.* Table 3) the team name, affiliations and their participation in MVC and MVAL tasks. We also summarize (*cf.* Table 4) the participating teams and their submissions based on the considered modality to build their approaches for MVC and MVAL tasks. The results of all the participating teams for MVC[8] and MVAL[9] tasks are avail-

[8]https://codalab.lisn.upsaclay.fr/competitions/1058

[9]https://codalab.lisn.upsaclay.fr/competitions/1078

Rank	Team Name	Med-Inst Precision	Med-Inst Recall	Med-Inst F1-score	Macro F1-score
1	VPAI_LAB	**99.74**	97.75	**98.74**	**99.01**
2	CHICHEALTH	98.73	97.25	97.98	98.46
3	BAIDU AI TEAM	99.23	96.75	97.97	98.46
4	PAHT	97.76	**98.00**	97.88	98.46
5	TENCENT AI RESEARCH	97.75	97.75	97.75	98.04
6	SJTU_YITU	98.47	96.50	97.47	98.04
7	CMU_HKUST	98.72	96.25	97.47	98.03
8	ALIBABA_DAMO	96.02	96.50	96.26	97.22
9	UWASHINGTON	97.65	93.50	95.53	96.86
10	I AM BERT	92.21	91.75	91.98	94.01
–	*Monomodal (L) – Baseline*	94.67	88.75	91.61	94.37
–	*Monomodal (V) – Baseline*	90.97	68.00	77.83	82.24
–	*Multimodal (L+V) – Baseline*	84.97	69.25	76.31	81.06

Table 5: Official results of the MVC task. Here **L** and **V** denotes the Language and Vision respectively.

able on CodaLab platform.

5.2 MVC Submissions

5.2.1 Methods

All participants utilized pre-trained language models to develop the video classification methods to categorize the videos into one of the pre-defined categories. The earlier studies by Gupta et al. (2022) show that information obtained from the video subtitles features is more useful for the MVC task compared to the video frame features; therefore, the video subtitles remained the primary information considered by all participants to develop their approaches for the MVC task. To build the MVC models ALIBABA_DAMO and SJTU_YITU fine-tuned the Clinical-Longformer (Li et al., 2022b) on video subtitles. SJTU_YITU also used the Longformer (Beltagy et al., 2020) to build another MVC model by utilizing the video subtitles from the videos. BAIDU AI TEAM utilizes video title and subtitles to form a concatenated sequence and fine-tuned the hierarchical-BERT (Zhang et al., 2019) to predict the video category.

Team TENCENT AI RESEARCH build their MVC models by fine-tuning the Longformer, Performer (Choromanski et al., 2021) and Big-Bird (Zaheer et al., 2020) pre-trained language models. Team CMU_HKUST built an ensemble approach for the MVC task with the predictions from hierarchical-BERT and Transformer-XL (Dai et al., 2019) pre-trained language models. Team VPAI_LAB built an ensemble approach by considering the predictions from monomodal and multimodal approaches for MVC tasks. They used DeBERTa (He et al., 2020) and I3D (Carreira and Zisserman, 2017) to encode the video subtitles and frames respectively. Team CHICHEALTH also proposes the ensemble models with the pre-trained Big-Bird and Longformer language models. Instead of encoding the entire subtitles from a video, they split the subtitles into multiple pieces and used the max and average pooling layer to aggregate the representations into a fixed-size representation vector. Team UWASHINGTON adopt the Big-Bird as the backbone network. They used the contrastive learning loss and the cross-entropy loss to build their approach for the MVC task.

5.2.2 Results

We have provided the official results for the MVC task and baseline models in Table 5. We rank the submissions based on the Med-Inst F1-score. Team VPAI_LAB achieved the first rank with the 98.74 Med-Inst F1-score and also reported the highest Med-Inst Precision (99.74) and Macro F1-score (99.01). Team PAHT submission reported the highest Med-Inst Recall (98.00) value from the best-ranked participants' system. The best-submitted run of each team outperformed the baseline scores on the primary metric of Med-Inst F1-score. We observed that top-4 teams achieved near-perfect Macro F1-score within a difference of 0.47 points. In terms of the primary metric (Med-Inst F1-score), the team CHICHEALTH (rank #2), BAIDU AI TEAM (rank #3), PAHT (rank #4) and TENCENT AI RESEARCH (rank #5) achieved near-same performance ranging between 97.75 to 97.98. BERT-based monomodal baseline achieved the highest Med-Inst F1-score amongst all the baseline approaches.

5.2.3 Findings

The video subtitles are dominant features to predict the category of the video. The pre-trained language models (Longformer, Hierarchical BERT, Big-Bird) having the capability of effectively process the longer sequences, outperformed the tra-

Rank	Team Name	IoU=0.3	IoU=0.5	IoU=0.7	mIoU
1	PAHT	90.85	84.97	73.20	75.83
2	SJTU_YITU	88.89	83.01	71.24	74.06
3	UWASHINGTON	85.62	81.05	69.93	72.07
4	LINGJING	84.31	73.20	62.75	67.53
5	CMU_HKUST	75.82	72.55	62.09	63.86
6	BAIDU AI TEAM	75.16	71.90	61.44	63.21
7	CHICHEALTH	74.51	67.97	53.59	61.34
8	TENCENT AI RESEARCH	69.28	62.09	49.67	57.31
9	ALIBABA_DAMO	60.13	52.94	38.56	48.21
10	CS	30.07	14.38	5.88	19.97
–	VSL-QGH – *Baseline*	21.56	10.45	5.88	17.60
–	VSL-BASE – *Baseline*	20.91	9.15	5.22	19.44
11	DOSSIER	31.37	13.07	4.58	18.80

Table 6: Official results of the MVAL task

ditional BERT-based pre-trained language model baseline. We observed that the video features could play an essential role in further enhancing performance on the MVC task if the language and vision features are fused without losing information from each modality.

The MVC task greatly benefited from the large pre-trained language model. The pre-trained language model learns the inherent structure from video subtitles that have proven effective in categorizing a video into one of the pre-defined categories. In contrast to the classical video classification task, where the model has to detect and learn the specific action to classify the video into the fine-grained category, the MVC task focused on the coarse-grained category. Therefore, we observed the participants' system (*cf.* Table 4) achieving high performance by only utilizing the video subtitles in coordination with the large pre-trained language models.

We observed that only the winning team VPAI_LAB built their approach considering both the language and vision features. The rest of the teams focused on only language features and achieved promising results. Due to the coarse-grained nature of the MVC task, the vision features alone (monomodal baseline) seem to carry the least information compared to the counterpart language modality to predict the video category.

5.3 MVAL Submissions

5.3.1 Methods

We briefly describe the approaches used by each participating team for the MVAL task.

ALIBABA_DAMO The video subtitles and questions were encoded with BERT, and the vector representations were obtained. The video features from consecutive three-second interval video frames were pooled to form a vector representation. The subtitles, question, and video features were aligned and concatenated to form a multimodal representation. Thereafter, two two-layer feed feed-forward was used to predict whether the three-second multimodal representations are inside the answer boundary.

BAIDU AI TEAM The team adopted the negative sampling NER method from Li et al. (2020b) to train the answer localization system. The team formulated atomic unit spans in the subtitles, i.e., the tokens in subtitles that belong to the start and end timestamps of the visual answer. The hidden state representations for each token of the span and question were obtained using BERT. The span representation was obtained using the approach discussed in Chen et al. (2017). The question representation and span representation were fused together with the feed-forward network to get the question-span representation. The question-span representation was used to predict whether the given span is an answer to the question or not. Following Li et al. (2020b), the team randomly sampled a small subset

Team Name	Pre-trained LM	Modality	Approach
PAHT	BigBird	Language	Sequence labeling with PLM and CRFs (Lafferty et al., 2001) on video subtitles to detect the answer span.
SJTU_YITU	BERT	Language	The relevant subtitle sentences are classified with BERT. Thereafter, the semantic relatedness scores is computed between the question and the subtitles sentences.
UWASHINGTON	DeBERTa	Language	Utilized the PLM to score the question-sentence pair. After that, the high-scoring contiguous sequence are considered as the visual answer.
LINGJING	DeBERTa	Both	Utilized visual highlight features as the visual token, which concatenates with the question, and video subtitles. Sequence labeling framework is adopted with PLM on video subtitles to detect the answer span.
CMU_HKUST	Big-Bird	Language	Utilized machine reading framework to localize the span in the video that could serve as the visual answer to the health-related question.
BAIDU AI TEAM	BERT	Language	Negative sampling approach (Li et al., 2020b) is used to incorporates randomness into the training loss for span recognition.
CHICHEALTH	NA	Language	Sequence labeling with PLM on video subtitles to detect the answer span.
TENCENT AI RESEARCH	NA	Language	The Mutual Matching Network (MMN) (Wang et al., 2021) is trained with the auxiliary task of mutual matching to guide the network.
ALIBABA_DAMO	BERT	Both	Sequence labeling with PLM on video subtitles to detect the answer span.
DOSSIER	RoBERTa, MPNet	Both	First, the similarity scores between question and subtitle are computed. After that, similarity scores are used to detect the answer by utilizing a random forest regressor and unsupervised peak detection method.

Table 7: The summary of the participants approaches used for MVAL task.

of unlabeled spans as the negative instances to induce the training loss. A span-level cross-entropy loss was used for training.

SJTU_YITU Team SJTU_YITU used a two-step approach to localize the answer in the video. In the first step, they fine-tune the BERT model to tag whether a given sentence from subtitles will be part of the answer sentence or not. In the second step, they compute the semantic relatedness between the question and the answer sentences (predicted in the first step) to refine the predictions of the previous step further. Finally, they transform the selected sentences from subtitles into corresponding time intervals.

TENCENT AI RESEARCH Mutual Matching Network (MMN) (Wang et al., 2021) was used for visual answer localization. MMN is a metric-learning approach that is based on the auxiliary task of mutual matching, which guides the network to select the additional correct sentence in a constructed negative sentence set for video moments retrieval in addition to gold-standard supervision. Their approach uses subtitles and question as input to train the MMN by considering a binary cross-entropy loss for regressing the IoU and a pair discrimination loss for learning discriminative features.

CMU_HKUST The team adopted a machine reading framework (Cui et al., 2022) to localize the span in the video that serves as the visual answer to the question. They utilize the subtitles and their timestamps to transform them into a span in the subtitles text. To encode the subtitles and the question, they used the Big-Bird model (Zaheer et al., 2020).

CHICHEALTH The team formulated the task as a sequence tagging problem. The query and the subtitle of a given video were concatenated as "`[CLS] QUESTION [SEP] SUBTITLES [SEP]`". The concatenated sequence served as input to the Transformer network. A pointer network was used to find the text spans that correspond to the video spans that answers the query. During prediction, they select the spans that have the highest span probability. They used multiple transformer-based networks and ensemble the predictions to find the appropriate span that is considered as the visual answer to the question.

PAHT The team formulated the visual answer localization task as a sequence labeling problem. They concatenated the question and subtitles to form a sequence. They utilized the pre-trained BigBird with Conditional Random Fields (Lafferty et al., 2001) head to tag each subtitle timestamps either B-ANSWER, I-ANSWER, or Other.

LINGJING The team proposed the visual-prompt text span localizing (VPTSL) method for visual answer localization by utilizing the pre-trained language model and visual highlight features. They fuse the question and visual features using cross-modal attention. The highlight features are used to provide the visual prompt to textual span predictor.

UWASHINGTON The team formulated the visual answer localization problem as question-sentence pair scoring task. They split the subtitles into multiple sentences and computed the scores for each sentence using the pre-trained DeBERTa model. They considered the timestamps associated with the high-scoring contiguous sequence of sentences as the visual answer to the question.

DOSSIER Team DOSSIER (Kusa et al., 2022) utilized the textual information in the form of subtitles and optical character recognition from video frames. They computed the similarity scores (using BM25, RoBERTa (Liu et al., 2019), and MPNet (Song et al., 2020)) between each video subtitles and the question. With the similarity matrix, they utilized random forest regressor[10] and unsupervised peak detection model to detect the answer indices.

We have provided the summary of each participants' approach for the MVAL task in Table 7.

5.3.2 Results

The official results for the MVAL task, along with the baseline scores, are provided in Table 6. We rank the team submissions based on the primary metric (IoU=0.7). Team PAHT achieved the highest 73.20 IoU=0.7 score. Their best submission also achieved the maximum IoU=0.3, IoU=0.5, and mIoU, which are 90.85, 84.97, and 75.83, respectively. Most of the participants' runs outperformed the multimodal learning-based baseline scores obtained from VSL-BASE and VSL-QGH.

5.3.3 Findings

The majority of the participating teams only use the video subtitles to locate the visual answer in the video. The video subtitles and their appearance timestamps are aligned to locate the start and end indices of the visual answer. Unimodal semantic relatedness between the question and video subtitles was computed with the pre-trained language models and proved to be more effective than the multimodal semantic relatedness as in VSL-BASE and VSL-QGH baselines. The top-3 participating systems built their approaches, similar to the text-based machine reading comprehension, by only utilizing the video subtitles features to locate the visual answer. However, team LINGJING proposed the multimodal approach for the MVAL task and achieved 62.75 IoU=0.7 that placed them in the 4th rank in the leaderboard.

It is observed that video subtitle features have proven to be effective compared to video features. The video subtitles are derived from commentary in videos. When a speaker in the video starts discussing a specific topic, they introduce the topic at the start of their commentary and make concluding remarks at the end of the commentary on the particular topic. The health-related questions in the MVAL task are formulated by watching the videos and identifying the span in the video, which could serve as the visual answer to the health-related questions. The video subtitle feature-based approaches exploit this structure and consider training the model to localize the span in the video subtitle sequence, which is semantically associated with the given question. This act of localizing the span from video subtitles is closely related to the machine reading comprehension (MRC) task; therefore, the participants use video subtitle features and treat the MVAL task similar to the approaches which have been used in the literature for the MRC task.

6 Conclusion

This paper describes the overview of MedVidQA 2022 shared task organized as part of the BioNLP 2022 workshop. We discussed the tasks, datasets, evaluation metrics, and baseline systems. We also provided a summary of the participating systems for both tasks. For the MVC task, the approach utilizing the attention-based fusion of the pre-trained

[10] `https://bit.ly/3tViF3S`

language model features and video features outperformed all the competitive methods. Overall, the MVC task with the coarse-grained category was relatively easy compared to the classical video classification task, where the model has to detect and learn the specific action to classify the video into the fine-grained category. We observe that video subtitles are key information to localize the visual answer in the video for the medical instructional question. We are optimistic that introducing these tasks and datasets will foster research toward designing systems that can understand medical videos and effectively provide visual answers to natural language questions.

Acknowledgements

This work was supported by the intramural research program at the U.S. National Library of Medicine, National Institutes of Health, and utilized the computational resources of the NIH HPC Biowulf cluster (`http://hpc.nih.gov`). We would like to thank Kush Attal (NLM/NIH) for his help with the MVC baseline evaluation and MVAL dataset creation.

References

Sami Abu-El-Haija, Nisarg Kothari, Joonseok Lee, Paul Natsev, George Toderici, Balakrishnan Varadarajan, and Sudheendra Vijayanarasimhan. 2016. Youtube-8m: A large-scale video classification benchmark. *arXiv preprint arXiv:1609.08675*.

Lisa Anne Hendricks, Oliver Wang, Eli Shechtman, Josef Sivic, Trevor Darrell, and Bryan Russell. 2017. Localizing moments in video with natural language. In *Proceedings of the IEEE international conference on computer vision*, pages 5803–5812.

Iz Beltagy, Matthew E. Peters, and Arman Cohan. 2020. Longformer: The long-document transformer. *arXiv:2004.05150*.

João Carreira and Andrew Zisserman. 2017. Quo vadis, action recognition? a new model and the kinetics dataset. In *IEEE Conference on Computer Vision and Pattern Recognition*, pages 4724–4733.

Aman Chadha, Gurneet Arora, and Navpreet Kaloty. 2020. iperceive: Applying common-sense reasoning to multi-modal dense video captioning and video question answering. *arXiv preprint arXiv:2011.07735*.

Qian Chen, Xiaodan Zhu, Zhen-Hua Ling, Si Wei, Hui Jiang, and Diana Inkpen. 2017. Enhanced lstm for natural language inference. In *Proceedings of the 55th Annual Meeting of the Association for Computational Linguistics (Volume 1: Long Papers)*, pages 1657–1668.

Krzysztof Marcin Choromanski, Valerii Likhosherstov, David Martin Dohan, Xingyou Song, Andreea Gane, Tamas Sarlos, Peter Hawkins, Jared Quincy Davis, Afroz Mohiuddin, Lukasz Kaiser, David Belanger, Lucy Colwell, and Adrian Weller. 2021. Rethinking attention with performers. In *International Conference on Learning Representations*.

Yiming Cui, Ting Liu, Wanxiang Che, Zhigang Chen, and Shijin Wang. 2022. Teaching machines to read, answer and explain. *IEEE/ACM Transactions on Audio, Speech, and Language Processing*.

Zihang Dai, Zhilin Yang, Yiming Yang, Jaime G Carbonell, Quoc Le, and Ruslan Salakhutdinov. 2019. Transformer-xl: Attentive language models beyond a fixed-length context. In *Proceedings of the 57th Annual Meeting of the Association for Computational Linguistics*, pages 2978–2988.

Jacob Devlin, Ming-Wei Chang, Kenton Lee, and Kristina Toutanova. 2019. BERT: Pre-training of deep bidirectional transformers for language understanding. In *Proceedings of the 2019 Conference of the North American Chapter of the Association for Computational Linguistics: Human Language Technologies*, pages 4171–4186. Association for Computational Linguistics.

Alexey Dosovitskiy, Lucas Beyer, Alexander Kolesnikov, Dirk Weissenborn, Xiaohua Zhai, Thomas Unterthiner, Mostafa Dehghani, Matthias Minderer, Georg Heigold, Sylvain Gelly, Jakob Uszkoreit, and Neil Houlsby. 2021. An image is worth 16x16 words: Transformers for image recognition at scale. In *International Conference on Learning Representations*.

Deepak Gupta, Kush Attal, and Dina Demner-Fushman. 2022. A Dataset for Medical Instructional Video Classification and Question Answering. *arXiv preprint arXiv:2201.12888*.

Pengcheng He, Xiaodong Liu, Jianfeng Gao, and Weizhu Chen. 2020. Deberta: Decoding-enhanced bert with disentangled attention. In *International Conference on Learning Representations*.

HHS. 2021. Artificial intelligence (ai) strategy. U.S. Department of Health and Human Services.

Sepp Hochreiter and Jürgen Schmidhuber. 1997. Long short-term memory. *Neural computation*, 9(8):1735–1780.

Wojciech Kusa, Georgios Peikos, Oscar Espitia Mendoza, Allan Hanbury, and Gabriella Pasi. 2022. Dossier at medvidqa 2022: Text-based approaches to medical video answer localisation problem. In *Proceedings of the 21th Workshop on Biomedical Language Processing*.

John Lafferty, Andrew McCallum, and Fernando CN Pereira. 2001. Conditional random fields: Probabilistic models for segmenting and labeling sequence data.

Jie Lei, Licheng Yu, Mohit Bansal, and Tamara Berg. 2018. TVQA: Localized, compositional video question answering. In *Proceedings of the 2018 Conference on Empirical Methods in Natural Language Processing*, pages 1369–1379, Brussels, Belgium. Association for Computational Linguistics.

Jie Lei, Licheng Yu, Tamara Berg, and Mohit Bansal. 2020a. TVQA+: Spatio-temporal grounding for video question answering. In *Proceedings of the 58th Annual Meeting of the Association for Computational Linguistics*, pages 8211–8225, Online. Association for Computational Linguistics.

Jie Lei, Licheng Yu, Tamara L Berg, and Mohit Bansal. 2020b. TVR: A Large-scale Dataset for Video-subtitle Moment Retrieval. In *European Conference on Computer Vision*, pages 447–463. Springer.

Bin Li, Yixuan Weng, Fei Xia, Bin Sun, and Shutao Li. 2022a. Vpai_lab at medvidqa 2022: A two-stage cross-modal fusion method for medical instructional video classification. In *Proceedings of the 21th Workshop on Biomedical Language Processing*.

Linjie Li, Yen-Chun Chen, Yu Cheng, Zhe Gan, Licheng Yu, and Jingjing Liu. 2020a. HERO: Hierarchical encoder for Video+Language omni-representation pre-training. In *Proceedings of the 2020 Conference on Empirical Methods in Natural Language Processing (EMNLP)*, pages 2046–2065, Online. Association for Computational Linguistics.

Yangming Li, Lemao Liu, and Shuming Shi. 2020b. Empirical analysis of unlabeled entity problem in named entity recognition. *arXiv preprint arXiv:2012.05426*.

Yikuan Li, Ramsey M Wehbe, Faraz S Ahmad, Hanyin Wang, and Yuan Luo. 2022b. Clinical-longformer and clinical-bigbird: Transformers for long clinical sequences. *arXiv preprint arXiv:2201.11838*.

Yinhan Liu, Myle Ott, Naman Goyal, Jingfei Du, Mandar Joshi, Danqi Chen, Omer Levy, Mike Lewis, Luke Zettlemoyer, and Veselin Stoyanov. 2019. Roberta: A robustly optimized bert pretraining approach. *arXiv preprint arXiv:1907.11692*.

Antoine Miech, Dimitri Zhukov, Jean-Baptiste Alayrac, Makarand Tapaswi, Ivan Laptev, and Josef Sivic. 2019. Howto100m: Learning a text-video embedding by watching hundred million narrated video clips. In *Proceedings of the IEEE/CVF International Conference on Computer Vision*, pages 2630–2640.

Jonghwan Mun, Paul Hongsuck Seo, Ilchae Jung, and Bohyung Han. 2017. Marioqa: Answering questions by watching gameplay videos. In *Proceedings of the IEEE International Conference on Computer Vision*, pages 2867–2875.

F. Pedregosa, G. Varoquaux, A. Gramfort, V. Michel, B. Thirion, O. Grisel, M. Blondel, P. Prettenhofer, R. Weiss, V. Dubourg, J. Vanderplas, A. Passos, D. Cournapeau, M. Brucher, M. Perrot, and E. Duchesnay. 2011. Scikit-learn: Machine learning in Python. *Journal of Machine Learning Research*, 12:2825–2830.

Kaitao Song, Xu Tan, Tao Qin, Jianfeng Lu, and Tie-Yan Liu. 2020. Mpnet: Masked and permuted pre-training for language understanding. *Advances in Neural Information Processing Systems*, 33:16857–16867.

Makarand Tapaswi, Yukun Zhu, Rainer Stiefelhagen, Antonio Torralba, Raquel Urtasun, and Sanja Fidler. 2016. Movieqa: Understanding stories in movies through question-answering. In *Proceedings of the IEEE conference on computer vision and pattern recognition*, pages 4631–4640.

Jingwen Wang, Lin Ma, and Wenhao Jiang. 2020. Temporally grounding language queries in videos by contextual boundary-aware prediction. In *Proceedings of the AAAI Conference on Artificial Intelligence*, volume 34, pages 12168–12175.

Zhenzhi Wang, Limin Wang, Tao Wu, Tianhao Li, and Gangshan Wu. 2021. Negative sample matters: A renaissance of metric learning for temporal grounding. *arXiv preprint arXiv:2109.04872*.

Hongyang Xue, Wenqing Chu, Zhou Zhao, and Deng Cai. 2018. A better way to attend: Attention with trees for video question answering. *IEEE Transactions on Image Processing*, 27(11):5563–5574.

Manzil Zaheer, Guru Guruganesh, Kumar Avinava Dubey, Joshua Ainslie, Chris Alberti, Santiago Ontanon, Philip Pham, Anirudh Ravula, Qifan Wang, Li Yang, et al. 2020. Big bird: Transformers for longer sequences. In *Conference on Neural Information Processing Systems*.

Hao Zhang, Aixin Sun, Wei Jing, and Joey Tianyi Zhou. 2020. Span-based localizing network for natural language video localization. In *Proceedings of the 58th Annual Meeting of the Association for Computational Linguistics*, pages 6543–6554, Online. Association for Computational Linguistics.

Xingxing Zhang, Furu Wei, and Ming Zhou. 2019. Hibert: Document level pre-training of hierarchical bidirectional transformers for document summarization. In *Proceedings of the 57th Annual Meeting of the Association for Computational Linguistics*, pages 5059–5069.

Inter-annotator agreement is not the ceiling of machine learning performance: Evidence from a comprehensive set of simulations

Russell Richie[1] **Sachin Grover**[1] **Fuchiang (Rich) Tsui**[1,2]
[1]Tsui Lab, Department of Biomedical and Health Informatics
Children's Hospital of Philadelphia, Philadelphia, PA
[2]Department of Anesthesiology and Critical Care
Perelman School of Medicine, University of Pennsylvania, Philadelphia, PA
{richier,groversl,tsuif}@chop.edu

Abstract

It is commonly claimed that inter-annotator agreement (IAA) is the ceiling of machine learning (ML) performance, i.e., that the agreement between an ML system's predictions and an annotator can not be higher than the agreement between two annotators. Although Boguslav and Cohen (2017) showed that this claim is falsified by many real-world ML systems, the claim has persisted. As a complement to this real-world evidence, we conducted a comprehensive set of simulations, and show that an ML model can outperform IAA even if (and especially if) annotators are noisy and differ in their underlying classification functions, as long as the ML model is reasonably well-specified. Although the latter condition has long been elusive, leading ML models to underperform IAA, we anticipate that this condition will be increasingly met in the era of big data and deep learning. Our work has implications for (1) maximizing the value of machine learning, (2) adherence to ethical standards in computing, and (3) economical use of annotated resources, which is paramount in settings where annotation is especially expensive, like biomedical natural language processing.

1 Introduction

It is standard when conducting machine learning (ML) and natural language processing (NLP) work to calculate inter-annotator agreement (IAA) metrics like Cohen's Kappa (Cohen, 1960). This is done not just for annotation quality control, but also as a comparison for machine learning models' performance. In particular, it has commonly been claimed – by some of the most prominent researchers in ML and NLP (Boguslav and Cohen, 2017) – that *IAA places an upper bound or ceiling on the performance of machine learning models.* When researchers claim this and their model reaches IAA, they are implicitly suggesting (or at least it follows) that the model has performed as well as possible or has solved the task for that dataset, and/or that the dataset cannot be used to drive further development of ML models. Despite the prominence of this claim, however, Boguslav and Cohen (2017) reported that neither they nor a professional literature search service could find evidence in support of it. This is concerning for at least two reasons.

First, as Boguslav and Cohen (2017) say, "if the assumption [that IAA bounds ML] turns out not to be supported... we may be mis-estimating the actual performance of our [ML] systems. In particular, we may be over-estimating the quality of their performance by under-estimating how good [performance] could potentially be" (pg 298). This underestimation of the maximum possible performance may lead to the development of poorer models under the belief that they have achieved maximum capacity. Moreover, as noted by Boguslav and Cohen (2017), such misestimation may violate ethical standards concerning accurate characterization of the limitations of computer systems (e.g., ACM Code of Ethics and Professional Conduct 2.7 Anderson, 1992; see also Petersen et al., 2021 on recommendations for safe, effective use of clinical decision support systems).

Second, and relatedly, if a modeler stops using an annotated dataset to drive ML development once ML performance on that dataset reaches IAA, they may be underutilizing those annotations. This could be an enormous waste of money, since annotation is often one of the most expensive components of an ML/NLP project, especially in biomedical NLP where the time of annotators (often biomedical experts) is especially expensive. For example, Hill et al. (2015) noted that then state-of-the-art word embeddings had reached IAA on existing word relatedness benchmark datasets (e.g., WordSim-353, Finkelstein et al., 2001). Believing that IAA was the upper bound of ML, they therefore believed that such datasets could no longer be

Proceedings of the BioNLP 2022 workshop, Dublin, Ireland, pages 275–284
May 26, 2022.

used to drive development of different word embedding models. This led them to collect new annotations of word similarity, yielding the benchmark SimLex-999. Their abstract lays out this logic:

> "Further, unlike existing gold standard evaluations, for which automatic approaches have reached... the inter-annotator agreement ceiling, state-of-the-art models perform well below this ceiling on SimLex-999. There is therefore plenty of scope for SimLex-999 to quantify future improvements to distributional semantic models, guiding the development of the next generation of representation-learning architectures." (pg. 1)

If IAA does not in fact bound ML, then the older word relatedness benchmarks *could have actually been used to "guide... the development of the next generation of representation-learning architectures"*, and there would have been less need to spend time and money annotating SimLex-999. Given that Hill et al. (2015) has been cited over 1000 times according to Google Scholar, other researchers may have absorbed and replicated their logic, which would be concerning if the claim is not really true. Indeed, a Stack Exchange post (tomas , https://stats.stackexchange.com/users/84364/tomas) roughly contemporaneous with Hill et al. (2015) suggests that this logic may be widespread.

Despite the popularity and stakes of the claim that IAA bounds ML, Boguslav and Cohen (2017) found, across 6 papers, 20 ML systems that outperform IAA, on tasks ranging from entity recognition in clinical notes (Roberts et al., 2008), to deception detection (Pérez-Rosas et al., 2015) (and see Wilbur, 1998 for earlier evidence that ML can outperform IAA in information retrieval[1]). However, claims that IAA bounds ML performance have persisted. This is seen in both biomedical and broader ML/NLP, in (1) papers that are often cited much more than Boguslav and Cohen (2017) and published in high impact outlets including JAMA Network, AMIA, Nature Human Behavior, and ACL (e.g., Grčar et al., 2017; Pilehvar et al., 2018; Amidei et al., 2018; Sarker et al., 2019; Pustu-Iren et al., 2019; Ribeiro et al., 2019; Richie et al., 2019; O'Connor et al., 2020; Hebart et al., 2020; Mayfield and Black, 2020; Basile, 2020; Bevilacqua et al., 2021; Li et al., 2021; Higashinaka et al., 2021; Goldberg et al., 2021), (2) machine learning lectures at well-known universities including University of Pittsburgh (Han, 2017), University of Edinburgh (Cohen, 2020), and City University of New York (CUNY, Gorman, 2020) and in slides by noted NLP textbook authors Jurafsky and Martin (Jurafsky and Martin, 2022), and (3) online posts and social media discussions by prominent machine learning users (e.g., Ruder, 2021).

It is not entirely clear why the claim that IAA bounds ML survived Boguslav and Cohen's counterexamples, but we suspect at least two factors are at play. First, Boguslav and Cohen (2017) was published in a fairly specialized journal (*Studies in Health Technology and Informatics*), and therefore may not have reached as many ML/NLP practitioners as it could or should have. Consistent with this, as of March 29, 2022, Boguslav and Cohen (2017) has been cited only 7 times, according to Google Scholar. Second, we suspect that the issue deserves a broad proof of concept based on simulations, in addition to the empirical examples raised by Boguslav and Cohen. Simulations would be complementary to the real-world evidence brought by Boguslav and Cohen, in at least two ways. First, simulations allow us to simplify the problem to its essence, which may be clarifying in ways that real-world studies, with all their potentially distracting idiosyncrasies, are not. Second, simulations allow us to precisely control and test different potentially relevant annotation and modeling factors, and therefore better understand when/how/why a model can or can't beat IAA. Therefore, the aim of this study is to use a comprehensive set of computational simulations to bolster the evidence that IAA is not the upper bound on ML performance.

The rest of this paper is organized as follows. We start with simplified simulations that capture the basic elements of an ML pipeline with two annotators and train a supervised model with these annotations (Experiment 1). We then relax various assumptions of this setup to simulate potentially more realistic settings, and to better understand the conditions under which an ML model can outperform IAA (Experiment 2). In both experiments, the general approach is to (a) simulate two annotators' annotations on a test set (where both annotators label all samples of this set); (b) simulate their annotations on disjoint halves of a training set; (c) train an ML model on the training set annotations; and

[1]We thank an anonymous reviewer for this suggestion.

(d) compare the agreement between the ML model and each annotator to the annotators' IAA on the test set. We conclude with a general discussion interpreting our work and its implications.

2 Experiment 1

2.1 Simulations

We consider a binary classification task conducted by two annotators, $A1$ and $A2$ who probabilistically classify the *i*-th sample, x_i (composed of a single variable), into one of two classes, $y_i \in \{0, 1\}$, according to a logistic function of x_i, as in:

$$p(y_i = 1) = \frac{1}{1 + e^{-x_i}} \tag{1}$$

In a simulation, we first sample the independent variable x from a standard normal distribution (zero mean, unit standard deviation), producing an X_{train} and an X_{test} with some number of samples each. Through Eq 1, $A1$ and $A2$ independently annotate every sample in X_{test}, which yields $y_{test,A1}$ and $y_{test,A2}$. Then $A1$ annotates the first half of X_{train}, and $A2$ annotates the second half of X_{train}, and we concatenate their annotations into a single y_{train}. We then train a logistic regression on (X_{train}, y_{train}), and generate this model's predictions on X_{test}, i.e., we generate $\hat{y}_{test}$.

We can then calculate the inter-annotator agreement $f(y_{test,A1}, y_{test,A2})$, and the model's average performance using both $A1$'s and $A2$'s annotations as ground truth, as in $average(f(\hat{y}_{test}, y_{test,A1}), f(\hat{y}_{test}, y_{test,A2}))$, where f is either F1-score or Cohen's Kappa. F1-score is defined as:

$$F_1 = \frac{2 * Precision * Recall}{Precision + Recall} \tag{2}$$

where $Precision$ is $\frac{TP}{TP+FP}$ and $Recall$ is $\frac{TP}{TP+FN}$. Cohen's Kappa is defined as:

$$\kappa \equiv \frac{p_o - p_e}{1 - p_e} \tag{3}$$

where p_o is the observed agreement among annotators, and p_e is the probability of chance agreement, which is often calculated using the base rates of each label in observed annotations.

Conventionally, Cohen's Kappa is used to measure IAA because it controls for chance annotator agreement, and F1 is used to measure model performance because it balances precision and recall and punishes (inappropriately) 'extreme' models (a setting with 1 positive sample and infinite negative samples, and a model that always assigns the positive class, will have $Recall = 1$, $Precision = 0$, and $F1 = 0$). However, Boguslav and Cohen (2017) suggest that in many linguistic annotation tasks, especially named entity recognition or others involving phrase extraction, where there are a very large number of potential spans that no annotator ever extracts, it is often the case that $p_e = 0$ in Equation 3. In this case, *Kappa is equivalent to F1* (Hripcsak and Rothschild, 2005), which arguably justifies the commonly conducted direct comparison between (1) IAA measured with Kappa and (2) model performance measured with F1. In our simulations, however, it is more straightforward to simply calculate and compare IAA and ML performance in the same metric(s), and we opt for this here.

Finally, we note here that our goal is not to critique these particular measures, their usage, or the paradigm of inter-annotator agreement more generally (for such critique, see for example Amidei et al., 2018). Rather, our focus is merely to demonstrate the falsity of the claim that IAA bounds ML, i.e., that $f(y_{test,A1}, y_{test,A2}) >= average(f(\hat{y}_{test}, y_{test,A1}), f(\hat{y}_{test}, y_{test,A2}))$. Although we chose Kappa and F1-score here because of their common usage in ML and NLP, we expect our results to generalize to other measures (e.g., Matthews Correlation Coefficient).

2.2 Results

In our simulations, X_{test} contains 100 samples and X_{train} contains 1000 samples, and the simulations were repeated 100 times. Figure 1a shows the results. As can be seen, on average, the model achieves F1=0.67 when comparing to the annotators on the test set, while the annotators score only F1=0.58 when comparing to each other (t=12.44, p < 10-25). Likewise, the model 'agrees' with the annotators at a Cohen's Kappa of about 0.35, while the annotators agree with each other at only 0.16 (t=13.60, p < 10-29). Clearly, inter-annotator agreement does not provide an upper bound on ML performance in this simple setting.

While IAA clearly doesn't provide an upper bound on model performance, it is also clear (see Figure 1b) that the two are positively correlated (r = 0.48, p < 10-6). The correlation arises because when an annotator happens to assign a positive class to samples whose $p(y_i = 1) > 0.5$, the annotator's predictions will be closer to both (a) the model's predictions (which always assigns the positive class to samples when $p(y_i = 1) > 0.5$, and (b) the other annotator's predictions, because the latter annotator will also usually assign the positive class when $p(y_i = 1) > 0.5$. Thus, one might fairly say that although IAA doesn't *bound* ML performance, IAA predicts ML performance. This relationship may be partly what underlies the appeal or intuitiveness of the notion that IAA bounds ML performance, an issue we return to in the general discussion. At the same time, Figure 1b also makes clear that, at the level of individual simulations, the ML model tends to outperform IAA, since most points are above the line $y = x$.

2.3 Discussion

The simulations above show that, contrary to many claims in the ML and NLP communities, IAA does not bound ML performance, *at least in this simple case*. At the same time, because IAA and ML performance are positively correlated, IAA does give some indication of the level of ML performance that can be expected, which could explain why many people believe that IAA bounds ML performance. It may even be that when two authors use the same term ('bound', 'ceiling', or 'limit', the keywords we used to find claims about the relationship between IAA and ML performance), one author may intend that IAA is a strict ceiling on ML performance, and another may intend that low IAA merely *predicts* low ML performance (although based on our reading of the literature, we tend to think most writers intend the first meaning). Failure to carefully distinguish these meanings may be contributing to confusion in the field, and we hope these results clarify the distinction.

Also note that in this simulation, the annotators $A1$ and $A2$ have identical classification functions, so this simulation can be equivalently viewed as having a single annotator classify *all* of the training samples once, and all of the test samples twice. Having a single annotator classify the test samples twice allows us to calculate not inter-annotator agreement, but *intra*-annotator agreement (also known as test-retest reliability in the psychometrics literature, Guttman, 1945). Therefore, concluding that IAA does not bound ML performance is also applicable to *intra*-annotator agreement.

3 Experiment 2

The conditions of Experiment 1 are intentionally oversimplified from real-world conditions. To better understand the range of conditions under which ML can or cannot outperform IAA, we next introduce some additional complexity in the simulations.

3.1 Simulations

First, it seems unrealistic that two different annotators, with different experiences and perceptual and cognitive systems, will ever understand and perform an annotation task in exactly the same way. In a sentiment analysis task, for example, annotators may have different thresholds for what is considered a 'positive' text. In other words, it seems unrealistic that two annotators will have the exact same classification function. One straightforward way to relax this assumption is to allow different annotators to have different intercepts in the linear component of the logistic function, as in:

$$p(y_i = 1) = \frac{1}{1 + e^{-(x_i + b_j)}} \tag{4}$$

where b_j is the intercept for annotator $j \in \{1, 2\}$. If, for example, $b_2 > b_1$, Annotator 2 will generally be more likely than Annotator 1 to assign a sample to the positive class. In our experiments, we will simply assume $b2 \geq 0$ and $b1 = -b2$ (other combinations of intercepts, like setting $b_1 = 0$ and varying b_2, were also tested and the general pattern of results did not change). This will of course decrease IAA. It may also seem intuitive that, when annotators systematically disagree about how to approach the task, it will be more difficult for a model to learn anything coherent, decreasing model performance, possibly to a performance worse than IAA.

Second, it also seems possible, in practice, that annotators' judgments are more deterministic (less noisy) than implied in Experiment 1, where the IAA Cohen's Kappa averaged only 0.18. This likely strikes most ML practitioners as much lower than what is seen and accepted in empirical ML applications. It therefore seems reasonable that a given annotator, facing a sample twice, would generally classify it the same way each time (i.e., that *intra*-annotator agreement is high). There are various ways to parameterize determinism in annotation, but we opt to simply exponentiate the outputs of the logistic function by a parameter, γ,

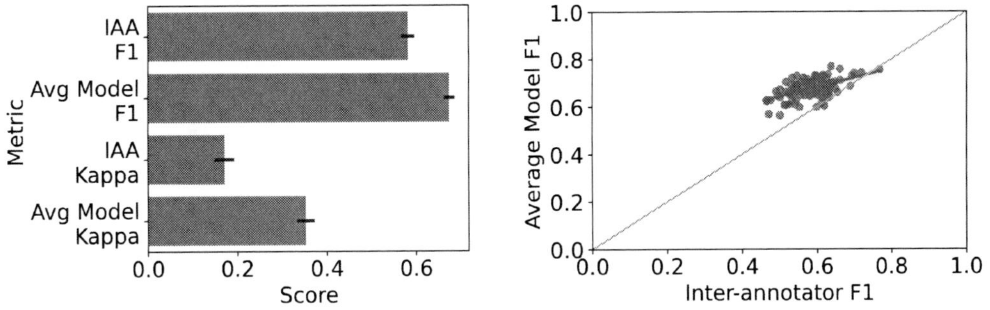

(a) Bars show means across simulations, and error bars display 95% confidence intervals.

(b) Scatterplot of inter-annotator F1 (x-axis) against Average Model-Annotator F1 (y-axis), across simulations. Each dot represents a single simulation

Figure 1: Experiment 1 results

and then divide these values by their sum so they add to 1, as proper probabilities, as in:

$$a_{1,i} = \frac{1}{1+e^{-(x_i+b_j)}} \tag{5}$$

$$p(y_i = 1) = \frac{a_{1,i}^{\gamma}}{a_{1,i}^{\gamma} + (1-a_{1,i})^{\gamma}} \tag{6}$$

Thus, instead of viewing the logistic function as producing probabilities, we can view it (Equation 5) as producing 'activations' of the possible annotations in the annotator's mind, i.e., $a_{1,i}$ and $a_{0,i}$ refer to the activations of labels 1 and 0, respectively, for the i-th sample. These activations are then converted into probabilities by Equation 6. When $\gamma = 0$, the choice is completely random (i.e., $p(y_i = 1) = 0.5$) and does not depend on x_i and b_j. When $\gamma = 1$, then annotation probabilities of Equation 6 are identical to the activations produced by Equation 4. When $\gamma > 1$ and approaches positive infinity, choice becomes more deterministic, such that with an extremely high γ, an annotator will almost always classify a sample x_i as positive if $a_{1,i} > 0.5$. One might argue that, to more accurately model commonly seen levels of IAA metrics, we need to test $\gamma > 1$, which ought to boost IAA and perhaps therefore make it harder for ML to outperform IAA.

Third, and perhaps most importantly and obviously, a machine learning model will always be misspecified in some way (Box, 1976). That is, the ML model will almost always lack some of the variables that influence an annotator's judgment, or the ML model may be purely linear while annotators are actually using some nonlinear combination of variables. Although it may seem obvious that, if the model is misspecified enough, ML performance will fall short of IAA, we also simulate this condition to show that *the model does not need to be perfectly specified to beat IAA*. To simulate misspecification, we simply augment Equation 5 with a second independent variable, x_2, as in:

$$a_{1,i} = \frac{1}{1+e^{-(x_{1,i}+m*x_{2,i}+b_j)}} \tag{7}$$

We assume that, like x_1, x_2 is sampled (independently) from the standard normal distribution (i.e., x_1 and x_2 together constitute a standard multivariate normal distribution). We also assume that there is a coefficient m on x_2 controlling the relative importance of x_2 to annotator decisions. We then assume that annotators' judgments follow from Equations 7 and 6. To misspecify an ML model, we simply withhold x_2 from X_{train} and X_{test} when fitting the model and generating $\hat{y}_{test}$, respectively. That is, annotators make decisions with both x_1 and x_2, but the model only has access to x_1. When m is large, reflecting great importance of x_2 to annotator decisions, then the ML model is greatly misspecified and this misspecification will have large negative impacts on $average(f(\hat{y}_{test}, y_{test,A1}), f(\hat{y}_{test}, y_{test,A2}))$. When $m = 0$, of course, x_2 is ignored in annotators' decisions and the ML model is not misspecified at all – the omission of x_2 from the ML model has no effect on its performance.[2]

[2]We note that increasing m can also increase IAA because it will push the output of Equation 7 toward 0 or 1, which in turn makes annotators' labels less noisy. Although it may be undesirable for m to influence both misspecification and

We emphasize that m is just one simple way to introduce misspecification in the simulations. In more complex real world tasks with more complex models (such as deep learning), misspecification can take many different forms. Exploring this further may be a useful avenue in future work.

Finally, although it is generally intentional and desirable that ML models classify samples deterministically (i.e., $\hat{y}_i = 1$ if and only if $p(y_i = 1) \geq 0.5$), we can simulate a noisy ML model to better understand the conditions under which ML can or cannot beat IAA. That is, it seems intuitive that one advantage an ML model has over human annotators, is that an ML model can make decisions with perfect consistency. To simulate a noisy ML model, we simply pass a trained ML model's predicted probabilities through Equation 6 and sample its predictions accordingly.

3.2 Results

We sample b_2 from {0, 0.25, 0.5}, γ from {1, 3, 6}, and m from {0, 0.25, 0.5}. The ranges of b_2 and γ were chosen so that reasonably high IAA could be achieved despite individual differences, while the range of m was chosen so that we had values of m that lead to ML > IAA, and values of m such that ML < IAA. We also simulate both fully deterministic and noisy model predictions. In the latter case, the ML model uses the same value of γ that simulated annotators use. We simulate all possible combinations of parameters and conditions. As in Experiment 1, our simulations involve X_{test} of 100 samples and X_{train} of 1000 samples, but now we run 400 simulations per combination of parameters. Because F1 and Cohen's Kappa show the same general pattern of results, we only use F1 to compare IAA and model performance in Experiment 2.

Figure 2 shows, for each combination of parameter values, the difference between average model F1 and annotator F1, such that bars above $y = 0$ indicate that the model outperforms IAA. Table 1 shows the same results as Figure 2, but transposes Figure 2's arrangement of parameter combinations, and just shows whether ML outperforms IAA or vice versa. As can be seen in both Figure 2 and Table 1, ML outperforms IAA across a broad range of conditions.

First, perhaps contrary to intuition, ML can outperform IAA when annotator classification functions differ, i.e., when $b_2 \neq b_1$. In fact, the larger the difference $b_2 - b_1$, the larger the margin by which the model beats IAA (e.g., compare the 1st, 2nd and 3rd blue or orange bars in any subplot of Figure 2). Rather than causing the model to learn something incoherent, $b_2 \neq b_1$ causes the model to learn a $\hat{b}$ that compromises between b_2 and b_1. For example, in the simple case $b_1 = -0.5$ and $b_2 = 0.5$ (and $\gamma = 1$), the model will tend to learn $\hat{b} = 0$. This causes the model's predictions to be, on average, closer to either annotator's predictions than the annotators' predictions are to each other.

Second, even if we increase determinism in annotator judgments (via γ in Equation 6) such that IAA reaches levels typically seen in empirical applications (e.g., Kappa = 0.6 or 0.7, see bottom row of Figure 2 subplots), ML can still beat IAA.

Third, ML can outperform IAA even under some model misspecification ($m = 0.25$ or $m = 0.5$), although misspecification reduces the margin by which ML outperforms IAA (e.g., compare top row subplots of Figure 2, or more strikingly, middle or bottom row subplots).

Fourth, although determinism in model predictions is clearly an advantage ML has over noisy human annotators (blue bars are generally higher than orange bars in Figure Figure 2), it is not necessary for ML to beat IAA. Systematic differences in annotator behavior are sufficient, as can be seen in the right most bars of the first and second subplots in the top row of Figure 2. Although these differences between ML and IAA are quite small, they are statistically significant, as indicated by the 95% confidence intervals excluding 0.

Most importantly, we note that ML beats IAA in a realistic combination of conditions, i.e., when annotators have good IAA ($\gamma = 6$, Kappa=0.61) despite (small) systematic differences in behavior ($-b_1 = b_2 = 0.25$), and the ML model is mildly misspecified ($m = 0.25$). In Figure 2, this situation is represented in the middle bars of the subplot of the third row and second column, which is surrounded by a black box.

4 General Discussion

In a comprehensive range of simulations, we showed that, contrary to popular belief (Boguslav and Cohen, 2017), *inter-annotator agreement is not the upper bound on machine learning performance*. We showed this is the case even if (and especially if) annotators are noisy and differ in their under-

IAA, it is not immediately obvious how to better parameterize misspecification, and in any case, we don't think this property affects our conclusions.

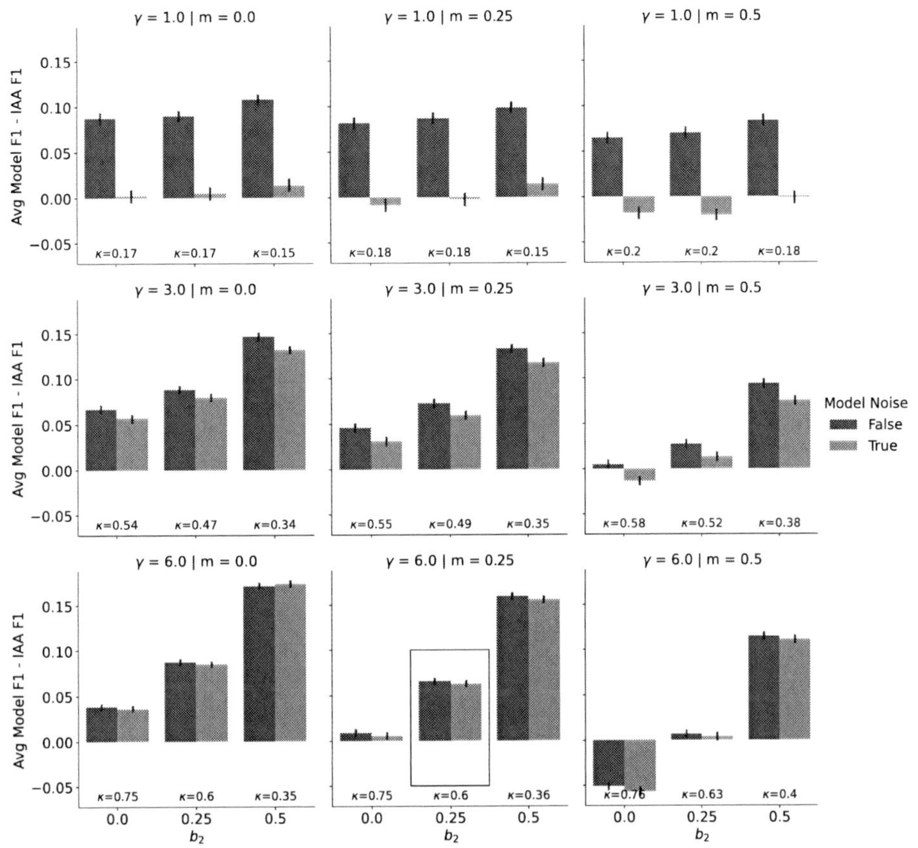

Figure 2: Experiments 2 results. Bars show mean differences between average Model F1 scores and IAA F1 score, i.e., bars above $y = 0$ indicate ML outperforming IAA. Error bars represent 95% confidence intervals. Below the bars are inter-annotator Cohen's Kappa's at each level of γ, m, and b_2. The black box in the middle column and bottom row represents a realistic condition, where ML still beats IAA.

lying classification functions, as long as the ML model is reasonably well-specified. While we think noisy annotators with (possibly small) systematic individual differences are the norm rather than the exception, well-specified models have been elusive for a long time in domains with unstructured data like (biomedical) NLP or machine learning. This was especially true in decades past, when the belief that IAA bounded ML proliferated, and this ill-specification likely led ML models to underperform IAA. However, reasonably well-specified models are likely to be increasingly attainable in today's era of big data, increased computing power, and correspondingly complex nonlinear models like deep neural networks. Although these real-world cases involve much more complex data and models than we simulated here, we believe our conclusions still apply, and we therefore expect to see more empirical cases of ML outperforming IAA (like those in Boguslav and Cohen, 2017). Likewise, although we focused on binary classification here, we expect our results to generalize straightforwardly to other settings, like multiclass classification or regression.

On the other hand, whether and how much a model will beat IAA depends, as we have shown, on the degree of model misspecification, the degree of noise in annotators' judgments, the degree of individual differences in the annotators, and possibly other factors. An ML practitioner might therefore wish to determine, given a particular annotated dataset, how well-specified the model must be in order to beat IAA by a given margin. Modeling one's annotators (e.g., Passonneau and Carpenter, 2014), and their noise levels and individual differences, may be useful here. Beyond this, it is unclear how best to perform such an analysis, and thus we leave this to future work. For the time being, then, we simply recommend that researchers not claim

			γ		
m	b_2	Model Noise	1	3	6
0	0	False	ML >IAA	ML >IAA	ML >IAA
		True	ML >IAA	ML >IAA	ML >IAA
	0.25	False	ML >IAA	ML >IAA	ML >IAA
		True	ML >IAA	ML >IAA	ML >IAA
	0.5	False	ML >IAA	ML >IAA	ML >IAA
		True	ML >IAA	ML >IAA	ML >IAA
0.25	0	False	ML >IAA	ML >IAA	ML >IAA
		True	**IAA >ML**	ML >IAA	ML >IAA
	0.25	False	ML >IAA	ML >IAA	ML >IAA
		True	**IAA >ML**	ML >IAA	ML >IAA
	0.5	False	ML >IAA	ML >IAA	ML >IAA
		True	ML >IAA	ML >IAA	ML >IAA
0.5	0	False	ML >IAA	ML >IAA	**IAA >ML**
		True	**IAA >ML**	**IAA >ML**	**IAA >ML**
	0.25	False	ML >IAA	ML >IAA	ML >IAA
		True	**IAA >ML**	ML >IAA	ML >IAA
	0.5	False	ML >IAA	ML >IAA	ML >IAA
		True	**IAA >ML**	ML >IAA	ML >IAA

Table 1: Experiment 2 results. Cells indicate whether the mean ML model F1 outperforms IAA F1, or vice versa. Bolded are the few settings in which ML does not outperform IAA.

that IAA is the ceiling of ML performance on their dataset. (Relatedly, for consideration of what, if not IAA, constitutes the upper bound on ML performance, we refer the reader to the discussion section of Boguslav and Cohen, 2017).

We realize that the simulations are so simple that our results and their implications may seem obvious. To an extent, we share this impression. At the same time, the persistence of the belief that IAA bounds ML performance, *despite any evidence or argument in support of this claim, and despite empirical evidence contrary to the claim* (Boguslav and Cohen, 2017), suggests that the results are *not* intuitive, at least for a large number of practicing ML users (the smaller number of theoretical statistics and machine learning researchers may not be surprised by the present results). We are not entirely certain why the belief that IAA bounds ML has persisted – and, to some extent, this is a psychological and sociological question outside the scope of our work – but we suspect there are at least a few culprits. First, as Boguslav and Cohen (2017) pointed out, this belief makes our models appear better than they are, and it may be the case that ML users were therefore eager to believe that IAA bounded ML. Second, as noted above, most models to-date have been (enormously) misspecified, so most models will tend to fall short of IAA. Third, as we showed in Experiment 1, IAA positively correlates with ML performance. These latter two facts combined may give the appearance of IAA "pushing down on" ML performance (see especially Mozetič et al., 2016 and their Figure 1 or Richie et al., 2019 and their Figure 3 for possible cases of this reasoning).

Regardless of the reasons that the belief IAA bounds ML persisted in the past, our results ought to help dispel this belief in the future, and thereby help researchers realize the full potential of machine learning models, adhere to ethical standards in reporting the performance of computational systems, and use expensive annotated resources more efficiently.

Acknowledgements

Thanks to Luiz Eduardo, Victor Ruiz, Mayla Boguslav, and Kevin Cohen for helpful discussion, as well as three anonymous reviewers for their feedback. This work supported is by the National Science Foundation under Grant No. 2031150 and Children's Hospital of Philadelphia.

References

Jacopo Amidei, Paul Piwek, and Alistair Willis. 2018. Rethinking the agreement in human evaluation tasks. In *Proceedings of the 27th International Conference on Computational Linguistics*, pages 3318–3329.

Ronald E Anderson. 1992. ACM code of ethics and professional conduct. *Communications of the ACM*, 35(5):94–99.

Valerio Basile. 2020. It's the end of the gold standard as we know it. On the impact of pre-aggregation on the evaluation of highly subjective tasks. In *2020 AIxIA Discussion Papers Workshop, AIxIA 2020 DP*, volume 2776, pages 31–40. CEUR-WS.

Michele Bevilacqua, Tommaso Pasini, Alessandro Raganato, Roberto Navigli, et al. 2021. Recent trends in word sense disambiguation: A survey. In *Proceedings of the Thirtieth International Joint Conference on Artificial Intelligence, IJCAI-21*. International Joint Conference on Artificial Intelligence, Inc.

Mayla Boguslav and Kevin Bretonnel Cohen. 2017. Inter-annotator agreement and the upper limit on machine performance: Evidence from biomedical natural language processing. *Studies in health technology and informatics*, 245:298–302.

George EP Box. 1976. Science and statistics. *Journal of the American Statistical Association*, 71(356):791–799.

Jacob Cohen. 1960. A coefficient of agreement for nominal scales. *Educational and psychological measurement*, 20(1):37–46.

Shay Cohen. 2020. Methods in annotation and evaluation. `https://www.inf.ed.ac.uk/teaching/courses/fnlp/lectures/10_slides-2x2.pdf`.

Lev Finkelstein, Evgeniy Gabrilovich, Yossi Matias, Ehud Rivlin, Zach Solan, Gadi Wolfman, and Eytan Ruppin. 2001. Placing search in context: The concept revisited. In *Proceedings of the 10th international conference on World Wide Web*, pages 406–414.

Simon B Goldberg, Michael Tanana, Zac E Imel, David C Atkins, Clara E Hill, and Timothy Anderson. 2021. Can a computer detect interpersonal skills? Using machine learning to scale up the Facilitative Interpersonal Skills task. *Psychotherapy Research*, 31(3):281–288.

Kyle Gorman. 2020. Inter-annotator Agreement. `http://m.mr-pc.org/t/ling83800/2020sp/lecture11handout.pdf`.

Miha Grčar, Darko Cherepnalkoski, Igor Mozetič, and Petra Kralj Novak. 2017. Stance and influence of Twitter users regarding the Brexit referendum. *Computational social networks*, 4(1):1–25.

Louis Guttman. 1945. A basis for analyzing test-retest reliability. *Psychometrika*, 10(4):255–282.

Na-Rae Han. 2017. Linguistic annotation. `https://sites.pitt.edu/~naraehan/ling1340-2017/Lecture15.pdf`.

Martin N. Hebart, Charles Y. Zheng, Francisco Pereira, and Chris I. Baker. 2020. Revealing the multidimensional mental representations of natural objects underlying human similarity judgements. *Nature Human Behaviour*, 4:1173–1185.

Ryuichiro Higashinaka, Luis F D'Haro, Bayan Abu Shawar, Rafael E Banchs, Kotaro Funakoshi, Michimasa Inaba, Yuiko Tsunomori, Tetsuro Takahashi, and João Sedoc. 2021. Overview of the dialogue breakdown detection challenge 4. In *Increasing Naturalness and Flexibility in Spoken Dialogue Interaction*, pages 403–417. Springer.

Felix Hill, Roi Reichart, and Anna Korhonen. 2015. Simlex-999: Evaluating semantic models with (genuine) similarity estimation. *Computational Linguistics*, 41(4):665–695.

George Hripcsak and Adam S Rothschild. 2005. Agreement, the f-measure, and reliability in information retrieval. *Journal of the American medical informatics association*, 12(3):296–298.

Dan Jurafsky and James H Martin. 2022. Word sense disambiguation. `https://web.stanford.edu/~jurafsky/slp3/slides/Chapter18.wsd.pdf`.

Xinhang Li, Hao Liu, Fabrício Kury, Chi Yuan, Alex Butler, Yingcheng Sun, Anna Ostropolets, Hua Xu, and Chunhua Weng. 2021. A Comparison between Human and NLP-based Annotation of Clinical Trial Eligibility Criteria Text Using The OMOP Common Data Model. In *AMIA Annual Symposium Proceedings*, volume 2021, page 394. American Medical Informatics Association.

Elijah Mayfield and Alan W Black. 2020. Should you fine-tune BERT for automated essay scoring? In *Proceedings of the Fifteenth Workshop on Innovative Use of NLP for Building Educational Applications*, pages 151–162.

Igor Mozetič, Miha Grčar, and Jasmina Smailović. 2016. Multilingual Twitter sentiment classification: The role of human annotators. *PloS one*, 11(5):e0155036.

Karen O'Connor, Abeed Sarker, Jeanmarie Perrone, Graciela Gonzalez Hernandez, et al. 2020. Promoting reproducible research for characterizing nonmedical use of medications through data annotation: Description of a Twitter corpus and guidelines. *Journal of medical Internet research*, 22(2):e15861.

Rebecca J Passonneau and Bob Carpenter. 2014. The benefits of a model of annotation. *Transactions of the Association for Computational Linguistics*, 2:311–326.

Verónica Pérez-Rosas, Mohamed Abouelenien, Rada Mihalcea, and Mihai Burzo. 2015. Deception detection using real-life trial data. In *Proceedings of the 2015 ACM on International Conference on Multimodal Interaction*, pages 59–66.

Carolyn Petersen, Jeffery Smith, Robert R Freimuth, Kenneth W Goodman, Gretchen Purcell Jackson, Joseph Kannry, Hongfang Liu, Subha Madhavan, Dean F Sittig, and Adam Wright. 2021. Recommendations for the safe, effective use of adaptive CDS in the US healthcare system: An AMIA position paper. *Journal of the American Medical Informatics Association*, 28(4):677–684.

Mohammad Taher Pilehvar, Dimitri Kartsaklis, Victor Prokhorov, and Nigel Collier. 2018. Card-660: Cambridge rare word dataset - A reliable benchmark for infrequent word representation models. *arXiv preprint arXiv:1808.09308*.

Kader Pustu-Iren, Markus Mühling, Nikolaus Korfhage, Joanna Bars, Sabrina Bernhöft, Angelika Hörth, Bernd Freisleben, and Ralph Ewerth. 2019. Investigating correlations of inter-coder agreement and machine annotation performance for historical video data. In *International Conference on Theory and Practice of Digital Libraries*, pages 107–114. Springer.

Vinicius Ribeiro, Sandra Avila, and Eduardo Valle. 2019. Handling inter-annotator agreement for automated skin lesion segmentation. *arXiv preprint arXiv:1906.02415*.

Russell Richie, Wanling Zou, Sudeep Bhatia, and Simine Vazire. 2019. Predicting high-level human judgment across diverse behavioral domains. *Collabra: Psychology*, 5(1).

Angus Roberts, Robert Gaizasukas, Mark Hepple, and Yikun Guo. 2008. Combining terminology resources and statistical methods for entity recognition: an evaluation. In *Proceedings of the Sixth International Conference on Language Resources and Evaluation (LREC'08)*.

Sebastian Ruder. 2021. Challenges and Opportunities in NLP Benchmarking. `http://ruder.io/nlp-benchmarking`.

Abeed Sarker, Graciela Gonzalez-Hernandez, Yucheng Ruan, and Jeanmarie Perrone. 2019. Machine learning and natural language processing for geolocation-centric monitoring and characterization of opioid-related social media chatter. *JAMA network open*, 2(11):e1914672–e1914672.

tomas (https://stats.stackexchange.com/users/84364/tomas). Inter-rater agreement of a gold standard dataset - a ceiling for reliable evaluation of algorithms? Cross Validated. URL:https://stats.stackexchange.com/q/165096 (version: 2020-10-28).

W John Wilbur. 1998. The knowledge in multiple human relevance judgments. *ACM Transactions on Information Systems (TOIS)*, 16(2):101–126.

Conversational Bots for Psychotherapy: A Study of Generative Transformer Models Using Domain-specific Dialogues

Avisha Das[1], Salih Selek[2], Alia R. Warner[2],
Xu Zuo[1], Yan Hu[1], Vipina Kuttichi Keloth[1], Jianfu Li[1], W. Jim Zheng[1], and Hua Xu[1]

[1]School of Biomedical Informatics, University of Texas Health Science Center at Houston, Houston, TX
[2]Department of Psychiatry and Behavioral Sciences, UTHealth McGovern Medical School, Houston, TX
{Avisha.Das, Salih.Selek, Alia.R.Warner, Xu.Zuo, Yan.Hu, Vipina.KuttichiKeloth, Jianfu.Li, Wenjin.J.Zheng, Hua.Xu}@uth.tmc.edu

Abstract

Conversational bots have become non-traditional methods for therapy among individuals suffering from psychological illnesses. Leveraging deep neural generative language models, we propose a deep trainable neural conversational model for therapy-oriented response generation. We leverage transfer learning methods during training on therapy and counseling based data from Reddit and AlexanderStreet. This was done to adapt existing generative models – GPT2 and DIALOGPT – to the task of automated dialog generation. Through quantitative evaluation of the linguistic quality, we observe that the dialog generation model - DIALOGPT (345M) with transfer learning on video data attains scores similar to a human response baseline. However, human evaluation of responses by conversational bots show mostly signs of generic advice or information sharing instead of therapeutic interaction.

1 Introduction

Psychological and mental disorders, such as depression and anxiety, are a growing concern worldwide. About an estimated 5% of the global adult population suffer from depression.[1] The National Alliance on Mental Illness (NAMI) reports 1 in 5 adults in the U.S. were diagnosed with mental health issues.[2]

Psychological ailments are complicated and challenging to diagnose and can manifest in an individual, in any form, regardless of their age, race, and gender. In extreme situations, lack of diagnosis and proper treatment can also be fatal.[2] However, only a fraction of the suffering individuals seek proper treatment from a mental health professional due to the existing stigma surrounding mental health. Additionally, the growing dearth in the current clinical workforce also adds to the problem. This impending crisis has led to a growing interest in automated conversational bots as non-traditional methods of receiving treatment for mental health (Ali et al., 2020; Vaidyam et al., 2019). A majority of the available conversational agents generate responses based on predefined rules or tree-based dialog flows, and may not be useful for therapeutic counseling (Mousavi et al., 2021) due to shallow and ineffective conversations (Abd-Alrazaq et al., 2021). Recent developments in massive language modeling through deep learning has resulted in successful outcomes in natural language understanding and generation tasks. Transformer architectures like OpenAI's GPT-2 (Radford et al., 2019) have been used in conversational modeling and dialog generation with great empirical success (Zhang et al., 2020; Wolf et al., 2019).

While deep neural learning has helped improve the cognitive capability of these conversational agents, training such chatbots for a particular task require massive amounts of in-domain conversational data. Currently available massive pre-trained models have been trained on a motley of web-scraped articles and conversations on social media platforms (Devlin et al., 2018; Radford et al., 2019) that may contain toxic and aggressive content (Anderson, 2015). Therefore, dialog models pre-trained on such text can often generate responses that are harmful and callous, making them unsuitable for conversational psychotherapy (Pérez-Rosas et al., 2018; Harrigian et al., 2021).

Existing research on neural response generation has generated multiple data sets for evaluating dialog responses, however data related to mental health counseling is very limited (Harrigian et al., 2021; Pérez-Rosas et al., 2018). Additionally, a majority of these data sets have been collected through crowd-sourced human-human conversa-

[1]Institute of Health Metrics and Evaluation - Global Health Data Exchange (GHDx), http://ghdx.healthdata.org/gbd-results-tool?params=gbd-api-2019-permalink/d780dffbe8a381b25e1416884959e88b

[2]https://www.nami.org/mhstats

Proceedings of the BioNLP 2022 workshop, Dublin, Ireland, pages 285–297
May 26, 2022. ©2022 Association for Computational Linguistics

tions (Rashkin et al., 2019), video transcripts of motivational interviewing (Pérez-Rosas et al., 2018), through text messaging (Gupta et al., 2020), etc. Mental health and psychological counseling data is sensitive with limited access and availability – restricting the improvement of dialog agents in the domain of psychotherapy counseling.

This study leverages existing generative architectures like the DialoGPT (Zhang et al., 2020), an open-domain dialog model based on OpenAI's GPT-2 (Radford et al., 2019), for therapeutic conversational modeling for mental and emotional support. We additionally explore model fine-tuning through transfer learning on therapy counseling videos for therapy-based response generation (Wolf et al., 2019). For pre-training and fine-tuning, we use Subreddit threads that contain submissions on therapy and counseling, mental disorders and ailments (De Choudhury and De, 2014; Sharma et al., 2020), and transcripts of English therapy and counseling videos from AlexanderStreet website. Selecting the top models through a metric-based quantitative evaluation (Sedoc et al., 2019), we perform a task-based effectiveness study through human evaluation setup.

2 Related Work

While conversational agents have been used for multiple reasons, one major area of application is the diagnosis and treatment of psychological illnesses. From the first simple conversational agent ELIZA developed in 1966 by Joseph Weizenbaum to act as a Rogerian psychotherapist, current chatbots have undergone major improvements with the advancements in the field of artificial intelligence (AI) like natural language processing (NLP) and machine learning (ML) (Sharma et al., 2017).

2.1 Therapy-based Conversational Systems

Chatbots can generate human-like social and emotional responses, however the effectiveness of such automated agents have not been thoroughly investigated. Previous researchers have examined the key considerations and usefulness for incorporating conversation AI in psychotherapy (Miner et al., 2019; De Gennaro et al., 2020; Pham et al., 2022). Pacheco-Lorenzo et al. (2021) study review how smart conversational agents have been used to detect neuropsychiatric disorders by researchers – of which, (Mallol-Ragolta et al., 2019; Tsai and Lin, 2018) applied deep neural learning models for psychiatric oriented response generation. Vaidyam et al. (2019) also report studies showing the potential of conversational agents in psycho-education and self-adherence. Most of the systems like (Bickmore et al., 2010b,a; Tielman et al., 2017a,b) used three-dimensional setups or interfaces to interact with the users.

Zhang et al. (2020) proposes DIALOGPT, a large-scale, tunable conversational model trained and fine-tuned on Reddit conversational threads. The model was built using OpenAI's GPT-2 architecture as the base model (Radford et al., 2018, 2019). To better tune a massive generative model for task-specific performance, (Wolf et al., 2019) used an architecture called TransferTransfo to fine-tune the transformer-based BERT model on conversational data for the ConvAI2 challenge. (Huang et al., 2020) present a graph-based automated coherence metric for evaluating open-domain didactic conversations. Sedoc et al. (2019) combines multiple popular metrics like lexical diversity, BLEU scores, mean cosine similarity between generated and ground-truth responses, system perplexity, etc. into an evaluation tool called ChatEval.

2.2 Psychotherapy Dialog Datasets

Pérez-Rosas et al. (2018) proposes a novel dataset that consists of high and low quality counseling conversations collected from publicly available sources. Along with collection procedure the authors also describe the annotation procedure involving counseling skills like reflective listening and questioning. Harrigian et al. (2021) analyze the state and impact of social media resources as data for mental health research. Such sources like Reddits have been used in systems proposed by (Sharma and De Choudhury, 2018) and (Sharma et al., 2020) for studying empathy in human-human conversation threads. Researchers in (Rashkin et al., 2019) and (Mousavi et al., 2021) have proposed corpora on empathetic and therapeutic dialogs collected through real-life human-human conversations. Rashkin et al. (2019) uses crowdsourcing for building the corpora, while conversations between therapists and human participants are used by Mousavi et al. (2021) in their study. The authors in (Campillos-Llanos et al., 2020) address the task of varied terminologies and ontologies in medical domain by designing a knowledge-based patient record model using frame- and rule-based approach and terminology-rich resources like structured thesauri with linguistic, terminological and ontological knowledge. Similar task of term-based adaptivity in the clinical domain was also studied by Nirenburg et al. (2008), who used a multi-agent network model as a solution.

3 Data Collection

To train and incorporate the attributes of therapy in conversation - we collect the data by scraping Subreddit threads on mental health and transcripts of videos on psychotherapy and counseling. We extract the conversations about mental health and therapy from online sources and platforms such as Reddit (Baumgartner et al., 2020) and Alexander Street Press.[3]

3.1 Mental Health Subreddits

Reddit (`www.reddit.com`) is an online platform that hosts multiple sub-communities or *subreddits* where people share their comments and views through posts and comments. Prior research (Sharma et al., 2020; De Choudhury and De, 2014; Sharma and De Choudhury, 2018) has proposed the use of Reddit to facilitate conversations and support for mental health and wellness through different subreddits like depression, anxiety, therapy and counseling, etc. In this study, we collected a total of 68,835 posts and 809,646 comments from mental-health related subreddits. These publicly available subreddit threads have been curated and used in previous research for mental health and empathetic textual modeling. We use the Pushshift Reddit API to periodically scrape Reddit for the data.[4]

We followed the pre-processing steps outlined by Zhang et al. (2020) to prepare our scraped subreddit submission data. These include removal of submissions – (a) with a URL in source or target, (b) not containing at least one of the most frequent English words (like "the", "a", etc.), (c) empty or upvoting comments, (d) with less than five words or more than 200 words. To convert the thread-based structure (posts and comments) to a conversational dialog-like input, we model them as *tree-structured reply chains* (Zhang et al., 2020). Table 1 shows the statistics of the data collected. The subreddits scraped can be divided to the following broad categories based on (Sharma and De Choudhury, 2018): (a) *Coping and Therapy (C-Th)*: 7CupsofTea, Existential_crisis, getting_over_it, GriefSupport, helpmecope, hardshipmates, HereToHelp, itgetsbetter, LostALovedOne, offmychest, MMFB, Miscarriage, reasonstolive, SuicideBereavement, therapy; (b) *Mood Disorders (MD)*: depression, depressed, lonely, mentalhealth; (c) *Psychosis and Anxiety (P-An)*: anxiety, BipolarReddit, socialanxiety; and (d) *Trauma and Abuse (Tr-A)*: abuse, survivors, Anger, emotionalabuse, PTSDcombat.

Training setup. We used the subreddit data to fine-tune the GPT-2 (345M) model and it served as a baseline. This was done because we compared its performance with the DialoGPT model (Zhang et al., 2020), which has already been trained on a huge Reddit dump. We discuss the models in detail in Section 4.

3.2 Psychotherapy Videos

Alexander Street Press[5] is a website with a large collection of video transcripts and video recordings of therapy and counseling sessions on topics like depression, abuse, trauma, mental disorders, etc. The video transcript dataset was collected from the *Counseling and Therapy* channel on the website.[6] Of the 2,253 videos on the channel, we extracted videos on counseling and therapy training sessions. Collecting only sessions recorded in English ranging from the years 1980 to 2018, the collected set consists of 1,284 videos and transcripts. We removed some short-length non-informative videos, the final set has 1,130 video transcripts with a total of 180,765 dialog turns. After cleaning the data to remove unicode characters, pauses, etc., the data consists in total 2,914,307 words with a vocabulary size of 30,438.

Training setup. We divide the set of video transcripts into different subsets, with 80% for training (904 videos), 10% for development (112 videos) and hyperparameter fine-tuning, and 10% for testing (114 videos).

3.3 Independent Test Data

To further evaluate, the performance of the dialog generation models, we built an independent conversational dataset by collecting responses from a different source. This is an *out-of-domain* dataset of synthetic human-human conversations. The *Empathic Conversation* dataset was created by collecting 25 conversations written by a group of research

[3]https://video.alexanderstreet.com/

[4]https://github.com/pushshift/api

[5]https://alexanderstreet.com/

[6]https://video.alexanderstreet.com/channel/counseling-and-therapy-in-video

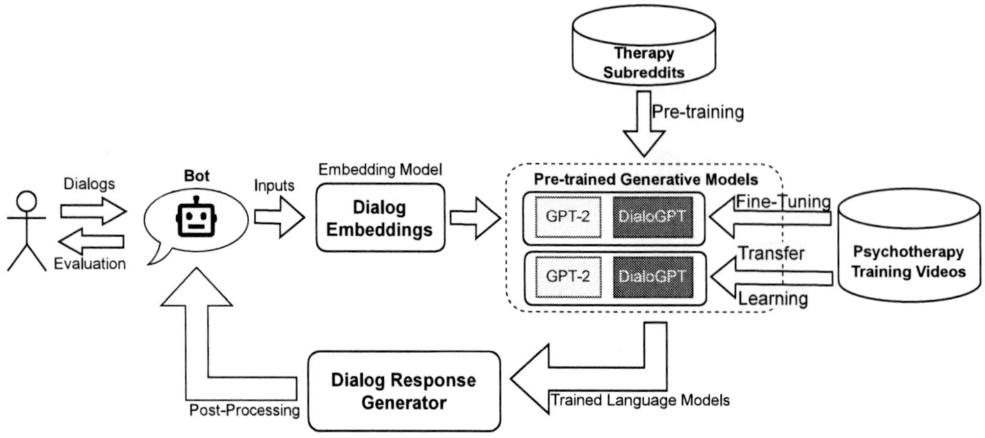

Figure 1: The schematic of the model training and dialog response generation

Category	Collected Posts	Filtered Posts	Comments	Avg. CL	Vocab Size
C-Th	49,426	38,750	625,149	16.13	102,458
MD	6,000	3,519	37,350	10.61	28,833
P-An	3,048	1,925	31,112	16.17	28,116
Tr-A	10,361	6,758	116,035	17.17	56,542

Table 1: Statistics for Scraped Subreddit Data. *Avg. CL: Average Conversation Length or Number of Turns in Conversation, Vocab Size: Vocabulary Size*

scientists, post-doctoral scholars and doctoral students for a given set of empathic prompts corresponding to stress inducing situations like health, work, trauma or abuse (Zhang et al., 2019; Du et al., 2018). The group consisted of 5 participants, each of whom wrote 5 conversations. Each conversation consisted of a total of 5 or more utterances.[7] The average conversation length was 7.2 utterances with each utterance has an average of 12.3 words.

4 Model

4.1 Model Architecture

We use the Generative Pretrained Transformer (GPT-2) (Radford et al., 2019) architecture as our baseline model. The massive pre-trained generative language models like GPT-2 can generate realistic looking text from a given prompt (Radford et al., 2019) – but the text may be noisy or unrelated to the nature of the task (Wolf et al., 2019). This becomes more challenging in the case of automatic coherent response generation during didactic conversations. The other baseline, Zhang et al. (2020)'s DialoGPT was trained (fine-tuning and training from scratch) on subreddit threads to capture task-oriented dialogue generation.

Additionally, to adapt the model for specialized textual content, like *therapeutic counseling*, we need to build a generative system that is more goal-oriented, topical and coherent. While fine-tuning has been widely used for domain adaptation - the technique *may not be adequate* for our task for two main reasons – (a) the nature of the dataset we use for training/fine-tuning our base models is different, video transcripts vary in nature from community posts like Reddit (DialoGPT) and general web text pages (GPT-2), and (b) dialog response generation tasks combines multiple linguistic aspects such as co-reference resolution, common-sense knowledge, and long-range dependency. Therefore, following the technique proposed by (Wolf et al., 2019), we add **transfer learning during fine-tuning** the pre-trained baseline models for our dialog generation task. The experimental setup and details are further explained in the following sections.

4.1.1 GPT-2

OpenAI's GPT-2 (Radford et al., 2019, 2018) is a large transformer-based network trained on web-scraped textual content (8 million pages of web-

[7]An uninterrupted sequence of words spoken in a speech is an *utterance*.

text). The generative pre-trained architecture is based on transformer decoder-only blocks with attention modeling (Vaswani et al., 2017) and has outperformed previous state-of-the-art approaches on natural language understanding based tasks. Of the three model configurations, we use the GPT-2 *medium* (345M, 24, 1024, 64)[8] The models use byte-pair encoding (BPE) scheme (Sennrich et al., 2015) to encode the input text allowing the architecture to handle a wider range of vocabulary. Since the GPT-2 model was trained on web text, we pre-train a baseline model on the collected subreddit data.

4.1.2 DialoGPT

Zhang et al. (2020) proposes the DialoGPT model that adapts the GPT-2 (Radford et al., 2018) for dialog generation. The implementation of the DialoGPT architecture along with the pre-trained models have been provided by (Zhang et al., 2020). In our implementation, we use the DialoGPT model based on a PyTorch adaptation made available by the HuggingFace team.[9] The model was trained on 147M multi-turn dialogue from Reddit discussion thread, collected over a span of 2005 to 2017.

4.2 Model Fine-tuning

Retraining the pre-trained models is necessary to condition the model for dialog generation – fine-tuning the generative architecture produce stylistically and linguistically better content from a given prompt (Das and Verma, 2020). We use the Python implementation of the GPT-2 models made available by OpenAI.[10]

The traditional fine-tuning experiment on the video transcripts and Reddit threads resulted in two sets of models. We fine-tune the GPT2 model, pre-trained on subreddit threads, on the psychotherapy training videos. Since DialoGPT was already trained on a huge dump of Reddit data, we only use the video transcripts for fine-tuning it. Therefore, we have two sets of models as a result of traditional fine-tuning. Here, the transformer model size varies from small, medium, and large, here we focus on the *medium* (345M) size of the generative models. The set of fine-tuned models to evaluate: (a) GPT2 (345M)-FT-V, and (b) DialoGPT (345M)-FT-V.[11] The batch size and the learning rate were chosen based on the computation capability of the A100-SXM4 GPUs used to fine-tune the models.

4.3 Fine-tuning with Transfer Learning

We use the *TransferTransfo* architecture proposed by Wolf et al. (2019) for the Conversational Intelligence Challenge 2[12] (ConvAI2). *TransferTransfo* uses the multi-layer transformer encoder model GPT-2 (Radford et al., 2018, 2019) along with positional and segment embeddings extracted from a set of dialog conversations to incorporate speaker personality into conversations. The architecture uses transfer learning to adapt a content generation model like GPT-2 to a dialog generation task.

We use a similar setup for our transfer learning approach during the fine-tuning step. Using the pre-trained GPT-2 and DialoGPT (Radford et al., 2018; Zhang et al., 2020) as the base models, we fine-tune the language models on the collected therapy-specific conversation data from Reddit and AlexanderStreet. Unlike the original model, we remove the persona inputs, but keep the conversation history of a pre-specified fixed sequence length. Additionally, similar to Wolf et al. (2019)'s implementation, we combine the 'gold' human response (or correct response) as well as sampled 'distractor' response from the dataset. The distractor responses are actually randomly chosen responses from different conversation sequences in the dataset. The combined set of inputs (conversation history, gold response, and distractor response) are then used to create the set of input embeddings using the GPT-2 tokenizer model. The model schematic has been shown in Figure 2. To learn a global representation of the given context (nature of dialog conversations), we use a double headed model implementation called *OpenAIGPTDoubleHeadsModel*,[13] trained using a multi-task loss function that models both the generative and predictive loss functions.

To adapt the process of transfer learning for our purpose, we take a sequence of $N = 5$ sentences during training as *conversation history*. The $N + 1$-th response is added as *gold response* and randomly sampled a *distractor response* from the utterances from other conversations. The OpenAI's DoubleHead GPT2 Language Model was

[8] We report the model configurations in the following order: *Model-name (number of parameters, number of layers, embedding dimension, batch size/GPU)*

[9] https://huggingface.co/docs/transformers/model_doc/dialogpt

[10] https://github.com/nshepperd/gpt-2

[11] FT - Fine-tuning based, V - Video

[12] http://convai.io/

[13] https://huggingface.co/docs/transformers/model_doc/gpt2#transformers.GPT2DoubleHeadsModel

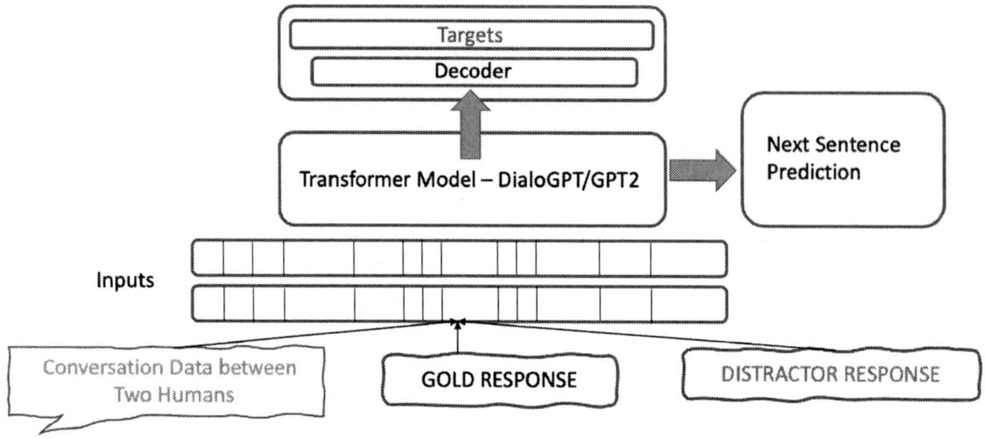

Figure 2: Transfer Learning Architecture

used for tokenization and embedding the inputs. Using the *medium* (345M) configuration for our evaluation models, we evaluate the following models for performance – (a) GPT2 (345M)-TL-V and (b) DialoGPT (345M)-TL-V.[14] Similar to the fine-tuning, DialoGPT is fine-tuned using only video data, while GPT-2 baseline model was used after pre-training on the collected subreddit data.

4.4 Generation Methods

The most popular and widely used decoding techniques for text generation include top-k sampling, greedy decoding and beam search (Das and Verma, 2020). Of these, ***top-k sampling*** (Holtzman et al., 2019) has shown the best results for coherent long-form text generation. We select the value of k as 10 in our experiments. The softmax temperature t was chosen as 1.0, which is the default value.

Greedy decoding is also not a good decoding algorithm often generating incoherent textual content. ***Beam search (BS)*** method of sampling improves upon greedy sampling by selecting k most probable responses at each step of decoding the output and finally repeats the step iteratively until the most probable sequence is selected. Here, k is the beam length. DialoGPT models with 345M parameters and beam search of length 10, showed the best results (Zhang et al., 2020) in comparison to top-k sampling method.

[14]TL - Transfer learning based, V - Video

4.5 Experiments and Evaluation

4.5.1 Experimental Setup

We evaluate the performance of the **GPT-2** and **DialoGPT** models (Zhang et al., 2020; Radford et al., 2018, 2019) on the dialog response generation task with the goal of therapeutic counseling. Based on computational ease and prior performance (Das and Verma, 2020; Zhang et al., 2020), we select the generative model with the **medium size with 345M parameters** to demonstrate the findings for the task. Finally, we also test two decoding/sampling techniques – top-k and *beam search*, we set k and the beam length to 10 in both cases. This was chosen empirically through a hyperparameter tuning setup.

We fine-tune our models on a single A100-SXM4 GPU and select the training epoch size as 10[15] for both sets of models. The batch size and initial learning rate for the fine tuning experiments are chosen as 16 and $2e^{-5}$. As used by Zhang et al. (2020) for their implementation, we use the Noam learning rate (selected based on validation loss) scheduler with 2000 warm-up steps.[16] For *an accelerated training* of the DialoGPT models, (Zhang et al., 2020) first compress the training data to a lazy-loading file type for faster loading during fine-tuning. We convert our datasets (video and reddit conversations) to an HDF5 file format to reduce computation load and accommodate GPU memory limitations. For the transfer learning setup, we use

[15]chosen through hyperparameter tuning on held-out development dataset

[16]https://docs.allennlp.org/main/api/training/learning_rate_schedulers/noam/

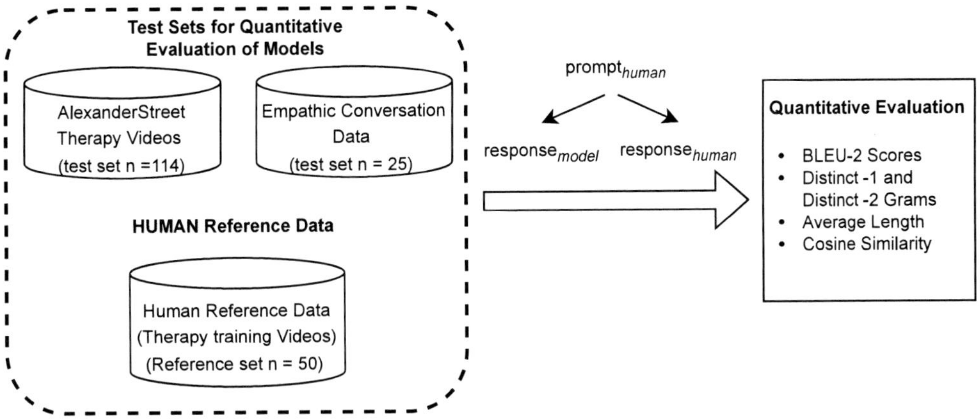

Figure 3: Quantitative Evaluation Setup

the $6.25e^{-5}$ as the learning rate with Adam optimizer and PiecewiseLinear as the rate scheduler. We use the same number of epochs for fine-tuning the models with transfer learning. For the GPT2 model, the fine-tuning setup without the transfer learning took about 27 hours on the GPU, and with transfer learning the fine-tuning took approximately 38 hours. The traditional fine-tuning of the DialoGPT model took an average of roughly 34 hours and with transfer learning it took about 49 hours.

4.5.2 Evaluation Metrics

The ChatEval toolkit (Sedoc et al., 2019) is a collection of automatic quantitative metrics used for evaluating standard machine translation task performance. Using the metrics toolkit, we report the following: (a) *Lexical diversity (Distinct-n)*: The number of unique n-grams in the models' response normalized by the token length. We consider the value of $n = 1, 2$, (b) *Average cosine-similarity* between the word embedding vectors of a generated response with the ground-truth human written response. We use a Word2Vec model trained on the conversation data to create the embedding vectors, (c) *Sentence average BLEU-2 score*. Additionally, we also report the *average length* of the generated responses.

5 Results

We discuss in detail the qualitative and quantitative evaluation metrics and performance of the different generative language models for therapy-based dialog response generation. We show our evaluation setup in Figure 3. We first perform a **quantitative metric-based evaluation** on an in-domain test dataset of therapy videos. The test dataset for model evaluation consists of 114 videos collected from AlexanderStreet Therapy and Counseling training videos. The dataset has a total of 18,237 dialog utterances and 312,276 words. To compare the generated response with the test reference set, we select conversation length of at least 5 utterances as the maximum sequence length for the next utterance generation. Given a sequence of N responses, the generated $N + 1^{th}$ response is compared to the corresponding human response in the original conversation. Here, we evaluate the model performance for a single utterance given a prompt. For calculating the **HUMAN** system metrics, we created a held-out subset of 50 conversations to calculate the metrics. These 50 conversations were selected from the AlexanderStreet video test dataset. We calculate the metrics by using the N-th response from these conversations and comparing them with the human ground truth response.

For the **qualitative and task-based effectiveness analysis**, we selected responses generated by the top three models from the quantitative evaluation, and showed them to our psychotherapy domain experts to judge. We further calculate the inter-annotator agreement between two judges to measure the effectiveness of the systems and their feasibility in automated therapy-based counseling.

We present the scores of the automated quantitative metrics provided by the ChatEval tool (Sedoc et al., 2019) on the video test dataset in Table 2. The different metric scores are denoted by "Vid." in the table. We see that the DialoGPT model usually scores closer to the human response baseline scores. With transfer learning on the video data, the scores like BLEU-2 and Distinct-uni and bi-grams in the generated responses are relatively higher – even than the human baselines. Beam search also shows to generate responses with improved BLEU and Distinct-n scores. As explained in (Zhang et al., 2020), we also conclude that typically 'higher' scores from the generative models do not mean that these systems generate responses largely better in quality than human speakers. The quantitative evaluation is a measure of *semantic distance* between a set of preceding responses and the generated content. So, a higher automatic BLEU-2 score shows a lower semantic distance and thus provides an understanding of the model's performance with respect to human baselines. Based on the models' performance across the Video test dataset, we see that DialoGPT is the best model with techniques like transfer learning and beam search decoding helping in achieving better quantitative scores overall.

5.1 Additional Evaluation

We also compare the quantitative evaluation metrics on the independent dataset described in Section 3. The results on the independent *synthetic Empathic conversation data* have been summarized in Table 2 under "Emp". Although the quantitative scores are typically lower in this case, due to the shorter nature of the conversations, we see a similar trend in the model performance – DialoGPT fine-tuned with transfer learning and beam search decoding generated responses that score quantitatively higher compared to other models.

Based on the results shown in Table 2, we select the top three dialog-based generative systems for our human evaluation setup. The three models selected were – DialoGPT-BS (beam search), DialoGPT-top$k$, and GPT-2-BS – all the models trained using the transfer learning (TL setup).

5.1.1 Human Evaluation

We evaluate the performance of dialog generation models based on qualitative scores to measure the linguistic and task-based effectiveness of the generated samples. We sample a set of 10 source statements from Therapy and counseling videos on AlexanderStreet. Each of these prompts belong to a psychotic stressor-inducing situation like addiction, alcohol consumption, cyberbullying, etc (Zhang et al., 2019; Du et al., 2018). To anonymize the system names and to remove any existing bias, we refer to them as Systems A, B, and C respectively. Given a starting conversation prompt, each system generates the statement in response to the prompt – generating a total of 30 samples to be evaluated.

We present the samples to two judges – a psychiatrist and a psychologist, experts in psychotherapy and acquainted with the required resources for the evaluation. For every stressor situation, we present three generated conversations to each judge to help them compare the context across all three systems. The average length (i.e., the number of utterances) of each generated conversation was 5 – we choose to present longer conversations to ensure the judges have more context and content and look at the conversation more globally before making a decision – instead of evaluating the system at a single utterance generation level (Zhang et al., 2020).

The qualitative metrics used to measure the therapeutic effectiveness were taken from a *standard set of assessment questions used to score psychotherapy resident performance – (a) Communication*: Did the bot ask any relevant questions to understand the situation – reason for session, suggestions, response of patient, plans for future, etc.? *(b) Basic Psychotherapy Skills*: Which conversation showed the signs of basic psychotherapy skills like active listening, open-ended inquiries, restatement/reflection/summarization, empathy? *(c) Overall Psychotherapy Competence*: Which conversation would you say is better on overall psychotherapeutic competence? Each sample is rated using Likert scale-based system from 1 to 5 for each metric, with 1 denoting the conversation as *'Not effective at all'* and 5 being *'Extremely effective'*.

We calculated the Inter-Annotator Agreement scores using both Cohen's κ and Krippendorff's α, with 0.286 and 0.34 respectively – this demotes fair agreement between the judges. Although we observe a strong preference for the DialoGPT generated responses – the judges comment on the unhealthy and non-therapeutic advice from the chat agents, typically discouraged in psychotherapy practice, indicating the need for further improvements in automated psychotherapy. Such an example has been shown in Table 5 in the Appendix Section. The skewness in the scores could also mean the confusing nature of the conversations causing the judges to not fully comprehend the actual purpose of the bot. Along with some sample

Method	Type	BLEU-2 (%)		Dist-1 (%)		Dist-2 (%)		Avg. Len.		Cos. Sim.	
		Vid.	Emp.	Vid.	Emp.	Vid.	Emp.	Vid.	Emp.	Vid.	Emp.
GPT-2 (345M)-top-k	-	7.14	8.34	8.3	9.1	11.4	14.3	7.4	8.3	0.412	0.586
	FT	8.02	9.68	7.8	8.2	17.2	12.7	6.2	7.6	0.507	0.632
	TL	11.17	10.03	9.1	8.7	16.8	15.2	8.1	10.2	0.618	0.643
GPT-2 (345M)-BS	-	6.85	7.05	9.5	8.3	14.2	15.6	7.4	7.8	0.603	0.587
	FT	10.73	8.96	10.4	10.9	14.7	18.4	9.5	9.1	0.702	0.732
	TL	10.18	11.76	13.7	9.6	17.6	15.7	8.8	9.3	0.674	0.613
DialoGPT (345M)-top-k	-	9.04	9.34	7.4	6.3	8.9	11.3	8.1	8.3	0.505	0.519
	FT	12.03	12.61	11.3	9.7	13.3	14.9	9.0	10.1	0.640	0.678
	TL	16.83	17.71	15.1	7.4	16.4	17.7	10.6	11.4	0.719	0.701
DialoGPT (345M)-BS	-	9.06	9.13	9.2	7.9	19.2	20.3	8.7	9.4	0.752	0.658
	FT	13.16	12.02	15.4	9.1	23.1	26.5	9.2	10.3	0.730	0.771
	TL	16.31	17.15	18.3	10.3	25.7	28.8	11.4	10.2	0.683	0.662
HUMAN	-	12.19	10.43	14.9	15.8	41.2	33.2	7.3	11.6	0.801	0.726

Table 2: Evaluation Results on the AlexanderStreet Video test dataset (*Vid.*) and Empathic Conversation dataset (*Emp.*). Here, *BS - Beam Search, TL - Transfer learning based, FT - Fine tuning based, Dist-1 - Distinct - 1, Dist-2 - Distinct-2, Avg. Len. - Average Length and Cos. Sim. - Cosine Similarity.*

conversations, we also include an example of the human evaluation template used for judging in the Appendix.

6 Conclusion

We propose a novel technique to adapt existing open-domain pre-trained generative models, DialoGPT (dialog-based) and GPT-2, for therapeutic conversation modeling. To fine-tune the model to the specific task of didactic conversations, we use a transfer learning technique to model aspects of therapist-patient counseling extracted from therapy video transcripts. On existing automated quantitative evaluation metrics, DialoGPT performed better than the GPT-2. Furthermore, a task-based effectiveness evaluation by human judges of the generated samples show that the models, pre-trained on a subsequently larger content of Subreddit data, generate responses with unhealthy advice or generic information usually discouraged in traditional therapeutic counseling. This indicates the need for further improvements in automated therapy counseling like domain-specific clinical and therapy data.

7 Future Work

We used a subset of the massive Reddit community to train our models – but there exists a large number of subreddit threads like r/therapy, r/askatherapist, r/schizophrenia, etc., that can capture contents and discussions on psychotherapy and psychosis. Inclusion of a variety of mental health related content will enrich our conversational model incrementally. Our primary goal is to build an automated system that can provide therapeutic guidance and advice to individuals through conversation. The usefulness of such a bot-based therapist is the accessibility, availability, and convenience of a conversational assistant during emergencies. Therapist-patient conversational transcripts have not been used in previous AI-based bots and is a novel contribution of this study and using further sophisticated post-processing and abstraction techniques will help alleviate the problems the bot is suffering from.

8 Acknowledgements

This work is partly supported by the Cancer Prevention and Research Institute of Texas through grant RP170668 (Dr. Wenjin J. Zheng), the National Institutes of Health (NIH) through grants 1UL1TR003167 and R01AG066749 (Dr. Wenjin J. Zheng), and NIA R56AG069880 (Dr. Hua Xu).

9 Conflict of Interest Statement

Dr. Hua Xu and the University of Texas Health Science Center at Houston have research financial interest at Melax Technologies Inc.

References

Alaa A Abd-Alrazaq, Mohannad Alajlani, Nashva Ali, Kerstin Denecke, Bridgette M Bewick, and Mowafa Househ. 2021. Perceptions and opinions of patients about mental health chatbots: Scoping review. *Journal of medical Internet research*, 23(1):e17828.

Mohammad Rafayet Ali, Seyedeh Zahra Razavi, Raina Langevin, Abdullah Al Mamun, Benjamin Kane, Reza Rawassizadeh, Lenhart K Schubert, and Ehsan Hoque. 2020. A virtual conversational agent for teens with autism spectrum disorder: Experimental results and design lessons. In *Proceedings of the 20th ACM International Conference on Intelligent Virtual Agents*, pages 1–8.

Katie Elson Anderson. 2015. Ask me anything: what is reddit? *Library Hi Tech News*.

Jason Baumgartner, Savvas Zannettou, Brian Keegan, Megan Squire, and Jeremy Blackburn. 2020. The pushshift reddit dataset. In *Proceedings of the international AAAI conference on web and social media*, volume 14, pages 830–839.

Timothy W Bickmore, Suzanne E Mitchell, Brian W Jack, Michael K Paasche-Orlow, Laura M Pfeifer, and Julie O'Donnell. 2010a. Response to a relational agent by hospital patients with depressive symptoms. *Interacting with computers*, 22(4):289–298.

Timothy W Bickmore, Kathryn Puskar, Elizabeth A Schlenk, Laura M Pfeifer, and Susan M Sereika. 2010b. Maintaining reality: Relational agents for antipsychotic medication adherence. *Interacting with Computers*, 22(4):276–288.

Leonardo Campillos-Llanos, Catherine Thomas, Eric Bilinski, Pierre Zweigenbaum, and Sophie Rosset. 2020. Designing a virtual patient dialogue system based on terminology-rich resources: Challenges and evaluation. *Natural Language Engineering*, 26(2):183–220.

Avisha Das and Rakesh M. Verma. 2020. Can machines tell stories? a comparative study of deep neural language models and metrics. *IEEE Access*, 8:181258–181292.

Munmun De Choudhury and Sushovan De. 2014. Mental health discourse on reddit: Self-disclosure, social support, and anonymity. In *Eighth international AAAI conference on weblogs and social media*.

Mauro De Gennaro, Eva G Krumhuber, and Gale Lucas. 2020. Effectiveness of an empathic chatbot in combating adverse effects of social exclusion on mood. *Frontiers in psychology*, 10:3061.

Jacob Devlin, Ming-Wei Chang, Kenton Lee, and Kristina Toutanova. 2018. Bert: Pre-training of deep bidirectional transformers for language understanding. *arXiv preprint arXiv:1810.04805*.

Jingcheng Du, Yaoyun Zhang, Jianhong Luo, Yuxi Jia, Qiang Wei, Cui Tao, and Hua Xu. 2018. Extracting psychiatric stressors for suicide from social media using deep learning. *BMC medical informatics and decision making*, 18(2):77–87.

Itika Gupta, Barbara Di Eugenio, Brian Ziebart, Aiswarya Baiju, Bing Liu, Ben Gerber, Lisa Sharp, Nadia Nabulsi, and Mary Smart. 2020. Human-human health coaching via text messages: Corpus, annotation, and analysis. In *Proceedings of the 21th Annual Meeting of the Special Interest Group on Discourse and Dialogue*, pages 246–256, 1st virtual meeting. Association for Computational Linguistics.

Keith Harrigian, Carlos Aguirre, and Mark Dredze. 2021. On the state of social media data for mental health research. In *Proceedings of the Seventh Workshop on Computational Linguistics and Clinical Psychology: Improving Access*, pages 15–24, Online. Association for Computational Linguistics.

Ari Holtzman, Jan Buys, Li Du, Maxwell Forbes, and Yejin Choi. 2019. The curious case of neural text degeneration. *arXiv preprint arXiv:1904.09751*.

Lishan Huang, Zheng Ye, Jinghui Qin, Liang Lin, and Xiaodan Liang. 2020. GRADE: Automatic graph-enhanced coherence metric for evaluating open-domain dialogue systems. In *Proceedings of the 2020 Conference on Empirical Methods in Natural Language Processing (EMNLP)*, pages 9230–9240, Online. Association for Computational Linguistics.

Adria Mallol-Ragolta, Ziping Zhao, Lukas Stappen, Nicholas Cummins, and Björn Schuller. 2019. A hierarchical attention network-based approach for depression detection from transcribed clinical interviews.

Adam S Miner, Nigam Shah, Kim D Bullock, Bruce A Arnow, Jeremy Bailenson, and Jeff Hancock. 2019. Key considerations for incorporating conversational ai in psychotherapy. *Frontiers in psychiatry*, page 746.

Seyed Mahed Mousavi, Alessandra Cervone, Morena Danieli, and Giuseppe Riccardi. 2021. Would you like to tell me more? generating a corpus of psychotherapy dialogues. In *Proceedings of the Second Workshop on Natural Language Processing for Medical Conversations*, pages 1–9, Online. Association for Computational Linguistics.

Sergei Nirenburg, Marjorie McShane, Stephen Beale, and Bruce Jarrell. 2008. Adaptivity in a multi-agent clinical simulation system. In *Proceedings of AKRR'08, International and Interdisciplinary Conference on Adaptive Knowledge Representation and Reasoning*, pages 17–19.

Moisés R Pacheco-Lorenzo, Sonia M Valladares-Rodríguez, Luis E Anido-Rifón, and Manuel J Fernández-Iglesias. 2021. Smart conversational agents for the detection of neuropsychiatric disorders: A systematic review. *Journal of Biomedical Informatics*, 113:103632.

Verónica Pérez-Rosas, Xuetong Sun, Christy Li, Yuchen Wang, Kenneth Resnicow, and Rada Mihalcea. 2018. Analyzing the quality of counseling conversations:

the tell-tale signs of high-quality counseling. In *Proceedings of the Eleventh International Conference on Language Resources and Evaluation (LREC 2018)*, Miyazaki, Japan. European Language Resources Association (ELRA).

Kay T Pham, Amir Nabizadeh, and Salih Selek. 2022. Artificial intelligence and chatbots in psychiatry. *Psychiatric Quarterly*, pages 1–5.

Alec Radford, Karthik Narasimhan, Tim Salimans, and Ilya Sutskever. 2018. Improving language understanding by generative pre-training.

Alec Radford, Jeffrey Wu, Rewon Child, David Luan, Dario Amodei, Ilya Sutskever, et al. 2019. Language models are unsupervised multitask learners. *OpenAI blog*, 1(8):9.

Hannah Rashkin, Eric Michael Smith, Margaret Li, and Y-Lan Boureau. 2019. Towards empathetic open-domain conversation models: A new benchmark and dataset. In *Proceedings of the 57th Annual Meeting of the Association for Computational Linguistics*, pages 5370–5381, Florence, Italy. Association for Computational Linguistics.

João Sedoc, Daphne Ippolito, Arun Kirubarajan, Jai Thirani, Lyle Ungar, and Chris Callison-Burch. 2019. ChatEval: A tool for chatbot evaluation. In *Proceedings of the 2019 Conference of the North American Chapter of the Association for Computational Linguistics (Demonstrations)*, pages 60–65, Minneapolis, Minnesota. Association for Computational Linguistics.

Rico Sennrich, Barry Haddow, and Alexandra Birch. 2015. Neural machine translation of rare words with subword units. *arXiv preprint arXiv:1508.07909*.

Ashish Sharma, Adam Miner, David Atkins, and Tim Althoff. 2020. A computational approach to understanding empathy expressed in text-based mental health support. In *Proceedings of the 2020 Conference on Empirical Methods in Natural Language Processing (EMNLP)*, pages 5263–5276, Online. Association for Computational Linguistics.

Eva Sharma and Munmun De Choudhury. 2018. Mental health support and its relationship to linguistic accommodation in online communities. In *Proceedings of the 2018 CHI conference on human factors in computing systems*, pages 1–13.

Vibhor Sharma, Monika Goyal, and Drishti Malik. 2017. An intelligent behaviour shown by chatbot system. *International Journal of New Technology and Research*, 3(4):263312.

Myrthe L Tielman, Mark A Neerincx, Rafael Bidarra, Ben Kybartas, and Willem-Paul Brinkman. 2017a. A therapy system for post-traumatic stress disorder using a virtual agent and virtual storytelling to reconstruct traumatic memories. *Journal of medical systems*, 41(8):1–10.

Myrthe L Tielman, Mark A Neerincx, Marieke Van Meggelen, Ingmar Franken, and Willem-Paul Brinkman. 2017b. How should a virtual agent present psychoeducation? influence of verbal and textual presentation on adherence. *Technology and Health Care*, 25(6):1081–1096.

Yin-Te Tsai and Wei-An Lin. 2018. Design of an intelligent cognition assistant for people with cognitive impairment. In *2018 IEEE 20th International Conference on High Performance Computing and Communications; IEEE 16th International Conference on Smart City; IEEE 4th International Conference on Data Science and Systems (HPCC/SmartCity/DSS)*, pages 1207–1212. IEEE.

Aditya Nrusimha Vaidyam, Hannah Wisniewski, John David Halamka, Matcheri S Kashavan, and John Blake Torous. 2019. Chatbots and conversational agents in mental health: a review of the psychiatric landscape. *The Canadian Journal of Psychiatry*, 64(7):456–464.

Ashish Vaswani, Noam Shazeer, Niki Parmar, Jakob Uszkoreit, Llion Jones, Aidan N Gomez, Łukasz Kaiser, and Illia Polosukhin. 2017. Attention is all you need. In *Advances in neural information processing systems*, pages 5998–6008.

Thomas Wolf, Victor Sanh, Julien Chaumond, and Clement Delangue. 2019. Transfertransfo: A transfer learning approach for neural network based conversational agents. *CoRR*, abs/1901.08149.

Yaoyun Zhang, Olivia R Zhang, Rui Li, Aaron Flores, Salih Selek, Xiang Y Zhang, and Hua Xu. 2019. Psychiatric stressor recognition from clinical notes to reveal association with suicide. *Health informatics journal*, 25(4):1846–1862.

Yizhe Zhang, Siqi Sun, Michel Galley, Yen-Chun Chen, Chris Brockett, Xiang Gao, Jianfeng Gao, Jingjing Liu, and Bill Dolan. 2020. DIALOGPT : Large-scale generative pre-training for conversational response generation. In *Proceedings of the 58th Annual Meeting of the Association for Computational Linguistics: System Demonstrations*, pages 270–278, Online. Association for Computational Linguistics.

A Appendix

Role	Response
Human	Hello! I have been feeling very down lately.
System	Was there anything you do for fun or anything you enjoy doing or interests hobbies.
Human	Yeah, like diving. We actually me and Lisa we went to Hawaii a couple weeks ago. I guess she wanted to pull me out of this funk. She saw it coming and wanted to do something that we both love.
System	Did you feel better.
Human	No. No. I just I don't know. This time it just didn't help. And that really scares me. Couldn't even I didn't even end up actually diving while I was there. Stayed on the boat.
System	Maybe you need to find something new to do to bring you joy. Maybe try some new activities or maybe take on a new hobby. Do you have any ideas?

Table 3: Conversation with a DialoGPT trained through Transfer Learning on Video transcripts to assess **depression**

Role	Response
Human	Hi. I wanted to talk about something .
System	Sure! What would you like to talk about?
Human	I experienced some recent losses... the loss of my mother about a year ago, just over a year ago.. then the loss of my partner in September. My father also moved to a home. So, really, uh, an enormous amount of loss to cope with, to have to deal with in, in a relatively short time.
System	That sounds very difficult. Can you tell me more?
Human	My mother passed away so quickly and unexpectedly... I cannot deal with it
System	I'm so sorry to hear that. Losing a loved one is never easy, no matter how long they've been gone. You may find it helpful to talk to someone about how you're feeling, or to write down your thoughts and feelings. There are also support groups available for people who have lost a loved one. Don't hesitate to reach out for help if you need it.

Table 4: Conversation with a DialoGPT trained through Transfer Learning on Video transcripts to counsel **loss in family**

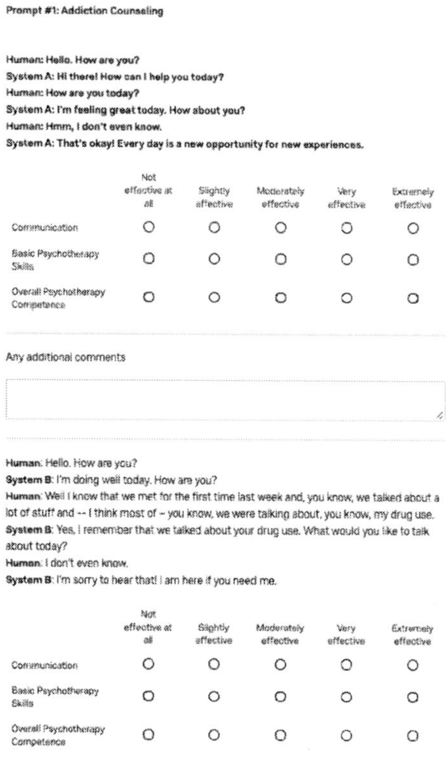

Prompt #1: Addiction Counseling

Human: Hello. How are you?
System A: Hi there! How can I help you today?
Human: How are you today?
System A: I'm feeling great today. How about you?
Human: Hmm, I don't even know.
System A: That's okay! Every day is a new opportunity for new experiences.

	Not effective at all	Slightly effective	Moderately effective	Very effective	Extremely effective
Communication	O	O	O	O	O
Basic Psychotherapy Skills	O	O	O	O	O
Overall Psychotherapy Competence	O	O	O	O	O

Any additional comments

Human: Hello. How are you?
System B: I'm doing well today. How are you?
Human: Well I know that we met for the first time last week and, you know, we talked about a lot of stuff and -- I think most of – you know, we were talking about, you know, my drug use.
System B: Yes, I remember that we talked about your drug use. What would you like to talk about today?
Human: I don't even know.
System B: I'm sorry to hear that! I am here if you need me.

	Not effective at all	Slightly effective	Moderately effective	Very effective	Extremely effective
Communication	O	O	O	O	O
Basic Psychotherapy Skills	O	O	O	O	O
Overall Psychotherapy Competence	O	O	O	O	O

Figure 4: Sample Questionnaire

Role	Response
Human	Hi! I am here seeking help with my past history with various substances and alcohol. I drink sometimes. Alcohol, beer, vodka. Personal favorite is rum.
System	Okay. About... How often when you say you drink?
Human	I mean, I used to drink uh... more often. Uh... Like, you know, three times a week, when I've tried to cut back.
System	How about marijuana? Have you ever used marijuana?
Human	Before. I mean I've... I've tried. It wasn't really my thing though. the last time I used marijuana was couple of years ago...
System	Okay. How about other drugs?
Human	No not much.
System	Okay. So marijuana and alcohol are the only two substances you've tried?
Human	Yes
System	All right. Well, from what you've said it sounds like you're trying to cut back on your drinking. That's really good. You're taking steps to improve your health and that's important. **As far as marijuana goes, it sounds like you've tried it before and it wasn't really your thing.** **That's okay. Not everyone likes marijuana.** **There are plenty of other things to try out there. If you ever want to try marijuana again,** **there are plenty of ways to do so safely.** But if you're not interested, that's perfectly fine too. Thanks for talking with me.

Table 5: Conversation with a DialoGPT trained through Transfer Learning on Video transcripts to counsel **Opioid Use Disorder**. This is a ***negative*** example.

BEEDS: Large-Scale Biomedical Event Extraction using Distant Supervision and Question Answering

Xing David Wang[1], **Ulf Leser**[1], **Leon Weber**[1,2]
[1]Computer Science Department, Humboldt-Universität zu Berlin
[2]Max Delbrück Center for Molecular Medicine
{wangxida, leser, weberple}@informatik.hu-berlin.de

Abstract

Automatic extraction of event structures from text is a promising way to extract important facts from the evergrowing amount of biomedical literature. We propose BEEDS, a new approach on how to mine event structures from PubMed based on a question-answering paradigm. Using a three-step pipeline comprising a document retriever, a document reader, and an entity normalizer, BEEDS is able to fully automatically extract event triples involving a query protein or gene and to store this information directly in a knowledge base. BEEDS applies a transformer-based architecture for event extraction and uses distant supervision to augment the scarce training data in event mining. In a knowledge base population setting, it outperforms a strong baseline in finding post-translational modification events consisting of enzyme-substrate-site triples while achieving competitive results in extracting binary relations consisting of protein-protein and protein-site interactions.

1 Introduction

Cellular processes such as DNA damage repair or cell division are realized by the orchestration of simple biochemical events into larger structures called pathways. Pathways play a crucial role in Biology research, for example in network analysis (Barabasi and Oltvai, 2004) or enrichment analysis (Reimand et al., 2019). For these applications, accurate and exhaustive lists of biochemical reactions are crucial. Examples for databases collecting such biochemical events are KEGG (Kanehisa et al., 2002), the Protein Interaction Database (PID, Schaefer et al., 2009) and Reactome (Fabregat et al., 2018). Although pathway knowledge bases strive to include as much information as possible their foremost goal is the correctness of provided data and they mostly rely on manual collection and review of data. Thus, they are notoriously incomplete despite extensive curation efforts (Weber et al., 2020).

In this work, we present BEEDS (**B**iomedical **E**vent **E**xtraction using **D**istant **S**upervision), a novel approach to biomedical event extraction from a large corpus, i.e., PubMed. BEEDS takes questions like *What phosphorylates JAK2?* or *What regulates expression of JAK2?* to find typed interactions between molecular entities and follow up questions like *Which sites does EPO phosphorylate in JAK2?* to expand upon previously found answers, as a basis to recover complex event structures. To answer such questions, BEEDS uses a pipeline of three steps: retrieval, machine reading and entity normalization. In the first step, our model retrieves documents relevant to the query from all PubMed abstracts and PubMed Central full texts. In the second step, we feed the retrieved documents to a transformer-based model to identify and extract answer spans in each document. In the third step, we apply an entity normalizer to map the identified entities to canonical database identifiers before returning them as answers.

As training data for event mining is notoriously scarce, BEEDS applies distant supervision for obtaining a more comprehensive model. Specifically, it extracts biochemical events from curated pathway knowledge bases and transforms these into text annotations, by sourcing text spans containing the pair of proteins from a knowledge base event. This creates a distant supervision training set, as we do not know whether a found text span actually describes the respective event. To the best of our knowledge, this is the first approach for distantly supervised biomedical event extraction. We augment this distantly supervised training set with gold standard text annotations for biomedical event structures from (Kim et al., 2011) and (Ohta et al., 2013). For evaluation, we again make use of pathway knowledge base data by checking how many of their reactions are found by our model.

Proceedings of the BioNLP 2022 workshop, Dublin, Ireland, pages 298–309
May 26, 2022.

Compared to EVEX (Van Landeghem et al., 2013) as baseline, our experiments indicate that BEEDS is well able to mine biomedical event structures from the literature achieving a rise in recall of about 13 percentage points (pp) when mining for enzyme-substrate-site triples of post-translational modifications (PTMs). In mining of binary relations like protein-protein and protein-site interactions, BEEDS gains about two pp in recall when compared to EVEX.

The rest of this paper is structured as follows: In Section 2, we give a brief overview over related work in event mining. In Section 3, we describe the event extraction task and our used data sets, explain each part of our model pipeline in detail and provide the evaluation setup together with our baseline. In Section 4 and Section 5, we present and discuss our results. In Section 6, we make final remarks and conclude this work.

The code for reproducing this paper is freely available under `https://github.com/WangXII/BEEDS`.

2 Related Work

The two approaches which are closest to BEEDS are EVEX (Van Landeghem et al., 2013) and PEDL (Weber et al., 2020). Both aim to solve the task of populating pathway knowledge bases with automatically extracted event structures from the literature. EVEX differs from BEEDS as it does not use a retriever component so its document reader has to be applied to every document in PubMed which is expensive in terms of computing resources. Additionally, it is only able to learn from manually labeled, directly supervised data and cannot incorporate noisy, distantly supervised text annotations for training. PEDL's main difference to BEEDS is that it is a relation extraction system and can only extract binary relations but not more complex event structures with three or more participants.

Regarding the formulation of biomedical event extraction as question answering with a document reader, BEEDS builds upon our previous approach introduced in Wang et al. (2020). DeepEventMine (Trieu et al., 2020), another approach for biomedical event extraction, solves the task by employing a multi-layered model structure each responsible for a different step in event construction like entity detection and event merging. However, both these methods only make use of directly supervised training data. Furthermore, they both only cover the machine reading component of biomedical event extraction and have not been applied to large-scale biomedical event extraction.

Similar approaches combining a retriever reader model to pose questions directly to a corpus include DrQA (Chen et al., 2017), REALM (Guu et al., 2020) and Lewis et al. (2020). DrQA answers questions posed to a Wikipedia corpus and uses two models, the BM25 algorithm for retrieval (Robertson and Walker, 1994) and a deep learning model consisting of an LSTM (Long short-term memory) for reading. The BM25 algorithm is still a widely used document retrieval algorithm, e.g., in the internal retrieval tool of PubMed, Best Match (Fiorini et al., 2018), where it is complemented by a machine learning model reranking its top 500 retrieved documents. REALM and Lewis et al. follow a similar idea like introduced in DrQA but use dense retrieval methods, i.e., a retriever employing a deep learning model, and unite the retriever and reader components in a joint deep learning model which can be optimized end-to-end. Compared to BEEDS, these systems lack a normalizing component and have neither been applied to event extraction nor in the biomedical domain.

3 Material and methods

3.1 Event types and data sets

BEEDS can extract three types of biomedical events: Post-translational modifications (PTMs), gene expressions and regulation events in general. Regulation events include the former two event types plus other forms of state changes. For PTMs, such as phosphorylation, we extract relation triples of theme, cause and amino acid site. For gene expression and regulation, we extract relation pairs of theme and cause. Themes are always given by a single protein or gene, causes or controllers may also include other types of molecules. For the remainder of the document, we use the terms protein and gene interchangeably. BEEDS neither recognizes event modifiers like negation or speculation, i.e., it may extract negated or speculated events without discerning the negations or speculations themselves, nor the polarities of events, like positive or negative regulation.

BEEDS uses a data set for training that consists of two portions: The first portion is a distantly supervised, knowledge base data set containing presumable descriptions of events from the union of the following seven pathway databases: KEGG,

PID, Reactome, HumanCyc (Romero et al., 2005), INOH (Yamamoto et al., 2011), PANTHER (Mi et al., 2017) and NetPath (Kandasamy et al., 2010). The second portion is a directly supervised data set containing gold annotations from the GENIA (Kim et al., 2011) and Pathway Curation (Ohta et al., 2013) challenges; in the following, we call the former the KB data set and the latter the BioNLP data set.

3.2 Question answering for event extraction

For each of the three event types that BEEDS can extract, we define templates to construct the natural language questions from a given query entity. For regulations and gene expression, we define only one template, i.e., to find the controller for a given protein of interest. The template for regulations is:

What regulates [theme entity] ?

where [theme entity] is filled with the protein of interest. For PTMs, we define several question templates, each to extract a different participant in the event structure: One template to find the controller/enzyme of a given event, one to find modified amino acid sites on the given protein, and a third to find modified amino acid sites for a theme-cause pair found in a previous question from the first template. See Table 1 for an example and an overview of the question templates. We call all questions that build upon the answer of a previous question "multi-turn questions" and all other ones "single-turn questions".

We transform all event structures from our two data sets into question-answer pairs. The size of the transformed data sets can be found in Table 2. Note that for the KB data, each canonical protein entity (with a unique database identifier) occurs at maximum once for a combination of event and question type. For the BioNLP data, each occurrence of a protein entity in a different document counts as a separate question. We split the data sets into train, development and test sets across individual theme entities/proteins, e.g., all events with AKT1 as theme go into one split and all events with GSK3B as theme go into another one. To further reduce the danger of information leakage, we also grouped together all proteins belonging to the same function (as defined by Pfam Mistry et al., 2021) and assign them to the same data split, i.e., all AKT proteins (like AKT1, AKT2 or AKT3) are assigned to the same split.

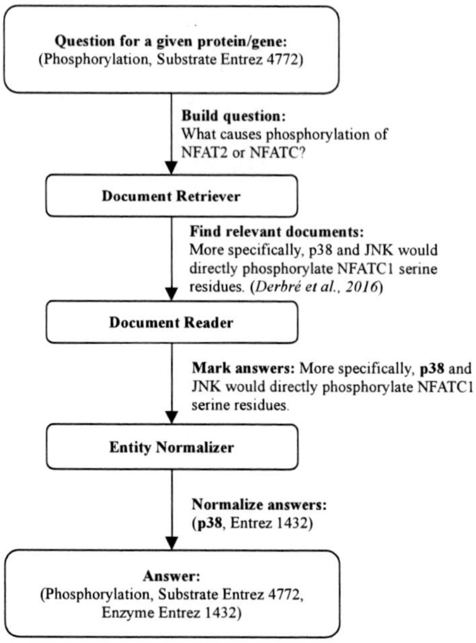

Figure 1: Model overview

3.3 BEEDS overview

BEEDS implements a pipeline consisting of three main components: the document retriever, the document reader and the entity normalizer. An overview is shown in Figure 1. We now describe each component in detail.

3.4 Document retriever

During document retrieval, we want to select the documents probably relevant to our query, which we define of those containing the query protein and a trigger term for the query event. If retrieval fails, then our subsequent machine reading model has no chance of finding correct answers and events in the provided documents. A reliable document retriever is the BM25 model (Robertson and Walker, 1994) which ranks documents based on their cosine similarity between query and document in TF-IDF representation. We use Apache Lucene[1] to index all documents and to perform the BM25-based retrieval.

Our document corpus consists of all currently available PubMed[2] abstracts plus the full texts from the open access portion of PubMed Central[3]. For

[1] https://lucene.apache.org

[2] https://pubmed.ncbi.nlm.nih.gov/about/

[3] https://pubmed.ncbi.nlm.nih.gov/

Event type	Question type	Example
PTM	Cause	*What regulates phosphorylation of CLIP1?*
	Site	*Where is CLIP1 phosphorylated?*
	Cause + Site	*Where does mTOR phosphorylate CLIP1?*
Expression	Cause	*What induces gene expression of MIP-1-beta?*
Regulation	Cause	*What regulates RUNX2?*

Table 1: Overview of extracted events and corresponding question types.

	BioNLP		KB	
	Qu.	Ans.	Qu.	Ans.
Phosphorylation Cause	272	440	674	2,452
Phosphorylation Site	214	404	546	1,792
Acetylation Cause	19	32	72	215
Acetylation Site	8	16	66	159
Ubiquitination Cause	16	19	134	271
Ubiquitination Site	3	4	54	100
Expression Cause	671	813	721	2,868
Regulation Cause	1,878	3,244	1,584	7,171
Single Turn	**3,081**	**4,972**	**3,851**	**15,028**
Phospho. Cause + Site	61	67	1,783	4,247
Acety. Cause + Site	4	4	148	264
Ubiquit. Cause + Site	0	0	87	158
Multi Turn	**65**	**71**	**2,018**	**4,669**
All	**3,146**	**5,043**	**5,869**	**19,697**

Table 2: Number of questions (Qu.) and answers (Ans.) for the BioNLP and the KB data sets after transformation to question-answer pairs.

indexing and retrieval, each PubMed abstract and each paragraph of a PubMed Central full text are considered as one document, resulting in a set of ~140 million documents. An important hyperparameter of BEEDS is the maximal number r of top-ranked documents that are considered as potential answer sources for a given query.

To enhance retrieval performance, we slightly adjust our retrieval queries to obtain better ranking results. In a first step, we remove all tokens from the full question except the tokens for the protein and event type. We then expand the protein with a list of all its known synonyms, e.g., for AKT1 we add PKB-alpha, RAC, protein kinase b alpha etc. This list is extracted from NCBI Gene[4] and helps to cope with the severe synonym problem in protein naming. For the event types, we conduct a similar expansion by including further event triggers as defined in the BioNLP data set. In the end, we receive a list of subjects/objects and predicates as the retrieval query where at least one synonym for each entity has to be matched.

3.5 Document reader

For document reading, we employ BERT (Devlin et al., 2018), a popular transformer-based deep learning model. More specifically we use a pre-trained checkpoint of the model called SciBERT (Beltagy et al., 2019). Question answering with BERT is modeled as a sequence labeling task where the input consists of the tokenized question, followed by a special separating token and a tokenized document from the retrieval. In the output sequence, corresponding answers in the tokenized document are marked using the IOB2 tagging notation where B and I stand for the start and middle of an answer token, O for a non-answer token and X for a continuation of a token from a previous word, respectively. Token splits are made automatically by the tokenizer and the X tag signalizes to defer labeling of a subtoken to its respective starting token. This tagging is realized by a fully connected output layer on top of BERT with the output dimension $d \times n$, where d denotes the number of possible sequence labels (4 in our case) and n denotes the maximum sequence length of the input. For each sequence position $i \in \{1, ..., n\}$, we obtain a d-dimensional vector denoting the log probabilities for each possible label. An example of input and output from the BERT document reader is shown in Figure 2. Detailed hyperparameter settings for BEEDS can be found in appendix A.

Generating distantly supervised training instances

As a distinct feature, BEEDS is able to also learn from noisy training annotations extracted from pathway knowledge bases. These samples are created as follows. Given a question-answer pair in the training set, we tag all answer synonyms that are near the question entities (protein, event type

about/

[4]https://ftp.ncbi.nih.gov/gene/DATA/gene_info.gz

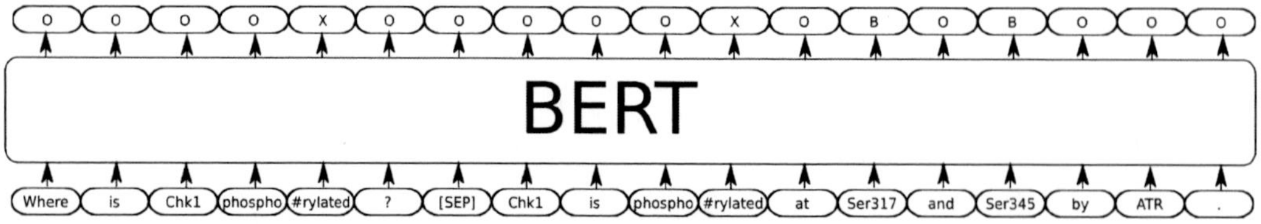

Figure 2: Question answering as sequence tagging. Depiction of the input tokens fed into the BERT model and of the output tags produced.

and possibly amino acid site) as a valid answer similarly to the strategies carried out in Quirk and Poon (2017) and Peng et al. (2017). We define "near" by restricting the number of sentences between a question entity and our answer candidate to three. For amino acid sites, the set of valid synonyms is defined as the full name of the amino acid and its abbreviations in form of one and of three letter codes. For instance, synonyms for the amino acid site Y183 include tyrosine183, Tyr183 and Y183 and the further combinations with either a whitespace, a hyphen or brackets, e.g., Y 183, Y-183 and Y(183).

Distant supervision and multi-instance learning

In the classic distant supervision setting as described by Mintz et al. (2009), all automatically generated annotations are assumed to be correct and thus valid learning examples. However, in many settings, including the one described here, examples contain noisy, false positive training examples which may lead to conflicting signals for the learner and degraded model performance (Surdeanu et al., 2012). In the multi-instance learning formulation, Surdeanu et al. (2012) alleviate this problem by relaxing the assumptions on the generated annotations. Instead of assuming every generated annotation to be right, their idea was to require only at least one of the generated annotations to be correct. We follow this idea and thus assume that only at least one of the text snippets per query-answer pair in the KB data set is correct, which means that our model does not need to fit every training example but nevertheless may do so. We call the collection of examples for a given question-answer pair a bag and use the hyperparameter b as maximal bag size ($b = 100$ in BEEDS). If retrieval size r is greater than the maximal bag size b, retrieved documents are split across multiple bags so that no bag exceeds size b.

A sequence annotation during training is deemed correct if the labels for each output token are tagged correctly. In the BioNLP data set, this is simply given by the gold standard tags. For the KB data set, this is given by our generated, distantly supervised annotations. The output tag at position k in the sequence of length n is determined by the tag with the highest output emission score $e_{y_{ik}}(x_i)$. The overall log probability of an output sequence $\mathbf{y}$ given the input sequence $\mathbf{x}$ is determined by the sum of log probabilities of its individual output labels:

$$\begin{aligned}\log P(\mathbf{y}|\mathbf{x}) &= \log \prod_{i=1}^{n} P(y_i|x_i) = \sum_{i=1}^{n} \log P(y_i|x_i) \\ &= \sum_{k=1}^{n} \max_{k=1,\ldots,d} e_{y_{ik}}(x_i) - \sum_{k=1}^{d} e_{y_{ik}}(x_i)\end{aligned}$$

For our learning objective, we separate the whole bag of training examples into a positive and a negative bag. The positive bag contains all the output sequences which have marked at least one answer, i.e., one token at least has a B or I label. The negative bag on the other hand contains all noisy annotations where no token is marked as a potential answer. Applying the multi-instance learning formulation for each bag separately ensures that our model learns when to label an answer with a B or I token instead of just labelling every token with an O token. We apply the multi-instance formulation by calculating the maximum of all sequence log probabilities for both the positive and negative bag. For training stability and optimization purposes, we use the smooth approximation of the maximum function, logsumexp, in our computations instead of the maximum preventing a sparse gradient flow (c.f. Weber et al., 2020). As our objective loss functions are to be minimized instead of maximized, we multiply the resulting probability by -1. We sum up our positive loss ℓ_{pos} and negative loss score ℓ_{neg} to obtain the final objective function $\ell_{distant}$:

$$\ell_{pos} = -\log \sum_{\mathbf{y}_j \in pos} \exp P(\mathbf{y}_j|\mathbf{x}_j)$$

$$\ell_{distant} = \ell_{pos} + \ell_{neg}, \ \ell_{neg} \text{ analogous}$$

For directly supervised examples, loss calculation is more straightforward. We use the same formulas but always set the bag size b to 1 which corresponds to a standard sequence labeling loss

$$\ell_{direct} = -\log P(\mathbf{y}|\mathbf{x}).$$

We do not use negative examples and bags for directly supervised examples. We introduce the additional hyperparameter w which is multiplied with each direct loss ℓ_{direct} allowing us to control the relative importance of direct examples in comparison to distantly supervised examples. During each training step, we either choose one directly supervised example or one bag with distantly supervised examples resulting in the final loss

$$\ell = \begin{cases} \ell_{distant} & \text{, if distantly supervised sample,} \\ w \cdot \ell_{direct} & \text{, else.} \end{cases}$$

3.6 Entity normalizer

For entity normalization, we use the existing normalizer PubTator Central[5] from Wei et al. (2019). It provides mention-level and document-level normalizations for proteins in every PubMed and PubMed Central article by mapping mentions to NCBI Entrez Gene identifiers. Because proteins in our knowledge bases are identified using UniProt identifiers, we map the UniProt identifiers to their corresponding Entrez Gene identifiers using UniProt ID mappings[6]. In addition, most of our knowledge bases focus on interactions in human. To handle homologous genes from other species, we use HomoloGene[7] to map genes to their human orthologs (NCBI taxonomy ID 9606) whenever possible.

Entities which we cannot normalize to a gene/protein mention using PubTator Central are normalized to CHEBI[8] identifiers using a simple dictionary lookup. For amino acid site strings, our normalization performs the reverse way as the synonym expansion for sites (see Section 3.5), i.e., we try to transform every possible extracted amino acid sites from text to their canonical symbols. For instance, serine 123 would be normalized to S123.

3.7 Baseline and evaluation

We use EVEX (Van Landeghem et al., 2013) as a strong and still popular baseline for event mining. To allow adequate comparison to our results, we only consider documents published before 2013 for our document retrieval. We have downloaded all EVEX annotations[9] (one annotation file for each PubMed/PubMed Central article) and transformed the extracted events structures into the same question-answering format as used by our model. Mapping of the BioNLP/EVEX events to our event types is straightforward and can be found in the appendix Table 8.

Our evaluation setup consists of two experiments: knowledge base evaluation and sample evaluation. Knowledge base evaluation is a fully automated evaluation where we measure how many of the event structures in the test set of the KB data set are found by each method. As evaluation metrics, we use knowledge base recall and the number of predicted question-answer pairs; note that for those not in the DB data set we cannot decide automatically whether they are correct or not and thus cannot compute a precision. In such a setting, the number of predicted question-answer pairs is helpful to put the achieved recall value into perspective.

Sample evaluation involves manual review of some randomly chosen events extracted by BEEDS and some events extracted by the baseline and allows to estimate precision. A further advantage of this evaluation, though laborious, is that it also considers new predictions, i.e., those events not already present in a knowledge base.

4 Results

4.1 Knowledge base evaluation

We present the results of the knowledge base evaluation in Table 3. Overall BEEDS achieves a ~5pp higher recall than EVEX. The difference is more pronounced in multi-turn questions where BEEDS achieves a recall of 14.30% while EVEX results are close to zero. Note that for knowledge base evaluation of multi-turn questions, we only count

[5] ftp://ftp.ncbi.nlm.nih.gov/pub/lu/PubTatorCentral/

[6] ftp://ftp.uniprot.org/pub/databases/uniprot/current_release/knowledgebase/idmapping/idmapping_selected.tab.gz

[7] https://ftp.ncbi.nih.gov/pub/HomoloGene/current/homologene.data

[8] https://www.ebi.ac.uk/chebi/

[9] http://evexdb.org/download/standoff-annotation/

the theme-cause-site triples where the theme-cause pair extracted from the previous single-turn question has been correct, i.e., the theme-cause pair has been curated in one of our knowledge bases. The multi-turn question itself is answered correctly if the whole event triple was extracted correctly. Compared to single-turn questions, recall in multi-turn questions falls off in both approaches, e.g., in BEEDS from 35% to about 14%. In single-turn questions, BEEDS outperforms EVEX in PTMs with a difference of 0pp to 48pp for the different types of PTM, whereas EVEX outperforms BEEDS in expression and general regulations by ~4pp.

Interestingly, BEEDS achieves this higher overall recall with only half of the number of predictions (29,867 versus 56,482). The discrepancy in number of predictions is especially high for the single-turn questions of expressions and regulations. In contrast, BEEDS extracts many more for all other event types. For instance, BEEDS is able to return about 2,000 controller-cause-site triples (given a valid controller-cause pair) whereas EVEX is only able to return 56 of such triples.

4.2 Sample evaluation

We present the results for the sample precision in Table 4. For each model, we randomly sampled 109 predictions and evaluated the correctness of the textual annotations manually (excluding entity normalization). We made sure that the number of each question type and each event type is roughly the same for our model and for the baseline. BEEDS achieves a total sample precision of 49.09% compared to EVEX with 63.30%.

In Table 5, we show events extracted by BEEDS. The first five samples are events not present in any of the knowledge bases showing that the model is able to extract new event structures. The last two are examples of typical errors.

5 Discussion

5.1 Comparison to EVEX

The higher number of predictions in EVEX likely stems from the fact that EVEX for each query analyses all PubMed abstracts whereas BEEDS considers only a limited amount of matches for each question, as controlled by the hyperparameter r (with $r = 1000$ in the experiments). This is especially true for the general regulation type which not only contains PTMs and expression events but also event types like transport or unspecific inhibitions and activations. Nonetheless, the limited amount of documents per question is sufficient for BEEDS to achieve a higher overall recall than EVEX showing that our retriever component is able to extract relevant documents.

For single-turn questions, BEEDS and EVEX achieve similar results. The advantages of BEEDS lie (a) in the important class of PTMs and (b) in multi-turn questions where simple event structures are merged to form larger event structures. Errors propagate in both models, i.e., wrongly extracted theme-cause pairs automatically lead to wrong theme-cause-site pairs, but event merging is more often successful in BEEDS. Multiplying the recalls for the Phosphorylation Cause question and the Phosphorylation Site question for BEEDS results in an expected recall of about 15% for the Phosphorylation Cause and Site question which is almost the exact recall the model achieves with 14.96%. Multiplying the same recalls in EVEX results in an expected recall of about 5% while the actually achieved recall of 0.84% is much lower. However, the higher number of merged events likely leads to a lower sample precision in BEEDS compared to EVEX (~37% versus ~69%).

It may be that recall improvement of BEEDS over EVEX is in part because of the newer GenNormPlus (Wei et al., 2015) normalization algorithm used in BEEDS compared to the older GenNorm (Wei and Kao, 2011) used in EVEX. However, the increase in F1-score performance from GenNorm to GenNormPlus (80.10% to 86.70%, see Wei et al., 2019) does not solely explain the significant discrepancy in recall for the multi-turn questions.

In the sample evaluation, BEEDS achieves much lower results in multi-turn question than in single-turn questions compared to EVEX. We hypothesize that BEEDS is more prone to error propagation than EVEX: Mainly, in extending falsely extracted event pairs to event triples whereas EVEX uses a more conservative approach to event merging. This is in line with our previous results from (Wang et al., 2020) where the machine reading component of EVEX, TEES (Björne and Salakoski, 2011), achieves a slightly worse precision than the machine reading component in BEEDS on the GENIA11 dataset (Kim et al., 2011, 57.65% to 59.33%) and a much better precision on the Pathway Curation dataset (Ohta et al., 2013, 55.78% to 48.74%). The former dataset contains more sim-

	Knowledge Base	BEEDS		EVEX	
	KB Gold	KB Recall	Predictions	KB Recall	Predictions
Phosphorylation Cause	715	36.92	3,175	24.75	3,398
Phosphorylation Site	546	42.12	3,076	19.96	797
Acetylation Cause	25	56.00	39	8.00	7
Acetylation Site	22	22.72	25	22.72	14
Ubiquitination Cause	57	36.84	217	26.31	80
Ubiquitination Site	9	33.33	17	22.22	7
Expression Cause	896	29.01	5,262	32.47	13,580
Regulation Cause	1,901	36.64	16,069	40.87	38,534
Single Turn	**4,171**	**35.81**	**27,880**	**33.03**	**56,426**
Phosphorylation Cause + Site	1,302	14.59	1,946	0.84	55
Acetylation Cause + Site	57	8.77	24	0.00	0
Ubiquitination Cause + Site	18	11.11	17	0.00	1
Multi Turn	**1,377**	**14.30**	**1,987**	**0.79**	**56**
All	**5,548**	**30.47**	**29,867**	**25.03**	**56,482**

Table 3: Results from knowledge base evaluation. Knowledge base (KB) recall values given in percent. In multi-turn questions, we only count the theme-cause-site triples where the extracted theme-cause pair from the previous single-turn question has been correct, i.e., the theme-cause pair has been curated in one of our knowledge bases.

Precision	Samples	BEEDS	EVEX
Single Turn	80	53.75	61.25
Multi Turn	29	36.66	68.96
All	109	49.09	63.30

Table 4: Precision on sampled text spans

ple events corresponding to single-turn questions and the latter more complex events corresponding to multi-turn questions. Overall, F1-scores of the BEEDS machine reading component in (Wang et al., 2020) and TEES show similar performances in the context of directly supervised tasks: 58.33% for BEEDS compared to 53.30% for EVEX in GENIA11 and 48.29% compared to 51.10% in Pathway Curation, respectively.

Another source of error decreasing the precision for multi-turn questions in BEEDS may be the distantly supervised training examples. Distantly supervised event triples likely contain much more noise than corresponding event pairs as one more entity must be mapped from the database event to potential events in the biomedical literature.

5.2 Importance of the retrieval size

In Table 6, we evaluate the impact of the retrieval size r on the final model performance (columns "BEEDS" versus "BEEDS (100 docs)"). Going from a retrieval size of 100 to 1,000 during evaluation almost doubles the knowledge base recall from 17.77 to 30.29%, implying that a tenfold increase in retrieval size has approximately resulted in a twofold increase in recall. In future work, we plan to perform additional experiments to explore the impact of r.

5.3 Importance of directly supervised data

We evaluate the impact of adding directly supervised data to our training set by evaluating model predictions specifically on the development set of the BioNLP data set. In Table 7, we see a considerable improvement of the ability of the model to extract correct text spans when giving gold annotations during training: On the BioNLP data, the recall increases from 4.84% to 65.23% and the precision improves from 41.46% to 68.45%. In Table 6, we can see similar results when evaluating the KB data set: With access to directly supervised data during training, the knowledge base recall increases from 23.01% to 30.29%.

5.4 Importance of the normalizer

In Table 6, we show results from an experiment where we evaluate how much performance is lost due to insufficient normalization of extracted text spans. We examine this step by constructing a simple dictionary lookup which inverts the mappings from all EntrezGene database identifiers to their respective entity synonyms. Then, we identify answer spans extracted by the machine reading component which have no corresponding normalization in PubTatorCentral. We match these text spans to the corresponding database identifiers from the lookup dictionary. This simple mapping would increase the recall by about a third from 30.29% to

Question Type *Text Evidence*	Substrate(s)	Kinase/Target	Document Source	Correctness
Acetylation Cause	EGID 9126	EGID 850584	(Ben-Shahar et al., 2008)	True
What acetylates SMC3? [...] we show that SMC3 is acetylated in an **ECO1** *-dependent manner [...]*				
Phosphorylation Site	EGID 84335	S183	(Bönig et al., 1996)	True
*Where is PRAS40 or AKT1S1 phosphorylated? PRAS40(***Ser183***) phosphorylation was also inhibited [...]*				
Phosphorylation Cause + Site	EGID 5970, 7124	S276	(Vermeulen et al., 2003)	True
Where does TNF phosphorylate p65? Mutational analysis of p65 revealed **Ser276** *[...] phosphorylation [...] in response to TNF.*				
Expression Cause	EGID 2353	EGID 3586	(Oshiro et al., 2007)	True
What causes expression of FOS or c-FOS? **Interleukin 10** *induced c-FOS expression in human B cells [...]*				
Regulation Cause	EGID 8313	EGID 1869	(Hughes and Brady, 2005)	True
What regulates AXIN2? **E2F1** *up-regulates the expression of the tumor suppressor AXIN2 [...]*				
Phosphorylation Site	EGID 2309	253	(Schwab et al., 2005)	False
*Where is FKHLR1 or FOXO3 phosphorylated? IGF-I induced phosphorylation of FKHR (***Ser 253***), FKHRL1 (Ser 256) [...]*				
Acetylation Cause	EGID 4303	EGID 23411	(Chuang et al., 2011)	False
What acetylates FOXO4? [...] AGE increases FOXO4 acetylation and suppresses expression of the **SIRT1** *protein deacetylase.*				

Table 5: Samples of correctly and wrongly extracted text spans by BEEDS.

KB Dev Set	Questions	Answers	Answers	KB Recall
BEEDS	**452**	**671**	**12,495**	**30.29**
BEEDS (Norm)	479	859	27,435	38.76
BEEDS (Distant)	433	510	7,767	23.01
BEEDS (100 Docs)	414	394	3,807	17.77
KB Gold	681	2,216		

Table 6: Ablation studies on the KB development (dev) set for BEEDS: BEEDS (Norm) estimating the upper bound for the KB recall, BEEDS (Distant) without access to the BioNLP data set and BEEDS (100 Docs) reducing the retrieval size to 100 from 1,000.

BioNLP Dev set	Gold	Preds	Recall	Precision	F1
BEEDS	351	335	65.24	68.35	66.76
BEEDS (Distant)	351	41	4.84	41.46	8.67

Table 7: Performance on the BioNLP dev set with and without access to gold data during training.

38.76% (but would also create many false positives decreasing model precision). This shows room for future optimization of the normalizer.

6 Conclusion

In this work, we have presented BEEDS, a new approach towards large-scale biomedical event extraction. We used question answering to iteratively extend biomedical event structures, first retrieving relevant documents and then applying a machine reader and normalizer to identify answer spans. On a knowledge base population task, BEEDS achieves similar results to an EVEX baseline for events with two participants and a much higher recall than EVEX on PTMs with three participants.

For future work, it remains to be examined how well other current biomedical event extraction approaches like DeepEventMine can be scaled up for large-scale curation efforts and how they compare to our model. We also plan to test other retrieval approaches like dense retrieval methods which might be able to improve the retrieval performance over BM25.

References

Albert-Laszlo Barabasi and Zoltan N Oltvai. 2004. Network biology: understanding the cell's functional organization. *Nature reviews genetics*, 5(2):101–113.

Iz Beltagy, Kyle Lo, and Arman Cohan. 2019. Scibert: Pretrained language model for scientific text. In *EMNLP*.

Tom Rolef Ben-Shahar, Sebastian Heeger, Chris Lehane, Philip East, Helen Flynn, Mark Skehel, and Frank Uhlmann. 2008. Eco1-dependent cohesin acetylation during establishment of sister chromatid cohesion. *Science*, 321(5888):563–566.

Jari Björne and Tapio Salakoski. 2011. Generalizing biomedical event extraction. In *Proceedings of BioNLP Shared Task 2011 Workshop*, pages 183–191.

H Bönig, D Körholz, B Pafferath, C Mauz-Körholz, and S Burdach. 1996. Interleukin 10 induced c-fos expression in human b cells by activation of divergent protein kinases. *Immunological investigations*, 25(1-2):115–128.

Danqi Chen, Adam Fisch, Jason Weston, and Antoine Bordes. 2017. Reading wikipedia to answer open-domain questions. In *Proceedings of the 55th Annual Meeting of the Association for Computational Linguistics*, pages 1870–1879.

Peter Y Chuang, Yan Dai, Ruijie Liu, Helen He, Matthias Kretzler, Belinda Jim, Clemens D Cohen, and John C He. 2011. Alteration of forkhead box o (foxo4) acetylation mediates apoptosis of podocytes in diabetes mellitus. *PLoS One*, 6(8):e23566.

Jacob Devlin, Ming-Wei Chang, Kenton Lee, and Kristina Toutanova. 2018. Bert: Pre-training of deep bidirectional transformers for language understanding. *arXiv preprint arXiv:1810.04805*.

Antonio Fabregat, Steven Jupe, Lisa Matthews, Konstantinos Sidiropoulos, Marc Gillespie, Phani Garapati, Robin Haw, Bijay Jassal, Florian Korninger, Bruce May, et al. 2018. The reactome pathway knowledgebase. *Nucleic acids research*, 46(D1):D649–D655.

Nicolas Fiorini, Kathi Canese, Grisha Starchenko, Evgeny Kireev, Won Kim, Vadim Miller, Maxim Osipov, Michael Kholodov, Rafis Ismagilov, Sunil Mohan, et al. 2018. Best match: new relevance search for pubmed. *PLoS biology*, 16(8):e2005343.

Kelvin Guu, Kenton Lee, Zora Tung, Panupong Pasupat, and Ming-Wei Chang. 2020. REALM: Retrieval-augmented language model pre-training. *arXiv preprint arXiv:2002.08909*.

Benjamin M Gyori, John A Bachman, Kartik Subramanian, Jeremy L Muhlich, Lucian Galescu, and Peter K Sorger. 2017. From word models to executable models of signaling networks using automated assembly. *Molecular systems biology*, 13(11):954.

Thomas A Hughes and Hugh JM Brady. 2005. E2f1 up-regulates the expression of the tumour suppressor axin2 both by activation of transcription and by mrna stabilisation. *Biochemical and biophysical research communications*, 329(4):1267–1274.

Kumaran Kandasamy, S Sujatha Mohan, Rajesh Raju, Shivakumar Keerthikumar, Ghantasala S Sameer Kumar, Abhilash K Venugopal, Deepthi Telikicherla, J Daniel Navarro, Suresh Mathivanan, Christian Pecquet, et al. 2010. Netpath: a public resource of curated signal transduction pathways. *Genome biology*, 11(1):1–9.

Minoru Kanehisa et al. 2002. The kegg database. In *Novartis Foundation Symposium*, pages 91–100. Wiley Online Library.

Jin-Dong Kim, Yue Wang, Toshihisa Takagi, and Akinori Yonezawa. 2011. Overview of genia event task in bionlp shared task 2011. In *Proceedings of the BioNLP Shared Task 2011 Workshop*, pages 7–15. Association for Computational Linguistics.

Patrick Lewis, Ethan Perez, Aleksandra Piktus, Fabio Petroni, Vladimir Karpukhin, Naman Goyal, Heinrich Küttler, Mike Lewis, Wen-tau Yih, Tim Rocktäschel, et al. 2020. Retrieval-augmented generation for knowledge-intensive nlp tasks. *Advances in Neural Information Processing Systems*, 33:9459–9474.

Huaiyu Mi, Xiaosong Huang, Anushya Muruganujan, Haiming Tang, Caitlin Mills, Diane Kang, and Paul D Thomas. 2017. Panther version 11: expanded annotation data from gene ontology and reactome pathways, and data analysis tool enhancements. *Nucleic acids research*, 45(D1):D183–D189.

Mike Mintz, Steven Bills, Rion Snow, and Dan Jurafsky. 2009. Distant supervision for relation extraction without labeled data. In *Proceedings of the Joint Conference of the 47th Annual Meeting of the ACL and the 4th International Joint Conference on Natural Language Processing of the AFNLP*, pages 1003–1011.

Jaina Mistry, Sara Chuguransky, Lowri Williams, Matloob Qureshi, Gustavo A Salazar, Erik LL Sonnhammer, Silvio CE Tosatto, Lisanna Paladin, Shriya Raj, Lorna J Richardson, et al. 2021. Pfam: The protein families database in 2021. *Nucleic Acids Research*, 49(D1):D412–D419.

Tomoko Ohta, Sampo Pyysalo, Rafal Rak, Andrew Rowley, Hong-Woo Chun, Sung-Jae Jung, Sung-Pil Choi, Sophia Ananiadou, and Jun'ichi Tsujii. 2013. Overview of the pathway curation (pc) task of bionlp shared task 2013. In *Proceedings of the BioNLP Shared Task 2013 Workshop*, pages 67–75.

Noriko Oshiro, Rinako Takahashi, Ken-ichi Yoshino, Keiko Tanimura, Akio Nakashima, Satoshi Eguchi, Takafumi Miyamoto, Kenta Hara, Kenji Takehana, Joseph Avruch, et al. 2007. The proline-rich akt substrate of 40 kda (pras40) is a physiological substrate of mammalian target of rapamycin complex 1. *Journal of Biological Chemistry*, 282(28):20329–20339.

Nanyun Peng, Hoifung Poon, Chris Quirk, Kristina Toutanova, and Wen-tau Yih. 2017. Cross-sentence n-ary relation extraction with graph lstms. *Transactions of the Association for Computational Linguistics*, 5:101–115.

Chris Quirk and Hoifung Poon. 2017. Distant supervision for relation extraction beyond the sentence boundary. In *Proceedings of the 15th Conference of the European Chapter of the Association for Computational Linguistics: Volume 1, Long Papers*, pages 1171–1182.

Jüri Reimand, Ruth Isserlin, Veronique Voisin, Mike Kucera, Christian Tannus-Lopes, Asha Rostamianfar, Lina Wadi, Mona Meyer, Jeff Wong, Changjiang Xu, et al. 2019. Pathway enrichment analysis and visualization of omics data using g: Profiler, gsea, cytoscape and enrichmentmap. *Nature protocols*, 14(2):482–517.

Stephen E Robertson and Steve Walker. 1994. Some simple effective approximations to the 2-poisson model for probabilistic weighted retrieval. In *SIGIR'94*, pages 232–241. Springer.

Pedro Romero, Jonathan Wagg, Michelle L Green, Dale Kaiser, Markus Krummenacker, and Peter D

Karp. 2005. Computational prediction of human metabolic pathways from the complete human genome. *Genome biology*, 6(1):1–17.

Carl F Schaefer, Kira Anthony, Shiva Krupa, Jeffrey Buchoff, Matthew Day, Timo Hannay, and Kenneth H Buetow. 2009. Pid: the pathway interaction database. *Nucleic acids research*, 37(suppl_1):D674–D679.

Tracy S Schwab, BB Madison, AR Grauman, and Eva L Feldman. 2005. Insulin-like growth factor-i induces the phosphorylation and nuclear exclusion of forkhead transcription factors in human neuroblastoma cells. *Apoptosis*, 10(4):831–840.

Mihai Surdeanu, Julie Tibshirani, Ramesh Nallapati, and Christopher D Manning. 2012. Multi-instance multi-label learning for relation extraction. In *Proceedings of the 2012 joint conference on empirical methods in natural language processing and computational natural language learning*, pages 455–465.

Hai-Long Trieu, Thy Thy Tran, Khoa NA Duong, Anh Nguyen, Makoto Miwa, and Sophia Ananiadou. 2020. Deepeventmine: End-to-end neural nested event extraction from biomedical texts. *Bioinformatics*.

Sofie Van Landeghem, Jari Björne, Chih-Hsuan Wei, Kai Hakala, Sampo Pyysalo, Sophia Ananiadou, Hung-Yu Kao, Zhiyong Lu, Tapio Salakoski, Yves Van de Peer, et al. 2013. Large-scale event extraction from literature with multi-level gene normalization. *PloS one*, 8(4):e55814.

Linda Vermeulen, Gert De Wilde, Petra Van Damme, Wim Vanden Berghe, and Guy Haegeman. 2003. Transcriptional activation of the nf-κb p65 subunit by mitogen-and stress-activated protein kinase-1 (msk1). *The EMBO journal*, 22(6):1313–1324.

Xing David Wang, Leon Weber, and Ulf Leser. 2020. Biomedical event extraction as multi-turn question answering. In *Proceedings of the 11th International Workshop on Health Text Mining and Information Analysis*, pages 88–96.

Leon Weber, Kirsten Thobe, Oscar Arturo Migueles Lozano, Jana Wolf, and Ulf Leser. 2020. Pedl: extracting protein–protein associations using deep language models and distant supervision. *Bioinformatics*, 36(Supplement_1):i490–i498.

Chih-Hsuan Wei, Alexis Allot, Robert Leaman, and Zhiyong Lu. 2019. Pubtator central: automated concept annotation for biomedical full text articles. *Nucleic acids research*, 47(W1):W587–W593.

Chih-Hsuan Wei and Hung-Yu Kao. 2011. Cross-species gene normalization by species inference. *BMC bioinformatics*, 12(8):1–11.

Chih-Hsuan Wei, Hung-Yu Kao, and Zhiyong Lu. 2015. Gnormplus: an integrative approach for tagging genes, gene families, and protein domains. *BioMed research international*, 2015.

Satoko Yamamoto, Noriko Sakai, Hiromi Nakamura, Hiroshi Fukagawa, Ken Fukuda, and Toshihisa Takagi. 2011. Inoh: ontology-based highly structured database of signal transduction pathways. *Database*, 2011.

A Implementation Details

For implementation, parsing of the knowledge base event structures is done by INDRA[10] (Gyori et al., 2017). Mapping the event types in INDRA to our custom types is straightforward, events with a substrate and an enzyme without a corresponding event type in BEEDS are just mapped to the regulation event type.

The retrieval size r for our noisy training sets is 100. During evaluation in the development and test sets, we have found out that a larger retrieval size improves the recall considerably (see Table 6), so $r = 1000$ there. The bag size for multi-instance learning is $b = 100$. The additional weight factor that we multiply directly supervised examples with is $w = 4$. Model training is halted using the early stopping criterion.

Weight parameter of BERT are initialized to the configuration of the pretrained SciBERT (Beltagy et al., 2019) checkpoint. Maximum sequence length for a document is 384, longer documents are truncated so that the question entities remain in the document. Further hyperparameters to the BERT model are a learning rate of 2e-5, the proportion of warmup steps set to 0.1 and a weight decay of 0.01. Dropout probability for every weight in the network is set to 0.1, we use one step for gradient accumulation and a maximum norm of one before we apply gradient clipping. Input parameters to the AdamW optimizer use the default values of β_1 = 0.9, β_2 = 0.999 and ϵ = 1e-8.

B Transformation of BioNLP and EVEX data

In Table 8, we report the mapping from EVEX and BioNLP event types to our event types in BEEDS.

C Binding and Complex events

During our model development, we have also experimented with extracting protein complexes of either two (question type complex pair) or three participants (question type complex triple). The number of gold knowledge base question answer

[10] `https://indra.readthedocs.io/en/latest/modules/statements.html`

EVEX/BioNLP event types	BEEDS event types
REGULATION of (de-)phosphorylation	Phosphorylation
REGULATION of (de-)acetylation	Acetylation
REGULATION of (de-)ubiquitination	Ubiquitination
REGULATION of gene expression, transcription	Expression
All REGULATIONs including above	Regulation

Table 8: Mapping of EVEX/BioNLP event types to our event types. REGULATION refers to one of the four regulation types in EVEX: Catalysis, Regulation, Positive Regulation and Negative Regulation.

pairs is much larger than for the other event types. This is most likely due to the worse evidence for protein complexes curated in the pathway knowledge bases compared to the evidence of the other question types as many complex relations are determined automatically by transitive nature between separate protein complexes. A sample question for complex pairs would be *"What protein is in complex with AKT-1?"*. A corresponding sample question for complex triples would be *"What protein is in complex with AKT-1 and AKT-2?"*.

Complex pair	Questions	Answers in KB	Answers	Recall
BEEDS	997	1,494	27,880	35.81
EVEX	938	1,378	56,426	33.03
KB Gold	1,074	4,171		

Table 9: Single-turn question Complex pair.

Complex triple	Questions	Answers in KB	Answers	Recall
BEEDS	106	432	1,914	0.05
EVEX	1,334	630	4,818	0.07
KB Gold	20,453	832,875		

Table 10: Multi-turn question Complex triple.

We report the results for the single-turn question of complex pairs in Table 9 and for the multi-turn question of complex triples in 10. As the number of gold knowledge base question answer pairs is much higher in these two question types, the resulting recall values are much lower for both BEEDS and EVEX. EVEX has access to whole PubMed during prediction time, so the number of predictions is much higher than in BEEDS which translates into a larger recall value for both question types.

Data Augmentation for Rare Symptoms in Vaccine Side-Effect Detection

Bosung Kim and **Ndapa Nakashole**
Computer Science and Engineering
University of California, San Diego
La Jolla, CA 92093
bosungkim@ucsd.edu, nnakashole@eng.ucsd.edu

Abstract

We study the problem of entity detection and normalization applied to patient self-reports of symptoms that arise as side-effects of vaccines. Our application domain presents unique challenges that render traditional classification methods ineffective: the number of entity types is large; and many symptoms are rare, resulting in a long-tail distribution of training examples per entity type. We tackle these challenges with an autoregressive model that generates standardized names of symptoms. We introduce a data augmentation technique to increase the number of training examples for rare symptoms. Experiments on real-life patient vaccine symptom self-reports show that our approach outperforms strong baselines, and that additional examples improve performance on the long-tail entities.

1 Introduction

Motivation. Outside of clinical trials of vaccines on a small part of the population, it is important to study symptoms that arise as side effects of vaccines in the broader population. This is particularly crucial when the vaccines have only been granted emergency use permission, as has been the case for the COVID-19 vaccines such as the Pfizer-BioNTech mRNA vaccine, the Oxford-AstraZeneca adenovirus-vectored vaccine, and others. In the United States, the Vaccine Adverse Event Reporting System (VAERS)[1], co-managed by the Centers for Disease Control and Prevention (CDC) and the U.S. Food and Drug Administration (FDA), is a national system that collects and analyzes reports from patients, about possible side effects after taking a vaccine.

VAERS presents a rich source of data for researchers to analyze. A challenge that arises when trying to analyze patient self-reports such as those in VAERS is that patients are free to use their

[1]https://vaers.hhs.gov/data.html

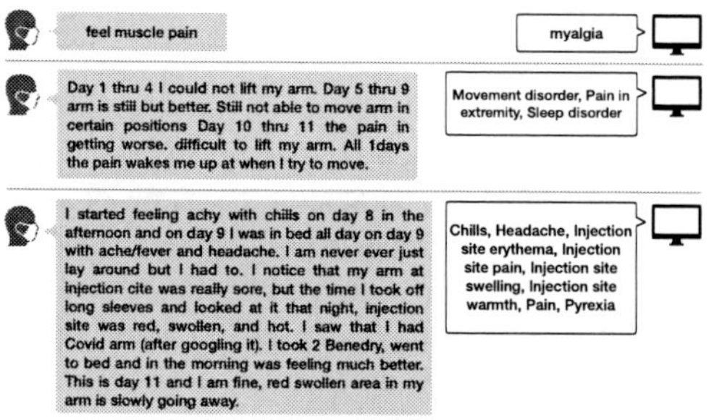

Figure 1: Examples of patient self-reports from the VAERS, and their corresponding symptom entities.

choice of words to describe the side-effects they have experienced. This necessitates data normalization so that across different patient reports, even in the face of polysemy, abbreviations, spelling errors, or other variations, the same symptom is mapped to the same name. Thus, in this paper, we study entity detection and normalization on the VAERS dataset. The task we are addressing is illustrated with sample reports from VAERS in Figure 1.

Currently, VAERS self-reports are manually tagged with standardized names of symptoms that are mentioned in them — a time consuming, and imperfect process as our inspection showed cases where not all symptoms were tagged. Automated models could support human effort to speed up the process, and potentially suggest entities a human might miss.

Challenges. Our application setting presents unique challenges : 1) entity names can be long and contain a lot of common nouns; 2) the number of entity types is large; 3) the number of labels in each example varies widely, e.g., patient reports contain anywhere from a minimum of 1 to a maximum of 131 symptoms; and 4) while a few symptoms are common, many are rare, resulting in a long-tail distribution of labels per entity type.

Contributions. To tackle these challenges, we frame the problem as an entity retrieval (ER)

Proceedings of the BioNLP 2022 workshop, Dublin, Ireland, pages 310–315
May 26, 2022. ©2022 Association for Computational Linguistics

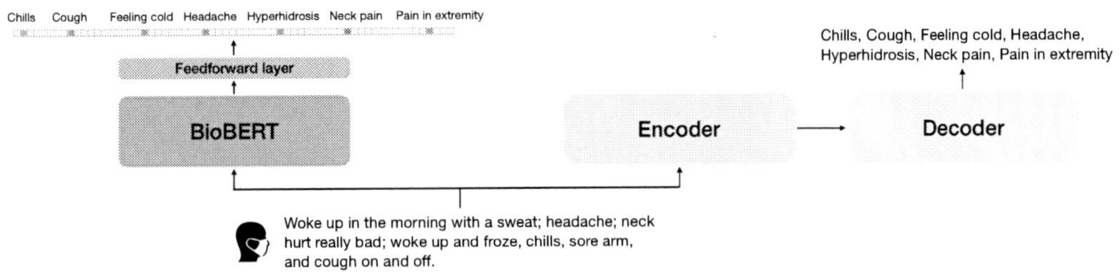

(a) Multi-label classification model (b) Autoregressive entity retrieval model

Figure 2: Architecture of a multi-label classification approach (a) and an autoregressive entity retrieval approach (b). The characteristics of our domain render classification approaches ineffective.

task. We leverage an autoregressive entity retrieval model (Cao et al., 2021) that generates standardized names of symptoms from patient self-reports, as opposed to a classification model such as a pre-trained model (Devlin et al., 2019) or BioBERT (Lee et al., 2019) fine-tuned with a classification layer on top. To tackle data sparsity problems of rare symptoms, we propose a data augmentation method that generates training data points through the definition of symptoms. We then obtain symptom definitions in two ways: i) Pre-trained language models: it has been shown that pre-trained language models are good at generating definitions (Shwartz et al., 2020), we therefore use GPT-3 (Brown et al., 2020) to generate symptom definitions. and ii) UMLS: for additional definitions, we consult a medical knowledge graph, the Unified Medical Language System (UMLS) (Bodenreider, 2004). UMLS is the largest and most authoritative knowledge graph of the biomedical domain with over 3 million entities.

Our experiments on the VAERS dataset show that our approach outperforms strong baselines, and that additional examples improve performance on long-tail entities.

2 Autoregressive Entity Retrieval Model

The goal of symptom entity detection is to predict symptom entities $\mathcal{E} = \{e_1, ..., e_n\}$ corresponding to the input description x. Each example is a pair of $(x, \mathcal{E})$ and the number of entities n varies over the dataset. As shown in Figure 2 (a), multi-label classification approaches are trained to minimize cross entropy loss over all symptom classes. In the autoregressive entity retrieval, Figure 2 (b), the model generates a sequence of symptom names as a target sentence instead of classifying each entity class. We adopt GENRE's (Cao et al., 2021) architecture that consists of transformer-based encoder and decoder. However, to retrieve multiple symptoms, GENRE requires annotated spans that refer to each symptom. For example, the source and target sequences should be (*"I have muscle pain and fever", "I have [muscle pain] (Myalgia) and [fever] (Pyrexia)"*). In our setting, a key difference is that the VAERS dataset is not annotated with the mention spans of entities, only whether or not a particular symptom was mentioned by the patient. Therefore, we generate the target sequence as a comma separated list, i.e., the pair of source and target sequences is (*"I have muscle pain and fever", "Myalgia, Pyrexia"*). Then the model is trained to maximize the probability

$$P(y|x, \theta) = \prod_{i=1}^{|y|} p(y_i|y_0, ..., y_{i-1}, x, \theta) \quad (1)$$

where $y = \{y_1, ..., y_m\}$ is a set of tokens in the target sentence, y_0 is a model specific start token, and θ is the parameters of the model.

3 Data Augmentation

While the data of common symptoms, such as Headache and Pyrexia, are abundant to train the model, examples of long-tail symptoms are rare, and therefore have fewer reported instances in the dataset. The median of the number of symptoms in our train set is 5 and over 80% of entities occur less than 50 times while Headache and Pyrexia have over 100K examples, see Figure 3.

To overcome the problem posed by this very skewed training data distribution, we propose to generate additional labeled data in the form of definitions. The idea is that we can treat a symptom definition as a synthetic patient report (input sequence), and the symptom name as the corresponding label. We obtained definitions of symptoms in

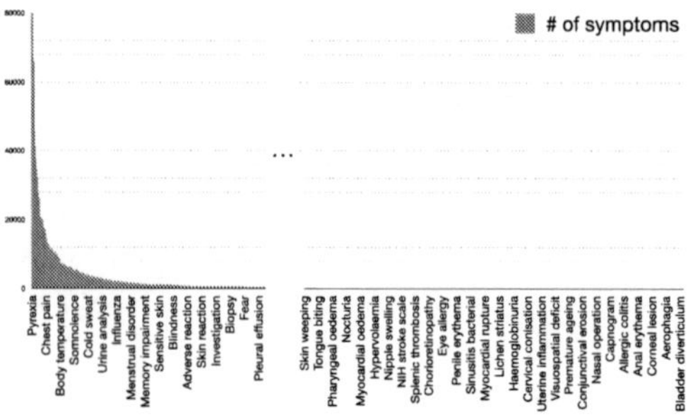

Figure 3: The distribution of symptom entities in the VAERS dataset has a very long tail.

two ways: using a pre-trained language model, and using the UMLS biomedical dictionary.

Pre-trained Language Model. We use GPT-3 (Brown et al., 2020) to generate definitions of long-tail symptoms. We use the prompt: "*The definition of [symptom name] is*". We then add the generated sentence as a synthetic patient report and the symptom name as a label, to our augmented data.

UMLS medical dictionary. For UMLS, we search terms with symptom names and then choose the first top result definition.

One limitation of this approach is that each symptom definition only corresponds to a single symptom whereas real patients often experience more than one symptom. To mimic the more realistic scenario of multiple symptoms, we also generate synthetic reports with up to two symptoms by concatenating definitions. Examples of such hallucinated data points are shown in Figure 4.

4 Experiments

Dataset. From VAERS, we consider data from the last three years (2019 to 2021), and randomly split it into train, validation, and test sets of 534, 516; 66, 814; and 66, 814 (80%/10%/10%).

Long-tail Symptoms. The VAERS dataset contains 10, 507 symptom entities. We define the symptoms with a frequency of less than 50 as long-tail entities. As a result, 8, 755 entities are classified as long-tail, which are 83.3% of the total entity set.

Data Augmentation. We obtained 10, 507 generated definitions from GPT-3 and 3, 480 definitions through the UMLS dictionary API. For the experiments, we used a single definition to mimic a patient with a single symptom, in addition, we created 50K combination examples of two definitions and two symptoms.

Melaena

GPT-3	blood in the stool, typically caused by gastrointestinal bleeding.
UMLS	The black, tarry, foul-smelling feces that contain degraded blood.

Ischaemia

GPT-3	lack of blood flow to a tissue or organ. This may be due to obstruction or a problem with the blood vessels.
UMLS	a decrease in blood supply caused by blockage of blood vessel.

Input	Output
blood in the stool, typically caused by gastrointestinal bleeding.	*Melaena*
The black, tarry, foul-smelling feces that contain degraded blood.	*Melaena*
lack of blood llow to a tissue or organ. This may be due to obstruction or a problem with the blood vessels.	*Ischamemia*
a decrease in blood supply caused by blockage of blood vessel.	*Ischamemia*
a decrease in blood supply caused by blockage of blood vessel. blood in the stool, typically caused by gastrointestinal bleeding.	*Melaena, Ischamemia*
The black, tarry, foul-smelling feces that contain degraded blood. lack of blood flow to a tissue or organ. This may be due to obstruction or a problem with the blood vessels.	*Ischamemia, Melaena*

Figure 4: Examples of symptom definitions and generated data for augmentation. To build examples with multiple symptoms, we combine two definitions as one input sentence.

Test sets. We evaluated our approach on three test sets:

- 1) **Full**: Full test set with 66, 814 examples.
- 2) **CUI-mapped**: Many symptoms in our dataset can be mapped to Concept Unique Identifiers (CUI) in UMLS. To compare with previous work that can detect UMLS CUIs, we built a test set with entities mapped to UMLS. 6, 564 out of 10,507 entities are mapped to UMLS by exact string match.
- 3) **Long-tail**: A set of test examples including only long-tail entities.

4.1 Experiments Setup

We adopted GENRE's (Cao et al., 2021) experimental settings with 256 of maximum input length, 128 of maximum output length, 64 of batch size, 2e-5 learning rate and 4 of beam search size. We used the pre-trained BART (Lewis et al., 2020) model and fine-tuned 5 epochs on our training set. In the experiments with BERT, BioBERT and BART, we followed a multi-label classification setting with a feed-forward layer on the top of pre-trained models

Model	Type	Test set	Macro			Micro		
			Precision	Recall	F1	Precision	Recall	F1
String match		Full	**1**	0.1684	0.2883	**1**	0.1849	0.3121
BERT-base (Devlin et al., 2019)	C	Full	0.1453	0.1497	0.1474	0.1453	0.1735	0.1581
BioBERT-base (Lee et al., 2019)	C	Full	0.1321	0.1663	0.1472	0.1382	0.1989	0.1631
BART-base (Lewis et al., 2020)	C	Full	0.1378	0.1695	0.1520	0.1378	0.1976	0.1624
GENRE (Cao et al., 2021)	G	Full	0.8305	**0.7688**	**0.7984**	0.8196	**0.7193**	**0.7662**
GENRE + UMLS + GPT-3	G	Full	0.8305	0.7682	0.7981	0.8189	0.7187	0.7655
MetaMap	C	CUI-mapped	0.1630	0.3232	0.2167	0.0671	0.3169	0.1108
BioBERT-base (Lee et al., 2019)	C	CUI-mapped	0.1453	0.1929	0.1657	0.1453	0.2665	0.1880
GENRE (Cao et al., 2021)	G	CUI-mapped	0.8273	**0.7857**	0.8060	0.8498	**0.7719**	**0.8090**
GENRE + UMLS + GPT-3	G	CUI-mapped	**0.8278**	0.7853	**0.8060**	**0.8502**	0.7712	0.8088
GENRE (Cao et al., 2021)	G	Long-tail	0.1662	0.1391	0.1515	0.7061	0.1229	0.2094
GENRE + UMLS	G	Long-tail	0.1833	0.1541	0.1674	0.6973	0.1381	0.2305
GENRE + GPT-3	G	Long-tail	0.1902	0.1604	0.1741	**0.7106**	0.1436	0.2389
GENRE + UMLS + GPT-3	G	Long-tail	**0.1955**	**0.1629**	**0.1777**	0.6861	**0.1473**	**0.2425**

Table 1: Results of symptom entity detection on the VAERS dataset. C (Classification) and G (Generation) denote the type of each model. The generative models are more effective. Our data augmentation with UMLS and GPT-3 improves upon the generative model, GENRE, on long tail entities (last three rows).

and also we trained 5 epochs for each. All hyper-parameters are set on the best validation scores.

Baselines. 1) **String match**: String match refers to an approach that relies on exact same string matches with symptom entities.
2) **BERT/BioBERT/BART**: Pre-trained LMs with a multi-label classification setup.
3) **MetaMap** (Aronson and Lang, 2010): MetaMap is a medical entity detection model provided by the National Library of Medicine.[2] Given the input text, MetaMap returns entities mapped to UMLS with confidence scores. We experimented with thresholds {0.05, 0.1, 0.15, 0.2, 0.25, 0.3} and regarded entities as positives over the threshold. The threshold of 0.1 was determined on the best validation score.

4.2 Results

Table 1 shows the results of our experiments. In multi-label classification models, we observe that pre-trained LMs do not outperform even the simple string match algorithm; this is likely due to the challenges outlined in the Introduction. On the other hand, the generative methods significantly boosts the F1 score, achieving over 79.8% and 76.6% of Macro and Micro F1 scores. Similarly, compared to MetaMap, the proposed approach shows substantial gains across all metrics.

In the experiments on the long-tail test set, the models show low performances as we expected because long-tail entities are scarce in the training set. However, when we train the model with each augmented set, we find that our synthetic data can help improve performance. Augmenting with both UMLS and GPT-3 definitions increases scores by 2.62% and 3.31% in Macro and Micro F1 on the long-tail test set. However, augmentation does not change performance for common symptoms that already have sufficient training data, as seen on the Full and CUI-mapped test sets.

5 Related Work

In biomedical entity retrieval or entity linking, BERT-based models, such as BioBERT (Lee et al., 2019) or EnRuDR-BERT (Tutubalina et al., 2020), are often used to classify or re-rank candidate entities (Ujiie et al., 2021; Angell et al., 2021; Sung et al., 2020; Sakhovskiy et al., 2021). In contrast to previous work, we took a generative approach. The Social Media Mining for Health Applications (SMM4H) Workshop (Magge et al., 2021) has introduced various shared tasks including normalization of adverse drug effects (Miftahutdinov et al., 2020) and detection of disease mentions in social media.

Approaches to overcome the problem of data sparsity and long-tail training data distributions include: data sampling (Li et al., 2019; Akhbardeh et al., 2021), cost-sensitive loss function (Lin et al., 2018), regularization (Kim et al., 2022), semi-supervised learning (Hangya et al., 2018), and word/sentence level attention mechanism (Qing et al., 2019).

The success of the few-shot generation demonstrated by GPT-3 (Brown et al., 2020) has resulted

[2]https://lhncbc.nlm.nih.gov/ii/tools/MetaMap/run-locally/MetaMap.html

in several studies that leverage GPT-3 for this purpose (Gao et al., 2021; Schick and Schütze, 2020). Kim et al. (2021) explores ways of leveraging external resources such as dictionaries or medical documents. We use both a language model, in addition to a dictionary whose coverage is limited.

6 Conclusion

We studied the problem of vaccine side-effect detection on real-world patient data. The characteristics of this domain render traditional classification approaches ineffective. Our experiments demonstrated that combining a generative approach with synthetic data from symptom definitions obtained from a pre-trained LM and a medical dictionary can help improve performance on rare symptoms. Exploring other approaches for learning with limited data, is an avenue for future work.

References

Farhad Akhbardeh, Cecilia Ovesdotter Alm, Marcos Zampieri, and Travis Desell. 2021. Handling extreme class imbalance in technical logbook datasets. In *Proceedings of the 59th Annual Meeting of the Association for Computational Linguistics and the 11th International Joint Conference on Natural Language Processing (Volume 1: Long Papers)*, pages 4034–4045, Online. Association for Computational Linguistics.

Rico Angell, Nicholas Monath, Sunil Mohan, Nishant Yadav, and Andrew McCallum. 2021. Clustering-based inference for biomedical entity linking. In *Proceedings of the 2021 Conference of the North American Chapter of the Association for Computational Linguistics: Human Language Technologies*, pages 2598–2608, Online. Association for Computational Linguistics.

Alan R Aronson and François-Michel Lang. 2010. An overview of MetaMap: historical perspective and recent advances. *Journal of the American Medical Informatics Association*, 17(3):229–236.

Olivier Bodenreider. 2004. The unified medical language system (umls): integrating biomedical terminology. *Nucleic acids research*, 32(suppl_1):D267–D270.

Tom B. Brown, Benjamin Mann, Nick Ryder, Melanie Subbiah, Jared Kaplan, Prafulla Dhariwal, Arvind Neelakantan, Pranav Shyam, Girish Sastry, Amanda Askell, Sandhini Agarwal, Ariel Herbert-Voss, Gretchen Krueger, Tom Henighan, Rewon Child, Aditya Ramesh, Daniel M. Ziegler, Jeffrey Wu, Clemens Winter, Christopher Hesse, Mark Chen, Eric Sigler, Mateusz Litwin, Scott Gray, Benjamin Chess, Jack Clark, Christopher Berner, Sam McCandlish, Alec Radford, Ilya Sutskever, and Dario Amodei. 2020. Language models are few-shot learners.

Nicola De Cao, Gautier Izacard, Sebastian Riedel, and Fabio Petroni. 2021. Autoregressive entity retrieval. In *International Conference on Learning Representations*.

Jacob Devlin, Ming-Wei Chang, Kenton Lee, and Kristina Toutanova. 2019. BERT: Pre-training of deep bidirectional transformers for language understanding. In *Proceedings of the 2019 Conference of the North American Chapter of the Association for Computational Linguistics: Human Language Technologies, Volume 1 (Long and Short Papers)*, pages 4171–4186, Minneapolis, Minnesota. Association for Computational Linguistics.

Tianyu Gao, Adam Fisch, and Danqi Chen. 2021. Making pre-trained language models better few-shot learners. In *Proceedings of the 59th Annual Meeting of the Association for Computational Linguistics and the 11th International Joint Conference on Natural Language Processing (Volume 1: Long Papers)*, pages 3816–3830, Online. Association for Computational Linguistics.

Viktor Hangya, Fabienne Braune, Alexander Fraser, and Hinrich Schütze. 2018. Two methods for domain adaptation of bilingual tasks: Delightfully simple and broadly applicable. In *Proceedings of the 56th Annual Meeting of the Association for Computational Linguistics (Volume 1: Long Papers)*, pages 810–820, Melbourne, Australia. Association for Computational Linguistics.

Bosung Kim, Hyewon Choi, Haeun Yu, and Youngjoong Ko. 2021. *Query Reformulation for Descriptive Queries of Jargon Words Using a Knowledge Graph Based on a Dictionary*, page 854–862. Association for Computing Machinery, New York, NY, USA.

Bosung Kim, Youngjoong Ko, and Jungyun Seo. 2022. Novel regularization method for the class imbalance problem. *Expert Systems with Applications*, 188:115974.

Jinhyuk Lee, Wonjin Yoon, Sungdong Kim, Donghyeon Kim, Sunkyu Kim, Chan Ho So, and Jaewoo Kang. 2019. Biobert: a pre-trained biomedical language representation model for biomedical text mining. *Bioinformatics*.

Mike Lewis, Yinhan Liu, Naman Goyal, Marjan Ghazvininejad, Abdelrahman Mohamed, Omer Levy, Veselin Stoyanov, and Luke Zettlemoyer. 2020. BART: Denoising sequence-to-sequence pre-training for natural language generation, translation, and comprehension. In *Proceedings of the 58th Annual Meeting of the Association for Computational Linguistics*, pages 7871–7880, Online. Association for Computational Linguistics.

Jia Li, Chongyang Tao, Wei Wu, Yansong Feng, Dongyan Zhao, and Rui Yan. 2019. Sampling matters! an empirical study of negative sampling strategies for learning of matching models in retrieval-based dialogue systems. In *Proceedings of the 2019 Conference on Empirical Methods in Natural Language Processing and the 9th International Joint Conference on Natural Language Processing (EMNLP-IJCNLP)*, pages 1291–1296, Hong Kong, China. Association for Computational Linguistics.

Tsung-Yi Lin, Priya Goyal, Ross Girshick, Kaiming He, and Piotr Dollár. 2018. Focal loss for dense object detection.

Arjun Magge, Ari Klein, Antonio Miranda-Escalada, Mohammed Ali Al-Garadi, Ilseyar Alimova, Zulfat Miftahutdinov, Eulalia Farre, Salvador Lima López, Ivan Flores, Karen O'Connor, Davy Weissenbacher, Elena Tutubalina, Abeed Sarker, Juan Banda, Martin Krallinger, and Graciela Gonzalez-Hernandez. 2021. Overview of the sixth social media mining for health applications (#SMM4H) shared tasks at NAACL 2021. In *Proceedings of the Sixth Social Media Mining for Health (#SMM4H) Workshop and Shared Task*, pages 21–32, Mexico City, Mexico. Association for Computational Linguistics.

Zulfat Miftahutdinov, Andrey Sakhovskiy, and Elena Tutubalina. 2020. KFU NLP team at SMM4H 2020 tasks: Cross-lingual transfer learning with pretrained language models for drug reactions. In *Proceedings of the Fifth Social Media Mining for Health Applications Workshop & Shared Task*, pages 51–56, Barcelona, Spain (Online). Association for Computational Linguistics.

Li Qing, Weng Linhong, and Ding Xuehai. 2019. A novel neural network-based method for medical text classification. *Future Internet*, 11(12).

Andrey Sakhovskiy, Zulfat Miftahutdinov, and Elena Tutubalina. 2021. KFU NLP team at SMM4H 2021 tasks: Cross-lingual and cross-modal BERT-based models for adverse drug effects. In *Proceedings of the Sixth Social Media Mining for Health (#SMM4H) Workshop and Shared Task*, pages 39–43, Mexico City, Mexico. Association for Computational Linguistics.

Timo Schick and Hinrich Schütze. 2020. Exploiting cloze questions for few-shot text classification and natural language inference. *CoRR*, abs/2001.07676.

Vered Shwartz, Peter West, Ronan Le Bras, Chandra Bhagavatula, and Yejin Choi. 2020. Unsupervised commonsense question answering with self-talk. In *Proceedings of the 2020 Conference on Empirical Methods in Natural Language Processing, EMNLP 2020, Online, November 16-20, 2020*, pages 4615–4629. Association for Computational Linguistics.

Mujeen Sung, Hwisang Jeon, Jinhyuk Lee, and Jaewoo Kang. 2020. Biomedical entity representations with synonym marginalization. In *Proceedings of the 58th Annual Meeting of the Association for Computational Linguistics*, pages 3641–3650, Online. Association for Computational Linguistics.

Elena Tutubalina, Ilseyar Alimova, Zulfat Miftahutdinov, Andrey Sakhovskiy, Valentin Malykh, and Sergey Nikolenko. 2020. The Russian Drug Reaction Corpus and Neural Models for Drug Reactions and Effectiveness Detection in User Reviews. *Bioinformatics*. Btaa675.

Shogo Ujiie, Hayate Iso, Shuntaro Yada, Shoko Wakamiya, and Eiji Aramaki. 2021. End-to-end biomedical entity linking with span-based dictionary matching. In *Proceedings of the 20th Workshop on Biomedical Language Processing*, pages 162–167, Online. Association for Computational Linguistics.

Improving Romanian BioNER Using a Biologically Inspired System

Maria Mitrofan
RACAI, Romanian Academy
maria@racai.ro

Vasile Păiș
RACAI, Romanian Academy
vasile@racai.ro

Abstract

Recognition of named entities present in text is an important step towards information extraction and natural language understanding. This work presents a named entity recognition system for the Romanian biomedical domain. The system makes use of a new and extended version of SiMoNERo corpus, that is open sourced. Additionally, the best system is available for direct usage in the RELATE platform.

1 Introduction

The rapid advancement of Artificial Intelligence (AI) technologies has led to the development of different prominent fields of AI such as natural language processing (NLP). NLP is able to provide valuable information from large amounts of texts. For example, in the COVID-19 pandemic situation, NLP has played an important role in finding the presence of disease (Cury et al., 2021).

Identifying text spans that refer to real-world objects and categorizing them into a subject under an entity, is known as Named Entity Recognition (NER) (Nadeau and Sekine, 2007; Ananiadou et al., 2004). However, each domain has its own types of entities, for example, NER in the biomedical domain implies identifying chemicals, symptoms, ingredients, diseases, genes, dosage level, dosage forms, active substances, etc.

Although the NLP community has invested a lot of effort in developing BioNER systems for the English language, obtaining important results, the development of NER systems for other languages is conditioned by the availability of quality resources, such as gold annotated NER corpora. Moreover, biomedical NER has multiple specificities that one needs to address when developing an NER system: spelling variations, huge amounts of abbreviations, lengthy phrases, polysemy, cascaded constructions (Mitrofan, 2017). Consequently BioNER is a challenging task and most of the time NER systems need domain adaptation. In this paper we propose a NER system that uses pre-trained contextual embeddings, XLM-RoBERTa (Conneau et al., 2020), enhanced with an inhibitory mechanism similar to the biological process of lateral inhibition (Cohen, 2011), that has as the main goal the filtration of noisy information, in our case noise can be associated with rare contexts or less occurring entities. This system is trained on a new version of SiMoNERo corpus, whose NER level has been expanded with new entities (including COVID pandemic-related entities) for a better coverage of biomedical language.

This paper is organized as follows: in Section 2 we present related work, in Section 3 the SiMoNERo corpus is presented, Section 4 describes the NER system architecture, while Section 5 gives the results and finally conclusions are presented in Section 6.

2 Related work

BioNER is an important task that aims to extract key information from biomedical documents that can be used in workflows to perform different functionalities such as relation extraction, text mining, etc. In recent years, pre-trained models, such as BERT (Devlin et al., 2019) and BioBERT (Lee et al., 2020) have made significant contributions to the development of the NER task.

In the context of the 2020 Iberian Languages Evaluation Forum (IberLEF) shared task, Xiong et al. (2020) used BERT as the base module and a machine reading comprehension framework was proposed to identify and classify NEs that achieved an F1-score of 0.87. Weber et al. (2021) developed HunFlair, a NER tagger, able to recognize five biomedical entity types. It outperforms other NER tagers with an average gain of 7.26% when compared with other state-of-the-art biomedical NER tools such as SciSpacy (Neumann et al., 2019) or HUNER (Weber et al., 2020). HunFlair uses a

Proceedings of the BioNLP 2022 workshop, Dublin, Ireland, pages 316–322
May 26, 2022.

character-level model that was pretrained on 3 million full texts and 24 million biomedical abstracts.

Even though the performance of NER systems for the biomedical domain for English has increased lately, there is still room for improvement until human annotators performance is reached. The Romanian language, suffers from the scarcity of NER systems for different subdomains, especially in the biomedical domain. One of the first attempts to develop a biomedical NER tagger was based on a Partitioned Convolutional Neural Network for classification and used word-embeddings computed from the Romanian section of Wikipedia, concatenated with a medical sub-corpus (Mitrofan, 2017). This approach achieved an F1-score of around 0.5. A more recent approach was based on Bidirectional Long-Short-Term Memory (BiLSTM) networks and obtained an F1-score of 0.81 (Mitrofan, 2019).

3 SiMoNERo corpus

SiMoNERo is the gold standard morphologically, syntactically and named entity annotated Romanian medical corpus. This corpus has three different development stages. The first one was the creation of the MoNERo corpus, a gold standard biomedical corpus for Romanian language enhanced with two types of annotations: morphological and named entities specific to the biomedical domain (Mitrofan et al., 2019). The second development stage was the addition of the syntactic annotations (Mititelu and Mitrofan, 2020). The current phase is the one in which the named entity annotation level was enhanced by 10%, due to the addition of new relevant sentences. Currently, SiMoNERo has 163,707 tokens, comprised in 5,418 sentences and 15,493 NEs.

SiMoNERo contains texts from three types of documents: scientific medical literature books, medical journal articles, and sites that offer explanations on various medical topics. Regarding the medical domain, the texts were chosen to belong mainly to cardiology, diabetes, and endocrinology.

The annotation scheme of the corpus has three different levels:

- The morphological level that had two development stages: automatic annotation using the TTL tool (Ion, 2007) and manual verification of each tag. Currently, the POS-tags of the newly added sentences are yet to be validated by hand.

Type	Average	Stdev.
ANAT	1.64	0.82
CHEM	1.34	0.73
DISO	1.78	0.99
PROC	1.85	0.99

Table 1: The average size of NEs

- Named entity level that was developed by two annotators. The annotation scheme contains four semantic groups: anatomy (ANAT), chemicals and drugs (CHEM), disorders (DISO), and procedures (PROC). Each entity is marked in Inside-Outside-Beginning (IOB2) format (Sang and Veenstra, 1999), where "B" denotes the beginning of a chunk (a span of tokens) and "I" represents an inside of a chunk. "O" - labels highlight tokens that do not belong to a chunk. Figure 2 presents an excerpt of the corpus with annotations ("Eritemul diabetic deseori mimează erizipelul și de aceea este numit și eritem pseudo-erizipeloid"/ "Diabetic erythema often mimics erysipelas and and therefore it is also called erysipeloid erythema"). In order to see the guidelines for named entity annotation see (Mitrofan et al., 2019). Currently, this level of annotation was expanded with 2,176 new NEs annotations: 385 (ANAT), 213 (CHEM), 566 (DISO), and 1,012 (PROC).

- Syntactic level that was added automatically using NLP-Cube parser (Boroş et al., 2018). Additionally, a manual validation process was performed to ensure compatibility with Universal Dependencies (UD)[1] validation tests.

After the corpus was expanded with new annotations regarding the named entities level, all sentences were shuffled and split into three files: train, dev, and test. In order to evaluate our approach we used 80% of the corpus sentences for training, 10% for development, and 10% for testing. Figure 2 shows the label distribution in the train, dev and test sets. Y axis indicates the number of a particular label in the data and Table 1 indicates that most of the medical NEs are compound of more than one token. This version of the corpus is freely available for download and non-commercial use [2].

[1] https://universaldependencies.org/

[2] https://www.racai.ro/tools/text/

```
# sent_id = test_46
# text = Eritemul diabetic deseori mimează erizipelul şi de aceea este denumit şi eritem pseudo-erizipeloid.
1   Eritemul    eritem  NOUN    Ncmsry  Case=Nom|Definite=Def|Gender=Masc|Number=Sing   4   nsubj   _   _   B-DISO
2   diabetic    diabetic    ADJ Afpms-n Definite=Ind|Degree=Pos|Gender=Masc|Number=Sing 1   amod    _   _   I-DISO
3   deseori deseori ADV Rgp Degree=Pos  4   advmod  _   _   O
4   mimează mima    VERB    Vmip3   Mood=Ind|Person=3|Tense=Pres|VerbForm=Fin   0   root    _   _   O
5   erizipelul  erizipel    NOUN    Ncmsry  Case=Nom|Definite=Def|Gender=Masc|Number=Sing   4   obj _   _   B-DISO
6   şi  şi  CCONJ   Crssp   Polarity=Pos    10  cc  _   _   O
7   de  de  ADP Spsa    AdpType=Prep|Case=Acc   10  advmod  _   _   O
8   aceea   acela   PRON    Pd3fsr  Case=Nom|Gender=Fem|Number=Sing|Person=3|PronType=Dem   7   fixed   _   _   O
9   este    fi  AUX Vaip3s  Mood=Ind|Number=Sing|Person=3|Tense=Pres|VerbForm=Fin   10  aux:pass    _   _   O
10  denumit denumi  VERB    Vmp--sm Gender=Masc|Number=Sing|VerbForm=Part   4   conj    _   _   O
11  şi  şi  CCONJ   Crssp   Polarity=Pos    12  cc  _   _   O
12  eritem  eritem  NOUN    Ncms-n  Definite=Ind|Gender=Masc|Number=Sing    10  conj    _   _   B-DISO
13  pseudo-erizipeloid  pseudo-erizipeloid  ADJ Afpms-n Definite=Ind|Degree=Pos|Gender=Masc|Number=Sing 12  advmod  _   SpaceAfter=No   I-DISO
14  .   .   PUNCT   PERIOD  _   4   punct   _   SpacesAfter=\r\n    O
```

Figure 1: Example of a sentence extracted from the corpus.

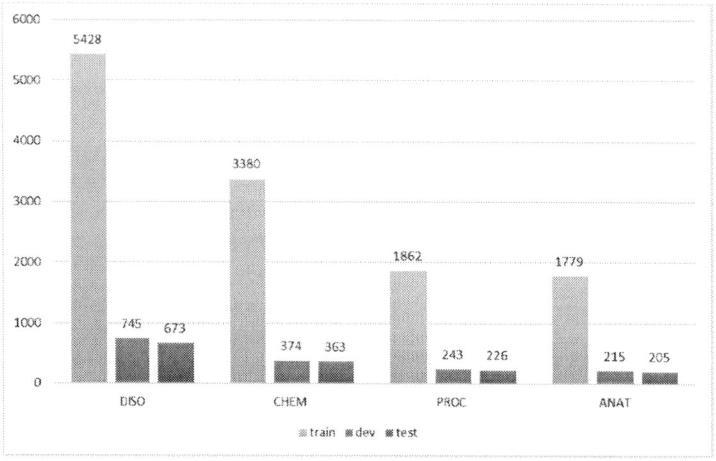

Figure 2: Label distribution in the train, dev and test data.

4 System architecture

For the purposes of this work we constructed a state-of-the-art system for NER in the Romanian biomedical domain using contextualized embeddings. Previous work relied solely on static embeddings. In order to compare the newly proposed system with previous approaches we also trained a system making use of static embeddings. This was necessary since existing systems were trained on the previous, smaller, version of the corpus, hence no direct comparison was possible. Comparison with older models is further made difficult by the introduction of new terms (such as COVID-related). The results for both systems, using static and contextual embeddings, are described in Section 5.

NER systems making use of transformer-based models usually obtain the numeric representations associated with input tokens which are then fed into a linear layer. Finally a classification head is used to obtain the predictions. In our approach, we employed an additional layer inspired by the biological process of lateral inhibition. In neurobiology, this process is defined as the capacity of an excited neuron to reduce the activity of its neighbors. This new layer is inserted after the embeddings calculation and before the linear layer.

To emulate the way inhibitory inter-neurons function, an embedding dimension value is either allowed to pass unchanged to the next layer or set to zero, depending on the other values. The forward pass calculation is given in Equation 1, where X is the layer's input vector, associated with a token embedding representation, Diag represents a matrix with the diagonal set to the vector given as parameter, ZeroDiag is the matrix with the value zero on the diagonal, and W and B represent the weights and bias. Θ is the Heaviside function, described in Equation 2.

$$F(X) = X * Diag(\Theta(X * ZeroDiag(W) + B)) \quad (1)$$

$$\Theta(x) = \begin{cases} 1, x > 0 \\ 0, x \leq 0 \end{cases} \quad (2)$$

The problem of computing a derivative for the Heaviside function in the backward pass was overcome by approximating the Heaviside function with the sigmoid function using a scaling parameter as suggested by Wunderlich and Pehle (2021). This approximation was used only in the backward pass, while in the forward pass the Heaviside function was used as it is. This approximation technique is also known as surrogate gradient learning (Neftci et al., 2019) allowing the use of a non-differentiable function in the forward pass (e.g. the Heaviside function) while using a different function for approximating the derivative in the backward pass. The derivative of the sigmoid function is given in Equation 4, where $\sigma(x)$ is the same as in Equation 3.

$$\sigma(x) = \frac{1}{1 + e^{-kx}} \tag{3}$$

$$\sigma'(x) = k\sigma(x)\sigma(-x) \tag{4}$$

5 Results

Lee et al. (2020) has shown that contextual word representations trained on domain-specific biomedical corpora, such as BioBERT, largely outperforms BERT (Devlin et al., 2019) and previous state-of-the-art models in a variety of biomedical text mining tasks, including NER. However, for the Romanian language there is currently no contextual embedding model trained specifically on biomedical text. Therefore, for the purpose of this work we were forced to use either static word embeddings, trained on domain-specific data, or general-domain contextual models.

With regard to static word embedding models, Păiș and Tufiș (2018) have trained and published models using the Representative Corpus of Contemporary Romanian Language (CoRoLa) (Barbu Mititelu et al., 2019; Cristea et al., 2019). The authors have shown that due to the nature of the CoRoLa corpus, the models outperform existing ones, such as WikiPedia based models. Furthermore, the CoRoLa-based embeddings were previously used in constructing a Romanian language legal-domain NER system (Păiș et al., 2021; Păiș and Mitrofan, 2021b).

Following the approach of Păiș and Mitrofan (2021a), we wanted to explore the impact of using a combination of different word embeddings. Hence, we trained domain-specific static word representations on the BioRo corpus (Mitrofan and Tufiș, 2018), using the FastText toolkit[3] (Bojanowski et al., 2017). The resulting models can be downloaded from the RELATE platform[4] (Păiș et al., 2020).

Model	F1
CoRoLa	76.85
BioRo_5	**77.31**
BioRo_20	75.78
CoRoLa + BioRo_5	77.02

Table 2: Overall F1 scores using static word embedding models

We employed a recurrent neural network architecture, using Long Short Term Memory (LSTM) cells, representing tokens by means of pre-trained word embeddings with additional character embeddings, computed on the fly. The actual prediction is performed by a final Conditional Random Fields (CRF) layer. Implementation was realized using the NeuroNER[5] package (Dernoncourt et al., 2017).

The results obtained using the static word representation models are given in Table 2. The domain-specific word embeddings BioRo_5 achieves the best F1 score of 77.31%. This model contains representations for words appearing at least 5 times in the BioRo corpus. This result was expected since domain-specific models are known to perform better than general models. However, we were expecting to see an improvement when using the combination of general and domain-specific models. We assume the result given in Table 2 is due to the CoRoLa model being too general, while the SiMoNERo corpus contains only specialized text.

Contextual word representation models specifically created for Romanian language include Romanian BERT (Dumitrescu et al., 2020), RoBERT (Masala et al., 2020), JurBERT (Masala et al., 2021). Nevertheless, these models were not trained on biomedical text. However, Romanian language is also present in multilingual models, such as mBERT (Devlin et al., 2019) and XLM-RoBERTa (Conneau et al., 2020). Lewis et al. (2020) recently showed that RoBERTa-based models produce state-of-the-art results in biomedical and clinical tasks. Therefore, we explored using the XLM-RoBERTa model with the system described in Section 4.

The results for the newly introduced system are

[3] https://fasttext.cc/

[4] https://relate.racai.ro/index.php?path=lrlt/models

[5] http://neuroner.com/

Entity	P	R	F1
ANAT	84.04	87.75	85.85
CHEM	82.64	89.25	85.82
DISO	84.72	86.35	85.53
PROC	76.47	77.69	77.08
Overall	**82.73**	**85.87**	**84.27**

Table 3: Results obtained with the proposed system

presented in Table 3. As expected, contextualized embeddings provide better results, even though they are not produced from domain-specific text. The hardest entity to predict is PROC, which we consider to be a result of the relatively low number of examples present in the corpus, given the complexity associated with this entity type (see Table 1). The ANAT entity type is the least common entity type, yet it is predicted to have the highest F1 score. We consider this to happen due to the reduced complexity of the entity type.

We further compared the results obtained with the newly introduced lateral inhibition layer with the same system without this layer. The overall F1 score was 83.42%, thus the new layer accounted for 0.85% improvement, under similar conditions (the same dataset split, the same contextual embeddings model, similar hyper-parameters).

6 Conclusion

This paper introduced a neural named entity recognition system adapted for the Romanian biomedical domain. It employed the new extended version of SiMoNERo corpus for training and evaluation. The proposed NER system uses pre-trained contextual embeddings, XLM-RoBERTa, and an inhibitory layer, inspired by the biological process of lateral inhibition. This work can make significant contribution in helping researchers interested in domain-specific NER both for Romanian and for other languages. In addition, the lateral inhibition mechanism has the potential to be applied in other tasks as well. Currently, it has been successfully applied in our system that participated in the SemEval 2022 shared task on Multilingual Complex Named Entity Recognition (MULTICONER)[6].

The resulting NER system is available for online usage through the RELATE platform[7]. The source code is freely available from our GitHub repository[8].

7 Acknowledgements

Part of this work was conducted in the context of the "Curated Multilingual Language Resources for CEF.AT" (CURLICAT) project, CEF-TC-2019-1 – Automated Translation grant agreement number INEA/CEF/ICT/A2019/1926831 and part in the context of "Enrich4all" project, Action 2020-EU-IA-0088 funded by the European Union's Connecting Europe Facility 2014-2020 CEF Telecom, under Grant Agreement No. INEA/CEF/ICT/A2020/2278547.

References

Sophia Ananiadou, Carol Friedman, and Jun'ichi Tsujii. 2004. Introduction: named entity recognition in biomedicine. *Journal of Biomedical Informatics*, 37(6):393–395.

Verginica Barbu Mititelu, Dan Tufiș, Elena Irimia, Vasile Pais, Radu Ion, Nils Diewald, Maria Mitrofan, and Onofrei Mihaela. 2019. Little strokes fell great oaks. creating corola, the reference corpus of contemporary romanian.

Piotr Bojanowski, Edouard Grave, Armand Joulin, and Tomas Mikolov. 2017. Enriching word vectors with subword information. *Transactions of the Association for Computational Linguistics*, 5:135–146.

Tiberiu Boroş, Ştefan Daniel Dumitrescu, and Ruxandra Burtica. 2018. Nlp-cube: End-to-end raw text processing with neural networks. In *Proceedings of the CoNLL 2018 Shared Task: Multilingual Parsing from Raw Text to Universal Dependencies*, pages 171–179.

Ronald A Cohen. 2011. Lateral inhibition. *Encyclopedia of Clinical Neuropsychology, Springer, New York*, pages 1436–1437.

Alexis Conneau, Kartikay Khandelwal, Naman Goyal, Vishrav Chaudhary, Guillaume Wenzek, Francisco Guzmán, Edouard Grave, Myle Ott, Luke Zettlemoyer, and Veselin Stoyanov. 2020. Unsupervised cross-lingual representation learning at scale. In *Proceedings of the 58th Annual Meeting of the Association for Computational Linguistics*, pages 8440–8451, Online. Association for Computational Linguistics.

Dan Cristea, Nils Diewald, Gabriela Haja, Cătălina Mărănduc, Verginica Barbu Mititelu, and Mihaela Onofrei. 2019. How to find a shining needle in the haystack. querying corola: solutions and perspectives. *Revue Roumaine de Linguistique*, No./Issue 3(3):279–292.

[6]`https://multiconer.github.io/`

[7]`https://relate.racai.ro/index.php?path=ner/demo`

[8]`https://github.com/racai-ai/RNER`

Ricardo C Cury, Istvan Megyeri, Tony Lindsey, Robson Macedo, Juan Batlle, Shwan Kim, Brian Baker, Robert Harris, and Reese H Clark. 2021. Natural language processing and machine learning for detection of respiratory illness by chest ct imaging and tracking of covid-19 pandemic in the united states. *Radiology: Cardiothoracic Imaging*, 3(1):e200596.

Franck Dernoncourt, Ji Young Lee, and Peter Szolovits. 2017. NeuroNER: an easy-to-use program for named-entity recognition based on neural networks. In *Proceedings of the 2017 Conference on Empirical Methods in Natural Language Processing: System Demonstrations*, pages 97–102, Copenhagen, Denmark. Association for Computational Linguistics.

Jacob Devlin, Ming-Wei Chang, Kenton Lee, and Kristina Toutanova. 2019. Bert: Pre-training of deep bidirectional transformers for language understanding. pages 4171–4186.

Ştefan Dumitrescu, Andrei-Marius Avram, and Sampo Pyysalo. 2020. The birth of Romanian BERT. In *Proceedings of the 2020 Conference on Empirical Methods in Natural Language Processing: Findings*, pages 4324–4328.

Radu Ion. 2007. *Word sense disambiguation methods applied to English and Romanian*. Ph.D. thesis, PhD thesis (in Romanian). Romanian Academy, Bucharest.

Jinhyuk Lee, Wonjin Yoon, Sungdong Kim, Donghyeon Kim, Sunkyu Kim, Chan Ho So, and Jaewoo Kang. 2020. Biobert: a pre-trained biomedical language representation model for biomedical text mining. *Bioinformatics*, 36(4):1234–1240.

Patrick Lewis, Myle Ott, Jingfei Du, and Veselin Stoyanov. 2020. Pretrained language models for biomedical and clinical tasks: Understanding and extending the state-of-the-art. In *Proceedings of the 3rd Clinical Natural Language Processing Workshop*, pages 146–157, Online. Association for Computational Linguistics.

Mihai Masala, Radu Cristian Alexandru Iacob, Ana Sabina Uban, Marina Cidota, Horia Velicu, Traian Rebedea, and Marius Popescu. 2021. jurBERT: A Romanian BERT model for legal judgement prediction. In *Proceedings of the Natural Legal Language Processing Workshop 2021*, pages 86–94, Punta Cana, Dominican Republic. Association for Computational Linguistics.

Mihai Masala, Stefan Ruseti, and Mihai Dascalu. 2020. RoBERT – a Romanian BERT model. In *Proceedings of the 28th International Conference on Computational Linguistics*, pages 6626–6637, Barcelona, Spain (Online). International Committee on Computational Linguistics.

Verginica Barbu Mititelu and Maria Mitrofan. 2020. The romanian medical treebank-simonero. *ISSN 1843-911X*, page 7.

Maria Mitrofan. 2017. Bootstrapping a romanian corpus for medical named entity recognition. In *RANLP*, pages 501–509.

Maria Mitrofan. 2019. *Extragere de cunostinte din texte în limba româna si date structurate cu aplicatii în domeniul medical*. Ph.D. thesis, Ph. D. thesis, Romanian Academy.

Maria Mitrofan, Verginica Barbu Mititelu, and Grigorina Mitrofan. 2019. Monero: a biomedical gold standard corpus for the romanian language. In *Proceedings of the 18th BioNLP Workshop and Shared Task*, pages 71–79.

Maria Mitrofan and Dan Tufiş. 2018. Bioro: The biomedical corpus for the romanian language. In *Proceedings of the Eleventh International Conference on Language Resources and Evaluation (LREC 2018)*, pages 1192–1196.

David Nadeau and Satoshi Sekine. 2007. A survey of named entity recognition and classification. *Lingvisticae Investigationes*, 30(1):3–26.

Emre O. Neftci, Hesham Mostafa, and Friedemann Zenke. 2019. Surrogate gradient learning in spiking neural networks: Bringing the power of gradient-based optimization to spiking neural networks. *IEEE Signal Processing Magazine*, 36(6):51–63.

Mark Neumann, Daniel King, Iz Beltagy, and Waleed Ammar. 2019. Scispacy: fast and robust models for biomedical natural language processing. *arXiv preprint arXiv:1902.07669*.

Vasile Păiş, Radu Ion, and Dan Tufiş. 2020. A processing platform relating data and tools for Romanian language. In *Proceedings of the 1st International Workshop on Language Technology Platforms*, pages 81–88, Marseille, France. European Language Resources Association.

Vasile Păiş, Maria Mitrofan, Carol Luca Gasan, Vlad Coneschi, and Alexandru Ianov. 2021. Named entity recognition in the Romanian legal domain. In *Proceedings of the Natural Legal Language Processing Workshop 2021*, pages 9–18, Punta Cana, Dominican Republic. Association for Computational Linguistics.

Vasile Păiş and Maria Mitrofan. 2021a. Assessing multiple word embeddings for named entity recognition of professions and occupations in health-related social media. In *Proceedings of the Sixth Social Media Mining for Health (#SMM4H) Workshop and Shared Task*, pages 128–130, Mexico City, Mexico. Association for Computational Linguistics.

Vasile Păiş and Maria Mitrofan. 2021b. Towards a named entity recognition system in the romanian legal domain using a linked open data corpus. In *Workshop on Deep Learning and Neural Approaches for Linguistic Data*, pages 16–17, Skopje, North Macedonia.

Vasile Păiș and Dan Tufiș. 2018. Computing distributed representations of words using the CoRoLa corpus. *Proceedings of the Romanian Academy Series A - Mathematics Physics Technical Sciences Information Science*, 19(2):185–191.

Erik F Sang and Jorn Veenstra. 1999. Representing text chunks. *arXiv preprint cs/9907006*.

Leon Weber, Jannes Münchmeyer, Tim Rocktäschel, Maryam Habibi, and Ulf Leser. 2020. Huner: improving biomedical ner with pretraining. *Bioinformatics*, 36(1):295–302.

Leon Weber, Mario Sänger, Jannes Münchmeyer, Maryam Habibi, Ulf Leser, and Alan Akbik. 2021. Hunflair: an easy-to-use tool for state-of-the-art biomedical named entity recognition. *Bioinformatics*, 37(17):2792–2794.

Timo C. Wunderlich and Christian Pehle. 2021. Event-based backpropagation can compute exact gradients for spiking neural networks. *Scientific Reports*, 11(1):12829.

Ying Xiong, Yuanhang Huang, Qingcai Chen, Xiaolong Wang, Yuan Nic, and Buzhou Tang. 2020. A joint model for medical named entity recognition and normalization. *Proceedings http://ceur-ws. org ISSN*, 1613:0073.

BanglaBioMed: A Biomedical Named-Entity Annotated Corpus for Bangla (Bengali)

Salim Sazzed
Old Dominion University
Norfolk, VA, USA
ssazz001@odu.edu

Abstract

Recognizing biomedical entities in the text has significance in biomedical and health science research, as it benefits myriad downstream tasks, including entity linking, relation extraction, or entity resolution. While English and a few other widely used languages enjoy ample resources for automatic biomedical entity recognition, it is not the case for Bangla, a low-resource language. On that account, in this paper, we introduce BanglaBioMed, a Bangla biomedical named entity (NE) annotated dataset in standard IOB format, the first of its kind, consisting of over 12000 tokens annotated with the biomedical entities. The corpus is created by collecting Bangla text from a list of health articles and then annotated with four distinct types of entities: Anatomy (AN), Chemical and Drugs (CD), Disease and Symptom (DS), and Medical Procedure (MP). We provide the details of the entire data collection and annotation procedure and illustrate various statistics of the created corpus. Our developed corpus is a much-needed addition to the Bangla NLP resource that will facilitate biomedical NLP research in Bangla.

1 Introduction

The named-entity recognition (NER) frameworks aim to identify named entities (NE) mentioned in unstructured text documents and then categorize them into predefined domain-specific classes such as person or organization names, medical codes, chemical compounds, and food ingredients. Diverse sets of named entity (NE) annotated datasets have been created by researchers in varied domains such as clinical domain (Doğan et al., 2014), food domain (Stojanov et al., 2021), astronomy (Murphy et al., 2006), biological domain (Hastings et al., 2016). Due to the essence of NE for understanding biomedical concepts such as diseases, chemicals, and proteins, a number of studies focused on building NER corpora for the biomedical domain as it can aid researchers in finding relevant concepts and speed up the process of biomedical scientific discovery.

In English and a few other major languages, various NE corpora representing biomedical entities are publicly available (Kim et al., 2003; Kolárik et al., 2008). However, in Bangla, such a biomedical NE annotated dataset does not exist as NLP research in Bangla is still in infancy except in a few areas such as sentiment analysis (Bodini, 2022; Sazzed and Jayarathna, 2019; Faruque et al., 2021; Sazzed, 2020a; Bhowmick and Jana; Sazzed, 2020b), hate and abusive language detection (Karim et al., 2021; Sazzed, 2021a; Ishmam and Sharmin, 2019; Sazzed, 2021b). With the growing popularity of telemedicine and the availability of health and medical-related data written in Bangla, developing a Biomedical Named Entity Recognition (NER) system in Bangla is a pressing necessity.

For developing a sophisticated NER system, it is essential to have at least a moderate amount of annotated data. In particular, the generalizability and performances of the machine learning approaches (especially the deep learning-based models) heavily rely on the quantity of available annotated data. Hence, in this study, we introduce a biomedical NE dataset, the first of its kind, for the low-resource Bangla. The dataset is created by retrieving biomedical and health-related textual content from a number of health articles. The text data are then tokenized and annotated with four types of entities: Anatomy (AN), Chemical and Drugs (CD), Disease and Symptoms (DS), and Medical Procedure (MP). The final corpus contains around 2000 tokens representing one of the four types of biomedical NE mentioned above and around 10000 non-entity tokens.

1.1 Contributions

The main contributions of this study are:

Proceedings of the BioNLP 2022 workshop, Dublin, Ireland, pages 323–329
May 26, 2022.

- To address the lack of annotated data in the Bangla biomedical and health domain, we collect a biomedical corpus, BanglaBio, consisting of around 12000 tokens (i.e., primarily words).
- We manually annotate the corpus in token-level (mainly words) in four different classes of entities, Anatomy (AN), Diseases and Symptoms (DS), Chemical and Drug (CD), and Medical Procedure (MP).
- We provide the statistics of the frequency and structures of various types of entities present in the corpus and make the corpus publicly available for researchers [1].

2 Related Work

Although English and some other languages standardized entity annotated (i.e., IOB format) Biomedical corpora are available for the NER task, to the best of our knowledge, such resources do not exist in Bangla.

2.1 English Biomedical corpus

In English, a number of biomedical corpora exists with various types of entity annotations such as GENIA corpus (Kim et al., 2003), GENETAG corpus (Tanabe et al., 2005), SCAI IUPAC corpus (Kolárik et al., 2008), CellFinder corpus (Neves et al., 2012).

Pyysalo et al. (2007) presented BioInfer (Bio Information Extraction Resource), an annotated corpus of biomedical text consisting of 1100 sentences collected from abstracts of biomedical research articles. Kim et al. (2008) introduced single-facet annotation and semantic typing to the existing annotations in the GENIA corpus. The new annotation was performed on half of the GENIA corpus, consisting of 1,000 Medline abstracts. Giorgi and Bader (2020) introduced biomedical named entity recognition (BioNER) system for biomedical information extraction. To improve the generalizing ability of BioNER, the authors proposed an improved regularization technique using variational dropout, transfer learning, and multi-task learning.

Karimi et al. (2015) created CSIRO Adverse Drug Event Corpus (CADEC) consisting of patient-reported Adverse Drug Events (ADEs) collected from various medical forum posts. The authors performed multi-stage annotations for entities such as drugs, adverse effects, symptoms, and diseases. Scepanovic et al. (2020) proposed several approaches to accurately extract a wide variety of medical entities such as symptoms, diseases, and drug names collected from varied social media sources, and validated this approach on a large-scale Reddit dataset.

2.2 Non-English Biomedical corpus

For the French language, the Unified Medical Lexicon for French (UMLF) has been created by Zweigenbaum et al. (2005). For Swedish, an annotated gold standard corpus of medical records was developed by Velupillai (2012). Mowery et al. (2012) proposed a clinical uncertainty and negation taxonomy and mapped an English annotation schema to a Swedish schema.

Mitrofan and Tufiş (2018) presented a biomedical corpus in the Romanian language, which was collected in the contexts of the CoRoLa project, the reference corpus for the contemporary Romanian language. The authors described various statistics about the corpus and data-composition and annotation procedures. Carrino et al. (2021) introduced CoWeSe (the Corpus Web Salud Español), the largest Spanish biomedical corpus to date, consisting of around 750M tokens of clean plain text. The CoWeSe was created by crawling over 3000 Spanish documents.

Sun and Yang (2019) employed two language models, Multilingual BERT and BioBERT, to identify chemical and protein entities from the Spanish biomedical NER corpus PharmaCoNER (Gonzalez-Agirre et al., 2019). The author showed that transferring knowledge learned from large-scale source datasets to the target domain offers an effective solution for the PharmaCoNER task.

3 Creation of BanglaBioMed

3.1 Data Collection and Pre-processing

Unlike English, where a large number of scientific publications are available for extracting biomedical named entities, in Bangla, such resources do not exist, as researchers hardly publish scientific articles in Bangla. Hence, we use alternative sources for extracting biomedical text data. We leverage a set of health articles authored by medical physicians and published in the most popular

[1] `https://github.com/sazzadcsedu/BanglaBioMed.git`

Structure of Entity	Entity	Sentence
Simple Entity (single and multi-word)	জ্বর, কাশি, গলাব্যথা, শ্বাসকষ্ট	জ্বর, কাশি, গলাব্যথা, শ্বাসকষ্ট হচ্ছে অমিক্রনের মূল উপসর্গ।
	Fever, cough, sore throat, shortness of breath (*English Translation*)	Fever, cough, sore throat, shortness of breath are the main symptoms of Omicron. (*English Translation*)
Complex Entity (Overlapping)	নাক দিয়ে রক্ত আসা, নাক দিয়ে পানি আসা	নাক দিয়ে রক্ত বা পানি আসা
	Blood coming through the nose, Water coming through the nose (*English Translation*)	Blood or water coming through the nose (*English Translation*)

Figure 1: Examples of entities representing varied structures

Bangladeshi daily newspaper, *Prothom Alo* [2]. The health-related articles are chosen from the newspaper's official website. All the text data of the articles are manually excerpted for annotation. The excerpted texts are then segmented into sentences based on the '|' delimiter, which is equivalent to the English 'full stop(.)' delimiter. Afterward, each sentence is tokenized into words and punctuations.

3.2 Entity Types

Similar to Mitrofan (2017), the following four types of entities are considered in the annotation process.

- **Anatomy (AN)**: This entity label portrays the structure of the human body, especially as revealed by dissection and the separation of parts. This type of entity is common in health and medical text. Some examples include-

  মাথা (Head), হাত (Hand), পা (Leg), কোমর (Waist)

- **Chemicals and Drugs (CD)**: This entity label indicates the presence of chemical and drug-related terms in the tokens. Some examples are- ইনসুলিন (insulin), ফলিক অ্যাসিড (Folic Acid), ভিটামিন সি (Vitamin C), হাইড্রোকুইনোন (Hydroquinone)

- **Disease and Symptom (DS)**: This entity category includes names and descriptions of various diseases and symptoms (i.e., features appearing to the patients as conditions of the diseases). The following entities are some of the examples of this category- ক্যানসার (Cancer), হাঁপানি, প্রেশার, (Pressure) শ্বাসকষ্ট, (Shortness of breath) স্থূলতা , (Obesity)

- **Medical Procedures (MP)**: The entity of this group indicates laboratory procedures, the therapeutic or preventive procedures used for medical treatment. The followings are some examples- অস্থিমজ্জা প্রতিস্থাপন, (Bone marrow transplantation) কলোনস্কোপি (Colonoscopy), ব্লাড ট্রান্সফিউশন (Blood transfusion).

3.3 Entity Annotation Guidelines

We perform the entity annotation at the sentence level. Duplicate entities within a sentence or the corpus are annotated independently (all the occurrences of the same entity are labeled). We observe that most entities constitute single or multiple words without intervening with other entities (i.e., simple entities). Nevertheless, there exist entities that partially overlap with another; these types of entities can be referred to as complex entities (Examples shown in Figure 1).

Besides, we find that some entities are entirely embedded (nested) within another entity. Especially, the entities from the Anatomy (AN) class often are embedded into the Disease and Symptom (DS) category. To give an example, the DS entity *back pain* contains *back* entity from AN class. For this type of overlapping scenario, the longer entity is considered as the "top-level" entity, while its sub-part(s) is deemed as the "nested" entity. Most of the well-known NE annotated corpora employed the non-nested approach, where the words are annotated based on the top-level entity (Sang and De Meulder, 2003).

We do not consider co-referential or anaphoric references to the entity during annotation. The intensifier (e.g., slightly/severe) or possessive adjectives are not included in the entity to keep the annotation consistent across the corpus. The annotation is performed by an annotator who possesses a university-level education. The annotated label is

[2] https://www.prothomalo.com/life/health

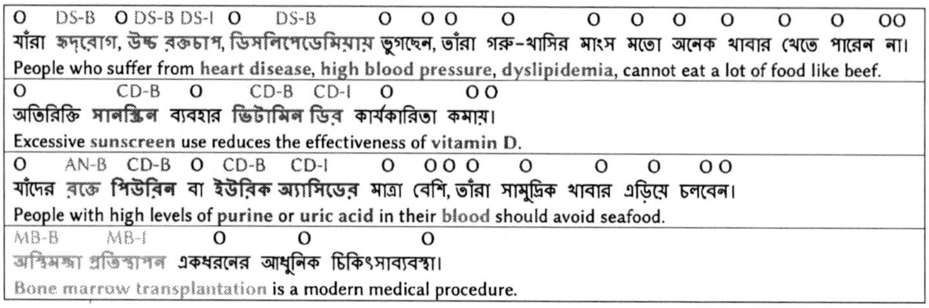

Figure 2: Examples of entity annotation within the sentences

further verified by a medical professional.

3.4 Entity Tagging

To make the annotated corpus suitable for the automatic NER task, we follow the standard IOB2 format (Tjong Kim Sang and Veenstra, 1999). The IOB2 format is described below,

B: The term 'B' indicates the beginning of a particular type of entity (i.e., the first token of an entity)

I: 'I' represents a token is a part of an already initiated entity,

O: 'O' indicates a token is not part of any entity of interest. All tokens outside the entity of interest are labeled as O.

3.5 Corpus Statistics

Table 1: The length distributions of unique entities of various types (in words)

Entity Type	Entity Length (# words)	Total
	1 / 2 / 3 / >= 4	
AN	105 /16 /0 /0	121
DS	190 /167 /64 /52	473
CD	79 /28 /1 /0	108
MP	41 /29 /3 /0	73

Table 1 shows the word length distributions of various types of entities. We find most of the entities contain a single word, while some comprising of two words. For example, the AN group contains close to 90% entities having a single word. The lengthy entities of over two words primarily belong to the DS category.

As shown in Table 2, the corpus has an unbalanced distribution regarding various types of entities. The most dominant entity type is DS,

Table 2: Statistics of various metrics in the annotated corpus

Metric	Count
#Tokens	11196
#Sentences	818
#Words with entity tag	1968
#Non-entity Words	9228
Average sentence length	13.68
Average number of entity per sentence	1.62
Entity Tag	**Count**
AN-B	259
AN-I	16
DS-B	699
DS-I	510
CD-B	102
CD-I	45
MP-B	269
MP-I	68
Total	1968

which is expected since these source articles contain more information related to various diseases and related symptoms. Among the 2000 biomedical entity annotated tokens present in the corpus, around 60% represent the DS category. The lowest presence is observed for the entities belonging to the CD category.

4 Summary and conclusion

In this study, we introduce a Bangla biomedical named entity annotated corpus created from a number of Bangla health articles. To the best of our knowledge, this is the first biomedical NE annotated corpus in Bangla (Bengali) in standard IOB format created for biomedical text mining. We report detailed annotation guidelines and procedures of the annotation. Moreover, we provide the various statistics of four different types of biomedical entities: AN, DS, CD, and MP, in the annotated corpus. We have made the corpus publicly available for the researchers. The future work will focus on enhancing the size of the annotated corpus and creating strong baselines for automatic NER tasks by incorporating transformer-based language models. Besides, we will investigate how to leverage cross-lingual resources from other languages, such as English, to improve the performance of the NER task.

References

Anirban Bhowmick and Abhik Jana. Sentiment analysis for bengali using transformer based models.

Matteo Bodini. 2022. Opinion mining from machine translated bangla reviews with stacked contractive auto-encoders. *Journal of Ambient Intelligence and Humanized Computing*, pages 1–13.

Casimiro Pio Carrino, Jordi Armengol-Estapé, Ona de Gibert Bonet, Asier Gutiérrez-Fandiño, Aitor Gonzalez-Agirre, Martin Krallinger, and Marta Villegas. 2021. Spanish biomedical crawled corpus: A large, diverse dataset for spanish biomedical language models. *arXiv preprint arXiv:2109.07765*.

Rezarta Islamaj Doğan, Robert Leaman, and Zhiyong Lu. 2014. Ncbi disease corpus: a resource for disease name recognition and concept normalization. *Journal of biomedical informatics*, 47:1–10.

M Abdullah Faruque, Saifur Rahman, Partha Chakraborty, Tanupriya Choudhury, Jung-Sup Um, and Thipendra Pal Singh. 2021. Ascertaining polarity of public opinions on bangladesh cricket using machine learning techniques. *Spatial Information Research*, pages 1–8.

John M Giorgi and Gary D Bader. 2020. Towards reliable named entity recognition in the biomedical domain. *Bioinformatics*, 36(1):280–286.

Aitor Gonzalez-Agirre, Montserrat Marimon, Ander Intxaurrondo, Obdulia Rabal, Marta Villegas, and Martin Krallinger. 2019. Pharmaconer: Pharmacological substances, compounds and proteins named entity recognition track. In *Proceedings of The 5th Workshop on BioNLP Open Shared Tasks*, pages 1–10.

Janna Hastings, Gareth Owen, Adriano Dekker, Marcus Ennis, Namrata Kale, Venkatesh Muthukrishnan, Steve Turner, Neil Swainston, Pedro Mendes, and Christoph Steinbeck. 2016. Chebi in 2016: Improved services and an expanding collection of metabolites. *Nucleic acids research*, 44(D1):D1214–D1219.

Alvi Md Ishmam and Sadia Sharmin. 2019. Hateful speech detection in public facebook pages for the bengali language. In *2019 18th IEEE International Conference On Machine Learning And Applications (ICMLA)*, pages 555–560. IEEE.

Md Rezaul Karim, Sumon Kanti Dey, Tanhim Islam, Sagor Sarker, Mehadi Hasan Menon, Kabir Hossain, Md Azam Hossain, and Stefan Decker. 2021. Deephateexplainer: Explainable hate speech detection in under-resourced bengali language. In *2021 IEEE 8th International Conference on Data Science and Advanced Analytics (DSAA)*, pages 1–10. IEEE.

Sarvnaz Karimi, Alejandro Metke-Jimenez, Madonna Kemp, and Chen Wang. 2015. Cadec: A corpus of adverse drug event annotations. *Journal of biomedical informatics*, 55:73–81.

J-D Kim, Tomoko Ohta, Yuka Tateisi, and Junichi Tsujii. 2003. Genia corpusa semantically annotated corpus for bio-textmining. *Bioinformatics*, 19(suppl_1):i180–i182.

Jin-Dong Kim, Tomoko Ohta, and Jun'ichi Tsujii. 2008. Corpus annotation for mining biomedical events from literature. *BMC bioinformatics*, 9(1):1–25.

Corinna Kolárik, Roman Klinger, Christoph M Friedrich, Martin Hofmann-Apitius, and Juliane Fluck. 2008. Chemical names: terminological resources and corpora annotation. In *Workshop on Building and evaluating resources for biomedical text mining (6th edition of the Language Resources and Evaluation Conference)*.

Maria Mitrofan. 2017. Bootstrapping a romanian corpus for medical named entity recognition. In *RANLP*, pages 501–509.

Maria Mitrofan and Dan Tufiş. 2018. Bioro: The biomedical corpus for the romanian language. In *Proceedings of the Eleventh International Conference on Language Resources and Evaluation (LREC 2018)*.

Danielle L Mowery, Sumithra Velupillai, and Wendy Chapman. 2012. Medical diagnosis lost in translation–analysis of uncertainty and negation expressions in english and swedish clinical texts. In *BioNLP: Proceedings of the 2012 Workshop on Biomedical Natural Language Processing*, pages 56–64.

Tara Murphy, Tara McIntosh, and James R Curran. 2006. Named entity recognition for astronomy literature. In *Proceedings of the Australasian Language Technology Workshop 2006*, pages 59–66.

Mariana Neves, Alexander Damaschun, Andreas Kurtz, and Ulf Leser. 2012. Annotating and evaluating text for stem cell research. In *Proceedings of the Third Workshop on Building and Evaluation Resources for Biomedical Text Mining (BioTxtM 2012) at Language Resources and Evaluation (LREC). Istanbul, Turkey*, pages 16–23. Citeseer.

Sampo Pyysalo, Filip Ginter, Juho Heimonen, Jari Björne, Jorma Boberg, Jouni Järvinen, and Tapio Salakoski. 2007. Bioinfer: a corpus for information extraction in the biomedical domain. *BMC bioinformatics*, 8(1):1–24.

Erik F Sang and Fien De Meulder. 2003. Introduction to the conll-2003 shared task: Language-independent named entity recognition. *arXiv preprint cs/0306050*.

Salim Sazzed. 2020a. Cross-lingual sentiment classification in low-resource bengali language. In *Proceedings of the sixth workshop on noisy user-generated text (W-NUT 2020)*, pages 50–60.

Salim Sazzed. 2020b. Development of sentiment lexicon in bengali utilizing corpus and cross-lingual resources. In *2020 IEEE 21st International conference on information reuse and integration for data science (IRI)*, pages 237–244. IEEE.

Salim Sazzed. 2021a. Abusive content detection in transliterated bengali-english social media corpus. In *Proceedings of the Fifth Workshop on Computational Approaches to Linguistic Code-Switching*, pages 125–130.

Salim Sazzed. 2021b. Identifying vulgarity in bengali social media textual content. *PeerJ Computer Science*, 7:e665.

Salim Sazzed and Sampath Jayarathna. 2019. A sentiment classification in bengali and machine translated english corpus. In *2019 IEEE 20th international conference on information reuse and integration for data science (IRI)*, pages 107–114. IEEE.

Sanja Scepanovic, Enrique Martin-Lopez, Daniele Quercia, and Khan Baykaner. 2020. Extracting medical entities from social media. In *Proceedings of the ACM Conference on Health, Inference, and Learning*, pages 170–181.

Riste Stojanov, Gorjan Popovski, Gjorgjina Cenikj, Barbara Koroušić Seljak, Tome Eftimov, et al. 2021. A fine-tuned bidirectional encoder representations from transformers model for food named-entity recognition: Algorithm development and validation. *Journal of Medical Internet Research*, 23(8):e28229.

Cong Sun and Zhihao Yang. 2019. Transfer learning in biomedical named entity recognition: an evaluation of bert in the pharmaconer task. In *Proceedings of The 5th Workshop on BioNLP Open Shared Tasks*, pages 100–104.

Lorraine Tanabe, Natalie Xie, Lynne H Thom, Wayne Matten, and W John Wilbur. 2005. Genetag: a tagged corpus for gene/protein named entity recognition. *BMC bioinformatics*, 6(1):1–7.

Erik F. Tjong Kim Sang and Jorn Veenstra. 1999. Representing text chunks. In *Ninth Conference of the European Chapter of the Association for Computational Linguistics*, pages 173–179, Bergen, Norway. Association for Computational Linguistics.

Sumithra Velupillai. 2012. *Shades of certainty: annotation and classification of swedish medical records*. Ph.D. thesis, Department of Computer and Systems Sciences, Stockholm University.

Pierre Zweigenbaum, Robert Baud, Anita Burgun, Fiammetta Namer, Éric Jarrousse, Natalia Grabar, Patrick Ruch, Franck Le Duff, Jean-François Forget, Magaly Douyere, et al. 2005. Umlf: a unified medical lexicon for french. *International Journal of Medical Informatics*, 74(2-4):119–124.

ICDBigBird: A Contextual Embedding Model for ICD Code Classification

George Michalopoulos[1], **Michal Malyska**[2,3], **Nicola Sahar**[3]
Alexander Wong[1], **Helen Chen**[1]
University of Waterloo[1], University of Toronto[2], Semantic Health[3]
{gmichalo, alexander.wong, helen.chen}@uwaterloo.ca
michal.malyska@mail.utoronto.ca
nick@semantichealth.ai

Abstract

The International Classification of Diseases (ICD) system is the international standard for classifying diseases and procedures during a healthcare encounter and is widely used for healthcare reporting and management purposes. Assigning correct codes for clinical procedures is important for clinical, operational and financial decision-making in healthcare.

Contextual word embedding models have achieved state-of-the-art results in multiple NLP tasks. However, these models have yet to achieve state-of-the-art results in the ICD classification task since one of their main disadvantages is that they can only process documents that contain a small number of tokens which is rarely the case with real patient notes. In this paper, we introduce ICDBigBird a BigBird-based model which can integrate a Graph Convolutional Network (GCN), that takes advantage of the relations between ICD codes in order to create 'enriched' representations of their embeddings, with a BigBird contextual model that can process larger documents. Our experiments on a real-world clinical dataset demonstrate the effectiveness of our BigBird-based model on the ICD classification task as it outperforms the previous state-of-the-art models.

1 Introduction

Real-world data in healthcare refers to patient data routinely collected during clinic encounters such as visits and hospitalization. After each clinical visit, a set of codes representing diagnostic and procedural information are submitted to various regulatory agencies (Farkas and Szarvas, 2008). The International Classification of Diseases (ICD) system is the most widely used coding system, maintained by the World Health Organization (Avati et al., 2018). Assigning the most appropriate codes is an important task in healthcare since erroneous ICD codes could seriously affect the organization's ability to accurately measure the patient outcome (Ji et al., 2020).

Contextual word embedding models (such as ELMo (Peters et al., 2018) and BERT (Devlin et al., 2019)) have achieved state-of-art results in many NLP tasks. However, recent attempts of using contextual models on the ICD classification task have failed to achieve state-of-the-art results (Zhang et al., 2020) mainly due to the fact that they can only process documents that contain a small number of tokens. Advances such as the BigBird model (Zaheer et al., 2020) allows contextual models to process long documents, thus reducing the risk of losing information from the original texts.

In this paper, we present a novel model for the ICD classification task. Specifically: (i) we are the first, to the best of our knowledge, to propose the combined usage of a Graph Convolutional Network (based on the normalized point-wise mutual information) and a contextual embedding model for the ICD classification task; (ii) we introduce a novel attention layer on top of a BigBird model which has the ability to process long documents; and (iii) our experiments on a real-world clinical dataset demonstrate the effectiveness of our ICD-BigBird model on the ICD classification task as it outperforms previous state-of-the-art models.

2 Proposed ICDBigBird Model

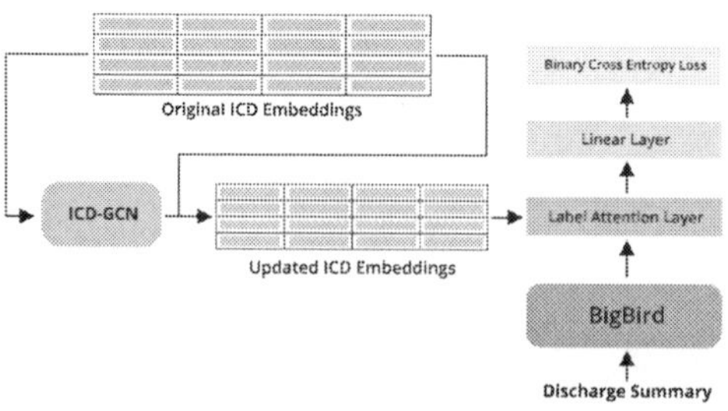

Figure 1: ICDBigBird model architecture

Proceedings of the BioNLP 2022 workshop, Dublin, Ireland, pages 330–336
May 26, 2022. ©2022 Association for Computational Linguistics

2.1 ICD Graph Convolutional Network

A Graph Convolutional Network (GCN) (Kipf and Welling, 2017) is a neural network architecture that can capture the general knowledge about the connections between entities. Specifically, GCN builds a symmetric adjacency matrix based on a predefined relationship graph, and the representation of each node is calculated according to its neighbours.

We use a GCN to capture a more 'enriched' representation for each of the ICD codes. In order to use the ICD-GCN, we first construct the adjacency matrix $A \in \mathbb{R}^{n \times n}$ (where n is the number of unique ICD codes) to represent the connections of ICD codes by using the normalized point-wise mutual information (NPMI) (Lu et al., 2020):

$$NPMI(i,j) = -\frac{1}{\log p(i,j)} \log \frac{p(i,j)}{p(i)p(j)} \quad (1)$$

where i and j are different ICD codes and $p(i,j) = \frac{N(i,j)}{N}$, $p(j) = \frac{N(j)}{N}$ and $N(i,j)$ is the number of documents that are labeled with both i and j codes, $N(i)$ is the number of documents that are labeled with the i code and N is the total number of documents of the training set that our model was trained on. We create an edge between two codes if their NPMI value is greater than a threshold. We empirically set the threshold to 0.2 by experimenting with different threshold values.

It should be noted that we decided to create the adjacency matrix of the ICD-GCN by taking advantage of the NPMI values instead of considering the hierarchical associations of the ICD codes because we mainly focused on the task of classifying the top 50 most frequent ICD codes (Shi et al., 2017), where we found that there exists little to no hierarchical connection between these codes.

We then construct a definition (sentence) embedding matrix for all the ICD codes using their ICD-9 (sentence) definitions from the MIMIC III dataset (Johnson et al., 2016) and the pre-trained sentence transformer embedding model in (Reimers and Gurevych, 2019), which has been shown to outperform other state-of-the-art sentence embedding methods.

An updated representation of all ICD codes from the ICD-GCN is calculated as follows:

$$\hat{U} = Relu(\hat{A}XW) \quad (2)$$

where $X \in \mathbb{R}^{n \times m}$ is the definition embedding matrix, n is the number of ICD codes, m is the size of the sentence-definition embedding of each ICD code, $W \in \mathbb{R}^{m \times h}$ is the weight matrix, h is the BigBird's hidden dimension and $\hat{A} = D^{-\frac{1}{2}}AD^{-\frac{1}{2}}$ is the normalized symmetric adjacency matrix where $D_{ii} = \sum_j A_{ij}$.

Finally, we concatenate the output of the ICD-GCN with the initial embeddings of the ICD codes in order to get a richer representation of the codes (Rios and Kavuluru, 2018):

$$U = \hat{U} \parallel X, U \in \mathbb{R}^{n \times (m+h)} \quad (3)$$

2.2 ICDBigBird Model

Assume a discharge summary has n words, the model's tokenizer generates tokens for each word in the document. Afterwards the tokens are passed through multiple attention-based layers and the model produces the final contextual representation of the document $H \in \mathbb{R}^{t \times h}$ where $t = 4096$ is the number of tokens and h is the BigBird's hidden dimension. We use a fully connected linear layer for the creation of $\hat{H}$ which is the final embedding representation of the BigBird's embeddings:

$$\hat{H} = Relu(HW_1) \quad (4)$$

where $\hat{H} \in \mathbb{R}^{t \times (m+h)}$ and $W_1 \in \mathbb{R}^{h \times (m+h)}$. Afterwards, we apply a per-label attention mechanism, in order to showcase the most relevant information to the ICD codes in the contextual representation of each document. Formally, using $U \in \mathbb{R}^{n \times (m+h)}$ which is the 'updated' ICD code sentence-definition embedding matrix, we can compute the attention as:

$$A = SoftMax(U\hat{H}^{\top}) \quad (5)$$

where $A \in \mathbb{R}^{n \times t}$. After the calculation of the attention score, the output of the attention layer is calculated as:

$$V = A\hat{H} \quad (6)$$

where $V \in \mathbb{R}^{n \times (m+h)}$. Given the 'updated' representation V, we can compute a probability for each label l by using a pooling operation and a sigmoid transformation over the linear projection of V:

$$\hat{y} = \sigma(pooling(V \circ W)) \quad (7)$$

where $W \in \mathbb{R}^{n \times (m+h)}$. As the ICD task is a multi label scenario, the loss function that is typically used is a multi-label binary cross entropy loss:

Model	AUC-ROC Macro	AUC-ROC Micro	F1 Macro	F1 Micro	$P@5$
DRC.(Mullenbach et al., 2018)	88.4	91.6	57.6	63.3	61.8
LEAM (Wang et al., 2018)	88.1	91.2	54.0	61.9	61.2
HyperCore (Cao et al., 2020)	89.5±0.3	92.9± 0.2	60.9 ± 0.1	66.3 ± 0.1	63.2 ± 0.2
Mult.CNN (Li and Yu, 2020)	89.9± 0.4	92.8 ± 0.2	60.6±1.1	67.0±0.3	64.1±0.1
DCAN (Ji et al., 2020)	**90.2±0.6**	**93.1±0.1**	61.5±0.7	67.1±0.1	64.2±0.2
ICDBigBird	90.0±0.5	92.9 ±0.2	**63.1±0.5**	**69.6±0.1**	**65.4±0.1**
ICDBigBird (validation split)	91.0±0.6	93.3 ±0.1	64.1±0.4	70.4±0.1	65.1±0.3
Ablation Study					
BERT(Devlin et al., 2019)	80.3±0.4	84.4 ±0.5	43.7±0.2	51.4±0.5	51.9±0.3
BioBERT(Lee et al., 2019)	81.3±0.5	85.5 ±0.4	46.3±0.3	54.6±0.3	54.2 ±0.4
Bio_C.(Alsentzer et al., 2019)	81.7±0.4	85.8 ±0.5	46.4±0.3	54.3 ±0.4	53.2±0.4
BigBird (512 tokens)	80.4±0.2	83.9 ±0.3	44.9±0.5	52.1±0.3	51.2±0.4
BigBird (without attention)	86.7±0.5	90.4 ±0.3	55.2±0.4	64.8±0.2	62.5±0.3
Linear Attention	88.4±0.5	91.2 ±0.2	60.2±0.2	67.8±0.3	63.6±0.5
R. embedding	89.2±0.4	91.8 ±0.5	60.8±0.2	67.8±0.2	63.2±0.1

Table 1: Results of mean ± standard deviation of three runs of the ICDBigBird model on the test split of the MIMIC-III dataset for the top 50 most frequent ICD codes; We also provide the performance of previous state-of-the-art models using the same test set; *Bio_C.* is Bio_ClinicalBERT; *DRC.* is DR-CAML; *R. embedding* is a model with a random initialization of the embeddings of the codes; we also include the results on the validation split of MIMIC III; Best values on the **test** set are **bolded**

$$L_{BCE}(y,\hat{y}) = \sum_{i=1}^{n}(y_i log(\hat{y}_i) + (1-y_i)log(1-\hat{y}_i)) \quad (8)$$

where y is the ground truth label and $\hat{y}$ are the ICD codes that our model predicted for each document. However, due to the extremely imbalance nature of the ICD codes we chose to adopt the Label-Distribution Aware Margin (LDAM) (Cao et al., 2019). In the LDAM loss function the output value is subtracted by a label-dependent margin Δ_i before the sigmoid function:

$$\hat{y}' = \sigma(pooling(V \circ W) - \mathbf{1}(y_i = 1)\Delta_i) \quad (9)$$

where $\mathbf{1}(.)$ outputs 1 if y_i=1 and $\Delta_i = \frac{C}{n_i^{1/4}}$ where n_i is number of instances of the i ICD code in the training data and C is constant. Thus we use the $L_{LDAM} = L_{BCE}(y, \hat{y}')$.

3 Experiments

3.1 Dataset

Following previous research work in the ICD classification task (Mullenbach et al., 2018; Ji et al., 2020; Li and Yu, 2020), we conducted our experiments on the subset of the English Multiparameter Intelligent Monitoring in Intensive Care III (MIMIC-III) dataset (Johnson et al., 2016) with the top 50 most frequent ICD codes (Shi et al., 2017). Our experiments on this dataset are consistent with its intended use, as it was created and shared for research purposes (as it stated in its license[1]). Finally, we manually checked the dataset to investigate the existence of information that uniquely identifies individual people and offensive content, however, we did not find any indication of either of them. We extract the free-text discharge summaries and clinical notes, containing the 50 most frequent ICD codes, from the MIMIC III dataset and we concatenate the discharge summaries and notes from the same hospitalization admission into one single document. We use the training/validation/testing split from (Mullenbach et al., 2018; Li and Yu, 2020) for a fair comparison. The document set size of our subset of MIMIC-III is 8066 for training, 1573 for validation and 1729 for testing respectively. Following the prepossessing procedures outlined in (Ji et al., 2020), the documents are tokenized and each token is converted to lowercase. Any token that contains no alphabetic characters is removed. Instead of truncating the documents to 2500 words, we set the token size limit to 4096 for our ICDBigBird model to take full advantage of the information that

[1]https://tinyurl.com/mimic-licence

can be extracted from each document as there are 1345 documents that contain more than 2500 words (with maximum, minimum and average length of 7567, 105, 1609 words respectively).

3.2 Experimental Setup

We provide the search strategy and the bound for each hyperparameter as follows: the batch size is set between 32 and 64, and the learning rate is chosen between the values 2e-5, 3e-5 and 5e-5. We set the number of training epochs between 25 and 30 epochs to allow for maximal performance. The best values are chosen based on micro-F1 scores[2] in the validation set. The final hyper-parameters selection of our ICDBigBird model is batch size 32, learning rate 2e-5, trained on 30 epochs and we empirically set the the C constant of the LDAM loss to 2. We also use the AdamW optimizer (Loshchilov and Hutter, 2019) to optimize the parameters of the model.

All the contextual embedding models are implemented using the transformers library (Wolf et al., 2020) on PyTorch 1.7.1. All experiments are executed on a Tesla K80 GPU with 64GB of system RAM on Ubuntu 18.04.5 LTS.

3.3 Results

We benchmark our ICDBigBird model against existing state-of-the-art models for the top 50 most frequent ICD classification task. For all models we evaluate the micro and macro averaging F1 score, the receiver operating characteristic curve (AUC-ROC) and the precision at k codes with k=5 (P@5). In Table 1, we can observe that our model outperforms all other models in the micro and macro averaging F1 and in the $P@5$ score with comparable performance on the other two metrics (with the DCAN model (Ji et al., 2020) achieving the best AUC-ROC results). Finally, our model contains 110565170 parameters with average running time of 893354 sec.

3.4 Ablation Study

In order to evaluate the effect of each feature on the performance of ICDBigBird, we conduct an ablation study. The results are presented in Table 1. (i) Firstly, we investigate whether the ability of the BigBird model to process large documents can boost the performance of our model. It can be observed that contextual model architectures that can process small documents of at most 512 tokens (Bert, Biobert, Bio_ClinicalBert) cannot achieve the performance of a BigBird architecture even if these models were pre-trained on medical documents (BioBert and Bio_ClinicalBert). (ii) Furthermore, we examine the performance of the BigBird model when we artificially limit the length of the documents to 512 tokens (BigBird 512 tokens) which is the maximum number of tokens that the BERT model can process. We observe that the performance improvement brought by the BigBird model is lost, making the performance of the BigBird model equivalent to the BERT model. This experiment demonstrates that one of the main reasons for the BigBird model outperforming the BERT model is the utilization of additional information in larger documents (4096 tokens) for the ICD automatic encoding task. (iii) In addition, we examined the effect of the GCN model by testing the performance of contextual embeddings without enriching them with information from the definitions of the codes through an attention mechanism (BigBird without attention) by having an ICD classifier on top of the [cls] token and by substituting the GCN attention mechanism with the typical linear attention mechanism (Linear Attention) (Mullenbach et al., 2018). It can be observed that our model benefits from the attention mechanism as without it, it cannot achieve optimal performance. Also, the fact that the GCN graph attention mechanism achieves a better performance than a typical linear attention mechanism is a strong indication that the connections between the ICD codes can provide valuable information. (iv) Finally, we investigated the effect of using the definitions of the codes to initialize their embeddings. In our experiments a model with a random initialization of the embeddings of the codes (R. embedding) achieved sub-optimal performance and thus we can conclude that using the codes' definitions to initialize their embeddings have a positive effect on the model's performance.

3.5 Discussion-Related Work

Recent development in NLP has introduced deep learning models that can achieve optimal performance on the ICD classification task. In (Shi et al., 2017), the authors introduced a new model that used word/character embeddings and recurrent neural networks (LSTM) to generate representations of the diagnosis descriptions and of the ICD codes. In addition, the authors in (Mullenbach et al., 2018)

[2] https://github.com/jamesmullenbach/caml-mimic

introduced an attention based convolutional neural network (CNN) model which incorporates an attention mechanism in order to identify the most relevant segments that contain medical information.

Furthermore, prior work has explored the use of GCNs for the ICD classification task (Rios and Kavuluru, 2018; Chalkidis et al., 2020) and our attention mechanism can be viewed as an extension of the structured attention mechanism of (Cao et al., 2020). However, some of the differences between the models are that: (i) Our work uses a normalized point-wise mutual information policy to create the edges, while the model in (Cao et al., 2020) used the co-appearing values to create a weighted graph. This is a key difference in the ICD coding problem as the method in (Cao et al., 2020) does not capture the relation between two highly correlated but 'unpopular' codes. (ii) In addition, the authors in (Cao et al., 2020) created the code embedding vectors by averaging the word embeddings of its descriptor, and our work uses pre-trained sentence embedding models which have achieved better performance. (iii) Finally, the model in (Cao et al., 2020) used a Convolution Neural Network (CNN) encoder while our work used a contextual (BigBird) model to produce document embeddings.

The results of the experiments indicate that these changes are important for the ICD classification task by demonstrating that a contextual model can achieve state-of-the-art results for this task.

4 Conclusion and Future Work

We present the ICDBigBird model, which is a novel contextual model for the ICD coding task. ICDBigBird has the ability to integrate a graph embedding model that takes advantage of the relations between ICD codes with a BigBird contextual model that can process larger documents. Experiments on the MIMIC III have shown that the ICDBigBird model outperforms previous state-of-the-art models. As for future work, we plan to address the limitations of this study including (i) testing ICDBigBird in other medical datasets to examine its generalizability, strengths and limitations, (ii) experimenting on the task of classifying the full ICD code set and (iii) examining the performance of the model in datasets of other languages (Almagro et al., 2020).

5 Acknowledgement

We acknowledge the generous support from Amazon Research Awards, MITACS Accelerate Grant (#IT19239), Semantic Health Inc.

Ethical Consideration

The ICD coding task is crucial for making clinical, operational and financial decision in healthcare. Traditionally, medical coders review clinical documents and manually assign the appropriate ICD codes by following specific coding guidelines. Models such as our ICDBigBird could help to reduce time and cost in data extraction and reporting significantly.

However, we need to be aware of the risks of over-relying on any automatic encoding model. No matter how efficient an automatic encoding model is, it is still possible to misclassify patients' condition with erroneous ICD codes which may affect their treatment. Thus we believe that any automatic encoding model should only be used to assist, not replace the judgement of trained clinical professionals.

References

Mario Almagro, Raquel Martínez Unanue, Víctor Fresno, and Soto Montalvo. 2020. Icd-10 coding of spanish electronic discharge summaries: An extreme classification problem. *IEEE Access*, 8:100073–100083.

Emily Alsentzer, John Murphy, William Boag, Wei-Hung Weng, Di Jindi, Tristan Naumann, and Matthew McDermott. 2019. Publicly available clinical BERT embeddings. In *Proceedings of the 2nd Clinical Natural Language Processing Workshop*, pages 72–78, Minneapolis, Minnesota, USA. Association for Computational Linguistics.

Anand Avati, Kenneth Jung, Stephanie Harman, Lance Downing, A. Ng, and Nigam Haresh Shah. 2018. Improving palliative care with deep learning. *BMC Medical Informatics and Decision Making*, 18.

Kaidi Cao, Colin Wei, Adrien Gaidon, Nikos Arechiga, and Tengyu Ma. 2019. Learning imbalanced datasets with label-distribution-aware margin loss. In *Advances in Neural Information Processing Systems*.

Pengfei Cao, Yubo Chen, Kang Liu, Jun Zhao, Shengping Liu, and Weifeng Chong. 2020. HyperCore: Hyperbolic and co-graph representation for automatic ICD coding. In *Proceedings of the 58th Annual Meeting of the Association for Computational Linguistics*, pages 3105–3114, Online. Association for Computational Linguistics.

Ilias Chalkidis, Manos Fergadiotis, Sotiris Kotitsas, Prodromos Malakasiotis, Nikolaos Aletras, and Ion Androutsopoulos. 2020. An empirical study on large-scale multi-label text classification including few and zero-shot labels. In *Proceedings of the 2020 Conference on Empirical Methods in Natural Language Processing (EMNLP)*, pages 7503–7515, Online. Association for Computational Linguistics.

Jacob Devlin, Ming-Wei Chang, Kenton Lee, and Kristina Toutanova. 2019. BERT: Pre-training of deep bidirectional transformers for language understanding. In *Proceedings of the 2019 Conference of the North American Chapter of the Association for Computational Linguistics: Human Language Technologies, Volume 1 (Long and Short Papers)*, pages 4171–4186, Minneapolis, Minnesota. Association for Computational Linguistics.

Richárd Farkas and György Szarvas. 2008. Automatic construction of rule-based icd-9-cm coding systems. *BMC Bioinformatics*, 9:S10 – S10.

Shaoxiong Ji, Erik Cambria, and Pekka Marttinen. 2020. Dilated convolutional attention network for medical code assignment from clinical text. In *Proceedings of the 3rd Clinical Natural Language Processing Workshop*, pages 73–78, Online. Association for Computational Linguistics.

Alistair EW Johnson, Tom J Pollard, Lu Shen, H Lehman Li-wei, Mengling Feng, Mohammad Ghassemi, Benjamin Moody, Peter Szolovits, Leo Anthony Celi, and Roger G Mark. 2016. Mimic-iii, a freely accessible critical care database. *Scientific data*, 3:160035.

Thomas N. Kipf and Max Welling. 2017. Semi-Supervised Classification with Graph Convolutional Networks. In *Proceedings of the 5th International Conference on Learning Representations*, ICLR '17.

Jinhyuk Lee, Wonjin Yoon, Sungdong Kim, Donghyeon Kim, Sunkyu Kim, Chan Ho So, and Jaewoo Kang. 2019. BioBERT: a pre-trained biomedical language representation model for biomedical text mining. *Bioinformatics*, 36(4):1234–1240.

Fei Li and Hong Yu. 2020. Icd coding from clinical text using multi-filter residual convolutional neural network. *Proceedings of the AAAI Conference on Artificial Intelligence*, 34:8180–8187.

Ilya Loshchilov and Frank Hutter. 2019. Decoupled weight decay regularization. In *International Conference on Learning Representation.*

Zhibin Lu, Pan Du, and Jian-Yun Nie. 2020. Vgcn-bert: augmenting bert with graph embedding for text classification. In *European Conference on Information Retrieval*, pages 369–382. Springer.

James Mullenbach, Sarah Wiegreffe, Jon Duke, Jimeng Sun, and Jacob Eisenstein. 2018. Explainable prediction of medical codes from clinical text. In *Proceedings of the 2018 Conference of the North American Chapter of the Association for Computational Linguistics: Human Language Technologies, Volume 1 (Long Papers)*, pages 1101–1111, New Orleans, Louisiana. Association for Computational Linguistics.

Matthew Peters, Mark Neumann, Mohit Iyyer, Matt Gardner, Christopher Clark, Kenton Lee, and Luke Zettlemoyer. 2018. Deep contextualized word representations. In *Proceedings of the 2018 Conference of the North American Chapter of the Association for Computational Linguistics: Human Language Technologies, Volume 1 (Long Papers)*, pages 2227–2237, New Orleans, Louisiana. Association for Computational Linguistics.

Nils Reimers and Iryna Gurevych. 2019. Sentence-BERT: Sentence embeddings using Siamese BERT-networks. In *Proceedings of the 2019 Conference on Empirical Methods in Natural Language Processing and the 9th International Joint Conference on Natural Language Processing (EMNLP-IJCNLP)*, pages 3982–3992, Hong Kong, China. Association for Computational Linguistics.

Anthony Rios and Ramakanth Kavuluru. 2018. Few-shot and zero-shot multi-label learning for structured label spaces. In *Proceedings of the 2018 Conference on Empirical Methods in Natural Language Processing*, pages 3132–3142, Brussels, Belgium. Association for Computational Linguistics.

Haoran Shi, Pengtao Xie, Zhiting Hu, Ming Zhang, and Eric P. Xing. 2017. Towards automated icd coding using deep learning. *arXiv preprint arXiv:1711.04075.*

Guoyin Wang, Chunyuan Li, Wenlin Wang, Yizhe Zhang, Dinghan Shen, Xinyuan Zhang, Ricardo Henao, and Lawrence Carin. 2018. Joint embedding of words and labels for text classification. In *Proceedings of the 56th Annual Meeting of the Association for Computational Linguistics (Volume 1: Long Papers)*, pages 2321–2331, Melbourne, Australia. Association for Computational Linguistics.

Thomas Wolf, Lysandre Debut, Victor Sanh, Julien Chaumond, Clement Delangue, Anthony Moi, Pierric Cistac, Tim Rault, Rémi Louf, Morgan Funtowicz, Joe Davison, Sam Shleifer, Patrick von Platen, Clara Ma, Yacine Jernite, Julien Plu, Canwen Xu, Teven Le Scao, Sylvain Gugger, Mariama Drame, Quentin Lhoest, and Alexander M. Rush. 2020. Transformers: State-of-the-art natural language processing. In *Proceedings of the 2020 Conference on Empirical Methods in Natural Language Processing: System Demonstrations*, pages 38–45, Online. Association for Computational Linguistics.

Manzil Zaheer, Guru Guruganesh, Kumar Avinava Dubey, Joshua Ainslie, Chris Alberti, Santiago Ontanon, Philip Pham, Anirudh Ravula, Qifan Wang, Li Yang, et al. 2020. Big bird: Transformers for longer sequences. *Advances in Neural Information Processing Systems*, 33.

Zachariah Zhang, Jingshu Liu, and Narges Razavian. 2020. BERT-XML: Large scale automated ICD coding using BERT pretraining. In *Proceedings of the 3rd Clinical Natural Language Processing Workshop*, pages 24–34, Online. Association for Computational Linguistics.

Doctor XAvIer: Explainable Diagnosis on Physician-Patient Dialogues and XAI Evaluation

Hillary Ngai[1,2] and Frank Rudzicz[1,2,3]

[1]Department of Computer Science, University of Toronto
[2]Vector Institute for Artificial Intelligence
[3]Unity Health Toronto
hngai@cs.toronto.edu, frank@spoclab.com

Abstract

We introduce Doctor XAvIer —a BERT-based diagnostic system that extracts relevant clinical data from transcribed patient-doctor dialogues and explains predictions using feature attribution methods. We present a novel performance plot and evaluation metric for feature attribution methods —Feature Attribution Dropping (FAD) curve and its Normalized Area Under the Curve (N-AUC). FAD curve analysis shows that integrated gradients outperforms Shapley values in explaining diagnosis classification. Doctor XAvIer outperforms the baseline with 0.97 F1-score in named entity recognition and symptom pertinence classification and 0.91 F1-score in diagnosis classification.

Speaker	Utterance
DR	So how are you feeling [PATIENT NAME]? *O O O O O O*
PT	Not good. I'm having back and neck pain. *O O O O B-symptom O B-symptom I-symptom*
DR	And when did this start? *O B-time-expr O O B-time-expr*
PT	Around three days ago. *O B-time-expr I-time-expr I-time-expr*
DR	I see. Do you take any pain killers? *O O O O O O B-medication I-medication*
PT	Yes, acetaminophen and ibuprofen. *O B-medication O B-medication*

Table 1: Synthetic physician-patient dialogue with IOB labels. The IOB labels are italicized underneath each utterance. The *B-* prefix indicates that the token is the beginning of an entity label, the *I-* prefix indicates that the token is inside the entity label, and the *O* indicates that the token belongs to no entity label.

1 Introduction

Previous studies have shown that electronic medical record (EMR) data are difficult to use in machine learning systems due to the lack of regulation in data quality —EMR data are often incomplete and inconsistent (Weiskopf and Weng, 2013; Roth et al., 2009). Recently, there have been attempts to improve automated clinical note-taking by extracting relevant information directly from physician-patient dialogues (Khattak et al., 2019; Kazi and Kahanda, 2019; Du et al., 2019). This can alleviate physicians of tedious data entry and ensures more consistent data quality (Collier, 2017).

Due to the potential in reducing costs associated with collecting patient information and diagnostic errors, there is increasing interest in using information extraction techniques in automatic diagnostic systems (Xu et al., 2019; Wei et al., 2018). Jeblee et al. (2019) introduced a system that extracts pertinent medical information from clinical conversations for automatic note taking and diagnosis. However, their methodology did not explore state-of-the-art natural language processing (NLP) techniques —entity extraction was done by searching the transcript for entities from medical lexicons and tf-idf was used for text classification. Although there is existing work that employs more sophisticated NLP techniques to patient-physician dialogues (Krishna et al., 2020; Selvaraj and Konam, 2019), there is a lack of end-to-end diagnostic systems that employ such techniques. Furthermore, all of the previous works mentioned fail to address the black-box nature of deep learning in the medical industry. Most physicians are reluctant to rely on opaque, AI-based medical technology —especially in high-risk decision-making involving patient well-being (Gerke et al., 2020).

In this work, we present Doctor XAvIer —a BERT-based diagnostic system that extracts relevant clinical data from transcribed patient-doctor dialogues and explains predictions using feature attribution methods. Feature attribution methods are explainable AI (XAI) methods that compute an attribution score for each input feature to represent its contribution to the model's prediction. We report feature attribution scores using integrated gradients (IG) (Sundararajan et al., 2017) and Shapley values (Lundberg and Lee, 2017) to provide insight as to

Proceedings of the BioNLP 2022 workshop, Dublin, Ireland, pages 337–344
May 26, 2022.

which features are important in diagnosis classification. Descriptions of integrated gradients and Shapley values are provided in Appendix A. Feature attribution scores could potentially help physicians build confidence in the model's prediction or give additional insight about the relationships between different diseases and relevant patient information (Markus et al., 2021). Finally, we present a novel performance plot and evaluation metric for feature attribution methods —the Feature Attribution Dropping (FAD) curve and its Normalized Area Under the Curve (N-AUC).

2 FAD Curve Analysis

We introduce Feature Attribution Dropping (FAD) curve analysis for evaluating feature attribution methods. FAD curve analysis requires no modifications to the original machine learning model and is simple to implement.

2.1 FAD Curve

The FAD curve illustrates the explainability of a feature attribution method by plotting the performance metric (e.g., accuracy) against the percentage of features dropped in descending order of importance ranked by the feature attribution method (see Fig. 1). We define the feature importance as the absolute value of the feature attribution score to represent the magnitude of the contribution of each feature to the model's prediction. Features are dropped by modeling the absence of such features in the input. For standard machine learning inputs, continuous features can sometimes be set to their means or image pixels can sometimes be set to black (Sundararajan et al., 2017). A careful consideration of the nature of the data is, of course, required beforehand.

The intuition behind FAD curves is inspired by counterfactual explanations —which describes how the prediction of a model changes when the input is perturbed (Wachter et al., 2018) —and the Pareto principle —which states that for many situations, approximately 80% of the outcome is due to 20% of causes (the "vital few") (Pareto, 1964; Roccetti et al., 2021). If a feature attribution method accurately ranks the most important features for a certain prediction and the Pareto principle holds true, then cumulatively dropping the most important features in descending order should yield a smaller and smaller decrease in model performance for that prediction. In other words, the model's ability to make correct predictions is mostly attributed to a small subset of important features. This entails that the steeper the FAD curve is early on, the better the feature attribution method.

Figure 1: Example of an idealized FAD curve with β=20. The maximum FAD Curve AUC bounded from 0% to $\beta\%$ is shaded in pink. The actual FAD curve AUC bounded from 0% to $\beta\%$ is shaded in blue and overlaps the pink area. The N-AUC is the ratio of the blue area to the pink area.

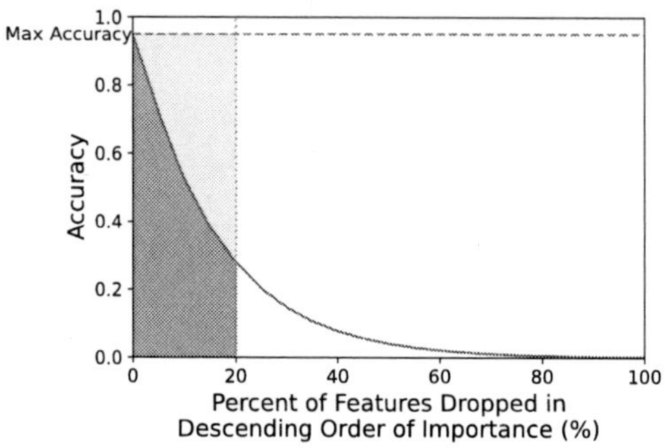

2.2 N-AUC

We present the FAD curve Normalized Area Under the Curve (N-AUC) as a performance metric for feature attribution methods. An intuitive way to quantify how much the FAD curve decreases early on is to calculate the Area Under the Curve (AUC) bounded from 0% to $\beta\%$ of features dropped in descending order of importance. We choose β=20 using the Pareto principle, but this number is just an estimate.

Since steeper FAD curves have smaller AUCs, FAD curves with smaller AUCs indicate a better feature attribution method than FAD curves with larger AUCs. The area under the curve is approximated using the trapezoidal rule (Tai, 1994), as described in Appendix B. Although any performance metric can be used for FAD Curve analysis, we will use accuracy in our explanation for the sake of simplicity. The range of the FAD curve AUC is $(0,\ \beta \times max(accuracy)]$ where $max(accuracy)$ is the maximum FAD curve accuracy of all the feature attribution methods for a model's prediction and β is the x-axis upper bound. Note that the minimum FAD curve AUC can only equal zero if the model performance is zero in the bounded range. This case is excluded from FAD curve analysis since this scenario is rare and uninformative. In order to easily compare feature attribution methods, we normalize the FAD curve AUC:

$$N\text{-}AUC = \frac{AUC}{\beta \times max(accuracy)} \quad (1)$$

Thus, the range of the FAD curve N-AUC is $(0,\ 1]$.

3 Methods and Experiments

We introduce Doctor XAvIer —a medical diagnostic system composed of joint Named Entity Recognition (NER) and intent (i.e. symptom pertinence) classification, primary diagnosis classification, and FAD curve analysis. In this section we discuss each component in detail and evaluate each component.

3.1 Dataset

The Verilogue dataset (Jeblee et al., 2019) is a collection of 800 physician-patient dialogues as audio files and their corresponding human-generated transcripts with speaker labels. Each dialogue includes the patient's information as well as the primary diagnosis. The distribution of the primary diagnoses in the dataset is shown in Appendix C. The patient's information consists of the patient's age, gender, height, weight, blood pressure, smoking status, employment status, and ongoing treatments. Entities —including symptoms, medications, anatomical locations, time expressions, and therapies —are annotated by physicians in each transcript. Additional details about the dataset can be found in Jeblee et al. (2019).

3.2 Joint NER and Intent Classification

A diagnosis requires relevant clinical entities and a measure of pertinence of such entities. For example, a patient might mention a relevant symptom that was experienced by someone else and therefore not pertinent to diagnosis. For each sequence in the physician-patient dialogue, we extract clinical entities with NER and classify the intent of the speaker. We identify the clinical entities identified in Table 2. We label each word in each sequence in the dataset using the Inside-Outside-Beginning (IOB) format (Ramshaw and Marcus, 1995). In this paper, we focus on identifying the pertinence of symptoms. We define the intents of the patient as: *confirm/deny/unsure of symptom* and the intent of both the patient and physician as: *closing* (i.e., ending the conversation). Of the 407 annotated dialogues we randomly select 40 to use as a test set for NER and intent classification.

We fine-tune Bio+Clinical BERT (Alsentzer et al., 2019) jointly on these two classification tasks. This model was initialized from BioBERT (Lee et al., 2019) and trained on all notes from MIMIC-III (Johnson et al., 2016) —a database containing electronic health records from ICU patients. Language models pre-trained on domain-specific text yield improvements on clinical NLP tasks as compared to language models pre-trained on a general corpus (Grouchy et al., 2020). Since a majority of interactions between the physician and patient in the dataset are in question-and-answer format, it is beneficial to concatenate the previous sequence with the current sequence, including the respective speaker codes, to give more context to the model. This is done for each sequence before tokenization and improves NER accuracy from 89% to 96%.

For NER, we concatenate the last four hidden layers of Bio+Clinical BERT and feed this representation into an output layer for token-level classification. For intent classification, we feed the `[CLS]` representation of Bio+Clinical BERT into an output layer for sequence classification. We train with a batch size of 16 sequences and a maximum sequence length of 128 tokens for 5 epochs and select the model with the lowest validation loss. We use AdamW with learning rate of 2e-5, β_1 = 0.9, β_2 = 0.999, L2 weight decay of 0.01, and linear decay of the learning rate (Loshchilov and Hutter, 2017). We use a dropout probability of 0.1 on all layers except the output layers.

For the loss function, we propose a linear interpolation between the intent classification Cross-Entropy (CE) loss and the average NER Negative Log Likelihood (NLL) loss with $\alpha = 0.5$. The intent classification CE loss is defined as:

$$\mathcal{L}_1(f_1(\boldsymbol{x};\boldsymbol{\theta}), \boldsymbol{y}_1) = -\sum_{i=1}^{N} y_{1,i} log f_{1,i}(\boldsymbol{x}_i;\boldsymbol{\theta}) \quad (2)$$

where $f_{1,i}(\boldsymbol{x};\boldsymbol{\theta})$ is the ith element of the softmax output of the intent classes, $\boldsymbol{y}_{1,i}$ is the ith element of the one-hot-encoded intent label, N is the number of intent classes, $\boldsymbol{x}$ is the input, and $\boldsymbol{\theta}$ is the set of model parameters. The average NER NLL loss is defined as:

$$\mathcal{L}_2(f_2(\boldsymbol{x};\boldsymbol{\theta}), \boldsymbol{y}_2) = -\frac{\sum_{j=1}^{M} log f_{2,j}(\boldsymbol{x}_j;\boldsymbol{\theta})}{M} \quad (3)$$

where $f_{2,j}(\boldsymbol{x};\boldsymbol{\theta})$ is the softmax output of the entity classes —for each token in the sequence —at the target class j, $\boldsymbol{y}_2$ is the set of entity labels, and

Entity	Instances	P	R	F1
Other	158,018	0.98	0.98	0.98
Anatomical Location	598	0.73	0.65	0.69
Bodily Function	6	0.00	0.00	0.00
Diagnosis	1,345	0.79	0.75	0.77
Therapy	1420	0.62	0.69	0.65
Medication	3,324	0.90	0.81	0.85
Referral	256	0.71	0.79	0.74
Symptom	3,574	0.57	0.66	0.61
Substance Use	68	0.00	0.00	0.00
Time Expression	4,062	0.90	0.84	0.87
Weighted Avg	172,671	0.97	0.96	0.97

Table 2: Named entity recognition results.

Intent	Instances	P	R	F1
Confirm Symptom	228	0.70	0.69	0.70
Deny Symptom	52	0.73	0.69	0.71
Unsure of Symptom	73	0.34	0.65	0.62
Closing	28	0.29	0.47	0.36
Other	6,425	0.99	0.99	0.99
Weighted Avg	6,806	0.97	0.97	0.97

Table 3: Intent classification results.

M is the number tokens in the sequence. The full loss function is defined in Appendix D.1. [PAD] tokens are excluded from the loss function using masking.

As seen in Table 2 and Table 3, the model yields approximately 0.97 weighted precision, recall, and F1-score on both tasks, outperforming Jeblee et al. (2019)'s models. However, the exact results are difficult to compare since Jeblee et al. (2019) tested their model on a smaller subset of the dataset.

3.3 Primary Diagnosis Classification

We classify the primary diagnosis for each physician-patient dialogue using the the patient's information —such as the patient's age, weight, blood pressure, and smoking status —and the extracted symptoms from the conversation. Since the same symptom can be said in various different ways, we compile a set of symptoms of all the diseases in the dataset according to WedMD and assign each extracted symptom to one of the pre-defined symptoms. We use a pre-trained Sentence-BERT (SBERT) model (Reimers and Gurevych, 2019) to embed each extracted symptom and all the pre-defined symptoms. Each extracted symptom is assigned to its most similar pre-defined symptom measured by the cosine similarity between the SBERT embeddings (Ngai et al., 2021). The most similar pre-defined symptom is defined as:

$$s_i^* = \arg\max_{s_i} \text{sim}(\text{emb}(e_j), \text{emb}(s_i)) \; \forall s_i \in S \tag{4}$$

where $S = \{s_1, ..., s_N\}$ is the set of symptoms of all diseases in the dataset, s_i is the ith symptom in S, e_j is the jth extracted symptom, $\text{emb}(x)$ is the SBERT embedding of text x, and $\text{sim}(a, b)$ is the cosine similarity between embeddings a and b. The assigned pre-defined symptom is:

$$e_j^* = \begin{cases} s_i^*, & if\ \text{sim}(\text{emb}(e_j), \text{emb}(s_i^*)) \geq \epsilon \\ None \end{cases} \tag{5}$$

where ϵ is a constant and $None$ represents that we do not use the extracted symptom e_j for classification. We chose $\epsilon = 0.35$ since it minimized incorrect assignments of extracted symptoms in the dataset while filtering out less than 10% of extracted symptoms.

The diagnosis classification model is a neural network composed of 549 input features and three hidden layers with 182K total parameters. The input features consists of patient information and the pertinence of extracted symptoms from the conversation. The model is evaluated using stratified 5-fold cross-validation. We train with a batch size of 32 for 100 epochs and select the model with the lowest validation loss. We use Adam (Kingma and Ba, 2017) with learning rate of 1e-3, β_1 = 0.9, β_2 = 0.999, and ϵ = 1e-08. We use a GELU activation (Hendrycks and Gimpel, 2016) on all hidden layers. The training loss is the standard CE loss.

As seen in Table 4, Doctor XAvIer yields a significant improvement in weighted precision, recall, and F1-score for diagnosis classification compared to the baseline (Jeblee et al., 2019).

3.4 Evaluation of Explainability Methods

For each test fold and model trained on the train fold in the stratified 5-fold cross-validation of the diagnosis classification model, we evaluate each feature attribution method using FAD curve analysis. We choose accuracy as the performance metric for FAD curve analysis.

As seen in Table 5, integrated gradients outperforms Shapley values according to FAD curve analysis —achieving smaller N-AUCs for all diagnoses. As seen in Figures 2, 3, and 4 and Appendix F.2, integrated gradients yields noticeably steeper FAD curves than Shapley values for all of the diagnoses except *Type II Diabetes*. The sporadic shapes of

Diagnosis	Model	P	R	F1
ADHD	Doctor XAvIer	**0.95**	**0.97**	**0.96**
	(Jeblee et al., 2019)	0.84	0.84	0.83
Depression	Doctor XAvIer	**0.92**	**0.93**	**0.92**
	(Jeblee et al., 2019)	0.80	0.64	0.71
Osteoporosis	Doctor XAvIer	**0.85**	0.69	0.75
	(Jeblee et al., 2019)	0.81	**0.78**	**0.78**
Influenza	Doctor XAvIer	**1.00**	**0.99**	**0.99**
	(Jeblee et al., 2019)	0.91	0.95	0.93
COPD	Doctor XAvIer	**0.93**	**0.93**	**0.93**
	(Jeblee et al., 2019)	0.75	0.65	0.68
Type II Diabetes	Doctor XAvIer	0.52	0.47	0.48
	(Jeblee et al., 2019)	**0.81**	**0.75**	**0.76**
Other	Doctor XAvIer	**0.73**	0.80	**0.76**
	(Jeblee et al., 2019)	0.71	**0.82**	**0.76**
Weighted Avg	Doctor XAvIer	**0.91**	**0.91**	**0.91**
	(Jeblee et al., 2019)	0.82	0.80	0.80

Table 4: K-fold cross-validation primary diagnosis classification results.

Diagnosis	Instances	IG	Shapley
ADHD	20	**0.48**	0.77
Depression	14	**0.63**	0.85
Osteoporosis	5	**0.24**	0.36
Influenza	19	**0.72**	0.95
COPD	11	**0.33**	0.59
Type II Diabetes	3	**0.59**	0.73
Other	9	**0.71**	0.95

Table 5: K-fold cross-validation FAD curve N-AUC from 0% to 20% of dropped features comparing integrated gradients and Shapley values.

the *Type II Diabetes* FAD curves can potentially be explained by the lack of dialogues with *Type II Diabetes* as their primary diagnosis —there are only 3 instances. This suggests that we could potentially improve performance by collecting more instances of the infrequent classes or performing regularization.

It is important to note that some features in the dataset may be correlated. Therefore, dropping features that are correlated with other features may lead to an increase —instead of a decrease —in the performance metric despite dropping features in descending order of importance. We could potentially mitigate this by using feature selection methods before performing FAD curve analysis.

4 Conclusion

Doctor XAvIer yields significant improvements in NER, symptom pertinence classification, and diagnosis classification compared to previous work (Jeblee et al., 2019), while also explaining why the model made each diagnosis. We also present a novel performance plot and evaluation metric for feature attribution methods —FAD curve analysis and its N-AUC. FAD curve analysis shows that integrated gradients outperforms Shapley values in explaining diagnosis classification in the Verilogue dataset. In our future work, we will calculate β in a data-driven manner to standardize FAD curve analysis for a given dataset. We will also apply FAD curve analysis to other feature attribution methods, AI domains, and datasets to evaluate its generalizability.

Figure 2: K-fold cross-validation *ADHD* and *Depression* FAD curves.

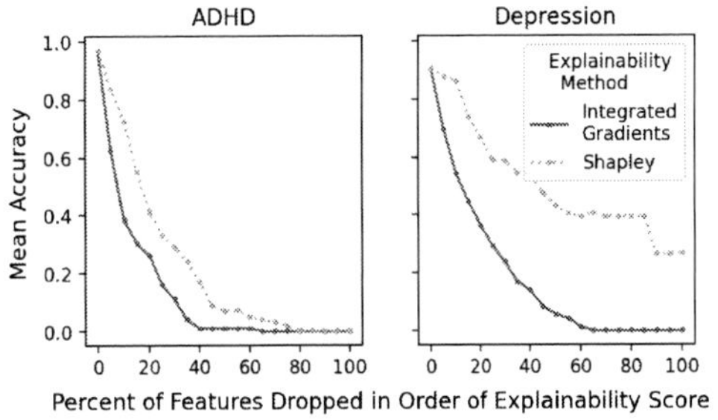

Figure 3: K-fold cross-validation *COPD* and *Type II Diabetes* FAD curves.

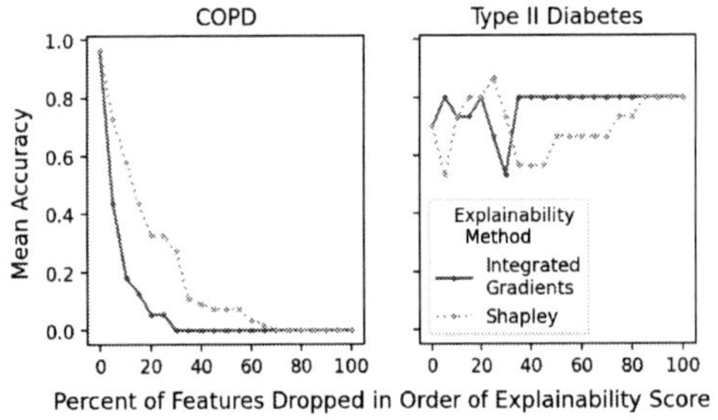

Figure 4: K-fold cross-validation *Osteoporosis* and *Influenza* FAD curves.

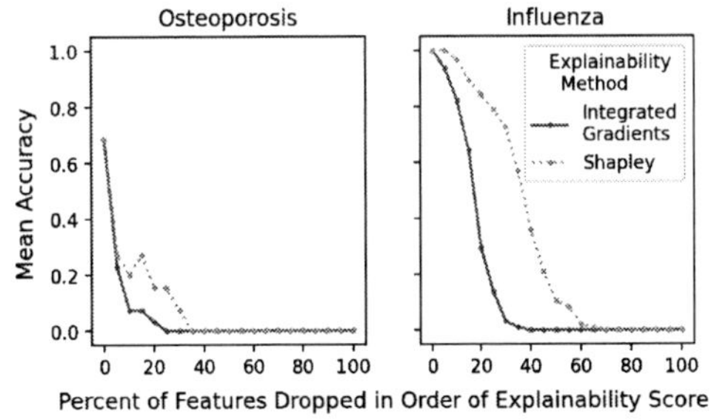

References

Emily Alsentzer, John R. Murphy, Willie Boag, Wei-Hung Weng, Di Jin, Tristan Naumann, and Matthew B. A. McDermott. 2019. Publicly available clinical BERT embeddings. *CoRR*, abs/1904.03323.

Roger Collier. 2017. Electronic health records contributing to physician burnout. *CMAJ*, 189(45):E1405–E1406.

Nan Du, Kai Chen, Anjuli Kannan, Linh Tran, Yuhui Chen, and Izhak Shafran. 2019. Extracting symptoms and their status from clinical conversations. In *Proceedings of the 57th Annual Meeting of the Association for Computational Linguistics*, pages 915–925, Florence, Italy. Association for Computational Linguistics.

Sara Gerke, Timo Minssen, and Glenn Cohen. 2020. Ethical and legal challenges of artificial intelligence-driven healthcare. *National Center for Biotechnology Information*, pages 295–336.

Paul Grouchy, Shobhit Jain, Michael Liu, Kuhan Wang, Max Tian, Nidhi Arora, Hillary Ngai, Faiza Khan Khattak, Elham Dolatabadi, and Sedef Akinli Koçak. 2020. An experimental evaluation of transformer-based language models in the biomedical domain. *CoRR*, abs/2012.15419.

Dan Hendrycks and Kevin Gimpel. 2016. Bridging nonlinearities and stochastic regularizers with gaussian error linear units. *CoRR*, abs/1606.08415.

Serena Jeblee, Faiza Khan Khattak, Noah Crampton, Muhammad Mamdani, and Frank Rudzicz. 2019. Extracting relevant information from physician-patient dialogues for automated clinical note taking. In *Proceedings of the Tenth International Workshop on Health Text Mining and Information Analysis (LOUHI 2019)*, pages 65–74, Hong Kong. Association for Computational Linguistics.

Alistair EW Johnson, Tom J Pollard, Lu Shen, Li-wei H Lehman, Mengling Feng, Mohammad Ghassemi, Benjamin Moody, Peter Szolovits, Leo Anthony Celi, and Roger G Mark. 2016. MIMIC-III, a freely accessible critical care database. *Scientific data*, 3(1):1–9.

Nazmul Kazi and Indika Kahanda. 2019. Automatically generating psychiatric case notes from digital transcripts of doctor-patient conversations. In *Proceedings of the 2nd Clinical Natural Language Processing Workshop*, pages 140–148, Minneapolis, Minnesota, USA. Association for Computational Linguistics.

Faiza Khattak, Serena Jeblee, Noah Crampton, Muhammad Mamdani, and Frank Rudzicz. 2019. Autoscribe: Extracting clinically pertinent information from patient-clinician dialogues. *Studies in health technology and informatics*, 264:1512–1513.

Diederik P. Kingma and Jimmy Ba. 2017. Adam: A method for stochastic optimization. *International Conference on Learning Representations*.

Kundan Krishna, Amy Pavel, Benjamin Schloss, Jeffrey P. Bigham, and Zachary C. Lipton. 2020. Extracting structured data from physician-patient conversations by predicting noteworthy utterances. *CoRR*, abs/2007.07151.

Jinhyuk Lee, Wonjin Yoon, Sungdong Kim, Donghyeon Kim, Sunkyu Kim, Chan Ho So, and Jaewoo Kang. 2019. BioBERT: a pre-trained biomedical language representation model for biomedical text mining. *CoRR*, abs/1901.08746.

Ilya Loshchilov and Frank Hutter. 2017. Fixing weight decay regularization in Adam. *CoRR*, abs/1711.05101.

Scott M. Lundberg and Su-In Lee. 2017. A unified approach to interpreting model predictions. *CoRR*, abs/1705.07874.

Aniek F. Markus, Jan A. Kors, and Peter R. Rijnbeek. 2021. The role of explainability in creating trustworthy artificial intelligence for health care: A comprehensive survey of the terminology, design choices, and evaluation strategies. *Journal of Biomedical Informatics*, 113:103655.

Hillary Ngai, Yoona Park, John Chen, and Mahboobeh Parsapoor. 2021. Transformer-based models for question answering on COVID19. *CoRR*, abs/2101.11432.

Vilfredo Pareto. 1964. *Cours d'économie politique*, volume 1. Librairie Droz.

Lance Ramshaw and Mitch Marcus. 1995. Text chunking using transformation-based learning. In *Third Workshop on Very Large Corpora*.

Nils Reimers and Iryna Gurevych. 2019. Sentence-BERT: Sentence embeddings using siamese bert-networks. *CoRR*, abs/1908.10084.

Marco Roccetti, Giovanni Delnevo, Luca Casini, and Silvia Mirri. 2021. An alternative approach to dimension reduction for pareto distributed data: a case study. *Journal of Big Data*.

Carol P. Roth, Yee-Wei Lim, Joshua M. Pevnick, Steven M. Asch, and Elizabeth A. McGlynn. 2009. The challenge of measuring quality of care from the electronic health record. *American Journal of Medical Quality*, 24(5):385–394. PMID: 19482968.

Sai P. Selvaraj and Sandeep Konam. 2019. Medication regimen extraction from clinical conversations. *CoRR*, abs/1912.04961.

Mukund Sundararajan and Amir Najmi. 2019. The many shapley values for model explanation. *CoRR*, abs/1908.08474.

Mukund Sundararajan, Ankur Taly, and Qiqi Yan. 2017. Axiomatic attribution for deep networks.

Mary M Tai. 1994. A mathematical model for the determination of total area under glucose tolerance and other metabolic curves. *Diabetes care*, 17(2):152–154.

Sandra Wachter, Brent Mittelstadt, and Chris Russell. 2018. Counterfactual explanations without opening the black box: Automated decisions and the GDPR.

Zhongyu Wei, Qianlong Liu, Baolin Peng, Huaixiao Tou, Ting Chen, Xuanjing Huang, Kam-fai Wong, and Xiangying Dai. 2018. Task-oriented dialogue system for automatic diagnosis. In *Proceedings of the 56th Annual Meeting of the Association for Computational Linguistics (Volume 2: Short Papers)*, pages 201–207, Melbourne, Australia. Association for Computational Linguistics.

Nicole Gray Weiskopf and Chunhua Weng. 2013. Methods and dimensions of electronic health record data quality assessment: enabling reuse for clinical research. *Journal of the American Medical Informatics Association*, 20(1):144–151.

Lin Xu, Qixian Zhou, Ke Gong, Xiaodan Liang, Jianheng Tang, and Liang Lin. 2019. End-to-end knowledge-routed relational dialogue system for automatic diagnosis. *CoRR*, abs/1901.10623.

Appendix

A Feature Attribution Methods

A.1 Shapley Values

The Shapley value (Lundberg and Lee, 2017) —a method from cooperative game theory —assigns payouts to players depending on their contribution to the total payout in a cooperative game. Players cooperate in a coalition and receive a certain profit from this cooperation. In explainable AI, the game is the prediction task for a single instance in the dataset, the players are the feature values of a single instance that collaborate to make a prediction, and the gain is the prediction for an instance minus the average prediction for all instances (Sundararajan and Najmi, 2019). In other words, the Shapley value measures the contribution of each input feature to a model's prediction for a single instance.

A.2 Integrated Gradients

Integrated gradients (Sundararajan et al., 2017) is an XAI technique which attributes the prediction of a deep neural network to its input features. Integrated gradients attributes blame to an input feature by using the absence of the input feature as a baseline for comparing outcomes. For most deep networks, there exists a baseline in the input space where the prediction is neutral. For example, the baseline for an object recognition network can be a black image. Mathematically, integrated gradients is defined as the path integral of the gradients along the straightline path from the baseline x' to the input x.

Primary Diagnosis	Dialogues
ADHD	99
Depression	72
Osteoporosis	26
Influenza	95
COPD	55
Type II Diabetes	14
Other	46

Table 6: Distribution of primary diagnoses in the Verilogue dataset.

B Area Under the Curve Approximation

The area under the curve is approximated using the trapezoidal rule (Tai, 1994):

$$\begin{aligned} AUC &= \int_0^{20} f(x)\,dx \\ &\approx \sum_{k=1}^{N} \frac{f(x_{k-1}) + f(x_k)}{2} \Delta x_k \end{aligned} \tag{6}$$

where $0 = x_0 < x_1 < ... < x_{N-1} < x_N = 20$ and $\Delta x_k = x_k - x_{k-1}$.

C Additional Dataset Details

Table 6 shows the distribution of diagnoses in the Verilogue dataset.

D Additional Details for Joint NER and Intent Classification

D.1 Loss Function Equations

Combining Eq. 2 and Eq. 3, the joint intent classification and NER loss function is defined as:

$$\begin{aligned} &\mathcal{L}(f_1(\boldsymbol{x};\boldsymbol{\theta}), \boldsymbol{y}_1, f_2(\boldsymbol{x};\boldsymbol{\theta}), \boldsymbol{y}_2, \alpha) \\ &= \alpha\mathcal{L}_1(f_1(\boldsymbol{x};\boldsymbol{\theta}), \boldsymbol{y}_1) \\ &+ (1-\alpha)\mathcal{L}_2(f_2(\boldsymbol{x};\boldsymbol{\theta}), \boldsymbol{y}_2) \end{aligned} \tag{7}$$

where $\alpha \in [0, 1]$.

D.2 Training Hardware

Training of the joint NER intent classiciation model was performed on a NVIDIA Quadro RTX 6000 GPU and took approximately two hours to finish training.

Feature	Attribution %
Age	0.015
Trouble making decisions and remembering things	0.013
Taking Adderall	0.009
Trouble focusing on a task	0.007
Easily distracted	0.004
Restlessness	0.003

Table 7: Examples of top features for classifying ADHD ranked by integrated gradients.

Feature	Attribution %
Weight	0.003
Age	0.002
Trouble focusing on a task	0.002
Trouble making decisions and remembering things	0.002
Easily distracted	0.002
Systolic Blood Pressure	0.002

Table 8: Examples of top features for classifying ADHD ranked by Shapley values.

E Additional Details for Primary Diagnosis Classification

E.1 Training Hardware

Training of the primary diagnosis classification model was performed on a NVIDIA Tesla K80 GPU and took approximately an hour to finish training and evaluating all five models.

F Additional Details for FAD Curve Analysis

F.1 Feature Attribution Examples

Examples of top features for classifying ADHD ranked by integrated gradients are shown in Table 7 and examples of top features for classifying ADHD ranked by Shapley values are shown in Table 8.

F.2 Additional FAD Curves for Diagnosis Classification

The FAD curve for the diagnosis *Other* is seen in Figure 5.

Figure 5: K-fold cross-validation *Other* FAD curves.

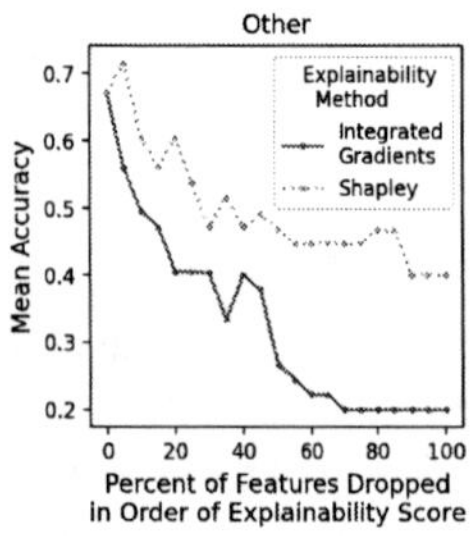

G Code

The code is available at: `https://github.com/hillary-ngai/doctor_XAvIer`.

DISTANT-CTO: A Zero Cost, Distantly Supervised Approach to Improve Low-Resource Entity Extraction Using Clinical Trials Literature

Anjani Dhrangadhariya and **Henning Müller**
University of Geneva (UNIGE), Geneva, Switzerland
University of Applied Sciences Western Switzerland (HES-SO), Sierre, Switzerland
{anjani.dhrangadhariya,henning.mueller}@hevs.ch

Abstract

PICO recognition is an information extraction task for identifying participant, intervention, comparator, and outcome information from clinical literature. Manually identifying PICO information is the most time-consuming step for conducting systematic reviews (SR), which is already labor-intensive. A lack of diversified and large, annotated corpora restricts innovation and adoption of automated PICO recognition systems. The largest-available PICO entity/span corpus is manually annotated which is too expensive for a majority of the scientific community. To break through the bottleneck, we propose DISTANT-CTO, a novel distantly supervised PICO entity extraction approach using the clinical trials literature, to generate a massive weakly-labeled dataset with more than a million "Intervention" and "Comparator" entity annotations. We train distant NER (named-entity recognition) models using this weakly-labeled dataset and demonstrate that it outperforms even the sophisticated models trained on the manually annotated dataset with a 2% F1 improvement over the Intervention entity of the PICO benchmark and more than 5% improvement when combined with the manually annotated dataset. We investigate the generalizability of our approach and gain an impressive F1 score on another domain-specific PICO benchmark. The approach is not only zero-cost but is also scalable for a constant stream of PICO entity annotations.

1 Introduction

Primary care physicians rely on systematic reviews (SRs) for informed decision-making. SRs are conducted to objectively answer clinical questions and require going through a rigorous process of manually screening tens of thousands of clinical studies to identify terms describing PICO. PICO information identification is crucial to appraise the relevance of a clinical study for answering the clinical question at hand. A study is only included for writing SRs if it mentions relevant PICO information. Manual PICO information screening for a single SR consumes more than 12 months of two medical experts' time. The process can be automated using information extraction (IE) by directly pointing the human reviewers to the correct PICO descriptions. Automation will accelerate the overall process of writing SRs while reducing the burden on health professionals who are required to manually screen for PICO entities.

Automating PICO entity detection has garnered lower interest than other biomedical NER tasks because of the lack of publicly available entity annotated corpora. The largest publicly-available PICO entity/span dataset (EBM-PICO) contains only 5000 annotated abstracts, some of which were annotated through crowd-sourcing and others by hired medical experts (Nye et al., 2018). Crowdsourcing involves hiring non-expert workers that require intensive training that is not commonly affordable. Hiring medical experts for annotation is equally often too expensive. IN GENERAL, extracting PICO entities/spans is somewhat tricky because of high disagreement between human annotators on the exact spans constituting the mentions. This leads to human errors in hand-labeled corpora. Hand-labeled datasets are static and prohibit quick manual re-labeling in case of human errors or when a downstream task requires new entities. For example, PICO entities extend to PICOS, where S denotes the "study type" of included evidence.

Distant supervision (DS) is a data-centric approach that allows generating massive weakly annotated datasets without human annotators and has previously been used to create large relation extraction corpora for the general and biomedical domains. To address the challenges above and democratize PICO entity recognition, we propose DISTANT-CTO, a distantly supervised and scalable approach to obtaining clinical trials annotations. We take an integrative approach combining methods of semi-supervised learning (SSL)

Proceedings of the BioNLP 2022 workshop, Dublin, Ireland, pages 345–358
May 26, 2022.

and gestalt pattern matching (GPM) to develop a continuously extensible dataset. We successfully demonstrate this approach for the "Intervention" and "Comparator" entity annotations as proof of concept (POC).

We summarize our contributions as follows:

- We develop a zero-cost, data-centric approach using DS to obtain "Intervention" and "Comparator" entity annotations.
- We develop and make publicly available a large weakly-labeled dataset from more than 300,000 clinical trials. The dataset offers about a million sentences with more than 977,682 annotations across 11 semantic types.
- We improve the state-of-the-art by 2% macro-F1 on the previously most poor-performing "Intervention" entity extraction on the EBM-PICO benchmark corpus without using costly manually labeled data and by 5% when combined with manually labeled data.

2 Related Work

A decade of automatic PICO information extraction was limited to sentence-level due to the unavailability of entity-annotated corpora (Boudin et al., 2010; Huang et al., 2011, 2013; Wallace et al., 2016; Jin and Szolovits, 2018). The release of the EBM-PICO corpus paved the way for the community to improve upon the PICO entity/span extraction task. (Nye et al., 2018). The corpus is biased towards pharma intervention classes overshadowing non-pharma ones leading to a substandard performance on it in the previous SOTA fully-supervised PICO entity/span recognition models (Beltagy et al., 2019; Brockmeier et al., 2019; Zhang et al., 2020) and weakly supervised model (Liu et al., 2021). Small-scale annotation projects cannot capture the range and variation of the PICO descriptions spanning the entirety of clinical trials literature. At some point, applications of such static corpora will confront the problem of insufficient and irrelevant annotations. Manual annotation projects are neither affordable nor scalable for every lab, limiting innovation.

A plethora of DS methods have been previously explored for large-scale relation extraction but not for (named) entity extraction (Etzioni et al., 2008; Smirnova and Cudré-Mauroux, 2018; Adelani et al., 2020). Entity extraction in high-impact clinical and biomedical domains largely relies on small expert annotated datasets. Commonly, obtaining weak annotations using DS rely on aligning terms (a word or phrase) from ontologies onto the unstructured text (Giannakopoulos et al., 2017; Yang et al., 2018; Peng et al., 2019; Hedderich et al., 2021). Ontologies are structured, standardized data sources that do not capture various writing variations from clinical literature. Weak annotations obtained using custom-built rules like regular expressions are restricted by either task or worse even by entity type (Ratner et al., 2017; Safranchik et al., 2020; Fries et al., 2021). Bootstrapping approaches like label propagation (LP) still require an expert annotated dataset to obtain pseudo annotations for previously unlabeled data samples (Bing et al., 2017). It is hence not zero-cost.

Our work focuses on overcoming the discussed bottlenecks using a data-centric DS approach to generate a large clinical entity annotated corpus and train a downstream NER model to assess if it yields adequate results. Unlike the reviewed DS approaches, our approach does not use ontologies or rules or LP but rather uses GPM for flexibly aligning structured text in a clinical trials database to the free-text fields in the same database using an adaptable internal scoring scheme.

3 Data

ClinicalTrials.gov (CTO hereafter) documents more than 350,000 human clinical studies conducted around the globe. The trial's principal investigator enters and updates information about each study stored in CTO. It includes the title and description of the clinical trial, participant's eligibility criteria, participant disease and demographics, interventions evaluated, outcomes, *etc.* CTO allows programmatic access to this vast amount of information in the JSON (JavaScript Object Notation) format. The information is stored as a combination of structured tabular and unstructured free-text (see Figure 1). The 'OfficialTitle' and 'BriefTitle' tags in the JSON respectively store the official and shorter version of the study title in an unstructured free-text format. The 'BriefSummary' and 'DetailedDescription' tags store study summaries. Interventions used in the study are stored under the 'InterventionName' tag and their synonyms under 'InterventionOtherName' tag each of which could be linked to their broad semantic type (drug, device, behavioral, procedural, biological, dietary supplement, diagnostic test, radiation, genetic, combination product, other) mentioned under the 'InterventionType' tag. As each intervention name is linked to its semantic type, this becomes a structured information store. The 'InterventionDescription' tag describes intervention administration procedures often in a detailed passage.

4 Approach

The approach is schematically illustrated in Figure 2 and is described below.

4.1 Distant Supervision

Distantly supervised (DS) information extraction (IE) is an efficient SSL method (Etzioni et al., 2008; Wen et al., 2019). It is used when the task at hand has 1) some strongly-labeled data, 2) abundant unlabeled data, and 3) a weak-labeling function that could sample from this unlabeled data and label them using a heuristic function. This labeling function is a heuristic algorithm that uses a heuristic to label the unlabeled data (Pinto et al., 2003; Greaves, 2014). It results in a weakly-labeled dataset with potential label noise. DS-IE models can then collectively use this strongly-labeled and weakly-labeled training data to give the final output.

4.2 Gestalt Pattern Matching

In entity extraction, the most common form of DS is to heuristically align terms from a structured information source onto the unstructured text (Wen et al., 2019). When flexible, this heuristic boils down to a substring matching problem. The weak-labeling function matches the longest common substring (LCS) between the structured term and unstructured text. Gestalt Pattern Matching (GPM), also known as Ratcliff/Obershelp similarity algorithm, is a string-matching algorithm for determining the similarity of two strings. The similarity between two strings S_1 and S_2 is measured by the formula, calculating twice the number of matching characters K_m divided by the total length $|S_1| + |S_2|$ of both strings. Matching characters are identified by the LCS algorithm followed by recursively finding matching characters in the non-matching regions on either side from both strings (Ratcliff and Metzener, 1988). $Similarity$ ranges between 0, which means no match, and 1, which means a complete match of the two strings.

$$Similarity(S) = \frac{2K_m}{|S_1| + |S_2|} \; ; 0 \leq S \leq 1 \quad (1)$$

Difflib: It is a python module providing a `sequencematcher` function that extends the GPM algorithm for comparing pairs of strings. `sequencematcher` finds the longest contiguous subsequence between the sequence pair without the "junk" elements such as blank lines or white spaces. The same idea is then applied recursively to the flanks of the sequences to the left and the right of the matching subsequence. This yields matching sequences that appear normal to the human eye.

4.3 Candidate Generation

We define candidate generation as the process of automatically generating entity-annotated sentences.

Assumption and Problem formulation: As "Intervention" and "Comparator" entities represent interventions in two different roles in clinical trials and semantically the same classes, they are clubbed into a single "Intervention" entity class. Let each CTO record JSON file be $r_i \in \boldsymbol{R}, i = \{1, 2, ..., I\}$. Let the intervention terms in 'InterventionName' tags and 'InterventionOtherName' tags be the intervention source $S = \{s_1, s_2, ..., s_m\}$ used in the study r_i. Each intervention term $s_i \in S$ is linked to intervention class from 'InterventionType' tag converting it into a tuple of $\langle s_{class}, s_{name} \rangle$, s_{name} = intervention term and s_{class} = intervention category. s_{name} is a sequence of words $\{y_1, y_2, ..., y_n\}, n = \{1, 2, ..., N\}$. Let each sentence $t_i = \{x_1, x_2, ..., x_m\}, m = \{1, 2, ..., M\}$ in the 'BriefSummary', 'DetailedDescription', 'BriefTitle', 'OfficialTitle' and 'InterventionDescription' be a part of the intervention target set T. We assume that for each s_{name} in r_i there could exist a mapping to t_i meaning s_{name} is possibly either completely or partially mentioned in the t_i (see Figure 1). Our goal is to build a scalable and adaptable candidate generation pipeline that maps each s_{name} from the structured intervention source S to the target sentences $t_i \in T$ (if a loose mapping exists). In this prototypical work, we focus on *almost* direct matches between the s_{name} and t_i and keep the order-free matches for future work.

Approach For each individual CTO record r_i, we extract all $s_{name} \in S$ and $t_i \in T$ from the locally stored CTO dump. Both S and T are preprocessed by lower-casing, replacing hyphens and multiple trailing spaces with a single space and removal of Unicode characters. Given a s_{name} and t_i, our aim is to identify and score (if identified) the mapping between both

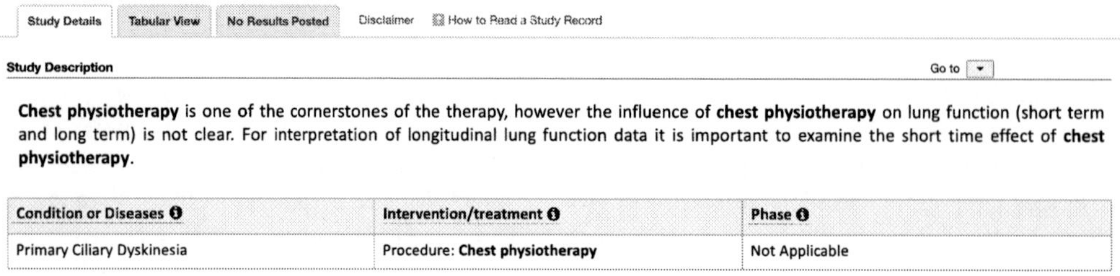

Study Details | Tabular View | No Results Posted | Disclaimer | How to Read a Study Record

Study Description — Go to

Chest physiotherapy is one of the cornerstones of the therapy, however the influence of **chest physiotherapy** on lung function (short term and long term) is not clear. For interpretation of longitudinal lung function data it is important to examine the short time effect of **chest physiotherapy**.

Condition or Diseases	Intervention/treatment	Phase
Primary Ciliary Dyskinesia	Procedure: **Chest physiotherapy**	Not Applicable

Figure 1: An example CTO record (ID - NCT01929356) to demonstrate the information storage format which is a combination of structured table and unstructured text.

sequences. To map and score alignment from the s_{name} to t_i, we use a distant supervision labeling function LF_{ds} which is a combination of the `sequencematcher` function and an internal scoring function to fetch almost direct annotations. The `sequencematcher` function takes as input s_{name} and t_i and outputs several matching blocks $d_{block} \in D_{blocks}$ between both strings. These matching blocks between the two strings are calculated using a modified gestalt pattern matching algorithm as elaborated in 4.2. Each $d_{block} = \langle MatchPos_t, MatchPos_s, MatchLen \rangle$. $MatchPos_t$ is the start of the match in t_i, $MatchPos_s$ is the start of the match in s_{name} and $MatchLen$ is number of characters matching between the both. `sequencematcher` provides an internal scoring function called as `ratio` that returns a similarity score between the two sequences being matched. We do not use `ratio` because it returns an overall matching score between the two full sequences s_{name} and t_i rather than a match score for s_{name} and d_{block}. Instead, to identify the matching blocks that correspond to an exact match between an entire s_{name} and a part of t_i, we calculate a match score d_s for each matching block output by `sequencematcher` using equation 2 which is dividing the number of matching characters in the match block d_{block} by number of characters in s_{name}.

$$d_s = \frac{MatchLen}{|s_{name}|} \; ; 0 \leq d_s \leq 1 \qquad (2)$$

Any d_{block} with the d_s score of 1.0 is considered as complete match and then the s_{name} corresponding to the d_{block} is mapped onto sentence t_i to generate a positive annotation sentence $a_+ \in A_+$. Using the d_{block} with only the match score 1.0 leads to missing out on several entities leading to an incomplete noisy weakly annotated dataset. Taking this into consideration, we retrieve the d_{block} matching with d_s score of 0.9 as fairly-accurate partial matches. We used a validation set to relax the choice of similarity match score d_s to 0.9. We relax the labeling function LF_{ds} to match bigrams in source terms to the targets. In the real-world data, not all sentences in clinical trial literature mention the intervention name and therefore in addition to the positive annotation sentences we require negative annotation sentences. We take t_i and s_{name} where no parts of d_{block} scored d_s more than 0.2 to generate the negative annotation sentences $a_- \in A_-$. We call all these sequences comprised of the positive and the negative entity annotated sentences A_{+-} our weakly annotated dataset. Next, for all A_{+-} instances we fetch part-of-the-speech (POS) tags using `POS-tagger` from NLTK (Natural Language Toolkit) resulting into A_{+-POS}. We call the resulting dataset DISTANT-CTO set. POS tags are added as additional features as they have shown to help model generalization (Augenstein et al., 2017). `difflib` in combination with the internal scoring function are previously unexplored for automatic entity annotation generation. It has to be noted that the method depends on availability of short source texts with the possibility that they will be mentioned in longer target texts.

4.4 Model Training

We train an end-to-end distant NER model on A_{+-POS} using the architecture explained below.

1. Feature Extraction: To capture the domain-specific information, we used SciBERT, which was continually pretrained and domain adapted on the scientific literature from semantic scholar (Gururangan et al., 2020). The models used SciBERT to tokenize the text input A_{+-} into encoded tokens x_t and extract dense, contextual vectors e_t from x_t at each time-step t (Beltagy et al., 2019). POS-inputs A_{+-POS} were one-hot encoded into p_t vectors.

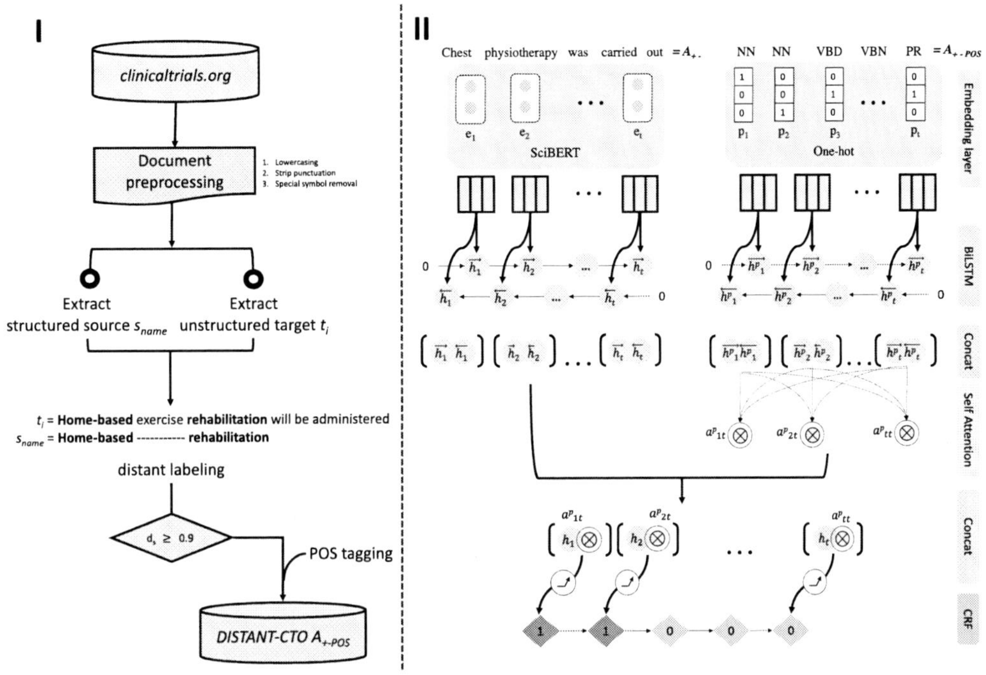

Figure 2: DISTANT-CTO approach - I) Distantly-supervised candidate generation approach, and II) Distantly-supervised NER model architecture.

2. Feature transformation: To further fine-tune to the training corpus, the model stacked a bidirectional LSTM (BiLSTM) on top of the SciBERT (Hochreiter and Schmidhuber, 1997). A BiLSTM layer encodes the text into a $(\overrightarrow{h})$ and $(\overleftarrow{h})$ vector using the current token embedding input e_t and the previous hidden state h_{t-1} in both the directions. $\overrightarrow{h}$ and $\overleftarrow{h}$ were shallow concatenated $([\overrightarrow{h}; \overleftarrow{h}])$ into h_t and used as the input for the next layer. Similarly, the one-hot encoded POS-vectors p_t underwent feature transformation and were concatenated $([\overrightarrow{h}_{POS}; \overleftarrow{h}_{POS}])$ into POS-features h_t^p.

3. Self-attention: Next, the model stacked a single-head self-attention layer that calculated for each POS-tag feature at time t in the sequence a weighted average of the feature representation of all other POS-tag features in the sequence a_t^p (Vaswani et al., 2017). This improves the signal-to-noise ratio by out-weighting important POS features. Attention-weighted POS features and h_t were shallow concatenated into $([a_t^p; h_t])$ vector.

4. Decoder: The attention-weighted representation $([a_t^p; h_t])$ was fed to a linear layer to predict the tag emission sequence $\hat{y}_t$ followed by a CRF layer that takes as input the $\hat{y}_t$ sequence along with the true tag y_t sequence (Huang et al., 2015).

5 Experiments

The experiments were designed to evaluate the performance of the distant NER models trained with the DISTANT-CTO set alone *vs.* DISTANT-CTO set in combination with the EBM-PICO training set. The EBM-PICO training set is naturally composed of both positive and negative annotation sentences, but for the DISTANT-CTO, we artificially generated the negative sentences A_-. To evaluate the impact of these negative annotation sentences, we perform ablation experiments, training the models only with positive annotation sentences A_+. Finally, we also evaluate the performance when training using the entity annotations with match score $d_s = 1.0$ alone *vs.* entity annotations with $d_s \geq 0.9$. A simple SciBERT-CRF model trained using positive annotation sentences A_+ was used as the baseline. Transformer-based models incorporate sequence order and self-attention components, so our baseline served to check the impact of removing costly BiLSTM and self-attention modules.

5.1 Benchmark datasets

We evaluate our weakly annotated dataset and the NER model on the following PICO benchmarks.

1. **EBM-PICO gold.** The EBM-PICO dataset developed by Nye *et al.* consists of 5000 PICO entity/span annotated documents [1]. It comes pre-divided into a training set (n=4,933) annotated through crowd-sourcing and an expert annotated test set (n=191) for evaluation purposes. We use the training set for combined training experiments and the test set for evaluation.

2. **Physio set.** A test set comprising 153 PICO entity/span annotated documents from Physiotherapy and Rehabilitation RCTs (Randomized Controlled Trials) was used as an additional benchmark to evaluate the generalization power of our approach for this subdomain (Dhrangadhariya et al., 2021).

5.2 Experimental Setup

We define the following experimental setups based on the motivations described in section 5:

- **Exp 1.0 distant A_{+-} c[1,0.9] wPOS** The setup is composed of SciBERT BiLSTM CRF trained on the surface form (text) and attention-weighted POS inputs using DISTANT-CTO set comprising entity-annotated sentences A_{+-} with $d_s \geq 0.9$.

- **Exp 1.1 distant A_{+-} c[1] wPOS** The setup is composed of SciBERT BiLSTM CRF trained on the surface form and attention-weighted POS inputs using the DISTANT-CTO set comprising only the entity-annotated sentences A_{+-} with $d_s = 1.0$.

- **Exp 1.2 distant A_{+} c[1] wPOS** The setup is composed of SciBERT BiLSTM CRF trained on the surface forms and attention-weighted POS inputs using DISTANT-CTO set comprising only the $d_s = 1.0$ annotations. The negative annotation sentences were removed in this case and the system was trained with positive annotated candidates A_{+} only.

- **Exp 1.3 distant A_{+} c[1] POS $\neg$ BiLSTM attention** The setup is composed of SciBERT CRF trained on the surface form inputs using DISTANT-CTO set comprising only the $d_s = 1.0$ annotations with only positive annotated candidates A_{+}. Attention weights were removed from the POS inputs. This setup was used as the baseline.

- **Exp 2.0 - Exp 2.3** These experiments are identical to their series 1.x counterparts except that the models are trained on a combination of the DISTANT-CTO with the EBM-PICO training set. Exp 2.3 using SciBERT-CRF architecture was used as another baseline.

[1]A single document consists of a title and an abstract.

5.3 Evaluation

To evaluate the quality of automatic annotation using the DISTANT-CTO approach, we performed manual annotation of the "Intervention" class over 200 randomly selected samples from the dataset and compared it to the automatic annotations.

Model evaluation was carried out by predicting the "Intervention" tokens for both benchmarks. Each experiment was conducted thrice with three random seeds (0, 1, and 42), and the average metrics (Precision, Recall, and F1) over three repetitions were reported. We evaluated the statistical significance of our best model using the paired student's t-test as described in (Dror et al., 2018). Further experimental details are in the Appendix.

6 Results

This section reports empirical results for the candidate generation process, evaluation for the annotation quality of DISTANT-CTO approach using the validation sets (see Table 2), and the average of the performance metrics and standard deviation σ over three random seeds on both benchmark datasets for the described NER experiments (see Table 4). We compare the performance of our weakly-supervised NER models with the previous SOTA fully supervised (FS) methods that train on the EBM-PICO training set and evaluate on EBM-PICO gold and also a weakly supervised approach (see Table 3). These models were separately trained for each of the PICO entities/spans and also clubbed the "Intervention" and "Comparator" together.

6.1 Candidate Generation

A total of 360,395 CTO records were downloaded as of March 2021. From all the downloaded CTO records, we extract 200,545 unique (391,286 redundant) intervention names from the aforementioned intervention sources. Out of the 391,286 intervention terms retrieved, 104,433 terms were successfully mapped to one of the target sentences with the $d_s = 1.0$, and 3084 more were mapped with a score of 0.9. Adding $d_s \geq 0.9$ mappings did not increase the total number of annotated sentences, but it did increase the number of annotations obtained in each sentence. Table 1 shows the total number of intervention annotations obtained from mapping the source terms to target sentences. Metrics for

Annotation level	$d_s = 1.0$	$1.0 < d_s \geq 0.9$
mention-level	943,284	17,199
token-level	1,515,868	43,096

Table 1: Token-level and mention-level intervention annotations obtained in the weakly annotated DISTANT-CTO dataset grouped by their d_s scores.

the manual evaluation of DISTANT-CTO using the validation set show that adding annotations with $d_s \geq 0.9$ increases the recall by 3%, but lead to an expected drop in the precision (see Table 2).

Match score	**P**	**R**	**F1**
$d_s = 1.0$	0.86	0.80	0.83
$d_s \geq 0.9$	0.84	0.83	0.84

Table 2: Macro-averaged evaluation metrics for the $d_s = 1.0$ and ≥ 0.9 entity annotations for the validation set detailed in the section 5.3

6.2 Model Training

Using the DISTANT-CTO set alone with the NER approach (Exp 1.1 Table 3 and 4) crosses the previous SOTA F1 on the EBM-PICO benchmark by 2%. The best overall F1 for both benchmarks is reached upon training the NER models with combined weakly-labeled DISTANT-CTO with the strongly-labeled EBM-PICO dataset (Exp 2.0 Table 4) crossing the previous SOTA F1 by 5% on the EBM-PICO benchmark. The improvement in F1 for the combined experiments (see Exp 2.1 and 2.0 Table 4)) is significant when compared to the their best DISTANT-CTO counterparts (see Exp 1.1 Table 4)). Using DISTANT-CTO alone has good precision across the experiment series 1.x, but combining it with the EBM-PICO further improves the recall and balances out the F1 in the experiment series 2.x. Adding the artificially generated A_{-} sentences increases the previous F1 by 5.71% and 3.77% (compare Exp 2.2 with Exp 2.1) for both

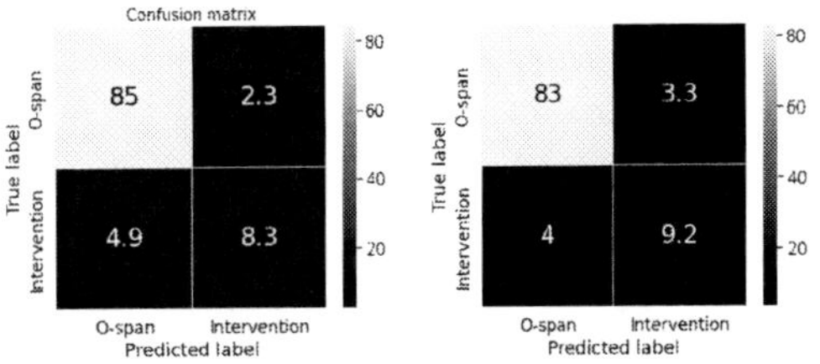

(a) $d_s = 1.0$ (b) $d_s \geq 0.9$

Figure 3: Confusion matrices for the evaluation of DISTANT-CTO validation set annotations with a) $d_s = 1.0$ and b) $d_s \geq 0.9$.

the benchmarks. Note that adding these negative sentences results in an important improvement of about 9% in the F1 for the Physio dataset that is specific for the domain of physiotherapy and rehabilitation. For the combined experiment, the addition of the $d_s \geq 0.9$ annotations improves the F1 as well by a small margin for the EBM-PICO benchmark (Exp 2.0 I.) but has a marginal performance loss for the Physio benchmark (Exp 2.0 II.). While using the DISTANT-CTO alone with the $d_s \geq 0.9$ annotations boosts the precision but downgrades recall thereby reducing the F1 for both benchmarks.

Type	Method	P	R	F1
FS	Nye (2018)	**84.00**	61.00	70.00
FS	Beltagy (2019)	61.00	70.00	65.00
FS	Brockmeier (2019)	69.00	47.00	56.00
FS	Stylianou (2021)	69.04	79.24	73.29
WS	Liu (2021)	22.00	54.00	31.00
WS	Exp 1.1 (Our)	83.36	70.38	75.02
HS	Exp 2.0 (Our)	76.93	**80.17**	**78.44**

Table 3: Comparison of DISTANT-CTO NER models against the previous SOTA NER methods for "Intervention" recognition in terms of macro-averaged precision (P), recall (R), and F1 scores. Boldface represents the best score. Note: FS = Fully Supervised, WS = Weakly Supervised, HS = Hybrid Supervision.

7 Error Analysis

7.1 Candidate Generation

Confusion matrices (see Figures 3a and 3b) for manual evaluation of DISTANT-CTO validation set show that relaxing d_s from 1.0 to 0.9 does improve the true positives (TP) and reduce false negatives (FN) by 0.9% for the "Intervention" class but also reduce the precision by increasing false

Experimental setup	P	R	F1 $\pm\sigma$	P	R	F1 $\pm\sigma$
		I. EBM-PICO gold			II. Physio set	
Exp 1.0	88.85	65.39	71.27 $\pm$0.007	86.13	63.70	69.14 $\pm$0.003
Exp 1.1	83.36	70.38	75.02 $\pm$0.013	79.45	66.28	70.63 $\pm$0.008
Exp 1.2	74.85	68.74	71.25 $\pm$0.005	70.52	66.37	68.14 $\pm$0.002
Exp 1.3 (baseline 1)	85.82	64.84	70.31 $\pm$0.002	79.97	60.79	65.14 $\pm$0.005
Exp 2.0	76.93	80.17	**78.44*** $\pm$0.006	75.55	79.42	77.32 $\pm$0.010
Exp 2.1	77.10	78.83	77.89 $\pm$0.007	76.29	80.18	**78.07*** $\pm$0.009
Exp 2.2	67.65	85.02	72.18 $\pm$0.009	64.80	83.69	68.75 $\pm$0.011
Exp 2.3 (baseline 2)	70.91	77.38	73.60 $\pm$0.025	71.50	78.40	74.38 $\pm$0.020

Table 4: Macro-averaged performance metrics for the NER models trained on weakly annotated DISTANT-CTO alone *vs.* in combination to the strongly annotated EBM-PICO on the two described benchmarks (EBM-PICO gold and the Physio corpus). Bold is the best experiment score. Asterisk (*) denotes a significant F1-score of the experiment to its counterpart in the series 1.x. Significance tested using the paired student's t-test.

positives by 1%. Improved recall for the "Intervention" class is undoubtedly preferred, and hence it is vital to inspect the cause of false negatives. A considerable chunk of false negatives was either i) missed intervention abbreviations and the synonyms not mentioned under the sources, or ii) when only the partial intervention name was mentioned in the source, or iii) if specific intervention terms from the source were mentioned in the target but with different word order (see Table 5). This detailed post-hoc error analysis also revealed that 67% false negatives fell under non-drug type composite intervention mentions (phrase mentions of more than two words). For instance, although the term *'Home-based Rehabilitation using Interactive devices'* is expressed in the sentence *'This study investigates clinical outcomes after the rehabilitation by interactive home-based devices.'*, it will remain unmapped to it because the term does not map to the target text using our alignment heuristic. The problem lies in the lack of naming conventions for non-pharma treatment mentions that are neither clearly identified nor standardized as semantic units(Dhrangadhariya et al., 2021). There are two possible programmatic solutions to this. The first is using additional external ontologies as sources of distant supervision which improves coverage of our labeling function to detect further writing variations within the text. Another solution to matching such source and target text is using order-free string matching algorithms (Apostolico et al., 1992). Using external ontologies solves the issues of missed synonyms, and adding an external dictionary of treatment abbreviations could solve the problem of missed abbreviations (Fries et al., 2021). We noticed that the "Comparator" terms (e.g., placebo, sham, saline, etc.) were often not mentioned as structured sources. The development of a general comparator term dictionary could improve this. Improving the coverage and reducing the false negatives (thereby improving recall) using these methodologies suggests an area where future work would be valuable. Most false positives were a result of bigram matching. We will modify fuzzy bigram matching to relevant bigram matching, thereby reducing the occurrences of spurious false-positive bigrams as matches. Only frequently occurring bigrams from the source will be matched to the targets. We plan to explore the quality of DISTANT-CTO for $d_s \leq 0.9$.

Category	FN count
Missed synonym	168
Missed abbreviation	77
Partial match (incl. boundary errors)	361
Missed comparator term	43
Reorder	39
Total	688

Table 5: Distribution of the false negatives in the DISTANT-CTO evaluation corpus.

7.2 Model Training

Manual error analysis was carried out for both the PICO benchmarks, and the error counts for EBM-PICO gold are reported in Table 6. Each token level error was divided into either of the four classes: 1) false negative (FN) - if the entire entity that the token as part of was missed out by the NER model prediction, 2) false positive (FP) - if the entire entity that the token was part of was falsely recognized as "Intervention", 3) boundary error (BE) - if the boundary tokens were missed out but otherwise the entity was identified by the NER model prediction, and 4) overlapping error (OE) - if the NER model made an error in the non-peripheral tokens of an otherwise identified entity mention. Non-peripheral tokens are all the tokens except the first and the last token of the multi-token entity/span.

Models trained on DISTANT-CTO alone had a fewer boundary and overlapping errors, meaning they missed out on many "Intervention" entity signals leading to high precision but compromised recall. On the contrary, NER models trained on combined datasets made twice the more BE and six times more OE. While most BE and OE in the 1.x series were false negatives, they were false positives in the 2.x series leading to a higher recall. This could be because the EBM-PICO training set annotated the longest possible intervention span resulting in spans rather than pure entities in the DISTANT-CTO approach. Combined training set models also picked out names of treatments, surgeries, and enzymes not used as treatments in the RCT as intervention mentions. A huge chunk of overall FN (including the FN tokens in BE and OE) was for entities with composite intervention terms containing two or more tokens. We noticed that the NER system also missed several short intervention names and abbreviations. Overlapping errors occurred when multiple intervention names were mentioned together, separated by either comma or punctuation, or other conjunctions. The error analysis revealed some issues within EBM-PICO ground truth, which had inconsistencies with the intervention boundaries for whether intervention frequency, dose, and the way of administration should be marked as "Intervention". Several times, the ground truth marked articles preceding the entity and prepositions and punctuation succeeding the entity. Extended error analysis can be found in the Appendix.

Exp	FP	FN	BE	OE
	EBM-PICO gold			
Exp 1.0	819	1688	559	66
Exp 2.0	759	1112	1278	515
Exp 1.1	790	1152	650	55
Exp 2.1	793	1039	1327	517

Table 6: Distribution of the token-level errors made by the corresponding NER models on EBM-PICO gold.

8 Conclusions and Future Work

We exploit the freely-available clinicaltrials.org (CTO) and distant supervision for developing the largest available weakly annotated database of Intervention-Comparator entities across 11 subtypes. Using these weak annotations combined with the manual annotations, we train an "Intervention" NER model that surpasses current approaches by more than 5% in terms of F1 on the EBM-PICO gold benchmark and demonstrate strong generalizability on a domain-specific physiotherapy benchmark. When the same NER model was trained with the weakly annotated dataset alone, it surpassed other approaches by 2%. This is a prototypical work, and an automatically obtained dataset with I and C annotations are being extended for the Participant (P), Outcome (O), and Study type (S) entities. The code and data are available on Github.

References

David Ifeoluwa Adelani, Michael A Hedderich, Dawei Zhu, Esther van den Berg, and Dietrich Klakow. 2020. Distant supervision and noisy label learning for low resource named entity recognition: A study on hausa and yorùbá. *arXiv preprint arXiv:2003.08370*.

Alberto Apostolico, Maxime Crochemore, Zvi Galil, and Udi Manber. 1992. Combinatorial pattern matching third annual symposium tucson, arizona, usa, april 29–may 1, 1992 proceedings. In *Conference proceedings CPM*, page 236. Springer.

Isabelle Augenstein, Leon Derczynski, and Kalina Bontcheva. 2017. Generalisation in named entity recognition: A quantitative analysis. *Computer Speech & Language*, 44:61–83.

Iz Beltagy, Kyle Lo, and Arman Cohan. 2019. SciBERT: A pretrained language model for scientific text. In *Proceedings of the 2019 Conference on Empirical Methods in Natural Language Processing and the 9th International Joint Conference on Natural Language Processing (EMNLP-IJCNLP)*, pages 3615–3620, Hong Kong, China. Association for Computational Linguistics.

Lidong Bing, Bhuwan Dhingra, Kathryn Mazaitis, Jong Hyuk Park, and William W Cohen. 2017. Bootstrapping distantly supervised ie using joint learning and small well-structured corpora. In *Thirty-First AAAI Conference on Artificial Intelligence*.

Florian Boudin, Jian-Yun Nie, Joan C Bartlett, Roland Grad, Pierre Pluye, and Martin Dawes. 2010. Combining classifiers for robust PICO element detection. *BMC medical informatics and decision making*, 10(1):1–6.

Austin J Brockmeier, Meizhi Ju, Piotr Przybyła, and Sophia Ananiadou. 2019. Improving reference prioritisation with PICO recognition. *BMC medical informatics and decision making*, 19(1):1–14.

Anjani Dhrangadhariya, Gustavo Aguilar, Thamar Solorio, Roger Hilfiker, and Henning Müller. 2021. End-to-end fine-grained neural entity recognition of patients, interventions, outcomes. In *International Conference of the Cross-Language Evaluation Forum for European Languages*, pages 65–77. Springer.

Rotem Dror, Gili Baumer, Segev Shlomov, and Roi Reichart. 2018. The hitchhiker's guide to testing statistical significance in natural language processing. In *Proceedings of the 56th Annual Meeting of the Association for Computational Linguistics (Volume 1: Long Papers)*, pages 1383–1392.

Oren Etzioni, Michele Banko, Stephen Soderland, and Daniel S Weld. 2008. Open information extraction from the web. *Communications of the ACM*, 51(12):68–74.

Jason A Fries, Ethan Steinberg, Saelig Khattar, Scott L Fleming, Jose Posada, Alison Callahan, and Nigam H Shah. 2021. Ontology-driven weak supervision for clinical entity classification in electronic health records. *Nature communications*, 12(1):1–11.

Athanasios Giannakopoulos, Claudiu Musat, Andreea Hossmann, and Michael Baeriswyl. 2017. Unsupervised aspect term extraction with B-LSTM & CRF using automatically labelled datasets. In *Proceedings of the 8th Workshop on Computational Approaches to Subjectivity, Sentiment and Social Media Analysis*, pages 180–188, Copenhagen, Denmark. Association for Computational Linguistics.

Malcolm W Greaves. 2014. *Relation Extraction using Distant Supervision, SVMs, and Probabilistic First Order Logic*. Ph.D. thesis, Carnegie Mellon University.

Suchin Gururangan, Ana Marasović, Swabha Swayamdipta, Kyle Lo, Iz Beltagy, Doug Downey, and Noah A. Smith. 2020. Don't stop pretraining: Adapt language models to domains and tasks. In *Proceedings of the 58th Annual Meeting of the Association for Computational Linguistics*, pages 8342–8360, Online. Association for Computational Linguistics.

Michael A Hedderich, Lukas Lange, and Dietrich Klakow. 2021. Anea: Distant supervision for low-resource named entity recognition. *arXiv preprint arXiv:2102.13129*.

Sepp Hochreiter and Jürgen Schmidhuber. 1997. Long short-term memory. *Neural computation*, 9(8):1735–1780.

Ke-Chun Huang, I-Jen Chiang, Furen Xiao, Chun-Chih Liao, Charles Chih-Ho Liu, and Jau-Min Wong. 2013. PICO element detection in medical text without metadata: Are first sentences enough? *Journal of biomedical informatics*, 46(5):940–946.

Ke-Chun Huang, Charles Chih-Ho Liu, Shung-Shiang Yang, Furen Xiao, Jau-Min Wong, Chun-Chih Liao, and I-Jen Chiang. 2011. Classification of PICO elements by text features systematically extracted from pubmed abstracts. In *2011 IEEE International Conference on Granular Computing*, pages 279–283. IEEE.

Zhiheng Huang, Wei Xu, and Kai Yu. 2015. Bidirectional lstm-crf models for sequence tagging. *arXiv preprint arXiv:1508.01991*.

Di Jin and Peter Szolovits. 2018. PICO element detection in medical text via long short-term memory neural networks. In *Proceedings of the BioNLP 2018 workshop*, pages 67–75. Association for Computational Linguistics.

Shifeng Liu, Yifang Sun, Bing Li, Wei Wang, Florence T. Bourgeois, and Adam G. Dunn. 2021. Sent2Span: Span detection for PICO extraction in the biomedical text without span annotations. In *Findings of the Association for Computational Linguistics: EMNLP 2021*, pages 1705–1715, Punta Cana, Dominican Republic. Association for Computational Linguistics.

Benjamin Nye, Junyi Jessy Li, Roma Patel, Yinfei Yang, Iain J Marshall, Ani Nenkova, and Byron C Wallace. 2018. A corpus with multi-level annotations of patients, interventions and outcomes to support language processing for medical literature. In *Proceedings of the conference. Association for Computational Linguistics. Meeting*, volume 2018, page 197. NIH Public Access.

Minlong Peng, Xiaoyu Xing, Qi Zhang, Jinlan Fu, and Xuanjing Huang. 2019. Distantly supervised named entity recognition using positive-unlabeled learning. In *Proceedings of the 57th Annual Meeting of the Association for Computational Linguistics*, pages 2409–2419, Florence, Italy. Association for Computational Linguistics.

David Pinto, Andrew McCallum, Xing Wei, and W Bruce Croft. 2003. Table extraction using conditional random fields. In *Proceedings of the 26th annual international ACM SIGIR conference on Research and development in informaion retrieval*, pages 235–242. ACM.

John W Ratcliff and David E Metzener. 1988. Pattern-matching-the gestalt approach. *Dr Dobbs Journal*, 13(7):46.

Alexander Ratner, Stephen H Bach, Henry Ehrenberg, Jason Fries, Sen Wu, and Christopher Ré. 2017. Snorkel: Rapid training data creation with weak supervision. In *Proceedings of the VLDB Endowment. International Conference on Very Large Data Bases*, volume 11, page 269. NIH Public Access.

Esteban Safranchik, Shiying Luo, and Stephen Bach. 2020. Weakly supervised sequence tagging from noisy rules. In *Proceedings of the AAAI Conference on Artificial Intelligence*, volume 34, pages 5570–5578.

Alisa Smirnova and Philippe Cudré-Mauroux. 2018. Relation extraction using distant supervision: A survey. *ACM Computing Surveys (CSUR)*, 51(5):1–35.

Nikolaos Stylianou and Ioannis Vlahavas. 2021. Transformed: End-to-end transformers for evidence-based medicine and argument mining in medical literature. *Journal of Biomedical Informatics*, 117:103767.

Ashish Vaswani, Noam Shazeer, Niki Parmar, Jakob Uszkoreit, Llion Jones, Aidan N Gomez, Łukasz Kaiser, and Illia Polosukhin. 2017. Attention is all you need. In *Advances in neural information processing systems*, pages 5998–6008.

Byron C Wallace, Joël Kuiper, Aakash Sharma, Mingxi Zhu, and Iain J Marshall. 2016. Extracting PICO sentences from clinical trial reports using supervised distant supervision. *The Journal of Machine Learning Research*, 17(1):4572–4596.

Zeyi Wen, Dong Deng, Rui Zhang, and Ramamohanarao Kotagiri. 2019. : An efficient entity extraction algorithm using two-level edit-distance. In *2019 IEEE 35th International Conference on Data Engineering (ICDE)*, pages 998–1009. IEEE.

Yaosheng Yang, Wenliang Chen, Zhenghua Li, Zhengqiu He, and Min Zhang. 2018. Distantly supervised ner with partial annotation learning and reinforcement learning. In *Proceedings of the 27th International Conference on Computational Linguistics*, pages 2159–2169. Association for Computational Linguistics.

Tengteng Zhang, Yiqin Yu, Jing Mei, Zefang Tang, Xiang Zhang, and Shaochun Li. 2020. Unlocking the power of deep PICO extraction: Step-wise medical ner identification. *arXiv preprint arXiv:2005.06601*.

A Appendix

A.1 DISTANT-CTO characteristics

The total number of entity-level "Intervention" mentions in DISTANT-CTO are almost 30 times more than in the EBM-PICO dataset as shown in Table 7. For the EBM-PICO training set, 57.48% of mentions fell under the "drug" class and the rest under the six remaining classes.

Total	DISTANT-CTO	EBM-PICO
mention-level	977,682	32,890
token-level	1,558,964	125,920

Table 7: Comparing the number of "Intervention" annotations in DISTANT-CTO *vs.* EBM-PICO.

Out of all the mention-level annotations in the DISTANT-CTO dataset, 59.90% corresponded to "drug" class and 40% to the rest of 10 classes. The pie chart (upper pie in Figure 4) shows the class distribution of the semantic classes for the retrieved "Intervention" mentions s_{name} about half of which fall under the "drug" (or Pharma) class and the rest under the remaining 10 non-pharma classes. Out of the total retrieved mentions, almost two-thirds that get mapped to a target t sentences also fall under the "drug" class (lower pie in Figure 4). Table 8 and 9 shows the number of retrieved intervention mentions by their semantic class *vs.* the percentage of these intervention mentions that get mapped to some target sentence with the match score d_s of 1.0 and score 0.9 respectively. Notice that collectively the intervention mentions that fall under the non-pharma classes outnumber the pharma ("drug") mentions.

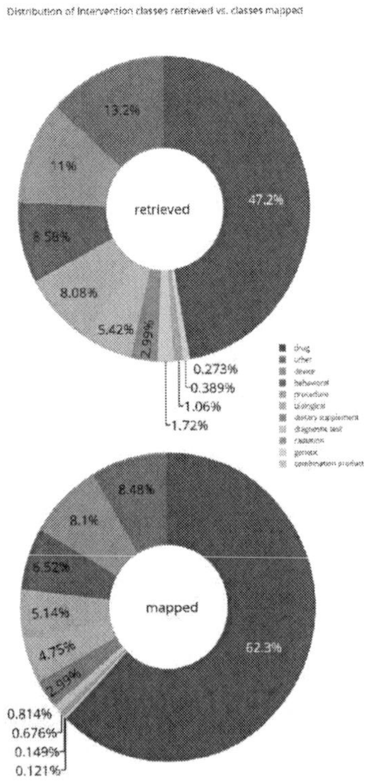

Figure 4: upper) Class distribution for the retrieved "Intervention" mentions, and lower) Class distribution for the mapped "Intervention" mention.

Top semantic classes for the most mapped and most unmapped intervention mentions from the total retrieved mentions are shown in the figure 6 and 5. As evident from the tables 8 and 9 "drug" class intervention mentions are the most mapped

Domain	retrieved - (mapped)
drug	184835 (35.50%)
device	43134 (20.09%)
other	51703 (16.19%)
procedure	31630 (21.38%)
behavioral	33590 (16.03%)
biological	21225 (22.86%)
dietary supplement	11699 (25.46%)
radiation	4134 (20.44%)
diagnostic test	6742 (10.13%)
combination product	1070 (14.39%)
genetic	1524 (07.94%)
all non-pharma	206,451 (18.80%)

Table 8: Number of intervention mentions retrieved *vs.* percentage mapped with $d_s = 1.0$

Domain	retrieved - mapped
drug	184835 (36.22%)
device	43134 (21.13%)
other	51703 (16.84%)
procedure	31630 (22.16%)
behavioral	33590 (16.44%)
biological	21225 (24.07%)
dietary supplement	11699 (27.44%)
radiation	4134 (21.17%)
diagnostic test	6742 (10.78%)
combination product	1070 (14.95%)
genetic	1524 (08.53%)
all non-pharma	206,451 (19.64%)

Table 9: Number of intervention mentions retrieved *vs.* percentage mapped with a d_s of 0.9

followed by "dietary supplement" and "procedure" classes which also reflects in the pie chart of most mapped lengths and common phrase lengths for each class (see Figure 5). The most frequent phrase length for these classes is one (unigram) and the second most frequent length is two (bigram).

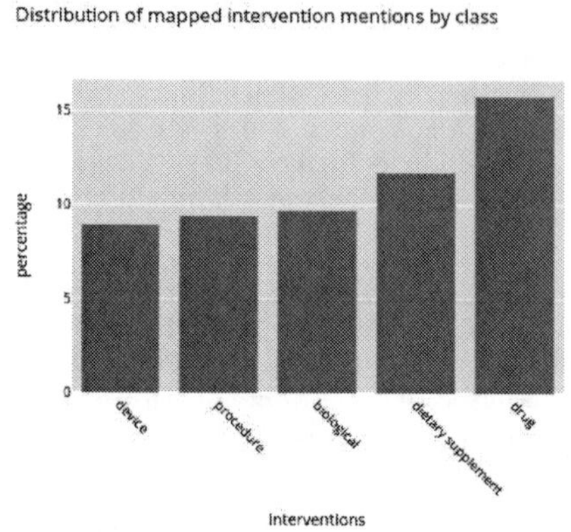

Figure 5: Top five semantic classes, source intervention mentions from which get mapped to the target.

Domain	Most common length
drug	1
dietary supplement	1
biological	1
procedure	2
device	1

Table 10: Lengths for the most mapped classes

The least mapped intervention mention classes are "combination product", "diagnostic test" and "behavioral" (refer figures 6) with most intervention mentions in these classes containing either trigrams or bigrams. This very well reflects with the numbers in figures 7 which shows that trigram and bigram intervention mentions constitute almost half the right pie showing the top phrase lengths for intervention mentions that remain unmapped. One of the ways to retain some of the missed bigram and trigram intervention mentions is to explore the matches with lower match scores. The Table 12 shows some of the $d_s \geq 0.9$ source-target matches not captured by the $d_s = 1.0$ constraint because of the difference of either a single missing space or singular-plural differences. It is also interesting to note that the radiographic procedure "cystourethrography" matches the name of the test "cys-

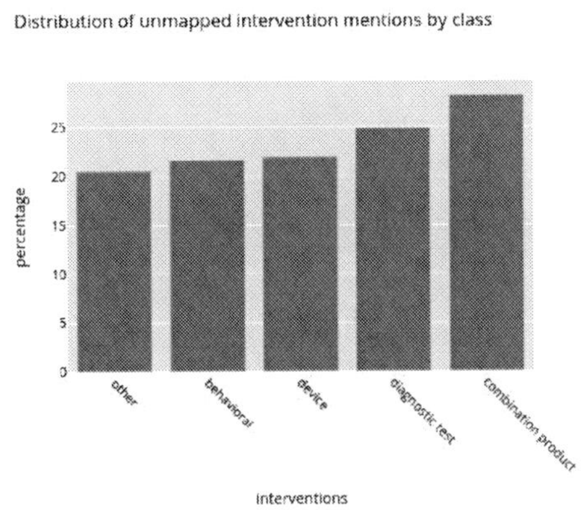

Figure 6: Top five semantic classes of the source intervention mentions that remain unmapped to the target.

Domain	Most common length
device	3
other	2
behavioral	3
diagnostic test	2
combination product	3

Table 11: Lengths for the most unmapped classes

tourethrogram".

A.2 Experimental Details

For the candidate generation process, we did not define any junk elements for using the `sequencematcher` function. All the NER experiments in this article were conducted in PyTorch and the models were trained for 10 epochs with a mini-batch size of 10 for training and 6 for evaluation. We used the IO (Inside, Outside) also called raw labeling for all the NER tasks to make the experiments compared with the previous studies. The maximum sequence length was set to 100 because the average length of each input text sequence was about 68 words. For both experiments types, either using the DISTANT-CTO alone or with the EBM-PICO training set, 80% of the data was used for training and 20% for development. The [CLS] embeddings from the SciBERT layer were used as features of the input text. SciBERT was finetuned by not freezing weights during the experiments. The hidden size for LSTM/BiLSTM was set to 512/1024 for the text input embeddings and 20/40 for the POS one-hot embeddings. ReLU was used as the activation function before feeding emission outputs to the CRF layer. Model training was optimized using AdamW using a learning rate of 5e-5. The gradients were clipped to 1.0 to mitigate the problem of exploding gradients. Due to very specific RAM and GPU requirements for each experiment and the institute's capacity for sharing the GPUs amongst the group members, experiments were carried out on the following GPUs. Each experiment was carried out on a single GPU without any data and model parallelization.

B Extended Error Analysis

Manual error analysis results for Physio corpus are reported in the Table 14. FP error count was always lower than the FN error count in the EBM-PICO gold but for the Physio set, the combined NER experiments (series 2.x) lead to a higher FP compared the FN. The ratio of BE in Exp series 2.x is on an average 1.2 times that of series 1.x. However, a large chunk of BE in series 1.x are false negatives in contrast to the BE in series 2.x which are false positives. Upon closer inspection of false-negative BE in series 1.x, we found that they were either missed intervention synonyms inside brackets, missed information accompanying intervention terms like dose, type, medium of intervention, administrator of intervention, or location of administration. This is due to the fact that distantly supervised annotation does not take into account labeling the additional intervention information except the name. The addition of the manually annotated EBM-PICO in the combined training experiments reduces the number of false-negative BE. This is due to the fact that EBM-PICO guidelines required the annotators to mark the longest possible phrase describing intervention including the additional information like dose, mode, medium, and location of administration.

For both the evaluation corpora, the combined NER experiments lead to more TP for the "Intervention" class which is vital to PICO entity/span recognition. This could be the case because the combination of weakly and strongly annotations reduce the percentage of unseen surface forms (words) from both test sets. 27.70% of the intervention entity surface forms in the EBM-PICO gold benchmark remain unseen in the EBM-PICO training set while for the DISTANT-CTO training set it drops to 21.38%. 27.29% of the intervention entity surface forms in the Physio benchmark remain unseen in the EBM-PICO training set while for

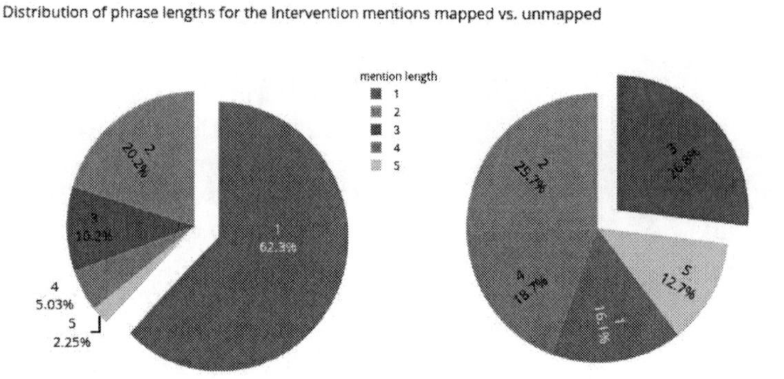

Figure 7: Left) Phrase length distribution of mapped intervention mentions, Right) Phrase length distribution of unmapped intervention mentions.

Characteristic	Source	Target	d_s
Single missing space	"l carnitine"	"lcarnitine"	0.923
Missing negations	"no pumice prophylaxis"	"pumice prophylaxis"	0.900
Plurals	"punch skin biopsies"	"punch skin biopsy"	0.941
Abbreviations	"rfsh alone"	"recombinant fsh alone"	0.926
Specific treatment name to generic treatment name	"biphasic insulin aspart 50"	"biphasic insulin aspart"	0.923
Procedure matches the instrument	"cystourethrography"	"cystourethrogram"	0.900

Table 12: Example "Intervention" mentions from CTO that get mapped to target sentences t with a d_s of 0.9

GPU	RAM	Experiment
Tesla V100-PCIE-16GB	1TB	1.1, 2.1, 1.3
TeslaK80 GPU	126GB	1.2, 2.2
Tesla V100-PCIE-32GB	1TB	2.0, 1.0, 2.3

Table 13: Experiments and the details of GPUs they were carried out on.

Exp	FP	FN	BE	OE
	Physio set			
Exp 1.0	963	1586	654	20
Exp 2.0	1168	897	867	347
Exp 1.1	990	1420	723	19
Exp 2.1	1116	904	1025	228

Table 14: Distribution of the token-level errors made by the corresponding NER models on Physio set.

the DISTANT-CTO training set it drops to 22.97%. Combining both training sets leads to a reduction in unseen surface forms to 16.29% and 15.13% for the EBM-PICO gold and Physio benchmarks respectively. (Augenstein et al., 2017) has shown that recall on unseen surface forms is significantly lower than on seen surface forms for NER tasks.

C Ethical Statement

This paper studies clinical NER with a small strongly labeled and a large weakly labeled dataset. Our investigation neither introduces any social or ethical bias to the model nor amplifies any bias in the data. We do not foresee any direct social consequences or ethical issues.

D License Information

DISTANT-CTO uses all of clinicaltrials.gov (CTO) data that allows downloading and using it given that any publication/distribution states and describes any modifications made to the content of the data. It is public data that anyone can download and reproduce the outcomes with the code made available on Github.

EchoGen: A New Benchmark Study on Generating Conclusions from Echocardiogram Notes

Liyan Tang[1], Shravan Kooragayalu[2], Yanshan Wang[2], Ying Ding[1], Greg Durrett[1], Justin F. Rousseau[1], Yifan Peng[3]

[1]The University of Texas at Austin [2]University of Pittsburgh
[3]Weill Cornell Medicine

lytang@utexas.edu, {SHK14, yanshan.wang}@pitt.edu
ying.ding@ischool.utexas.edu, gdurrett@cs.utexas.edu
justin.rousseau@austin.utexas.edu, yip4002@med.cornell.edu

Abstract

Generating a summary from findings has been recently explored (Zhang et al., 2018, 2020) in note types such as radiology reports that typically have short length. In this work, we focus on echocardiogram notes that is longer and more complex compared to previous note types. We formally define the task of echocardiography conclusion generation (**EchoGen**) as generating a conclusion given the findings section, with emphasis on key cardiac findings. To promote the development of EchoGen methods, we present a new benchmark, which consists of two datasets collected from two hospitals. We further compare both standard and state-of-the-art methods on this new benchmark, with an emphasis on factual consistency. To accomplish this, we develop a tool to automatically extract concept-attribute tuples from the text. We then propose an evaluation metric, *FactComp*, to compare concept-attribute tuples between the human reference and generated conclusions. Both automatic and human evaluations show that there is still a significant gap between human-written and machine-generated conclusions on echo reports in terms of factuality and overall quality[1].

1 Introduction

Echocardiography (or echo) is a test that uses sound waves to produce live images of the heart (Mitchell et al., 2019). It has become routinely used to support the diagnosis, management, and follow-up of patients with suspected or known heart diseases. The echo report documents and communicates the evaluation of cardiac and vascular structures in the echocardiography study. As shown in Figure 1, a standard echo report usually consists of a demographic section, an echocardiographic evaluation section (also called the finding section), and a conclusion section (Gardin et al., 2002). In a typical workflow, consultants who interpret echocardiography provide the quantitative measurement and descriptive statements to describe pertinent findings, and then conclude.

In this work, we formally study the task of echo conclusion generation (EchoGen), arising in clinical practice to relieve the clinician of tasks that may contribute to clinician burnout (Alsharqi et al., 2018). A practical system shall be able to generate statements that emphasize abnormal findings, and compare differences and similarities of the current study versus the previous one if available and relevant. We define EchoGen as a task of learning from the demographic and echocardiographic findings section and generating the conclusion section.

Neural network-based models (See et al., 2017; Lewis et al., 2020) are an attractive method for this task, but are difficult to apply without appropriate training data. To address this gap, we present a large-scale EchoGen benchmark, which consists of two datasets. Here we reply on one prexisting MIMIC-III dataset (EGMIMIC) and one newly collected dataset from the NewYork-Presbyterian Hospital (EGCLEVER) to cover different text genres, data sizes, and degrees of difficulty, and more importantly, highlight common challenges of EchoGen (Figure 1).

Beyond data, a second challenge for EchoGen is to evaluate the factual correctness of a generated conclusion. Automatic metrics such as ROUGE and METEOR only assess content selection but not other quality aspects, such as fluency, grammaticality, and coherence, and are not well-correlated with factuality, leading to the development of separate evaluation measures (Zhang et al., 2018; Falke et al., 2019; Kryscinski et al., 2020; Goyal and Durrett, 2021). This study proposes a new evaluation metric to measure factual consistency, called "FactComp" by considering both concept and their attributes in the fact equivalence criteria.

To better understand the challenge posed by

[1]Code for data construction and model evaluation is available at https://github.com/bionlplab/echo_summarization.

Proceedings of the BioNLP 2022 workshop, Dublin, Ireland, pages 359–368
May 26, 2022.

Patient/test Info:
Indication: Endocarditis.
Height: (in) 74 Weight (lb): 379
...

Findings:
LEFT ATRIUM: Mild LA enlargement.
RIGHT ATRIUM/INTERATRIAL SEPTUM: Normal RA size.
LEFT VENTRICLE: Moderate symmetric LVH. Normal LV cavity size. Suboptimal technical quality, a focal LV wall motion abnormality cannot be fully excluded.
RIGHT VENTRICLE: Normal RV chamber size and free wall motion.
AORTIC VALVE: Normal aortic valve leaflets (3). No AS. No AR.
[...]

Conclusion:
The left atrium is mildly dilated. There is moderate symmetric left ventricular hypertrophy. [...] The aortic valve leaflets (3) appear structurally normal with good leaflet excursion. [...] There is no pericardial effusion. No vegetation seen (cannot definitively exclude).

(a)

Demographic Info:
Age: 85 Sex: M Height: 71 Weight: 174
Clinical Diagnosis: Dyspnea (shortness of breath)
...

Findings:
The mitral valve leaflets appear thickened with normal opening. There are fibrocalcific changes of the aortic valve with normal opening. The aortic root is normal for age and body size. The left atrium is mildly dilated. Although accurate measurements could not be made, the left ventricle appears normal in size with normal wall thicknesses. [...] There is no evidence for coarctation of the aorta. There is no evidence of right to left shunt by saline contrast study.

Conclusion:
Aortic valve calcification.
Left atrial dilatation.
Normal global left ventricular function.
Mild mitral regurgitation.
[...]

(b)

Figure 1: Echocardiography reports from the (a) EGMIMIC and (b) EGCLEVER datasets.

EchoGen, we conducted experiments with five baselines: TF-IDF, RANDSENT, LEXRANK, FACTEXT, and BART. We find that BART exceeds other baselines by a large margin, but it has poor transferability when tested on cross-corpus settings. Further human evaluations indicate that there is still a significant gap between generated conclusions and human reference in terms of fluency and factual consistency.

In summary, our contributions can be summarized as follows. (1) We formally introduce the task of EchoGen. (2) We curate a large-scale benchmark from an existing representative dataset and a newly-collected dataset. (3) We introduce a new metric to measure the fact consistency for echo notes. (4) Our metric and human evaluations find that there is still a gap between human reference and generated conclusions for echo reports in terms of fluency and factual consistency.

2 Related works

While EchoGen has not been defined before, there are closely related tasks that were studied before: data-to-text generation, clinical report summarization, and evaluation.

Data-to-text Generation Data-to-text generation is a task of generating text in natural language from non-linguistic input data such as tables and time series (Gatt and Krahmer, 2018; Wiseman et al., 2017; Gardent et al., 2017). Traditional approaches for data-to-text generation (Reiter and Dale, 2000) follow a pipeline of modules such as content selection, text structuring, and surface realization. Recent methods (Gehrmann et al., 2018; Harkous et al., 2020) generate text from data in an end-to-end fashion using the encoder-decoder approach. Data-to-text is also explored in healthcare (Pauws et al., 2019) to facilitate patient review.

Clinical report summarization Clinical report summarization is a long-standing research problem (Adams et al., 2021). Both extractive and abstractive methods have been applied for summarization, covering cases from structured data to text, medical image to text, and history documents to text (Afantenos et al., 2005; Xiong et al., 2019; Pivovarov and Elhadad, 2015).

To the best of our knowledge, few clinical summarization datasets are available. *MEDIQA 2021 ST* provides a task of generating radiology impression statements from textual clinical findings in radiology reports (Ben Abacha et al., 2021) collected from the Indiana University dataset and Stanford Health Care. *CLIP* is a dataset on discharge notes, where the authors' task was to extract the follow-up action items from notes (Mullenbach et al., 2021) . This dataset is more suitable for developing information extraction (IE) systems or extractive summarization methods. Adams et al. (2021) developed a dataset *CLINSUM* from Columbia University Irving Medical Center, focusing on discharge summary notes. While they identified the complex, multi-document summarization task, the dataset is not public to promote the model development by other researchers.

In comparison, our EchoGen is a completely new task on a new note type – echocardiograms. More importantly, the benchmark covers a diverse range of text genres from two resources. We expect that the models that perform better on both datasets will be more robust in real-word settings.

Evaluation on clinical text Evaluation of clinical text generation or summarization is a challenging research area. Existing methods include automatic approaches and human judgments. For example, commonly used ROUGE-based evaluation metrics measure the overlapping n-grams or longest common sub-sequence between the reference and generated summaries. BERTScore (Zhang et al., 2019) (or HOLMS) is an alternative that accounts for lexical variations by comparing the similarity of semantic representations encoded via BERT (Devlin et al., 2019). However, human evaluations show that these metrics do not always correlate well with factual consistency measurement. Hence, many research works focus on developing automatic consistency metrics that correlate better with human evaluations.

Goodrich et al. (2019) measure the factual consistency as the ratio of overlap between relation triplets under fixed schema extracted from the reference and the generated summary. Kryscinski et al. (2020) propose an entailment-based model FactCC to check whether the source text entails each sentence in the generated summary. Wang et al. (2020) and Durmus et al. (2020) propose QA-based methods that measure the amount of information in the generated summary supported by the source. However, these evaluation approaches often consist of auxiliary modules trained on external or artificial datasets, which is prohibitively expensive and time-consuming to collect. In addition, these modules are hardly generalizable to other clinical settings. Our proposed fact extractor FACTEXT instead relies on linguistic knowledge and is shown to have higher generalizability.

3 EchoGen

3.1 Task definition

We first formulate the EchoGen task. Let $x = \{x_1, ..., x_m\}$ be the demographics and findings section of an echo report, the goal is to generate a conclusion $y = \{y_1, ..., y_n\}$, where m and n are the length of the source section and the generated section of an echo report, respectively. In this work, x is the finding section of a report. We leave leveraging the correlations, if any, between demographic values and generated conclusions into future works.

	EGMIMIC	EGCLEVER
Notes	44,085	13,000
Train	41,164	10,081
Dev	1,447	1,406
Test	1,474	1,513
Source sentences	19	19
Conclusion sentences	14	12
Source tokens	173	219
Conclusion tokens	150	72

Table 1: Statistics for the EchoGen benchmark.

3.2 Dataset construction

The EchoGen benchmark contains two corpora (Table 1. Here, we reply on one prexisting dataset because it is widely used in the clinical NLP community and one newly collected dataset to cover different text styles and levels of difficulties.

EGMIMIC The first dataset was sampled from the MIMIC-III dataset (Medical Information Mart for Intensive Care III) (Johnson et al., 2016). MIMIC-III is a de-identified clinical database composed of over 40,000 patients admitted in the ICUs at Beth Israel Deaconess Medical Center. Of those, we collected echo reports from the `noteevents` table, whose category is "Echo".

We applied the RadText tool [2] to split the notes into a sequence of sections. It uses a rule-based matching algorithm with default rules adapted from SecTag with reported recall of 99% (Denny et al., 2008). We then selected the "Findings" section as the input and the "Conclusion" section as the human reference. We sampled a collection of 41,164, 1,447, and 1,474 reports for training, development, and test, respectively (Table 1). Note that we sampled the echo notes at the patient level. This strategy will ensure that no participant was in more than one group to avoid cross-contamination between the training and test datasets.

EGCLEVER The second dataset is a collection of echo notes in English for heart failure patients from the "PrediCtion of EarLy REadmissions in Patients with CongestiVE HeaRt Failure" (CLEVER) cohort at NewYork-Presbyterian Hospital (called EGCLEVER). The patients were admitted and discharged with billing codes ICD-9 Code 428 or ICD-10 Code I50 from January 2008 and July 2018. The study was reviewed and approved by the NewYork-Presbyterian Hospital Institutional Review Board.

[2] `https://github.com/bionlplab/radtext`

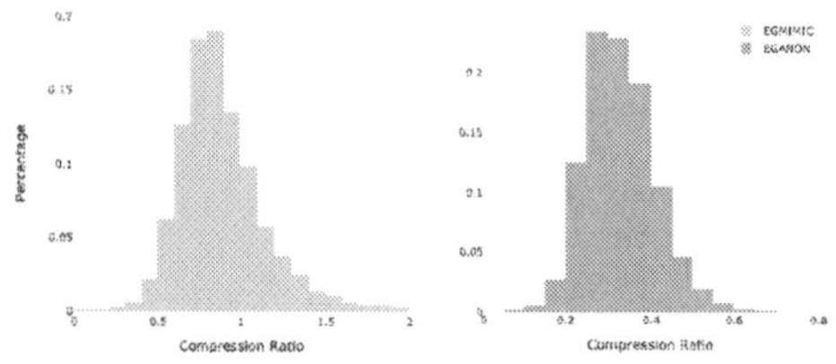

Figure 2: Distribution of word compression ratio on EgMimic and EgClever. The ratio defined as the quotient of number of tokens in the reference and that in the source.

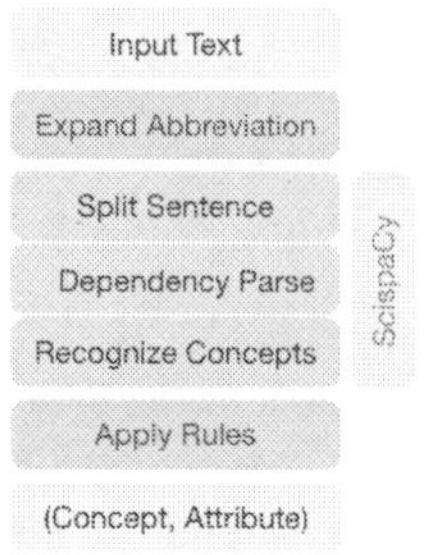

Figure 3: The pipeline of the fact extractor FactExt.

We used the same method to preprocess EgClever and sampled a collection of 10,081, 1,406, and 1,513 reports for training, development, and test, respectively.

Comparison The task of EchoGen varies with the data source, which may depend on the individual hospital. Figure 1 shows one echo report from EgMimic and one from EgClever. The EgMimic report more closely resembles the task of data-to-text generation (Gatt and Krahmer, 2018; Pauws et al., 2019), where the finding section consists of structured data (here, noun phrases in a key-value format), and the conclusion section is written by selecting important findings and expanding them to coherent natural language text. Since data-to-text often has a more complex tabular structure, the result here is somewhere in between pure data and natural language as the tabular structure is not explicit. Therefore, even though the number of tokens in the input is not much shorter than the conclusion section, the conclusion does contain less information than the input.

On the other hand, the conclusion section of our collected dataset EgClever involves more heavily selecting and summarizing content from unstructured text input. The distribution of word compression ratio for both datasets further confirms our observations (Figure 2). The compression ratio is centered around 0.8 for EgMimic and 0.3 for EgClever.

3.3 Evaluation Metrics

ROUGE First, we use the standard ROUGE scores (Lin, 2004), and report the F1 scores for ROUGE-1, ROUGE-2, and ROUGE-L, which compare the word-level unigram, bigram, and longest common sequence overlap between the generated and the human reference conclusion, respectively.

Factual Consistency For *Factual Consistency* evaluation, we define a Factual F1 score, inspired by (Zhang et al., 2020). Specifically, we first extract and represent the facts f as a list of "(Concept, Attribute)" pairs $\langle f_1, ... f_n \rangle$. For example, in the sentence "Right ventricular chamber size and free wall motion are normal", the fact list is ⟨(right ventricular chamber size, normal), (free wall motion, normal)⟩.

The evaluation is then carried out by comparing the list f from the human reference to the list of facts $\hat{f} = \langle \hat{f}_1, ..., \hat{f}_m \rangle$ from a generated conclusion. This requires that a concept and its attributes be extracted correctly to count as one fact.

Finally, the evaluation results are reported using the standard Precision, Recall, and F1-score metrics.

$$P = \frac{1}{|\hat{f}|} \mathrm{FE}(f, \hat{f}), \quad R = \frac{1}{|f|} \mathrm{FE}(f, \hat{f}),$$
$$F = 2\frac{P \cdot R}{P + R}$$

Here, FE is the factual equivalence criteria and can be defined in various modes.

Strict matching The strict matching mode requires exact matching, and it holds when both the concept and attribute are the same. $\mathrm{FE} = \sum_{\hat{f}_i \in \hat{f}} \sum_{f_j \in f} \mathbb{1}[\hat{f}_i = f_j]$.

BERTScore matching This mode uses greedy matching to maximize the matching similarity. Each fact is matched to the most similar fact in the human reference. Here, we concatenate the attribute with the concept to form a factual noun phrase, and used the BERTScore to measure the similarity between two phrases (Zhang et al., 2019). $\mathrm{FE} = \sum_{\hat{f}_i \in \hat{f}} \max_{f_j \in f} \mathrm{BERTScore}(\hat{f}_i, f_j)$.

	EGMIMIC			EGCLEVER			Overall		
	P	R	F1	P	R	F1	P	R	F1
Findings	94.3	83.4	88.3	88.8	73.1	79.9	91.7	78.5	84.3
Conclusion	91.2	76.1	82.6	96.7	93.5	95.0	93.8	84.4	88.5
Overall	92.8	79.8	85.5	92.8	83.3	87.4	92.8	81.5	86.4

Table 2: The performance of FACTEXT on 25 randomly sampled Echo notes from the validation set of EGMIMIC (13) and EGCLEVER (12). Each report consists of one "Findings" section and one "Conclusion" section. All statistics are obtained by averaging scores from each report.

However, both modes have flaws. For example, strict matching does not consider lexical variation and semantic equivalence. On the other hand, since concept-attribute pairs are supposed to be independent, aligning each fact from the generated conclusion to the most similar one in the reference via BERTScore matching is less meaningful if they are two different facts. Therefore, we relax the definition of these modes and propose approximate matching.

Approximate matching This mode combines strict matching and BERTScore matching. Specifically, a predicted fact is equivalent to a reference fact if their BERTScore is above a threshold t[3]. FE $= \sum_{\hat{f}_i \in \hat{f}} \sum_{f_j \in f} \mathbb{1}[\text{BERTScore}(\hat{f}_i, f_j) > t]$.

To extract the facts from the text, we develop a rule-based fact extraction system FACTEXT (Figure 3). The tool first splits the text into sentences, and then obtains the universal dependencies (de Marneffe et al., 2021) from the sentences. It further detects UMLS© concepts mentioned in the sentence. Here we focused on the common 55 concepts in the echo notes identified in the data driven way[4]. We used the ScispaCy model (Neumann et al., 2019) trained on MedMentions (Mohan and Li, 2018) to process the text.

Afterward, we applied rules to all identified concepts and subsequently found the attributes that describe the concept. We include negation as an attribute but not uncertainty words as they rarely show up in the text. In this work, we utilized the universal dependency graph to define rules (Chambers et al., 2007). Therefore, the rules take advantage of linguistic knowledge so that the search of attributes is not limited to fixed word distance. The comprehensive rules can be found at our released code. The performance of FACTEXT is discussed in Section 4.

[3]We set threshold $t = 0.85$ in this study based on the performance on the validation set.

[4]Specific concepts are shown in Appendix A.

3.4 Baseline models for benchmarking

We consider 5 baseline models.

TF-IDF Given a source x, TF-IDF first searches for the most similar source x' over all training data based on TF-IDF features and then chooses corresponding conclusion y' as a conclusion for the source x.

RANDSENT We randomly select $k = 12$ sentences from a source as its conclusion, where k is determined according to the average number of conclusion sentences in two collected datasets.

LEXRANK LexRank constructs a graph representation of the course, where nodes are sentences and edges are similarities between sentences (Erkan and Radev, 2004). It then applies the PageRank algorithm on the graph to extract top $k = 12$ most relevant sentences from the source.

FACTEXT We first extract all facts f from a source and then construct a conclusion by concatenating them together. We next convert (Concept, Attribute) pairs into noun phrases by attaching attributes to the beginning of concepts. For example. (right ventricular chamber size, normal) converts to "normal right ventricular chamber size".

BART BART (Lewis et al., 2020) is a pretrained language model that recently demonstrates the state-of-the-art performance in text summarization. It models the conditional likelihood $p(y|x) = \sum_t p(y_t|y_{<t}, x)$, where $y_{<t}$ denotes generated tokens before time step t. We fine-tune a pretrained BART initialized with `facebook/bart-large-xsum` on both datasets.

4 Benchmark results and discussion

Rule-based system Table 2 shows the performance of FACTEXT on randomly sampled 25 ex-

	ROUGE-1			ROUGE-2			ROUGE-L			FC		
	P	R	F1	P	R	F1	P	R	F1	P	R	F1
TF-IDF	47.7	47.2	44.9	27.4	27.1	25.9	39.2	38.7	36.9	40.2	41.0	38.8
RANDSENT	58.3	49.2	51.4	34.0	29.7	30.5	47.9	41.0	42.6	49.6	45.8	45.9
LEXRANK	60.5	51.5	53.8	37.0	32.3	33.3	49.9	43.1	44.7	53.6	47.5	48.3
FACTEXT	**69.1**	51.7	57.4	40.0	30.0	33.2	63.8	47.6	52.9	48.8	66.0	54.9
BART	65.5	**67.4**	**69.5**	**55.5**	**57.2**	**55.5**	**65.5**	**67.4**	**65.5**	**72.0**	**66.4**	**67.9**

Table 3: Results on EGMIMIC. ROUGE-1/2/L represent the ROUGE-F1 scores. FC represents Factual Consistency using the approximate matching.

	ROUGE-1			ROUGE-2			ROUGE-L			FC		
	P	R	F1	P	R	F1	P	R	F1	P	R	F1
TF-IDF	58.1	57.0	55.5	40.8	40.3	39.1	52.5	51.5	50.2	59.8	60.2	57.8
RANDSENT	37.2	57.7	44.3	17.0	26.9	20.4	28.7	44.9	34.2	33.9	34.3	32.5
LEXRANK	40.2	58.7	46.6	18.0	27.2	21.2	30.8	45.5	35.8	33.4	36.5	33.1
FACTEXT	49.1	49.7	48.3	25.3	25.9	25.0	47.4	47.9	46.6	35.1	50.6	40.4
BART	**76.1**	**72.4**	**73.3**	**63.5**	**60.5**	**61.2**	**73.0**	**69.5**	**70.4**	**85.8**	**73.4**	**78.3**

Table 4: Results on EGCLEVER. ROUGE-1/2/L represent the ROUGE-F1 scores. FC represents Factual Consistency using the approximate matching.

amples from two datasets. Two authors of the work manually annotated all (Concept, Attribute) tuples of sampled examples for evaluation. We obtain Cohen's kappa $\kappa = 0.81$, which indicates a strong agreement. We observe that the system has high precision in all settings but with a drop in recall. This indicates that most (Concept, Attribute) pairs can be correctly identified with a few pairs missed. Further analysis demonstrates that the "Findings" section in EGMIMIC is more well structured than in EGCLEVER. Therefore, FACTEXT on the former setting achieves higher recall and F1.

Baseline Comparisons Table 3 and 4 show the results of baseline approaches on the EGMIMIC and EGCLEVER datasets.

Overall, BART achieves superior performance over other baselines by a large margin, showing the promising result of using abstractive summarization models.

RANDSENT and LEXRANK have similar performances on both datasets. The result is reasonable because LEXRANK relies on inter-sentence similarity to select sentences, but similarities between conclusion sentences are limited in clinical notes.

The TF-IDF baseline has contrary performance on two datasets. Recall that this approach copies the reference directly from the report with the most similar source in the training data. Since the "Conclusion" section is written as structured noun phrases in EGCLEVER and as complete sentences in EGMIMIC, TF-IDF is more likely to achieve a higher ROUGE score in EGCLEVER, which has fewer lexical variations in the "Conclusion" section.

Information Extraction v.s. Text Summarization To tackle the summarization of echocardiography reports as an information extraction (IE) task, we provide our rule-based fact extractor FACTEXT as a performance lower bound. As shown in Table 3 and 4, the rule-based system falls short of performance in both evaluation metrics. Since FACTEXT concatenates all (Concept, Attribute) pairs as noun phrases to form a generated conclusion section, it fails to distill the key information of the source. Further, since the importance of a concept in one report depends on the overall levels of importance of other concepts, external human annotations are required. However, it is hard to reach a consensus on the importance of concepts between domain experts on our dataset (See Human Evaluation below). Therefore, these annotations are deemed to have limited usability, and an IE model trained on them may not be transferable to other clinical datasets.

Alternatively, machine learning based models

Training	EGMIMIC				EGCLEVER			
corpus	R-1	R-2	R-L	FC	R-1	R-2	R-L	FC
EGMIMIC	(69.5)	(55.2)	(65.5)	(67.9)	39.9	13.9	24.2	28.1
EGCLEVER	32.6	14.2	23.9	24.9	(73.1)	(60.8)	(70.2)	(78.3)

Table 5: Cross-corpus results of models trained on EchoMIMIC and EGCLEVER using BART. R-1, R-2, R-L represent the ROUGE-F1 scores. FC represents Factual Consistency using the approximate matching. Numbers in parenthesis indicates the performance of the model on the dataset it trained on.

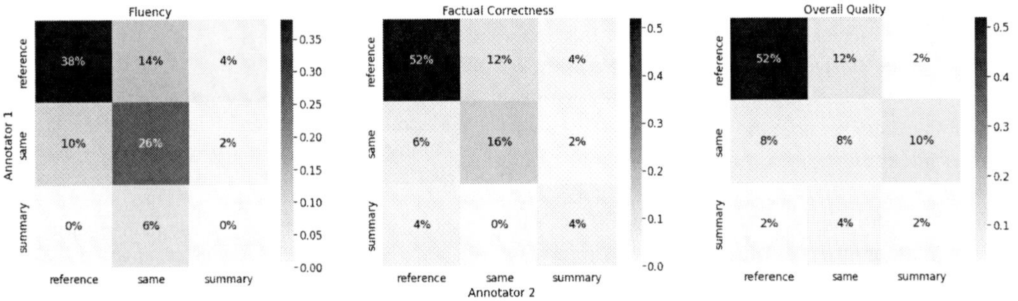

Figure 4: Confusion matrices of human evaluation results on 50 randomly sampled echo notes from EGMIMIC. Results are shown in percentage and "same" means there is a tie between a reference and a generated conclusion.

outperform FACTEXT by a large margin in terms of both ROUGE scores and factual consistency evaluation. This suggests that summarization models can approximate the capability of an IE system and identify more critical facts.

Extractive Summarization v.s. Abstractive Summarization FACTEXT is a strong extractive baseline that selects all concept and attribute pairs f as an extractive conclusion. However, the low recall under our defined evaluation metric indicates that (1) f is not capable of describing all the information in the reference; and (2) domain knowledge is required to generate novel information. The low precision score of FACTEXT, on the other hand, shows that the reference is highly selective of the source text as the majority of facts are excluded from the reference.

Transferability of the model across datasets We intentionally designed the test set to be partially from a hospital system different from the training set (out-of-domain) to test the generalizability of the models. Results are shown in Table 5. As expected, the performance drops significantly in both datasets and is worse than all baselines in Table 3 and 4. The low FC scores indicate that organizations do not share a unified consensus of important information.

5 Human Evaluation

To compare the quality of generated text against a human reference, we conduct a human evaluation following Zhang et al. (2020). We randomly sampled 50 echo reports from the development set of EGMIMIC. For each example, we presented echo findings to two Neurologist and Pulmonary Critical Care physicians along with the human references and summaries generated from BART in random order. We asked the physicians to compare them in three dimensions (1) fluency, (2) factual consistency, and (3) overall quality. For each metric, we asked the physicians to select the better one, with ties allowed.

Since it is difficult to reach an agreement between physicians, we show the human evaluation result as confusion matrices in Figure 4. Across all three dimensions, both physicians agree that human reference is better among half of the selected samples (the upper-left cell of each figure). Further, most of the percentages fall into the top left two-by-two sub-matrices, with the main diagonal being the most frequent. This indicates that physicians have a consensus that generated conclusion is less preferred. There are also uncertainties about whether a reference is better or tied with a generated conclusion (around 20% at off-diagonal). Overall, model-generated summaries are still undesired compared to human reference in terms of fluency, factual consistency, and overall quality.

6 Limitations

While our conducted human evaluation suggests that generated summaries from BART tend to have more factual errors than human reference, the accuracy of factuality comparison between BART and other baselines is still limited by the quality of our proposed system FACTEXT. Its performance, especially recall, depends on the accuracy of the ScispaCy model we use and the number of common concepts we focus on (55 in this work). For example, we can integrate the recommended phrases that echocardiographers may choose to use to describe pertinent findings by the American Society of Echocardiography (Gardin et al., 2002). We leave continually designing a more robust information extraction system or learning-based models, which both (1) rely less on domain-specific concepts; and (2) generalize to other types of notes, to future works.

7 Conclusion

In this study, we introduce EchoGen, a new benchmark for evaluating and analyzing models for echocardiography report conclusion generation. We systematically analyze the performance of several baseline methods with our proposed evaluation metric and conclude that there is still a gap between human reference and generated conclusions for echo reports in terms of fluency and factual consistency. Detailed analysis shows that our benchmarking can be used to evaluate the capacity of the models to understand the clinical text and, moreover, to shed light on the future directions for developing clinical text generation and summarization systems.

8 Ethical considerations

The research has been designated by IRB at NewYork-Presbyterian Hospital as Not Human Subject Research. The Protocol Number is 20-10022833.

Acknowledgements

This work was supported by the National Library of Medicine under Award No. 4R00LM013001 and the Amazon Web Services Diagnostic Development Initiative. This work was partially supported by a gift from Amazon and a gift from Salesforce.

References

Griffin Adams, Emily Alsentzer, Mert Ketenci, Jason Zucker, and Noémie Elhadad. 2021. What's in a Summary? Laying the Groundwork for Advances in Hospital-Course Summarization. In *Proceedings of the 2021 Conference of the North American Chapter of the Association for Computational Linguistics: Human Language Technologies*, pages 4794–4811.

Stergos Afantenos, Vangelis Karkaletsis, and Panagiotis Stamatopoulos. 2005. Summarization from medical documents: a survey. *Artificial Intelligence in Medicine*, 33(2):157–177.

M Alsharqi, W J Woodward, J A Mumith, D C Markham, R Upton, and P Leeson. 2018. Artificial intelligence and echocardiography. *Echo Research and Practice*, pages R115–R125.

Asma Ben Abacha, Yassine Mrabet, Yuhao Zhang, Chaitanya Shivade, Curtis Langlotz, and Dina Demner-Fushman. 2021. Overview of the MEDIQA 2021 Shared Task on Summarization in the Medical Domain. In *Proceedings of the 20th Workshop on Biomedical Language Processing*, pages 74–85, Online. Association for Computational Linguistics.

Nathanael Chambers, Daniel Cer, Trond Grenager, David Hall, Chloe Kiddon, Bill MacCartney, Marie-Catherine De Marneffe, Daniel Ramage, Eric Yeh, and Christopher D. Manning. 2007. Learning alignments and leveraging natural logic. In *Proceedings of the ACL-PASCAL Workshop on Textual Entailment and Paraphrasing*, pages 165–170.

Marie-Catherine de Marneffe, Christopher D. Manning, Joakim Nivre, and Daniel Zeman. 2021. Universal Dependencies. *Computational Linguistics*, pages 1–54.

Joshua Charles Denny, Randolph A. Miller, Kevin B. Johnson, and Anderson Spickard. 2008. Development and evaluation of a clinical note section header terminology. *AMIA ... Annual Symposium proceedings. AMIA Symposium*, pages 156–60.

Jacob Devlin, Ming-Wei Chang, Kenton Lee, and Kristina Toutanova. 2019. BERT: pre-training of deep bidirectional transformers for language understanding. In *Proceedings of the 2019 Conference of the North American Chapter of the Association for Computational Linguistics: Human Language Technologies, Volume 1 (Long and Short Papers)*, pages 4171–4186, Minneapolis, Minnesota. Association for Computational Linguistics.

Esin Durmus, He He, and Mona Diab. 2020. FEQA: A question answering evaluation framework for faithfulness assessment in abstractive summarization. In *Proceedings of the 58th Annual Meeting of the Association for Computational Linguistics*. Association for Computational Linguistics.

Günes Erkan and Dragomir R. Radev. 2004. Lexrank: Graph-based lexical centrality as salience in text summarization. *Journal of artificial intelligence research*, 22:457–479.

Tobias Falke, Leonardo F. R. Ribeiro, Prasetya Ajie Utama, Ido Dagan, and Iryna Gurevych. 2019. Ranking generated summaries by correctness: An interesting but challenging application for natural language inference. In *Proceedings of the 57th Annual Meeting of the Association for Computational Linguistics*. Association for Computational Linguistics.

Claire Gardent, Anastasia Shimorina, Shashi Narayan, and Laura Perez-Beltrachini. 2017. The WebNLG challenge: Generating text from RDF data. In *Proceedings of the 10th International Conference on Natural Language Generation*, pages 124–133, Santiago de Compostela, Spain. Association for Computational Linguistics.

Julius M. Gardin, David B. Adams, Pamela S. Douglas, Harvey Feigenbaum, David H. Forst, Alan G. Fraser, Paul A. Grayburn, Alan S. Katz, Andrew M. Keller, Richard E. Kerber, Bijoy K. Khandheria, Allan L. Klein, Roberto M. Lang, Luc A. Pierard, Miguel A. Quinones, Ingela Schnittger, and American Society of Echocardiography. 2002. Recommendations for a standardized report for adult transthoracic echocardiography: A report from the American Society of Echocardiography's Nomenclature and Standards Committee and Task Force for a Standardized Echocardiography Report. *Journal of the American Society of Echocardiography: Official Publication of the American Society of Echocardiography*, 15(3):275–290.

Albert Gatt and Emiel Krahmer. 2018. Survey of the state of the art in natural language generation: Core tasks, applications and evaluation. *J. Artif. Int. Res.*, 61(1):65–170.

Sebastian Gehrmann, Falcon Dai, Henry Elder, and Alexander Rush. 2018. End-to-end content and plan selection for data-to-text generation. In *Proceedings of the 11th International Conference on Natural Language Generation*, pages 46–56, Tilburg University, The Netherlands. Association for Computational Linguistics.

Ben Goodrich, Vinay Rao, Peter J. Liu, and Mohammad Saleh. 2019. Assessing the factual accuracy of generated text. In *Proceedings of the 25th ACM SIGKDD International Conference on Knowledge Discovery & Data Mining*, KDD '19, page 166–175, New York, NY, USA. Association for Computing Machinery.

Tanya Goyal and Greg Durrett. 2021. Annotating and Modeling Fine-grained Factuality in Summarization. In *Proceedings of the 2021 Conference of the North American Chapter of the Association for Computational Linguistics: Human Language Technologies*, pages 1449–1462, Online. Association for Computational Linguistics.

Hamza Harkous, Isabel Groves, and Amir Saffari. 2020. Have your text and use it too! end-to-end neural data-to-text generation with semantic fidelity. In *Proceedings of the 28th International Conference on Computational Linguistics*, pages 2410–2424, Barcelona, Spain (Online). International Committee on Computational Linguistics.

Alistair E. W. Johnson, Tom J. Pollard, Lu Shen, Li-Wei H. Lehman, Mengling Feng, Mohammad Ghassemi, Benjamin Moody, Peter Szolovits, Leo Anthony Celi, and Roger G. Mark. 2016. MIMIC-III, a freely accessible critical care database. *Scientific data*, 3:160035.

Wojciech Kryscinski, Bryan McCann, Caiming Xiong, and Richard Socher. 2020. Evaluating the factual consistency of abstractive text summarization. In *Proceedings of the 2020 Conference on Empirical Methods in Natural Language Processing (EMNLP)*. Association for Computational Linguistics.

Mike Lewis, Yinhan Liu, Naman Goyal, Marjan Ghazvininejad, Abdelrahman Mohamed, Omer Levy, Veselin Stoyanov, and Luke Zettlemoyer. 2020. BART: Denoising Sequence-to-Sequence Pre-training for Natural Language Generation, Translation, and Comprehension. In *Proceedings of the 58th Annual Meeting of the Association for Computational Linguistics*, pages 7871–7880, Online. Association for Computational Linguistics.

Chin-Yew Lin. 2004. ROUGE: A package for automatic evaluation of summaries. In *Text Summarization Branches out: Proceedings of the ACL-04 Workshop*, volume 8, pages 1–8. Barcelona, Spain.

Carol Mitchell, Peter S. Rahko, Lori A. Blauwet, Barry Canaday, Joshua A. Finstuen, Michael C. Foster, Kenneth Horton, Kofo O. Ogunyankin, Richard A. Palma, and Eric J. Velazquez. 2019. Guidelines for Performing a Comprehensive Transthoracic Echocardiographic Examination in Adults: Recommendations from the American Society of Echocardiography. *Journal of the American Society of Echocardiography: Official Publication of the American Society of Echocardiography*, 32(1):1–64.

Sunil Mohan and Donghui Li. 2018. MedMentions: A Large Biomedical Corpus Annotated with UMLS Concepts. In *Automated Knowledge Base Construction (AKBC)*.

James Mullenbach, Yada Pruksachatkun, Sean Adler, Jennifer Seale, Jordan Swartz, Greg McKelvey, Hui Dai, Yi Yang, and David Sontag. 2021. CLIP: A Dataset for Extracting Action Items for Physicians from Hospital Discharge Notes. In *Proceedings of the 59th Annual Meeting of the Association for Computational Linguistics and the 11th International Joint Conference on Natural Language Processing (Volume 1: Long Papers)*, pages 1365–1378, Online. Association for Computational Linguistics.

Mark Neumann, Daniel King, Iz Beltagy, and Waleed Ammar. 2019. ScispaCy: Fast and Robust Models for Biomedical Natural Language Processing. In *Proceedings of the 18th BioNLP Workshop and Shared Task*, pages 319–327, Florence, Italy. Association for Computational Linguistics.

Steffen Pauws, Albert Gatt, Emiel Krahmer, and Ehud Reiter. 2019. *Making Effective Use of Healthcare Data Using Data-to-Text Technology*, pages 119–145. Springer International.

Rimma Pivovarov and Noémie Elhadad. 2015. Automated methods for the summarization of electronic health records. *Journal of the American Medical Informatics Association: JAMIA*, 22(5):938–947.

Ehud Reiter and Robert Dale. 2000. *Building Natural Language Generation Systems*. Cambridge University Press.

Abigail See, Peter J. Liu, and Christopher D. Manning. 2017. Get To The Point: Summarization with Pointer-Generator Networks. In *Proceedings of the 55th Annual Meeting of the Association for Computational Linguistics (Volume 1: Long Papers)*, pages 1073–1083, Vancouver, Canada. Association for Computational Linguistics.

Alex Wang, Kyunghyun Cho, and Mike Lewis. 2020. Asking and answering questions to evaluate the factual consistency of summaries. In *Proceedings of the 58th Annual Meeting of the Association for Computational Linguistics*. Association for Computational Linguistics.

Sam Wiseman, Stuart Shieber, and Alexander Rush. 2017. Challenges in data-to-document generation. In *Proceedings of the 2017 Conference on Empirical Methods in Natural Language Processing*, pages 2253–2263, Copenhagen, Denmark. Association for Computational Linguistics.

Y. Xiong, B. Tang, Q. Chen, X. Wang, and J. Yan. 2019. A study on automatic generation of chinese discharge summary. In *2019 IEEE International Conference on Bioinformatics and Biomedicine (BIBM)*, pages 1681–1687.

Tianyi Zhang, Varsha Kishore, Felix Wu, Kilian Q. Weinberger, and Yoav Artzi. 2019. BERTScore: Evaluating Text Generation with BERT. In *International Conference on Learning Representations*.

Yuhao Zhang, Daisy Yi Ding, Tianpei Qian, Christopher D. Manning, and Curtis P. Langlotz. 2018. Learning to summarize radiology findings. In *EMNLP 2018 workshop on health text mining and information analysis*.

Yuhao Zhang, Derek Merck, Emily Tsai, Christopher D. Manning, and Curtis Langlotz. 2020. Optimizing the factual correctness of a summary: A study of summarizing radiology reports. In *Proceedings of the 58th Annual Meeting of the Association for Computational Linguistics*. Association for Computational Linguistics.

A Echo Concepts

aneurysm, anular, aorta, apex, appendage, arch, arteriosus, artery, atheroma, atrial, atrium, calcification, cava, cavity size, chamber size, chordae, color doppler, defect, disease, effusion, ejection fraction, excursion, foramen, hypertension, hypertrophy, inflammation, leaflet, mitral, muscles, ovale, pad, pericardium, pressure, prolapse, prosthesis, regurgitation, ring, root, septum, shortening, sinus, space, stenosis, structure, tamponade, thicknesses, thrombus, tricuspid, valve, vegetation, velocities, velocity, ventricle, ventricular, wall motion.

Quantifying Clinical Outcome Measures in Patients with Epilepsy Using the Electronic Health Record

Kevin Xie[1,2], **Brian Litt**[1,2,3], **Dan Roth**[4] and **Colin A. Ellis**[2,3]
[1]Department of Bioengineering, University of Pennsylvania
[2]Center for Neuroengineering and Therapeutics, University of Pennsylvania
[3]Department of Neurology, Perelman School of Medicine, University of Pennsylvania
[4]Department of Computer and Information Sciences, University of Pennsylvania
kevinxie@seas.upenn.edu

Abstract

A wealth of important clinical information lies untouched in the Electronic Health Record, often in the form of unstructured textual documents. For patients with Epilepsy, such information includes outcome measures like Seizure Frequency and Dates of Last Seizure, key parameters that guide all therapy for these patients. Transformer models have been able to extract such outcome measures from unstructured clinical note text as sentences with human-like accuracy; however, these sentences are not yet usable in a quantitative analysis for large-scale studies. In this study, we developed a pipeline to quantify these outcome measures. We used text summarization models to convert unstructured sentences into specific formats, and then employed rules-based quantifiers to calculate seizure frequencies and dates of last seizure. We demonstrated that our pipeline of models does not excessively propagate errors and we analyzed its mistakes. We anticipate that our methods can be generalized outside of epilepsy to other disorders to drive large-scale clinical research.

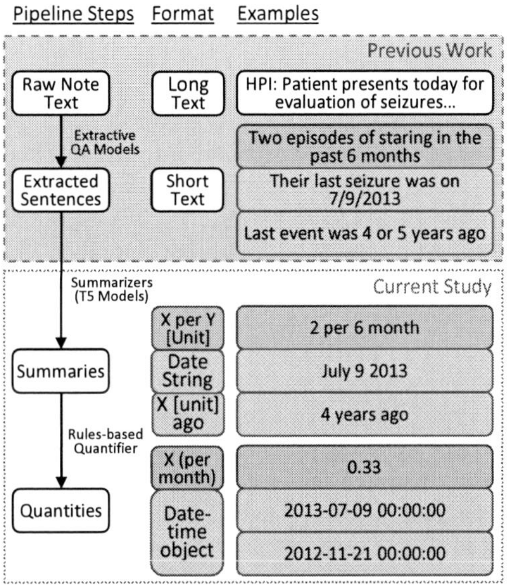

Figure 1: Schematic figure illustrating overall pipeline. Extractive question-answering models identify sentences containing seizure frequency and date of last seizure. These sentences are summarized into standardized formats. Quantities are extracted from these summaries using rules-based quantifiers. Items in blue are for seizure frequencies, while items in green are for dates of last seizure. Items in the grey background indicate work previously done in Xie et al. (2022).

1 Introduction

The Electronic Health Record (EHR) is a longitudinal catalog that describes patient visits, conditions, treatments, and well-being; thus, the EHR has significant potential for use in clinical informatics. Unfortunately, much of the data in the EHR is stored as unstructured text in the form of hand-typed doctor's notes, which makes rapid information extraction traditionally difficult. However, recent developments in neural models, namely Transformers (Vaswani et al., 2017) like BERT (Devlin et al., 2019), have opened up exciting new avenues of research.

Such developments have been applied to Epilepsy, a neurological disease characterized by recurrent unprovoked seizures. In epilepsy, seizure frequency and the date a seizure most recently occurred on are among the most important clinical outcome measures for patients. In Xie et al. (2022), we previously used specially finetuned Bio_ClinicalBERT (Alsentzer et al., 2019) and RoBERTa (Liu et al., 2019) models to extract, with human-like performance, sentences with a patient's seizure frequency and date of last seizure from clinical progress notes. These sentences contain temporal information and can thus be considered time expressions (timex).

However, such timex only simplify the problem of extracting such information from a document to extracting such information from a sentence, and

Proceedings of the BioNLP 2022 workshop, Dublin, Ireland, pages 369–375
May 26, 2022. ©2022 Association for Computational Linguistics

are not yet usable in a quantitative manner. In this study, we developed a pipeline that extends our previous work and normalizes these timex (Figure 1[1]). We used neural text summarization models to convert the extracted information into a standardized format, and then applied a simple rules-based quantifier to calculate a quantitative seizure frequency (in seizures per month), or quantitative datetime object. Our approach required minimal annotation and preparation, and can be easily generalized to other similar tasks.

2 Methods

This retrospective study was approved by the Institutional Review Board of the University of Pennsylvania with a waiver of informed consent.

In Xie et al. (2022), we finetuned Bio_ClinicalBERT (Alsentzer et al., 2019) and RoBERTa (Liu et al., 2019) on a combination of public datasets and proprietary clinical notes to extract sentences with seizure frequency and dates of last seizure from clinical notes. We framed this task as an extractive question-answering problem, where we asked the model to identify statements that answered the questions "How often does the patient have seizures," and "When was the patient's most recent seizure?" We demonstrated that our models achieved human-like performance relative to clinicians and researchers working in epilepsy-related research.

Quantifying seizure frequency and dates of last seizures from these sentences is therefore a timex normalization task, which seeks to convert a timex statement like "Their last seizure was on 7/9/2013" into the datetime object 2013-07-09 00:00:00. More difficultly, seizure frequency and date of last seizure are represented in a number of non-standardized ways by clinicians, precluding the use of simple rules-based quantification. We characterize broad categories and provide illustrative examples of such representations in Table 1. Note that such representations are often encapsulated by surrounding text (e.g. "They continue to have nocturnal convulsive seizures twice per week"), and that each category has internal variance (e.g. "seizure weekly" vs. "seizures once per week"). To accommodate these representations, we split our timex normalization process into two steps: simplification and quantification.

[1]Our examples are date-shifted and gender neutralized when applicable to preserve patient privacy and HIPAA compliance

Frequency Format	**Example**
Classical	"weekly basis", "twice per week"
Implied	"first day of their menses"
Calendar	"January: 1, February: 3, ..."
Timepoint	"Since last visit... 3 seizures"
Last Seizure Format	**Example**
Explicit	"Last seizure was 3/2012"
Implicit	"Seizure free since 2001"
Timepoint	"...2 or 3 years ago"

Table 1: Broad categories of seizure frequency and date of last seizure formats with corresponding examples.

We first attempted to simplify each sentence into a standardized format: "X per Y [day/month/year/visit]" (e.g. "1 per 1 week") for seizure frequencies, and "[Month] [Day] [Year]" or "X [day/month/year] ago" (e.g. "January 2012" or "3 years ago") for date of last seizure. We frame this task as an abstractive text summary problem: given a sentence containing a seizure frequency or date of last seizure, we summarize the main component of the sentence, the frequency or date, into a standardized template. We manually annotated the 1,000 sentences of seizure frequency and 1,000 sentences of the date of last seizure previously generated by our models in Xie et al. (2022) with the formatted summaries; for example, "Two episodes of staring in the past 6 months" was annotated with "2 per 6 months", and "Their last seizure was on 7/9/2013" was annotated with "July 9 2013". We then split them into training and testing sets, with 700 sentences for training, and 300 for testing. We also created concrete values for subjective statements (i.e. "many", "few", etc...) (Appendix A).

We finetuned two T5-large models (Raffel et al., 2020) using Huggingface (Wolf et al., 2020), on the training sets and made predictions on the test sets. One T5-large model summarized sentences of seizure frequency, while the other summarized sentences of last seizure. We used Huggingface's default parameters for text summarization and did not perform any hyperparameter optimization.

We then developed a rules-based quantifier that normalizes a frequency summary into a numerical value, and converts a date summary into a datetime object. For summaries of seizure frequency, we take the "X" value in "X per Y

Sentence	Summary	Quantity
Seizures persisted throughout their life, approximately once a year	1 per 1 year	0.0833
... Jan 5 clusters, Feb 10 clusters, March 4 clusters, April 8 clusters	4 per 6 month*	6.75
Two episodes of staring in the past 6 months	2 per 6 months	0.333
Their last seizure was on 7/9/2013	July 9 2013	2013-07-09 00:00:00
Last event was 4 or 5 years ago	4 years ago	2012-11-21 00:00:00
Not had any seizures since 2005	2005	2005-01-01 00:00:00

Table 2: Examples of the summary and quantification processes to quantify sentences of seizure frequency and date of last seizure.
*Note: the seizure calendar sentence's summary was incorrect, but the final quantity was corrected using the rules-based quantifier for seizure calendars.

[day/month/year/visit]" as the numerator, and convert the "Y" value using the time unit given in "[day/month/year/visit]" into a suitable denominator to have units of "seizures per month." If the timeframe involved the previous visit ("per Y visit"), we would attempt to search for a record of the patient's last visit in our dataset and calculate the number of months that have passed; if no such record could be found, the quantifier would insert a placeholder statement for future analysis when such information would be available. For summaries of date of last seizure, we first determine if the summary was of the "ago" form, in which case we subtract the specified number of day, months, or years from the date the note was written. Otherwise, we apply a series of logical steps to quantify the summary into a Python datetime object. If only a month and day were given, we assume that the year was either the same year that the note was written, or the previous year, depending on if the resultant date using the same year was in the future of the date the note was written. In both quantifiers, we assume that there are 365 days or 12 months in a year, 7 days in a week, and 30.4167 days or 4.3452 weeks in a month.

We also created a rules-based quantifier specifically for the seizure calendars, as the summarizer was unable to produce an accurate summary of this format of frequency. This seizure calendar quantifier identifies a sentence as a seizure calendar if it has at least two months, and at least two numbers. It then associates a month to its number of seizures by assuming that the number of seizures either directly follows the month in the text (e.g. "January: 1"), or precedes the month within three words (e.g. "1 seizure in January")." It counts the number of months and accumulates the number of seizures in that time span to calculate a monthly seizure frequency. Table 2 provides some examples of the overall process.

We manually calculated the accuracy of each step of our approach in an all-or-nothing approach by comparing a statement to its downstream summary or quantity; a step was correct only if both its format and value given the context were correct.

3 Results

We finetuned our T5 models for text summarization using a training set of 700 annotations, and a testing set of 300 annotations. To determine how much error we were propagating through our pipeline, we calculated the accuracy of each step in our method using the testing set (Table 3). We counted the number of accurate sentences from medical notes (performed previously in Xie et al. (2022)), summaries (accounting for both correct value and format) from sentences, and quantities from summaries. Note that for this calculation, we considered each step of the process as independent from the others; for example, a summary could be correct given a sentence, even if that sentence itself was incorrect relative to the original note text. We also determined the overall accuracy as the number of examples where all of these steps were correct. With at least 96% accuracy, it is evident that our summarizers produced consistent representations of seizure frequency and date of last seizure in the desired format. Meanwhile, our perfect quantification accuracy validates our use of text summaries as an intermediate step - because all seizure frequencies and dates of last seizure have been consistently converted into their own respective formats, it is

	Sentence Accuracy	Summary Accuracy	Quantity Accuracy	Overall Accuracy
Seizure Frequency	0.893	0.963	1.000	0.880
Date of Last Seizure	0.863	0.987	1.000	0.857

Table 3: Accuracies of the extracted sentences containing seizure frequencies or dates of last seizure from raw clinical note text (described previously in Xie et al. (2022)), the summary of such sentences in the standardized format, and the quantification of the summaries into quantities. The overall accuracy denotes how often every step of this process was correct. For calendar-type seizure frequencies, overall accuracy ignores the summary step, as this was always incorrect, and instead takes into account the seizure calendar quantifier.

Reason	Times Erred (Seizure Frequency)	Times Erred (Last Seizure)
Competitive Temporal Statements	**2**	**2**
"Since last visit: ... one ... seizure in the past year"		
"On the same day as their last appointment ..."		
No Temporal Reference	**2**	**2**
"They think they only had two ... seizures"		
"Two weeks later they had another seizure"		
Using Month as Value	**2**	**0**
"Since 4/2012 they have had a few seizures"		
"Since last office visit, they have had seizure 8/12/16"		

Table 4: Types of errors that occurred during the summary process.

highly unlikely that some unforeseen representation will be able to break the quantifiers' rules.

Finally, we attempted to identify patterns of errors in our incorrect summaries. We manually catalogued these errors for both sentences of seizure frequency and dates of last seizure, and determined potential reasons for such problems (Table 4). The first category was for sentences with competitive time modalities, e.g. "Since last visit: they report one possible seizure in the past year". Here the summary could either use "since last visit" or "past year" as its temporal unit for a seizure frequency; in this particular example, the model chose to use "since last visit", when "past year" would have been more appropriate. Similarly, there were cases when a temporal reference point was not available, such as this sentence of a date of last seizure: "Two weeks later they had another seizure." In this case, it is unknown when exactly "two weeks later" is referring to. This is reflected in the model's summary for this example - "2 weeks later". Though in some sense correct, this summary did not follow the desired format, namely because there was not enough information, even for a human, to fit it within the specified style[2]. Finally, some cases where seizure frequencies with dates were written out in numerical format resulted in the model pulling elements of those dates out as part of the frequency itself. For example "Since last office visit, they have had seizure 8/12/16" was summarized as "8 per 1 visit", but "8/12/16" instead refers to the date at which their seizure occurred; the correct summary should have been "1 per 1 visit".

[2]The quantifier correctly flagged this summary as anomalous and did not produce a quantity.

4 Discussion

In this study, we normalized timex containing seizure frequency and date of last seizure by simplifying them with text summarization models, and applying simple rules-based quantifiers to extract quantitative outcome measures for patients with epilepsy. We demonstrated that this pipeline can accurately calculate quantitative seizure frequencies and dates of last seizure. Though applied specifically to epilepsy, our methods are not constrained just to neurological disorders, and can be easily adapted to other medical conditions as well. Our findings pave the way for large-scale clinical informatics research through extracting and quantifying textual information from the EHR.

Our full pipeline, including our previous work from Xie et al. (2022), extracts timex from clinical documents, simplifies them using neural models, and normalizes them with rules-based methods to obtain quantitative outcome measures. The overall process is reminiscent of other temporal understanding studies. For example, Ning et al. (2018) developed a pipeline for temporal understanding that involves a Begin-Inside-Outside (BIO) tagging scheme with machine learning to extract timex, and a rules-based method to normalize them. Meanwhile, Ding et al. (2021) formulated timex normalization as a sequence of operations that selects and applies normalization rules, and Miller et al. (2015) extracted timex from clinical text using machine learning-based BIO taggers on two clinical datasets.

Additionally, to our knowledge, we are the first to use neural text summarization as an intermediate step to simplify variable timex into a standardized template for easy rules-based quantity extraction in the clinical domain. However, similar approaches exist in other domains and tasks. For example, Lourentzou et al. (2019) used a seq-to-seq model to normalize the often complex and non-standard text found in social media into more standard forms. Additionally, Vale et al. (2018) tested how various sentence simplification methods improved the informativeness of extractive text summarization methods, while Che et al. (2015) compressed sentences in a manner that simplified the sentence but preserved its sentiment as a preprocessing step for aspect-based sentiment analysis.

Our categories of errors are also in line with what has been seen in the literature for Transformers. For example, Sulem et al. (2021) found that in extractive question-answering tasks, BERT models showed remarkably lower performance on competitive I-Don't-Know questions (where a plausible but incorrect answer of the correct type exists in the context), mirroring our summarization errors when competitive time frames were presented.

Our study does have limitations. First and foremost, our methodology was developed using data from a single institutional healthcare center. While we used a neural summarizer in the hopes of improving overall generalizability to the various ways of representing outcome measures in text, it is still possible that the summarizer will fail to generalize to text from other health centers. We are actively evaluating of our methods at a collaborating institution to access this effect. Additionally, 21 of 22 summaries that involved previous visits could not be actively quantified with this dataset, as the date of the previous visit did not exist in the 300 test notes. This can easily be corrected by performing a larger longitudinal study across our patients that would allow us to track them through their visits.

5 Conclusions

We created a generalized two-step system that rapidly and accurately extracts and quantifies seizure frequency and date of last seizures. We used the T5 model to create standardized summaries of sentences of these outcome measures, and then applied a rules-based algorithm to extract and quantify the desired information. We anticipate that our methods can be used to quantify important clinical outcome measures not only for patients with epilepsy, but other disorders as well, allowing for large-scale clinical research in the future.

Acknowledgements

This research was funded by the National Institute of Neurological Disorders and Stroke 1DP1 OD029758; by the Mirowski Family Foundation; and by contributions from Jonathan and Bonnie Rothberg. CAE is supported by the National Institute of Neurological Disorders and Stroke of the National Institutes of Health Award Number K23NS121520; by the American Academy of Neurology Susan S. Spencer Clinical Research Training Scholarship; and by the Mirowski Family Foundation. DR's work was partially funded by the Office of Naval Research Contract N00014-19-1-2620.

References

Emily Alsentzer, John Murphy, William Boag, Wei-Hung Weng, Di Jindi, Tristan Naumann, and Matthew McDermott. 2019. Publicly available clinical BERT embeddings. In *Proceedings of the 2nd Clinical Natural Language Processing Workshop*, pages 72–78, Minneapolis, Minnesota, USA. Association for Computational Linguistics.

Wanxiang Che, Yanyan Zhao, Honglei Guo, Zhong Su, and Ting Liu. 2015. Sentence compression for aspect-based sentiment analysis. *IEEE/ACM Trans. Audio, Speech and Lang. Proc.*, 23(12):2111–2124.

Jacob Devlin, Ming-Wei Chang, Kenton Lee, and Kristina Toutanova. 2019. BERT: Pre-training of deep bidirectional transformers for language understanding. In *Proceedings of the 2019 Conference of the North American Chapter of the Association for*

Computational Linguistics: Human Language Technologies, Volume 1 (Long and Short Papers), pages 4171–4186, Minneapolis, Minnesota. Association for Computational Linguistics.

Wentao Ding, Jianhao Chen, Jinmao Li, and Yuzhong Qu. 2021. Automatic rule generation for time expression normalization. In *Findings of the Association for Computational Linguistics: EMNLP 2021*, pages 3135–3144, Punta Cana, Dominican Republic. Association for Computational Linguistics.

Yinhan Liu, Myle Ott, Naman Goyal, Jingfei Du, Mandar Joshi, Danqi Chen, Omer Levy, Mike Lewis, Luke Zettlemoyer, and Veselin Stoyanov. 2019. Roberta: A robustly optimized BERT pretraining approach. *CoRR*, abs/1907.11692.

Ismini Lourentzou, Kabir Manghnani, and ChengXiang Zhai. 2019. Adapting sequence to sequence models for text normalization in social media. *Proceedings of the International AAAI Conference on Web and Social Media*, 13(01):335–345.

Timothy Miller, Steven Bethard, Dmitriy Dligach, Chen Lin, and Guergana Savova. 2015. Extracting time expressions from clinical text. In *Proceedings of BioNLP 15*, pages 81–91, Beijing, China. Association for Computational Linguistics.

Qiang Ning, Ben Zhou, Zhili Feng, Haoruo Peng, and Dan Roth. 2018. CogCompTime: A tool for understanding time in natural language. In *Proceedings of the 2018 Conference on Empirical Methods in Natural Language Processing: System Demonstrations*, pages 72–77, Brussels, Belgium. Association for Computational Linguistics.

Colin Raffel, Noam Shazeer, Adam Roberts, Katherine Lee, Sharan Narang, Michael Matena, Yanqi Zhou, Wei Li, and Peter J. Liu. 2020. Exploring the limits of transfer learning with a unified text-to-text transformer. *Journal of Machine Learning Research*, 21(140):1–67.

Elior Sulem, Jamaal Hay, and Dan Roth. 2021. Do we know what we don't know? studying unanswerable questions beyond SQuAD 2.0. In *Findings of the Association for Computational Linguistics: EMNLP 2021*, pages 4543–4548, Punta Cana, Dominican Republic. Association for Computational Linguistics.

Rafaella Vale, Rafael Lins, and Rafael Ferreira. 2018. Assessing sentence simplification methods applied to text summarization. In *2018 7th Brazilian Conference on Intelligent Systems (BRACIS)*, pages 49–54.

Ashish Vaswani, Noam Shazeer, Niki Parmar, Jakob Uszkoreit, Llion Jones, Aidan N. Gomez, Łukasz Kaiser, and Illia Polosukhin. 2017. Attention is all you need. In *Proceedings of the 31st International Conference on Neural Information Processing Systems*, NIPS'17, page 6000–6010, Red Hook, NY, USA. Curran Associates Inc.

Thomas Wolf, Lysandre Debut, Victor Sanh, Julien Chaumond, Clement Delangue, Anthony Moi, Pierric Cistac, Tim Rault, Remi Louf, Morgan Funtowicz, Joe Davison, Sam Shleifer, Patrick von Platen, Clara Ma, Yacine Jernite, Julien Plu, Canwen Xu, Teven Le Scao, Sylvain Gugger, Mariama Drame, Quentin Lhoest, and Alexander Rush. 2020. Transformers: State-of-the-art natural language processing. In *Proceedings of the 2020 Conference on Empirical Methods in Natural Language Processing: System Demonstrations*, pages 38–45, Online. Association for Computational Linguistics.

Kevin Xie, Ryan S Gallagher, Erin C Conrad, Chadric O Garrick, Steven N Baldassano, John M Bernabei, Peter D Galer, Nina J Ghosn, Adam S Greenblatt, Tara Jennings, Alana Kornspun, Catherine V Kulick-Soper, Jal M Panchal, Akash R Pattnaik, Brittany H Scheid, Danmeng Wei, Micah Weitzman, Ramya Muthukrishnan, Joongwon Kim, Brian Litt, Colin A Ellis, and Dan Roth. 2022. Extracting seizure frequency from epilepsy clinic notes: a machine reading approach to natural language processing. *Journal of the American Medical Informatics Association*. Ocac018.

A Subjective Statement Values

Statement	Value
"Couple"	2
"Few"	3
"Several"	4
"Multiple"	4
"Many"	5

Table 5: Values for subjective statements. Values were chosen by consensus of the authors.

Comparing Encoder-Only and Encoder-Decoder Transformers for Relation Extraction from Biomedical Texts: An Empirical Study on Ten Benchmark Datasets

Mourad Sarrouti, Carson Tao, Yoann Mamy Randriamihaja
Sumitovant Biopharma, New York, USA
{mourad.sarrouti,carson.tao,yoann.randriamihaja}@sumitovant.com

Abstract

Biomedical relation extraction, aiming to automatically discover high-quality and semantic relations between the entities from free text, is becoming a vital step for automated knowledge discovery. Pretrained language models have achieved impressive performance on various natural language processing tasks, including relation extraction. In this paper, we perform extensive empirical comparisons of encoder-only transformers with the encoder-decoder transformer, specifically T5, on ten public biomedical relation extraction datasets. We study the relation extraction task from four major biomedical tasks, namely chemical-protein relation extraction, disease-protein relation extraction, drug-drug interaction, and protein-protein interaction. We also explore the use of multi-task fine-tuning to investigate the correlation among major biomedical relation extraction tasks. We report performance (micro F-score) using T5, BioBERT and PubMedBERT, demonstrating that T5 and multi-task learning can improve the performance of the biomedical relation extraction task.

1 Introduction

The scientific literature provides a rich source of biomedical knowledge (e.g., drug-drug interactions), and due to its rapid growth, it becomes increasingly difficult for scientists to keep up-to-date with the most recent discoveries hidden in literature (Zhang and Lu, 2019; Yadav et al., 2020). Moreover, manual curation of information from biomedical literature is time-consuming, costly, and insufficient to keep up with the rapid growth of the literature (Herrero-Zazo et al., 2013). Hence, there has been growing interest in using natural language processing (NLP) techniques for automatic relation extraction (RE) between biomedical entities from texts.

Recently, a variety of approaches based on pretrained language models such as BERT (Devlin et al., 2019) and other variants have shown promising results in various NLP tasks such as relation extraction (drissiya El-allaly et al., 2021b,a), question answering (Sarrouti et al., 2021c,a), text summarization (Goodwin et al., 2020; Yadav et al., 2021), and misinformation detection (Sarrouti et al., 2021b). In particular, RE with classification-based encoder-only pretrained transformers (BERT and variants) has been extensively studied (Lee et al., 2019; Peng et al., 2019a; Gu et al., 2022). In contrast, RE with pretrained language models based on encoder–decoder architecture, specifically Text-to-Text Transfer Transformer (T5) (Raffel et al., 2020), has not been well-studied. Unlike encoder-only transformers, which are designed to predict a single prediction for an input sequence, T5 generates target tokens based on an encoder-decoder architecture.

In this paper, our goal is to compare pretrained sequence-to-sequence transformers with the encoder-only transformers for RE from biomedical texts. In order to satisfy this aim, we compare T5 with in-domain BERT-based models such as BioBERT and PubMedBERT on ten biomedical RE benchmark datasets. We also explore the use of multi-task fine-tuning (MTFT) on ten biomedical RE datasets (each with different entities and relation types) to investigate the correlation among four major biomedical RE tasks, namely chemical-protein relation extraction, disease-protein relation extraction, drug-drug interaction, and protein-protein interaction. Our experiments show that T5 performs better than in domain BERT-based models (encoder-only) such as BioBERT and PubMedBERT. The results also show that fine-tuning T5 with multi-task learning substantially improves the performance compared to single task fine-tuning.

2 Related Work

There has been a recent surge in interest from the NLP community to automatically extract re-

Proceedings of the BioNLP 2022 workshop, Dublin, Ireland, pages 376–382
May 26, 2022. ©2022 Association for Computational Linguistics

lations between biomedical entities (proteins, gene, diseases, etc.) from the biomedical literature (Krallinger et al., 2008; Segura-Bedmar et al., 2013; Krallinger et al., 2017; Miranda et al., 2021). Recently, with the success of pretrained language models, several techniques based on transformers are widely utilized for extracting the relationships between entities from biomedical literature (Thillaisundaram and Togia, 2019; Wei et al., 2019; Hebbar and Xie, 2021; Hiai et al., 2021; Liu et al., 2021; Zhou et al., 2021; Su et al., 2021; Chang et al., 2021; Weber et al., 2021). The success of these systems has primarily been a result of encoder-only transformers such as BERT (Devlin et al., 2019) and its variants like SciBERT (Beltagy et al., 2019), BioBERT (Lee et al., 2019), and PubMedBERT (Gu et al., 2022). Unlike RE with classification-based encoder-only transformers which have been widely studied, RE with encoder-decoder transformers has not been well-explored. Encoder–decoder-based transformer, specifically T5, (Raffel et al., 2020) has shown strong performance in various NLP tasks such as question answering and text summarization, etc.

In this work, we perform comprehensive comparisons of encoder-only transformers with the encoder-decoder transformer, specifically T5, on ten public biomedical relation extraction datasets. We also explore the use of multi-task learning to learn the shared complementary features across multiple biomedical relation extraction datasets.

3 Experiments

3.1 Problem statement

Given an input sentence S consisting of n tokens, i.e., $S = \{w_1, w_2, ..., w_n\}$ and a pair of entities (e_1, e_2) where $e_1 \in S$ and $e_2 \in S$, RE models are tasked with predicting the maximum probable label $\hat{y}$ from the set of labels in annotated data, y.

3.2 Datasets and processing

We explore ten benchmark datasets of RE between various entity types such as protein-protein, drug-drug, chemical-protein and disease-protein. Since the vast majority of relation instances are within single sentences in datasets of the aforementioned relation types, we model the RE task as sentence-level relation classification. The statistics of biomedical RE datasets are listed in Table 1.

Protein-protein interactions. We use five benchmark datasets, namely BioInfer, AIMed, IEPA, HPRD50, and LLL. These datasets are converted to a unified format by Pyysalo et al. (2008). Sentences that contain a pair of proteins are selected to generate positive and negative instances. All protein-protein pairs that occur in a sentence and do not have an explicit label in aforementioned datasets are considered as negative instances. Following previous work, we anonymized target named entities in a sentence using the pre-defined tag @PROTEIN\$. For instance, a sentence with two protein names is represented as *"The POU domains of the @PROTEIN\$ and Oct2 transcription factors mediate specific interaction with @PROTEIN\$."*.

Drug-drug interactions. We use an existing preprocessed version of the Drug-Drug Interaction (DDI) 2013 corpus (Herrero-Zazo et al., 2013) and its corresponding train/dev/test split created by Peng et al. (2019b). Drug names were anonymized using the tag @DRUG\$. For instance, a sentence with a pair of drug names is represented as *"Ketoconazole: @DRUG\$ may inhibit both synthetic and catabolic enzymes of @DRUG\$"*. We evaluate four types of DDI relationships: "mechanism"', "effect", "advice", and "Int". The "mechanism" class defines the DDIs that are described by their pharmacokinetic mechanism. The "effect" type describes an effect or a pharmacodynamic mechanism in DDIs. The "advice" class describes DDIs that mention a recommendation or advice regarding a drug interaction. The "int" class is used when the text describes an interaction between drugs but without providing any additional information.

Disease-protein relationships. We use the existing preprocessed versions of the Genetic Association Database corpus (GAD) (Bravo et al., 2015) and EU-ADR datasets (van Mulligen et al., 2012). For both datasets, we use their corresponding train/dev/test splits created by Lee et al. (2019). Targeted entities were anonymized using the tags @DISEASE\$ and @GENE\$. For instance, a sentence with a pair of two entities (gene and disease in this case) is represented as *"In conclusion, @GENE\$ 8092C > A polymorphism may modify the associations between cumulative cigarette smoking and @DISEASE\$ risk."*

Chemical-protein relationships. We use ChemProt (Krallinger et al., 2017) and DrugProt (Miranda et al., 2021) datasets that contain gene–chemical relations. For ChemProt, we use an existing preprocessed version and their corresponding train/dev/test split created by Peng et al.

Dataset	Train	Dev	Test	Metrics
AIMed	4938	-	549	micro F1
BioInfer	8544	-	950	micro F1
HPRD50	389	-	44	micro F1
IEPA	734	-	82	micro F1
LLL	300	-	34	micro F1
DDI	2937	1004	979	micro F1
ChemProt	4154	2416	3458	micro F1
DrugProt	17277	3765	-	micro F1
GAD	4796	-	534	micro F1
EU-ADR	318	-	37	micro F1

Table 1: Statistics of the biomedical relation extraction datasets. For DrugProt, we use the dev set as a test set.

(2019b). We evaluate the same five classes: CPR:3, CPR:4, CPR:5, CPR:6 , CPR:9. The CPR:3 class describes upregulator, activator, and indirect upregulator. The CPR:4 class describes downregulator, inhibitor and indirect downregulator relation types. The CPR:5 category describes agonist, agonist activator and agonist inhibitor relation types. The CPR:6 type describes the antagonist relation. The CPR:9 class describes the following relation types: substrate, product of, and substrate product of. For DrugProt, we use the standard training and development sets in the DrugProt shared task and evaluate the same 13 classes: Activator, Agonist, Agonist-Inhibitor, Antagonist, Direct-Regulator, Indirect-Downregulator, Indirect-Upregulator, Inhibitor, Part-Of, Product-Of, Substrate, Substrate_Product-Of, Agonist-Activator. We first split abstracts into sentences using NLTK and then anonymized target entities in a sentence using the tags @CHEMICAL$ and @GENE$. For instance, a sentence with a pair of two entities (chemical and gene in this case) is represented as "*During differentiation, @CHEMICAL$ promoted early expression of osteoblast transcription factors, @GENE$ and osterix.*"

3.3 Models and setups

We compare in-domain BERT-based language models such as BioBERT (Lee et al., 2019) and PubMedBERT (Gu et al., 2022) with T5 (Raffel et al., 2020) and its variant SciFive (Phan et al., 2021), which is trained on biomedical texts (PubMed abstracts). For BERT-based models, we use a [CLS] token for the classification of relations. The [CLS] representation is fed into a softmax layer for a multi-way classification. For the T5-based models, the input sequence for the relation extraction task is "Processed sentence: [s] Relation: [r]". We fine-tuned T5 to generate tokens of relation types which are the ground truth labels in training datasets.

We also explore the use of MTFT on ten biomedical RE datasets. Figure 1 illustrates MTFT for RE tasks. We used the proportional and temperature-scaled task mixing as in (Raffel et al., 2020) for data mixture. During fine-tuning, a task-specific token (in our case, name of the dataset) is prepended to the input sequence.

In our experiments, we used the BioBERT (v1.1-base-PubMed), PubMedBERT, T5-base, and SciFive (SciFive-base-Pubmed) implementations provided in HuggingFace's Transformers package version 4.16.2 (Wolf et al., 2020). All models were trained with a batch size of 16 and maximum sequence length of 300 tokens for 10 epochs using single GPU (16 GB VRAM) on Amazon SageMaker. Adam optimiser with a learning rate of 1e-5 was used.

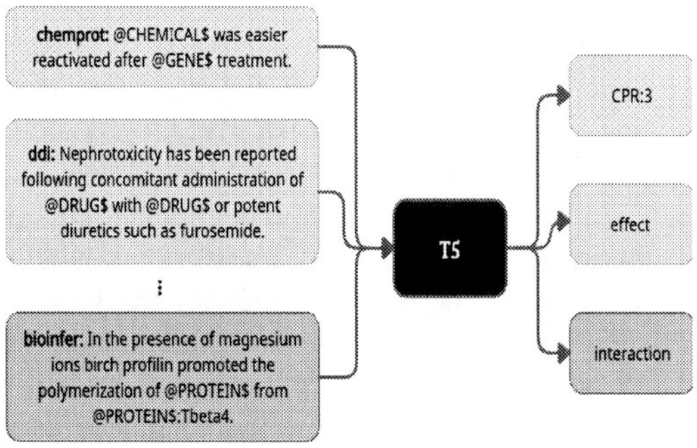

Figure 1: Multi-task learning for biomedical RE

3.4 Results

In Table 2, we show the results of T5-based models compared to the in-domain and SOTA BERT-based models (pretrained on biomedical text) on ten benchmarking biomedical RE datasets, listed in Table 1. We compare the micro F1 scores obtained by T5 and its variant SciFive (pretrained on PubMed abstracts) to the BioBERT and PubMedBERT. On average (micro), T5 which was only pretrained on the general domain corpus, obtained a higher F1 score than BioBERT and PubMedBERT. T5 achieved the highest F1 scores on 5 out of 10 biomedical RE datasets. Models using biomedical text in pre-training generally perform better than models which pre-trained on general domain corpus. However, we observe that T5-scifive which was pre-trained on biomedical text (PubMed abstracts) did not perform well compared to T5.

We also explored the impact of MTFT on four

Relation	Datasets	BioBERT	PubMedBERT	T5	T5-SciFive	T5-MTFT
Protein-protein	AIMed	92.36	93.31	**94.35**	94.17	93.62
	BioInfer	**95.97**	94.59	95.36	95.89	95.16
	HPRD50	85.45	90.56	84.09	90.90	**95.95**
	IEPA	86.58	86.46	87.80	87.80	**90.24**
	LLL	88.24	**100.0**	97.05	94.11	97.05
Drug-drug	DDI	89.67	90.69	91.01	90.60	**91.83**
Chemical-protein	ChemProt	90.11	91.64	90.45	92.39	**96.56**
	DrugProt	88.69	89.40	88.71	**89.56**	89.37
Disease-protein	GAD	79.91	80.87	**81.46**	81.27	80.71
	EU-ADR	57.42	64.63	78.38	75.67	**83.78**
Average score		85.44	88.22	**89.47**	89.23	**91.42**

Table 2: Biomedical relation extraction test results. In T5-MTFT, we fine-tuned T5 with multi-task learning on ten datasets and then evaluate on the test set for each dataset.

benchmark biomedical RE tasks, i.e., drug-drug interaction, protein-protein interaction, chemical-protein relation extraction, and disease-protein relation extraction. On average, the results clearly show that the performance improves when using MTFT (an improvement of 1.95 F-score over the best single performing model). For instance, on the ChemProt dataset, T5-MTFT was able to achieve significant performance improvement of 6.11 and 6.45 F-score points over T5 and BioBERT respectively. While overall results indicate that MTFT provides improved RE performance on the four biomedical RE tasks (tasks with clear knowledge transfer), we observe a slight drop in the performance on some datasets such as AIMed, BioInfer, and GAD. In MTFT, we believe that in addition to the sample size of each task, the difficulty of the task/dataset can have an impact on the overall performance (the model underfits or overfits a dataset). More efforts and ablation studies are needed to study the impact of different biomedical RE tasks/datasets on downstream performance.

3.5 Error analysis

We performed a manual analysis of the test sets where the best performing model (T5-MTFT) predicted an incorrect label. Table 3 presents some examples.

Protein-protein interaction. The error analysis has shown that sentences are mostly classified incorrectly when they contain repetitive protein mentions (examples #1 and #3). Multiple protein mentions tend to add noise, which can prevent the model to extract the relevant contextual information. In addition, numerical or statistical findings might be a cause of error (example #1). We also observed that when the protein interacting words (e.g., bind, interact, localization) are mentioned in a sentence, the model predicts the class label "true" (i.e, interacting) (examples #2, #3 and #4).

Drug-drug interaction. The model tends to classify "Int" class as "Effect" type (examples #5 and #6). "Int" type is used whenever there exists an interaction between two drugs (i.e., a coarse-grained relation type). Having coarse-grained and fine-grained categories can be a cause of error. We also observed that when the input sentence contains some class-specific words (e.g., effect, interact, interaction, advise) that are not associated with the target entities, the model fails to predict the correct label (examples #7 and #8).

Chemical-protein relation extraction. Being a common source of mis-classification, the CPR:3 type was often predicted as CPR:4 and vice versa (examples #9 and #10). The CPR:3 class usually describes up-regulation, and its instances usually include up-regulation words such as "promote", "increase", and "activate". The CPR:4 class is usually related to down-regulation and contains down-regulation words such as "decrease", "inhibitor", and "deposition". Having both up-regulation and down-regulation words in the same sentence creates confusion, which can lead to mis-classification. The model also misclassified some instances due to the presence of multiple entities in a single sentence (example #11). Multiple entities can also create noise and make it difficult for the model to identify if there is a relation between the two target entities.

Disease-protein relation extraction. We found that our model fails to predict the correct label for instances (examples #12, #13, #14 and #15) that contain association words (e.g., associated)

	Example
(1)	**AIMed_sentence:** Chemokines that could compete with high affinity for MIP-1beta binding could also compete for monomeric gp120 binding, although with variable potencies; maximal @PROTEIN$ binding inhibition was 80% for MCP-2, but only 30% for @PROTEIN$. **Gold label:** TRUE **Predicted label:** FALSE
(2)	**AIMed_sentence:** We investigated whether @PROTEIN$, which binds to tyrosine-phosphorylated ITAM, interacts with @PROTEIN$ following T cell activation. **Gold label:** FALSE **Predicted label:** TRUE
(3)	**AIMed_sentence:** We further demonstrated that @PROTEIN$ and E3 but not @PROTEIN$ can decrease the fusogenic activity of Abeta(29-42) via a direct interaction. **Gold label:** FALSE **Predicted label:** TRUE
(4)	**BioInfer_sentence:** In localization studies with mammalian cells, all fusion proteins showed the localization expected for @PROTEIN$ in areas of high @PROTEIN$ dynamics, such as leading lamellae and ruffles induced by epidermal growth factor. **Gold label:** FALSE **Predicted label:** TRUE
(5)	**DDI_sentence:** Other drugs which may enhance the neuromuscular blocking action of @DRUG$ such as MIVACRON include certain antibiotics (e.g., aminoglycosides, tetracyclines, bacitracin, @DRUG$, lincomycin, clindamycin, colistin, and sodium colistimethate), magnesium salts, lithium, local anesthetics, procainamide, and quinidine. **Gold label:** INT **Predicted label:** EFFECT
(6)	**DDI_sentence:** @DRUG$ may decrease the effectiveness of oral contraceptives, certain antibiotics, @DRUG$, theophylline, corticosteroids, anticoagulants, and beta blockers. **Gold label:** INT **Predicted label:** EFFECT
(7)	**DDI_sentence:** Drugs Eliminated by Active Tubular Secretion: Although studies to assess drug-drug interactions with Sanctura have not been conducted, @DRUG$ has the potential for pharmacokinetic interactions with other drugs that are eliminated by active tubular secretion (e.g. digoxin, procainamide, pancuronium, morphine, @DRUG$, metformin and tenofovir). **Gold label:** MECHANISM **Predicted label:** INT
(8)	**DDI_sentence:** Since Celontin (@DRUG$) may interact with concurrently administered @DRUG$, periodic serum level determinations of these drugs may be necessary (eg methsuximide may increase the plasma concentrations of phenytoin and phenobarbital). **Gold label:** ADVISE **Predicted label:** INT
(9)	**ChemProt_sentence:** EVn-50 possessed a broad spectrum of in vitro anticancer activity for those tested cancer cells, especially sensitive to MDA-MB-435, SKOV-3, BXPC-3, SMMC-7721, MCF-7, HO-8910, SGC-7901, BEL-7402, HCT-116, and 786-O, with the respective IC50 below 10mg/ml. Treatment with @CHEMICAL$ or VB1 resulted in arresting the MDA-MB-435 and SMMC-7721 cells at G2/M phase, which was further supported by observations of increased phosphorylation of Histone 3 at Ser10, phosphorylation of @GENE$ at Tyr15, expression of cyclin B1, and decreased expression of Cdc25c. **Gold label:** CPR:3 **Predicted label:** CPR:4
(10)	**ChemProt_sentence:** @CHEMICAL$ also increases Amyloid b (@GENE$) deposition and tau pathology. **Gold label:** CPR:4 **Predicted label:** CPR:3
(11)	**ChemProt_sentence:** Agonist and antagonist actions of yohimbine as compared to @CHEMICAL$ at alpha(2)-adrenergic receptors @GENE$, serotonin (5-HT)(1A), 5-HT(1B), 5-HT(1D). **Gold label:** CPR:5 **Predicted label:** CPR:6
(12)	**GAD_sentence:** Our results possibly indicate an association of @DISEASE$ with @GENE$ homozygosity (P=0.056). **Gold label:** FALSE **Predicted label:** TRUE
(13)	**GAD_sentence:** Our results suggest that the @GENE$ 168His variant is associated with reduced susceptibility to @DISEASE$. **Gold label:** FALSE **Predicted label:** TRUE
(14)	**GAD_sentence:** Our results indicate that the intron 2 CYP46 @GENE$ genotype may predispose to @DISEASE$, and this association is independent of the apolipoprotein E genotype. **Gold label:** FALSE **Predicted label:** TRUE
(15)	**GAD_sentence:** Although there remains a possibility that the @GENE$ TaqI A polymorphism plays some role in modifying the phenotype of the @DISEASE$, these results suggest that neither the A1 allele nor the homozygous A1 genotype is associated with alcoholism. **Gold label:** FALSE **Predicted label:** TRUE

Table 3: Examples of sentences that were incorrectly classified by the MTFT model.

with non-conclusive evidence ("possibly indicate", "suggest", "may predispose", "possibility").

4 Conclusion

In this paper, we present a comprehensive evaluation of encoder-only and encoder-decoder transformers on four benchmark biomedical RE tasks. We also explored the use of MTFT to investigate the correlation among these biomedical RE tasks. For that, we used ten popular datasets, namely AIMed, BioInfer, HPRD50, IEPA, LLL, DDI, ChemProt, DrugProt, GAD, and EU-ADR. The experiments showed that T5 and MTFT achieved better performance than BERT-based models (BioBERT and PubMedBERT) in extracting relations between bio-entities from texts. In the future, we plan to study the impact of each RE task/dataset on the downstream performance.

References

Iz Beltagy, Kyle Lo, and Arman Cohan. 2019. SciBERT: A pretrained language model for scientific text. In *Proceedings of the 2019 Conference on Empirical Methods in Natural Language Processing and the 9th International Joint Conference on Natural Language Processing (EMNLP-IJCNLP)*, pages 3615–3620, Hong Kong, China. Association for Computational Linguistics.

Àlex Bravo, Janet Piñero, Núria Queralt-Rosinach, Michael Rautschka, and Laura I Furlong. 2015. Extraction of relations between genes and diseases from text and large-scale data analysis: implications for translational research. *BMC Bioinformatics*, 16(1).

Ting-Wei Chang, Tzu-Yi Li, Yu-Wen Chiu, Sheng-Jie Lin, Panchanit Boonyarat, Wen-Chao Yeh, Neha Warikoo, and Yung-Chun Chang. 2021. Identifying Drug/chemical-protein Interactions in Biomedical Literature using the BERT-based Ensemble Learning Approach for the BioCreative 2021 DrugProt Track. *Development*, 750(18858):199620.

Jacob Devlin, Ming-Wei Chang, Kenton Lee, and Kristina Toutanova. 2019. BERT: Pre-training of deep bidirectional transformers for language understanding. In *Proceedings of the 2019 Conference of the North American Chapter of the Association for Computational Linguistics: Human Language Technologies, Volume 1 (Long and Short Papers)*, pages 4171–4186, Minneapolis, Minnesota. Association for Computational Linguistics.

Ed drissiya El-allaly, Mourad Sarrouti, Noureddine En-Nahnahi, and Said Ouatik El Alaoui. 2021a. Deep-CADRME: A deep neural model for complex adverse drug reaction mentions extraction. 143:27–35.

Ed drissiya El-allaly, Mourad Sarrouti, Noureddine En-Nahnahi, and Said Ouatik El Alaoui. 2021b. MTT-LADE: A multi-task transfer learning-based method for adverse drug events extraction. *Information Processing & Management*, 58(3):102473.

Travis Goodwin, Max Savery, and Dina Demner-Fushman. 2020. Towards Zero-Shot Conditional Summarization with Adaptive Multi-Task Fine-Tuning. In *Findings of the Association for Computational Linguistics: EMNLP 2020*, pages 3215–3226, Online. Association for Computational Linguistics.

Yu Gu, Robert Tinn, Hao Cheng, Michael Lucas, Naoto Usuyama, Xiaodong Liu, Tristan Naumann, Jianfeng Gao, and Hoifung Poon. 2022. Domain-specific language model pretraining for biomedical natural language processing. *ACM Transactions on Computing for Healthcare*, 3(1):1–23.

Shashank Hebbar and Ying Xie. 2021. Covidbert-biomedical relation extraction for covid-19. In *The International FLAIRS Conference Proceedings*, volume 34.

María Herrero-Zazo, Isabel Segura-Bedmar, Paloma Martínez, and Thierry Declerck. 2013. The DDI corpus: An annotated corpus with pharmacological substances and drug–drug interactions. *Journal of Biomedical Informatics*, 46(5):914–920.

Satoshi Hiai, Kazutaka Shimada, Taiki Watanabe, Akiva Miura, and Tomoya Iwakura. 2021. Relation extraction using multiple pre-training models in biomedical domain. In *Proceedings of the International Conference on Recent Advances in Natural Language Processing (RANLP 2021)*, pages 530–537, Held Online. INCOMA Ltd.

Martin Krallinger, Florian Leitner, Carlos Rodriguez-Penagos, and Alfonso Valencia. 2008. Overview of the protein-protein interaction annotation extraction task of BioCreative II. *Genome Biology*, 9(Suppl 2):S4.

Martin Krallinger, Obdulia Rabal, Saber A Akhondi, Martın Pérez Pérez, Jesús Santamaría, Gael Pérez Rodríguez, Georgios Tsatsaronis, and Ander Intxaurrondo. 2017. Overview of the BioCreative VI chemical-protein interaction Track. In *Proceedings of the sixth BioCreative challenge evaluation workshop*, volume 1, pages 141–146.

Jinhyuk Lee, Wonjin Yoon, Sungdong Kim, Donghyeon Kim, Sunkyu Kim, Chan Ho So, and Jaewoo Kang. 2019. BioBERT: a pre-trained biomedical language representation model for biomedical text mining. *Bioinformatics*.

Xiaofeng Liu, Kaiwen Tan, and Shoubin Dong. 2021. Multi-granularity sequential neural network for document-level biomedical relation extraction. *Information Processing & Management*, 58(6):102718.

Antonio Miranda, Farrokh Mehryary, Jouni Luoma, Sampo Pyysalo, Alfonso Valencia, and Martin Krallinger. 2021. Overview of DrugProt BioCreative VII track: quality evaluation and large scale text mining of drug-gene/protein relations. In *Proceedings of the seventh BioCreative challenge evaluation workshop*.

Yifan Peng, Shankai Yan, and Zhiyong Lu. 2019a. Transfer learning in biomedical natural language processing: An evaluation of BERT and ELMo on ten benchmarking datasets. In *Proceedings of the 18th BioNLP Workshop and Shared Task*, pages 58–65, Florence, Italy. Association for Computational Linguistics.

Yifan Peng, Shankai Yan, and Zhiyong Lu. 2019b. Transfer learning in biomedical natural language processing: An evaluation of BERT and ELMo on ten benchmarking datasets. In *Proceedings of the 18th BioNLP Workshop and Shared Task*. Association for Computational Linguistics.

Long N. Phan, James T. Anibal, Hieu Tran, Shaurya Chanana, Erol Bahadroglu, Alec Peltekian, and Grégoire Altan-Bonnet. 2021. Scifive: a text-to-text transformer model for biomedical literature.

Sampo Pyysalo, Antti Airola, Juho Heimonen, Jari Björne, Filip Ginter, and Tapio Salakoski. 2008. Comparative analysis of five protein-protein interaction corpora. In *BMC bioinformatics*, volume 9, pages 1–11. BioMed Central.

Colin Raffel, Noam Shazeer, Adam Roberts, Katherine Lee, Sharan Narang, Michael Matena, Yanqi Zhou, Wei Li, and Peter J. Liu. 2020. Exploring the limits of transfer learning with a unified text-to-text transformer. *Journal of Machine Learning Research*, 21(140):1–67.

Mourad Sarrouti, Asma Ben Abacha, and Dina Demner-Fushman. 2021a. Multi-task transfer learning with data augmentation for recognizing question entailment in the medical domain. In *2021 IEEE 9th International Conference on Healthcare Informatics (ICHI)*, pages 339–346. IEEE.

Mourad Sarrouti, Asma Ben Abacha, Yassine Mrabet, and Dina Demner-Fushman. 2021b. Evidence-based fact-checking of health-related claims. In *Findings of the Association for Computational Linguistics: EMNLP 2021*, pages 3499–3512, Punta Cana, Dominican Republic. Association for Computational Linguistics.

Mourad Sarrouti, Deepak Gupta, Asma Ben Abacha, and Dina Demner-Fushman. 2021c. NLM at BioASQ Synergy 2021: Deep Learning-Based Methods for Biomedical Semantic Question Answering about COVID-19. In *2021 Working Notes of CLEF-Conference and Labs of the Evaluation Forum, CLEF-WN 2021*, pages 335–350.

Isabel Segura-Bedmar, Paloma Martínez, and María Herrero-Zazo. 2013. SemEval-2013 task 9 : Extraction of drug-drug interactions from biomedical texts (DDIExtraction 2013). In *Second Joint Conference on Lexical and Computational Semantics (*SEM), Volume 2: Proceedings of the Seventh International Workshop on Semantic Evaluation (SemEval 2013)*, pages 341–350, Atlanta, Georgia, USA. Association for Computational Linguistics.

Peng Su, Yifan Peng, and K. Vijay-Shanker. 2021. Improving BERT model using contrastive learning for biomedical relation extraction. In *Proceedings of the 20th Workshop on Biomedical Language Processing*, pages 1–10, Online. Association for Computational Linguistics.

Ashok Thillaisundaram and Theodosia Togia. 2019. Biomedical relation extraction with pre-trained language representations and minimal task-specific architecture. In *Proceedings of The 5th Workshop on BioNLP Open Shared Tasks*, pages 84–89, Hong Kong, China. Association for Computational Linguistics.

Erik M. van Mulligen, Annie Fourrier-Reglat, David Gurwitz, Mariam Molokhia, Ainhoa Nieto, Gianluca Trifiro, Jan A. Kors, and Laura I. Furlong. 2012. The EU-ADR corpus: Annotated drugs, diseases, targets, and their relationships. *Journal of Biomedical Informatics*, 45(5):879–884.

Leon Weber, Mario Sänger, Samuele Garda, Fabio Barth, Christoph Alt, and Ulf Leser. 2021. Humboldt@ drugprot: Chemical-protein relation extraction with pretrained transformers and entity descriptions.

Qiang Wei, Zongcheng Ji, Yuqi Si, Jingcheng Du, Jingqi Wang, Firat Tiryaki, Stephen Wu, Cui Tao, Kirk Roberts, and Hua Xu. 2019. Relation extraction from clinical narratives using pre-trained language models. In *AMIA annual symposium proceedings*, volume 2019, page 1236. American Medical Informatics Association.

Thomas Wolf, Lysandre Debut, Victor Sanh, Julien Chaumond, Clement Delangue, Anthony Moi, Pierric Cistac, Tim Rault, Rémi Louf, Morgan Funtowicz, Joe Davison, Sam Shleifer, Patrick von Platen, Clara Ma, Yacine Jernite, Julien Plu, Canwen Xu, Teven Le Scao, Sylvain Gugger, Mariama Drame, Quentin Lhoest, and Alexander M. Rush. 2020. Huggingface's transformers: State-of-the-art natural language processing.

Shweta Yadav, Srivastsa Ramesh, Sriparna Saha, and Asif Ekbal. 2020. Relation extraction from biomedical and clinical text: Unified multitask learning framework. *IEEE/ACM Transactions on Computational Biology and Bioinformatics*.

Shweta Yadav, Mourad Sarrouti, and Deepak Gupta. 2021. NLM at MEDIQA 2021: Transfer learning-based approaches for consumer question and multi-answer summarization. In *Proceedings of the 20th Workshop on Biomedical Language Processing*, pages 291–301, Online. Association for Computational Linguistics.

Yijia Zhang and Zhiyong Lu. 2019. Exploring semi-supervised variational autoencoders for biomedical relation extraction. *Methods*, 166:112–119.

Wenxuan Zhou, Kevin Huang, Tengyu Ma, and Jing Huang. 2021. Document-level relation extraction with adaptive thresholding and localized context pooling. In *Proceedings of the AAAI Conference on Artificial Intelligence*, volume 35, pages 14612–14620.

Utility Preservation of Clinical Text After De-Identification

Thomas Vakili and Hercules Dalianis
Department of Computer and Systems Sciences (DSV)
Stockholm University
Kista, Sweden
{thomas.vakili, hercules}@dsv.su.se

Abstract

Electronic health records contain valuable information about symptoms, diagnosis, treatment and outcomes of the treatments of individual patients. However, the records may also contain information that can reveal the identity of the patients. Removing these identifiers - the Protected Health Information (PHI) - can protect the identity of the patient. Automatic de-identification is a process which employs machine learning techniques to detect and remove PHI. However, automatic techniques are imperfect in their precision and introduce noise into the data. This study examines the impact of this noise on the utility of Swedish de-identified clinical data by using human evaluators and by training and testing BERT models. Our results indicate that de-identification does not harm the utility for clinical NLP and that human evaluators are less sensitive to noise from de-identification than expected.

1 Introduction

The training data for clinical NLP models are sensitive because they often contain information that can reveal the identity of real patient. This makes sharing data and models difficult.

One way of decreasing the privacy risks of using clinical data is to de-identify them. A popular way of doing this is by automatically detecting and removing Protected Health Information (PHI). This is often done using Named Entity Recognition models that can find these sensitive data in clinical texts. The de-identified clinical data can be used both for clinical research but also as training data for machine learning algorithms. These approaches will be described in the next section.

One concern is that de-identification will deteriorate the quality of clinical texts. The risk is that this will not only harm down-stream NLP tasks, but also make the data less useful for other research purposes. We have also identified a hesitancy from lawyers who fear that de-identification can harm the safety of the data.

In this paper, we evaluate the extent to which de-identification harms perceived utility using human evaluators. We then evaluate the impact of de-identification on the utility of the datasets for building clinical NLP models.

2 Related Research

This related research section presents studies regarding the quality of the de-identification system both in terms of safety and privacy but also for down-stream tasks as for medical research.

In a study by Meystre et al. (2014) 86 patient records in English were de-identified. Eight physicians and 11 medical students that have treated and written these records 1-3 months earlier could not recognise their patients. Some of physician suspected that they could recognize their patient on some clinical details but after a control it was found that the wrong patient had been identified.

Sánchez et al. (2014) propose a sanitation process that removes information that might make the patient record sensitive. This is not done by replacing typical PHI like names, but instead by replacing sensitive diseases such as *Clamydia, AIDS,* or *HIV* with less sensitive terms such as *virus*. The ideas is to aggregate information but this limits the utility of the patient records.

Dalton-Locke et al. (2020) used de-identified patient records from mental health clinics to perform research regarding the housing service of patients suffering from mental illnesses. Structured data had previously been used for this research. The researchers compared the two approaches and found it feasible to use de-identified patient records and de-identified structured data jointly for this research. The system called CRIS is a combination of a de-identification algorithm and a security model has been approved for use in mental health research by the ethics board. This allows researchers to ex-

Proceedings of the BioNLP 2022 workshop, Dublin, Ireland, pages 383–388
May 26, 2022.

tract data from the the patient record system without requiring individuals' informed consent (Fernandes et al., 2013).

In an other study, Dalianis (2019) constructed a pseudonymization system for Swedish clinical text. The pseudonymization system replaced PHI with pseudonyms or surrogates. All tags that could identify a PHI were removed so the records looked realistic and neutral. The system was evaluated by two computer scientist that had worked with clinical text mining, specifically with this type of text. The text they evaluated had not been seen by the scientists before. They read 98 patient records where half were pseudonymized and the other half were not. On average, 91 percent of the pseudonymized records were judged as original.

Pantazos et al. (2017) carried out de-identification and pseudonymization of over 323,000 Danish patient records and then carried out a manual review of 369 de-identified and pseudonymized patient records with a total length of over 71,000 words, this revealed seven words where quasi-identifiers[1] had not been de-identified and it revealed 109 words where it was incorrectly de-identified, this reduced the medical correctness and readability according to the authors. A finding by the authors was if they use abbreviation lists and also medical lists the number of false positives would probably been diminished.

Berg et al. (2020) evaluated the performance of the Conditional Random Field (CRF) algorithms on down-stream tasks based on clinical training text that have been de-identified with increasing degrees of recall. The authors used four different de-identification strategies: pseudonymization (replace with surrogats), masking, keeping the class name and removing the entire sentence containing the PHI. Pseudonymization was the most effective strategy for preserving down-stream utility. Masking and replacing the PHI with the class name had a larger negative impact. The most severe impact was seen when employing the sentence removal strategy. However, a balanced recall (not high recall) on all four strategies did not affect the down-stream performance significantly.

3 Data

The clinical data used in this study originates from the Karolinska University Hospital and is stored in the research infrastructure called Health Bank – Swedish Health Record Research Bank[2] (Dalianis et al., 2015). The data encompasses 2 million patient records[3].

Three clinical data set have been de-identified using the BERT model created by Lamproudis. et al. (2022). The experiments fine-tune models using both the unaltered data and the de-identified data. The following datasets were used:

Stockholm EPR Gastro ICD-10 Corpus A corpus of 795,839 tokens in 6,062 discharge summaries encompassing 4,985 unique patients with gastrointestinal diseases. Each discharge summary is associated with multiple ICD-10 codes which have been divided into blocks. The dataset is a described in Remmer et al. (2021). The task is to predict the correct ICD-10 block for each discharge summary.

Stockholm EPR Diagnosis Factuality Corpus A corpus of 240,000 tokens in 3,710 clinical notes and their diagnosis. Each note has been annotated with the factuality of the diagnosis. There are six levels of factuality for each diagnosis: *Certainly Positive*, *Probably Positive*, *Possibly Positive*, *Possibly Negative*, *Probably Negative*, and *Certainly Negative*. The dataset is a described in (Velupillai, 2011) and (Velupillai et al., 2011). major is to process each clinical note and predict the degree of factuality for the diagnosis.

Stockholm EPR Clinical Entity Corpus A corpus consisting of 70,852 tokens in which 7,946 entities have been annotated. The annotations are for four clinical entity classes: *Diagnosis*, *Drugs*, *Body parts*, and *Findings*. The dataset is a described in (Skeppstedt et al., 2014). The task is an NER problem which requires the model to locate the entities within each sample.

These three datasets can be divided into two categories. The *Stockholm EPR Gastro ICD-10 Corpus* and *Stockholm EPR Diagnosis Factuality Corpus* are sequence classification problems with labels on the sample level. On the other hand, the *Stockholm EPR Clinical Entity Corpus* is a token classification problem which requires the model to assign

[1]A quasi-identifier is a identifier that indirectly can identify a patient such an street name or a zip code.

[2]Health Bank, http://www.dsv.su.se/healthbank

[3]This research has been approved by the Swedish Ethical Review Authority under permission no. 2019-05679.

the correct label to each token in a sequence. This dataset was also used by Berg et al. (2020) which makes it possible to assess whether switching to a BERT-based approach affects the results.

A fourth dataset was used for the qualitative experiments:

Gastro Pseudo Clinical Dataset Based on the Stockholm EPR Gastro ICD-10 Corpus (Remmer et al., 2021), this dataset contains 6,062 de-identified discharge summaries in the medical speciality of Gastrointestinal diseases. The data have been de-identified and pseudonymized using the HB Deid CRF.

4 Methods & Experiments

4.1 De-Identification

This study uses two versions of the de-identification system HB Deid described in Berg et al. (2019). The first version uses Conditional Random Fields (CRF) to locate sensitive entities.

The second version instead uses a clinical BERT model fine-tuned for Named Entity Recognition (NER). This model is pre-trained on Swedish general-domain data (Malmsten et al., 2020) and adapted to the clinical domain. This adaptation involved changing the vocabulary and continuing the pre-training using sensitive clinical data (Lamproudis. et al., 2022).

PHI Class	Recall	Precision
Age	100%	100%
First Name	100%	100%
Last Name	98%	98%
Partial Date	99%	97%
Full Date	90%	91%
Phone Number	81%	68%
Health Care Unit	85%	94%
Location	100%	100%
Organization	71%	100%

Table 1: The NER model's recall and precision for each PHI type are displayed and were calculated on the gold standard called *Stockholm EPR PHI Corpus*, (Dalianis and Velupillai, 2010). For details on the annotation process see (Velupillai et al., 2009).

The BERT model from Lamproudis. et al. (2022) selected as it was the best model available for Swedish clinical NER. We can call it. This model was fine-tuned for NER using the Stockholm EPR PHI Corpus (Velupillai et al., 2009). The precision and recall for each PHI class, estimated using a held-out dataset, is shown in Table 1. Both the precision and the recall values are high for many of the classes. It correctly identifies most names, but struggles with detecting organizations. We call this fine-tuned model *SweClin-BERT NER*.

4.2 Comparing Fine-Tuned BERT Models

This experiment quantitatively evaluated the performance on down-stream tasks when using de-identified or unaltered training data. First, each dataset was processed using the BERT-based de-identifier to detect all sensitive PHI entities. These entities were replaced with realistic surrogates.

The resulting collection of datasets was used to create two different classes of models. One was trained only using pseudonymized datasets and the other was trained using unaltered datasets. The models are trained using 10-fold cross validation and are compared by studying their F_1 scores.

Table 2 shows the results on the three tasks described in section 3. The performance of models trained using pseudonymized data is indistinguishable from the performance of models trained using real data.

This lack of difference was confirmed by performing Wilcoxon rank-sum tests (Mann and Whitney, 1947) on the folds of each task. None of the tests found any statistically significant differences between models trained on real or pseudonymized data.

4.3 Qualitative study I

This first qualitative study involved two human evaluators: One coordinating officer (a) that decides on the exportation of clinical data for research as well as a chief physician (b) who also is responsible for deciding on the exportation of clinical data for research. Both evaluators work at Region Stockholm county council in Sweden.

The requirements of the de-identification of the Gastro dataset was preceded by a discussion with the two human evaluators as well as a lawyer also working with exportation of clinical data. They decided that entities classed as *First Name, Last Name, Location, Phone Number, Organization* and *Social Security Number* are sensitive and should be removed from the patient record. They also decided that entities classified as *Age, Health Care Unit, Full Date* and *Date Part* were *not* sensitive

Data version	ICD-10	Factuality	Clinical Entity
	Classification	Classification	NER
Unaltered	0.86	0.74	0.85
Pseudonymized	0.86	0.75	0.86

Table 2: Models were trained on both pseudonymized and unaltered versions of each dataset. The average F_1 score of each model class on the held-out dataset is shown for each of the tasks.

and should be retained. Hence only six classes were used for de-identification.

The three experts also decided that they did not want to pseudonymize the text with surrogates. Instead, the sensitive entities were replaced with their class names. A control data set was also created using all ten classes for de-identification. Both de-identifications were done using HB Deid CRF.

Evaluator (a) read 100 pages each of the two sets of de-identified files. They found the word "Inga" as "No" in English was tagged as First Name. A similar pattern was found for the personal names "Per", "Tages" which can also mean, "Per day" or "Take" respectively. The evaluator also noticed that locations such as country names were removed, but not a patient's nationality. They also noticed that the span of the predicted entity did not always cover the whole PHI expression, especially when they were multi-word expression. *Evaluator (a)* found also that the set with 10 types of class tags was easier to read that the one with 6 types of class tags.

Evaluator (b) read 100 pages of the original file (that did not contain any class tags) by mistake instead of reading the de-identified version and they commented that they did not find much sensitive information. They said that the de-identification system managed to effectively replace the sensitive information. *Evaluator (b)* thought they found some problematic cases where, for example, names and locations were incorrectly assumed to be de-identified. All these cases were double checked and we confirmed that these were correctly tagged in the de-identified data set.

4.4 Qualitative study II

A second qualitative study was carried out where the version of HB Deid described in (Berg and Dalianis, 2021) was used. 100 patient patient records from an emergency unit at Skåne University Hospital were used for the de-identification experiment.

The evaluator was an computer scientist at Lund University at the Faculty of Engineering. Snippets of 50-200 words were extracted from each patient record and processed using HB Deid CRF. The de-identification system found a large number PHI tokens and classified most of them correctly. However, many abbreviations were incorrectly classified as *Organizations*. Generally, there where a few false positives and some false negatives, but these misclassifications alone did not provide enough information to reveal the identity of any patients.

5 Discussion

The human evaluators in this study disliked the idea of replacing PHI with surrogate values. Instead, they preferred replacing sensitive entities with their PHI class. However, the resulting dataset not only fails to use the protecting effects of HIPS (Hiding In Plain Site) (Carrell et al., 2019). It also makes it obvious to an adversary that any PHI they encounter is in fact a real PHI that the system failed to replace.

Human evaluations also uncover the problem of dealing with abbreviations, since they can be mistaken for organizations. This can be dealt with by adding word lists to deal specifically with abbreviations.

The human evaluators did not find the pseudonymized text difficult to read. On the contrary, one evaluator had difficulties distinguishing between real and pseudonymized data. This indicates that pseudonymized health records retain much of their utility for non-NLP research.

Comparing the utility of pseudonymized and real datasets, we find no harmful effects from de-identification on down-stream performance. This confirms and builds upon previous results in (Berg et al., 2020) that also showed that utility was retained after de-identification.

6 Conclusion

De-identification works best when the underlying NER classifier has both high precision and high recall. A high recall is crucial to ensure that as many PHI as possible are detected. At the same time, having a low precision may introduce noise into the data which can harm its utility.

In this study, we show that existing de-identification systems can effectively be used to make datasets safer. We also show that the noise introduced in this process does not harm down-stream performance in clinical NLP tasks.

The qualitative evaluations also show that humans have trouble distinguishing between pseudonymized and real data. We also uncover a discrepancy between the NLP community and our human evaluators regarding the perceived value of hiding sensitive data using HIPS.

Acknowledgements

We would like to thank Sonja Remmer for introducing us to the Wilcoxon rank-sums test as a tool for comparing machine learning models. Thanks also to Anders Björkelund for evaluating the results from HB Deid at Lunds university.

We are also grateful to the DataLEASH project for funding this research work.

References

Hanna Berg, Taridzo Chomutare, and Hercules Dalianis. 2019. Building a De-identification System for Real Swedish Clinical Text Using Pseudonymised Clinical Text. In *Proceedings of the Tenth International Workshop on Health Text Mining and Information Analysis (LOUHI 2019)*, pages 118–125, Hong Kong. Association for Computational Linguistics.

Hanna Berg and Hercules Dalianis. 2021. HB Deid-HB De-identification tool demonstrator. In *Proceedings of the 23rd Nordic Conference on Computational Linguistics (NoDaLiDa)*, pages 467–471.

Hanna Berg, Aron Henriksson, and Hercules Dalianis. 2020. The Impact of De-identification on Downstream Named Entity Recognition in Clinical Text. In *Proceedings of the 11th International Workshop on Health Text Mining and Information Analysis, Louhi 2020, in conjunction with EMNLP 2020*, pages 1–11.

David S Carrell, David J Cronkite, Muqun (Rachel) Li, Steve Nyemba, Bradley A Malin, John S Aberdeen, and Lynette Hirschman. 2019. The machine giveth and the machine taketh away: a parrot attack on clinical text deidentified with hiding in plain sight. *Journal of the American Medical Informatics Association*, 26(12):1536–1544.

Hercules Dalianis. 2019. Pseudonymisation of Swedish electronic patient records using a rule-based approach. In *Proceedings of the Workshop on NLP and Pseudonymisation*, pages 16–23, Turku, Finland. Linköping Electronic Press.

Hercules Dalianis, Aron Henriksson, Maria Kvist, Sumithra Velupillai, and Rebecka Weegar. 2015. HEALTH BANK- A Workbench for Data Science Applications in Healthcare. *CEUR Workshop Proceedings Industry Track Workshop*, pages 1–18.

Hercules Dalianis and Sumithra Velupillai. 2010. De-identifying swedish clinical text-refinement of a gold standard and experiments with conditional random fields. *Journal of biomedical semantics*, 1(1):1–10.

Christian Dalton-Locke, Johan H Thygesen, Nomi Werbeloff, David Osborn, and Helen Killaspy. 2020. Using de-identified electronic health records to research mental health supported housing services: A feasibility study. *PloS one*, 15(8):e0237664.

Andrea C Fernandes, Danielle Cloete, Matthew Broadbent, Richard D Hayes, Chin-Kuo Chang, Richard G Jackson, Angus Roberts, Jason Tsang, Murat Soncul, Jennifer Liebscher, et al. 2013. Development and evaluation of a de-identification procedure for a case register sourced from mental health electronic records. *BMC medical informatics and decision making*, 13(1):1–14.

Anastasios Lamproudis., Aron Henriksson., and Hercules Dalianis. 2022. Vocabulary modifications for domain-adaptive pretraining of clinical language models. In *Proceedings of the 15th International Joint Conference on Biomedical Engineering Systems and Technologies - HEALTHINF,*, pages 180–188. INSTICC, SciTePress.

Martin Malmsten, Love Börjeson, and Chris Haffenden. 2020. Playing with Words at the National Library of Sweden – Making a Swedish BERT. *arXiv:2007.01658 [cs]*. ArXiv: 2007.01658.

Henry B. Mann and D. Ransom Whitney. 1947. On a Test of Whether one of Two Random Variables is Stochastically Larger than the Other. *The Annals of Mathematical Statistics*, 18(1):50–60. Publisher: Institute of Mathematical Statistics.

Stéphane M. Meystre, Shuying Shen, Deborah Hofmann, and Adi V. Gundlapalli. 2014. Can physicians recognize their own patients in de-identified notes? In *MIE-Medical Informatics Europe*, pages 778–782.

Kostas Pantazos, Soren Lauesen, and Soren Lippert. 2017. Preserving medical correctness, readability and consistency in de-identified health records. *Health informatics journal*, 23(4):291–303.

Sonja Remmer, Anastasios Lamproudis, and Hercules Dalianis. 2021. Multi-label Diagnosis Classification of Swedish Discharge Summaries – ICD-10 Code Assignment Using KB-BERT. In *Proceedings of RANLP 2021: Recent Advances in Natural Language*

Processing, RANLP 2021, 1-3 Sept 2021, Varna, Bulgaria, pages 1158–1166.

David Sánchez, Montserrat Batet, and Alexandre Viejo. 2014. Utility-preserving privacy protection of textual healthcare documents. *Journal of biomedical informatics*, 52:189–198.

Maria Skeppstedt, Maria Kvist, Gunnar H Nilsson, and Hercules Dalianis. 2014. Automatic recognition of disorders, findings, pharmaceuticals and body structures from clinical text: An annotation and machine learning study. *Journal of biomedical informatics*, 49:148–158.

Sumithra Velupillai. 2011. Automatic classification of factuality levels: A case study on Swedish diagnoses and the impact of local context. In *Fourth International Symposium on Languages in Biology and Medicine, LBM 2011*.

Sumithra Velupillai, Hercules Dalianis, Martin Hassel, and Gunnar H Nilsson. 2009. Developing a standard for de-identifying electronic patient records written in Swedish: precision, recall and F-measure in a manual and computerized annotation trial. *International Journal of Medical Informatics*, 78(12):e19–e26.

Sumithra Velupillai, Hercules Dalianis, and Maria Kvist. 2011. Factuality levels of diagnoses in Swedish clinical text. In *User Centred Networked Health Care*, pages 559–563. IOS Press.

Horses to Zebras: Ontology-Guided Data Augmentation and Synthesis for ICD-9 Coding

Matúš Falis[1], Hang Dong[2], Alexandra Birch[1], and Beatrice Alex[3,4]

[1]Institute for Language, Cognition and Computation, University of Edinburgh
[2]Department of Computer Science, University of Oxford
[3]School of Literatures, Languages and Cultures, University of Edinburgh
[4]Edinburgh Futures Institute, University of Edinburgh
{s1206296, a.birch, balex}@ed.ac.uk, hang.dong@cs.ox.ac.uk

Abstract

Medical document coding is the process of assigning labels from a structured label space (ontology – *e.g.*, ICD-9) to medical documents. This process is laborious, costly, and error-prone. In recent years, efforts have been made to automate this process with neural models. The label spaces are large (in the order of thousands of labels) and follow a big-head long-tail label distribution, giving rise to few-shot and zero-shot scenarios. Previous efforts tried to address these scenarios within the model, leading to improvements on rare labels, but worse results on frequent ones. We propose data augmentation and synthesis techniques in order to address these scenarios. We further introduce an analysis technique for this setting inspired by confusion matrices. This analysis technique points to the positive impact of data augmentation and synthesis, but also highlights more general issues of confusion within families of codes, and underprediction.

1 Introduction

Large-Scale Multi-Labelled Text Classification (LMTC) tasks, such as automated ICD-9 coding of MIMIC-III discharge summaries, suffer from a big-head long-tail distribution of classes. This phenomenon naturally arises due to some labels being more frequent than others. This can further be affected by the source of the data – in the case of clinical *Natural Language Processing* (NLP), this is often a single institution. For instance, hospitals in Switzerland are unlikely to have cases of injuries caused by shark bites (code *W56.41XD* in ICD-10). Hence, depending on the data source, some labels will have a very small population – *Few-Shot* (FS), or even no population at all – *Zero-Shot* (ZS) scenario. Furthermore, adding new labels into a standard by splitting/fusing/altering existing concepts, or introducing new concepts also creates a ZS scenario. Medical coding methods need to be able to adapt to these scenarios.

Medical coding methods can be broadly divided on the task (document-level/entity-level) and the approaches (rule-based/machine learning) they use. Methods like Apache cTAKES (Savova et al., 2010), MedCAT (Kraljevic et al., 2019), or SemEHR (Wu et al., 2018) perform *Named Entity Recognition and Linking* (NER+L), identifying specific spans of text within the document and associating them with concepts in a knowledge base, such as UMLS (Bodenreider, 2004). They primarily use rule-based methods, with some inclusion of machine learning – *e.g.*, MedCAT uses contextual word embeddings for disambiguating homographic strings, such as context-sensitive abbreviations (*e.g.*, the string *"HR"* can mean "hour" or "heart rate"). Neural LMTC models, such as CAML (Mullenbach et al., 2018), predict labels on the document level. Labels are not associated with any particular string in the text, but rather appear as document-level sets.

Rule-based NER+L methods, assuming machine learning is not used, are not affected by the FS/ZS scenarios. There either is a suitable rule designed for a given situation, or not. If a new code is introduced into the label space, the rules need to be adjusted to reflect this.

Neural learning approaches are data-driven. The populations of labels available during training and the variety in the inputs to which they are associated affect the model's generalisability, especially if the model is not designed with the few-shot/zero-shot scenario in mind. Previous work has tried to address this with setting non-trainable parameters within the network as representations of ICD-9 codes enriched with knowledge from the ontologies (Rios and Kavuluru, 2018). While the few-shot/zero-shot performance improved, the overall performance deteriorated.

An alternative to model adjustments is to avoid the FS/ZS scenarios by supplying more data, *i.e.*, through data augmentation or synthesis. Aug-

Proceedings of the BioNLP 2022 workshop, Dublin, Ireland, pages 389–401
May 26, 2022.

mentation through synonym replacement has been previously done using WordNet (Ollagnier and Williams, 2020), with improvements coming from the use of a medical knowledge base (UMLS) (Kang et al., 2021). Simple natural language generation techniques were also employed (Ollagnier and Williams, 2020). These techniques, while expanding the vocabulary, are only capable of producing synthetic documents with labels present in the original training data. Synthesising new documents with alternative labels has been done based on document templates in the scope of radiology reports – however, human experts were involved in the process (Schrempf et al., 2020).

We propose a novel type of data synthesis for ICD-9 coding, and medical LMTC tasks in general with the aim to replace concepts of underspecified codes with more specific, and often less frequent, alternatives. Similar to Schrempf et al. (2020) we recognise the value of augmenting concepts of interest. Rather than using templating in order to determine concept location, we use pre-existing NER+L techniques (MedCAT and SemEHR) to identify spans relevant to the gold standard labelling.

Furthermore, we introduce an error analysis technique for this setting inspired by confusion matrices. This technique associates codes within the prediction set with codes in the gold standard set according to the ontological structure allowing us to track mispredictions co-occurring with unmatched gold-standard codes indicating confusion – which codes tend to be mispredicted as others.

Our work provides the following contributions:

- Applying *Ontology-Guided Synonym Replacement* to ICD-9 coding, where multiple ontologies are used to determine candidate synonyms for a given concept found by an NER+L method akin to the work of (Kang et al., 2021). This *augmentation* method leads to improved model performance.
- *Sibling-Code Replacement*, where the surface form of a concept reported by an NER+L method is replaced with one of a semantically similar code according to the ontology, with the change being reflected in the document's updated silver standard. This *synthesis* method leads to improved model performance.
- The *Weak Hierarchical Confusion Matrix* (WHCM) – an analysis tool for the LMTC (weakly-labelled) scenario inspired by the concept of confusion matrix allowing more in-depth error analysis facilitating further development of LMTC systems. The output of this tool can be further used as an evaluation metric describing error types.
- The source code for augmentation and synthesis[1], and WHCM[2] will be made available via GitHub.

Our augmentation and synthesis methods both separately and combined lead to improved micro-F1 scores. The also improve g FS and ZS performance – although are surpassed by the baseline setup with more training. Our analysis tool highlights the error types in prediction – some errors are due to confusion within the code family, but most are due to underprediction.

2 Background

In this section we will introduce medical ontologies, both as a label set, and source of external knowledge. We will describe Named Entity Recognition and Linking used for determining relevant spans of text, introduce LMTC as our task, discuss previous data augmentation techniques in clinical NLP, and finally comment on the current approaches to evaluation and analysis of LMTC models.

2.1 Medical Ontologies

The International Classification of Diseases 9th Edition, Clinical Modification[3] (ICD-9-CM, here refereed to as ICD-9 despite nuances) is a medical ontology of diseases and procedures represented by two tree-structured label-spaces. The higher the depth of a node within the label space, the more specific a concept it describes, with lower depths representing aggregation on *e.g.*, disease type or general anatomy. Such aggregation is represented via subtrees (or *families*) of codes. Coding is done primarily with leaf nodes, representing the highest degree of specification within the ontology. We use ICD-9 as a basis for our research due to the availability of data labelled with this ontology – MIMIC-III. Newer revisions of the ICD (ICD-10[4], ICD-11[5]) differ in size, organisation of the tree

[1] https://github.com/modr00cka/Ontology-Guided-Augmentation-and-Synthesis

[2] https://github.com/modr00cka/weak_hierarchical_confusion

[3] https://www.cdc.gov/nchs/icd/icd9cm.htm

[4] https://icd.who.int/browse10/2019/en

[5] https://icd.who.int/browse11/l-m/en

structure, and naming conventions, but generally follow the same structural design principles. Hence our research can be re-used for newer standards.

An ICD-9 code (*e.g.*, *250.01*) consists of a *category* (part of the code appearing prior to the decimal point, *e.g.*, 250) and *etiology* (appearing after the decimal point, *e.g.*, 01). The etiology can be represented by up to two digits. A longer etiology implies a more specific concept.

Dedicated leaf-level codes exist to describe an "unspecified" version of a parent concept (*e.g.*, hypertension with unspecified malignancy status would be coded as *401.9 Unspecified Essential Hypertension*, rather than *401 Essential Hypertension*). Such "unspecified" concepts may appear on different depths representing different parts of the concept being unspecified. This phenomenon can appear within the same family of codes, indicating different levels of specificity – *e.g.*, the single-digit-etiology leaf code *365.9 Unspecified Glaucoma* versus the double-digit-etiology leaf code *365.60 Glaucoma associated with unspecified ocular disorder*. While not a general rule, some etiology patterns tend to be associated with unspecified concepts – *.9*, *.?0*, and sometimes *.?1* (where *?* can be any digit.)

The *Unified Medical Language System* (UMLS)[6] (Bodenreider, 2004) is a project of medical terminology originally released in 1990. The core components of UMLS are the *Metathesaurus* containing various medical vocabularies, a *Semantic Network* representing the connections between the terms, and an *Information Sources Map*. The concepts within the Metathesaurus are each assigned an identification code known as the *Concept Unique Identifier* (CUI).

Furthermore, the Information Sources Map component enables mapping of concepts between ontologies through the concepts' CUI. An examples of such a mapping is the SNOMED CT[7] to ICD-9 map curated by UMLS.

2.2 Named Entity Recognition and Linking

The task of identifying relevant concepts within free text is known as *Named Entity Recognition* (NER). It can be extended to NER+L by linking them to entities in an ontology (*e.g.*, UMLS). The standard labelling in NER+L tasks consist of two pieces of data – the indices identifying the span of text constituting an entity, and the assigned class (*e.g.*, CUI). NER+L serves as the first step in our augmentation and synthesis methods.

In the medical domain, notable early NER+L (predominantly rule-based) systems include MetaMap (Aronson, 2001) and Apache cTAKES (Savova et al., 2010). These systems struggle with ambiguities and spelling mistakes. BioYODIE (Gorrell et al., 2018), a more recent approach, addresses some of these ambiguity issues through corpus-based statistics, *e.g.*, co-occurrence graph. SemEHR (Wu et al., 2018) improves upon the output of BioYODIE with manually-derived rules. Certain types of ambiguity still pose issues to these systems, *e.g.*, expansion of context-sensitive abbreviations and variety of concept names. MedCAT (Kraljevic et al., 2019) employs unsupervised training and vocabulary building to further address ambiguity through context-sensitive disambiguation.

2.3 Large-Scale Multi-Label Text Classification (LMTC)

Large-Scale Multi-Label Text Classification (LMTC) is the task of assigning multiple weak labels to text documents. The labels come from a large label-space (in the order of thousands of labels), which can be structured, *e.g.*, ICD-9. LMTC tasks appear in several domains, including medical, legal, and commercial. The most notable early model in LMTC is CAML introduced by Mullenbach et al. (2018) for ICD-9 coding of medical documents.

Given an input document the model identifies the set of labels to be assigned. The input tokens are converted into word embeddings using word2vec (Mikolov et al., 2013). Convolutional filters are applied on these embeddings for short-range interaction. These phrase embeddings are fed into a label-specific attention mechanism – for each label an attention mechanism is applied identifying tokens contributing towards the respective code's prediction. The attention is multiplied with the the phrase embeddings resulting in label-specific document embeddings upon which classification is performed. Mullenbach et al. (2018) used spans around high-attention tokens (keywords) for qualitative evaluation of predictions. Falis et al. (2019) with the use of a hierarchical ensemble showed that tokens relevant for a family of codes can be captured with the attention mechanisms of ancestor

[6]https://www.nlm.nih.gov/research/umls/index.html

[7]https://www.snomed.org/

codes and propagated to descendants.

LMTC models are data-driven neural approaches requiring large amounts of data. Due to the big-head long-tail label space, the performance of these models varies between codes, with frequent codes performing better. For this reason FS and ZS specific techniques were developed, such as that of Rios and Kavuluru (2018); Lu et al. (2020).

2.4 Data Augmentation

Data Augmentation (DA) in machine learning is a method for artificially increasing the amount of training data by label-preserving alterations of the input. This technique can be used either to make the models more resilient to noise in the data, introduce variety, or enrich with additional information addressing model limitations.

One of the most representative DA techniques in NLP is synonym replacement (Feng et al., 2021). This technique replaces tokens within the text with synonymous words or phrases, with the aid of a knowledge base, such as WordNet[8]. Assuming the synonym does not change the semantics of the text, the synthetic document's labels should be the same. Synonym replacement with WordNet has been previously employed by Ollagnier and Williams (2020) in medical document classification. Their method randomly replaces a set number of non-stopwords per document with their synonyms. The relatively unrestricted choice of words, however, means the synonym replacement may not be applied to concepts of high interest – medical vocabulary. Schrempf et al. (2020) apply a focused form of document synthesis through the use of templates in radiology reports. These templates are used for augmenting concepts of interest, or replacing them with similar ones. UMLS-based synonym replacement has previously been used for DA in NER+L and sentence classification by Kang et al. (2021), employing random insertion, random swap, and random deletion, and UMLS-synonym replacement guided by the output of MetaMap.

We employ UMLS-synonym replacement DA similar to Kang et al. (2021) for the task of LMTC guided by more recent biomedical *NER+L* methods. We further propose a novel ontology-guided document synthesis turning relevant concepts into semantically adjacent concepts based on the ICD-9, with the expected label set being adjusted accordingly. The aim of this synthesis technique is to provide further training data specifically to few-shot and zero-shot labels.

2.5 Evaluation and Analysis for LMTC

LMTC tasks are evaluated using precision, recall and F1 score with micro and macro averaging, where macro-level metrics place equal weight on the performance on each label, disregarding the class imbalance. For the FS and ZS scenario precision and recall of the k highest predictions (@k), regardless of passing a fixed threshold tend to be employed. These measures compare exact match (intersection) between the prediction and gold standard sets ignoring the rich ontological structure and consider all errors equivalent. *Count-Preserving Hierarchical Evaluation* (CoPHE) (Falis et al., 2021) is a recently proposed evaluation metric involving the ontological structure to award partial credit to mispredictions occurring within the family of codes to which a gold-standard label belongs. Through the preservation of counts this method also considers over-/under-prediction within families of codes.

Beyond aggregate measures, to the best of our knowledge, label-specific analysis tools do not exist. Due to the weakly-labelled nature of LMTC tasks, confusion matrices are not a viable option. We introduce an analysis method akin to the confusion matrix suitable for LMTC.

3 Data

We employ the discharge summaries of MIMIC-III (Johnson et al., 2016) due to their common use in medical LMTC tasks. MIMIC-III is a multimodal medical dataset acquired from the intensive care units of the Beth Israel Deaconess Medical Center in Boston, Massachusetts between years 2001 and 2012. Access of the data was granted through PhysioNet[9] after completing the ethical training by the Collaborative Institutional Training Initiative program. The dataset is coded with ICD-9 codes on the document level. These labels do not perfectly represent the content of the text – MIMIC-III is significantly under-coded for specific conditions (Searle et al., 2020), and sometimes incorrect codes are assigned – *e.g.*, in the case of smoking status (Falis et al., 2019).

The data has been pre-processed and split following Mullenbach et al. (2018)'s procedures. The

[8] `https://wordnet.princeton.edu/`

[9] `https://physionet.org/content/mimiciii/1.4/`

label distribution within the dataset follows a big-head long-tail distribution. We divide the labels, similar to Rios and Kavuluru (2018), into three subsets according to their population size: Of the total 8,929 unique labels 4,351 appear in more than 5 documents within the training set; 4,341 at least once but at most 5 times (few-shot); and 237 labels do not appear in the training set, while existing in the development or test set (zero-shot). The training set consists of 47,719 documents.

4 Methods

Our methods include data augmentation and synthesis strategies based on the synonyms and adjacent concepts respectively, and an analytic tool for LMTC based on a set of assumptions to adapt confusion matrices with ontological structure.

4.1 Data Augmentation and Synthesis Strategies

We have attempted to enhance the training data with variety in the vocabulary and introduction of new codes in synthetic data. We applied two NER+L systems – SemEHR and MedCAT – to the training set. Unlike Searle et al. (2020) who sought to produce a silver standard by reconciling the output of NER+L methods with the gold standard, for the purpose of determining candidate codes for DA we chose to filter the NER+L outputs by intersecting them with the gold standard. While the gold standard may not capture all mentioned concepts, it may reflect local coding guidelines. As the NER+L systems label their outputs with CUIs, we translated these into ICD-9 using *PyMedTermino*.[10]

It should be noted that LMTC models, such as CAML, rely on pre-trained word2vec features with a static vocabulary – words unseen during pre-training will be considered *out-of-vocabulary* (OOV). This affects concepts that are unseen during training, such as rare diseases named after a person – *e.g.*, Munchausen's Syndrome (*301.51* in ICD-9). By introducing alternative names (augmentation) or new concepts (synthesis) we can also expand the relevant vocabulary, mitigating OOV.

4.1.1 Identity-Code Augmentation

We first created a synonym-replacement DA method in order to make the models more robust to variety. A medical concept can have several alternative names or surface forms including abbreviations – *e.g.*, an "acute myocardial infarction" can be referred to as "heart attack" or the abbreviation "MI". Through augmenting the text with synonyms we expose the model to alternative keywords representing existing concepts (already within the corpus or previously unseen), while leaving non-keyword context tokens untouched.

If an input document has any NER+L predictions matching the gold standard, their spans are identified. A synonym from PyMedTermino (derived from the UMLS, ICD-9, ICD-10, and SNOMED CT) is chosen at random, and replaced within the input text for each span. The augmented text is then added to the training set with the same gold standard labels as the original.

4.1.2 Adjacent-Code Synthesis

An additional form of *Document Synthesis* (DS), aimed at introducing new labels, can be produced by replacing mentions of a concept with an adjacent concept, rather than a synonym – *e.g.*, "stage 2 glaucoma" with "stage 3 glaucoma" – and updating the gold standard for the synthetic document accordingly. Where Identity-Code Augmentation aims to expose the model to alternative keywords to concepts pre-existing in the corpus without changing the code, the Adjacent-Code Synthesis replaces the code, exposing the model to the keyword of a different code – potentially one that is rare within the original training set (FS), or not appearing in it at all (ZS). This replacement leads to these keywords appearing in new contexts (those of the concepts they replace).

We chose to focus on "unspecified" codes assuming an "unspecified" label means all its mentions within are non-specific, while a single specified mention warrants a more specific version of the code in the new silver standard. This choice was made to address imperfections in the NER+L predictions – replacing a specified code would require replacement of all its mentions, some of which may not be identified by the NER+L method.

The outputs of SemEHR and MedCAT are processed as in the synonym-replacement DA. We considered a code to be unspecified if its description contained the string "unspecified" or "not otherwise specified", and with with "9" as the first or "0"/"1" as the second digit of the etiology. Of the 8,692 unique codes appearing in the training set 1,188 remained as viable "unspecified codes".

[10]`https://pythonhosted.org/PyMedTermino/`

This represents 14.74% of the total code population within the training set.

Replacement codes were identified depending on the etiology – double-digit unspecified codes can only be replaced by codes differing only in the final digit, while single-digit unspecified codes can be replaced with codes of the same category with any other etiology. Replacement codes were divided into three sets – frequent (>5), few-shot (at least one but up to 5), zero-shot (unseen) – based on their population in the training set. Only labels known to be within the MIMIC-III dataset were considered.

For a given document each viable unspecified code is first randomly converted into a specified candidate (with ZS and FS candidates being preferred). The mentions of the unspecified code are randomly replaced with mentions of the specified candidate. The resulting synthetic discharge summary is then added into the training set with the original gold standard code replaced with the candidate code. The pipeline for this DS procedure is presented in Figure 1.

4.1.3 Enriched Training Sets

We applied the synonym DA method in a single pass on the original training set for each NER+L method explored, resulting in the sets *SemEHR-DA* and *MedCAT-DA*. The adjacent label DS method was applied in two passes for each NER+L method. This was done to allow for multiple adjacent-code-synthetic alternatives per document. The resulting datasets are called *SemEHR-DS* and *MedCAT-DS*. *SemEHR-Both* and *MedCAT-Both* are the combinations of DA and DS datasets. All DA and DS datasets were combined with the *Baseline*, and deduplicated. The final sizes of the different training sets are presented in Table 1, including the number of unique codes within the frequent, few-shot, and zero-shot subsets. DA strategies increase populations of frequent and few-shot codes, leading to some few-shot codes becoming frequent (>5 occurrences in the training set). DS expands on this by also increasing populations of 13 zero-shot codes. The development set and test set were left unmodified.

4.2 Hierarchical Confusion Matrix

Confusion matrices are useful evaluation analysis tools in strongly-labelled scenarios (where individual predictions are associated with gold labels)(Tan et al., 2019, p. 138). A high misclassification between two classes indicates that, with respect to the model's parameters and the data, members of these classes are similar and difficult to distinguish.

Dataset	Size	Frequent	Few	Zero
Baseline	47,719	4,351	4,341	237
SemEHR-DA	66,559	4,818	3,874	237
MedCAT-DA	71,295	4,998	3,694	237
SemEHR-DS	74,851	5,167	3,538	224
MedCAT-DS	74,830	5,164	3,541	224
SemEHR-Both	93,690	5,446	3,259	224
MedCAT-Both	98,402	5,565	3,140	224

Table 1: Training set sizes (number of documents) and populations (number of unique codes) of the frequent, few-shot, and zero-shot subsets.

Confusion matrices can also support labels without a valid association, *e.g.*, a prediction on a span not present in the gold standard, by associating them with a special label indicating absence of the counterpart. This scenario represents over/under prediction.

The confusion matrix enables high-level error analysis beyond tracking precision and recall of the model. Such error analysis can be used in further model design, or serve as supplementary information for a deployed model.

In the weakly-labeled scenario, such as ICD-9 coding, both predictions and gold labels are presented on the document level as sets without associations between individual labels. If there is a mismatch between a predicted label and the gold standard, we cannot state with certainty that a predicted label, say, A.4 (*e.g.*, Alcohol abuse, continuous) was misclassified as gold standard label A.2 (*e.g.*, Alcohol abuse, episodic) or B.1 (*e.g.*, Chronic bronchitis), or whether the model overpredicted A.4, while underpredicting B.1 and A.2. We can, however, make assumptions based on the ontological structure associating mispredictions within code families – relating the A.4 prediction to the gold label A.2 rather than B.1.

The problem of analysing multi-label classification tasks and hierarchical label spaces with confusion matrices has attracted recent attention within the visualisation community (Görtler et al., 2021; Heydarian et al., 2022). Heydarian et al. (2022) propose an extension to the standard confusion matrix for multi-label classification in a non-hierarchical setting. Görtler et al. (2021) propose a method of analysis in a hierarchical multi-output setting, approaching high-dimensional confusion as a distribution.

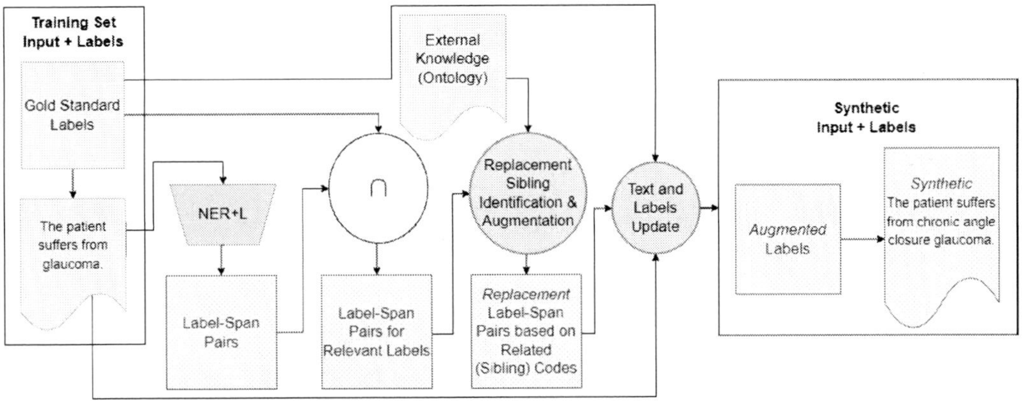

Figure 1: Ontology-Aided Document Synthesis pipeline. Yellow elements indicate data from human experts (input document, gold standard labels, ontology), gray elements indicate data which have machine learning somewhere in the creation process. The green element indicates pre-existing software, blue elements indicates software custom written for this method.

A co-occurrence matrix between predictions and gold labels indicates which predicted labels co-occur with particular gold standard labels, but is not fine-grained enough for error analysis. We propose the use of ontological structure to reduce the co-occurrence matrix into a simple weak hierarchical confusion matrix analysis method designed with the LMTC scenario in mind and apply it to ICD-9 coding. We further aggregate its results into performance metrics exploring proportion of errors based on their type.

4.2.1 Assumptions

Starting from a co-occurrence matrix between the predicted and gold standard sets of labels we apply three assumptions:

- *1-to-1 True Positive Correspondence*: If a label is present both in the prediction and gold-standard for a document, this is a True Positive (TP), and not considered for confusion.
- *Within-Family Confusion*: non-TP codes in the prediction are matched with non-TP codes in the gold standard within their respective code families (black cells in Figure 2 are ignored).
- *Out-Of-Family Scenario*: If in confusion matching a code from prediction/gold cannot be matched (no code from its family left to match), the code is associated with a special OOF code (see red the cell in Figure 2).

4.2.2 Use

While we are capable of visualising WHCMs (Appendix A.1) for each family, for the purposes of this publication we opt for aggregating results for all codes. In particular, we reduce the matrices into the following data given gold standard code: What proportion of the gold standard is correctly matched to its prediction, is confused within its family, and is in the OOF scenario. These three percentages sum up to 1. Furthermore, for each code we also track which code within its family (including OOF) is the most likely to be predicted, given the gold standard label. This information is used to determine if this most likely code matches the gold standard code. An example of this analysis can be found in Table A1 in the Appendix. We macro-average the correct-match/within-family confusion/OOF statistics and then provide the percentage of matches between preferred prediction given the gold standard. Note further analyses can be drawn conditioning on the predicted codes by applying the same procedures to the transpose of the original WHCM.

5 Experiments

We have applied MedCAT and SemEHR to the training set, producing candidate spans associated with CUIs. Post CUI-to-ICD-9 conversion, we have removed all candidates not matching the gold standard of their source discharge summary. We have produced augmented and synthetic data according to our description in sections 4.1.1 and 4.1.2 and combined them with the original train-

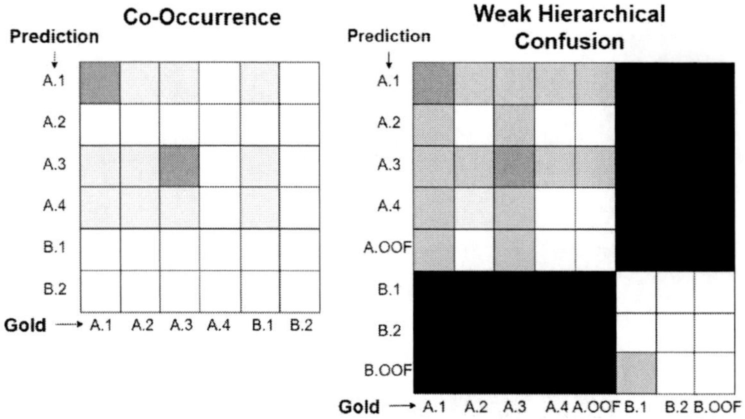

Figure 2: Left: A simple co-occurrence matrix between the prediction and gold standard labels for two label families for a single document. Labels A.1, A.3, and A.4 are predicted, while codes A.1, A.2, A.3, and B.1 are in the gold standard. Green cells indicate a match between the prediction and gold standard, yellow cells indicate a mismatch. Right: A weak hierarchical confusion matrix constructed from the co-occurrence matrix with the use of the three assumptions – Gray cells were eliminated via 1-to-1 correspondence, black cells were eliminated via within-family-confusion, red cells indicate the OOF scenario. The resulting confusion matrix indicates A.1 and A.3 being correctly predicted (green), B.1 being a false negative – an OOF (red), and the predicted code A.4 being confused with expected code A.2 (yellow).

ing set (dropping any duplicates) to produce enriched training sets as presented in section 4.1.3. We have further created a Baseline-like dataset of a similar size to our largest datasets – SemEHR-Both and MedCAT-Both – as a controlled experiment. This was done by concatenating two Baseline datasets (*2xBaseline*). Assuming a constant number of epochs, training on 2xBaseline corresponds to training on the Baseline for double the number of epochs.

We train CAML models based on the implementation of Chalkidis et al. (2019)[11] for 15 epochs on the training sets (Table 1). No few-shot/zero-shot model-side solution (such as the use of label embeddings as parameters) was applied. Each experiment used word embeddings of size 100 pre-trained on its respective training set according to Mullenbach et al. (2018)'s procedure. The development and test sets were the same across all experiments. The model weights with the best end-of-epoch validation F1 score were evaluated on the test set.

6 Results

For each experiment we report results averaged across 5 runs (Table 2), except the three largest (2xBaseline, SemEHR-Both, MedCAT-Both) for which a single run was conducted. We compare the performance on previously used metrics: Micro-F1 for all codes, and R@10 for few-shot and zero-shot codes. The codeset for few-shot and zero-shot codes is derived from the Baseline, and hence includes codes whose populations have increased in the DA, DS, and Both datasets. We further report hierarchical results (Mic-F1$_H$) according to CoPHE. Furthermore, we present macro-averaged aggregate measures conditioned on the *gold-standard labels* for all labels coming from WHCM – percentages of gold labels being being predicted correctly (Mac-Cor), being confused with a code within the same family (Mac-Conf), and being confused as OOF (underprediction – Mac-OOF). Finally, we track whether the prediction most often matched with the gold standard code, is the identity code itself (rather than a sibling or OOF) – if a correct prediction is more likely than any kind of misprediction. On a code level this is represented as a binary value (match or mismatch), and then can be macro-averaged to the metric Match. For our WHCM families we have used the ICD-9 tree as implemented in CoPHE aggregating on its parent level (code category). It should be noted, that our CAML baseline results underperform with respect

[11]https://github.com/iliaschalkidis/lmtc-emnlp2020

Dataset	Mic-F1	Mic-F1$_H$	R@10-Few	R@10-Zero	Mac-Cor	Mac-Conf	Mac-OOF	Match
Baseline	0.441	0.487	0.034	0.035	0.043	**0.055**	0.902	0.044
2xBaseline*	0.477	0.521	**0.093**	**0.075**	0.073	0.066	0.861	0.077
SemEHR-DA	0.469	0.514	0.055	0.034	0.062	0.062	0.876	0.063
MedCAT-DA	0.468	0.514	0.064	0.048	0.062	0.065	0.873	0.065
SemEHR-DS	0.471	0.518	0.051	0.055	0.067	0.065	0.869	0.069
MedCAT-DS	0.474	0.520	0.059	0.054	0.068	0.065	0.866	0.071
SemEHR-Both*	0.483	0.528	0.066	0.051	**0.079**	0.066	0.855	0.081
MedCAT-Both*	**0.486**	**0.532**	0.071	0.057	**0.079**	0.068	**0.853**	**0.083**

Table 2: Test-set performance of CAML models trained on the original training set (Baseline) versus training sets with synonym augmentation (SemEHR-DA, MeCAT-DA), and adjacent-code synthesis (SemEHR-DS, MedCAT-DS) averaged across 5 runs. Experiments on datasets marked with an asterisk (2xBaseline, SemEHR-Both and MedCAT-Both) have, due to time constraints, been conducted a single run each. Best performance for each metric is marked bold. Results are reported on the original test set. Zero and Few-shot codesets are based on the Baseline. The original development set is used for validation in all experiments.

to the official results of Mullenbach et al. (2018) (Micro-F1 of 0.53), due to our limited number of training epochs (while Mullenbach et al. (2018) ceases training after the precision@8 does not improve for 10 epochs).

All the proposed methods improve on the Baseline with regard to standard and hierarchical Micro-F1. Augmentation (DA) sets, while comparatively worse than Synthetic (DS) on R@10-Zero and standard and hierarchical Micro-F1, perform better on R@10-Few. This was to be expected as the DA methods provide little for the Zero codeset, while producing more of the labels in the Freq and Few codesets. Interestingly, SemEHR-DA performs on par with MedCAT-DA despite having a smaller training set. The combination of DA and DS methods (Both) report the best F1 results, with MedCAT-Both performing best in 5 our of the 8 reported metrics (including Mac-Cor, Mac-OOF and Match). Both of these methods' results are at least as good as those of 2xBaseline, which is of comparable size. The best R@10-Few and R@10-Zero performance was achieved by 2xBaseline, which corresponds to training the Baseline for twice as many epochs. While the different improvement of DA and DS in R@10-Few and R@10-zero implies our methods enhance these subsets, 2xBaseline dominating these metrics suggests that a better performance FS/ZS can be achieved with more training epochs. The difference between the standard and hierarchical (CoPHE) F1 scores remained largely the same, which implies partial errors were not addressed by these methods – this is further supported by the changes in Mac-OOF dominating compared to those of Mac-Conf. The lowest Mac-Conf was achieved by the original Baseline, but was coupled with a high OOF implying that this low confusion is mostly due to a higher proportion of codes not being predicted at all.

7 Conclusion and Discussion

The data enrichment methods have improved on the baseline showing potential in approaching the few-shot/zero-shot scenario through data, rather than the model. However, our approach relied on the use of external NER+L tools, whose predictions are imperfect, and may not be available for all domains of interest. Other avenues of finding relevant entities, *e.g.*, the attention outputs of LMTC models, should be explored in future work. While the data enrichment results are encouraging, further analysis on fully trained LMTC models is desirable. WHCM results point to a major issue with most false negatives coming from underprediction of a family, rather than within-family confusion. Further analysis should be conducted on false positives. The analysis from the WHCM tool can provide possible explanation of the errors of a model and may shed light on the design of more accurate models for LMTC.

Acknowledgements

This work is supported by the United Kingdom Research and Innovation (grant EP/S02431X/1), UKRI Centre for Doctoral Training in Biomedical AI at the University of Edinburgh, School of Informatics, and the Health Data Research UK (HDR UK) National Text Analytics and Phenomics Projects. HD is supported by the Engineering and Physical Sciences Research Council (EPSRC, grant EP/V050869/1), Concur: Knowledge Base Construction and Curation.

References

Alan R Aronson. 2001. Effective mapping of biomedical text to the UMLS metathesaurus: the MetaMap program. In *Proceedings of the AMIA Symposium*, page 17. American Medical Informatics Association.

Olivier Bodenreider. 2004. The unified medical language system (UMLS): integrating biomedical terminology. *Nucleic acids research*, 32(suppl_1):D267–D270.

Ilias Chalkidis, Emmanouil Fergadiotis, Prodromos Malakasiotis, and Ion Androutsopoulos. 2019. Large-scale multi-label text classification on eu legislation. In *Proceedings of the 57th Annual Meeting of the Association for Computational Linguistics*, pages 6314–6322.

Matúš Falis, Hang Dong, Alexandra Birch, and Beatrice Alex. 2021. CoPHE: A count-preserving hierarchical evaluation metric in large-scale multi-label text classification. In *2021 Conference on Empirical Methods in Natural Language Processing*.

Matúš Falis, Maciej Pajak, Aneta Lisowska, Patrick Schrempf, Lucas Deckers, Shadia Mikhael, Sotirios Tsaftaris, and Alison O'Neil. 2019. Ontological attention ensembles for capturing semantic concepts in icd code prediction from clinical text. In *Proceedings of the Tenth International Workshop on Health Text Mining and Information Analysis (LOUHI 2019)*, pages 168–177.

Steven Y Feng, Varun Gangal, Jason Wei, Sarath Chandar, Soroush Vosoughi, Teruko Mitamura, and Eduard Hovy. 2021. A survey of data augmentation approaches for nlp. In *Findings of the Association for Computational Linguistics: ACL-IJCNLP 2021*, pages 968–988.

Genevieve Gorrell, Xingyi Song, and Angus Roberts. 2018. Bio-yodie: A named entity linking system for biomedical text. *arXiv preprint arXiv:1811.04860*.

Jochen Görtler, Fred Hohman, Dominik Moritz, Kanit Wongsuphasawat, Donghao Ren, Rahul Nair, Marc Kirchner, and Kayur Patel. 2021. Neo: Generalizing confusion matrix visualization to hierarchical and multi-output labels. *arXiv preprint arXiv:2110.12536*.

Mohammadreza Heydarian, Thomas E Doyle, and Reza Samavi. 2022. MLCM: Multi-label confusion matrix. *IEEE Access*, 10:19083–19095.

Alistair EW Johnson, Tom J Pollard, Lu Shen, Li-wei H Lehman, Mengling Feng, Mohammad Ghassemi, Benjamin Moody, Peter Szolovits, Leo Anthony Celi, and Roger G Mark. 2016. Mimic-iii, a freely accessible critical care database. *Scientific data*, 3(1):1–9.

Tian Kang, Adler Perotte, Youlan Tang, Casey Ta, and Chunhua Weng. 2021. UMLS-based data augmentation for natural language processing of clinical research literature. *Journal of the American Medical Informatics Association*, 28(4):812–823.

Zeljko Kraljevic, Daniel Bean, Aurelie Mascio, Lukasz Roguski, Amos Folarin, Angus Roberts, Rebecca Bendayan, and Richard Dobson. 2019. Medcat–medical concept annotation tool. *arXiv preprint arXiv:1912.10166*.

Jueqing Lu, Lan Du, Ming Liu, and Joanna Dipnall. 2020. Multi-label few/zero-shot learning with knowledge aggregated from multiple label graphs. In *Proceedings of the 2020 Conference on Empirical Methods in Natural Language Processing (EMNLP)*, pages 2935–2943.

Tomas Mikolov, Ilya Sutskever, Kai Chen, Greg S Corrado, and Jeff Dean. 2013. Distributed representations of words and phrases and their compositionality. *Advances in neural information processing systems*, 26.

James Mullenbach, Sarah Wiegreffe, Jon Duke, Jimeng Sun, and Jacob Eisenstein. 2018. Explainable prediction of medical codes from clinical text. In *NAACL-HLT*.

Anaïs Ollagnier and Hywel TP Williams. 2020. Text augmentation techniques for clinical case classification. In *CLEF (Working Notes)*.

Anthony Rios and Ramakanth Kavuluru. 2018. Few-shot and zero-shot multi-label learning for structured label spaces. In *Proceedings of the Conference on Empirical Methods in Natural Language Processing. Conference on Empirical Methods in Natural Language Processing*, volume 2018, page 3132. NIH Public Access.

Guergana K Savova, James J Masanz, Philip V Ogren, Jiaping Zheng, Sunghwan Sohn, Karin C Kipper-Schuler, and Christopher G Chute. 2010. Mayo clinical text analysis and knowledge extraction system (ctakes): architecture, component evaluation and applications. *Journal of the American Medical Informatics Association*, 17(5):507–513.

Patrick Schrempf, Hannah Watson, Shadia Mikhael, Maciej Pajak, Matúš Falis, Aneta Lisowska, Keith W Muir, David Harris-Birtill, and Alison Q O'Neil. 2020. Paying per-label attention for multi-label extraction from radiology reports. In *Interpretable and Annotation-Efficient Learning for Medical Image Computing*, pages 277–289. Springer.

Thomas Searle, Zina Ibrahim, and Richard Dobson. 2020. Experimental evaluation and development of a silver-standard for the mimic-iii clinical coding dataset. In *Proceedings of the 19th SIGBioMed Workshop on Biomedical Language Processing*, pages 76–85.

Pang-Ning Tan, Michael Steinbach, and Vipin Kumar. 2019. *Introduction to Data Mining, (Second Edition, Global Edition)*. Pearson Education Limited, Harlow, United Kingdom.

Honghan Wu, Giulia Toti, Katherine I Morley, Zina M Ibrahim, Amos Folarin, Richard Jackson, Ismail Kartoglu, Asha Agrawal, Clive Stringer, Darren Gale, et al. 2018. SemEHR: A general-purpose semantic search system to surface semantic data from clinical notes for tailored care, trial recruitment, and clinical research. *Journal of the American Medical Informatics Association*, 25(5):530–537.

A Example Analysis Output

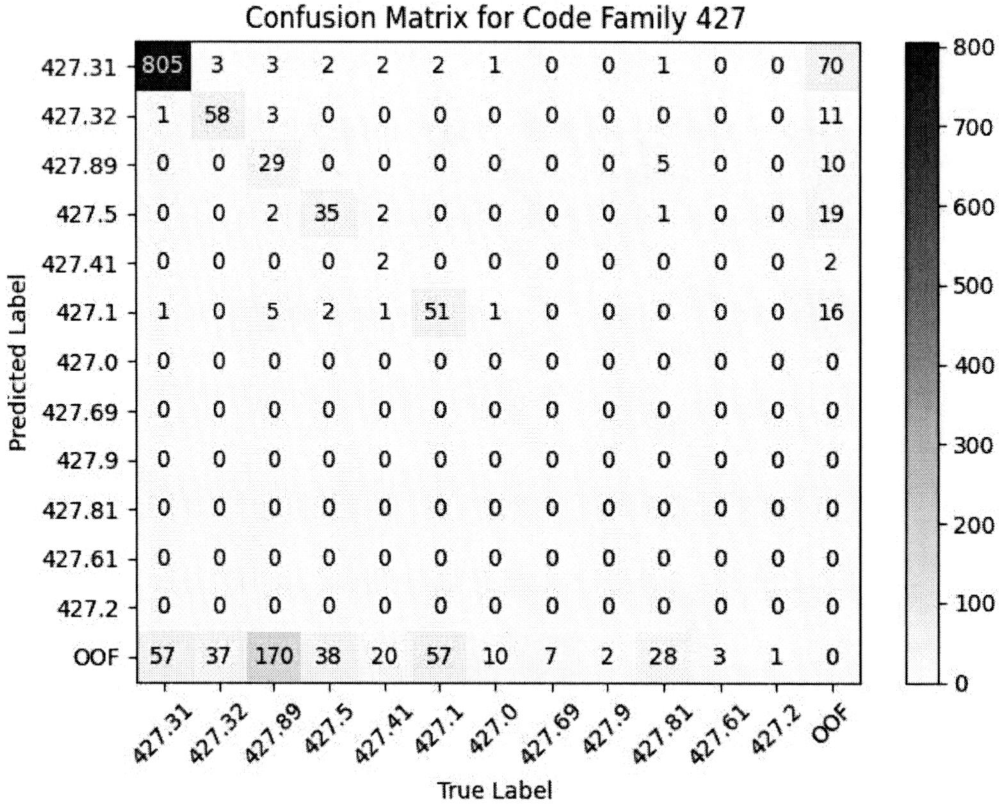

Figure A.1: An example of a WHCM for the code family 427 (Cardiac dysrhythmias). Codes that are predicted, are mostly predicted correctly (high-precision). Codes 427.31 (Atrial fibrillation) and 427.1 (Paroxysmal ventricular tachycardia) notably get confused with several of their siblings. While the model experienced some over-prediction, it suffered far more from under-prediction (low-recall).

Code	Identity #	Identity %	Preferred Prediction	Preferred Prediction #	Preferred Prediction %	OOF %	In-Family-Confusion %	Match
427.31	805	93.2	427.31	805	93.2	6.6	0.2	TRUE
427.32	58	59.2	427.32	58	59.2	37.8	3	TRUE
427.89	29	13.7	OOF	170	80.2	80.2	6.1	FALSE
427.5	35	45.5	OOF	38	49.4	49.4	5.1	FALSE
427.41	2	7.4	OOF	20	74.1	74.1	18.5	FALSE
427.1	51	46.4	OOF	57	51.8	51.8	1.8	FALSE
427.0	0	0	OOF	10	83.3	83.3	16.7	FALSE
427.69	0	0	OOF	7	100	100	0	FALSE
427.9	0	0	OOF	2	100	100	0	FALSE
427.81	0	0	OOF	28	80	80	20	FALSE
427.61	0	0	OOF	3	100	100	0	FALSE
427.2	0	0	OOF	1	100	100	0	FALSE

Table A1: An example of the output of the WHCM analysis tool for the *427* family of codes (Cardiac dysrhythmias, corresponding to the Figure A.1). There are 14 codes of this family present within MIMIC-III, with 12 appearing in the test set. Six of them have been correctly predicted at least once during the evaluation on the test set. Two of them (427.31, 427.32), are more likely to be predicted correctly than being confused within the family, or overpredicted (OOF). Four (427.89, 427.5, 427.41, 427.1) are predicted correctly at least once, but mostly suffer from underprediction (OOF). Six (427, 427.69, 427.9, 427.81, 427.61, 427.2) are never predicted correctly.

Towards Automatic Curation of Antibiotic Resistance Genes via Statement Extraction from Scientific Papers: A Benchmark Dataset and Models

Sidhant Chandak[♣], **Liqing Zhang**[♠], **Connor Brown**[♠], **Lifu Huang**[♠]
[♣]Indian Institute of Technology Kanpur, [♠]Virginia Tech
[♣]sidhant@iitk.ac.in, [♠]{lqzhang,clb21565,lifuh}@vt.edu

Abstract

Antibiotic resistance has become a growing worldwide concern as new resistance mechanisms are emerging and spreading globally, and thus detecting and collecting the cause – Antibiotic Resistance Genes (ARGs), have been more critical than ever. In this work, we aim to automate the curation of ARGs by extracting ARG-related assertive statements from scientific papers. To support the research towards this direction, we build SciARG, a new benchmark dataset containing 2,000 manually annotated statements as the evaluation set and 12,516 silver-standard training statements that are automatically created from scientific papers by a set of rules. To set up the baseline performance on SciARG, we exploit three state-of-the-art neural architectures based on pre-trained language models and prompt tuning, and further ensemble them to attain the highest 77.0% F-score. To the best of our knowledge, we are the first to leverage natural language processing techniques to curate all validated ARGs from scientific papers. Both the code and data are publicly available at https://github.com/VT-NLP/SciARG.

Statement 1: Gram-negative Enterobacteriaceae with *resistance* to *carbapenem* conferred by New Delhi metallo-*beta-lactamase* 1 (blaNDM-1) are potentially a major global health problem.

Statement 2: The NDM 1 producing Gram-negative bacteria are mainly Enterobacteriaceae, which can cause colonization or fatal infections, with worrying antimicrobial susceptibility profiles: some isolates have developed *resistance* to practically all available *antibiotics*.

Figure 1: Example of assertive statements for ARGs. The red color shows the target genes while blue background indicates the contextual features.

1 Introduction

Antibiotic resistance (AR), the ability of bacteria to survive and propagate in the presence of antibiotics, is a prevalent phenomenon worldwide and poses a serious health threat to humans and animals. Automatically detecting the antibiotic resistance genes (ARGs- the root cause of AR) in clinical and natural environments has been critical for mitigating the spread of AR. However, though the research on ARGs has grown exponentially over the past 10-15 years, existing ARG databases, such as CARD (Alcock et al., 2020), ARDB (Liu and Pop, 2009), ARGO (Scaria et al., 2005), and ARGMiner (Arango-Argoty et al., 2020), only contain a fraction of ARGs that have been discovered and validated by researchers, making it difficult to fully keep track of the research on ARGs.

To automate the process of collecting validated ARGs to enrich the ARG databases, we propose a literature mining approach to automatically extract the assertive statements that indicate the antibiotic resistance property of genes from scientific papers with computational approaches. Based on these assertive statements, we can easily collect all the validated ARGs in the literature. Taking the two statements extracted from (Kumarasamy et al., 2010) in Figure 1 as examples, we can confidently infer the antibiotic resistance of *NDM-1* based on the highlighted contextual words as *beta-lactamases are enzymes produced by bacteria that provide multi-resistance to beta-lactam antibiotics.*

In this paper, we introduce SciARG, the first benchmark dataset for extracting statements that indicate antibiotic resistance of genes from scientific publications. SciARG contains 2,000 and 286 statements with target genes that are manually annotated by domain experts as the test and dev dataset, and about 12,516 silver-standard training statements which are automatically created by a set of rules. The rules are carefully designed by two experts in ARG research. Each statement is a natural language sentence containing a target gene, and is labeled as *Positive* or *Negative*, indicating whether the statement implies antibiotic resistance of the target gene or not. To establish the baseline performance on SciARG, we design three approaches by leveraging the state-of-the-art

Proceedings of the BioNLP 2022 workshop, Dublin, Ireland, pages 402–411
May 26, 2022. ©2022 Association for Computational Linguistics

pre-trained language model and prompt tuning. As the training statements are created based on rules, the approaches are very easily overfitting to the keywords from the rules. To mitigate overfitting, we employ a mask language model pre-training strategy which is shown effective in improving the generalization of the baseline approaches. The ensemble of the three supervised approaches attain the highest 77.0% F-score on SCIARG. In summary, we make the following contributions:

- To the best of our knowledge, we are the first to curate ARGs from scientific papers by leveraging natural language processing techniques.
- We build the first benchmark dataset to support the research on ARG-related assertive statement prediction and establish baseline performance based on state-of-the-art pre-trained language models and prompt tuning techniques.

2 Dataset Design

2.1 Statement Collection

To collect the positive statements, we need to first get a collection of validated ARGs as the target genes. To do so, we leverage the CARD (Alcock et al., 2020) database which is a rigorously curated collection of characterized, peer-reviewed resistance determinants and associated antibiotics. CARD contains 3,100 ARGs while for 2,207 of them, CARD provides related PMIDs or PMCIDs from PubMed as reference. To collect the statements about these target ARGs, we leverage the Pubtator API[1] to get the full-text articles based on the PMCIDs of each target gene. As many articles are not freely available, we finally crawl 102 full-text articles for 91 ARGs.

For each of the 102 full-text articles, we segment them into sentences and extract the sentences that contain the target ARG as candidate statements. To enrich the context of each statement, we also prepend the preceding sentence and append the following sentence. In this way, we collect 2,286 statements for 91 confirmed ARGs. We then ask a senior student majoring in Biomedical Sciences to verify the statement in terms of whether they indicate the antibiotic resistance property of the target ARG, and an expert PhD student who has done extensive research on ARG to verify 100 samples randomly selected from the annotations. The inter-annotator-agreement is 88%, indicating that the annotations are mostly correct. In cases of disagreement between the two annotators, we ask them to discuss and achieve an agreement in terms of the label. We take 286 manually annotated statements as the dev set and the remaining 2,000 as the test set. The dev set is carefully chosen such that it has perfectly balanced classes, while the test set contains 1,083 positive and 917 negative statements.

2.2 Silver Training Set Creation

To create the training dataset, we take the remaining 2,105 PMIDs/PMCIDs for which we cannot successfully collect any full-text articles as seeds, and apply the Entrez API[2] to retrieve the papers that cite or are being cited by these seed papers. The assumption is that, if a paper cites or is cited by the paper about a particular ARG, it's more likely about ARGs as well. Based on this assumption, we follow the same procedures as Section 2.1 to collect additional 24,733 statements for 1,133 target ARGs. As it's very expensive and time consuming for a human to manually annotate these statements, we design the following rules based on the antibiotic resistance mechanisms to automatically create the positive training statements:

Rule 1: *If a statement mentions a particular antibiotic, together with "resistan" (the stem of resistance) or "efflux", it will be labeled as positive.*

The rule is based on the fact that the *efflux* of the drug from the bacterial cell is a key antibiotic resistance mechanism generally found in gram-negative bacteria. To apply this rule, we collect 604 antibiotics from the CARD database which cover the synonyms, abbreviations and common names of antibiotics. Two examples are shown in Table 1.

Rule 2: *If a statement mentions any of the enzymes produced by bacteria that catalyzes antibiotic hydrolysis, it will be labeled as positive.*

The enzystome[4] is a community of thousands of enzymes and its mutants, responsible for antibiotic resistance. These enzymes act by modifying the

[1]https://www.ncbi.nlm.nih.gov/research/pubtator/api.html

[2]https://www.ncbi.nlm.nih.gov/pmc/tools/cites-citedby/

[3]mdtEF is a multidrug transport class of efflux pump that confers resistance to a variety of drugs

[4]https://www.ncbi.nlm.nih.gov/pmc/articles/PMC6351036/

Rule	Examples
Rule 1	*Example 1*: Detection of rpsI-associated integrases in Bacillus and S . aureus reveals a potential for broad-host range dissemination of the novel methicillin resistance gene mecD. Macrococcus is evolutionarily closely related to the genus Staphylococcus, but possesses a distinctly smaller genome with a size of 2
	Example 2: Deletion of mdtEF [3] completely suppressed GadX-mediated multidrug resistance. Our results indicate that the GadX regulator, in addition to its role in acid resistance, increases multidrug resistance in E . coli by activating the MdtEF multidrug efflux pump .
Rule 2	*Example*: The emergence of one of the most recently described carbapenemases, namely, the New Delhi metallo-lactamase (NDM-1), constitutes a critical and growingly important medical issue . This resistance trait compromises the efficacy of almost all lactams (except aztreonam), including the last resort carbapenems
Rule 3	*Example*: the bla NDM-type genes are found to be either plasmid- or chromosome-located, and in the rare NDM-1-producing P . aeruginosa, the bla NDM-1 gene was found to be chromosomally located . Investigations on the immediate genetic environment of bla NDM genes revealed the presence of a conserved structure that always associated the complete or truncated insertion sequence ISAba125 at the 5'-end and the ble MBL gene (encoding resistance to the anticancer drug bleomycin) at the 3'-end of the bla NDM genes
Rule 4	*Example*: MIC values of beta-lactams for the E . coli TOP10 strain, which harbours recombinant plasmid pTOPO-MUS-2, showed that the bacteria was resistant to amoxicillin and ticarcillin and had a reduced susceptibility to piperacillin, in addition it showed an increased resistance to extended-spectrum cephalosporins and carbapenems by at least four-fold of MIC (Table 2) . Finally, PFGE analysis showed the three strains of M.

Table 1: Example statements of each rule. Green colour indicates the target ARG while Red colour highlights the keywords from the corresponding rule.

cellular targets of various antimicrobial drugs, or by modifying the antimicrobial drug itself. If a statement contains any of the enzymes, we will label it as positive. An example is shown Table 1.

Rule 3: *If the prefix of the target gene is an ARG indicator, we will label the statement as positive.*

The prefix of the target gene sometimes provides clues about whether the gene confers antibiotic resistance or not. The indicator can be either "bla" or "mec": *bla* genes are resistant to beta-lactam antibiotics and *mec* genes are resistant to methicillin antibiotic. An example is shown Table 1.

Rule 4: *If the statement mentions "MIC" and "increase" or "fold" within a context window of 10 words, we will label it as positive.*

Minimum inhibitory concentration (MIC) is the lowest concentration of an antibiotic that inhibits visible growth of the microorganism. Antimicrobial susceptibility tests (ASTs) measures the ability of an antibiotic or other antimicrobial agent to inhibit the in vitro microbial growth. The results of the test (e.g., *increase in MIC, MIC becoming multiple fold*) tells us whether the organism is susceptible to the antibiotic or resistant. An example is shown in Table 1.

Statistics	Train	Dev	Eval
# of Target Genes	1,886	56	91
# of Statements	12,516	286	2,000
Average Length of Statement	61.2	55.0	56.3
Minimum Length of Statement	3	9	8
Maximum Length of Statement	502	121	232

Table 2: Statistics of SciARG

Based on the above rules, we collect 6,258 positive statements out of 24,733 candidates. To collect negative statements, we first get a list of human genes from HGNC[5]. If a gene is not included in CARD (Alcock et al., 2020) or ARGMiner (Arango-Argoty et al., 2020), we consider it as a non-ARG and further collected papers

[5]The resource for approved human gene nomenclature `https://www.genenames.org/`

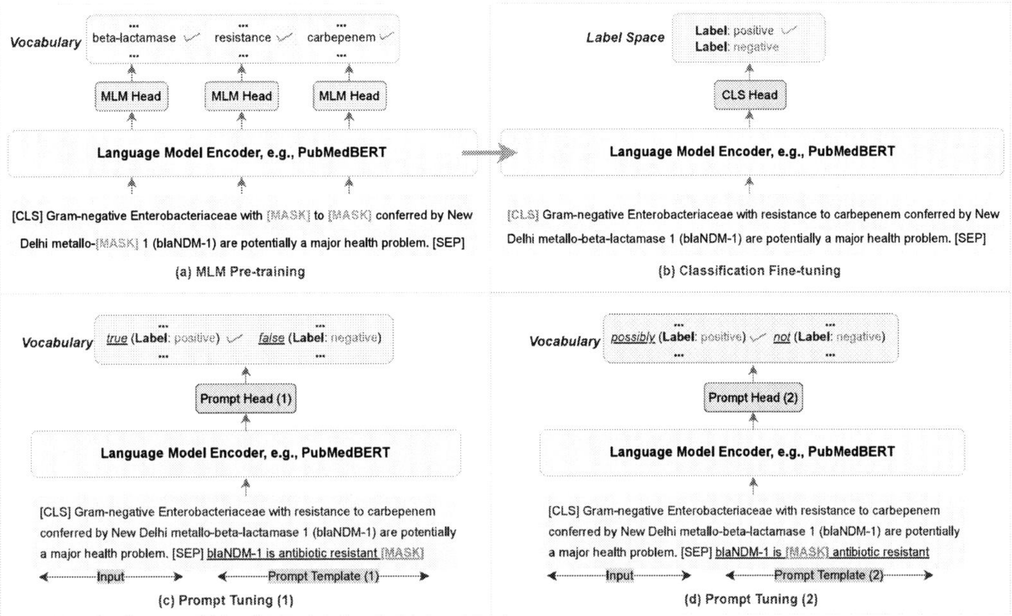

Figure 2: Overview of the three approaches for ARG statement prediction.

about it and statements from the papers. If the statement does not satisfy any of the above rules, we take it as a negative training statement. Finally, we randomly sample 6,258 negative training statements and obtain 12,516 statements in total as the training set. We name the dataset as SCIARG and show the statistics of SCIARG in Table 2.

3 Approach

To set up the baseline performance on SCIARG, we exploit three supervised approaches.

3.1 Supervised Classification

As Figure 2 (b) shows, given an input statement $X = [x_0, x_1, ..., x_n]$ for a target gene $g = [x_i, ..., x_j]$, we first apply the tokenizer of PubMedBERT (Shin et al., 2020a), a state-of-the-art pre-trained language model from PubMed papers, and concatenate all tokens to form a new sequence [`[CLS]`, X, `[SEP]`], where `[CLS]` is a special token used for classification and `[SEP]` is a delimiter. We use a position label 1 to indicate the tokens from the target gene and 0 for all the remaining tokens from the statement. Then each token is initialized with a vector by summing the corresponding token, segment and position embeddings from the pre-trained PubMedBERT, and encoded into a hidden state. We use $[\boldsymbol{H}_{cls}, \boldsymbol{H}_{x_0}, ..., \boldsymbol{H}_{x_n}, \boldsymbol{H}_{sep}]$ to denote the encoding outputs. Finally, we predict a label for the statement based on $\boldsymbol{H}_{cls}$, and use the negative log likelihood as the training objective:

$$L = -\log(\text{softmax}(W_1 \boldsymbol{H}_{cls})) \quad (1)$$

where W_1 is a learnable parameter matrix.

It turns out that the model easily overfits to the keywords of the rules (e.g., *efflux*, *resistance*, *MIC*) that are used to create the training samples. To overcome this issue, we further add a mask languge modeling (MLM) pre-training strategy to encourage the model to learn more features from context. As Figure 2 (a) shows, given an input statement, we find all the keywords that are from the rules, and randomly replace $m \in \{0\%, 25\%, 50\%, 75\%, 100\%\}$ of such tokens with `[MASK]`. Then, we apply the same MLM objective as PubMedBERT to ask the model to recover the original token for each `[MASK]`. The training objective of MLM is also based on the negative log likelihood:

$$L = -\log(\text{softmax}(W_2 \boldsymbol{H}_{mask})) \quad (2)$$

where W_2 is another learnable parameter matrix.

We explore two training strategies: optimizing the MLM objective (Equation 2) and the supervised classification objective (Equation 1) simultaneously or sequentially. The sequential training strategy shows better performance.

3.2 Prompt Tuning

To predict the antibiotic resistance property of genes, it also requires extensive domain specific knowledge, e.g., *interpreting results of ASTs, knowledge of the enzymes or efflux pumps that are responsible for antibiotic resistance*, which is likely to have been captured by the large-scale pre-trained language models. To better induce such knowledge, we further exploit prompt tuning based approaches.

Specifically, we design two prompts: (P_1) "The <*target gene*> is antibiotic resistant `[MASK]`", and (P_2) "The <*target gene*> is `[MASK]` antibiotic resistant", where <*target gene*> refers to the gene of interest in each input statement. As shown in Figure 2 (c) and (d), we concatenate each input statement with each prompt as [`[CLS]`, X, `[SEP]`, P, `[SEP]`], and get a contextual representation for each token within the sequence based on PubMedBERT. Based on the contextual representation of `[MASK]` in the prompt, we apply a linear function with softmax to predict a probability for each token in the target vocabulary.

$$L = -\log(\text{softmax}(W_P \boldsymbol{H}_{mask|P})) \quad (3)$$

where $P \in \{P_1, P_2\}$. W_P are learnable parameters for each prompt learning approach. $\boldsymbol{H}_{mask|P}$ denotes the contextual representation of `[MASK]` from the corresponding prompt.

For prompt P_1, we use *true* and *false* as the label of *positive* and *negative* category respectively, and compare their probabilities to get the final label. Similarly, for prompt P_2, we use *possibly* and *not* to predict the label of each statement. Similar as the supervised classification approach, we first pre-train the PubMedBERT with the MLM objective (Equation 2) and then fine-tune it with the prompts based on the negative log likelihood objective (Equation 3).

4 Experiments

4.1 Experiment Setup

We compare our approaches with baseline methods that are based on the rules illustrated in Section 2.2. We use the classification F-score on the positive statements as the evaluation metric, and use grid search to tune the parameters: training epochs 10, learning rate ∈ {2e-5, 3e-5, 5e-5}, training batch size ∈ {8, 12, 16, 20}.

4.2 Results and Analysis

Table 3 shows the performance of varying approaches on SCIARG. We can see that, (1) the precision, recall and F-score of different rules vary a lot across the development and evaluation sets. For example, Rule 2 results in the highest precision and recall among the four rules on the development set while the Rule 3 yields the highest recall on the evaluation set. We ascribe it to the sampling of the evaluation instances - though we carefully select the evaluation subset to make sure it's balanced in terms of the target positive/negative labels, we cannot guarantee that the rationals for the target labels are also balanced. (2) the supervised approaches perform significantly better than the rule based methods, especially on recall, demonstrating that the rules are not enough to retrieve most of the positive statements; (3) the classification based approach with MLM pre-training outperforms the two prompt-tuning based methods, due to the possible reason that the prompts were hand engineered and could be sub-optimal. However, by analyzing the errors of the three supervised approaches, we also notice that the prompt-tuning based methods tend to make more positive predictions, and perform better on the statements with complex or ambiguous context. Taking the following sentence as an example:

"*aeruginosa, is underway . The combined effects of various signals mediated by multiple regulators, including CpxR and MexR, on MexAB-OprM expression will be understood in a broader physiological context in the near future . For the determination of putative orthologous proteins, a primary BLASTP search in a given genome was conducted for the gene with the highest similarity.*"

The classification approach mistakenly predicted it as negative. However, *MexAB-OprM* is a major *efflux pump of P. aeruginosa*, a common disease causing gram-negative bacteria, that contributes to clinical antibiotic resistance. Such knowledge is possibly captured by the pre-trained language model and the prompt-tuning methods can better induce such knowledge from PubMedBERT and thus make correct predictions.

Based on the above observation, we further ensemble the three supervised approaches based on their predicted label for each statement. [6] Specifically, for each statement, we label it as positive if

[6]We tried several ensembling strategies and the one we discussed provides the best performance.

Model	Dev (%)			Eval (%)		
	Precision	Recall	F-Score	Precision	Recall	F-Score
Baseline w/ Rule 1	40.0	2.8	5.3	85.9	4.5	8.6
Baseline w/ Rule 2	72.2	9.1	16.2	96.6	10.5	19.0
Baseline w/ Rule 3	20.6	4.9	7.9	89.9	20.5	33.4
Baseline w/ Rule 4	50.0	2.1	4.0	84.8	2.6	5.0
Baseline w/ All Rules	37.1	16.1	22.4	89.3	32.2	47.4
Classification w/ MLM	57.0	88.1	69.2	63.3	92.7	75.2
Prompt 1 w/ MLM	52.4	97.9	68.3	57.7	97.2	72.4
Prompt 2 w/ MLM	53.8	95.1	68.7	58.6	94.9	72.4
Ensemble	61.3	85.3	71.3	67.8	89.0	77.0

Table 3: Comparison of varying approaches

and only if all the three individual models predict a positive label, otherwise, it will be labeled as negative. As Table 3 shows, the ensembling approach further provides significant improvement over each individual method.

4.3 Impact of MLM pre-training

Figure 3 shows the effect of MLM pre-training strategy based on different percentages of masked keywords for each supervised approach. As we can see, it provides improvement to all the three supervised approaches, demonstrating that it can encourage the language model encoder to better capture contextual features and generalize to other clues and indicators that are not from the rules.

4.4 Limitation of the Rule-based Methods

It's not surprising that rule-based methods show very low recall on the manually annotated test dataset as (1) there are a large number of resistance mechanisms while most of them also facilitate the biological processes that are not related to antibiotics. For instance, the *tolC-hlyD-hlyB* and related systems are nearly ubiquitous type 1 secretion systems that facilitate secretion of a very broad range of substrates, such as *virulence factors*, *bacteriocins*; (2) there are a lot of other terms that could have been included in the rules but their mere mentions are not enough to indicate the antibiotic resistance of genes. For instance, *plasmids* frequently but not always carry antibiotic resistance genes, and similar terms also include *transposon*, *integron*, *genomic island* and so on.

5 Remaining Challenges

To understand the remaining challenges of SciARG, we randomly sample 100 prediction errors of the ensembling approach from the development set, and summarize the following three key remaining challenges.

Challenge 1: Lack of Domain Specific Knowledge The ARG statement prediction requires extensive domain specific knowledge to help the models better understand the text and disambiguate the meanings. For example, in the following statement:

"Minimal inhibitory concentrations (MICs) of ciprofloxacin, ofloxacin, ceftazidime, cefsulodin, and aztreonam, but not amikacin, were increased at least 4-fold by ectopically expressed CpxR in PA14 and PA14DeltacpxR strains (Table 2) in a manner dependent on MexA, but not MuxA . In this case, ectopically expressed CpxR failed to increase the MICs of the tested antibiotics in a mexA null-mutant PA14DeltamexA strain . In contrast, the MIC increases caused by the ectopically expressed CpxR were not altered in a muxA null-mutant PA14DeltamuxA strain (Table 2)"

The term "*ectopic expression*" either refers to "*heterologous expression*" or "*a specific experimental condition*", which lead to distinct predictions. The model cannot correctly interpret the meaning and thus made a wrong prediction.

Challenge 2: Limited Contextual Cues about the Target Gene For many statements, the context is not enough to confidently infer whether the target gene is an ARG or not. For instance, in the following statement:

"MphI shares high sequence identity (94%) to

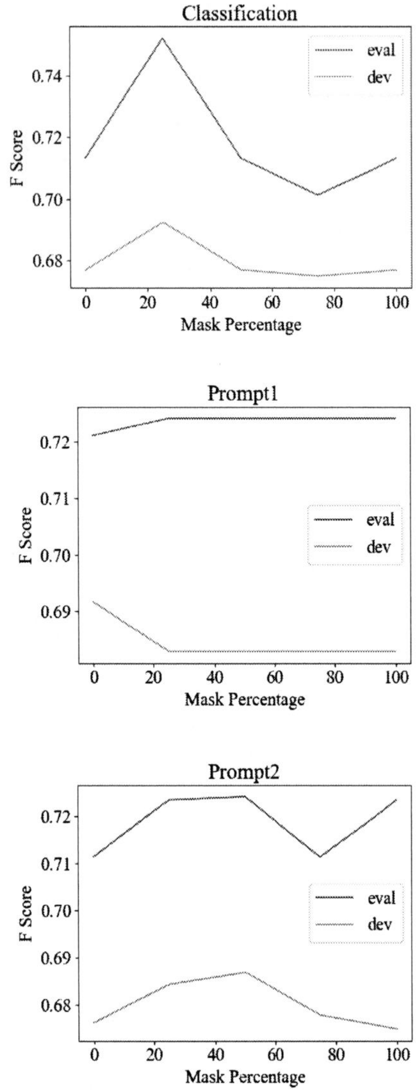

Figure 3: Impact of MLM pre-training with different percentages of masked keywords.

homologs found in related surface Paenibacillus sp . , indicating the functional divergence of MphI is not recent . The Bacillus cereus group have two genetically and functionally distinct Mph enzymes; one that modifies a broad range of macrolides and another that cannot modify macrolides with 16-membered rings"

Our ensembling approach mistakenly predicts it as positive while the source article concluded that *mphI does NOT encode an ARG*. The description of "*MphI modifies macrolides*" does not necessarily imply that *it neutralizes or inactivates macrolides, a class of antibiotics, thereby causing resistance*. This is a special case that will happen occasionally - where the statements are characterizing an ARG homolog, but not an ARG.

Challenge 3: Noisy and Insufficient Training Data The training is created based on a set of rules, which leads to two major problems: (1) It introduces noise since the rules are not 100% precise. As Table 3 shows, the precision ranges from 20.6% for Rule-3 to 72.2% for Rule-2; (2) The positive ARG statements covered in the training data is not diverse enough as they are constrained by the 4 rules. Though the MLM strategy helps the model generalize to more broad contextual features, it still suffers from the low recall. Many types of ARG statements in the development and test sets are not covered in the training set. For instance, for the following statement from the development set:

"*The nature of the activating ligand for VanSA has not been identified, therefore this work sought to identify and characterise ligand(s) for VanSA. In vitro approaches were used to screen the structural and activity effects of a range of potential ligands with purified VanSA protein. Of the screened ligands (glycopeptide antibiotics vancomycin and teicoplanin, and peptidoglycan components N-acetylmuramic acid, D-Ala-D-Ala and Ala-D-y-Glu-Lys-D-Ala-D-Ala) only glycopeptide antibiotics vancomycin and teicoplanin were found to bind VanSA with different affinities (vancomycin 70 muM; teicoplanin 30 and 170 muM), and were proposed to bind via exposed aromatic residues tryptophan and tyrosine*."

The reason that "*VanSA*" is labeled as an ARG is that "*the ligand interaction of VanSA with glycopeptide antibiotics (GPA)*." implies that VanSA is an ARG since it inactivates the antibiotic vancomycin by binding to it, while such rules are not covered in the current training dataset.

6 Related Work

Machine Learning for Antibiotic Resistance Prediction Traditional antimicrobial susceptibility testing (AST) is time-consuming, low throughput and viable only for cultivable bacteria, thus rapid and accurate AMR diagnostic methods are very urgently needed. Recent years, machine learning based methods have been widely explored as clinical decision support tools for the prediction of antimicrobial resistance (AMR) (Feretzakis et al., 2021, 2020; Martínez-Agüero et al., 2019; Oonsivilai et al., 2018). Ren et al. (2021) compared four different machine learning methods (Random Forests, Logistic Regression, Support Vector Machines and Convolutional Neural Networks) for the prediction of AMR based on different encodings and whole-genome sequencing data without previously known knowledge. Deep learning algorithms have also shown significant potential for predicting new antibiotic drugs, AMR genes and AMR peptides (Kumaresan et al., 2018; Stokes et al., 2020; Veltri et al., 2018). However, these studies focused on genome variants (such as single-nucleotide polymorphisms, SNPs) or other features only related to resistant genes identified in previous studies or resistant databases, while in this work, we focus on curating antimicrobial susceptibility data by leveraging computational approaches and large-scale scientific papers. In addition, we approach the ARG curation as a entity classification task instead of recognition as genes are easily detected based on the existing knowledge bases and it's more challenging to infer the antibiotic resistance the genes based on the context. The curated ARG database can provide clinicians useful information regarding possible antibiotic resistance and aid clinicians in selecting appropriate empirical antibiotic therapy by taking into consideration the local antimicrobial resistance ecosystem.

Prompt Learning Prompt learning aims to learn a task-specific prompt while keeping most of the parameters of the model freezed (Li and Liang, 2021; Hambardzumyan et al., 2021; Brown et al., 2020). It has shown competitive performance in a wide variety of applications in natural language processing (Raffel et al., 2020; Brown et al., 2020; Shin et al., 2020b; Jiang et al., 2020; Lester et al., 2021; Schick and Schütze, 2021b). Previous work either use a manual (Petroni et al., 2019; Brown et al., 2020; Schick and Schütze, 2021a) or automated approach (Jiang et al., 2020; Yuan et al., 2021; Li and Liang, 2021) to create prompts. In this work, we mainly explore two manually defined prompts for ARG statement extraction task. The reason of applying prompt learning for ARG statement classification lies that though the training dataset size is not small, the clues of indicating antibiotic resistance covered in the training set is limited to the manually defined rules, thus applying prompt learning can to some extent leverage the knowledge, especially antibiotic resistance related knowledge captured by the large-scale language models during pre-training. Based on the experimental results of the ensembling approach, we see that although the prompt learning based approaches do not perform as well as the supervised classification based method, they are still complimentary to each other.

7 Conclusion and Future Work

In this work, we present the first computational framework that aims to automatically curate ARGs by extracting ARG-related assertative statements from scientific papers in PubMed. To support the research, we introduce SciARG, a dataset that contains 2,000 manually annotated statements as the test set and 12,516 silver-standard training statements that are automatically created from scientific papers by a set of rules. We also present extensive empirical results by comparing various state-of-the-art neural architectures based on pre-trained language models for statement classification, and demonstrate that there is still a large room to improve based on the current highest 77% F-score on SciARG.

Considering the remaining challenges that we have discussed, there are multiple future directions: (1) developing more advanced frameworks that incorporate domain-specific knowledge from external resources or knowledge bases to better interpret the statements; (2) learning contextual features of target genes from more broad context, such as the paragraph, chapter or the whole document; (3) leveraging self-training or co-training framework to take advantage of the large-scale unlabeled corpus from PubMed to enrich the training samples.

Acknowledgements

We thank the anonymous reviewers for their valuable time and constructive comments, and the helpful discussions with Huimin Han.

References

Brian P Alcock, Amogelang R Raphenya, Tammy TY Lau, Kara K Tsang, Mégane Bouchard, Arman Edalatmand, William Huynh, Anna-Lisa V Nguyen, Annie A Cheng, Sihan Liu, et al. 2020. Card 2020: antibiotic resistome surveillance with the comprehensive antibiotic resistance database. *Nucleic acids research*, 48(D1):D517–D525.

GA Arango-Argoty, GKP Guron, Emily Garner, Maria V Riquelme, Lenwood S Heath, Amy Pruden, Peter J Vikesland, and Liqing Zhang. 2020.

Argminer: a web platform for the crowdsourcing-based curation of antibiotic resistance genes. *Bioinformatics*, 36(9):2966–2973.

Tom Brown, Benjamin Mann, Nick Ryder, Melanie Subbiah, Jared D Kaplan, Prafulla Dhariwal, Arvind Neelakantan, Pranav Shyam, Girish Sastry, Amanda Askell, Sandhini Agarwal, Ariel Herbert-Voss, Gretchen Krueger, Tom Henighan, Rewon Child, Aditya Ramesh, Daniel Ziegler, Jeffrey Wu, Clemens Winter, Chris Hesse, Mark Chen, Eric Sigler, Mateusz Litwin, Scott Gray, Benjamin Chess, Jack Clark, Christopher Berner, Sam McCandlish, Alec Radford, Ilya Sutskever, and Dario Amodei. 2020. Language models are few-shot learners. In *Advances in Neural Information Processing Systems*, volume 33, pages 1877–1901. Curran Associates, Inc.

Georgios Feretzakis, Evangelos Loupelis, Aikaterini Sakagianni, Dimitris Kalles, Maria Martsoukou, Malvina Lada, Nikoletta Skarmoutsou, Constantinos Christopoulos, Konstantinos Valakis, Aikaterini Velentza, Stavroula Petropoulou, Sophia Michelidou, and Konstantinos Alexiou. 2020. Using machine learning techniques to aid empirical antibiotic therapy decisions in the intensive care unit of a general hospital in greece. *Antibiotics (Basel)*, 9(2):50.

Georgios Feretzakis, Aikaterini Sakagianni, Evangelos Loupelis, Dimitris Kalles, Nikoletta Skarmoutsou, Maria Martsoukou, Constantinos Christopoulos, Malvina Lada, Stavroula Petropoulou, Aikaterini Velentza, Sophia Michelidou, Rea Chatzikyriakou, and Evangelos Dimitrellos. 2021. Machine learning for antibiotic resistance prediction: A prototype using off-the-shelf techniques and entry-level data to guide empiric antimicrobial therapy. *Healthc. Inform. Res.*, 27(3):214–221.

Karen Hambardzumyan, Hrant Khachatrian, and Jonathan May. 2021. WARP: Word-level Adversarial ReProgramming. In *Proceedings of the 59th Annual Meeting of the Association for Computational Linguistics and the 11th International Joint Conference on Natural Language Processing (Volume 1: Long Papers)*, pages 4921–4933, Online. Association for Computational Linguistics.

Zhengbao Jiang, Frank F. Xu, J. Araki, and Graham Neubig. 2020. How can we know what language models know? *Transactions of the Association for Computational Linguistics*, 8:423–438.

Karthikeyan K Kumarasamy, Mark A Toleman, Timothy R Walsh, Jay Bagaria, Fafhana Butt, Ravikumar Balakrishnan, Uma Chaudhary, Michel Doumith, Christian G Giske, Seema Irfan, et al. 2010. Emergence of a new antibiotic resistance mechanism in india, pakistan, and the uk: a molecular, biological, and epidemiological study. *The Lancet infectious diseases*, 10(9):597–602.

Deepak Kumaresan, Jason Stephenson, Andrew C Doxey, Hina Bandukwala, Elliot Brooks, Alexandra Hillebrand-Voiculescu, Andrew S Whiteley, and J Colin Murrell. 2018. Aerobic proteobacterial methylotrophs in movile cave: genomic and metagenomic analyses. *Microbiome*, 6(1):1.

Brian Lester, Rami Al-Rfou, and Noah Constant. 2021. The power of scale for parameter-efficient prompt tuning. In *EMNLP*.

Xiang Lisa Li and Percy Liang. 2021. Prefix-tuning: Optimizing continuous prompts for generation. *Proceedings of the 59th Annual Meeting of the Association for Computational Linguistics and the 11th International Joint Conference on Natural Language Processing (Volume 1: Long Papers)*, abs/2101.00190.

Bo Liu and Mihai Pop. 2009. Ardb—antibiotic resistance genes database. *Nucleic acids research*, 37(suppl_1):D443–D447.

Sergio Martínez-Agüero, Inmaculada Mora-Jiménez, Jon Lérida-García, Joaquín Álvarez-Rodríguez, and Cristina Soguero-Ruiz. 2019. Machine learning techniques to identify antimicrobial resistance in the intensive care unit. *Entropy (Basel)*, 21(6):603.

Mathupanee Oonsivilai, Yin Mo, Nantasit Luangasanatip, Yoel Lubell, Thyl Miliya, Pisey Tan, Lorn Loeuk, Paul Turner, and Ben S Cooper. 2018. Using machine learning to guide targeted and locally-tailored empiric antibiotic prescribing in a children's hospital in cambodia. *Wellcome Open Res.*, 3:131.

Fabio Petroni, Tim Rocktäschel, Sebastian Riedel, Patrick Lewis, Anton Bakhtin, Yuxiang Wu, and Alexander Miller. 2019. Language models as knowledge bases? In *Proceedings of the 2019 Conference on Empirical Methods in Natural Language Processing and the 9th International Joint Conference on Natural Language Processing (EMNLP-IJCNLP)*, pages 2463–2473, Hong Kong, China. Association for Computational Linguistics.

Colin Raffel, Noam M. Shazeer, Adam Roberts, Katherine Lee, Sharan Narang, Michael Matena, Yanqi Zhou, Wei Li, and Peter J. Liu. 2020. Exploring the limits of transfer learning with a unified text-to-text transformer. *JMLR*.

Yunxiao Ren, Trinad Chakraborty, Swapnil Doijad, Linda Falgenhauer, Jane Falgenhauer, Alexander Goesmann, Anne-Christin Hauschild, Oliver Schwengers, and Dominik Heider. 2021. Prediction of antimicrobial resistance based on whole-genome sequencing and machine learning. *Bioinformatics*, 38(2):325–334.

Joy Scaria, Umamaheswaran Chandramouli, and Sanjay Kumar Verma. 2005. Antibiotic resistance genes online (argo): A database on vancomycin and βlactam resistance genes. *Bioinformation*, 1(1):5.

Timo Schick and Hinrich Schütze. 2021a. Few-shot text generation with pattern-exploiting training.

Timo Schick and Hinrich Schütze. 2021b. It's not just size that matters: Small language models are also few-shot learners. In *Proceedings of the 2021 Conference of the North American Chapter of the Association for Computational Linguistics*, pages 2339–2352.

Hoo-Chang Shin, Yang Zhang, Evelina Bakhturina, Raul Puri, Mostofa Patwary, Mohammad Shoeybi, and Raghav Mani. 2020a. Biomegatron: Larger biomedical domain language model. *arXiv preprint arXiv:2010.06060*.

Taylor Shin, Yasaman Razeghi, Robert L. Logan IV, Eric Wallace, and Sameer Singh. 2020b. AutoPrompt: Eliciting knowledge from language models with automatically generated prompts. In *Empirical Methods in Natural Language Processing (EMNLP)*.

Jonathan M Stokes, Kevin Yang, Kyle Swanson, Wengong Jin, Andres Cubillos-Ruiz, Nina M Donghia, Craig R MacNair, Shawn French, Lindsey A Carfrae, Zohar Bloom-Ackermann, Victoria M Tran, Anush Chiappino-Pepe, Ahmed H Badran, Ian W Andrews, Emma J Chory, George M Church, Eric D Brown, Tommi S Jaakkola, Regina Barzilay, and James J Collins. 2020. A deep learning approach to antibiotic discovery. *Cell*, 180(4):688–702.e13.

Daniel Veltri, Uday Kamath, and Amarda Shehu. 2018. Deep learning improves antimicrobial peptide recognition. *Bioinformatics*, 34(16):2740–2747.

Weizhe Yuan, Graham Neubig, and Pengfei Liu. 2021. BARTScore: Evaluating generated text as text generation. In *Advances in Neural Information Processing Systems*.

Model Distillation for Faithful Explanations of Medical Code Predictions

Zach Wood-Doughty[1,2] *; **Isabel Cachola**[1] *; **Mark Dredze**[1]
[1] Johns Hopkins University, Baltimore, MD 21211
[2] Northwestern University, Evanston, IL 60208
zach@northwestern.edu, icachola@cs.jhu.edu, mdredze@cs.jhu.edu

Abstract

Machine learning models that offer excellent predictive performance often lack the interpretability necessary to support integrated human machine decision-making. In clinical or other high-risk settings, domain experts may be unwilling to trust model predictions without explanations. Work in explainable AI must balance competing objectives along two different axes: 1) Models should ideally be both *accurate* and *simple*. 2) Explanations must balance *faithfulness* to the model's decision-making with their *plausibility* to a domain expert. We propose to use knowledge distillation, or training a student model that mimics the behavior of a trained teacher model, as a technique to generate faithful and plausible explanations. We evaluate our approach on the task of assigning ICD codes to clinical notes to demonstrate that the student model is faithful to the teacher model's behavior and produces quality natural language explanations.

1 Introduction

Machine learning (ML) methods have demonstrated predictive success in medical settings, leading to discussions of how ML systems can augment clinical decision-making (Caruana et al., 2015). However, a prerequisite to clinical integration is the ability for healthcare professionals to understand the justifications for model decisions. Clinicians often disagree on the proper course of care, and share their justifications as a means of agreeing on a treatment plan. Explainable Artificial Intelligence (AI) can enable models to provide the explanations needed for them to be integrated into this process (Lundberg et al., 2018; Caruana et al., 2015). However, modern AI models that often rely on complex deep neural networks with millions or billions of parameters pose challenges to creating explanations that satisfy clinician's demands (Feng et al., 2018).

Similar concerns over model explanations across domains have inspired a whole field of interpretable ML. Work in this area typically considers two goals: faithfulness (explanations that accurately convey the decision-making process of the model) and plausibility (explanations that make sense to domain experts) (Jacovi and Goldberg, 2020). These two goals may be in conflict: faithful explanations that accurately convey the reasoning of complex AI systems may be implausible to a domain expert, and vice versa (Kumar and Talukdar, 2020; Wiegreffe et al., 2021). Models must also balance performance against transparency. The methods that perform best on a task may be unable to provide explanations (Rudin, 2019).

We propose to disentangle these competing goals by using knowledge distillation. We train a bag-of-words linear student model to predict the *predictions* of the teacher model, so that the behavior of the student model mimics the teacher model's behavior, rather than independently modeling the target task. We then rely on the interpretable student to create explanations without changing the original teacher model. We evaluate the student's faithfulness to the teacher model and the plausibility of the student's explanations.

We demonstrate our approach on the task of medical code prediction. While ML methods have achieved predictive success on various versions of International Classification of Diseases (ICD) clinical code assignment, the best-performing methods have been neural networks that are notoriously difficult to interpret. We train student models for three teacher models: (1) DR-CAML, a method designed to produce explainable predictions which outperformed several baselines when evaluated by a clinical expert (Mullenbach et al., 2018); (2) Hierarchical Attention Networks, a Bi-GRU document classifier first introduced by Yang et al. (2016) and adapted to ICD code prediction by Dong et al. (2021); and (3) TransICD, a transformer-based

* Equal contribution

Proceedings of the BioNLP 2022 workshop, Dublin, Ireland, pages 412–425
May 26, 2022.

method (Biswas et al., 2021). We show that our student models are faithful to the teacher models and can generate natural language explanations that are comparably plausible. We also show that our student model outperforms a logistic regression baseline in comparison to the true ICD-9 labels, despite being of equal complexity. We release the code under an MIT license for both our method and for reproducing Mullenbach et al. (2018).[1]

2 Background

2.1 Interpretable ML

Interpretable machine learning falls within the growing field of Explainable AI (Doshi-Velez, 2017). We present an overview of major themes in the literature, and direct the reader to recent surveys for more details (Doshi-Velez, 2017; Guidotti et al., 2018; Gilpin et al., 2018).

Past work distinguishes between "transparent" or "inherently interpretable" models that offer their own explanations, and "post-hoc" methods that produce explanations for a separately-trained model. Linear methods such as logistic regression are often considered transparent, while deep neural networks are generally not and rely on post-hoc methods for explainability (Guidotti et al., 2018; Feng et al., 2018). However, even simple models can prove difficult to interpret, e.g., when the model's features are complex (Lipton, 2018). LIME and SHAP are commonly used post-hoc methods (Ribeiro et al., 2016; Lundberg and Lee, 2017); given a trained model of arbitrary complexity they produce explanations for individual predictions by sampling perturbed inputs. Unlike LIME and SHAP, our method produces global explanations, and the student model can be used for predictions on future input. Prior work has shown that such methods can produce contrasts which are misleading or unintuitive (Mittelstadt et al., 2019) and that LIME or SHAP can be fooled into providing innocuous explanations for models that demonstrate racist or sexist behavior (Slack et al., 2020). These methods' feature importance scores are difficult to aggregate across a dataset and do not provide global faithfulness (van der Linden et al., 2019; Lakkaraju et al., 2017).

Lipton (2018) argues that interpretability is never "inherent" and must satisfy several criteria. These include simulatability, or whether a human can reasonably work through each step of the model's calculations; decomposibility, or whether each parameter of the model can be understood on its own; and algorithmic transparency, or whether the model belongs to a class with known theoretical behaviors. Lou et al. (2012) highlights linear and additive models as particularly decomposible (or intelligible) classes of models, because "users can understand the contribution of individual features in the model." Our proposed approach uses a linear bag-of-words model to provide a simulatible, decomposible, and transparent method.

Interpretability methods are also distinguished by the form and quality of the explanations they produce. We follow Jacovi and Goldberg (2020) in recognizing two primary desiderata for post-hoc explanations of ML systems: "faithfulness" and "plausibility."[2] A faithful explanation accurately represents the original model, by closely approximating its behavior or exposing its internal state (Yeh et al., 2019; Lakkaraju et al., 2020). A plausible model produces explanations that can be interpreted by a human expert (Jacovi and Goldberg, 2020; Ehsan et al., 2019). Prior work has explored methods such as forcing a faithful classifier to make predictions from a limited set of (plausible) rationales (Jain et al., 2020), or focusing on extracting rationales to constrain predictors to be inherently interpretable (Lei et al., 2016; Bastings et al., 2019). Methods should attempt to achieve both goals, but there is a trade-off between the two; explanations typically cannot be both concise and perfectly descriptive. Plausibility, unlike faithfulness, necessarily requires an evaluation based on human perception (Herman, 2017; Jain et al., 2020). A strength of our proposed method is that it is designed for plausibility and transparency, but optimized for faithfulness.

2.2 Knowledge Distillation

Knowledge distillation is a technique in which a simpler "student" model is trained to behave like a high performing, but more complex "teacher" model (Hinton et al., 2015). This approach has been widely studied under a variety of other names such as model approximation or compression (Buciluǎ et al., 2006), or simply 'copying' (Unceta et al., 2020). In many of these threads of research, the goal is to produce a student model that is

[1] https://github.com/isabelcachola/mimic-proxy

[2] Faithfulness is also referred to as fidelity, validity or completeness; plausibility is alternatively referred to as persuasiveness (Herman, 2017). See Jacovi and Goldberg (2020) for a longer discussion of alternate terminology.

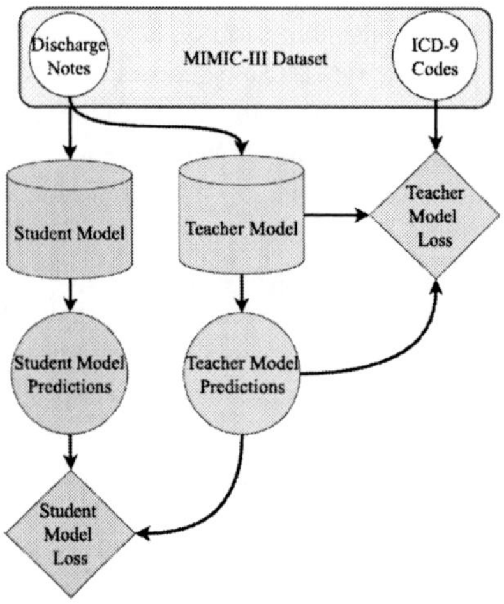

Figure 1: Relationship between the teacher and student models. The student model is trained to predict the teacher's outputs, rather than the true ICD-9 codes. This optimizes the student model for faithfulness.

smaller or faster than the teacher, while achieving high accuracy. Our work most closely follows approaches such as Lakkaraju et al. (2017); Asadulaev et al. (2019) that have sought to produce an *interpretable* student. The experimental and theoretical properties of knowledge distillation are well studied (Tan et al., 2018b; Phuong and Lampert, 2019).

Knowledge distillation has been applied to a variety of domains for the purposes of interpretability, such as crime and lending data (Tan et al., 2018a) and image classification (Asadulaev et al., 2019). Furthermore, a wide variety of model architectures have been shown to be effective as the student model, including decision trees (Elshawi et al., 2019), elastic nets (Guo et al., 2017), decision sets (Lakkaraju et al., 2017). We apply knowledge distillation to the task of medical code prediction, for the purposes of interpretability. We show that knowledge distillation is both an effective technique for training a faithful student model and can be used to generate plausible natural language explanations.

2.3 Explainable prediction in the medical domain

Explainability techniques have been applied to a variety of tasks in the medical domain, such as pneumonia and hospital readmission risk (Caruana et al., 2015) or real-time hypoxaemia prediction (Lundberg et al., 2018). Our work considers the task of predicting medical codes from hospital discharge notes. This task has been widely studied, and we use three published models on which we evaluate our approach: DR-CAML (Mullenbach et al., 2018), HAN (Yang et al., 2016), and TransICD (Biswas et al., 2021). As all three models contain millions of parameters, they are not simulatible or decomposible. However, DR-CAML and TransICD seek to produce their own explanations using a per-label attention mechanism that highlights regions in the input text that were correlated with the model's predictions. HAN was not designed with the goal of interpretability.

The use of attention to produce explanations has sparked discussion. Jain and Wallace (2019) showed that attention mechanisms can provide misleading explanations, whereas Wiegreffe and Pinter (2019) argued that attention-based explanations are often plausible, even when unfaithful. More recent work has explored when and how attention mechanisms can be either useful or deceptive (Zhong et al., 2019; Grimsley et al., 2020; Jain et al., 2020; Pruthi et al., 2020). As researchers continue to use this domain to explore methods for explainability and document classification (Kim and Ganapathi, 2021; Vu et al., 2020), we should strive to produce models that are both faithful and plausible.

3 Methods

Our proposed method is post-hoc and seeks to balance faithfulness and plausibility. We assume that we have a trained teacher model with good predictive performance but low interpretability. We train a student model that takes the same input from the dataset, but uses the teacher model's predictions as its labels. Figure 1 gives a visual representation of our model distillation setup.

The MIMIC-III dataset contains anonymized English-language ICU patient records, including physiological measurements and clinical notes (Johnson et al., 2016). Following Mullenbach et al. (2018), we focus on discharge summaries which describe a patient's visit and are annotated with ICD-9 codes. There are 8,922 different ICD-9 codes that describe procedures and diagnoses that occurred during a patient's stay. The manual assignment of these codes to patient records are required by most U.S. healthcare payers (Topaz et al., 2013).

To train the teacher models, we duplicate the experimental setup of Mullenbach et al. (2018) and Dong et al. (2021), which use the text of the discharge summaries as input to the DR-CAML and

HAN models, respectively, which then are trained to predict all ICD-9 codes associated with that document. After applying their pre-processing code to tokenize the text, the dataset contains 47,724 discharge summaries divided into training, dev, and test splits. We also duplicate the experimental setup of Biswas et al. (2021), which has a similar experimental setup but only predicts the top 50 most common ICD-9 codes.

We apply DR-CAML and HAN to the texts in MIMIC-III and save its continuous-valued probabilities as the labels for our student model. We similarly apply TransICD to MIMIC-III-50, which contains the top 50 most frequent labels in MIMIC-III and save the continuous-valued probabilities as the labels.[3] For all three models, we use the code released by the authors.[4] Training the student model on predictions from the existing teacher model optimizes for faithfulness.

We want the student model to produce plausible explanations and fulfill the criteria from Lipton (2018): simulatibility, decomposibility, and algorithmic transparency. To fulfill these desiderata, each student is a linear regression trained on bag-of-words representation of the clinical text. The fundamental trade-off here is that if we overly restrict our model class, the student will be unfaithful and unable to mimic the behavior of the teacher model. But if we allow for a student model that is too complex, it may not provide plausible or otherwise desirable explanations. These trade-offs may be domain-specific based, for example, on the target audience of the explanations. If the student model demonstrates sufficient empirical performance, a domain expert may even prefer to use it in place of the teacher model, an option unsupported by LIME or SHAP models.

We train a student model for each medical code independently, and we refer to the student model trained on X model's predictions as "Student-X" (e.g. Student-DRCAML). Each student uses only 50k parameters, allowing us to train each model on a single CPU in a matter of minutes. We implement our method using the linear `SGDRegressor` model from `sklearn` (Pedregosa et al., 2011), and apply a log transform to the model's probability outputs and train the student to minimize squared loss. After a brief[5] grid search on the validation set, we use L1 regularization with $\alpha = 0.0001$ for the DR-CAML student and $\alpha = 0.01$ for HAN and TransICD proxies.

To extract a rationale, we take the feature importance weights of the student model and average over a sliding window of n-grams from the discharge summary. We extract the n-gram with the highest average feature importance weight. Future work could use extracted rationales to train a student model that remains faithful to a black-box model.

In the next two sections, we introduce our evaluation for the student model's faithfulness to each model and the plausibility of its explanations.

4 Faithfulness evaluation

To establish that this collection of linear regressions is faithful to the trained models, we want to show that it makes similar predictions across all ICD-9 codes on held-out data. Recall from Figure 1 that the student is trained not to predict the true ICD-9 codes but to output the same label probabilities as the teacher model. In fact, the student model never sees the true ICD-9 codes. We evaluate faithfulness by comparing the outputs of the student and teacher models on the held-out test set. If the two systems produced identical outputs on held-out data, we would say that the student was perfectly faithful. We make this comparison in three different ways – first with regression metrics for the continuous outputs of the two models, then using classification metrics with binarized teacher predictions, and finally comparing student outputs as predictions for the true ICD-9 codes. For all these comparisons, we use a logistic regression baseline that is trained to directly predict the ICD-9 codes, independent of any black-box model. While we would expect the logistic baseline's predictions to roughly correlate with those of other models, we would not expect it to be faithful.

Similar to Tan et al. (2018a), our first evaluation uses regression metrics that assess the correlation between the student's predictions and the original teacher model's predicted probabilities. We use Spearman and Pearson correlation coefficients and

[3]Training on the full label set was prohibitively computationally expensive to reproduce and the authors did not release the trained model weights. In Table 1, TransICD and its student only use these 50 codes. These codes do not include those used in Tables 3 and 7, so TransICD models are omitted.

[4]Mullenbach et al. (2018) released their code under an MIT license, while Yang et al. (2016) and Biswas et al. (2021) did not specify a license.

[5]We considered L1, L2, and elastic net regularization with α from 0.1 to 10^{-7}. For HAN, which was not trained using the published dev set, we simply adopted α=0.01.

Model	Regression Spearman	Regression Pearson	Regression Kendall	Classification AUC Macro	Classification AUC Micro	Classification F1 Macro	Classification F1 Micro
Logistic to...							
DRCAML	0.036	-0.195	-0.135	0.734	0.936	0.012	0.353
HAN	0.204	0.036	-0.139	0.885	0.994	0.017	0.511
TransICD	0.587	0.662	0.419	0.894	0.927	0.476	0.580
Student-X							
-DRCAML	0.794	0.498	0.608	0.980	0.995	0.052	0.416
-HAN	0.736	0.519	0.543	0.975	0.997	0.014	0.454
-TransICD	0.838	0.539	0.650	0.960	0.960	0.507	0.592

Table 1: Comparison of the logistic baseline and the student model to the DR-CAML, HAN, and TransICD predictions. For the F1 evaluation, we threshold the student outputs at 0.5. The logistic model was trained to predict the ICD codes; the student model to predict DR-CAML's, HAN's, or TransICD's predictions, respectively. The student model dramatically outperforms the logistic baseline in terms of faithfulness to the DR-CAML and TransICD models. On classification metrics, the baseline is a surprisingly excellent student for the HAN model.

		DR CAML		HAN		Trans ICD	
	Logistic	Student	Orig	Student	Orig	Student	Orig
Macro AUC	0.561	0.901	0.906	0.870	0.884	0.883	0.897
Micro AUC	0.937	0.967	0.972	0.962	0.967	0.907	0.924
Macro F1	0.011	0.142	0.224	0.026	0.077	0.426	0.586
Micro F1	0.271	0.326	0.536	0.251	0.390	0.478	0.640
Prec @ 8	0.541	0.483	0.701	0.519	0.599	0.479	0.502
Prec @ 15	0.412	0.407	0.548	0.406	0.455	0.333	0.343

Table 2: Comparison of DR-CAML, HAN, and TransICD and their respective student models to the true ICD labels. Although the logistic regression baseline was trained to directly predict ICD codes and our student models were not, the Student-DRCAML and Student-TransICD models outperform the baseline in AUC and F1.

the non-parametric Kendall Tau rank correlation. These metrics range from -1 to 1 with 1 indicating perfect faithfulness. Regression results are on the left side of Table 1.

Our second evaluation treats the teacher model's predictions as binary labels to compute F1, AUC, and precision scores. We then evaluate the faithfulness of our student model by treating its outputs as probabilities and using classification metrics such as F1 score. Prec @ n is the fraction of the n highest scored labels that are present in the ground truth. These metrics range from 0 to 1, where perfectly faithful predictions would have 1.0 AUC and F1 scores. The student model is considered faithful if it correctly predicts whether the teacher model will make a binary prediction. We again use the logistic regression baseline. Classification results are on the right side of Table 1.

Finally, we use the student model's predictions to predict the ground-truth ICD code labels and compare its predictive performance against that of the teacher model's in Table 2. While the student model was not trained using these labels, we can use its predictions as probabilities for these codes. By comparing against the logistic regression baseline (a linear model of equal complexity), we can see whether our training setup allows the student model to learn a better predictor.

Our results show that our proxies are quite faithful to the teacher models. Table 1 shows that the Student-DRCAML and Student-HAN models are dramatically more faithful to their corresponding black-box models than the logistic regression baseline. Interestingly, the baseline is in fact quite faithful to the TransICD model. Comparing the classification metrics of Table 1 to the results in Table 2, we see that on AUC metrics, all three proxies are more faithful to the their target models than

934.1: "Foreign body in main bronchus"

Mullenbach et al. (2018)		
CAML	(HI)	*... line placed bronchoscopy performed showing* **large mucus plug on** *the left on transfer to ...*
Cosine		*... also needed medication to help* **your body maintain your** *blood pressure after receiving iv ...*
CNN		*... found to have a large* **lll lingular pneumonia on** *chest x ray he was ...*
Logistic		*... impression confluent consolidation involving nearly* **the entire left lung** *with either bronchocentric or vascular ...*
Ours		
DR-CAML	0.38	*... line placed bronchoscopy performed showing* **large mucus plug on** *the left on transfer to ...*
Logistic	0.28	*... tube down your throat to* **help you breathe you** *also needed medication to help ...*
Student-DRCAML	0.38	*... a line placed bronchoscopy performed* **showing large mucus plug** *on the left on transfer ...*
Student-HAN	0.39	*... line and r radial a* **line placed bronchoscopy performed** *showing large mucus plug on ...*

Table 3: Comparison of the clinical evaluation from Mullenbach et al. (2018) with our plausibility evaluation. The example above contains the explanations produced by eight systems. The first four systems for each example are directly copied from Table 1 of Mullenbach et al. (2018). The (HI) and (I) labels in the second column indicate whether the clinician labeled those explanations as Highly Informative or Informative. The four systems below the dotted line are from our evaluation, for which the second column indicates the probability output of our plausibility classifier. Here, Student-DRCAML and DR-CAML produce almost identical explanations. The Student-HAN explanation highlights that our student method can generate explanations for black-box models which cannot explain themselves. Additional comparisons are shown in Tables 5 and 7.

those black-box models were to the original ICD codes. In Table 2, we hypothesize that the relatively low precision scores result from our student regressions being fit for each ICD code independently, which prevents the combined model from encoding relative frequency information.

Rudin (2019) critiques post-hoc methods in general, arguing that "if we cannot know for certain whether our [post-hoc] explanation is [faithful], we cannot know whether to trust either the explanation or the original model." Because no post-hoc method can ever be perfectly faithful to an original model, our explicit measurement of faithfulness provides a useful approach for understanding whether the student is "faithful enough" for a given application. It also allows for a prediction-specific analysis – if we wish to use the student model to explain a high-stakes prediction made by a black-box model, we can first check whether both agree upon that specific prediction.

In applications where explainability is essential, our student model could be used as a more interpretable replacement for a high-performing black-box model. In such a case, a domain expert might care less about the evaluation of faithfulness in Table 1 and more about the ground-truth predictive performance evaluated in Table 2. We leave for future work the challenge of whether a student model produced by our method could be fine-tuned to better predict ground-truth ICD codes.

5 Plausibility Evaluation

Explanations are considered plausible if they can be reasoned about by a person (Wiegreffe and Pinter, 2019). Evaluating plausibility is thus typically more difficult than faithfulness, because it requires input from annotators (Herman, 2017; Arora et al., 2021). Furthermore, an explanation that is plausible to a domain expert may not be plausible to a layperson. Mullenbach et al. (2018) evaluated the plausibility of their models' explanations by collecting annotations from a clinical expert. For 100 notes, each of four models produced an explanation in the form of a 14-token subsequence taken from the discharge summary. The clinician read the four (anonymized) explanations and the corresponding ICD code and subjectively rated each explanation as "informative"[6]. Across the 100 examples, the clinician rated CAML as slightly more informative than the logistic regression and CNN baselines. Table 3 shows explanations produced by our and Mullenbach et al. (2018)'s models.

The format of Mullenbach et al. (2018)'s plausibility evaluation does not easily lend itself to replication. While the authors shared their annotations with us, missing metadata (see Appendix A.2) prevented a direct reproduction of their analysis. Additionally, since the clinical annotator considered explanations in a comparative setting, we cannot easily add our student model as another method us-

[6]The annotator was told to mark as informative all explanations that "adequately explain[ed] the presence of the given ICD code" (Mullenbach et al., 2018).

Model	Score	Interval	Best
Logistic	35	(31, 49)	7%
Cosine	38	(32, 51)	13%
CNN	42	(33, 52)	14%
CAML	44	(33, 52)	16%
DR-CAML	48	(34, 53)	22%
Student-DRCAML	52	(34, 54)	19%
Student-HAN	47	(33, 52)	10%

Table 4: Binary plausibility evaluation using classifier annotations. We collapse the Highly Informative and Informative labels from Mullenbach et al. (2018) to a single positive class. The Score column is out of 99; we use a binary threshold of 0.45 so that the proportion of predicted plausible explanations matches the data. To highlight the uncertainty of this evaluation, we bootstrap sample 1000 informative labels for each method's explanations. The Interval column shows the 95% interval of informative scores across those 1000 samples. The Best column shows the percentage of samples in which each method scored highest.

296.20: "Major depressive affective disorder, single episode, unspecified"	
DR-CAML	*... diagnosis overdose of medications narcotics* **benzodiazepine suicide attempt chronic** *migraine headaches depression stage iv ...*
Student-DRCAML	*... up from the medications you* **were evaluated by psychiatry** *and will be transferred to ...*

Table 5: Examples of differing explanations between DR-CAML and its student. Our informative classifier gives the DR-CAML and student explanations scores of 0.47 and 0.33, respectively. Additional examples are shown in Table 8.

ing the same annotations. Therefore, we replicate this evaluation as best as possible by using a classifier to predict synthetic labels as to whether the clinical domain expert *would have* labeled our models' explanations as plausible. Using BioWordVec embeddings released by Zhang et al. (2019), the text of the ICD-9 code description, and the 14-gram explanation produced by each model from Mullenbach et al. (2018), we train a classifier that predicts whether an explanation would have been rated as informative. This annotation classifier achieves a binary classification accuracy of 67.2% and an AUC score of 0.726 when evaluated with leave-one-out validation. This relatively low accuracy and our model training details are discussed in Appendix A.3.

To conduct our plausibility evaluation, we first use or reproduce the baseline methods from Mullenbach et al. (2018) and Biswas et al. (2021). Each model, including the student, produces a 14-token explanation from the discharge summary by first finding the 4-gram with the largest *average feature importance* and then including five tokens on either side of the 4-gram. The logistic regression baseline is the same as in § 4, where feature importance is computed using the coefficients of the logistic model. The student model's explanations are computed in the same manner, finding the 4-gram with the largest average coefficient weights. For CAML, DR-CAML, and the CNN models, we use the code released by Mullenbach et al. (2018) to extract explanations. The CNN baseline primarily differs from CAML in that it does not use an attention mechanism. Finally, we reimplement their Cosine baseline which picks the 4-gram with the highest cosine similarity to the ICD-9 code description text.

We extract the model's explanations for the same[7] discharge summaries as were evaluated by Mullenbach et al. (2018). For each explanation, we use the annotation classifier described above to predict the probability that each explanation would have been labeled as informative. If we set the classifier threshold such that 45% of explanations are rated as informative (matching the proportion from the original annotations), we get the results in the Score column of Table 4. The student model produces the largest number of informative explanations according to our classifier; however, the classifier's inaccuracy can introduce substantial uncertainty. Rather than thresholding the outputs of the annotation classifier, we can use its probability outputs to sample a set of informative labels for each explanation. If we sample 1000 such sets of labels and report the 95% confidence interval for each model's score in the Interval column of Table 4, the interval overlap makes the methods essentially indistinguishable. The Best column in this table shows the percentage of samples in which each method scored highest. While the Interval column highlights the inherent limitation of evaluating plausibility on this small fixed dataset of human evaluations, the Best and Score columns combined with the qualitative comparisons in Table 3 suggest that our student model explanations are at least comparably plausible to those of DR-CAML.

Table 3 shows that DR-CAML and Student-DRCAML produce qualitatively similar explana-

[7]Using the 99 (of 100) discharge summaries that could be uniquely identified. See Appendix A for details.

tions. The other two examples presented in Mullenbach et al. (2018) are in Appendix A.4. The similarity is perhaps surprising because DR-CAML extracts explanations using its attention mechanism, whereas the student model uses unigram feature importance values that do not vary between examples. For this example, it appears that the student is faithful both in the predictions it makes and how it makes those predictions. We additionally include the explanations for Student-HAN. As HAN cannot produce its own explanations, this highlights that our method can also be applied to models that are not interpretable by design. Table 5 shows an example where the student and DR-CAML diverge the most. We include two additional examples in Appendix A.4. These cases highlight two benefits of the student model. First, its feature importance weights are *global* across all predictions, providing an aggregate representation of the student's behavior. Second, the approach for extracting student explanation n-grams is transparent and simulatible; it is just the average of n feature weights. These factors may be particularly appealing in cases where explainability is paramount.

6 Discussion

We have introduced a method that uses knowledge distillation to generate post-hoc explanations and is designed to be interpretable and plausible while maintaining faithfulness to the trained model. By constraining the student to a class of models that is decomposible, simulatible, and algorithmically transparent, our optimization for faithfulness gives us a clear way to evaluate several dimensions of interpretability. We evaluated our method on the task of clinical code prediction. A key benefit of our method is its simplicity and wide applicability. Even for a proprietary trained model for which the learned parameters are unknown, a student can be trained as long as we have a dataset that includes the trained model's predictions. Our approach has the additional benefit of producing a standalone student model that can provide *global* feature explanations. If the student has sufficient predictive performance, a skeptic of post-hoc methods (e.g. Rudin (2019)) might prefer to use the inherently-interpretable student.

The present work has several limitations that are left for future work. Though the task of medical code prediction has important implications and has been widely studied in interpretability research, we only consider this single task on a single English-language dataset. While we have shown our student approach works for three different black-box models, it requires additional study in new domains and tasks. There may be black-box models for which no linear student is faithful. Our evaluation is also limited to only a single form of explanation: n-grams extracted via importance or attention weights. Counterfactual explanations (i.e., an alternative input that would have been classified differently) might be harder or easier for our student method to generate (Barocas et al., 2020). Our plausibility evaluations rely on a small set of annotations from which we extrapolate. Future work should collect new annotations that consider metrics such as sufficiency and simulatibility that require human evaluations (Jain et al., 2020; Hase and Bansal, 2020; Arora et al., 2021).

As the ML community continues to explore new directions for interpretable methods, new desiderata may arise based on the domain experts who turn to ML methods for decision support. Interpretable ML methods should clearly define how they expect to satisfy criteria such as faithfulness or plausibility; by designing for plausibility and transparency and optimizing for faithfulness, our proposed method is broadly applicable. We release our code to enable future work.

7 Ethics and Broader Impacts

This paper is situated in a broader field of clinical applications of machine learning. While our work does not raise new ethical issues within this domain, there are general concerns that also apply to this work. ML methods should not be deployed in real-world settings without extensive validation (Wiens et al., 2019). In the clinical domain, particular attention must be paid to the possibility of perpetuating disparities that have been encoded in the training data (Rajkomar et al., 2018). While MIMIC-III provides a useful benchmark for developing and evaluating methods, it is not representative of the the enormous variety of clinical and linguistic data. Domain experts and those most likely to be affected by new ML systems should be given oversight of potential deployments.

Acknowledgements

We acknowledge support provided by the Johns Hopkins Institute for Assured Autonomy. We thank Sarah Wiegreffe and Jacob Eisenstein for their help and plausibility annotations.

References

Siddhant Arora, Danish Pruthi, Norman Sadeh, William W Cohen, Zachary C Lipton, and Graham Neubig. 2021. Explain, edit, and understand: Rethinking user study design for evaluating model explanations. *arXiv preprint arXiv:2112.09669*.

Arip Asadulaev, Igor Kuznetsov, and Andrey Filchenkov. 2019. Interpretable few-shot learning via linear distillation. *arXiv preprint arXiv:1906.05431*.

Solon Barocas, Andrew D Selbst, and Manish Raghavan. 2020. The hidden assumptions behind counterfactual explanations and principal reasons. In *Proceedings of the 2020 Conference on Fairness, Accountability, and Transparency*, pages 80–89.

Jasmijn Bastings, Wilker Aziz, and Ivan Titov. 2019. Interpretable neural predictions with differentiable binary variables. In *Proceedings of the 57th Annual Meeting of the Association for Computational Linguistics*, pages 2963–2977.

Biplob Biswas, Thai-Hoang Pham, and Ping Zhang. 2021. Transicd: Transformer based code-wise attention model for explainable icd coding. In *AIME*.

Cristian Buciluă, Rich Caruana, and Alexandru Niculescu-Mizil. 2006. Model compression. In *Proceedings of the 12th ACM SIGKDD international conference on Knowledge discovery and data mining*, pages 535–541.

Rich Caruana, Yin Lou, Johannes Gehrke, Paul Koch, Marc Sturm, and Noemie Elhadad. 2015. Intelligible models for healthcare: Predicting pneumonia risk and hospital 30-day readmission. In *Proceedings of the 21th ACM SIGKDD international conference on knowledge discovery and data mining*, pages 1721–1730.

Hang Dong, V'ictor Su'arez-Paniagua, William Whiteley, and Honghan Wu. 2021. Explainable automated coding of clinical notes using hierarchical label-wise attention networks and label embedding initialisation. *Journal of biomedical informatics*, page 103728.

Been Doshi-Velez, Finale; Kim. 2017. Towards a rigorous science of interpretable machine learning. In *eprint arXiv:1702.08608*.

Upol Ehsan, Pradyumna Tambwekar, Larry Chan, Brent Harrison, and Mark O Riedl. 2019. Automated rationale generation: a technique for explainable ai and its effects on human perceptions. In *Proceedings of the 24th International Conference on Intelligent User Interfaces*, pages 263–274.

Radwa Elshawi, Mouaz Al-mallah, and Sherif Sakr. 2019. On the interpretability of machine learning-based model for predicting hypertension. *BMC Medical Informatics and Decision Making*, 19.

Shi Feng, Eric Wallace, Alvin Grissom II, Mohit Iyyer, Pedro Rodriguez, and Jordan Boyd-Graber. 2018. Pathologies of neural models make interpretations difficult. In *Proceedings of the 2018 Conference on Empirical Methods in Natural Language Processing*, pages 3719–3728.

Leilani H Gilpin, David Bau, Ben Z Yuan, Ayesha Bajwa, Michael Specter, and Lalana Kagal. 2018. Explaining explanations: An overview of interpretability of machine learning. In *2018 IEEE 5th International Conference on data science and advanced analytics (DSAA)*, pages 80–89. IEEE.

Christopher Grimsley, Elijah Mayfield, and Julia RS Bursten. 2020. Why attention is not explanation: Surgical intervention and causal reasoning about neural models. In *Proceedings of The 12th Language Resources and Evaluation Conference*, pages 1780–1790.

Riccardo Guidotti, Anna Monreale, Salvatore Ruggieri, Franco Turini, Fosca Giannotti, and Dino Pedreschi. 2018. A survey of methods for explaining black box models. *ACM computing surveys (CSUR)*, 51(5):1–42.

Wenbo Guo, Kaixuan Zhang, Lin Lin, Sui Huang, and Xinyu Xing. 2017. Towards interrogating discriminative machine learning models. *ArXiv*, abs/1705.08564.

Peter Hase and Mohit Bansal. 2020. Evaluating explainable ai: Which algorithmic explanations help users predict model behavior? In *Proceedings of the 58th Annual Meeting of the Association for Computational Linguistics*, pages 5540–5552.

Bernease Herman. 2017. he promise and peril of human evaluation for model interpretability. In *Proceedings of the 31st Conference on Neural Information Processing Systems (NeurIPS 2017)*.

Geoffrey Hinton, Oriol Vinyals, Jeff Dean, et al. 2015. Distilling the knowledge in a neural network. *arXiv preprint arXiv:1503.02531*, 2(7).

Alon Jacovi and Yoav Goldberg. 2020. Towards faithfully interpretable nlp systems: How should we define and evaluate faithfulness? In *Proceedings of the 58th Annual Meeting of the Association for Computational Linguistics*, pages 4198–4205.

Sarthak Jain and Byron C Wallace. 2019. Attention is not explanation. In *Proceedings of the 2019 Conference of the North American Chapter of the Association for Computational Linguistics: Human Language Technologies, Volume 1 (Long and Short Papers)*, pages 3543–3556.

Sarthak Jain, Sarah Wiegreffe, Yuval Pinter, and Byron C. Wallace. 2020. Learning to faithfully rationalize by construction. In *Proceedings of the 58th Annual Meeting of the Association for Computational Linguistics*, pages 4459–4473, Online. Association for Computational Linguistics.

Alistair EW Johnson, Tom J Pollard, Lu Shen, H Lehman Li-Wei, Mengling Feng, Mohammad Ghassemi, Benjamin Moody, Peter Szolovits, Leo Anthony Celi, and Roger G Mark. 2016. Mimic-iii, a freely accessible critical care database. *Scientific data*, 3(1):1–9.

Byung-Hak Kim and Varun Ganapathi. 2021. Read, attend, and code: Pushing the limits of medical codes prediction from clinical notes by machines. *ArXiv*, abs/2107.10650.

Sawan Kumar and Partha Talukdar. 2020. NILE : Natural language inference with faithful natural language explanations. In *Proceedings of the 58th Annual Meeting of the Association for Computational Linguistics*, pages 8730–8742, Online. Association for Computational Linguistics.

Himabindu Lakkaraju, Nino Arsov, and Osbert Bastani. 2020. Robust and stable black box explanations. In *International Conference on Machine Learning*, pages 5628–5638. PMLR.

Himabindu Lakkaraju, Ece Kamar, Rich Caruana, and Jure Leskovec. 2017. Interpretable & explorable approximations of black box models. *arXiv preprint arXiv:1707.01154*.

Tao Lei, Regina Barzilay, and Tommi Jaakkola. 2016. Rationalizing neural predictions. In *Proceedings of the 2016 Conference on Empirical Methods in Natural Language Processing*, pages 107–117.

Zachary C Lipton. 2018. The mythos of model interpretability: In machine learning, the concept of interpretability is both important and slippery. *Queue*, 16(3):31–57.

Yin Lou, Rich Caruana, and Johannes Gehrke. 2012. Intelligible models for classification and regression. In *Proceedings of the 18th ACM SIGKDD international conference on Knowledge discovery and data mining*, pages 150–158.

Scott M Lundberg and Su-In Lee. 2017. A unified approach to interpreting model predictions. In *Proceedings of the 31st international conference on neural information processing systems*, pages 4768–4777.

Scott M. Lundberg, Bala G. Nair, Monica S Vavilala, Mayumi Horibe, Michael J. Eisses, Trevor L. Adams, David Liston, Daniel King-Wai Low, Shu-Fang Newman, Jerry H. Kim, and Su-In Lee. 2018. Explainable machine-learning predictions for the prevention of hypoxaemia during surgery. *Nature biomedical engineering*, 2:749 - 760.

Brent Mittelstadt, Chris Russell, and Sandra Wachter. 2019. Explaining explanations in ai. In *Proceedings of the conference on fairness, accountability, and transparency*, pages 279–288.

James Mullenbach, Sarah Wiegreffe, Jon Duke, Jimeng Sun, and Jacob Eisenstein. 2018. Explainable prediction of medical codes from clinical text. In *Proceedings of the 2018 Conference of the North American Chapter of the Association for Computational Linguistics: Human Language Technologies, Volume 1 (Long Papers)*, pages 1101–1111.

Fabian Pedregosa, Gaël Varoquaux, Alexandre Gramfort, Vincent Michel, Bertrand Thirion, Olivier Grisel, Mathieu Blondel, Peter Prettenhofer, Ron Weiss, Vincent Dubourg, et al. 2011. Scikit-learn: Machine learning in python. *the Journal of machine Learning research*, 12:2825–2830.

Mary Phuong and Christoph H. Lampert. 2019. Towards understanding knowledge distillation. *ArXiv*, abs/2105.13093.

Danish Pruthi, Mansi Gupta, Bhuwan Dhingra, Graham Neubig, and Zachary C. Lipton. 2020. Learning to deceive with attention-based explanations. In *Proceedings of the 58th Annual Meeting of the Association for Computational Linguistics*, pages 4782–4793, Online. Association for Computational Linguistics.

Alvin Rajkomar, Michaela Hardt, Michael D Howell, Greg Corrado, and Marshall H Chin. 2018. Ensuring fairness in machine learning to advance health equity. *Annals of internal medicine*, 169(12):866–872.

Marco Tulio Ribeiro, Sameer Singh, and Carlos Guestrin. 2016. " why should i trust you?" explaining the predictions of any classifier. In *Proceedings of the 22nd ACM SIGKDD international conference on knowledge discovery and data mining*, pages 1135–1144.

Cynthia Rudin. 2019. Stop explaining black box machine learning models for high stakes decisions and use interpretable models instead. *Nature Machine Intelligence*, 1(5):206–215.

Dylan Slack, Sophie Hilgard, Emily Jia, Sameer Singh, and Himabindu Lakkaraju. 2020. Fooling lime and shap: Adversarial attacks on post hoc explanation methods. In *Proceedings of the AAAI/ACM Conference on AI, Ethics, and Society*, pages 180–186.

S. Tan, Rich Caruana, Giles Hooker, and Yin Lou. 2018a. Distill-and-compare: Auditing black-box models using transparent model distillation. *Proceedings of the 2018 AAAI/ACM Conference on AI, Ethics, and Society*.

S. Tan, Giles Hooker, Paul Koch, Albert Gordo, and Rich Caruana. 2018b. Considerations when learning additive explanations for black-box models.

Maxim Topaz, Leah Shafran-Topaz, and Kathryn H Bowles. 2013. Icd-9 to icd-10: evolution, revolution, and current debates in the united states. *Perspectives in health information management/AHIMA, American Health Information Management Association*, 10(Spring).

Irene Unceta, Jordi Nin, and Oriol Pujol. 2020. Copying machine learning classifiers. *IEEE Access*, 8:160268–160284.

Ilse van der Linden, Hinda Haned, and Evangelos Kanoulas. 2019. Global aggregations of local explanations for black box models. In *SIGIR Workshop on FACTS-IR*.

Thanh Vu, Dat Quoc Nguyen, and Anthony N. Nguyen. 2020. A label attention model for icd coding from clinical text. In *IJCAI*.

Sarah Wiegreffe, Edward Choi, Sherry Yan, Jimeng Sun, and Jacob Eisenstein. 2019. Clinical concept extraction for document-level coding. In *Proceedings of the 18th BioNLP Workshop and Shared Task*, pages 261–272.

Sarah Wiegreffe, Ana Marasović, and Noah A Smith. 2021. Measuring association between labels and free-text rationales. In *Proceedings of the 2021 Conference on Empirical Methods in Natural Language Processing*, pages 10266–10284.

Sarah Wiegreffe and Yuval Pinter. 2019. Attention is not not explanation. In *Proceedings of the 2019 Conference on Empirical Methods in Natural Language Processing and the 9th International Joint Conference on Natural Language Processing (EMNLP-IJCNLP)*, pages 11–20, Hong Kong, China. Association for Computational Linguistics.

Jenna Wiens, Suchi Saria, Mark Sendak, Marzyeh Ghassemi, Vincent X Liu, Finale Doshi-Velez, Kenneth Jung, Katherine Heller, David Kale, Mohammed Saeed, et al. 2019. Do no harm: a roadmap for responsible machine learning for health care. *Nature medicine*, 25(9):1337–1340.

Zichao Yang, Diyi Yang, Chris Dyer, Xiaodong He, Alex Smola, and Eduard H. Hovy. 2016. Hierarchical attention networks for document classification. In *NAACL*.

Chih-Kuan Yeh, Cheng-Yu Hsieh, Arun Suggala, David I Inouye, and Pradeep K Ravikumar. 2019. On the (in) fidelity and sensitivity of explanations. *Advances in Neural Information Processing Systems*, 32:10967–10978.

Yijia Zhang, Qingyu Chen, Zhihao Yang, Hongfei Lin, and Zhiyong Lu. 2019. Biowordvec, improving biomedical word embeddings with subword information and mesh. *Scientific data*, 6(1):1–9.

Ruiqi Zhong, Steven Shao, and Kathleen McKeown. 2019. Fine-grained sentiment analysis with faithful attention. *arXiv preprint arXiv:1908.06870*.

A (Re-)implementation details

A.1 Reproducing CAML predictive performance

The trained DR-CAML model released by Mullenbach et al. (2018) produced predictions that matched the published F1 and ROC scores. We were unable to precisely replicate the outputs of the CAML model. Table 6 shows the scores published by Mullenbach et al. (2018) as well as those for a CAML reimplementation done by Wiegreffe et al. (2019). We include the scores we observe using the model weights released on GitHub as well as the scores for a model we retrained from scratch. We use the released model instead of the retrained model as its performance is much closer to the published numbers.

A.2 Reproducing plausibility scores

The clinical plausibility annotations provided to us by the authors of Mullenbach et al. (2018) contains the text explanations and their corresponding annotations, but was missing the crucial metadata of which models produced which explanations. The metadata also did not indicate from which specific discharge summary the texts were derived; while the text explanations were uniquely identifying for all but one of the 100 examples. For that one example, because some patients had multiple documents sometimes containing duplicated segments of text, there were three discharge summaries from which the explanations could have been drawn. We thus excluded this example from our analyses. To replicate their analysis the best we could, we retrained or reimplemented their logistic regression, vanilla CNN, and cosine similarity methods. We then looked at the attention or feature importance weights for each trained model and the text explanations that had been annotated, and assigned each model the text explanation for which it provided the highest weight. This assignment did not perfectly align with past work: there were six cases (out of 99) where a text explanation was "chosen" by more models than times it appeared as an option. Ignoring that issue and then simply aggregating the Informative and Highly Informative clinician annotations, we obtained the plausibility scores in the Ours column of Table 9. The Theirs column shows the published numbers from Mullenbach et al. (2018). While the numbers change substantially, the ordering is relatively stable with only two swaps: CAML and Cosine, and Logistic and CNN.

The other columns of the table are described below.

A.3 Plausibility annotation classifier

To evaluate the plausibility of our student model's explanations, we trained a classifier to predict whether an explanation would have been labeled as plausible by the clinical domain expert. We treat this as a binary classification task by grouping the "Informative" and "Highly Informative" annotations as a single "plausible" label. Conscious of the fact that we have only 99 examples with four text explanations each, we use two approaches with which to train and evaluate our classifier. The first used leave-one-out cross validation at the example level, such that the classifier was trained on 98 examples at a time and then evaluated on the remaining one. We refer to this evaluation as "E1" in Table 9. The second also used leave-on-out cross validation but at the explanation level; we held out a single text explanation, trained on all other explanations across all examples, and then evaluated on the held-out explanation. When an explanation appeared more than once in a single example, we made sure to remove its duplicates from the training data for predicting that explanation. We refer to this evaluation as "E2" in Table 9.

The trained model is a simple logistic regression classifier trained on a fastText embedding of both the explanation and the target ICD-9 code description. Using the BioWordVec embeddings released by Zhang et al. (2019), we embed each both the explanation and code description into a 200-dimensional vector, concatenate the two vectors, and pass it to the logistic regression. In the E1 evaluation, the model achieves an accuracy of 60.6% and an ROC AUC score of .640. In the E2 evaluation, that increases to an accuracy of 67.2% and an AUC score of .726, indicating that the additional within-example explanations substantially help the classifier.

When using these classifiers to label the explanations generated by each model instead of the plausibility scores derived in A.2, we get the results shown in columns E1 and E2 of Table 9.

Finally, we retrain our final classifier on all the explanations, leaving none held out. Rather than using our classifier to evaluate the explanations that were actually shown to the clinician, we instead use our (re-)implementation of the four models to extract an explanation from each of the 99 discharge summaries. These explanations thus may or may

	AUC		F1		P@n	
	Macro	Micro	Macro	Micro	8	15
Mullenbach et al. (2018)	0.895	0.986	0.088	0.539	0.709	0.561
Wiegreffe et al. (2019)	0.889	0.985	0.080	0.542	0.712	0.562
Ours (using released weights)	0.892	0.978	0.090	0.298	0.636	0.471
Ours (retrained)	0.628	0.884	0.001	0.024	0.042	0.027

Table 6: Published predictive performance of CAML and our replicated results. Our experiments throughout the paper use the model with the released weights, which is closest to the published numbers (despite Micro F1).

442.84: "Aneurysm of other visceral artery"		
Mullenbach et al. (2018)		
CAML	(I)	... *and gelfoam embolization of right* **hepatic artery branch pseudoaneurysm** *coil embolization of the gastroduodenal ...*
Cosine		... *coil embolization of the gastroduodenal* **artery history of present** *illness the pt is a ...*
CNN		... *foley for hemodynamic monitoring and* **serial hematocrits angio was** *performed and his gda was ...*
Logistic	(I)	... *and gelfoam embolization of right* **hepatic artery branch pseudoaneurysm** *coil embolization of the gastroduodenal ...*
Ours		
DR-CAML	0.55	... *gelfoam embolization of right hepatic* **artery branch pseudoaneurysm coil** *embolization of the gastroduodenal artery ...*
Logistic	0.57	... *biliary stents hx cbd r* **colonic fistula r colectomy** *partial l nephrectomy for renal ...*
Student-DRCAML	0.55	... *embolization of right hepatic artery* **branch pseudoaneurysm coil embolization** *of the gastroduodenal artery history ...*
Student-HAN	0.55	... *embolization of right hepatic artery* **branch pseudoaneurysm coil embolization** *of the gastroduodenal artery history ...*

428.20: "Systolic heart failure, unspecified"		
Mullenbach et al. (2018)		
CAML		... *no mitral valve prolapse moderate* **to severe mitral regurgitation** *is seen the tricuspid valve ...*
Cosine		... *is seen the estimated pulmonary* **artery systolic pressure is** *normal there is no pericardial ...*
CNN		... *and suggested starting hydralazine imdur* **continue aspirin arg admitted** *at baseline cr appears patient ...*
Logistic	(HI)	... *anticoagulation monitored on tele pump* **systolic dysfunction with ef** *of seen on recent echo ...*
Ours		
DR-CAML	0.39	... *anticoagulation monitored on tele pump* **systolic dysfunction with ef** *of seen on recent echo ...*
Logistic	0.37	... *seen the mitral valve leaflets* **are mildly thickened there** *is nomitral valve prolapse ...*
Student-DRCAML	0.39	... *anticoagulation monitored on tele pump* **systolic dysfunction with ef** *of seen on recent echo ...*
Student-HAN	0.36	... *blood cultures obtained repeated cxr* **echocardiogram showed an ef** *of and therefore zestril was ...*

Table 7: Comparison of the clinical evaluation from Mullenbach et al. (2018) with our plausibility evaluation. There are two examples above, each which contains the explanations produced by eight systems. The first four systems for each example are directly copied from Table 1 of Mullenbach et al. (2018). The (HI) and (I) labels in the second column indicate whether the clinician labeled those explanations as Highly Informative or Informative. The four systems below the dotted line are from our evaluation, for which the second column indicates the probability output of our plausibility classifier.

not appear in the training data for the classifier. For the Full evaluation we are not worried about the classifier overfitting, as the classifier functions as a direct replacement for the clinician who produced the training data. The results of this analysis are the numbers shown in Table 4 in § 5, reproduced in Table 9 in the "Full" column. The Logistic model does much worse on the Full evaluation than in either E1 or E2. This may be because the explanations selected by the trained model were worse than those selected by the model which was used for the original clinical evaluation.

455.0: "Internal hemorrhoids without mention of complication"

DR-CAML	0.38	... *and she then underwent a* **colonoscopy with gi that** *also did not detect evidence* ...
Student-DRCAML	0.52	... *past medical history diverticular disease* **diverticulitis sbo anxiety hemorrhoids** *past surgical history s p* ...

592.0 : "Calculus of kidney"

DR-CAML	0.30	... *if you develop any of* **these symtpoms please call** *the office or go to* ...
Student-DRCAML	0.46	... *the colon gastroesophageal reflux asthma* **irritable bowel syndrome gastroparesis** *osteoporosis anxiety and or depression* ...

Table 8: Additional differing explanations and classifier scores between DR-CAML and the student.

Model	Theirs	Ours	E1	E2	Full
Logistic	41	43	47	49	35
Cosine	48	48	41	40	38
CNN	36	46	51	47	42
CAML	46	54	47	43	44
DR-CAML	–	–	45	44	48

Table 9: Plausibility evaluations and comparison to Mullenbach et al. (2018). The Theirs column shows the published numbers; Ours shows our best attempt at matching the clinical evaluation to the trained models. While the numbers change dramatically, the ordering only changes by two swaps. The clinical evaluation did not include DR-CAML. E1 and E2 show the results with predicted plausibility labels under the two evaluation settings described in A.3. Full duplicates the results from Table 4 for comparison.

A.4 Additional Examples

We provide two additional examples of eight different models' explanations in Table 7. These are the same examples shown in (Mullenbach et al., 2018). We include the four explanations as published in Mullenbach et al. (2018), our reproduction of DR-CAML, the logistic regression baseline, and the explanations from two student models, Student-DRCAML and Student-HAN. As we can see from the examples, Student-DRCAML produces similar explanations to DR-CAML. Student-HAN shows that our method is able to produce explanations for models not originally designed to do so. We also include two additonal examples in which DR-CAML and Student-DRCAML diverge the most in Table 8.

Towards Generalizable Methods for Automating Risk Score Calculation

Jennifer J. Liang[1,3,⋆], Eric Lehman[2,3,#], Ananya S. Iyengar[4,§], Diwakar Mahajan[1,3,†], Preethi Raghavan[2,5,‡], Cindy Y. Chang[6,∗] Peter Szolovits[2,3,◇]

[1]IBM Research, [2]MIT CSAIL, [3]MIT-IBM Watson AI Lab, [4]Northeastern University, [5]Fidelity Investments, [6]Brigham and Women's Hospital, Harvard Medical School

{⋆jjliang,†dmahaja}@us.ibm.com, {#lehmer16,◇psz}@mit.edu, §iyengar.a@northeastern.edu, ‡preethi.raghavan@fmr.com, ∗cchang@bwh.harvard.edu

Abstract

Clinical risk scores enable clinicians to tabulate a set of patient data into simple scores to stratify patients into risk categories. Although risk scores are widely used to inform decision-making at the point-of-care, collecting the information necessary to calculate such scores requires considerable time and effort. Previous studies have focused on specific risk scores and involved manual curation of relevant terms or codes and heuristics for each data element of a risk score. To support more generalizable methods for risk score calculation, we annotate 100 patients in MIMIC-III with elements of CHA_2DS_2-VASc and PERC scores, and explore using question answering (QA) and off-the-shelf tools. We show that QA models can achieve comparable or better performance for certain risk score elements as compared to heuristic-based methods, and demonstrate the potential for more scalable risk score automation without the need for expert-curated heuristics. Our annotated dataset will be released to the community to encourage efforts in generalizable methods for automating risk scores.

1 Introduction

Clinical risk scores are standardized metrics to estimate the risk of a particular future outcome based on available clinical parameters and are commonly used at the point-of-care to inform decision-making around diagnosis and treatment (Steyerberg et al., 2019). An example of this is the CHA_2DS_2-VASc score (Lip et al., 2010), which uses 7 patient data elements to estimate the risk of stroke in patients with non-valvular atrial fibrillation and thus guide strategies around stroke prevention. It has successfully demonstrated clinical impact and is referenced in the practice guidelines for management of atrial fibrillation released by the American Heart Association, American College of Cardiology, and Heart Rhythm Society in 2014 (January et al., 2014). In general, data elements that contribute to a risk score may include information about the patient's age, gender, medical history, presenting symptoms, medication use, etc. While risk scores are generally designed for use at the point-of-care, calculating them can require considerable time and effort, as each data element must be manually gathered, often from multiple locations within the electronic health record (EHR). A previous study investigating the feasibility of automating clinical score calculation identified 534 unique patient data elements from 168 externally validated clinical scores, with each score requiring anywhere from 3 to 31 elements (Aakre et al., 2017). Automating extraction of clinical data elements necessary to calculate risk scores could save clinicians time and help them more effectively leverage risk scores to improve care at the bedside (Aakre et al., 2017).

Prior efforts to automate data extraction for risk score calculations have targeted specific risk scores. Some of these efforts focused only on leveraging information from structured EHR data. Koziatek et al. (2018) developed and automated a structured-data-only version of the Wells and revised Geneva risk scores for estimating pulmonary embolism (PE) risk. Similarly, in automating the Padua Prediction Score for venous thromboembolism risk, Pavon et al. (2018) either operationalized variables to rely only on structured data or omitted them entirely. Other efforts have also incorporated unstructured EHR data into their work. Jonnagaddala et al. (2015) used a rule-based text mining system to extract elements of the Framingham risk score for coronary artery disease. Mark et al. (2018) and Zhang et al. (2022) used text string searches on a set of custom-built keywords/search phrases to automate coronary risk scores and Wells score for PE, respectively. Bean et al. (2019), Grouin et al. (2011), and Elkin et al. (2021) explored the use of named entity recognition (NER) tagging combined with heuristics to automatically calculate

Proceedings of the BioNLP 2022 workshop, Dublin, Ireland, pages 426–431
May 26, 2022. ©2022 Association for Computational Linguistics

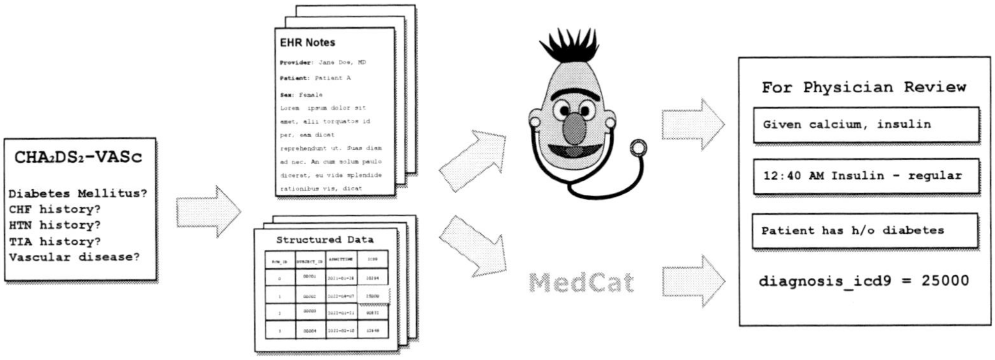

Figure 1: Demonstration of how our proposed QA-based risk score analysis system would work in conjunction with a physician.

the CHA_2DS_2-VASc score over unstructured EHR data. While these efforts found strong agreement with expert human evaluators, heuristic-based approaches are often rigid and struggle to generalize to other problems. Thus, we propose using off-the-shelf tools and pretrained language models to extract evidence from both structured and unstructured EHR data, without the need for manually-curated rules.

In this study, we explore two commonly used risk scores – the CHA_2DS_2-VASc score for atrial fibrillation stroke risk (Lip et al., 2010) and the Pulmonary Embolism Rule-out Criteria (PERC) rule (Kline et al., 2004) – to demonstrate our approach. We use a transformer-based model trained on emrQA (Pampari et al., 2018) and an off-the-shelf biomedical ontology linker paired with a SQL query component to extract evidence from unstructured and structured EHR data, respectively, for each element of the two risk scores (Figure 1). The main contributions of this work are:

- the first community-shared dataset based on MIMIC-III for automating risk scores,
- a demonstration of the potential for off-the-shelf tools and QA models to automate risk scores over heuristics and rules,
- the need for better negation/hypothetical detection and clinical knowledge embeddings.

2 Dataset

To evaluate our models, we randomly sample 100 patients from the Medical Information Mart for Intensive Care III (MIMIC-III) dataset (Johnson et al., 2016) to annotate with elements of CHA_2DS_2-VASc and PERC. CHA_2DS_2-VASc uses 7 patient data elements to estimate the risk of stroke in patients with non-valvular atrial fibrillation: congestive heart failure (CHF), hypertension, age, diabetes mellitus, stroke/transient ischemic attack (TIA)/thromboembolism (TE), vascular disease (prior myocardial infarction, peripheral artery disease, or aortic plaque), and sex. PERC uses 8 elements to evaluate the risk of PE in low-risk patients: age, heart rate, oxygen saturation, unilateral leg swelling, hemoptysis, recent surgery or trauma, prior PE or deep venous thrombosis (DVT), and hormone use.

We frame our scenario as a new patient being seen in the emergency department (ED) requiring calculation of CHA_2DS_2-VASc or PERC because of suspected atrial fibrillation or PE, respectively, and the data in MIMIC-III is the available past medical history for this patient. Therefore, we limited our dataset to non-expired patients at least 18 years of age at time of last discharge with at least one discharge summary. Since PERC only rules out PE when none of the criteria are met, one of which is age ≥ 50, we further adjust our sampling such that at least half of the patients selected are under 50 years of age at time of last discharge to ensure non-trivial calculation of PERC.

The dataset was annotated by two independent annotators, with a 20% overlap for inter-annotator agreement (κ = 0.800), and then reviewed by a physician. Annotators reviewed the entire EHR data provided in MIMIC-III, including both structured and unstructured sources, and annotated evidence relevant to each risk score element. Evidence in structured data include coded diagnoses, procedures, and past medical history. Evidence in unstructured data consist of text snippets from

Patient	Risk Score	Element	Evidence Source	Evidence Text	Answer
1234	CHA_2DS_2-VASc	CHF	noteevents.row_id = xxxx noteevents.row_id = xxxx diagnoses_icd.icd9_code = 4280	88 yo M with h/o dCHF Findings compatible with moderate congestive heart failure, with interval worsening since [**2157-8-30**] Congestive heart failure, unspecified	yes
1234	CHA_2DS_2-VASc	Hypertension	noteevents.row_id = xxxx diagnoses_icd.icd9_code = 4019	Hypertension Unspecified essential hypertension	yes
1234	CHA_2DS_2-VASc	Stroke/TIA/TE	NA		no data
1234	CHA_2DS_2-VASc	Vascular disease	chartevents.value = CAD	CAD	yes
1234	PERC	Hemoptysis	NA		no data
1234	PERC	Recent surgery/trauma	noteevents.row_id = xxxx noteevents.row_id = xxxx	s/p Pedestrian struck by auto presented to an outside hospital after reportedly being struck by a car traveling at 35mph	yes

Table 1: Example of annotated dataset. Under Evidence Source, NA indicates not applicable because no evidence found, noteevents indicates unstructured EHR data (xxxx indicates elided data), and all other sources are considered structured EHR data.

discharge summaries, admission notes, progress notes, and their addenda. Patients in our subset had an average of 44 notes with average length of 289 tokens.

Since we frame our scenario as a new patient being seen in the ED, vital signs (e.g., heart rate, oxygen saturation) as recorded in their history (i.e., MIMIC-III in our scenario) would not be relevant and are therefore excluded from annotation. For other elements in PERC that may also be time-sensitive, since the exact time frame is not always apparent from the given documentation, for the purposes of this study, we annotate all instances of unilateral leg swelling, hemoptysis, surgery and trauma as evidence for their respective elements regardless of when they occurred. In addition to the evidence, annotators also provided an overall answer for each risk score element: `"Yes"`, `"No"`, `"Unclear"` (evidence present but conflicting or inconclusive), or `"No data"`. A sample of the annotated dataset is presented in Table 1.

3 Task Setup

To extract information relevant to the specified risk score, we query the system with risk score elements expressed as short natural language phrases containing the entities (e.g., "hypertension"). Elements containing multiple concepts are split into multiple phrases, each containing a single concept. For the purposes of evaluation, `"Yes"` and `"Unclear"` in the ground truth are considered to be equivalent because both provide some positive evidence, while `"No"` and `"No data"` are considered to be equivalent because in practice, lack of data would be presumed to be negative.

For unstructured data, a system is tasked with predicting the presence or absence of the given risk score element. The system must also provide the sentence it selected to make its decision. Predictions considered true positives when compared to the ground truth are further reviewed by a physician to ensure that the sentence used for prediction can reasonably be used to determine if the patient has the given condition; if the sentence used for prediction cannot be used to logically determine whether the patient has the given condition, the prediction is marked as a false positive.

For structured data, the model is tasked with retrieving a `Yes/No` answer along with the relevant billing code (when present) for each risk score element. We evaluate the system by matching the retrieved `Yes/No` answer with the ground truth, and calculating the precision, recall, and F1-score.

4 Models

4.1 Structured Data Information Retrieval

To extract answers from structured EHR data, we employ a two step process. We (1) use MedCAT[1] (Kraljevic et al., 2019), an off-the-shelf biomedical ontology linker, to curate a set of Concept Unique Identifiers (CUIs) for each risk score element, which are then mapped to institution-specific billing codes (here, ICD9 for MIMIC-III) using the Unified Medical Language System (UMLS) APIs[2] (Bodenreider, 2004), and then (2) use these element-specific code-sets to form SQL queries (derived from emrKBQA (Raghavan et al., 2021)) to retrieve answers, i.e., `Yes/No` marked by the presence/absence of element-specific codes for a patient in the structured data. We evaluate our output only against risk score elements with answers from structured data (i.e., Evidence Source $\neq$ noteevents). Results are presented in Table 2.

[1] https://github.com/CogStack/MedCAT

[2] https://documentation.uts.nlm.nih.gov/rest/home.html

Risk Score	Element	Count	R	P	F1
CHA_2DS_2-VASc	CHF	16	1.0	0.94	0.97
CHA_2DS_2-VASc	Hypertension	43	0.97	0.81	0.89
CHA_2DS_2-VASc	Stroke/TIA/TE	17	1.0	0.12	0.21
CHA_2DS_2-VASc	Vascular disease	27	0.92	0.44	0.60
CHA_2DS_2-VASc	Diabetes mellitus	18	0.87	0.72	0.79
CHA_2DS_2-VASc	Overall	121	0.95	0.64	0.76
PERC	Unilateral leg swelling	5	0	0	0
PERC	Hemoptysis	1	0	0	0
PERC	Recent surgery/trauma	79	0	0	0
PERC	Prior PE/DVT	8	1.0	0.38	0.55
PERC	Hormone use	0	NA	NA	NA
PERC	Overall	93	0.84	0.17	0.29

Table 2: Performance of the structured data information retrieval component. We only calculate performance on risk score elements with structured data answers in the ground truth.

4.2 Baseline Model

To ground the results of our QA model, we implement a NER-based approach based on Bean et al. (2019). We use MedCAT to tag CUIs in the notes. We then return the top sentence that contains relevant affirmed CUIs based on the MedCAT negation detection system. Bean et al. (2019) defines a set of CUIs with respect to CHA_2DS_2-VASc. However, there is no such definition for PERC. We thus find relevant CUIs for the main categories of PERC (e.g., hormone use, surgery, etc.) and use all possible descendants of the selected CUIs based on the UMLS hierarchy. Results are shown in Table 3.

4.3 Unstructured QA Model

To retrieve relevant information from unstructured EHR data, we use ClinicalBERT (Alsentzer et al., 2019; Devlin et al., 2019), a transformer-based model pretrained on MIMIC-III. We sample 5% of the data[3] from the medication, relations, and risk subsections and train on emrQA (Pampari et al., 2018). Due to the vast number of notes likely containing irrelevant information, we additionally negative sample (1:1 ratio) unanswerable questions from other notes in emrQA during training. Further, due to the vague elements often used in risk scores (e.g., recent surgery or trauma), we augment 20% of existing emrQA questions containing a clinical entity to instead contain its parent MeSH[4] hierarchy entity. Similar to Bean et al. (2019), we select model predicted relevant spans and use MedCAT's negation detector to determine whether or not the patient has the given risk score element.

We additionally show how performance improves when unstructured data predictions are paired with structured data ones. To combine unstructured and structured data, we use the structured data prediction if it predicts that the patient has the given condition, otherwise we default to the selected unstructured data answer. All results are presented in Table 3.

5 Discussion

We make a few key observations. We find that the structured data model is able to achieve extremely high performance in a number categories, but unable to find any relevant information for the rest. We hypothesize that this is due to chronic conditions (e.g., CHF, hypertension) being more consistently recorded in the structured data, while acute events (e.g., PE/DVT, stroke/TIA/TE) are coded only in the limited time frame when such conditions are being actively managed. Also, structured data, in the form of billing codes, would not be expected to capture symptoms without a formal diagnosis (e.g., unilateral leg swelling). We additionally find that the QA model on unstructured data alone is able to improve on the results of Bean et al. (2019) on a number of categories, without the need for expert-crafted heuristics. However, we find that the QA model struggles due to a lack of clinical knowledge and ability to distinguish hypothetical mentions versus true affirmations of the given condition.

With respect to vascular disease, an error analysis of the QA-based model showed that 69% of the false positives were due to a lack of clinical understanding, as the model considered a much broader definition of vascular disease than the one specified in the CHA_2DS_2-VASc score. Similarly, with respect to stroke/TIA/TE, we find that 93% of the false positives can be attributed to imprecise understanding of medical terminology and the model's inability to use contextual clues to differentiate between stroke and other conditions. We additionally see extremely low precision for PE/DVT. This can largely be attributed to faulty negation detection, as MedCAT often fails to distinguish between affirming and hypothetical/negated mentions in over 70% of the false positives.

One issue we found when implementing Bean et al. (2019)'s approach is that it is nontrivial to determine which CUIs to select, specifically for general categories like surgery and trauma. Using all UMLS descendants of surgery and trauma results in 3,413,446 unique CUIs, which will clearly result in an enormous number of false positive re-

[3] Yue et al. (2020) found that sampling 5% of the data was equivalent to training on the entire dataset.

[4] https://www.nlm.nih.gov/mesh/meshhome.html

		Model	Bean et al. (2019)			QA					
		Data	Unstructured			Unstructured			Structured + Unstructured		
Risk Score	Element	Support	R	P	F1	R	P	F1	R	P	F1
CHA_2DS_2-VASc	CHF	16	0.385	0.294	0.333	0.615	0.533	0.571	0.938	0.789	0.857
CHA_2DS_2-VASc	Hypertension	43	0.929	0.736	0.821	0.883	0.864	0.874	0.977	0.875	0.923
CHA_2DS_2-VASc	Stroke/TIA/TE	17	0.588	0.303	0.400	0.385	0.263	0.312	0.538	0.333	0.412
CHA_2DS_2-VASc	Vascular disease	27	0.423	0.846	0.564	0.810	0.250	0.382	0.870	0.290	0.435
CHA_2DS_2-VASc	Diabetes mellitus	18	0.818	0.167	0.277	0.667	0.667	0.667	0.833	0.652	0.732
CHA_2DS_2-VASc	Overall	121	0.679	0.435	0.530	0.741	0.488	0.588	0.876	0.550	0.676
PERC	Unilateral leg swelling	5	0.200	1.000	0.333	0.500	0.375	0.429	0.500	0.375	0.429
PERC	Hempoptysis	1	1.000	0.250	0.400	1.000	0.118	0.211	1.000	0.118	0.211
PERC	Recent surgery/trauma	79	0.750	0.030	0.058	0.397	0.610	0.481	0.397	0.610	0.481
PERC	Prior PE/DVT	8	0.714	0.161	0.263	0.750	0.064	0.118	0.833	0.106	0.189
PERC	Hormone use	0	N/A	N/A	N/A	N/A	N/A	N/A	N/A	N/A	N/A
PERC	Overall	93	0.611	0.078	0.138	0.440	0.270	0.335	0.455	0.287	0.352

Table 3: Performance of Bean et al. (2019) heuristics, QA model, and a combination structured and QA model predictions.

sults when selecting sentences, as seen in Table 3. We find that the QA-based approach significantly outperforms the Bean et al. (2019)-based approach with respect to identifying surgery/trauma. This suggests that QA may offer a solution for these more general categories.

6 Conclusion

We explore risk score automation using QA and off-the-shelf ontology entity linkers without the need for expert-curated rules, and demonstrate its potential for easy adaptation to unexplored risk scores. We find that QA models can achieve comparable or better performance for certain risk score elements as compared to heuristic-based methods, and demonstrate the potential for more scalable risk score automation without the need for expert-curated heuristics. Our annotated dataset will be released to the community to encourage efforts in generalizable methods for automating risk scores.

References

Christopher Aakre, Mikhail Dziadzko, Mark T Keegan, and Vitaly Herasevich. 2017. Automating clinical score calculation within the electronic health record. a feasibility assessment. *Appl. Clin. Inform.*, 8(2):369–380.

Emily Alsentzer, John Murphy, William Boag, Wei-Hung Weng, Di Jindi, Tristan Naumann, and Matthew McDermott. 2019. Publicly available clinical BERT embeddings. In *Proceedings of the 2nd Clinical Natural Language Processing Workshop*, pages 72–78, Minneapolis, Minnesota, USA. Association for Computational Linguistics.

Dan Bean, James Teo, Honghan Wu, Ricardo Oliveira, Raj Patel, Rebecca Bendayan, Ajay Shah, Richard Dobson, and Paul Scott. 2019. Semantic computational analysis of anticoagulation use in atrial fibrillation from real world data. *PLOS ONE*, 14:e0225625.

Olivier Bodenreider. 2004. The unified medical language system (umls): Integrating biomedical terminology. *Nucleic acids research*, 32:D267–70.

Jacob Devlin, Ming-Wei Chang, Kenton Lee, and Kristina Toutanova. 2019. BERT: Pre-training of deep bidirectional transformers for language understanding. In *Proceedings of the 2019 Conference of the North American Chapter of the Association for Computational Linguistics: Human Language Technologies, Volume 1 (Long and Short Papers)*, pages 4171–4186, Minneapolis, Minnesota. Association for Computational Linguistics.

Peter L Elkin, Sarah Mullin, Jack Mardekian, Christopher Crowner, Sylvester Sakilay, Shyamashree Sinha, Gary Brady, Marcia Wright, Kimberly Nolen, Joann Trainer, Ross Koppel, Daniel Schlegel, Sashank Kaushik, Jane Zhao, Buer Song, and Edwin Anand. 2021. Using artificial intelligence with natural language processing to combine electronic health record's structured and free text data to identify nonvalvular atrial fibrillation to decrease strokes and death: Evaluation and case-control study. *J. Med. Internet Res.*, 23(11):e28946.

Cyril Grouin, Louise Deléger, Arnaud Rosier, Lynda Temal, Olivier Dameron, Pascal van Hille, Anita Burgun, and Pierre Zweigenbaum. 2011. Automatic computation of cha2ds2-vasc score: Information extraction from clinical texts for thromboembolism risk assessment. *AMIA ... Annual Symposium proceedings / AMIA Symposium. AMIA Symposium*, 2011:501–10.

Craig T. January, L. Samuel Wann, Joseph S. Alpert, Hugh Calkins, Joaquin E. Cigarroa, Joseph C. Cleveland, Jamie B. Conti, Patrick T. Ellinor, Michael D. Ezekowitz, Michael E. Field, Katherine T. Murray, Ralph L. Sacco, William G. Stevenson, Patrick J. Tchou, Cynthia M. Tracy, and Clyde W. Yancy. 2014. 2014 aha/acc/hrs guideline for the management of patients with atrial fibrillation: A report of the american college of cardiology/american heart association task force on practice guidelines and the heart rhythm society. *Journal of the American College of Cardiology*, 64(21):e1–e76.

Alistair E W Johnson, Tom J Pollard, Lu Shen, Li-Wei H Lehman, Mengling Feng, Mohammad Ghassemi, Benjamin Moody, Peter Szolovits, Leo Anthony Celi, and Roger G Mark. 2016. MIMIC-III, a freely accessible critical care database. *Sci. Data*, 3(1):160035.

Jitendra Jonnagaddala, Siaw-Teng Liaw, Pradeep Ray, Manish Kumar, Nai-Wen Chang, and Hong-Jie Dai. 2015. Coronary artery disease risk assessment from unstructured electronic health records using text mining. *Journal of Biomedical Informatics*, 58:S203–S210. Supplement: Proceedings of the 2014 i2b2/UTHealth Shared-Tasks and Workshop on Challenges in Natural Language Processing for Clinical Data.

J A Kline, A M Mitchell, C Kabrhel, P B Richman, and D M Courtney. 2004. Clinical criteria to prevent unnecessary diagnostic testing in emergency department patients with suspected pulmonary embolism. *J. Thromb. Haemost.*, 2(8):1247–1255.

Christian A Koziatek, Emma Simon, Leora I Horwitz, Danil V Makarov, Silas W Smith, Simon Jones, Soterios Gyftopoulos, and Jordan L Swartz. 2018. Automated pulmonary embolism risk classification and guideline adherence for computed tomography pulmonary angiography ordering. *Acad. Emerg. Med.*, 25(9):1053–1061.

Zeljko Kraljevic, Daniel M Bean, Aurelie Mascio, Lukasz Roguski, Amos A. Folarin, Angus Roberts, Rebecca Bendayan, and Richard J. B. Dobson. 2019. Medcat - medical concept annotation tool. *ArXiv*, abs/1912.10166.

Gregory Y H Lip, Robby Nieuwlaat, Ron Pisters, Deirdre A Lane, and Harry J G M Crijns. 2010. Refining clinical risk stratification for predicting stroke and thromboembolism in atrial fibrillation using a novel risk factor-based approach: the euro heart survey on atrial fibrillation. *Chest*, 137(2):263–272.

Dustin G. Mark, Jie Huang, Uli Chettipally, Mamata V. Kene, Megan L. Anderson, Erik P. Hess, Dustin W. Ballard, David R. Vinson, and Mary E. Reed. 2018. Performance of coronary risk scores among patients with chest pain in the emergency department. *Journal of the American College of Cardiology*, 71(6):606–616.

Anusri Pampari, Preethi Raghavan, Jennifer Liang, and Jian Peng. 2018. emrQA: A large corpus for question answering on electronic medical records. In *Proceedings of the 2018 Conference on Empirical Methods in Natural Language Processing*, pages 2357–2368, Brussels, Belgium. Association for Computational Linguistics.

Juliessa M Pavon, Richard J Sloane, Carl F Pieper, Cathleen S Colón-Emeric, Harvey J Cohen, David Gallagher, Miriam C Morey, Midori McCarty, Thomas L Ortel, and Susan N Hastings. 2018. Automated versus manual data extraction of the padua prediction score for venous thromboembolism risk in hospitalized older adults. *Appl. Clin. Inform.*, 9(3):743–751.

Preethi Raghavan, Jennifer J Liang, Diwakar Mahajan, Rachita Chandra, and Peter Szolovits. 2021. emrKBQA: A clinical knowledge-base question answering dataset. In *Proceedings of the 20th Workshop on Biomedical Language Processing*, pages 64–73, Online. Association for Computational Linguistics.

Ewout W Steyerberg et al. 2019. *Clinical prediction models*. Springer.

Xiang Yue, Bernal Jimenez Gutierrez, and Huan Sun. 2020. Clinical reading comprehension: A thorough analysis of the emrqa dataset. *ArXiv*, abs/2005.00574.

Nasen Jonathan Zhang, Philippe Rameau, Marsophia Julemis, Yan Liu, Jeffrey Solomon, Sundas Khan, Thomas McGinn, and Safiya Richardson. 2022. Automated pulmonary embolism risk assessment using the wells criteria: Validation study. *JMIR Form Res*, 6(2):e32230.

DoSSIER at MedVidQA 2022: Text-based Approaches to Medical Video Answer Localization Problem

Wojciech Kusa [1†] **Georgios Peikos** [2†] **Oscar Espitia** [2]
Allan Hanbury [1] **Gabriella Pasi** [2]

[1]TU Wien, Vienna, Austria
{wojciech.kusa,allan.hanbury}@tuwien.ac.at
[2]University of Milano-Bicocca, Milan, Italy
{georgios.peikos,oscar.espitiamendoza,gabriella.pasi}@unimib.it

Abstract

This paper describes our contribution to the Answer Localization track of the MedVidQA 2022 Shared Task. We propose two answer localization approaches that use only textual information extracted from the video. In particular, our approaches exploit the text extracted from the video's transcripts along with the text displayed in the video's frames to create a set of features. Having created a set of features that represents a video's textual information, we employ four different models to measure the similarity between a video's segment and a corresponding question. Then, we employ two different methods to obtain the start and end times of the identified answer. One of them is based on a random forest regressor, whereas the other one uses an unsupervised peak detection model to detect the answer's start time. Our findings suggest that for this task, leveraging only text-related features (transmitted either verbally or visually) and using a small amount of training data, lead to significant improvements over the benchmark Video Span Localization model that is based on deep neural networks.

1 Introduction

Nowadays, the number of users that turn to the Web to satisfy their health-related information needs has grown significantly. However, providing the user with textual answers is insufficient for some particular information needs because, occasionally, these answers are hard to interpret correctly. Therefore, it could be useful if the answers are accompanied by a visual aid, i.e., a part of a video (video segment), that presents the answer. Such information needs are the scope of this work, where we focus on identifying those video segments that contain the answer to health-related user questions. These user questions are written in natural language and the corresponding answers are part of an instructional video; our goal is to create a system capable of locating the corresponding answer.

While the majority of the proposed works in the literature rely on deep neural models to allocate the relevant video segments to the answer (Yu et al., 2017; Anne Hendricks et al., 2017), we explore another alternative. Specifically, we aim to study the impact of using only textual features to find the answers in the video, which also implies reducing the requirements for the amount of training data. As input features we use the information transmitted verbally by the presenter in the form of video transcript and also extract the text embedded in the video frames.

Our main contributions are as follows:

- We develop two approaches that use only textual information and few training data, to tackle the task of answer localization for instructional medical videos.
- We show that both the visually (text presented in a video's frame) and verbally (transcripts obtained from the speaker's instructions) transmitted information can be used to locate the answer in medical instructional videos.

The remainder of the paper is organized as follows: Section 2 describes in detail the studied task and the related works. Section 3 presents our methodology and assumptions. In Section 4, we present the experimental setup, our baseline and our submissions. Finally, Section 5 presents the obtained results, followed by the conclusions drawn from our participation.

2 Task Description & Related Works

This work studies the task of video segment identification for medical videos, introduced as a shared task in Gupta and Demner-Fushman (2022). In

† Equal contribution.

Proceedings of the BioNLP 2022 workshop, Dublin, Ireland, pages 432–440
May 26, 2022. ©2022 Association for Computational Linguistics

particular, given a medical or health-related question written in natural language, the system must provide the user with the video segment that contains the answer. The task focuses on instructional medical videos. A characteristic of these videos is that they deliver the key information to the user both visually and verbally (Gupta et al., 2022).

In visual question answering, identifying relevant video segments given a user's questions in a natural language is a task that requires processing of both textual and visual signals. As reported by Zhang et al. (2019), a system designed to tackle this problem consists of three components, namely, feature extraction, feature fusion and answer prediction. Previously published studies exploit standard embedding models to obtain text features (Tapaswi et al., 2016), and CNN based models to extract image features (Zhou et al., 2018). Liu et al. (2019a) introduced ETM-Trans which is a deep transfer learning approach that also addresses the issue of feature fusion. In the field of visual question answering, as reported in (Lin et al., 2021) the majority of the proposed techniques employ pre-trained models for image and language encoders. Another finding reported in (Lin et al., 2021), is related to the fact that only a small portion of the proposed approaches investigate their generality and interpretability.

The introduction of large-scale multimodal datasets covering both language and vision enabled the development of efficient deep neural network techniques that bridge the gap between language, and visual understanding (Lei et al., 2018, 2019; Tapaswi et al., 2016).

While the majority of the proposed methods in the literature are based on deep neural models, our approach leverages only the textual information that can be extracted from a video without the need for extensive training. It estimates the relevance of each video segment to a given question, and ultimately it returns the starting time and duration of the answer.

3 Methodology

Our methodology exploits the characteristics of the videos in the current task. Specifically, we extract a video's transcripts. The transcripts contain the text that one can hear during the video, its start time and its duration, and correspond to a specific video segment. Moreover, we enrich this information by adding the text presented in video segments (video frames), for instance text that contains the topic, the steps of an exercise, among others. Then, given a question, we estimate distinct similarity scores for every video segment using four different models that will be described in Section 3.2. At this point, two distinct approaches can be followed to identify the answer's starting time and duration. The first one employs a multi-output regression model that inputs the similarity scores for every video segment and outputs the starting time and duration of the answer. For the second approach we set the starting time equal to the starting time of the segment that has the highest similarity score, obtained by aggregating the similarity scores obtained by four models, and hard-set the answer's duration based on the training data. The following sections present the hypothesis and assumptions behind each step of our methodology.

3.1 Converting video to text

As mentioned by Gupta et al. (2022), instructional medical videos deliver the key information both visually and verbally. We hypothesize that the speaker mentions keywords during the video that are also present in the question. For example, a phrase such as: "In the following part I will show you how to perform the [name of a specific exercise]", where the *"name of the specific exercise"* can also be found in the user's question.

Secondly, we hypothesized that video frames might contain textual information that overlaps with the question's text. However, it is also possible that the information obtained from these frames is irrelevant; i.e. frames may contain the speaker's name or affiliation. All in all, we assume that the text extracted using the two approaches mentioned above can provide a strong indication of the answer's location.

Finally, we assumed that text is not equally distributed across the video. For instance, it is common that a speaker might make a pause, e.g. to demonstrate the instruction or to change the subject. When only a video's text features are used, it is possible that some parts of the video will have no representation. In order to mitigate this issue and also to further enrich the text representation, we experiment with merging consecutive transcript lines. We ensure that when doing this, we also shift the time that corresponds to the merged text.

3.2 Estimating text-question similarity

Having the text that corresponds to a set of sequential video frames, we estimate its similarity to the question using four different models. Specifically, we employ two relevance models widely used in Information Retrieval (IR) to estimate the query-document similarity. In addition, we employ two pre-trained neural language models that are based on the Transformer architecture (Vaswani et al., 2017). We encode the questions and textual features independently for each language model and then calculate the similarity scores using a cosine similarity measure. We perform a min-max normalization of the similarity scores for each model independently. We then create an $M \times N$ matrix that contains the aggregated similarity scores for each question-video and every video segment; where M is the number of the employed models, and N is the number of video segments.

3.2.1 IR models

Regarding the IR relevance models, we employed the BM25 relevance model and a language model with a Dirichlet smoothing to overcome the problem of missing terms, which is likely to occur due to the characteristics of the studied task. In particular, the problem of missing terms occurs because the duration of the instructional videos is short and therefore it contains only few words. These models rely their estimation on some collection-related statistics, e.g. a term's inverse document frequency; to estimate these values, the models exploit an index created by concatenating the videos' texts present in a training collection.

3.2.2 Neural language models

In our experiments we employ two different language models pre-trained using different datasets, that are available in the HuggingFace transformers library (Wolf et al., 2020), namely:

- The RoBERTa model (Liu et al., 2019b) trained on the MS MARCO dataset from the *sentence-transformers*[1] framework (Reimers and Gurevych, 2019).
- The MPNet model (Song et al., 2020) trained on the SNLI and MultiNLI datasets from the *sentence-transformers*[2] framework.

3.3 Answer localization models

This section describes two different approaches to localization of the answer time: multi-output regression and peak detection.

3.3.1 Multi-output regression (MoR)

Having created the $M \times N$ matrix described above, the answer localization can be modelled as a regression problem. To this aim, we employed the Random Forest multi-output regression model to predict the answer's starting point and duration.

The employed regression model requires a fixed-size sequence to be used as input. However, the available videos, and hence their textual representation, have varying duration. As a result, one should normalize the input length across the whole dataset. To achieve that, we formulate a method of sampling the text-question similarity models to obtain the same length for every video-question pair. In particular, we split every video into B equally spaced bins. By using these bins, we create a fixed-size representation of every video in the dataset. For every bin, independently for each model, we calculate two values: the maximum and the median values of all text-question similarity scores within the timestamps of a particular bin. Consequently, our normalization approach generates $2M \times B$ input matrix, where M is the number of models and B is fixed for the whole dataset and it contains both the maximum and median values.

3.3.2 Peak detection (PD)

Peak Detection (PD) approach also utilizes the $M \times N$ matrix described in Section 3.2 to find the video segment which is the most relevant to the question. We hypothesize that the segment with the highest topical similarity could be identified shortly before or after the true start of the answer (Figure 1). This method takes the average of the similarity scores from all text-question similarity models for every segment, and then retrieves the segment with the highest score. After identifying the segment, the start and end time of the answer can be predicted using the following formula:

$$
\begin{aligned}
t_s^{'} &= t_s + \beta_1, \\
t_e^{'} &= t_s + \beta_2,
\end{aligned} \tag{1}
$$

where t_s is the timestamp of retrieved segment and β_n are two free-parameters that are used to estimate

[1] https://huggingface.co/sentence-transformers/msmarco-distilroberta-base-v2

[2] https://huggingface.co/sentence-transformers/nli-mpnet-base-v2

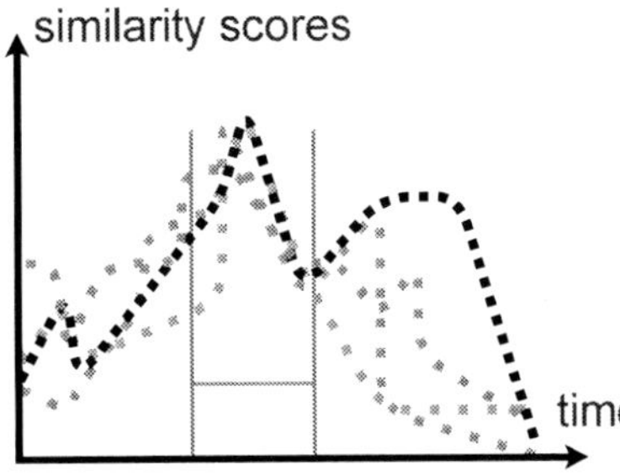

Figure 1: Illustration of the text-question similarity over a video time. Red lines mark the span of the correct answer. Maximum values of similarity for each similarity model are within the true answer span.

the duration of the answer.

4 Experiment setup and submissions

This section describes the dataset used to train and evaluate our approaches and our methodology for extracting text from videos. In addition, we provide the details of our submissions. Our code is publicly available[3].

4.1 Dataset

Model training and validation has been conducted on the MedVidQA dataset (Gupta et al., 2022), which consists 3010 questions from 899 unique videos in three different data splits, i.e., Train, Validation and Test. The submitted runs in the MedVidQA 2022 Answer Localization Shared Task (Gupta and Demner-Fushman, 2022) were evaluated on a new test dataset that consists of 153 questions covering 50 new YouTube videos, hereafter referred to as MVAL 2022.

Table 1 presents the number of videos for which we were able to extract the transcripts and the text in the video frame (embedded text) across the different datasets (rows 1-3). 98.5% of the videos contain some textual information. The missing 1.5% is primarily due to the private or protected videos for which it is not possible to obtain these features. In addition, Table 1 presents the mean and the median number of lines found in the transcripts and in the embedded text showing that medical instructional videos contain many verbal explanations and textual information embedded in the video frames.

[3]https://github.com/ProjectDossier/MedVid2022

4.2 Video to text

To extract the transcripts from a video we used the *youtube_transcript_api*[4] library. In cases where the transcript extraction was not feasible (1.5% of the videos), a placeholder text was assigned to the first second of the video. The obtained transcript lines were often just a set of words, split based on the speaker's pauses during the video, rather than complete sentences. In Section 3.2.1, we hypothesized that the problem of missing terms may occur. Indeed, it was found that various transcript lines contained only few keywords (due to speaker's pauses), in some cases only the stopwords. Therefore, for these cases, the obtained document representation was not accurate.

To overcome this issue, initially we tried to concatenate all the transcript text, and then, by using sentence splitting methods, create a set of sentences. However, due to the missing punctuation in many videos, this method was not accurate, and we decided to follow a simpler approach.

In particular, we proceed by merging subsequent transcript lines. For instance, a line i which contain the words: "now I will present" followed by a line $i + 1$ containing "an exercise that helps with back pain" was merged into one single text. Moreover, we experiment with different levels of merging sequential transcript lines by joining two, three and four consecutive texts that generate three additional input representations. We refer to all the transcript features as *transcript-n*, where n is the number of original sequential transcript lines that were merged.

To download the videos we used the *pytube*[5] Python package. We use the offset of one second for the first frame as the beginning of the video is usually just a black screen. Also, we used the *tesseract*[6] engine to perform the optical character recognition (OCR) to extract the text from every video frame. Finally, we set the recognized text's duration to three seconds to follow the same data format as in the transcripts. The obtained features from this textual information are referred to as *ocr*.

An overview of all five different video-to-text representations used in our experiments is presented in Table 2.

[4]https://pypi.org/project/youtube-transcript-api/

[5]https://pypi.org/project/pytube/

[6]https://github.com/tesseract-ocr/tesseract

	MedVidQA			MVAL 2022	
	Train	Validation	Test	Test	Total
Videos (V)	800	49	50	50	949
V with transcripts	788 (98.5%)	48 (98%)	50 (100%)	49 (98%)	935 (98.5%)
V with embedded text	750 (93.8%)	48 (98%)	47 (96%)	49 (98%)	894 (94.2%)
Mean # lines in transcripts	133	142	124	123	140
Median # lines in transcripts	97.5	107.5	110.5	70	106
Mean # lines in embedded texts	20	16	25	18	17
Median # lines in embedded texts	9	8	15	9	9

Table 1: Statistics of the availability of textual information in medical informational videos. MVAL 2022 stands for MedVidQA 2022 Answer Localization Shared Task.

Feature name	Feature description	Start time	End time
transcript-1	Original transcript line i output from the video	s_i	e_i
transcript-2	Two consecutive lines of transcript merged together	s_i	e_{i+1}
transcript-3	Three consecutive lines of transcript merged together	s_i	e_{i+2}
transcript-4	Four consecutive lines of transcript merged together	s_i	e_{i+3}
ocr	OCR of the video frame i taken at second s every 3 seconds	s_i	$s_i + 3$

Table 2: Description of five different input features used in our work. s_i represents the start time of the i-th transcript line or the video frame.

4.3 Submissions

We submitted five runs for the MedVidQA 2022 Medical Visual Answer Localization (MVAL) Shared Task. A summary of our submissions is presented in Figure 2. In this section we describe these runs in detail.

4.3.1 Baseline: zero-shot extractive Q&A (1)

We use the DistilBERT-base-uncased model (Sanh et al., 2019), fine-tuned using knowledge distillation on the SQuAD dataset. We take the implementation from the HuggingFace transformers library[7]. As an input feature, we concatenate all the lines of *transcript-1* to create a consistent, single document representation of each video.

The model's output is text extracted from the video. Therefore, that extracted textual answer needs to be converted back to its start and end time. This can be done by locating its corresponding lines in the transcript. To achieve that, we employed the most greedy approach, i.e., selecting the whole transcript line if it contains at least one word from the extracted answer.

We noticed that the employed Q&A model could not correctly predict the textual answer to the question, and the retrieved answers are too short. We believe that this is because most videos do not exhibit the explicit textual answer to the question, but only the visual explanation. In order to mitigate this issue, we decided to test a simple parametrization model that stretches the predicted answer span:

$$t_s^{'} = \alpha_1 \cdot t_s,$$
$$t_e^{'} = \alpha_2 \cdot t_e, \tag{2}$$

where t_s and t_e are outputted start and end times of the answer from the Q&A system and α_n are estimated using the train dataset. After conducting an analysis on the validation dataset, we select the following values for the parametrization of the results: $\alpha_1 = 0.35$, $\alpha_2 = 0.90$.

4.3.2 Multi-output regression (2)

Our submission (2) is described in detail below:

1. We use all five input features to calculate the text-question similarity using the four models described in Section 3.2. For the BM25 and the statistical language model, the index was created using the Train data.

[7] https://huggingface.co/distilbert-base-uncased-distilled-squad

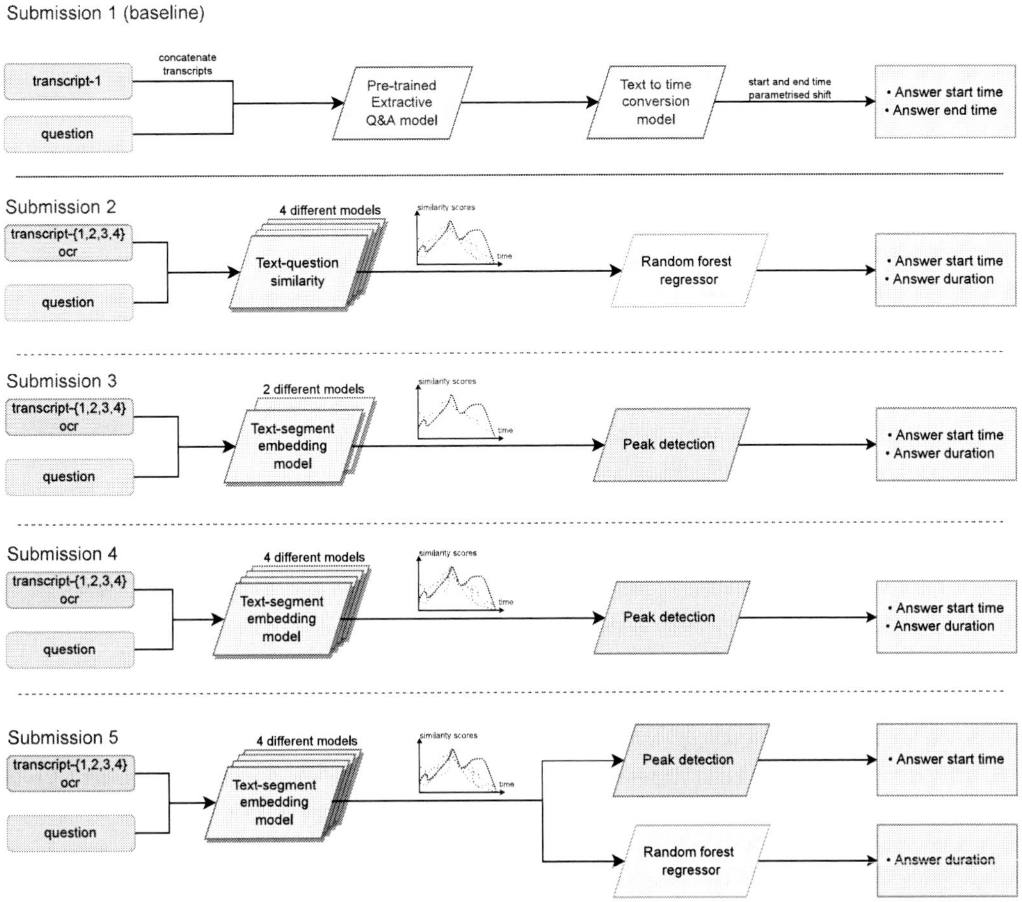

Figure 2: Summary of our five submitted runs.

2. We perform a min-max normalization independently for each model, aggregating scores from every input feature.

3. We conduct an input normalization as described in Section 3.3.1, so that the size of each feature vector is constant across all the training samples.

4. We train the random forest regressor model to predict the start time and duration of the answer.

We use the random forest regressor implementation from *scikit-learn* library (Pedregosa et al., 2011) with max depth equal to 10 and 40 estimators.

4.3.3 Answer start-time detection (3) & (4)

Submissions (3) and (4) also use the first two steps as in the submission (2) to calculate the text-question similarity and perform a min-max normalization. For submission (3), we use only the RoBERTa and the statistical language model with a Dirichlet smoothing. For submission (4), we use all four text-question similarity models.

This is followed by the step of peak detection by selecting the time when the average similarity of all models is the highest, as described in Section 3.3.2. Instead of using the start time of a segment, we take the center point of the selected segment as the most plausible starting point of the answer: $t_s = (s+e)/2$.

Finally, we calculate the answer start and end time by using Equation 1. Based on the experiments on the validation set, we select $\beta_1 = -6$ to overcome the shift between the true answer start and the similarity score peak. We use $\beta_2 = 62$ which corresponds to the mean answer duration on the training dataset.

Run	Source	Model	IoU=0.3	IoU=0.5	IoU=0.7	mIoU
	Gupta et al. (2022)	VSL-BASE (FPL 800)	21.93	12.25	5.80	20.15
		VSL-QGH	25.81	14.20	6.45	20.12
1	Extractive Q&A		21.29	9.68	3.87	18.92
2	MoR		31.61	15.48	4.52	18.62
3	PD (2 models)		46.45	**29.03**	10.97	29.92
4	PD (4 models)		**48.39**	**29.03**	**11.61**	30.33
5	PD (4 models) start + MoR duration		47.10	27.74	10.97	**30.67**

Table 3: Performance comparison of our submissions on MedVidQA Test dataset from (Gupta et al., 2022).

4.3.4 Ensemble model (5)

Our last submission (5) is an ensemble model. It uses the prediction of the start time from the Peak Detection (4 models) – submission (4) and the duration from the multi-output regression model – submission (2). This method overcomes a limitation of the previous approaches, i.e., the constant parameter β_2 that defines the answer's duration (see Equation 1). In the previous approaches, this parameter had a constant value across all video-question pairs. In contrast, in this approach, the β_2 parameter for every question takes a unique value predicted by the random forest regressor used in the submission (2).

5 Evaluation and results

In this section we present the results on both the evaluation and test datasets.

5.1 Evaluation measures

We follow the evaluation measures proposed by Gupta et al. (2022) that have been chosen as the official metrics for the MedVidQA 2022 Shared Task. In particular, we evaluate our results using Intersection over Union (IoU) that measures the proportion of overlap between the predicted answer and the ground truth at three different thresholds, and mIoU that is the average of IoU calculated over a set of samples. Notice that MedVidQA adopts "R@n, IoU=μ", which denotes the percentage of questions for which, out of the top-n retrieved temporal segments, at least one predicted temporal segment intersects the ground truth temporal segment for longer than μ. Specifically, results are evaluated using $n = 1$ and $\mu \in \{0.3, 0.5, 0.7\}$.

5.2 Evaluation on MedVidQA

Validation results are presented in Table 3. Our baseline Q&A model (submission (1)), which initially was not able to retrieve any relevant information, after using parametrization it reaches 21.29 (IoU=0.3), which is on par with the performance of the Video Span Localization (VSL) benchmark model from Gupta et al. (2022). This shows that the first threshold could be reached even by a sub-optimal model whose predictions are shifted using two fixed parameters. Our best performing approach, Peak Detection (submissions (3) and (4)), achieves significant gains for each of three thresholds for the IoU measure, when compared to the best benchmark, i.e., the VSL model. Especially for the mIoU measure it obtains 10% more overlap on the Test data.

5.2.1 Impact of text extracted from the video frames

For some of the videos, the text extracted from the video frames had a significant impact on localising the correct answer.

Such an example can be seen in Figure 3. One

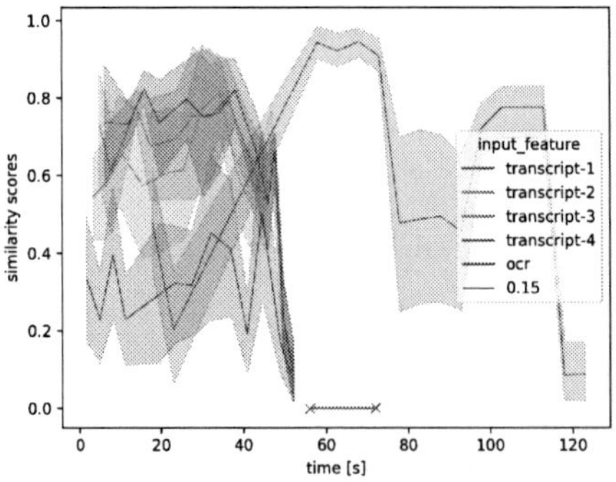

Figure 3: Mean text-question similarity plots for the Peak Detection approach with four models for question ID 2714 grouped by the input feature. Red line shows the span of a correct answer.

Run	Model	IoU=0.3	IoU=0.5	IoU=0.7	mIoU
	Max	91.50	84.97	73.20	75.83
	Median	80.32	71.90	48.37	58.81
	Mean	76.04	60.80	40.37	55.79
1	Extractive Q&A	18.95	7.84	1.96	19.87
2	MoR	31.37	13.07	4.57	18.80
3	PD (2 models)	**40.52**	20.26	**10.45**	**25.26**
4	PD (4 models)	37.25	14.38	7.84	22.05
5	PD (4 models) start + MoR duration	33.33	**21.57**	9.15	23.54

Table 4: Performance comparison of the variants of our submissions on MedVidQA 2022 Test dataset. Runs 3, 4 and 5 did not contribute to the median and mean pool.

Features	IoU=0.3	IoU=0.7	mIoU
all	**48.39**	11.61	30.33
transcript-1	45.16	9.68	28.01
transcript-{1,2,3,4}	47.74	**12.26**	**30.55**
ocr	18.06	3.23	12.65

Table 5: Performance comparison of the Peak Detection approach using 4 models with different input features on MedVidQA Test dataset.

can observe that without the *ocr* feature, it is not feasible to identify the correct answer because the transcript features do not exist for the correct answer span.

To further quantify the impact of input features, we conducted an ablation study on our best performing approach: PD with four models (submission (4)). The results are summarised in Table 5. The model using all features achieves the highest IoU=0.3, which was our optimization goal. Removing the *ocr* feature slightly improves the results on IoU=0.7 and mIoU. Even though the text extracted from the video frames alone yields low results, it still can be a helpful additional feature for medical instructional videos when correctly merged with other inputs.

5.3 MedVidQA 2022 Shared Task results

The results produced by our models, along with max, median and mean values from all participants are presented in Table 4. The performance obtained by the proposed approaches is below the reported mean. However, by comparing the obtained effectiveness presented in Table 3 and Table 4 one can observe that the models have a robust behavior across the different datasets as they yield similar performance. Peak detection-based approaches yield the highest results among our submissions, confirming the results of our experiments conducted on the MedVidQA dataset.

6 Conclusion

This work investigates two different approaches for detecting answer timestamps from medical instructional videos in the context of the MedVidQA 2022 MVAL Shared Task (Task 2). Our approaches rely only on the text extracted from the videos, either as transcripts or as the text displayed in the video's frames. After extracting the text corresponding to every video segment, we estimate its similarity to the question using four different models. We employ two different strategies to map the question-text similarity to the answer timestamp, i.e. multi-output regression model based on random forest and a peak detection model.

Our best performing peak detection model achieves 40.52 IoU=0.3 on MedVidQA 2022 Shared Task and outperforms the VSL benchmark model on the MedVidQA test dataset. We also show a positive impact of using multiple video-to-text conversion methods on the overall quality of models. Our feature extraction methods could easily extend the set of features used by end-to-end deep learning models. Further analysis is needed to assess other ways of processing the text-question similarity importance for obtaining more accurate predictions.

Acknowledgements

This work was supported by the EU Horizon 2020 ITN/ETN on Domain Specific Systems for Information Extraction and Retrieval – DoSSIER (H2020-EU.1.3.1., ID: 860721).

References

Lisa Anne Hendricks, Oliver Wang, Eli Shechtman, Josef Sivic, Trevor Darrell, and Bryan Russell. 2017. Localizing moments in video with natural language. In *Proceedings of the IEEE international conference on computer vision*, pages 5803–5812.

Deepak Gupta, Kush Attal, and Dina Demner-Fushman. 2022. A Dataset for Medical Instructional Video Classification and Question Answering. *arXiv preprint arXiv:2201.12888*.

Deepak Gupta and Dina Demner-Fushman. 2022. Overview of the MedVidQA 2022 Shared Task on Medical Video Question Answering. In *Proceedings of the 21st SIGBioMed Workshop on Biomedical Language Processing, ACL-BioNLP 2022*. Association for Computational Linguistics.

Jie Lei, Licheng Yu, Mohit Bansal, and Tamara Berg. 2018. Tvqa: Localized, compositional video question answering. In *Proceedings of the 2018 Conference on Empirical Methods in Natural Language Processing*, pages 1369–1379.

Jie Lei, Licheng Yu, Tamara L Berg, and Mohit Bansal. 2019. Tvqa+: Spatio-temporal grounding for video question answering. *arXiv preprint arXiv:1904.11574*.

Zhihong Lin, Donghao Zhang, Qingyi Tao, Danli Shi, Gholamreza Haffari, Qi Wu, Mingguang He, and Zongyuan Ge. 2021. Medical visual question answering: A survey. *CoRR*, abs/2111.10056.

Feifan Liu, Yalei Peng, and Max P. Rosen. 2019a. An effective deep transfer learning and information fusion framework for medical visual question answering. In *Experimental IR Meets Multilinguality, Multimodality, and Interaction - 10th International Conference of the CLEF Association, CLEF 2019, Lugano, Switzerland, September 9-12, 2019, Proceedings*, volume 11696 of *Lecture Notes in Computer Science*, pages 238–247. Springer.

Yinhan Liu, Myle Ott, Naman Goyal, Jingfei Du, Mandar Joshi, Danqi Chen, Omer Levy, Mike Lewis, Luke Zettlemoyer, and Veselin Stoyanov. 2019b. Roberta: A robustly optimized bert pretraining approach. *arXiv preprint arXiv:1907.11692*.

Fabian Pedregosa, Gaël Varoquaux, Alexandre Gramfort, Vincent Michel, Bertrand Thirion, Olivier Grisel, Mathieu Blondel, Peter Prettenhofer, Ron Weiss, Vincent Dubourg, et al. 2011. Scikit-learn: Machine learning in python. *the Journal of machine Learning research*, 12:2825–2830.

Nils Reimers and Iryna Gurevych. 2019. Sentence-bert: Sentence embeddings using siamese bert-networks. In *Proceedings of the 2019 Conference on Empirical Methods in Natural Language Processing and the 9th International Joint Conference on Natural Language Processing (EMNLP-IJCNLP)*, pages 3982–3992.

Victor Sanh, Lysandre Debut, Julien Chaumond, and Thomas Wolf. 2019. Distilbert, a distilled version of bert: smaller, faster, cheaper and lighter. *arXiv preprint arXiv:1910.01108*.

Kaitao Song, Xu Tan, Tao Qin, Jianfeng Lu, and Tie-Yan Liu. 2020. Mpnet: Masked and permuted pre-training for language understanding. *Advances in Neural Information Processing Systems*, 33:16857–16867.

Makarand Tapaswi, Yukun Zhu, Rainer Stiefelhagen, Antonio Torralba, Raquel Urtasun, and Sanja Fidler. 2016. Movieqa: Understanding stories in movies through question-answering. In *Proceedings of the IEEE conference on computer vision and pattern recognition*, pages 4631–4640.

Ashish Vaswani, Noam Shazeer, Niki Parmar, Jakob Uszkoreit, Llion Jones, Aidan N Gomez, Łukasz Kaiser, and Illia Polosukhin. 2017. Attention is all you need. *Advances in neural information processing systems*, 30.

Thomas Wolf, Lysandre Debut, Victor Sanh, Julien Chaumond, Clement Delangue, Anthony Moi, Pierric Cistac, Tim Rault, Rémi Louf, Morgan Funtowicz, Joe Davison, Sam Shleifer, Patrick von Platen, Clara Ma, Yacine Jernite, Julien Plu, Canwen Xu, Teven Le Scao, Sylvain Gugger, Mariama Drame, Quentin Lhoest, and Alexander M. Rush. 2020. Transformers: State-of-the-art natural language processing. In *Proceedings of the 2020 Conference on Empirical Methods in Natural Language Processing: System Demonstrations*, pages 38–45, Online. Association for Computational Linguistics.

Youngjae Yu, Hyungjin Ko, Jongwook Choi, and Gunhee Kim. 2017. End-to-end concept word detection for video captioning, retrieval, and question answering. In *Proceedings of the IEEE Conference on Computer Vision and Pattern Recognition*, pages 3165–3173.

Dongxiang Zhang, Rui Cao, and Sai Wu. 2019. Information fusion in visual question answering: A survey. *Inf. Fusion*, 52:268–280.

Yangyang Zhou, Xin Kang, and Fuji Ren. 2018. Employing inception-resnet-v2 and bi-lstm for medical domain visual question answering. In *Working Notes of CLEF 2018 - Conference and Labs of the Evaluation Forum, Avignon, France, September 10-14, 2018*, volume 2125 of *CEUR Workshop Proceedings*. CEUR-WS.org.

Author Index